Inhaltsverzeichnis. Contents

FORTSCHRITTE DER CHEMIE ORGANISCHER NATURSTOFFE

PROGRESS IN THE CHEMISTRY OF ORGANIC NATURAL PRODUCTS

BEGRÜNDET VON · FOUNDED BY

L. ZECHMEISTER

HERAUSGEGEBEN VON · EDITED BY

W. HERZ   H. GRISEBACH   G. W. KIRBY

TALLAHASSEE, FLA.   FREIBURG i. BR.   GLASGOW

VOL. 34

VERFASSER · AUTHORS

A. J. AASEN · D. P. CHAKRABORTY · C. R. ENZELL · D. GROSS
J. JACOB · K. H. OVERTON · D. J. PICKEN · A. R. PINDER
W. VOELTER · I. WAHLBERG

1977

WIEN · SPRINGER-VERLAG · NEW YORK

Mit 63 Abbildungen · With 63 Figures

© 1977 by Springer-Verlag/Wien
Softcover reprint of the hardcover 1st edition 1977
Library of Congress Catalog Card Number AC 39-1015

ISBN 978-3-7091-8478-3   ISBN 978-3-7091-8476-9 (eBook)
DOI 10.1007/978-3-7091-8476-9

Inhaltsverzeichnis. Contents   V

Mitarbeiterverzeichnis. List of Contributors

Aasen, Doc. Dr. A. J., Pharmaceutical Department, University of Oslo, Blindern, Post-books 1068, Oslo 6, Norway.

Chakraborty, D. P., Ph. D., D. Sc., Professor of Chemistry, Comparative Phytochemical Laboratory, Bose Institute, 93/1, Acharya Prafulla Ch. Road, Calcutta-700 009, India.

Enzell, Prof. Dr. C. R., Research Department, Swedish Tobacco Company, P. O. Box 17007, S-104 62 Stockholm 17, Sweden.

Gross, Dr. D., Institut für Biochemie der Pflanzen, Forschungszentrum für Molekular-biologie und Medizin der Akademie der Wissenschaften der DDR, DDR-401 Halle (Saale), Deutsche Demokratische Republik.

Jacob, Priv.-Doz. Dr. J., Biochemisches Institut für Umweltcarcinogene, Sieker Land-straße 19, D-2070 Ahrensburg/Holstein, Bundesrepublik Deutschland.

Overton, Prof. K. H., D. Sc., Department of Chemistry, The University of Glasgow, Glasgow G 128QQ, Scotland.

Picken, D. J., B. Sc., Department of Chemistry, The University of Glasgow, Glasgow G 128QQ, Scotland.

Pinder, Prof. Dr. A. R., Department of Chemistry and Geology, Clemson University, Clemson, SC 29631, U. S. A.

Voelter, Prof. Dr. W., Institut für Organische Chemie, Eberhard-Karls-Universität Tübin-gen, Auf der Morgenstelle 18, D-7400 Tübingen, Bundesrepublik Deutschland.

Wahlberg, Doc. Dr. I., Research Department, Swedish Tobacco Company, P. O. Box 17007, S-104 62 Stockholm 17, Sweden.

Isoprenoids and Alkaloids of Tobacco

By C. R. ENZELL, I. WAHLBERG, and A. J. AASEN, Chemical Research Department, Swedish Tobacco Company, Stockholm, Sweden

With 3 Figures

Contents

Acknowledgements: We are greatly indebted to Ms. Kerstin Karlsson, Ms. Ann-Marie Eklund, Ms. Barbro Bramberg, and Mr. Tommy Öhman, who assisted in the preparation of this manuscript.

I. Introduction

Apart from cotton and food-providing plants, tobacco is the major cultivated plant in the world. It is grown in some one hundred countries and the annual production is of the order of five million tons. In view of its economic importance both in producing and consuming countries, it is not surprising that the chemistry of tobacco has attracted the attention of many investigators. More recently, health aspects and the desire to produce a tobacco substitute to counter future tobacco shortages have given further impetus to chemical research on this plant.

The genus *Nicotiana,* named by Carl von Linné in memory of the French diplomat Jean Nicot who introduced the usage of tobacco into France, comprises more than fifty species but most of the tobacco products are prepared from one species only, *N. tabacum.* This species has, however, not yet been found in the wild state and its origin and evolution are therefore of great interest. The assumed genesis involving chromosome doubling after hybridisation of *N. sylvestris* with a species in the *tomentosae* section of *Nicotiana* has recently been confirmed by Gray *et al.* (*111*). These authors compared the polypeptide components of Fraction I protein from various species and concluded that the putative progenitors of *N. tabacum* (n = 24) are *N. sylvestris* ♀ (n = 12) and *N. tomentosiformis* ♂ (n = 12).

A recent phytochemical study by Reid *et al.* (*46, 225*) has demonstrated that *N. tomentosiformis* produces diterpenoids of the labdane type (I), whereas *N. sylvestris* produces diterpenoids of the thunbergane type (II). They also found that, in general, cultivars of *N. tabacum* synthesize either labdanes or thunberganes, although some cultivars produce trace amounts of the other group. In agreement with this both types are present in commercial Greek (*12, 19*) and Turkish tobaccos (*253*), whereas commercial Burley tobacco seems to produce thunberganoids exclusively (63, 234). While the biosyntheses of some of the tobacco alkaloids and of the cytoplasmic terpenoids, i.e. triterpenoids and sterols, have been studied in considerable detail, little is known about the biogenesis of the tobacco diterpenoids. This is due to the fact that the latter compounds are formed in the trichromes, a site virtually inaccessible to exogenous labelled precursors (225).

In addition to genetic factors and growth conditions, the processing of the leaves after harvest is of vital importance to the final product. During the curing process, which involves air, heat, fire or sun-drying and leads to the creation of the typical tobacco aroma, the tobacco constituents are subjected to various enzymatic, microbial, photochemical and oxidative reactions. Thus, air-curing, which occurs at ambient temperature and is used for Burley tobacco, yields a leaf low in polysaccharides and virtually devoid of monosaccharides, while flue-curing, the other major curing process which takes place at elevated temperatures and is used for Virginia tobacco, gives a leaf low in poly- but rich in monosaccharides. It has also been demonstrated by a recent examination of Virginia tobacco that the compositions of the volatile fractions are drastically affected by the post-harvest treatment (291). This is illustrated by the gas chromatograms shown in Figs. 1 and 2 (Tables 1 and 2), which reveal that curing and aging lead to the formation of a fair number of compounds. Nitrogen-containing constituents and isoprenoids such as carotenoids and diterpenoids are important precursors of the majority of the generated products, a finding particularly understandable in the case of the diterpenoids, since these are present in the gummy exudate of the leaf and hence highly susceptible to light and oxidation (225). The significance of degradation products for the tobacco flavour is reflected by the fact that a large number of them have been patented as tobacco additives. In view of this and despite the fact the volatile fractions contain compounds of widely different biogenetic origin, we have chosen to focus our discussion on tobacco isoprenoids, alkaloids and their degradation products. Since a large number of compounds belonging to these groups has been obtained from tobacco in recent years, it is now possible to forward some hypotheses as to the modes of formation of the nor-compounds; therefore this article is arranged according to the hypothetical biogenetic model. Our intention has been to give a comprehensive compilation of constituents of the groups selected; in

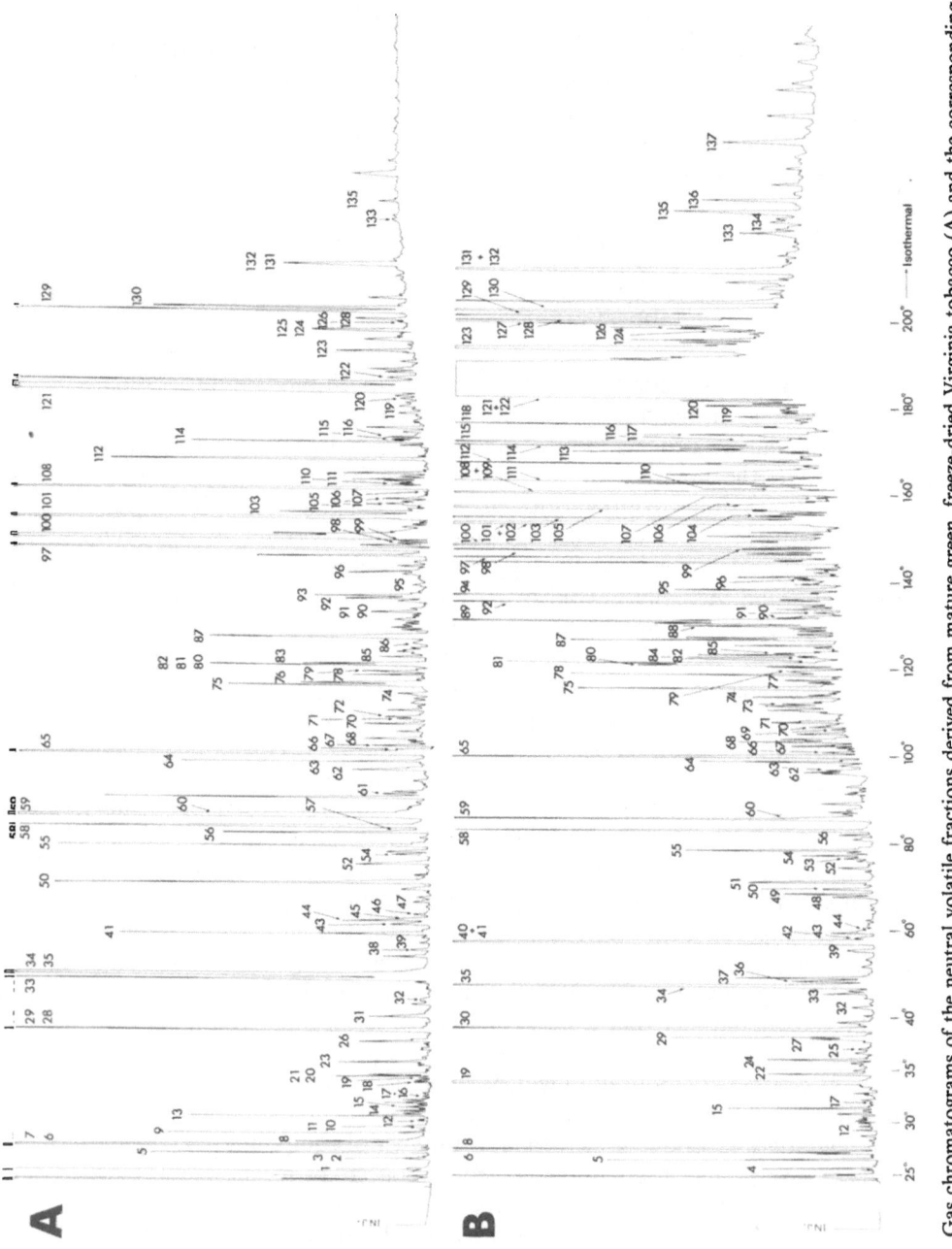

Fig. 1. Gas chromatograms of the neutral volatile fractions derived from mature green, freeze-dried Virginia tobacco (A) and the corresponding flue-cured tobacco aged for 24 months (B). [Glass capillary column (50 m × 0.25 mm) coated with HB 5100]

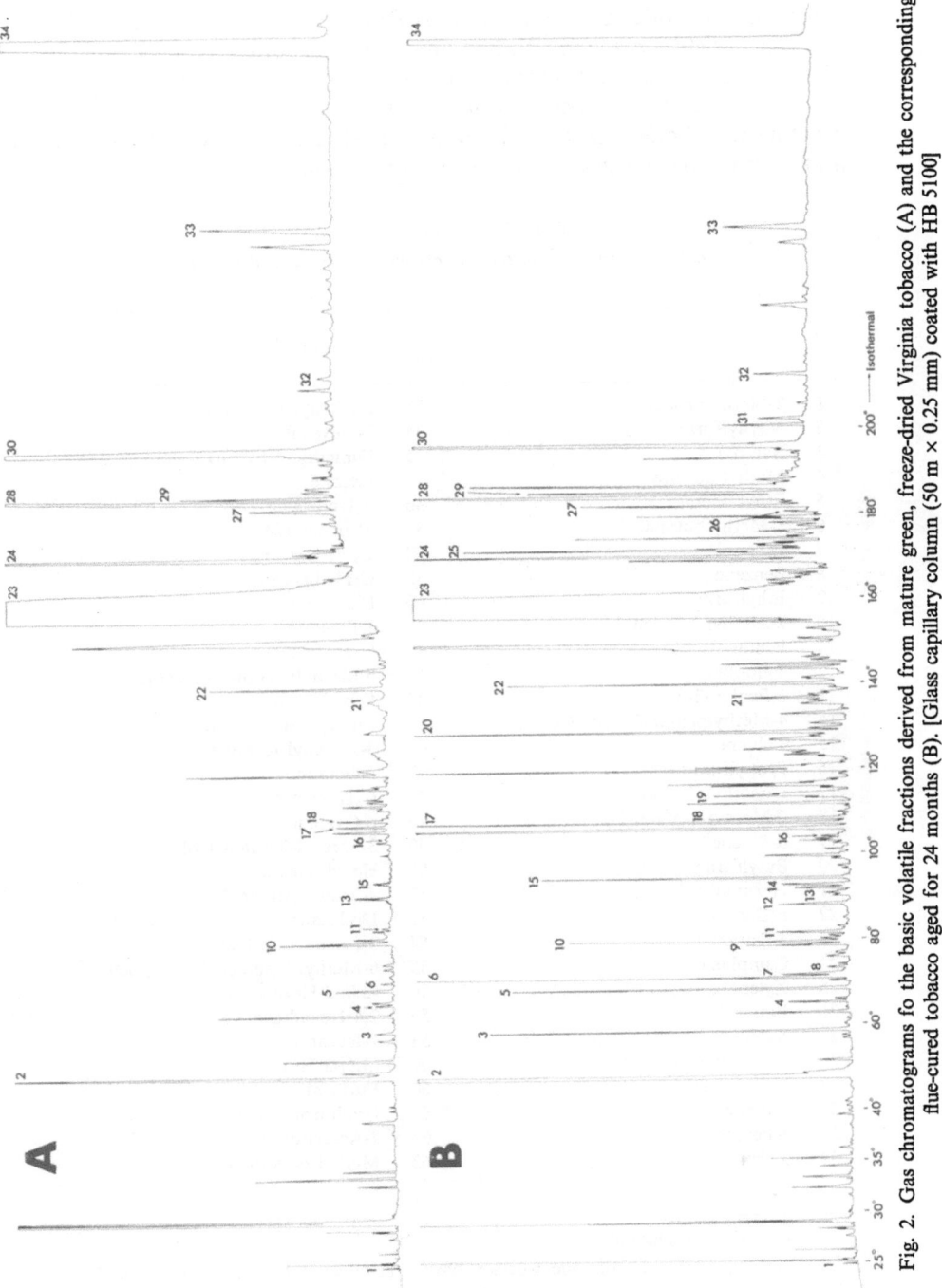

Fig. 2. Gas chromatograms fo the basic volatile fractions derived from mature green, freeze-dried Virginia tobacco (A) and the corresponding flue-cured tobacco aged for 24 months (B). [Glass capillary column (50 m × 0.25 mm) coated with HB 5100]

order to facilitate the presentation Arabic numerals are used for tobacco constituents and Roman numerals for other compounds. In view of the volume of relevant chemical literature it has been necessary to exclude certain aspects, *e. g.* spectroscopic studies, mammalian nicotine metabolism and isoprenoids and alkaloids of tobacco smoke, and to limit the discussion to certain other topics.

Table 1. *Neutral Compounds*
Peak numbers refer to the gas chromatograms shown in Fig. 1

Peak No.	Compound	Peak No.	Compound
1	2-Methylpropanal	33	2-Methyl-1-butanol
2	Methylfuran	34	2-Hexenal
3	Butanal	35	Dimethyl-4-hydroxybutanoic acid lactone
4	Methyl acetate	36	n-Undecane
5	Ethyl acetate	37	Myrcene (**24**)
6	3-Methylbutanal	38	C_3H_7-benzene
7	Ethanol	39	C_3H_7-benzene
8	Benzene	40	Limonene[b]
9	Ethylfuran	41	1-Pentanol
10	2-Methyl-1-buten-3-one	42	C_3H_7-benzene
11	Pentanal	43	6-Methylheptan-2-one (**113**)
12	Thiophene	44	Pentylfuran
13	1-Penten-3-one	45	Methyl 3-hexenoate
14	4-Methylpentan-2-one (**133**)	46	3-Hexenyl formate
15	Toluene	47	3-Octanone
16	Propylfuran	48	C_3H_7-benzene
17	Dimethyldisulfide	49	p-Cymene[b]
18	2-Methyl-2-buten-1-al	50	3-Methyl-2-buten-1-ol
19	α-Pinene[b]	51	Hexyl acetate
20	Butylfuran	52	4-Methylpentan-1-ol
21	Hexanal	53	Dodecane
22	n-Decane	54	3-Hexenyl acetate
23	3-Pentenal	55	6-Methyl-5-hepten-2-one (**110**)
24	Camphene[b]	56	(3 E)-3-Hexen-1-ol
25	Xylene	57	2,4-Hexadienal
26	Butanol	58	1-Hexanol
27	Xylene	59	(3 Z)-3-Hexen-1-ol
28	1-Penten-3-ol	60	Furfural
29	Pentyl acetate	61	Isophorone (**96**)
30	β-Pinene[b]	62	2-Acetylfuran
31	3-Pentenal	63	Methyl octanoate
32	Xylene		

[a] Tentative assignment.

[b] Probably derived from the wooden barrel, in which the tobacco was stored.

References, pp. 66—79

Table 1 (continued)

Peak No.	Compound	Peak No.	Compound
64	1-Heptanol	103	Damascone (**67**)
65	Benzaldehyde	104	2-Hydroxy-5-isopropyl-2-methyl-3-
66	1-Octen-3-ol		nonen-8-one (**175**)
67	2,4-Heptadienal	105	2-Acetylpyrrole (**337**)
68	2-Methyl-2-hepten-6-ol (**121**)	106	1-Decanol
69	Furfuryl acetate	107	Methyl dodecanoate
70	5-Isopropyl-3-hepten-2-one (**195**)	108	Geranylacetone (**109**)
71	Methylfurfural	109	2-Formylpyrrole (**336**)
72	3,5-Octadien-2-one	110	Solanofuran (**174**)
73	Benzonitrile (**382**)	111	8-Hydroxy-5-isopropyl-2-methyl-
74	2-Propylpyrrole[a]		1,3-nonadiene (**172**)
75	2-Methyl-3,6-heptadione (**208**)	112	β-Ionone (**38**)
76	3,5-Octadien-2-one	113	2-Formyl-5-methylpyrrole (**339**)
77	Methyl nonanoate	114	β-Ionon-5,6-epoxide (**39**)
78	Linalool (**25**)	115	1,3,7,7-Tetramethyl-9-oxo-2-oxa-
79	6-Methyl-3,5-heptadien-2-one (**120**)		bicyclo[4.4.0]-dec-5-ene (**59**)
80	Furfuryl alcohol	116	Methyl tridecanoate
81	Methyl benzoate	117	Megastigma-4,7,9-trien-3-one (**51**)
82	1-Octanol	118	1,3,7,7-Tetramethyl-9-oxo-2-oxa-
83	6-Hydroxy-2,2,6-trimethylcyclo-		bicyclo[4.4.0]-decane (**60**)
	hexanone (**88**)	119	1,1,3-Trimethyl-5-hydroxy-3-cyclo-
84	Acetophenone		hexen-2-one (**91**)
85	4-Methyl-4-hydroxy-5-hexenoic acid	120	2,3-Dimethyl-4-hydroxy-2,4-nona-
	lactone (**123**)		dienoic acid lactone
86	Benzyl formate	121	Neophytadiene (**16**)
87	β-Cyclocitral (**83**)	122	Methyl tetradecanoate
88	3-Cyanopyridine (**291**)	123	Megastigma-4,6,8-trien-3-one
89	Estragole[b]	124	Megastigma-4,6,8-trien-3-one
90	Methylfurfuryl alcohol	125	Nicotyrine (**285**)
91	Methyl decanoate	126	Methyl pentadecanoate
92	Benzyl acetate	127	Megastigma-4,6,8-trien-3-one
93	1-Nonanol	128	Hexahydrofarnesylacetone (**111**)
94	α-Terpineol (**270**)	129	Dihydroactinidiolide (**76**)
95	Methyl phenylacetate	130	Megastigma-4,6,8-trien-3-one
96	2-Phenylethyl formate	131	Methyl hexadecanoate
97	Solanone (**171**)	132	Phytofuran (**17**)
98	2-Phenylethyl acetate	133	Ethyl hexadecanoate
99	Methyl undecanoate	134	Phytol (**18**)
100	Benzyl alcohol	135	Farnesylacetone (**108**)
101	2-Phenylethanol	136	3-Hydroxy-damascone (**66**)
102	Damascenone (**65**)	137	3-Oxo-α-ionol (**49**)

[a] Tentative assignment.

[b] Probably derived from the wooden barrel, in which the tobacco was stored.

Table 2. *Basic Compounds*
Peak numbers refer to the gas chromatograms shown in Fig. 2

Peak No.	Compound	Peak No.	Compound
1	Dimethylamine	19	2-Acetylpyridine (**326**)
2	Pyridine (**300**)	20	3-Cyanopyridine (**291**)
3	α-Picoline (**301**)	21	Methyl nicotinate (**324**)
4	2-Methylpyrazine (**361**)	22	3-Acetylpyridine (**296**)
5	2,6-Dimethylpyridine (**306**)	23	Nicotine (**284**)
6	β-Picoline (**302**)	24	N-Methylmyosmine (**287**)[a]
7	γ-Picoline (**305**)	25	2-Formyl-5-methylpyrrole (**339**)
8	2-Ethylpyridine (**303**)	26	3,6,6-Trimethyl-5,6-dihydro-7H-2-
9	2,5-Dimethylpyrazine (**370**)		pyrindine-7-one (**100**)
10	2,6-Dimethylpyrazine (**366**)	27	Nornicotine (**289**)
11	2,3-Dimethylpyrazine (**364**)	28	Myosmine (**288**)
12	3-Ethylpyridine (**304**)	29	Nicotine-N-oxide (**299**)
13	2-Ethyl-6-methylpyrazine (**367**)	30	Nicotyrine (**285**)
14	2,3,6-Trimethylpyridine (**307**)	31	5,6,7,8-Tetrahydroindolizin-8-one
15	2,3,5-Trimethylpyrazine (**372**)		(**349**)
16	2-Ethyl-3,6-dimethylpyrazine (**374**)	32	2,3′-Dipyridyl (**311**)
17	Tetramethylpyrazine (**376**)	33	N-Methylnicotinamide (**297**)
18	3-(1-Propenyl)-pyridine (**323**)	34	Cotinine (**295**)

[a] Tentative assignment.

II. Carotenoids and Acyclic Isoprenoids
1. Carotenoids

Despite their now commonly accepted role as precursors of aroma constituents, the characterisation of the carotenoids in green and aged tobacco has been limited and to date only eleven have been claimed to have been identified. In addition to the four major carotenoids always found in green leaves – β-carotene (**4**), lutein (**6**), violaxanthin (**10**) and neoxanthin (**11**) – these comprise phytofluene (**1**), phytoene (**2**), α-carotene (**3**), cryptoxanthin (**5**), zeaxanthin (**7**), flavoxanthin (**8**) and antheraxanthin (**9**). The amounts and relative abundances of these compounds depend, in addition to several other factors such as leaf position and genetic background, upon the age of the tobacco. Maturation, senescence, curing and aging all involve a decrease which brings the level of the total carotenoids down from about 2000 ppm (w/w dry leaf) in the immature green leaf to about 100 ppm in the aged material (*233*). β-Carotene (**4**) and lutein (**6**) usually predominate and, according to a recent study by WHITFIELD and ROWAN (*301*), represent 15–30% and 40–60% respectively of the total carotenoids in

(1) (304)

(2) (304)

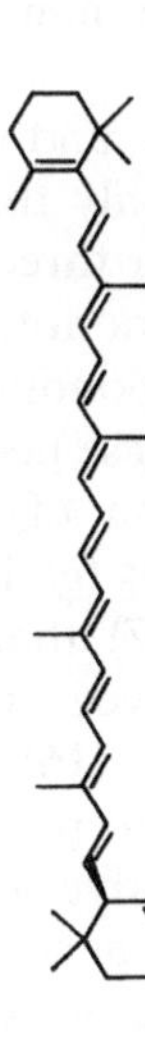
(3) (304)

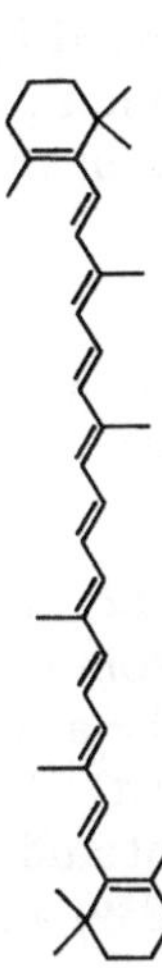
(4) (304)

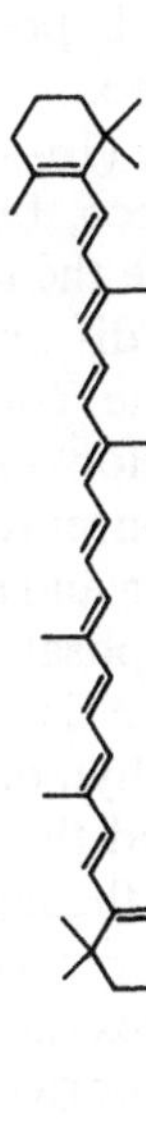
(5) (304)

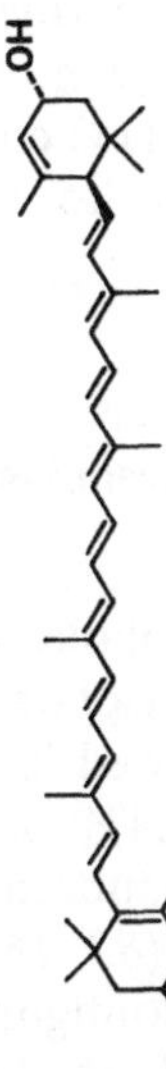
(6) (304)

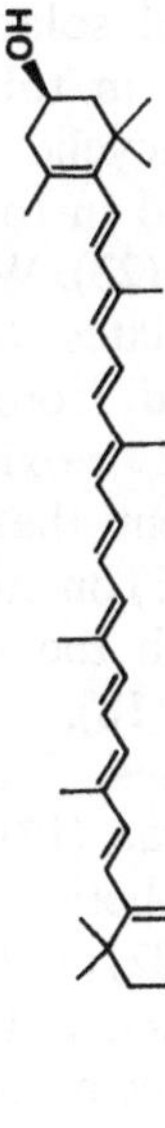
(7) (304)

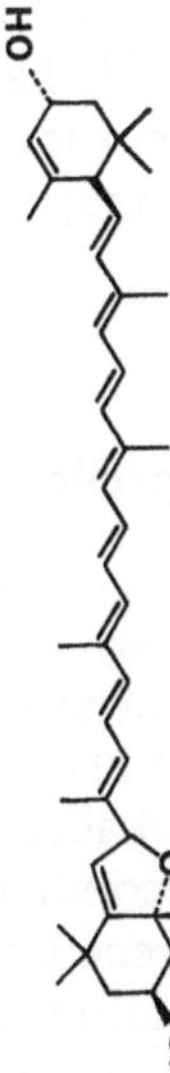
(8) (210)

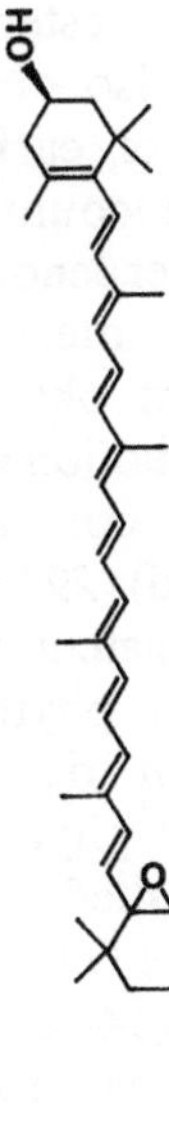
(9) (301)

(10) (279, 280)

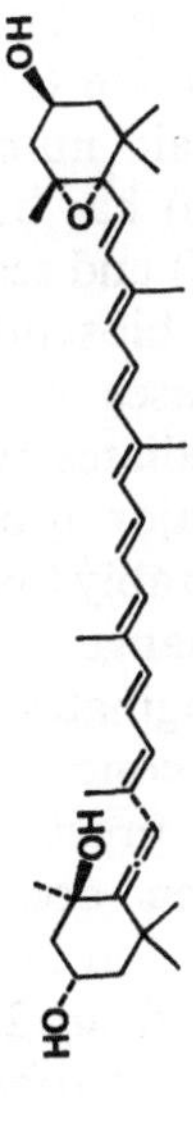
(11) (279, 280)

senescent tobacco leaves (Virginia Gold), while the epoxycarotenoids encountered (**9, 10, 11**) each account for 5–20%. Only one carotenoid, antheraxanthin (**9**), was found to be generated during this phase. In aged Burley tobacco, further, as yet unidentified carotenoids and several *cis*-isomers of known carotenoids have been observed by Wright *et al.* (*304*). Violaxanthin (**10**) and neoxanthin (**11**) could not be detected in this tobacco, although these were major xanthophylls in the corresponding fresh green material.

2. Acyclic Isoprenoids

The all-*trans* nonaprenyl alcohol solanesol (**12**), now regarded as a ubiquitous plant component, was first isolated from flue-cured tobacco by Rowland *et al.* (*241*), who also showed it to be a major constituent of green, cured and aged tobacco (0.4% dry wt). It was originally characterised as a decaprenyl alcohol, but subsequent studies demonstrated the presence of only nine isoprene units (*82, 163, 239*) a result confirmed by total synthesis (*244*). The all-*trans* configuration was established by ^{1}H-NMR (*84*) and X-ray diffraction (*244*), the latter method also showing that the two forms of solanesol observed in the solid state and differing in melting point and spectral characteristics are represented by a planar conformation (β) and one in which successive isoprene units are twisted relative to one another (α). The acetate (**13**) and a number of fatty acid esters of solanesol, predominantly the palmitate and linolate (*240*), also occur in tobacco.

A fair number of carboacyclic di- (**16—22**) and monoterpenoids (**24—30**) has been encountered in tobacco, but only three aliphatic tri- (**14—15**) and sesquiterpenoids (**23**). While the latter three compounds are regular biosynthetic intermediates, the di- and monoterpenoids are in most cases products obtained from the corresponding C_{20} and C_{10} intermediates by reduction and/or oxidation. Of these neophytadiene (**16**) is a major tobacco constituent the concentration of which increases considerably (fivefold) (*291, 303*) on curing and aging (*cf.* Fig. 1, peak 121). This change is consistent with the proposal (*237*) that neophytadiene (**16**) originates from phytol (**18**), which is derived from chlorophyll whose concentration decreases. Neophytadiene (**16**) is in turn an obvious precursor of phytofuran (**17**) and the corresponding conversion has been accomplished synthetically through dye-sensitized photo-oxygenation (*98*). The ketone (**22**) (*182*) is probably generated in a similar manner from 2,6,10,14-tetramethyl-10-hexadecene, a compound whose presence in tobacco has not yet been reported.

C$_{45}$

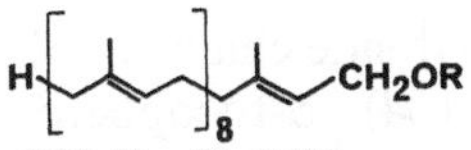

H$\left[\right]_8CH_2$OR

(12) R = H (*241*)
(13) R = COCH$_3$ (*239*)

C$_{30}$

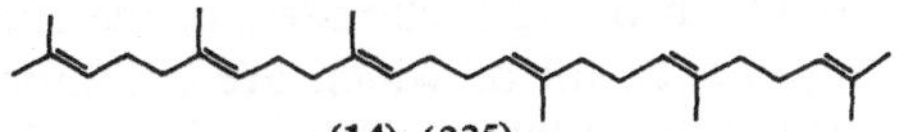

(14) (*225*)

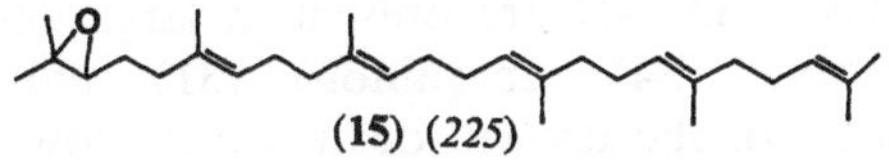

(15) (*225*)

C$_{20}$

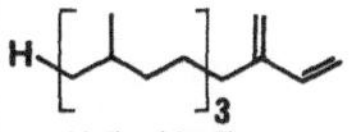

H$\left[\right]_3$

(16) (*237*)

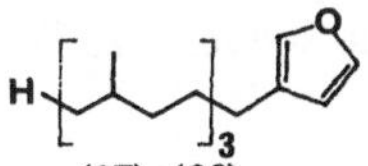

H$\left[\right]_3$

(17) (*98*)

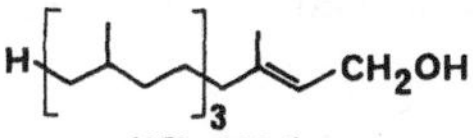

H$\left[\right]_3CH_2$OH

(18) (*291*)

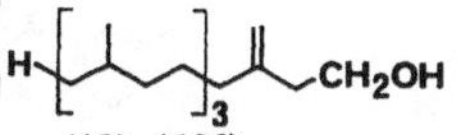

H$\left[\right]_3CH_2$OH

(19) (*182*)

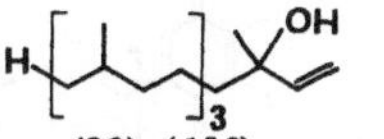

H$\left[\right]_3$

(20) (*182*)

H$\left[\right]_3$COOCH$_3$

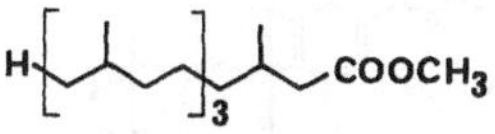

(21) (*151*)

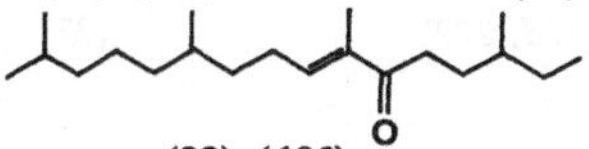

(22) (*182*)

C$_{15}$

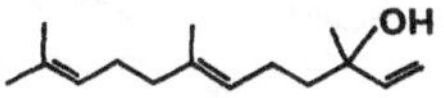

(23) (*182*)

C$_{10}$

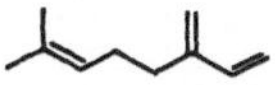

(24) (*182*)

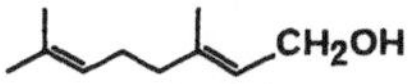

CH$_2$OH

(25) (*63*)

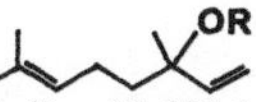

OR

(26) R = H (*234*)
(27) R = COCH$_3$ (*234*)

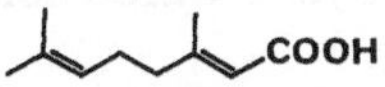

COOH

(28) (*64*)

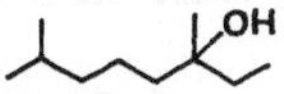

OH

(29) (*182*)

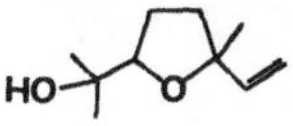

HO

(30) (*63*)

3. Meroterpenoids

Several meroterpenoids whose terpenoid portions are composed of ten (**31**), nine (**32, 33**) or four (**34—36**) all head-to-tail linked isoprene units have been isolated from tobacco. Moreover, GRIFFITHS *et al.* (*113*) have encountered two sets of monohydroxylated plastoquinones-9 (plastoquinones-C and -D) each comprising three representatives which all differ from plastoquinone-9 (**32**) in having an additional hydroxyl group inserted in one of the seven non-terminal isoprene units of the nonaprenyl side chain. Results obtained from tracer experiments support the view that solanesol (12) and the prenyl side chains of plastoquinone-9 (32), plastoquinone-C, phylloquinone (34), α-tocopherol (35) and α-tocopherolquinone (36) are biosynthesized from mevalonic acid in the chloroplast, while the prenyl side chain of ubiquinone-10 (31) is synthesized elsewhere, probably in the mitochondria of the leaf tissue (113). The levels of these components, all of which are commonly present in higher plants, increase as the green leaves mature. While plastoquinone-9 (32) and plastoquinones-C and -D are present in large amounts during senescence, phylloquinone (34), ubiquinone (31), chlorophyll and β-carotene (4) disappear. In the brown, dead leaves, however, the concentrations of the phylloquinones is reduced to less than half those observed in the senescent leaves (113).

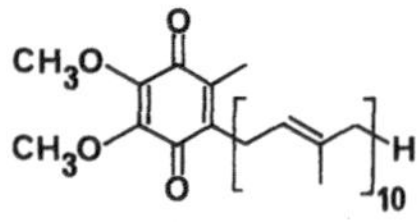

(31) (113) 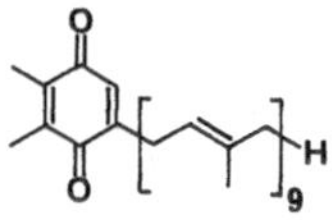   (32) (113, 277) 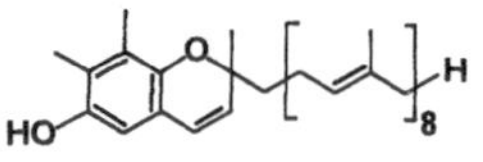   (33) (163, 238)

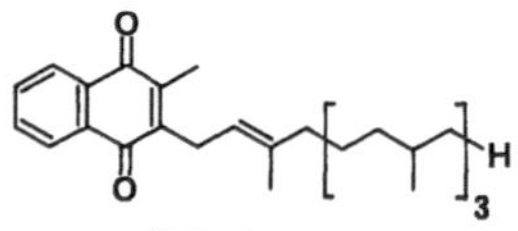

(34) (113) 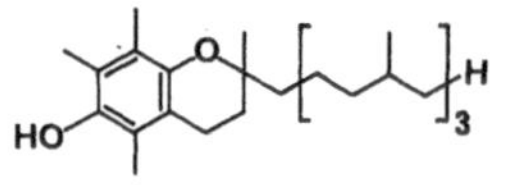   (35) (113, 238) 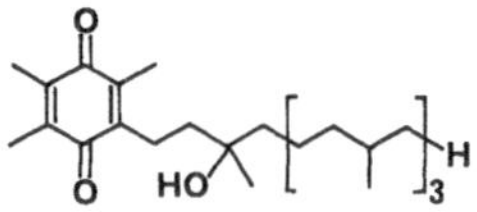   (36) (113)

4. Nor-Compounds Derived from Cyclic Carotenoids

a) General Considerations

As pointed out by Enzell (79) in 1969, a large number of the tobacco flavour constituents can be viewed as carotenoid degradation products formed by oxidative cleavage of the polyene chain. Subsequent work has lent support to this view. Thus an increasing number of new compounds conforming to the expected pattern has been found (cf. 37—99), reactions of this type have been shown to take place in plant material and chemical degradations of individual compounds have been carried out under well defined oxidative conditions.

An inverse relationship between flavour and carotenoid content of tobacco leaves has recently been demonstrated by Roberts et al. (233) and explained as being due to a more extensive degradation of carotenoids in the leaves which are richer in flavour – a finding in line with the earlier proposal (79) that tobacco carotenoids may provide a rationale for the use of leaf colour in evaluating leaf quality. More direct evidence has been presented by STEVENS (276), who has shown that the carotenoids in tomatoes are rapidly degraded enzymatically when the tissue is damaged and that there is a high correlation between the concentration of certain degradation products and the content of their presumed carotenoid precursors. SANDERSON et al. (246) were able to show that a tea enzyme preparation in the presence of tea flavanols converted ^{14}C-labelled β-carotene into β-ionone, the main product, and several other, unidentified volatile substances, some of which must be derived exclusively from the central part of the polyene chain. Furthermore, soy bean lipoxygenase is known to effect the conversion of violaxanthin (10) into 3-hydroxy-β-ionone epoxide (41) (86).

Singlet oxygen, which can be generated even with near-infrared irradiation in the presence of an appropriate sensitizer (201), is known to effect oxidative cleavage of a conjugated double bond. OHLOFF (206) and others (24, 40, 148) have suggested that such oxidations, which are proposed to involve shortlived intermediates such as A, B or C (Scheme 1), proceed in a manner similar to biological oxidations and hence have a bearing on this type of reaction.

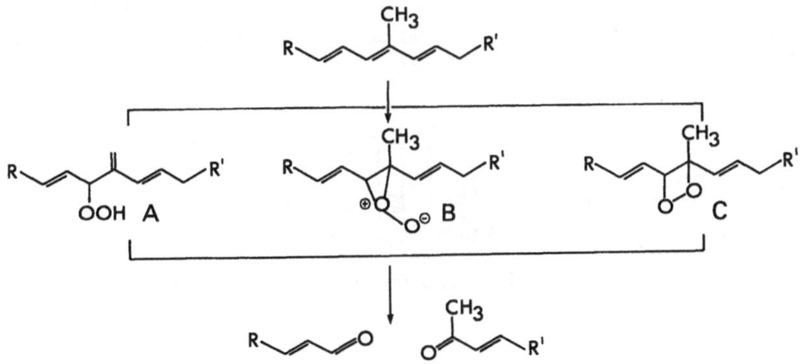

Scheme 1

It is thus of interest to note that exhaustive photo-oxygenation of β-carotene (4) yields desoxyxanthoxin (III), β-ionone (38), dihydroactinidiolide (76), 6-hydroxy-2,2,6-trimethylcyclohexanone (88) and geronic acid (IV), while the corresponding hydroxy derivative, zeaxanthin (7), when subjected to the same treatment, furnishes 3-hydroxy-β-ionon-5,6-epoxide (41), loliolide (77), isololiolide (V), 3-hydroxy-β-cyclocitral (VI), hydroxygeronic acid (VII) and 2,4-dihydroxy-2,6,6-trimethylcyclohexanone (VIII) (134). Under less drastic photo-oxidative conditions, β-carotene (**4**) gives only three of the above products (**38, 76** and **88**). These compounds have also been obtained, although in different proportions, on air oxidation of β-carotene (**4**) (*135*), a reaction also reported to afford β-ionon-5,6-epoxide (**39**) (*22*). Photo-oxygenation of violaxanthin (**10**) gives xanthoxin (**IX**), 3-hydroxy-β-ionon-5,6-epoxide (**41**) and loliolide (**77**) (*35*).

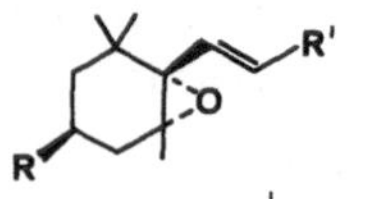

(**III**) R = H, R′ =

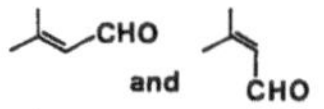

and

(**IX**) R = OH, R′ =

and

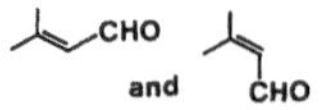

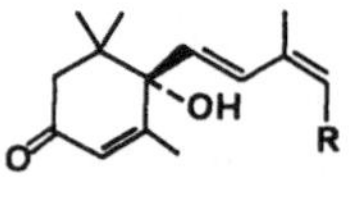

(**37**) R = COOH (*274*)
(**X**) R = CHO

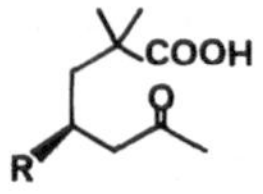

(**IV**) R = H
(**VII**) R = OH

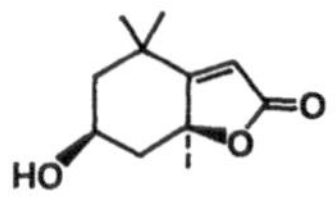

(**V**)

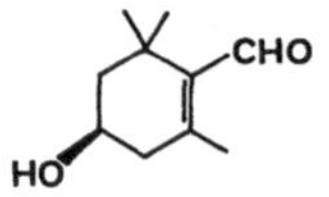

(**VI**)

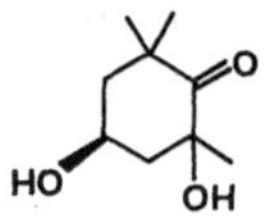

(**VIII**)

Consistent with this, virtually all tobacco constituents under consideration are derivable from cyclic carotenoids through cleavage of the 9,10, 8,9, 7,8 or 6,7 bonds (*cf*. Fig. 3).

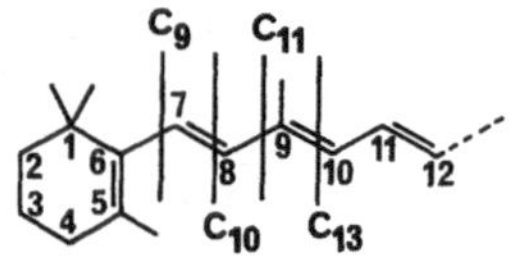

Fig. 3. Bond fragmentations commonly encountered in cyclic carotenoids

b) C₁₅ Constituent

The only compound encountered so far which comprises a larger fraction of the polyene chain is abscisic acid (**37**), which is more probably a genuine sesquiterpenoid (*193, 194*). Its formation from tobacco carotenoids is, however, by no means an improbable process. Thus oxidation of xanthoxin (**IX**), a potent growth inhibitor obtained by photo-oxidation of violaxanthin (**10**), yielded a mixture of abscisic aldehyde (**X**) and *trans*-abscisic aldehyde, which was converted to *trans*-abscisic acid and abscisic acid (**37**) on further oxidation (*35*).

c) C_{13} Constituents

All but two of the C_{13} ketones (**38—42**) derivable by simple oxidative cleavage of the 9,10 double bond of the tobacco carotenoids (**1—11**) have been detected in tobacco. The exceptions are the grasshopper ketone (**XI**), which comprises the allenic end group of neoxanthin (**11**), and 3-hydroxy-α-ionone (**XII**), which originates from lutein (**6**) and flavoxanthin (**8**). The absence of these two ketones (**XI, XII**) in tobacco may be fortuitous or, more likely, depend upon the fact that they are converted into more stable products at a higher rate than the others (**38—42**). Although alternative routes to some of the latter C_{13} ketones (**38—42**) are obvious, they may conveniently be regarded as primary degradation products and may be converted to a series of related tobacco constituents (**43—58**) by simple reactions such as reduction of double bonds, dehydration and interconversion of oxo and hydroxyl functions. Thus, β-ionone (**38**) could give rise to compounds (**43**) and (**44**) and 3-hydroxy-β-ionone (**40**) to the ketones (**45—47**). Similarly, 3-hydroxy-α-ionone (**XII**, *vide supra*) is a probable precursor of a group of ketones (**48—57**) which includes several flavour components (**51—55**). The concentrations of these flavour components are drastically increased during the post-harvest treatment of the tobacco (*cf.* Fig. 1, peaks 117, 123, 124, 127 and 130).

To the limited extent to which the absolute configurations of these compounds are established, they support the assumed precursor-product relationships*. Moreover, the stereochemistry of (3S,5S,6R)-5,6-epoxy-3-hydroxy-7-megastigmen-9-one (**58**) indicates that it is formed from (3R)-3-hydroxy-β-ionone (**40**) in an epoxidation reaction, which is either non-stereospecific or directed by the hydroxyl group. Alternatively, it may be derived from an appropriate carotenoid precursor which subsequently suffers a 9,10 bond cleavage.

* The frequent use of GC and GC-MS techniques for identification is the main reason for the lack of information regarding the absolute configurations of these compounds, despite the fact that in many cases they have been established for the reference materials. Further knowledge about the stereochemistry would greatly aid clarification of the genetic relationships.

Many of the C_{13} compounds are clearly formed by more complex routes which make the precursor-product relationships less obvious and, in the absence of isotopic experiments, leave the numerous chemical studies on nor-carotenoids as the sole source of information about these relationships. It is thus noteworthy that the pyran derivatives (59) and (60), whose concentrations are increased manifold during the post-harvest treatment (cf. Fig. 1, peaks 115 and 118), have been obtained by acid treatment of the alcohols (49) and (50) (236), their probable biological precursors. Similarly, the bicyclic dienone (61) has been produced by internal aldol condensation of the diketone (XIII) (230), the corresponding hydroxyketone (50) of which has been isolated from tobacco.

C₁₃

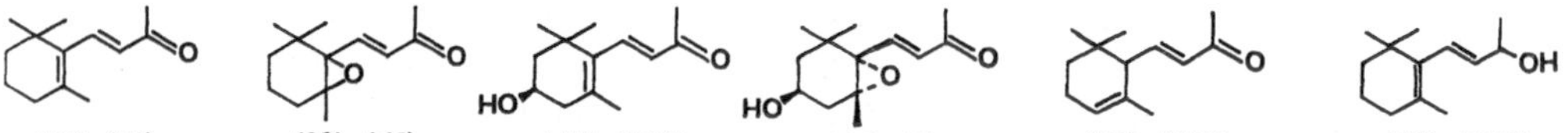

(38) (62)	(39) (62)	(40) (100)	(41) (1)	(42) (253)	(43) (182)
(44) (123)	(45) (63)	(46) (182)	(47) (233)	(48) (182)	(49) (14, 16)
(50) (7)	(51) (13, 234)	(52) (13, 234)	(53) (13, 234)	(54) (13, 234)	(55) (13, 234)
(56) (182)	(57) (233)	(58) (1)	(59) (63)	(60) (234)	(61) (234)
(62) (234)	(63) (151)	(64) (182)	(65) (62)	(66) (142)	(67) (62)
(68) (99)	(69) (142)	(70) (63)	(71) (63)	(72) (69)	(73 (234)

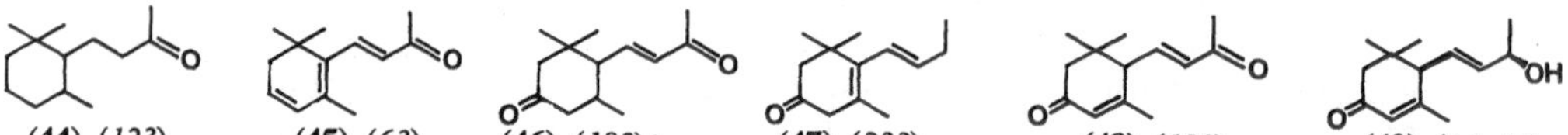

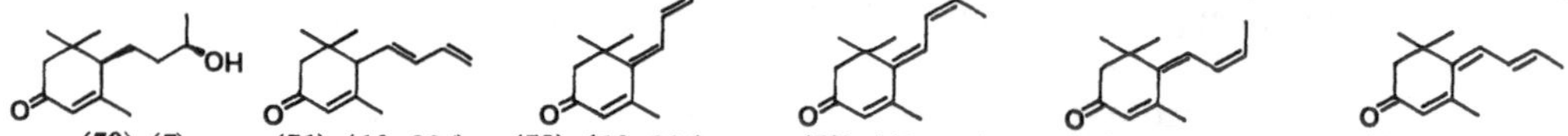

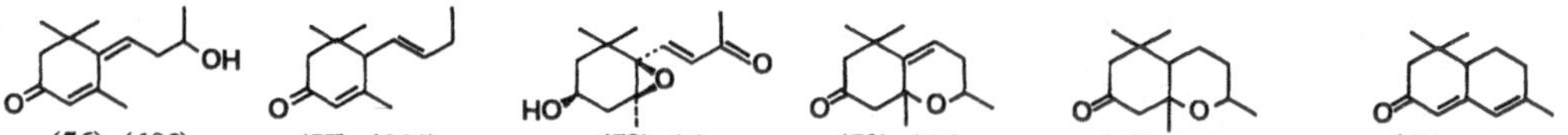

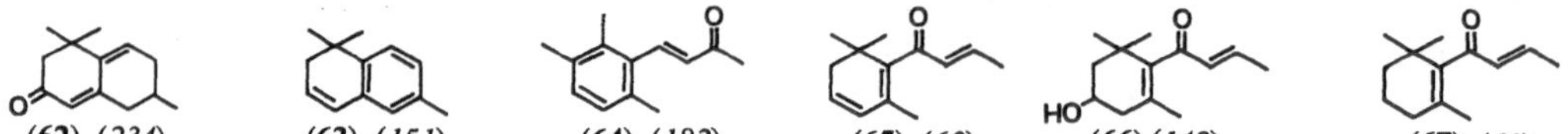

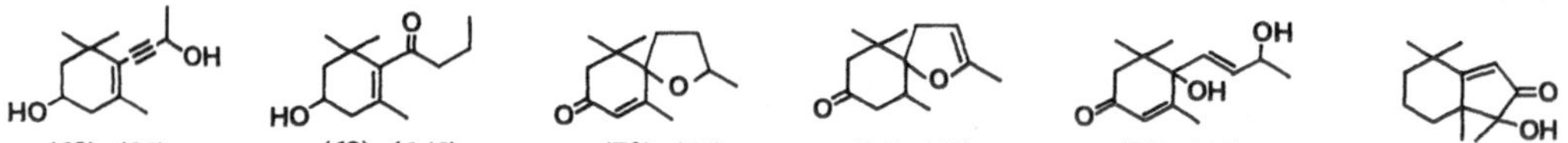

No close model has yet been presented for the formation of the isomeric bicyclic dienone (62) from any of the other known tobacco constituents, unless it merely represents a photochemically more stable isomer of the dienone (61). It may, however, be noted that (XIV), an experimentally well documented (222) intermediate in the photoconversion of 3,4-dehydro-β-ionone (45) to the dihydronaphthalene derivative (63), could also yield the dienone (62) by a reaction sequence comprising reduction to the corresponding alcohol (XV) and cyclisation of this to (**XVI**) followed by oxidation (Scheme 2). The dihydronaphthalene (**63**) has also been obtained from β-ionone epoxide (**39**) by the action of aqueous formic acid. Among other products, this reaction also furnished a rearranged compound, 4-(2,3,6-trimethylphenyl)-butan-2-one (**XVII**), a dihydro derivative of the very recently encountered tobacco constituent (**64**) (*275*). Further insight into the possible genesis of the latter compound (**64**) is provided by the observation that, in addition to the main product (**XVIII**), damascenone (**65**) afforded the isomer (**XIX**) on selenium dioxide oxidation (*250*). The phenylbutenone (**64**) may also be a degradation product of an aromatic carotenoid such as okenone (**XX**).

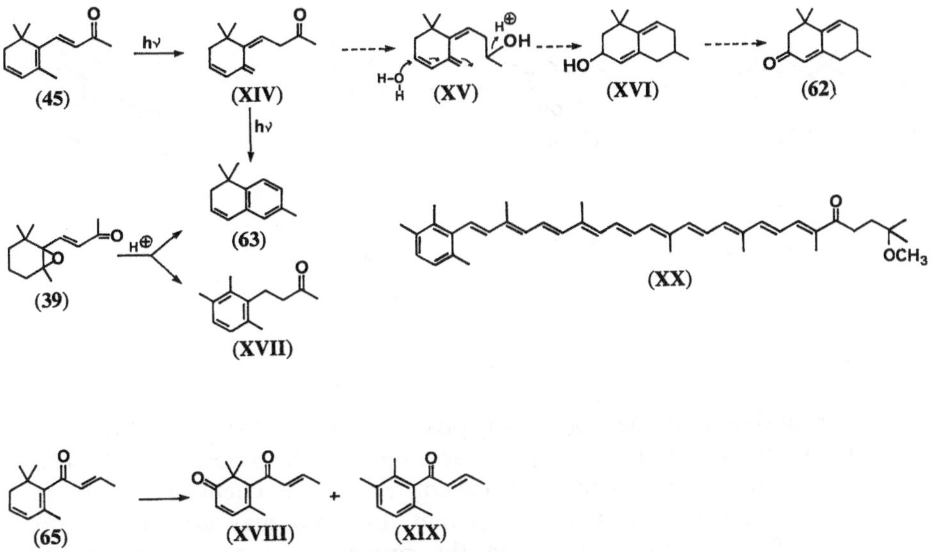

Scheme 2

On the basis of chemical studies and biogenetic consideration, OHLOFF et al. (*207*) and ISOE et al. (*137*) have shown that damascenone (**65**) as well as some of the other C$_{13}$ tobacco constituents (**66, 69**) can be regarded as derived from neoxanthin (**11**) *via* the grasshopper ketone (**XI**) and the allenic triol (**XXI**) (Scheme 3). Although none of these steps have been carried out experimentally, support is obtained from an analogous set of reactions involving the conversion of the allenic diol (**XXII**), obtained from β-ionol (**43**) by $^1O_2$ oxidation, to damascone (**67**) (*136*). An alternative route to the 7-oxo compounds, which is of considerable interest in view of the ready conversion of allenic carotenoids into their acetylenic counterparts (neoxanthin → diadinochrom) (*77*) and the recent isolation of the acetylenic diol (**68**) from tobacco, is that implied by OHLOFF's biomimetic synthesis. This involves the acid catalysed conversion of the trihydroxy-acetylene (**XXIII**) to damascenone (**65**) and 3-hydroxydamascone (**66**) (*207*). Hydride reduction of the latter furnished another tobacco consti-tuent, the hydroxyketone (**69**) (*207*). Damascenone (**65**) and damascone (**67**) are important flavour components, the levels of which are increased on curing and aging of tobacco (*cf.* Fig. 1, peaks 102 and 103) (*291*).

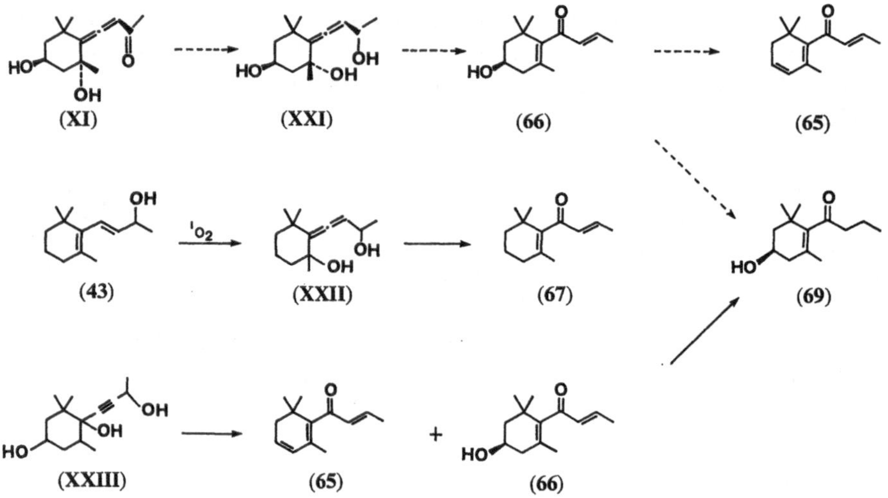

Scheme 3

In addition to the above 5,6-epoxides (**39, 41, 58**), only three C_{13} norcarotenoids having an oxygen substituent in the 6-position (**70—72**), have so far been found in tobacco. From the chemistry of these and related compounds, it is apparent that they may arise in several ways, of which those involving the primary C_{13} precursors are of particular interest. Thus Sarett oxidation (*197*) of 3-hydroxy-β-ionon-5,6-epoxide (**41**) and t-butyl chromate oxidation (*232*) of α-ionone (**42**) afford dehydrovomifoliol (**XXIV**), which contains a 6-hydroxyl group and is a genetically obvious and synthetically verified (*298*) precursor of blumenol A (**72**) (Scheme 4). Oxygenation of the 6-position can, however, also be effected by photo-oxygenation of related 5-enes and 3,5-dienes, the former yielding 4- and 5(5′)-en-6-ols *via* the corresponding 6-hydro-peroxides (*198, 251*) and the latter furnishing 3-hydroxy- or 3-oxo-4-en-6-ols *via* 3,6-epidioxides (*48, 198, 250*). It seems highly probable that blumenol A (**72**) can in turn give rise to blumenol B (**XXV**) and to the important flavour compound theaspirone (**70**). These transformations have been effected synthetically by catalytic reduction to blumenol B (**XXV**) and subsequent cyclisation of the corresponding secondary monomesylate, a reaction involving stereospecific inversion at C-9 (*298*). Theaspirone (**70**) has also been obtained from blumenol B (**XXV**) by dehydration (*116*). The other spiroether (**71**) may arise in a similar manner or possibly by a process analogous to the photochemical conversion of 6-hydroxy-α-ionone (**XXVI**) to the spiroether (**XXVII**) (*296*).

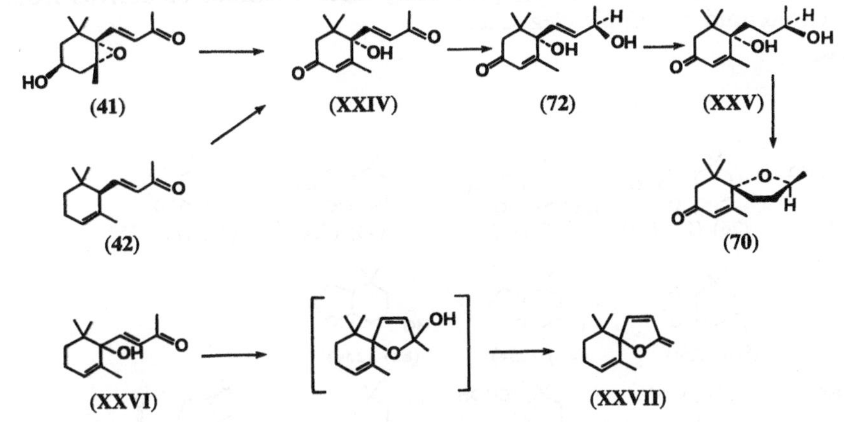

Scheme 4

The generation of the C_{13} ketol (**73**) from β-ionone (**38**) has been achieved by the biomimetic route indicated in Scheme 5 (*166*). Thus air oxidation of the pyran (**XXVIII**), produced on irradiation of β-ionone (**38**), yields the diol (**XXIX**) and, obviously by further oxidation of this, the aldehyde (**XXX**) and dihydroactinidiolide (**76**). Subsequent dehydration of the diol (**XXIX**) gives the ketol (**73**), probably by the mechanism given below (*166*).

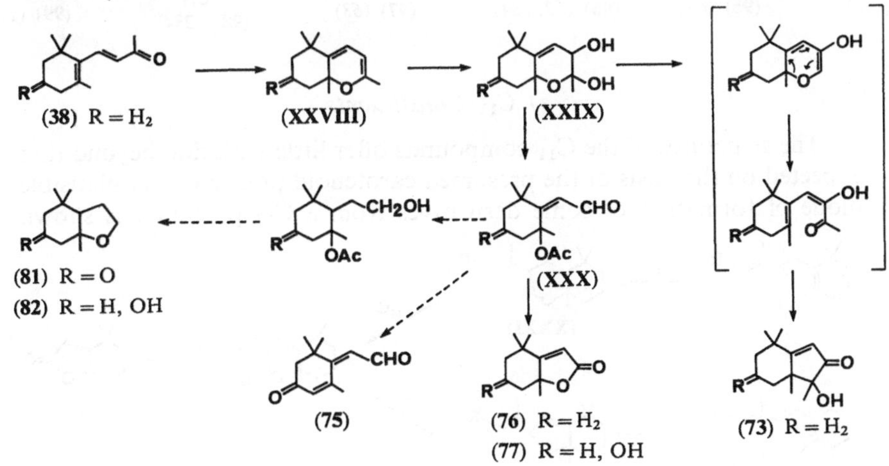

Scheme 5

2*

d) C_{12} Constituent

Only one C_{12} compound derivable from cyclic carotenoids, the cyclohexenone (74), has so far been obtained from tobacco. Although there is at present no indication as to its genesis, it seems likely that it is formed from a C_{13} precursor, since it cannot be derived from a carotenoid by simple cleavage.

C_{12}

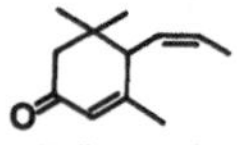

(74) (182)

C_{11}

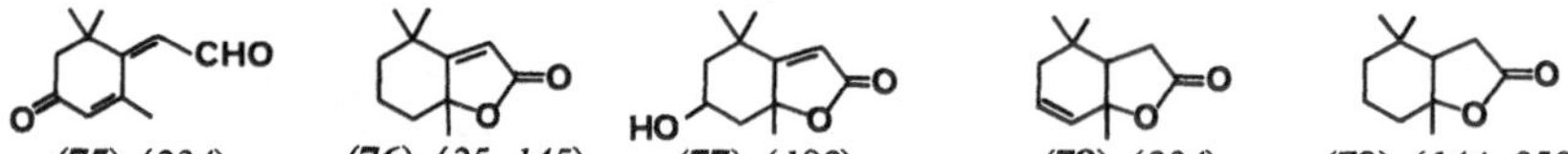

(75) (234)   (76) (25, 145)   (77) (182)   (78) (234)   (79) (144, 252)

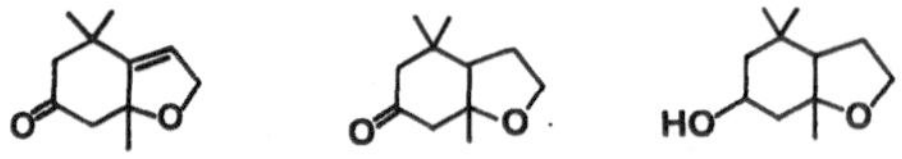

(80) (234)   (81) (234)   (82) (234)

C_{10}

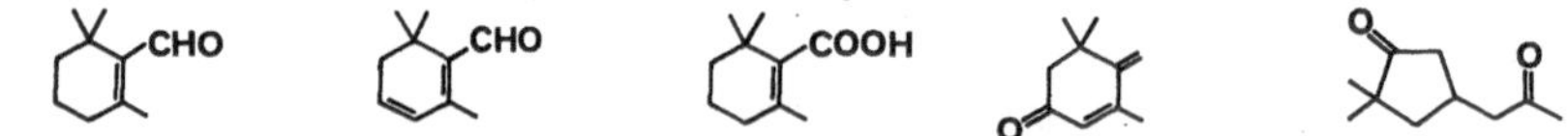

(83) (152)   (84) (152)   (85) (64)   (86) (234)   (87) (65)

C_{9}

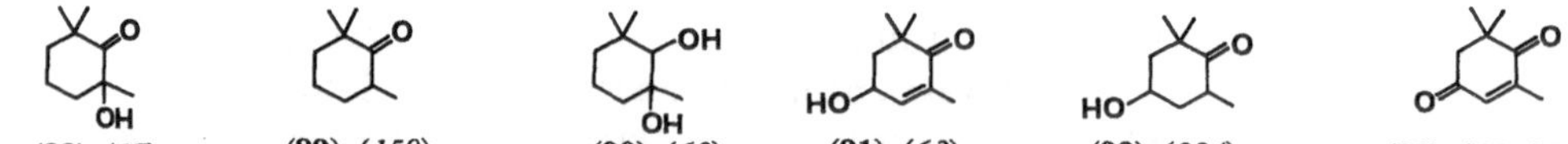

(88) (17)   (89) (150)   (90) (63)   (91) (63)   (92) (234)   (93) (63, 234)

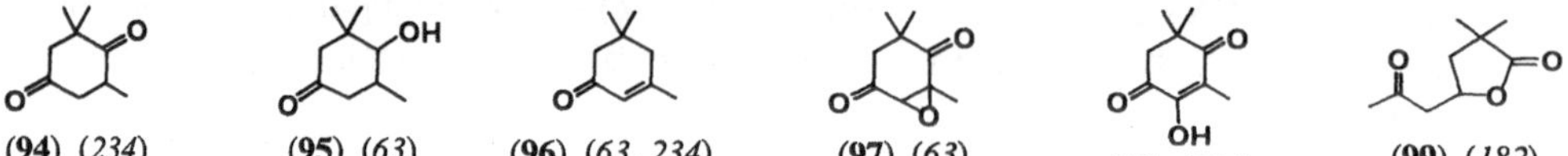

(94) (234)   (95) (63)   (96) (63, 234)   (97) (63)   (98) (234)   (99) (182)

e) C_{11} Constituents

The structures of the C_{11} compounds offer little variation beyond that expected on the basis of the presumed carotenoid precursors. A plausible mode of formation of some derivatives from a C_{13} precursor is shown

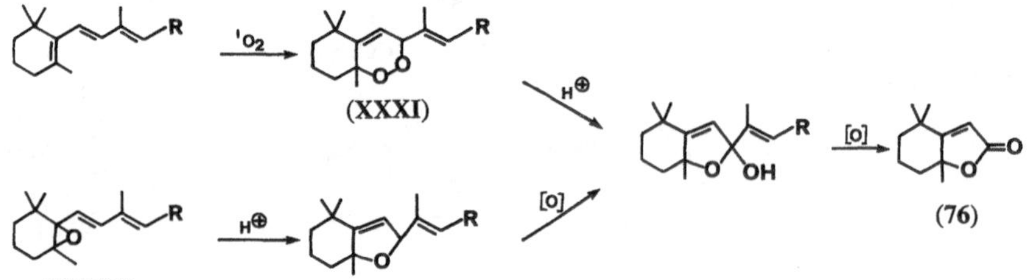

(XXXI)

(XXXII)   Scheme 6

(76)

in Scheme 5. However, alternative routes are likely to be of importance, e.g. via the initial formation of 5,8-epidioxides (**XXXI**) or 5,6-epoxides (**XXXII**) which can rearrange to 5,8-epoxides yielding C_{11} lactones on further oxidation (Scheme 6) (*88, 198, 255*).

f) C_{10} Constituents

Of the five C_{10} compounds (**83—87**) the genesis of only two (**86** and **87**) is ambiguous. The first of these (**86**), possessing an unaltered carbon skeleton, might be derived from the aldehyde (**84**), as illustrated in Scheme 7. The aldehyde is initially converted to the corresponding alcohol (**XXXIII**), which due to steric crowding may suffer a vinylogous allylic shift to the alcohol (**XXXIV**), yielding on subsequent oxidation the ketone (**86**). A route to the rearranged compound (**87**), which also accounts for the formation of the related tobacco alkaloid (**100**) and which involves acyclic intermediates, has recently been proposed by DEMOLE et al. (*65*) and is depicted in Scheme 7. This also details the proposed biogenesis of the other related tobacco alkaloid (**101**). Further variations on this theme are to be expected, since acyclic compounds closely related to the assumed intermediates have been encountered in a number of instances, e. g. on acid treatment as well as on irradiation of β-ionon-5,6-epoxide (**39**) (*258, 275, 295*) and 5,6-dihydro-β-ionol (*259, 296*), on heating of the 6,9-epidioxide of the pyran (**XXVIII**) (*83*) and on treatment of the 3,6-epidioxide of methyl β-ionylidenacetate with a ferrous salt (*198*).

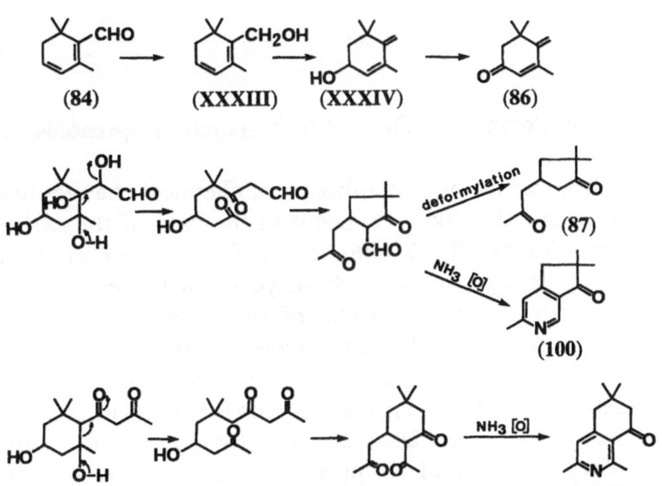

Scheme 7

g) C₉ Constituents

A major route to the C_9 nor-carotenoids (**88—99**) is indicated by the ready conversion of the hydroperoxides (**XXXVI**) and (**XXXVII**), derived from damascol (**XXXV**), to the cyclohexenones (**XXXVIII**) and (**XXXIX**) (*251*) (Scheme 8). Support is provided by the fact that stable 6-hydroperoxides have also been obtained from methyl β-ionylidenacetate and the corresponding 3-methoxy derivative (*198*). Subsequent reactions such as oxidation, reduction and hydration serve to explain the structural variation observed for the C_9 tobacco compounds. In contrast to the higher nor-carotenoids, the present group comprises two members oxygenated at C-4 (**97, 98**). Although these may arise by epoxidation of the corresponding 4-ene-3,6-dione (**93**), the fact that α-ionone (**42**) on $^1O_2$ oxidation followed by triphenyl phosphine reduction yields 4-hydroxy-α-ionone (*198*) implies that an alternative route to (**98**) involving two consecutive attacks of $^1O_2$ on a $\Delta^{4:5}$-precursor might merit consideration.

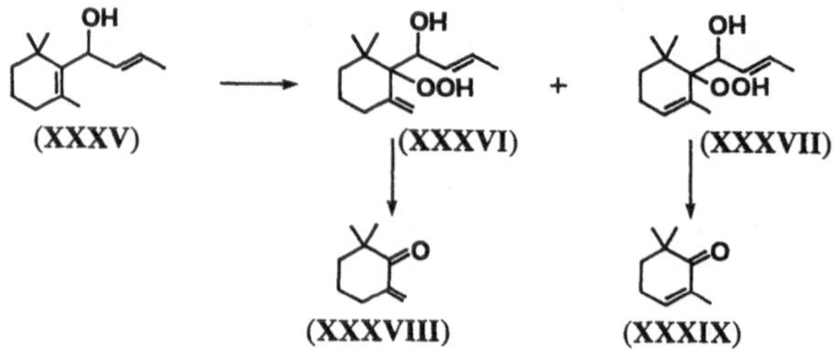

Scheme 8

5. Nor-Compounds Derived from Acyclic Isoprenoids

Oxidative fragmentation and subsequent chemical changes of the type discussed above are also likely to occur in the case of the carbo-acyclic tobacco isoprenoids (**1—2, 12—36**), and a fair number of the tobacco constituents may be regarded as carbo-acyclic nor-terpenoids (**102—144**) generated in this manner. The origin of those possessing less than ten carbon atoms may be questioned, however, since in most cases an appropriately placed methyl group represents the only criterion for the assumed genetic relationship. In view of this it should be noted that the C_8 compounds (**110, 113, 120—122**) and two of the C_7 compounds, (**123**) and (**124**), correspond to products generated by cleavage of a prenyl double bond in the normal position.

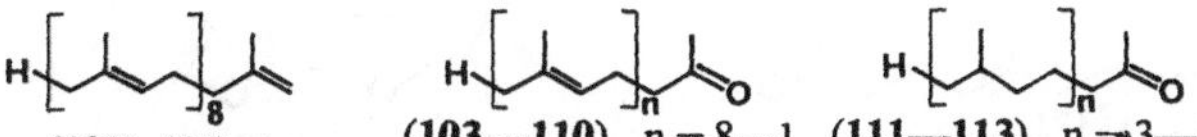

(102) *(80)*   **(103—110)**, n = 8—1 *(62, 139, 152)*   **(111—113)**, n ⇒ 3—1 *(62, 152)*

C_{16}

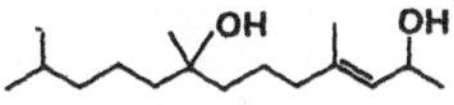

(114) *(182)*

C_{13} **(115)** *(152)*   **(116)** *(63)*   **(117)** *(21)*   **(118)** *(21)*

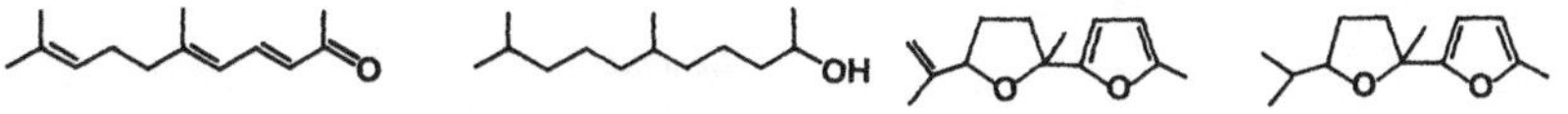

C_{11} **(119)** *(63)*

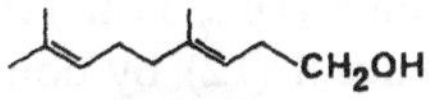

C_8 **(120)** *(152)*   **(121)** *(234)*   **(122)** *(253)*

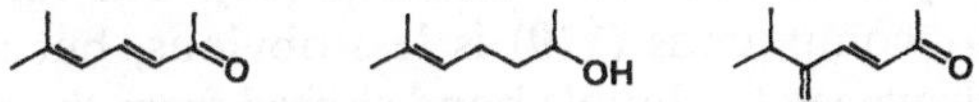

C_7 **(123)** *(234)*   **(124)** *(234)*   **(125)** *(234)*   **(126)** *(182)*

C_6 **(127)** *(63)*   **(128)** *(63)*   **(129)** *(102)*   **(130)** *(102)*   **(131)** *(50)*   **(132)** *(249)*   **(133)** *(39)*

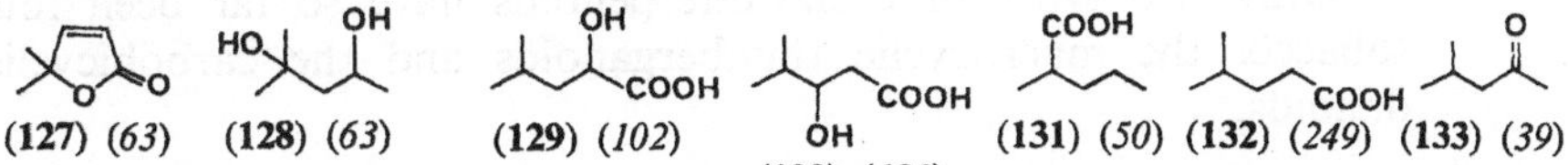

C_5 **(134)** R = H *(248)*   **(136)** *(63)*   **(137)** *(63)*   **(138)** *(253)*   **(139)** *(63)*   **(140)** *(63)*   **(141)** *(234)* 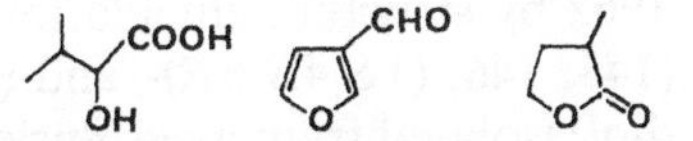
(135) R = CH₃ *(182)*
(142) *(102)*   **(143)** *(63)*   **(144)** *(182)*

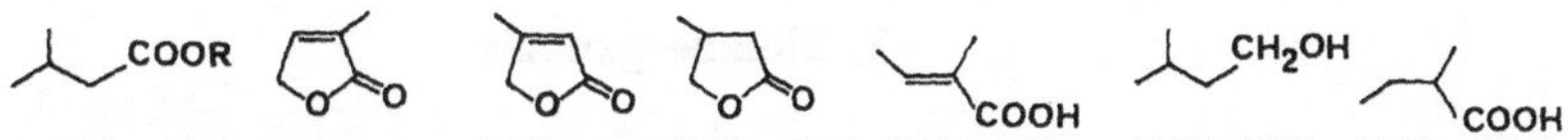

Of the carbo-acyclic nor-compounds containing more than one intact isoprene unit, the majority is derivable in an analogous fashion. Some of these **(103—109, 111)** correspond to the expected initial products while others **(112, 115—118)** have undergone subsequent transformations. Of the latter compounds, the furyl derivatives **(117)** and **(118)** are of particular interest since **(117)** can be viewed as being derived from pseudo-ionone **(115)** by initial $^1O_2$ attack on the isolated double bond followed by acid catalysed cyclisation as outlined in Scheme 9. In agreement with this, the *cis*- and *trans*-isomers of both this compound **(117)** and its dihydro-derivative **(118)** have been isolated from tobacco.

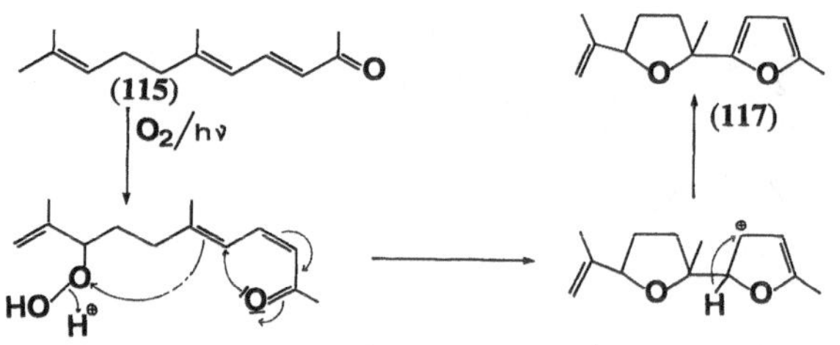

Scheme 9

Of the remaining three aliphatic nor-terpenoids (**102, 114, 119**), nor-solanesene (**102**) is obviously derived from solanesol (**12**) by conversion to the corresponding acid followed by decarboxylation, a route supported by the biomimetic total synthesis of nor-solanesene (*80*). The formation of the C_{16} (**114**) and C_{11} compounds (**119**) is less obvious, but may be explained by oxidative cleavage of a double bond shifted from the original position to the vicinal trisubstituted position as a result of $^1O_2$ attack.

III. Diterpenoids

Only two types of cyclic diterpenoids have so far been found in tobacco, the macrocyclic thunberganoids and the carbobicyclic lab-danoids.

1. Thunberganoids

The skeletal structure of the thunberganoids was first established in 1962 by structure elucidation of the α- and β-4,8,13-duvatriene-1,3-diols (**145, 146,** (1*S*,4*S*,6*R*)- and (1*S*,4*R*,6*R*)-2*E*,7*E*,11*E*-thunbergatriene-4,6-diol) isolated from aged Burley tobacco (*235*), and of cembrene (**147,**(1*S*)-2*E*,4*Z*,7*E*,11*E*-thunbergatetraene, Scheme 10), first obtained from *Pinus* species (*51, 159, 160*), and later found to be present in tobacco leaves when enzyme-blocking agents are administered (*224*). Subsequent studies by ROBERTS *et al.* (*234, 242, 243*) led to the isolation and structure elucidation of α- and β-4,8,13-duvatriene-1,5-diols (**148, 149**), three hydroxyethers (**150—152**) and one mono-ol (**153**) from the Burley tobacco. Recent investigations have revealed the presence of two diol ethers (**154, 155**) in Greek tobacco (*2, 19*), of an ether and a mono-ol (**156, 157**) in Virginia tobacco (*182*), and of (1*S*,4*S*)- and/or (1*S*,4*R*)-2*E*,7*E*,11*E*-thunbergatrien-4-ol-6-one (**158, 159**) as well as a *cis-trans* pair of the corresponding retro-aldol products (**160**) in dark-fired tobacco (*307*). Three further *seco*-terpenoids have been encountered in *N. tabacum*; two diastereomers of 6ξ,8ξ-dihydroxy-11ξ-isopropyl-4,8-dimethyl-14-oxo-4ξ, 9ξ-pentadecadienoic acid (**161**), probably differing at C(8), were isolated from Burley tobacco (*153*), and (10*S*)-3,7,13-trimethyl-10-isopropyl-2ξ, 6*E*,11*E*-13-tetradecatetraen-1-al (**162**) was obtained from tobacco flowers (*49*).

Except for cembrene (**147**), the hydroxyketoacid (**161**) and the diol ether (**154**), which have not been directly correlated with any of the others, the absolute configurations of these tobacco diterpenoids have been determined recently by way of degradation, LIS proton NMR and X-ray diffraction studies (*2, 10*). A most recent X-ray structure determination has revealed that a new growth inhibitor extracted from immature tobacco leaves is in all probability identical with β-4,8,13-duvatriene-1,3-diol (**146**) and possesses *trans*-oriented hydroxyl groups, a finding which provides the basis for the indicated configuration assigned to the 6-hydroxyl group of (**145**) and (**146**) (*264*). Moreover, ^{14}C-labelled cembrene, obtained from [2-^{14}C]-mevalonate in an inhibition biosynthesis, is converted to 4,8,13-duvatriene-1,3-diol when fed to tobacco leaf tissue (*224*), an observation which implies that the cembrene (**147**) obtained from tobacco has the same absolute configuration as that obtained from other sources (*72*) and constitutes a precursor of the oxygenated thunberganoids (**145, 146, 148—162**). Although there is at present no further direct evidence available and several routes for the formation of these compounds are possible, only one is considered in Scheme 10.

The common occurrence of compounds epimeric at C-4 suggests that the 4,5 double bond of cembrene (**147**) suffers an initial non-stereospecific oxidative attack yielding an epoxide (**XL**) or a similar species which is capable of furnishing, *via* an intermediate carbonium ion such as (**XLI**), five of the tobacco thunberganoids, *i. e.* the mono-ol (**157**) and the four diols (**145, 146, 148, 149**) (Scheme 11). The 4,6- and 4,8-diols are readily interconverted under mild chemical conditions (*242*), which indicates that they constitute the more stable of the possible stereoisomers and supports the mechanism outlined above. The observed preponderance of the 4,6-diols in green as well as in aged tobacco leaves (*225, 235, 242*) may be due to the fact that the equilibrium is strongly shifted towards the 4,6-diol side during their formation in tobacco. The 4,6-diols (**145**) and (**146**) are obvious and synthetically verified precursors of the hydroxyketones (**158**) and (**159**) respectively (*235*) and can also give rise to the mono-ol (**153**), a dehydration reaction recently carried out *in vitro*. Subsequent retro-aldol condensation of the hydroxyketones (**158, 159**) and 1,3-glycol cleavage of the 4,6-diols (**145, 146**) can generate the diketone (**160**) and the aldehyde (**162**) respectively, conversions which have been verified

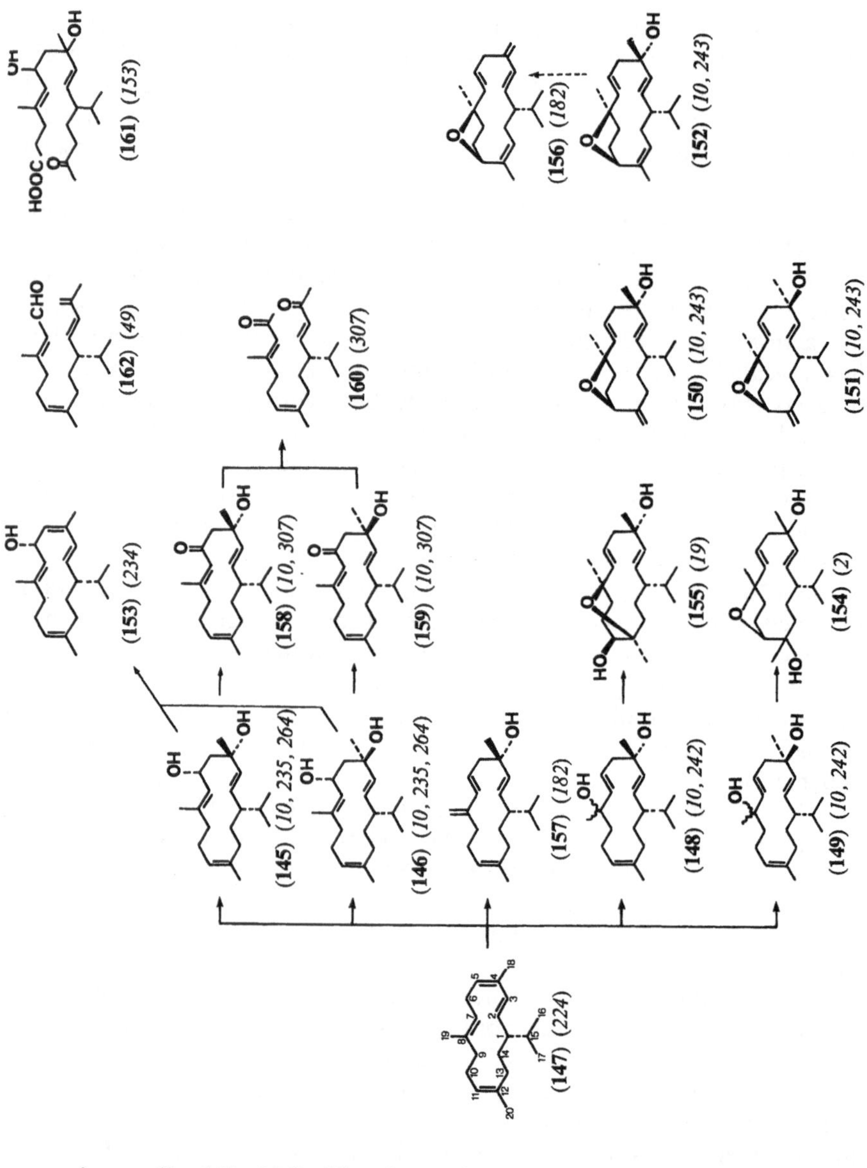

Scheme 10

experimentally (49, 235). The formation of the ketoacid (161) can be accounted for by oxidative cleavage of the 11,12 double bond in the 4,6-diols (145, 146) while the generation of the ethers (150—152, 154—156) may be explained in terms of an oxidative attack at the 11,12 double bond in the 4,6- and/or 4,8-diols (145, 146, 148, 149). Species such as the 11,12-epoxides (XLII, XLIII) are plausible intermediates in the latter

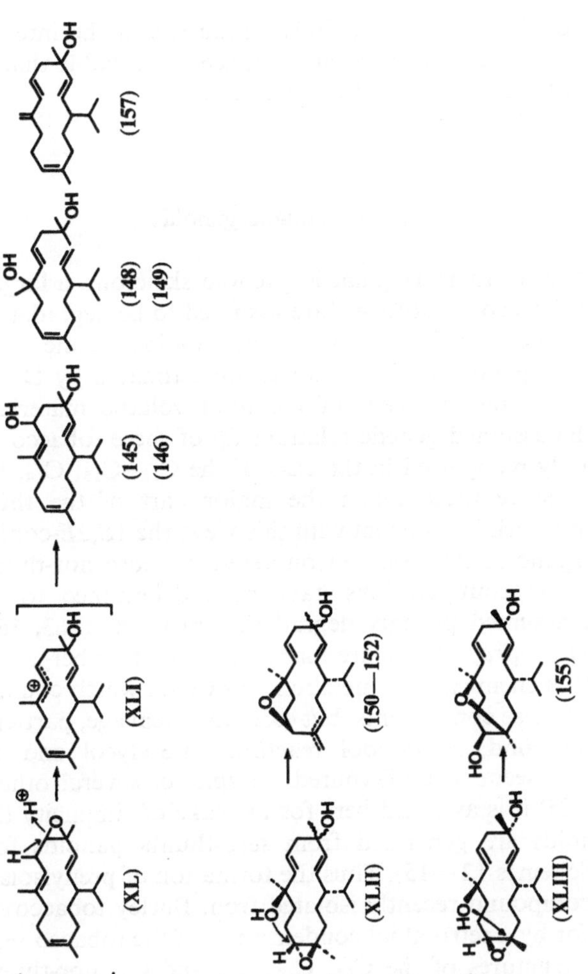

Scheme 11

reactions, which are likely to proceed by the mechanisms outlined in Scheme 11. Consonant with this, peracid oxidation of the 4S,6R-diol (145) has been shown to yield the corresponding 11,12-epoxide, which on mild acid treatment rearranges to the hydroxy ether (152) in quantitative yield (243). Since this reaction, which closely parallels the interconversion of the 4,6- and 4,8-diols, requires that the intermediate 4,8-diol epoxide has an $8R$ configuration, it seems probable that the tobacco diols (**148, 149**) also have $8R$ configurations.

2. Nor-Thunberganoids

On account of their irregular isoprenoid skeletons, a large number of the volatile tobacco constituents are assumed to be degradation products of thunberganoids. These nor-thunberganoids, some of which of considerable organoleptic importance, are estimated by Demole (*61*) to constitute some ten per cent of the total volatile material of Burley tobacco. The assumed genetic relationship of these tobacco constituents is most readily recognized in the case of the C_{18}, C_{15}, C_{14}, C_{13} and C_{12} compounds, since these retain the major part of the thunberganoid skeletal framework. Consistent with this view, the $1S,2E$-configuration of the thunberganoids has been encountered in those nor-thunberganoids whose absolute configurations have been determined to date. These include the assumed primary degradation products (**163, 164, 168, 171 and 179**) *(vide infra)*. There are many potential thunbergane precursors and numerous chemical and metabolic processes which could lead to the formation of these constituents, but oxidative cleavage, particularly of the 11,12-double bond, retro-aldol reactions, 1,3-glycol and 1,2-oxy-3-ol fragmentation seem to be favoured. In spite of several other proposals (*9, 43, 70, 225*) it is assumed here for the sake of simplicity that the nor-thunberganoids are generated from *seco*-thunberganoids found in tobacco (*cf.* Schemes 12—15). Thus the formation of prenylsolanone (**163**), a C_{18} nor-compound recently isolated from Burley tobacco (*70*), can be accounted for by a retro-aldol condensation of the tobacco *seco*-aldehyde (**162**). The structures of the C_{15}, C_{14}, C_{13} and C_{12} nor-thunberganoids suggest that their formation involves oxidative cleavage of the original 11,12 double bond and fragmentation of a further bond. Thus a plausible route to the C_{15} compounds (**164—166**), attractive since it also serves to explain the formation of the C_{14} compounds (**167, 168**), involves epoxidation of the *seco*-acid (**161**) or the rearranged isomer (**XLIV**) (Scheme 12). Acid catalysed fragmentation of the resulting epoxy acid (**XLV**) leads to the intermediate aldehydes which undergo subsequent oxidation or reduction.

Two additional compounds, the C_{16} ketone (**169**) and the C_{14} acid (**170**), are postulated to be nor-thunberganoids. While the C_{16} ketone must be generated by a complex multistep process (*6*), the formation of (**170**) is likely to occur by successive oxidations of the diketone (**160**).

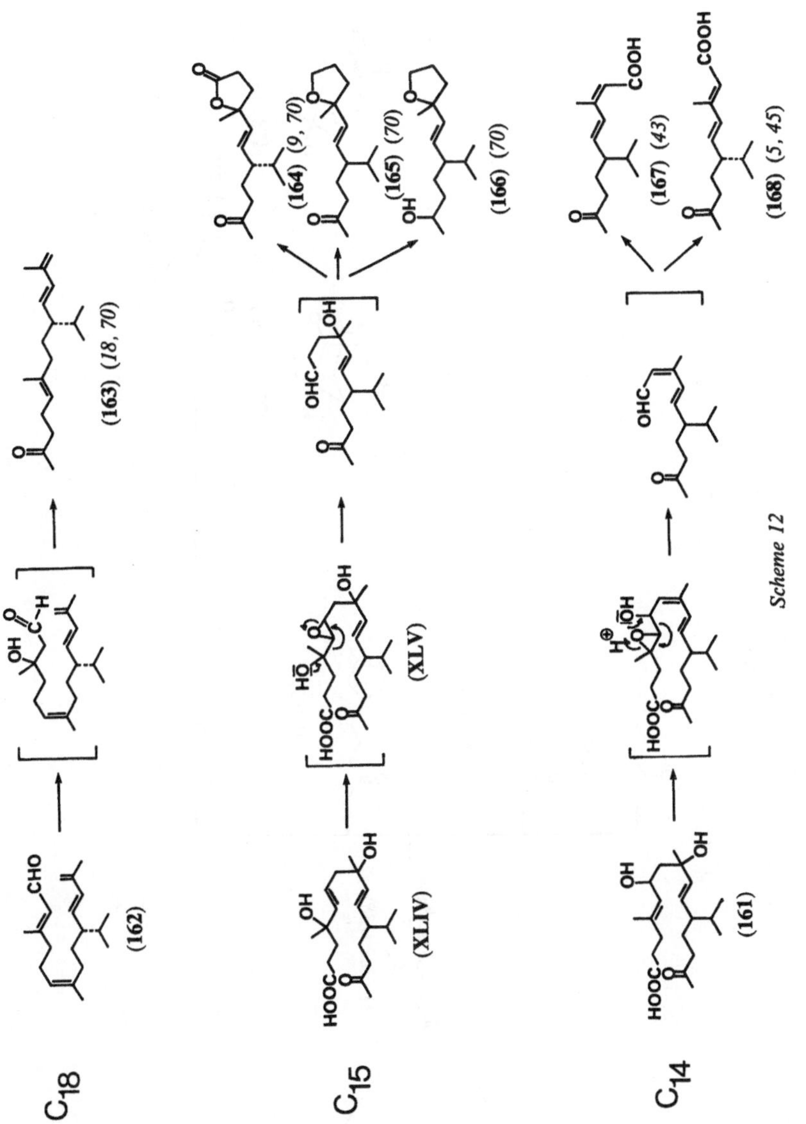

Scheme 12

Two acyclic precursors (161 and 162), each having undergone one of the two cleavage reactions required, have been proposed by KINZER *et al.* (*153*) and by DEMOLE (*70*) for the formation of the C_{13} nor-thunberganoids, typified by solanone (171). This compound (171), which is a major tobacco constituent (Fig. 1, peak 97), is likely to be a precursor of other C_{13} constituents (*cf.* Scheme 13). Thus compound (172), whose concentration

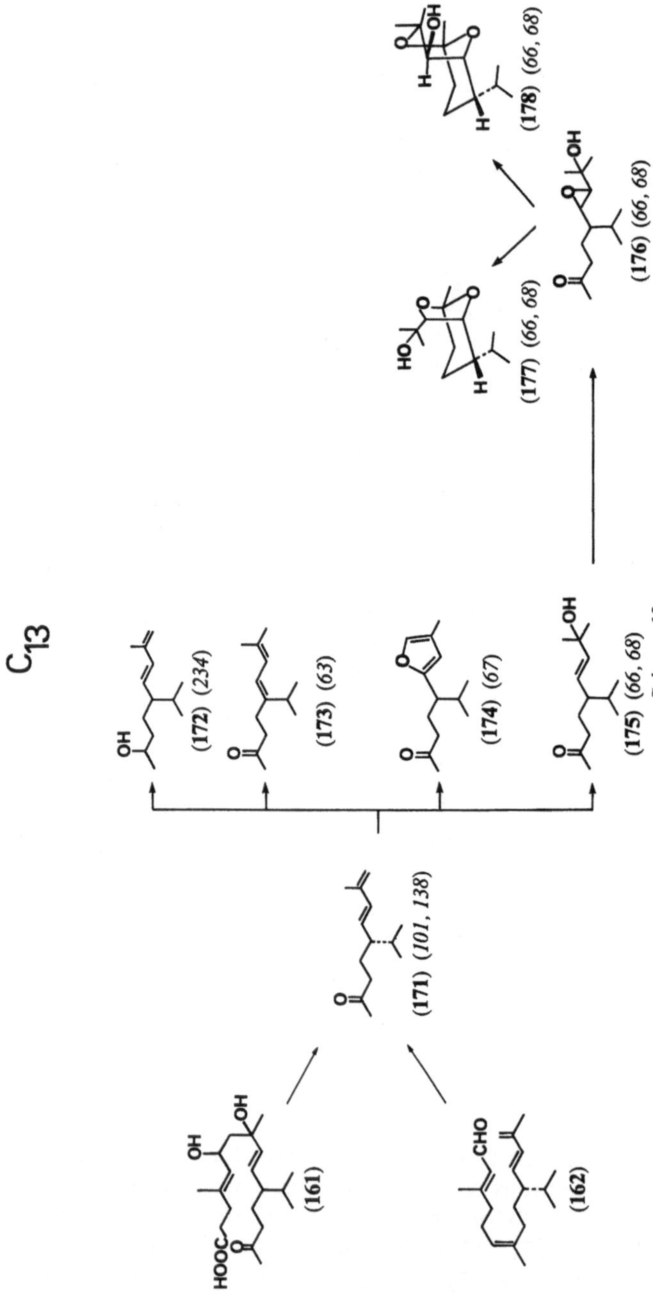

Scheme 13

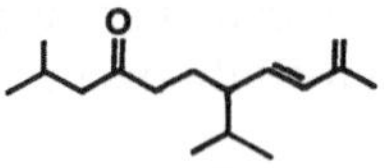

(169) (6) 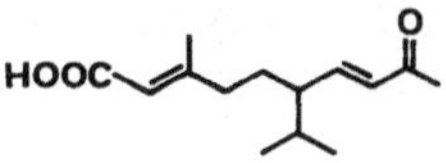 (170) (43)

is increased on curing and aging (*cf.* Fig. 1, peak 111) (*291*), can be viewed as having been generated by reduction, while the ketone (173) may arise by isomerization. The formation of compound (174) probably involves oxidative conversion of the diene group of solanone (171) to a furan moiety, a reaction which has been carried out synthetically (*67*). It also seems probable that solanone (171) is converted to the tertiary alcohol (175), and subsequently to the corresponding epoxide (176), which is capable of yielding the dioxabicyclo[3.2.1]octane (177) and the dioxabicyclo[3.3.1]nonane (178), as demonstrated experimentally by

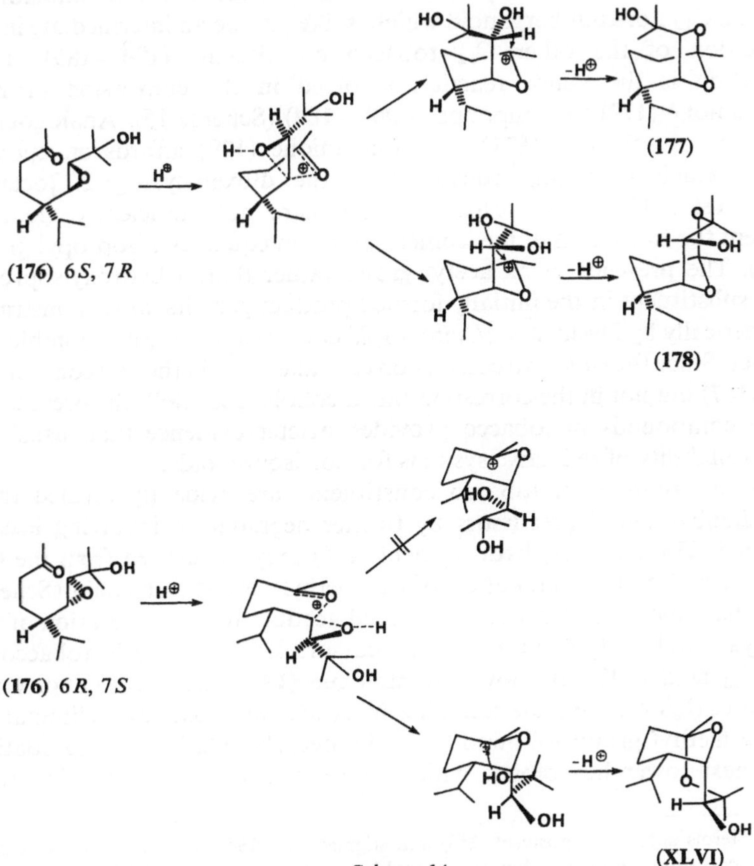

(176) 6S, 7R

(177)

(178)

(176) 6R, 7S

(XLVI)

Scheme 14

DEMOLE et al. (66). They also proposed a mechanism which accounts for the observed *endo* configuration of the naturally occurring dioxabicyclo-octane (177).

If (5S)-solanone (171) is accepted as the precursor, it can be advocated, as illustrated in Scheme 14, that of the two possible epoxides (176) only one, the 6S,7R-isomer, could lead to the formation of dioxabicyclo [3.2.1]octane (177), since steric hindrance between the isopropyl and hydroxyisopropyl groups in the 6R,7S-isomer would prevent bond formation between the oxygen atom of the epoxide group and C(2). Since recent findings by DEMOLE et al. (66) have shown that epoxidation of the racemic keto-alcohol (175) is highly stereospecific leading to the formation of E-threo epoxyketol (i.e. 176, 6S,7R), it can be inferred that the naturally occurring dioxybicyclo[3.3.1]nonane has the absolute configuration shown in formula (178) and not the stereo structure designated (XLVI).

The diketone (160) may be invoked as precursor of nor-solanadione (179), a C_{12} nor-thunberganoid which is likely to be an intermediate in the formation of the other C_{12} tobacco constituents (180—189). Thus reduction is the major reaction involved in the conversion of nor-solanadione (179) to compounds (180—184) (Scheme 15). Analogous to the case of solanone (171), nor-solanadione (179) affords an epoxide (185), which is readily converted to the dioxabicyclo[3.2.1]octanes (186—189). The steric restrictions are here not sufficiently great to prevent the formation of the isomer having an equatorial isopropyl group (186). The presence of an acetyl group rather than a hydroxyisopropyl *endo* substituent in the initially formed product permits, as demonstrated synthetically by DEMOLE et al. (66), rapid conversion to the more stable *exo* isomer. Since the *endo/exo* conversion can take place in the oxo compounds (186,187) but not in the corresponding alcohols (188, 189), the presence of these compounds in tobacco provides stricter evidence than usual for the availability of reductive systems for nor-isoprenoids.

A fair number of tobacco constituents are evidently formed from nor-thunberganoid precursors by further degradation involving loss of carbon*. Thus, the C_{12} hydroxy acid (190) may be derived from the C_{14} precursor (168) by degradation of the methyl ketone side-chain (Scheme 16). The same type of conversion would account for the formation of the epoxyalcohol (191) from a C-13 precursor. The presence in tobacco of the C_{12} acid (192), the lower homologue (193) and the alcohol (194) indicates that this degradation takes place *via* initial oxidative elimination of the methyl group followed either by decarboxylation or degradation to a next lower homologue which subsequently suffers reduction to a

---

* Sterols such as β-sitosterol (253) and stigmasterol (254), which have adequate side-chains, constitute alternative but less probable precursors of some of these compounds.

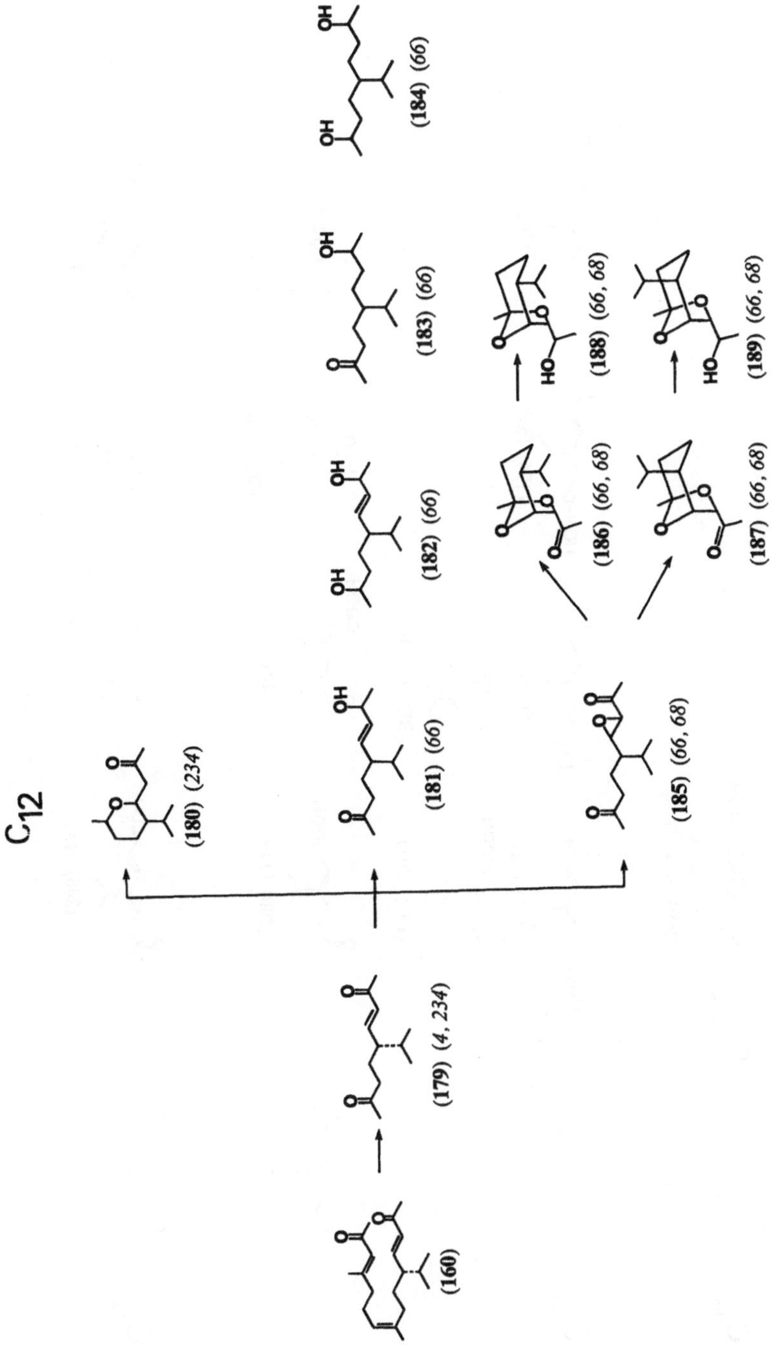

C₁₂

(184) (66)

(183) (66)

(188) (66, 68)

(189) (66, 68)

(182) (66)

(186) (66, 68)

(187) (66, 68)

(180) (234)

(181) (66)

(185) (66, 68)

Scheme 15

(179) (4, 234)

(160)

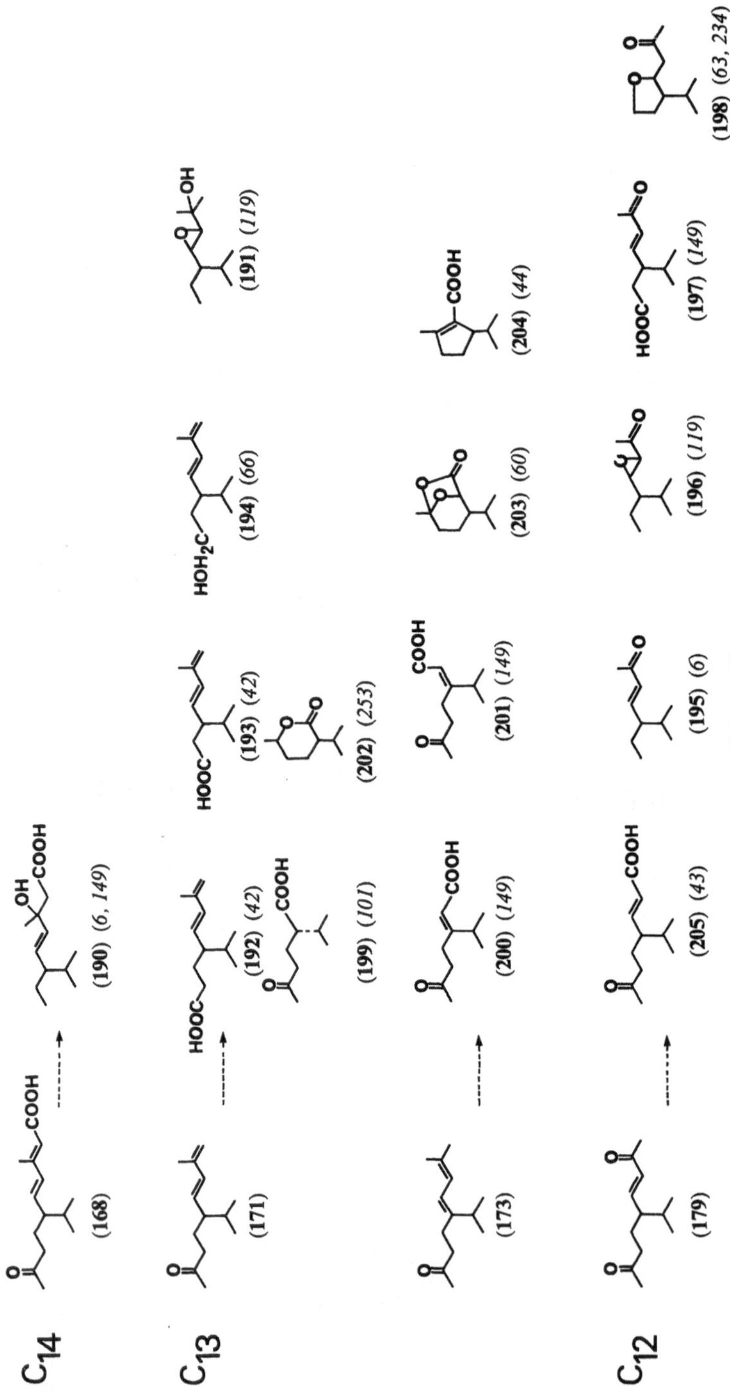

Scheme 16

product incorporating an ethyl sidechain. The compounds **(195—198)** are presumably derived in a similar manner from a C_{12} precursor. Oxidative cleavage of the olefinic bonds in solanone **(171)** and its isomer **(173)** is a plausible key step in the formation of the keto acids **(199—201)**, the lactones **(202, 203)** and the cyclised acid **(204)**. Although it is at present the sole example encountered, oxidative degradation of the unsaturated oxo grouping of the C_{12} precursor **(179)** appears to constitute a route to the acid **(205)**.

An account of the possible genesis of the C_{10} compounds has been given in Scheme 16 because of the presence in tobacco of adequate precursors and of analogous nor-thunberganoids which incorporate more than ten carbons. It should be inferred, however, that some of them, *e. g.* **(204)** *(44)* can arise from monoterpenoid precursors, an alternative not preferred here due to the low concentrations of these precursors. C_{10} constituents containing an isopropenyl group have been excluded since no thunberganoids possessing this group have been encountered. In view of the increasing number of possibilities, no comment is given regarding the possible genesis of C_8 and lower nor-isoprenoids **(206—214)**.

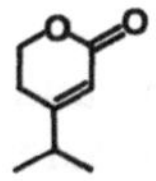

(206) *(63, 234)* 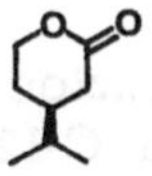   **(207)** *(143, 234)* 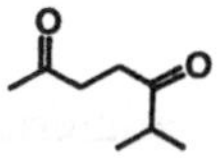   **(208)** *(234)* 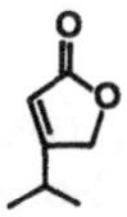   **(209)** *(63)* 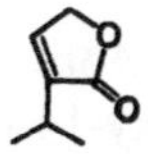   **(210)** *(182)*

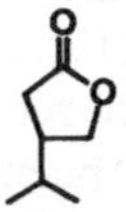

(211) *(234)* 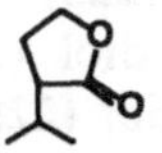   **(212)** *(234)* 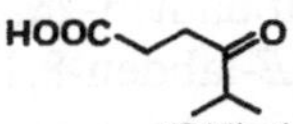   **(213)** *(149)* 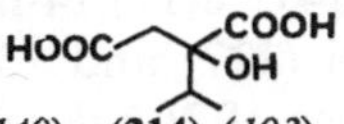   **(214)** *(103)*

3. Labdanoids and Nor-Labdanoids

The occurrence of labdanic diterpenoids in tobacco was disclosed in 1961—1963 by GILES *et al.* (*106, 107, 108*), who isolated α- and β-levantenolide **(215, 216)**, α2-levantanolide **(217)**, which is a dihydroderivative of **(215)**, and (12*S*,13*R*)-8,13-epoxy-14-labden-12-ol **(218)** from Turkish tobacco. The structures of the levantenolides **(215, 216)** have been verified by a biogenetic type of synthesis from *trans,trans*-farnesol (**XLVII**, Scheme 17). This was initially converted to the corresponding bromo derivative (**XLVIII**), which was reacted with 2,2′-di-3-methylfuryl mercury to yield the furan derivative (**XLIX**). Dye-sensitized photo-oxidation of the latter (**XLIX**) and subsequent oxidation of the resulting reaction mixture afforded the acyclic butenolide progenitor (**L**), which was cyclised using stannic chloride to a mixture of α- and β-levantenolide (**215, 216**) (*147, 283*).

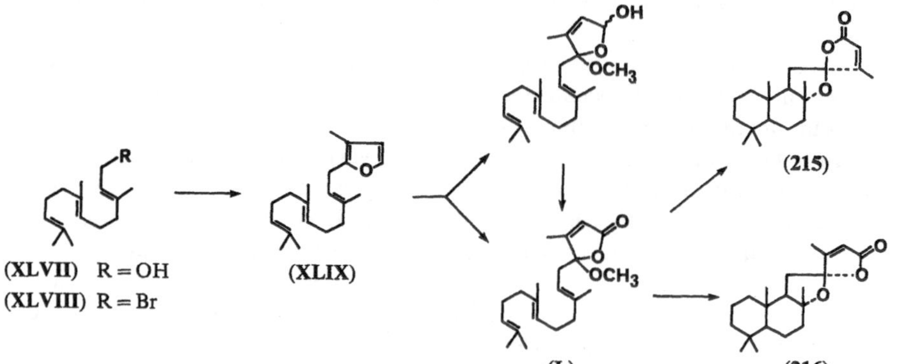

(**XLVII**) R = OH
(**XLVIII**) R = Br
(**XLIX**)
(**L**)
(**215**)
(**216**)

Scheme 17

More recently, two C-13 epimeric epoxyketones, 8,13- and 8,13β-epoxy-14-labden-12-one (**219, 220**), and four C-12,C-13 diasteromeric epoxyalcohols, 8,12ξ-epoxy-14-labden-13ξ-ols (**221**), have been obtained from Greek tobacco (*8, 12*), whereas a series of labdanols comprising abienol (**222**), 8(17),14-labdadien-13ξ-ol (**223**), 2-oxo-8(17),14-labdadien-13ξ-ol (**224**), 8(17),14-labdadien-2α,13ξ-diol (**225**), sclareol (**226**), 13-episclareol (**227**) and 13*E*-labden-8,15-diol (**228**) have been detected in green leaves of *N. tabacum* and other *Nicotiana* species (*46, 225*).

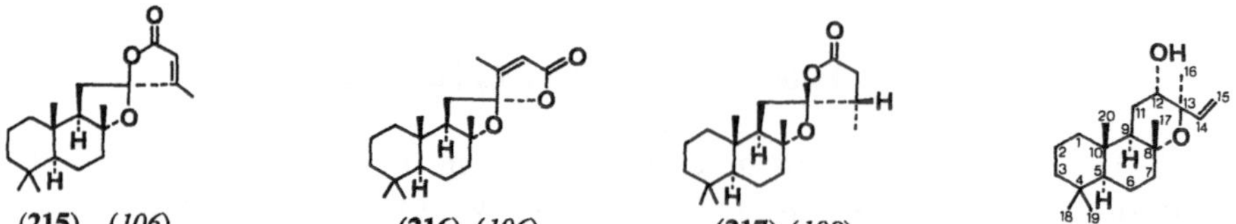

(**215**) (*106*) (**216**) (*106*) (**217**) (*108*) (**218**) (*107*)

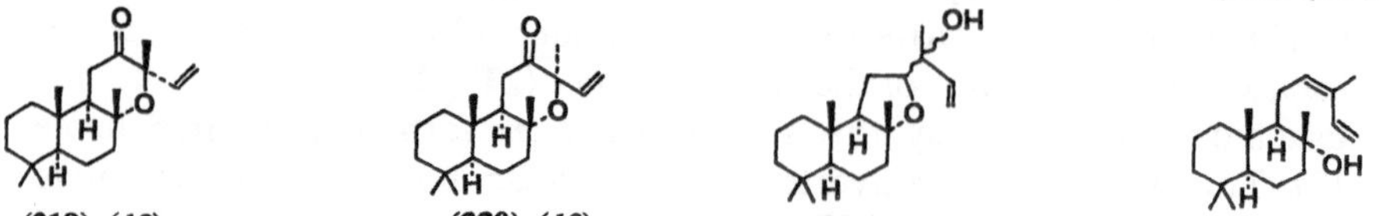

(**219**) (*12*) (**220**) (*12*) (**221**) (*8*) (**222**) (*46, 225*)

References, pp. 66—79

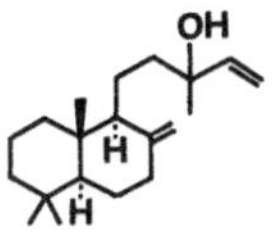

(223) (46, 225) 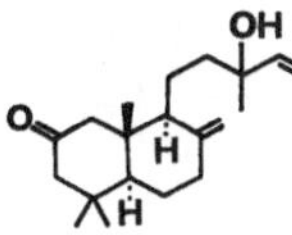 (224) (46, 225) 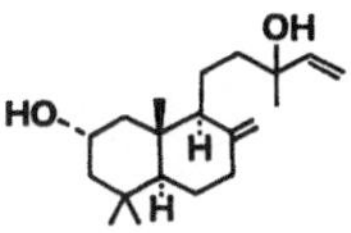 (225) (46, 225) 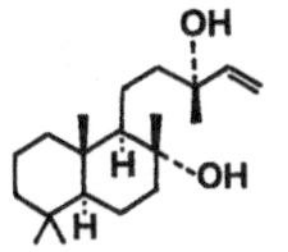 (226) (46, 225)

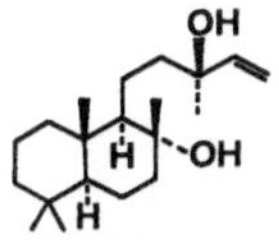

(227) (46, 225) 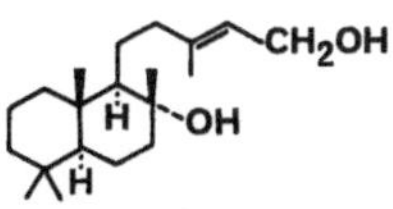 (228) (46, 225)

A number of compounds structurally closely related to the tobacco labdanoids but containing less than twenty carbon atoms has been encountered in Turkish (229—235) (253), Greek (230, 233) (121) and cigar (233) (141) tobacco. These include two C_{18} constituents, 8,13-epoxy-14,15-bisnor-12-labdene (229) and 14,15-bisnor-8-hydroxy-11E-labden-13-one (230). The latter is also a constituent of *Abies sibirica* (41). Of the two C_{17} constituents (231) and (232), ambreinolide (232) is a well-known degradation product of the triterpene ambrein. The C_{16} constituents are represented by norambreinolide (233), the corresponding 9,11-dehydro (234) and dihydro (235) derivatives. Most of these C_{16}–C_{18} compounds have been obtained synthetically from various labdanes by oxidative degradation (36, 37, 247, 278). Moreover, the unsaturated ketone (230) has been synthesized both from drimenol (LI) (121) and from ambreinolide (232) (212), as illustrated in Scheme 18. It is noteworthy that several of these C_{16}–C_{18} constituents have desirable flavour properties and give a cedar-like aroma to the smoke of Turkish tobacco (253).

While it is highly probable that the C_{16}–C_{18} tobacco constituents are formed by degradation of labdanic diterpenoids, the biogenetic origin of the C_{15} and C_{14} constituents (236—241) encountered in tobacco is

C_{18}

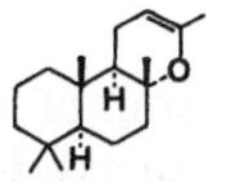

(229) (253) 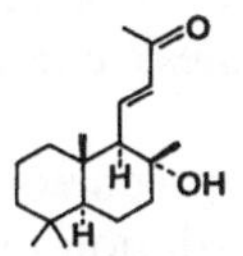 (230) (121) C_{17} (231) (253) 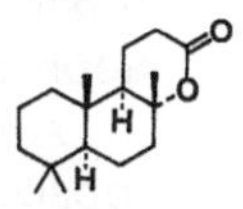 (232) (253)

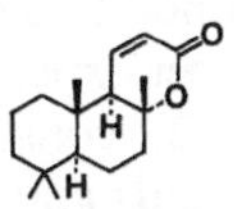

C_{16}

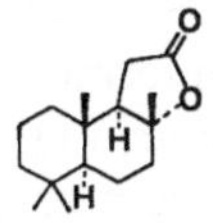

(233) (141) 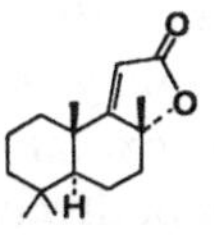 (234) (253) 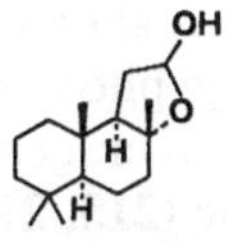 (235) (253)

C15

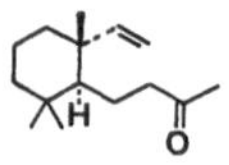

(236) (120) 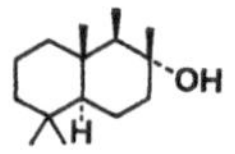   (237) (122) 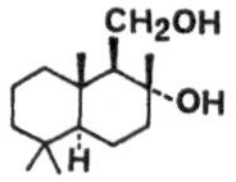   (238) (122) 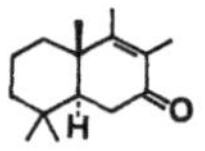   (239) (20)

C14

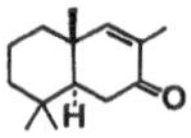

(240) (120) 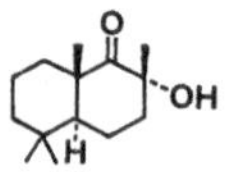   (241) (253, 3)

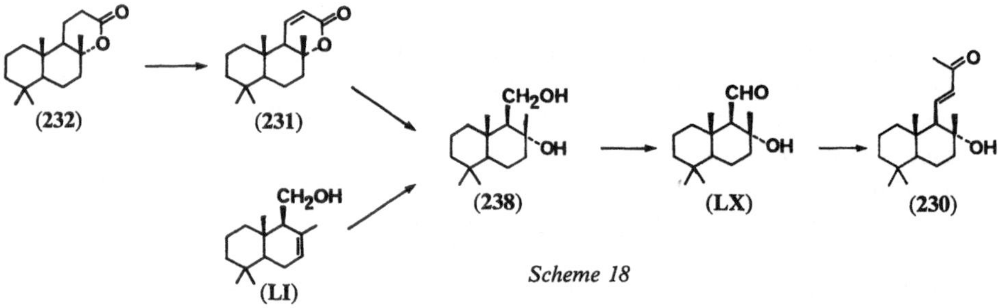

(232) — (231)

(238) — (LX) — (230)

(LI)

Scheme 18

less obvious. The latter (236—241), which have all been synthesized from drimenol (3, 20, 120, 122), may be genuine sesquiterpenoids of the drimane type or nor-derivatives thereof, biosynthesized from a farnesyl precursor. Consonant with this, the terminal monoepoxide of *trans, trans*-farnesyl acetate has been cyclised to drimane derivatives using a boron trifluoride ether complex or mineral acids (281, 282). However, tobacco contains fair amounts of appropriate labdane precursors as well as numerous nor-isoprenoids and the C_{14}–C_{15} constituents (236—241) conform structurally to the nor-labdane pattern. Furthermore, preliminary results indicate that nor-labdanes including the C_{15}-constituent (237), are formed on photo-oxygenation of abienol (222) (292). Therefore, neither of the two routes to compounds (236—241) can as yet be excluded in the case of tobacco and conclusive evidence of their genesis can only be obtained by appropriate tracer experiments.

Abienol (222) constitutes a plausible precursor of several of the tobacco labdanoids and their assumed degradation products, an assumption consonant with the observation that abienol is destroyed during the curing process (225). A hypothetical route for the genesis of some of the labdanes and nor-labdanes is shown in Scheme 19. The 1,3-diene moiety of abienol (222) may undergo photo-oxygenation yielding the intermediate peroxide (LII), which *via* rearrangement and oxidation to the acid (LIII) gives α- and β-levantenolide (215, 216). This pathway differs from earlier views on the biogenesis of these lactones in assuming that the precursor

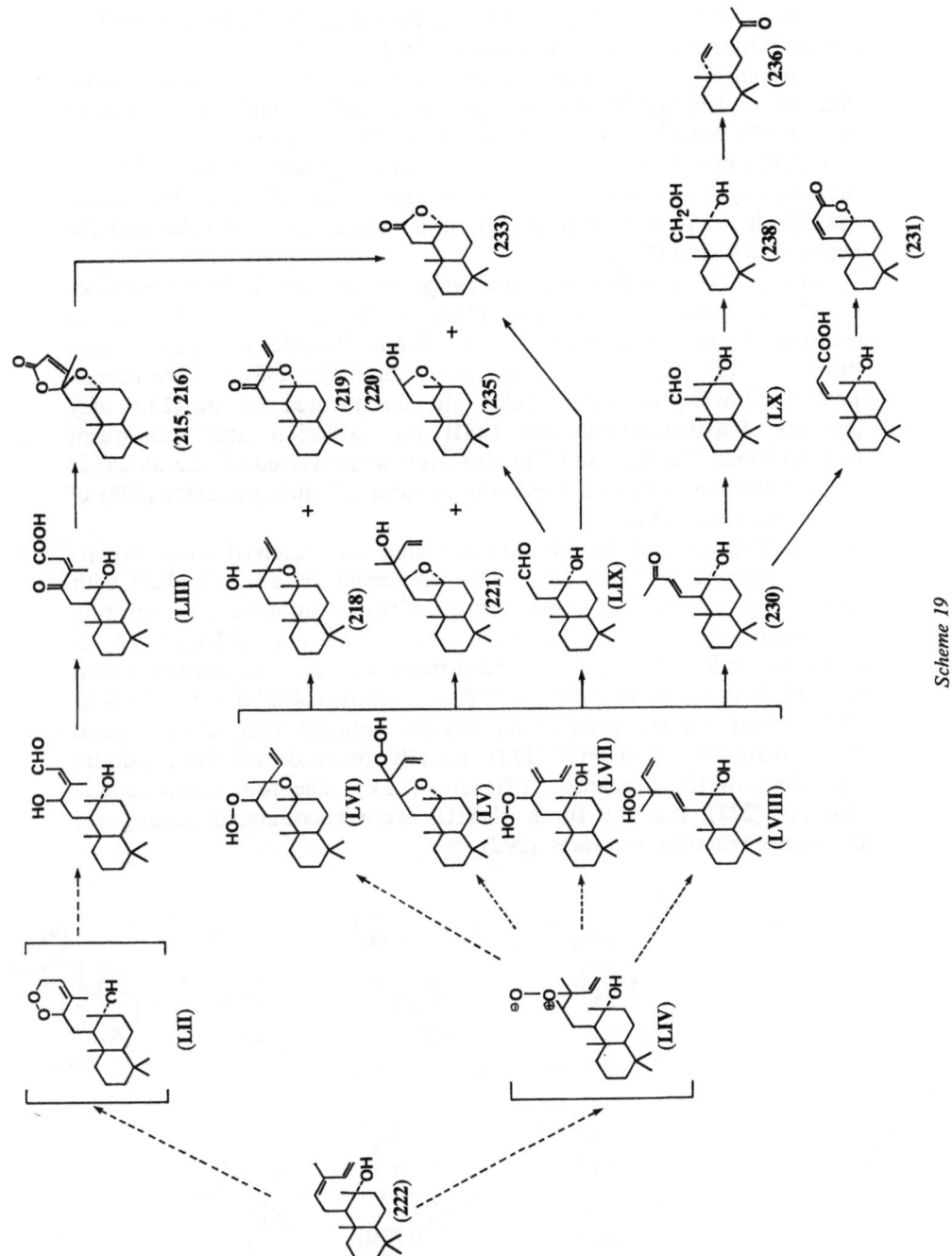

Scheme 19

is the bicyclic compound (**LIII**) and not the acyclic 3,7,11,15-tetramethyl-2,6,10,14-hexadecatetraenoic acid (*147, 283*).

Alternatively, photo-oxygenation of the 12,13 double bond of abienol (**222**) may yield the hypothetical peroxide (**LIV**), which by subsequent nucleophilic attack at C-12 and C-13 affords the hydroperoxides (**LV**) and (**LVI**) respectively. Reduction or rearrangement converts (**LV**) to the 8,12-oxygen-bridged tobacco constituents (**221, 233** and **235**), whereas (**LVI**) is the precursor of the 8,13-ether-bridged alcohol (**218**) and the ketones (**219**) and (**220**).

The peroxide (**LIV**) may rearrange to the allylic hydroperoxides (**LVII**) and (**LVIII**), which are stabilized by the formation of the nor-oxo products (**LIX**) and (**230**) respectively. While the aldehyde (**LIX**) constitutes an alternative precursor of norambreinolide (**233**) and the corresponding dihydro derivative (**235**), the unsaturated ketone (**230**) may produce dehydroambreinolide (**231**) on oxidation and subsequent lactonization. The ketone (**230**) can also be converted to the aldehyde (**LX**), which on reduction yields the obvious 1,3-diol precursor (**238**) of the vinylketone (**236**).

This hypothetical degradation sequence has received some experimental support (Scheme 20). Thus, treatment of abienol (**222**) with *m*-chloroperbenzoic acid gives a mixture of two compounds incorporating five-membered ether-bridged rings, (12*S*,13 *S*)- and (12*R*,13*R*)-8,12-epoxy-14-labden-13-ol (**221**), as well as two products having six-membered rings, (12*S*,13*S*)- (**LXI**) and (12*R*,13*R*)-8,13-epoxy-14-labden-12-ol (**LXII**) (*293*). Furthermore, preliminary results indicate that dye-sensitized photo-oxidation of abienol (**222**) actually proceeds *via* intermediates equivalent to the peroxides (**LII**) and (**LIV**). The four diastereomeric alcohols (**221**) and the furan (**LXIII**) were encountered among the numerous reaction products (*292*).

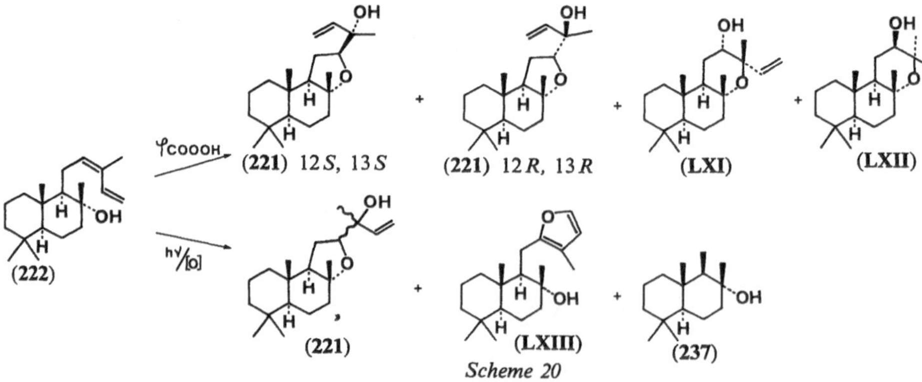

Scheme 20

IV. Triterpenoids and Steroids

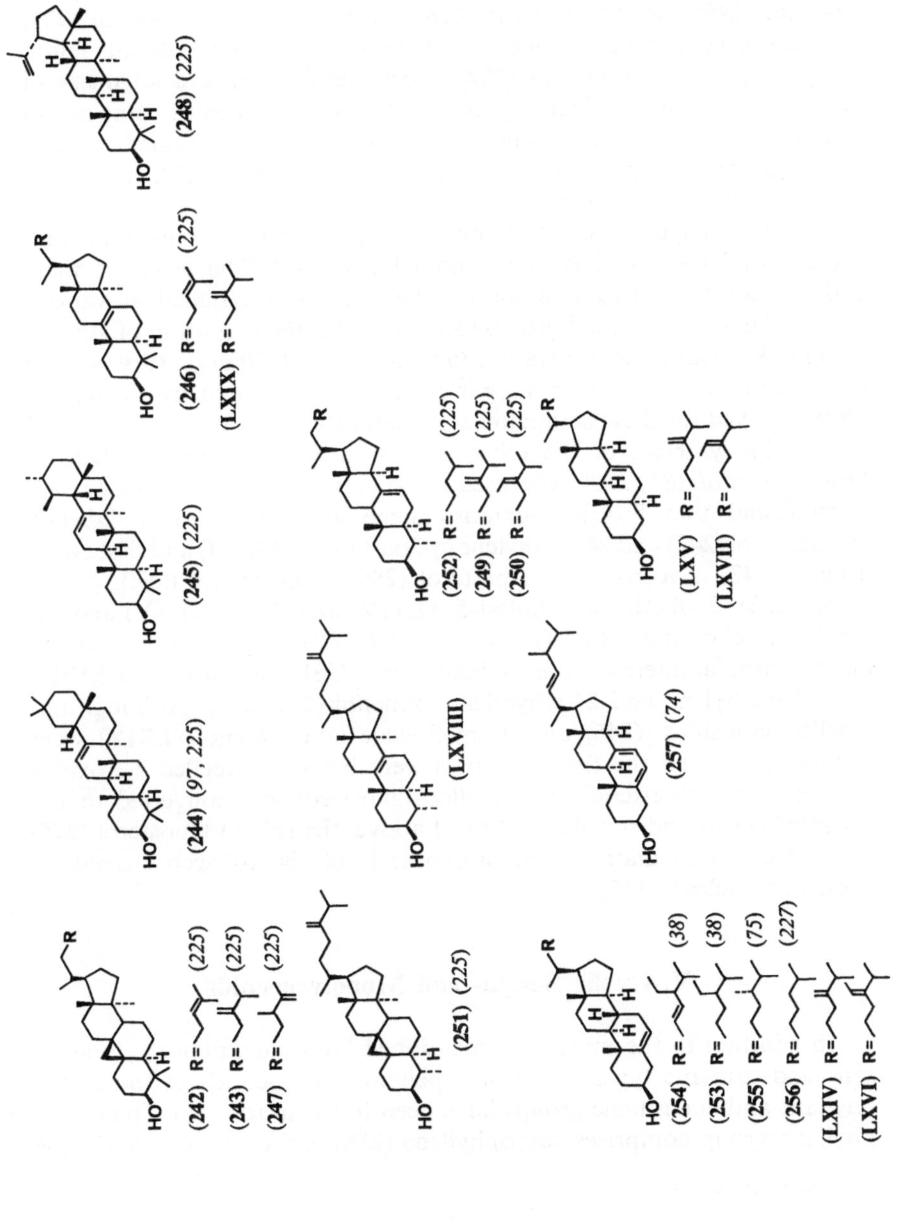

Studies on the chemistry and location of the major terpenoids of commercial tobacco varieties, *N. tabacum* (var. Virginia Gold) and *N. tabacum P. B.* (Bergerac) have shown that the cytoplasm contains

(248) (225)

(246) R= (225)
(LXIX) R=

(245) (225)

(252) R= (225)
(249) R= (225)
(250) R= (225)

(LXV) R=
(LXVII) R=

(244) (97, 225)

(LXVIII)

(257) (74)

(242) R= (225)
(243) R= (225)
(247) R= (225)

(251) (225)

(254) R= (38)
(253) R= (38)
(255) R= (75)
(256) R= (227)
(LXIV) R=
(LXVI) R=

triterpenoids, 4α-methylsteroids and steroids, while the cuticular extract consists mainly of thunberganoids and minor amounts of alkanes and steroids (225). Cycloartenol (242, 50%), 24-methylene cycloartanol (243, 20%) and β-amyrin (244, 10%) were major and α-amyrin (245), lanosterol (246), cyclolaudenol (247) and lupeol (248) minor components of the triterpenoid fraction. Of the 4α-methylsteroids, 24-methylene lophenol (249), 24-ethylidene lophenol (250) and cycloeucalenol (251) were major constituents, while one of the minor components was tentatively identified as lophenol (252). β-Sitosterol (253) and stigmasterol (254) were the predominant cytoplasmic steroids, whereas campesterol (255) and cholesterol (256), whose presence was not ascertained, were of low abundance (225). One further steroid, ergosterol (257), have been reported present in tobacco (74).

Esters and glucosides of stigmasterol, β-sitosterol, campesterol and cholesterol have also been encountered (29, 140). The levels of these and the corresponding free steroids have been determined at various stages of maturity of field grown tobacco. With the exception of stigmasterol (254), whose concentration increases from 0.50 mg/g (dry weight) in the immature leaf to 0.75 mg/g in the mature leaf, virtually no changes were observed for these compounds on maturation (114).

The biosynthesis of the tobacco phytosterols has been studied by Ourisson et al. (81, 118), who used tissue cultures of Nicotiana tabacum. They found that [^{14}C]-2,3-oxidosqualene was converted to labelled cycloartenol (242) and 24-methylene cycloartanol (243) (81) and that cycloartenol (242) gave rise to campesterol (255), stigmasterol (254), β-sitosterol (253), 24-methylene cholest-5- (LXIV) and -7-enol (LXV) and 24-ethylidene cholest-5- (LXVI) and -7-enol (LXVII) via the expected 4α-methylsteroidal intermediates cycloeucalenol (251), obtusifoliol (LXVIII) and 24-methylene and 24-ethylidene lophenol (249, 250). Although unlabelled lanosterol (246) and 24-methylene lanost-8-enol (LXIX) were encountered when the tissue cultures were fed with labelled 2,3-oxidosqualene or cycloartenol and labelled lanosterol was converted to the 4-methylsterols and sterols mentioned above, the role of lanosterol (246) as a true intermediate in the biosynthesis of the tobacco steroids is presently unclear (118).

V. Cyclic Sesqui- and Monoterpenoids

In addition to the previously mentioned farnesane, monocyclofarnesane and drimane types of sesquiterpenoids, compounds of the caryophyllane and eudesmane groups have been found in tobacco. The caryophyllane group comprises caryophyllene (258) and (−)-β-caryophyllene epoxide (**259**), while the eudesmane group is represented by 6-*epi*-α-cyperone (**260**), 1-hydroxy- (**261**) and 1-oxo-α-cyperone (**262**). Capsidiol (**263**), an eremophiloid closely related to the tobacco eudesmanes, was recently found to be produced in *N. tabacum* and *N. clevelandii* following infection with tobacco necrosis virus (*23*). Both capsidiol (**263**) and glutinosone (**LXX**), a nor-sesquiterpenoid obtained from leaves of *N. glutinosa* infected with tobacco mosaic virus (*34*), possess antifungal properties and are likely to represent part of the *Solanaceae* virus defence system.

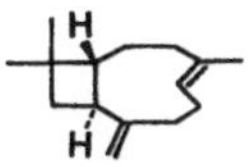

(258) (*63*) 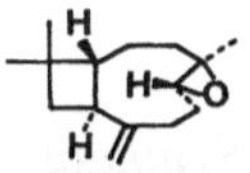   (259) (*123*) 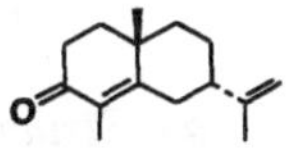   (260) (*182*) 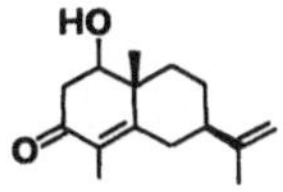   (261) (*182*)

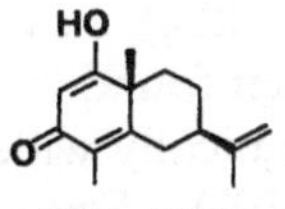

(262) (*231*) 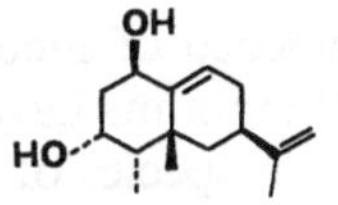   (263) (*23*) 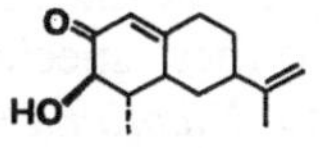   (LXX)

The skeletal variation within the group of tobacco monoterpenoids is quite limited and besides the linear compounds (**24—30**) mentioned earlier, only menthane, camphane and pinane derivatives have been encountered. Furthermore, there are only three representatives of the latter two groups, camphor (**264**), borneol (**265**) and myrtenol (**266**). The menthanes (**267—279**), however, display a greater structural variation and seem to have been exposed to reactions similar to those observed for the nor-carotenoids and nor-diterpenoids, an assumption supported by the fact that two nor-menthanes (**280, 281**) have been obtained from tobacco (*63*). The benzofuran derivatives (**282**) and (**283**) can be dissected into two isoprene units, which are linked in an irregular way (*11, 15*). Whether these compounds should be classified as monoterpenoids or not is a matter for conjecture.

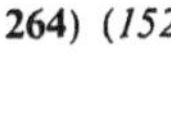

264) (*152*) 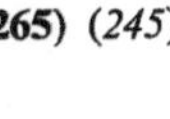   (265) (*245*) 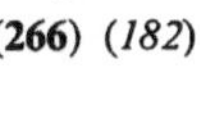   (266) (*182*) 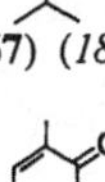   (267) (*182*) 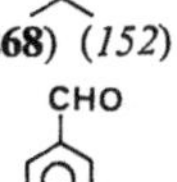   (268) (*152*) 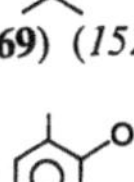   (269) (*152*) 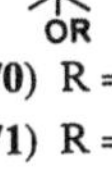   (270) R = H (*63*)
                                                                                                (271) R = COCH₃ (*182*)

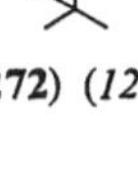

272) (*123*) 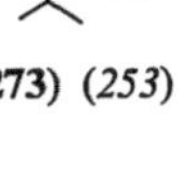   (273) (*253*) 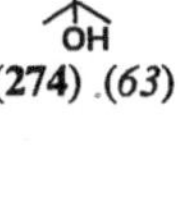   (274) (*63*) 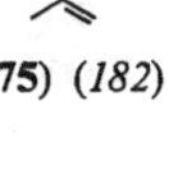   (275) (*182*) 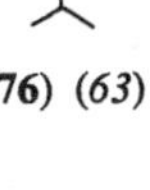   (276) (*63*) 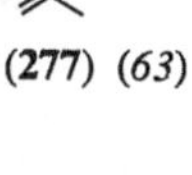   (277) (*63*)

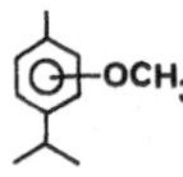

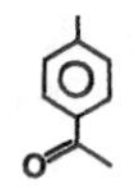

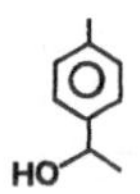

(278) (*151*) (279) (*182*) (280) (*63*) (281) (*63*)

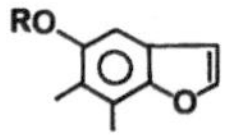

(282) R = H (*182*)
(283) R = CH₃ (*11, 15*)

VI. Pyridine Derivatives

1. Nicotine

a) Structure and Synthesis

Nicotine (284) is the major alkaloid of the cultivated tobacco species *N. tabacum* and *N. rustica,* in which its concentration ranges between 0.5 and 8%. The natural occurrence of nicotine (284), however, is not restricted to these species. Small amounts have been found in wild species of the genus *Nicotiana* and also in species of botanically widely different genera such as *Lycopodium* (*185, 188, 189, 190*), *Equisetum* (*184*), *Erythroxylum* (*85*), *Asclepias* (*186*), *Sedum* (*187*) and *Sempervivum* (*211*).

Nicotine (284) was first isolated in pure form by Posselt and Reimann in 1828 (*220*). The elucidation of its structure was based on degradation studies. Thus, oxidation of nicotine by chromic trioxide afforded nicotinic acid (*129*), which was decarboxylated to pyridine (*130*) and later identified as pyridine-3-carboxylic acid (*260*). It was therefore evident that an unidentified $C_5H_{10}N$ unit was attached to the β-position of the pyridine ring of nicotine. A clue to the nature of this unit was obtained when Pinner characterized the bromo derivatives dibromocotinine (LXXI) and dibromoticonine (LXXII) and showed that alkaline cleavage of (LXXI) yielded methylamine, oxalic acid and a third product, evidently 3-acetyl-pyridine, while dibromoticonine (LXXII) gave methylamine, malonic acid and nicotinic acid. This results established that nicotine (284) contains

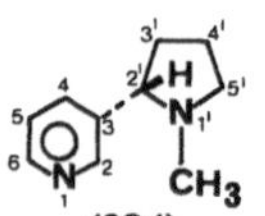

(284) 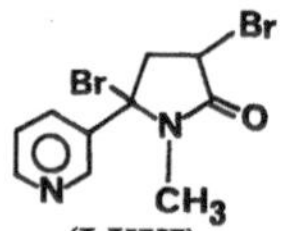   (LXXI) 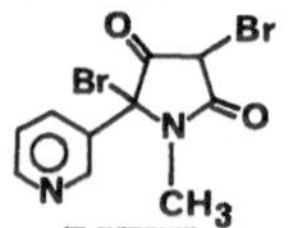   (LXXII)

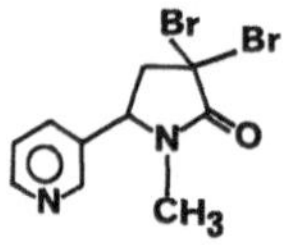

(LXXIII) 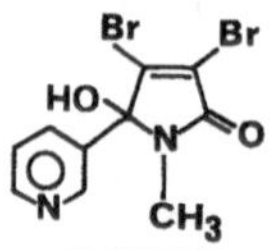   (LXXIV)

a *N*-methyl group and although later studies have necessitated a revision of the structures of dibromocotinine and dibromoticonine to (**LXXIII**) and (**LXXIV**) respectively (*73, 192*), it enabled PINNER in 1893 to propose the correct structure (**284**) for nicotine (*216, 218*).

The absolute configuration of nicotine (**284**) was determined by degradation to optically active hygric acid (**LXXVII**) (Scheme 21). Oxidation using ferricyanide and alkali of (**LXXV**), obtained by treatment of nicotine with methyl iodide and hydrogen iodide, afforded *N*-methylnicotone (**LXXVI**), which was converted to hygric acid (**LXXVII**) on oxidation with chromic trioxide. Since this compound (**LXXVII**) could be methylated to optically pure *L*-stachydrine (**LXXVIII**), which has the same configuration as *L*-proline and the natural amino acids, it was evident that *S*-nicotine is the naturally occurring enantiomer (*146*). This has later been confirmed by the correlation of (−)-nicotine (**284**) with *L*-serine (*131*).

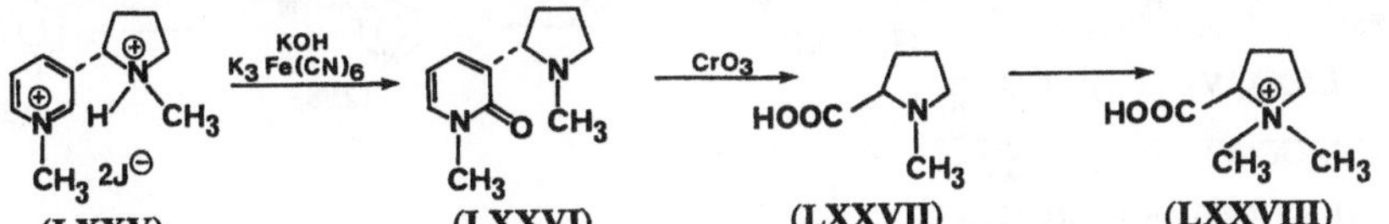

Scheme 21

The first synthesis of nicotine was published by PICTET and CRÉPIEUX in 1895 (*214*). It involves conversion of di-3-aminopyridine mucate (**LXXIX**) to 3-pyridyl-1′-pyrrol (**LXXX**), rearrangement of this to 3-pyridyl-2′-pyrrol (**LXXXI**) and *N*-methylation with the formation of nicotyrine (**285**). This compound, which was later found in tobacco, gave nicotine (**284**) on reduction (Scheme 22). The need for specifically labelled nicotine, nornicotine and myosmine derivatives as well as nicotine analogues for biosynthetic, metabolic and pharmacological studies has, however, stimulated the elaboration of a number of modern and efficient synthetic procedures.

One method, originally described by SPÄTH *et al.* (*266, 272*) involves condensation of ethyl nicotinate (**LXXXII**) with *N*-methylpyrrolidone (**286**) and treatment of the product (**LXXXIII**) with hydrochloric acid (Scheme 23). The *N*-methylmyosmine (**287**) thus formed is hydrogenated to yield nicotine

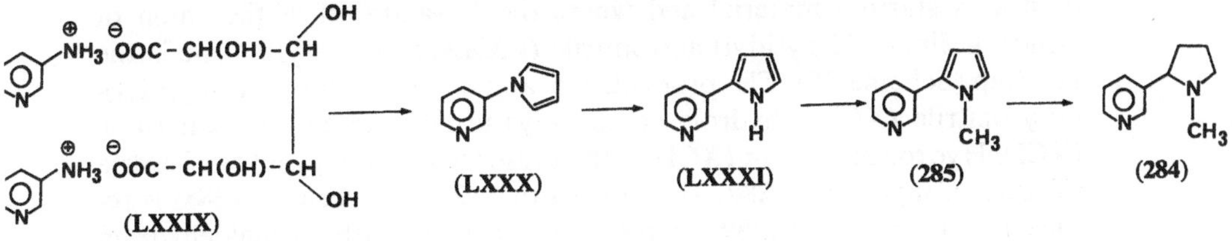

Scheme 22

(**284**) (*164*). The use of *N*-benzoylpyrrolidone (**LXXXIV**) (*270*), *N*-nicotinoyl-pyrrolidone (**LXXXV**) (*164*), 3-lithio-*N*-trimethylsilyl-2-pyrrolidone (**LXXXVI**) (*128*), or *N*-vinylpyrrolidone (**LXXXVII**) (*32*) as the ketonic unit in the condensation leads to the formation of myosmine (**288**). This minor tobacco alkaloid, first isolated from tobacco smoke (*300*), is converted to nornicotine (**289**) on reduction with sodium borohydride. [$2'$-^{14}C]-Nicotine (*59*), [$2'$-^{14}C]-myosmine (*128*) and [$2'$-^{14}C]-nornicotine (*128*) have been prepared from labelled ethyl (or methyl) nicotinate by this method.

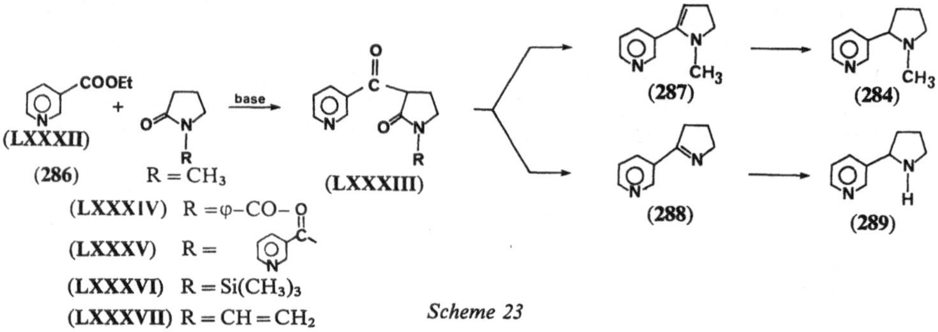

Scheme 23

Another route (Scheme 24), which has also found application in the preparation [$2'$-^{14}C]-nicotine (*47*), utilizes nicotinamide (**290**) as a starting material and proceeds *via* 3-cyanopyridine (**291**) to 3-pyridylcyclopropyl ketone (**LXXXVIII**). This is converted to nicotine (**284**) in a one-step process by treatment with *N*-methylformamide and magnesium chloride (*33*).

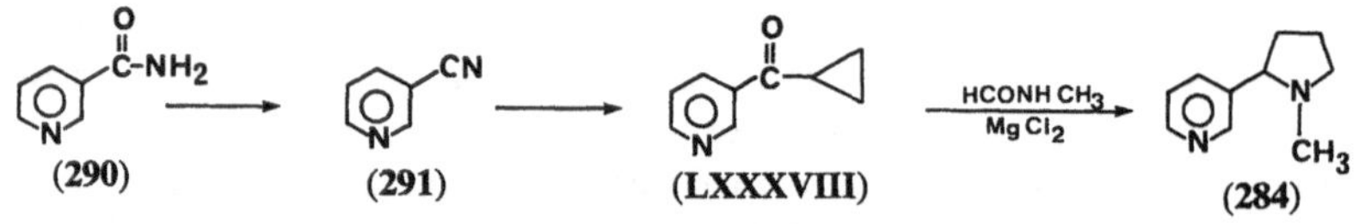

Scheme 24

Leete *et al.* (*178*) devised a method in which 3-formylpyridine (**292**) is used as a starting material and where the 1,4-addition of the anion of α-morpholino-α-(3-pyridyl)-acetonitrile (**LXXXIX**) to acrylonitrile is the key step (Scheme 25). The product, γ-cyano-γ-morpholino-γ-(3-pyridyl)-butyronitrile (**XC**) is hydrolysed to 3-cyano-1-(3-pyridyl)-propan-1-one (**XCI**). Hydrogenation of (**XCI**) in the presence of Raney nickel affords a mixture of myosmine (**288**) and nornicotine (**289**). Myosmine (**288**) is reduced to nornicotine (**289**), which when reacted with formaldehyde in formic acid yields nicotine (**284**). Addition of various agents such as

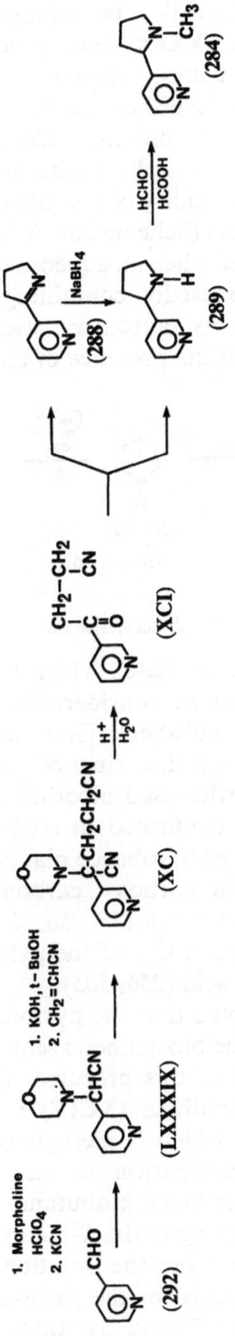

Scheme 25

crotonitrile, α-methylacrylonitrile, cinnamoylnitrile and methylvinyl
ketone to the carbanion (**LXXXIX**) gave products which were readily
converted to nicotine analogues, whereas use of [^{14}CHO]-3-formyl-
pyridine as starting material furnished nicotine labelled at C-2'.

A biomimetic synthesis of nicotine (**284**), also applicable to the
preparation of analogues, N-methylanabasine (**308**, *vide infra*) and
[2'-^{14}C]-nicotine, utilizes the ability of 1,4-dihydropyridines to react with
N-methyl-Δ^1-pyrrolinium salts (Scheme 26). A 21% radiochemical yield of
[2'^{14}C]-nicotine was obtained when an aqueous solution of glutaraldehyde
(**XCII**) and ammonia, supposedly containing small amounts of 1,4-
dihydropyridine (**XCIII**) or its tautomers, was stirred with [2'^{14}C]-N-
methylpyrrolinium acetate in the presence of air at pH 10.3 (*172*).

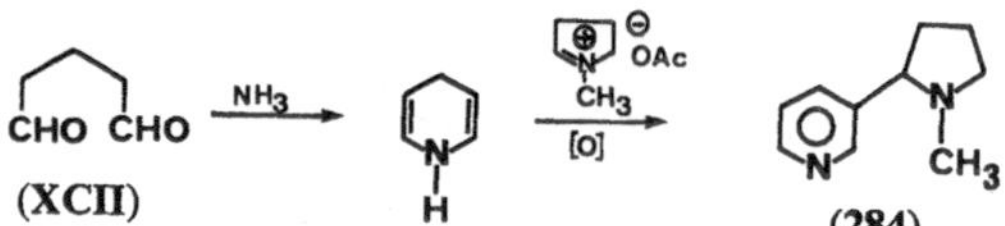

Scheme 26

b) Biosynthesis

The biosynthesis of nicotine (**284**), which takes place in the tobacco
roots (*54*), has been studied in considerable detail and the principal
route is now fairly well established. Thus, nicotinic acid (**293**) is an
efficient precursor of the pyridine ring of nicotine, a result initially
obtained by Dawson (*56*), who used nicotinic acids labelled in the ring
with ^{14}C or ^{3}H, and later confirmed in studies in which [2,3,7-^{14}C]-
nicotinic acid was administered to tobacco plants (*256, 305*). Experimental
data has also shown that the carboxyl carbon of nicotinic acid is not
incorporated in the nicotine molecule (*57, 256, 305*) and that the pyrrolidine
moiety is exclusively attached at C-3 of the pyridine ring, *i. e.* the site of
decarboxylation of nicotinic acid (*256, 305*).

It is now generally accepted that the pyrrolidine ring is derived from
ornithine (**XCIV**) and that the biosynthetic sequence shown in Scheme 27
accounts for the conversion of this precursor (**XCIV**) to the N-methylpyrrolinium salt (**XCV**). Ornithine (**XCIV**) is initially decarboxylated
yielding putrescine (**XCVI**), which is methylated to N-methylputrescine
(**XCVII**). On oxidative deamination of the primary amino group,
(**XCVII**) is converted to 4-methylaminobutanal (**XCVIII**), which undergoes spontaneous ring closure to give the N-methylpyrrolinium salt (**XCV**).

A number of tracer studies and the isolation of appropriate enzymes
constitute the experimental basis for the pathway outlined above. Thus,
early experiments by Leete (*167*) and Byerrum et al. (*71*) demonstrated

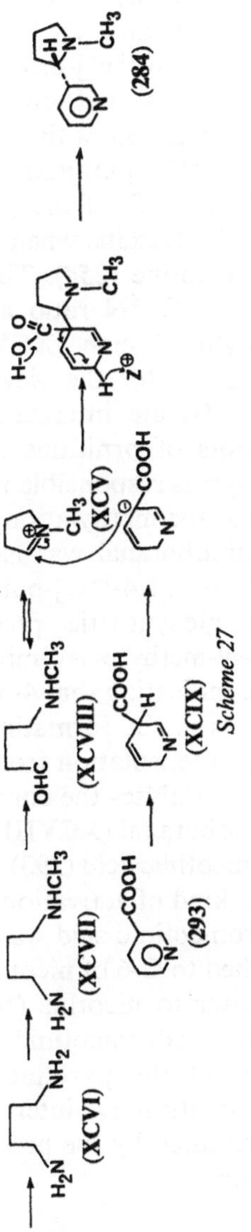

Scheme 27

that [2'-^{14}C]-ornithine is incorporated into the pyrrolidine ring of nicotine (284) and that C-2' and C-5' are labelled equally. Labelled nicotine is produced from [δ-^{15}N]-ornithine but not from [α-^{15}N]-ornithine, which demonstrates that only the δ-nitrogen of ornithine (XCIV) is utilized in the formation of the pyrrolidine ring (180).

Administration of [1,4-^{14}C]-putrescine afforded nicotine, which was labelled equally at C-2' and C-5' of the pyrrolidine ring (169), whereas [Me-^{14}C,1-^{15}N]-N-methylputrescine when fed to Nicotiana rustica plants gave rise to radioactive nicotine (254). This was labelled at the methyl group and had the same ^{14}C/^{15}N ratio as the administered precursor suggesting that N-methylputrescine (XCVII) is incorporated without degradation. Further support for the view that putrescine (XCVI) and N-methylputrescine (XCVII) are intermediates was obtained from the discovery in tobacco roots of ornithine decarboxylase and putrescine N-methyltransferase, enzymes responsible respectively for the decarboxylation of ornithine and for the methylation of putrescine (195).

Labelled 4-methylaminobutanal was isolated from tobacco roots to which [2-^{14}C]-ornithine or [1,4-^{14}C]-putrescine had been fed (154). Administration of the biosynthetic product, 4-methylaminobutanal (XCVIII) or of [2-^{14}C]-N-methylpyrrolinium chloride (170) to tobacco yielded labelled nicotine, indicating that 4-methylaminobutanal (XCVIII) is one of the intermediates in the formation of the N-methylpyrrolidine ring. This was verified by the isolation from tobacco roots of N-methylputrescine oxidase, which catalyses the conversion of N-methylputrescine (XCVII) to 4-methylaminobutanal (XCVIII) (196).

The condensation of nicotinic acid (293) with the N-methylpyrrolinium (XCV) salt requires some kind of activation. The possibility that this may take place via 1,6-dihydronicotinic acid was rendered likely by the observation that tritium attached to C-6 of nicotinic acid is not retained during the conversion of the latter to nicotine (56, 181). Most recently it has been suggested that 3,6-dihydronicotinic acid (XCIX) may constitute the immediate precursor of the pyridine ring (174). This compound (XCIX) was regarded as an attractive intermediate since the formation of a carbanion at C-3 is facilitated by the presence of the adjacent carboxyl group and the C=N group.

c) Metabolism in the Growing Plant

Nicotine (284) does not, however, represent an end product of the metabolism in the tobacco plant. Its N'-methyl group is readily lost by transmethylation (53, 132, 175, 285) to yield nornicotine (289), a minor component of N. tabacum but the major alkaloid of some Nicotiana species, e. g. N. glutinosa (262). This dealkylation proceeds in a fairly non-specific manner as evidenced by the fact that nornicotine (**289**) is produced from (−)- and (±)-nicotine and from (±)-ethylnornicotine in the leaves of *N. glutinosa* (*55*). It has also been reported that the conversion of (−)-nicotine (**284**) to nornicotine (**289**) involves a partial racemization (*156*).

In a recent study, LEETE fed a mixture of (−)-[2′-^{3}H]-nicotine and (±)-[2′-^{14}C]-nicotine to *N. glauca* and, as expected, isolated radioactive nornicotine as the major metabolite (*177*). This had the same ^{3}H/^{14}C ratio as the mixture of precursors administered, which implies that in this plant (±)- and (−)-nicotine are demethylated at similar rates, an observation which does not, however, agree with a previous finding for *N. tabacum* (*157*). The radioactive nornicotine consisted of 48% (−)-[2′-^{14}C,^{3}H]-nornicotine and 52% (+)-[2′-^{14}C]-nornicotine. This indicates that the conversion of (−)-nicotine to (+)-nicotine, if it occurs, involves the loss of hydrogen from C-2′ (Scheme 28).

[2′-^{14}C]-Myosmine was also obtained from the plants to which the mixture of the labelled nicotines was fed, whereas no activity was found in anabasine (**294**), cotinine (**295**), 3-acetylpyridine (**296**) or nicotinic acid (**293**) (*177*). Myosmine (**288**) is probably formed by dehydrogenation

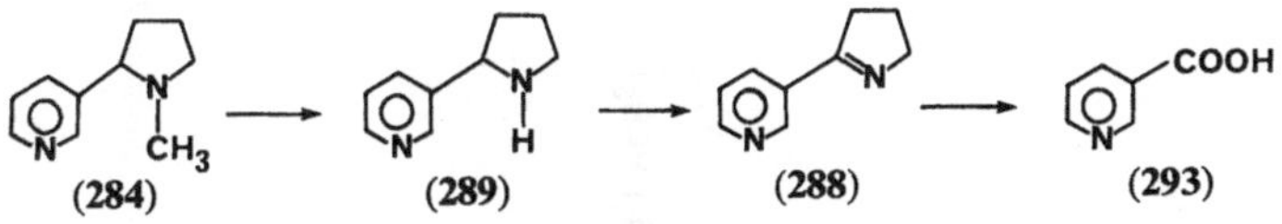

Scheme 28

of nornicotine (**289**) (*158*). The reaction sequence is, however, irreversible since [2-^{14}C]-myosmine failed to produce labelled nicotine or nornicotine. On the other hand, the isolation of nicotinic acid (**293**) labelled at the carboxyl group indicated that myosmine (**288**) is a direct precursor of this compound.

d) Post-Harvest Reactions

The post-harvest treatment of tobacco, *i. e.* curing, fermentation and aging, results in a certain loss of nicotine (**284**) and a concomitant accumulation of various degradation products. This degradation may be ascribed to the action of enzymes present in the tobacco leaf, of bacteria contaminating the leaf and of non-enzymic reactions. Thus, FRANKENBURG *et al.* (*91—94, 96*) isolated a series of nicotine transformation products (**290, 293, 295—299**) from a fermented Pennsylvania cigar filler tobacco whose nicotine content had undergone a marked reduction, 30—50%, as a result of the treatment. It was suggested that

4*

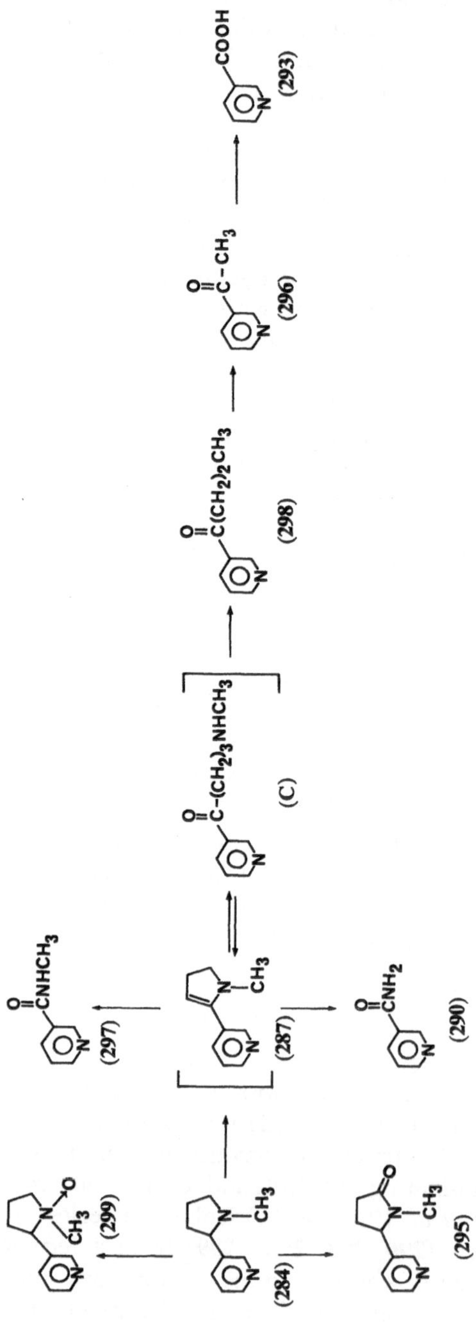

C. R. Enzell, I. Wahlberg, and A. J. Aasen:

Scheme 29

these compounds, which all retain the pyridine ring, are formed from nicotine (284) as illustrated in Scheme 29. Nicotine (284) is converted to the hypothetical intermediate N-methylmyosmine (287) (*117*) which is degraded to N-methylnicotinamide (297) and nicotinamide (290). The hydrated form of N-methylmyosmine, 3-pyridyl-3-methylaminopropyl ketone (pseudooxynicotine) (C) (*117*) gives rise to 3-pyridylpropyl ketone (298), which is degraded further with the formation of 3-acetylpyridine (296) and nicotinic acid (293). Nicotine (284) may also be converted to nicotine N'-oxide (299) and cotinine (295).

Support for this degradation pathway was obtained from experiments carried out *in vitro*. Thus, treatment of (−)-nicotine (284) with hydrogen peroxide afforded the same components as those encountered in the fermented tobacco. Nicotine N'-oxide (299), recently resolved into two diastereoisomers (*31*) of now known absolute configurations (*26*), and cotinine (295) were the major products, whereas N-methylnicotinamide (297), nicotinamide (290), 3-pyridylpropyl ketone (298), 3-acetylpyridine (296) and nicotinic acid (293) were minor components. Furthermore, the latter compounds (290, 293, 296—298) were also obtained when N-methyl-myosmine (287) was reacted with hydrogen peroxide. As expected, nicotine N'-oxide (299) and cotinine (295) were not detected in this case (*93, 94*).

Auto-oxidation of nicotine (284) resulted in the formation of nicotine N'-oxide (299), cotinine (295), nicotyrine (285), myosmine (288), nicotinic acid (293), methylamine and ammonia (*288*), whereas photo-oxygenation was found to proceed *via* an initial attack at the nitrogen of the pyrrolidine ring of nicotine (284). However, the products obtained were not fully characterised (*297*).

It should be noted, however, that the ultimate content of alkaloids is dependent on several factors such as type of tobacco and curing procedure. This is exemplified by a recent study on volatile constituents derived from a Virginia tobacco at different stages of flue-curing and aging (*291*). Thus, besides nicotine (284) a number of nitrogen-containing constituents are already present in green freeze-dried leaves of this tobacco. Of these N-methylmyosmine (287), myosmine (288), nicotyrine (285), N-methylnicotinamide (297) and cotinine (295) are fairly abundant (*cf.* Fig. 2). Moreover, flue-curing and aging lead to a reduction in the concentrations of all these compounds, whereas roughly constant levels are observed for certain constituents, *e. g.* nornicotine (289) and nicotine N'-oxide (299).

These results may still be in harmony with the degradation pathway suggested by FRANKENBURG (*93, 94*) if it is assumed that the intermediates are metabolised further during the curing and aging of the Virginia

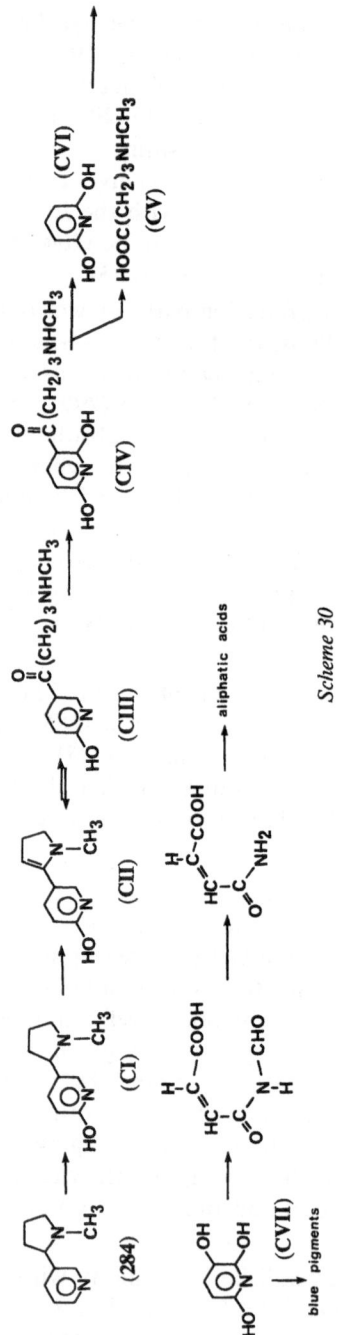

Scheme 30

tobacco. This view is corroborated by the fact that a number of structurally simple components accumulate in the aged tobacco.

Thus, higher concentrations of pyridine (300), α- and β-picoline (301, 302) are found in the flue-cured than in the green tobacco (291). Several constituents, not present in the green leaves, are formed during the curing and aging processes, e. g. the pyridine derivatives 3-cyano-pyridine (291), 2- and 3-ethylpyridine (303, 304), γ-picoline (305), 2,6-dimethylpyridine (306) and 2,3,6-trimethylpyridine (307). However, it must be emphasized that a thorough knowledge of the reactions which take place in the harvested leaf and give rise to these constituents can only be obtained by extensive tracer experiments.

e) Bacterial Degradation

A number of bacteria isolated from soil or tobacco leaves and seeds are able to utilize nicotine (284) as a source of carbon and nitrogen and thereby eventually give rise to products which do not retain the pyridine ring. Thus, detailed studies have shown that the degradation of nicotine (284) by the aerobic *Arthrobacter oxydans* proceeds as indicated in Scheme 30. Nicotine (284) is initially hydroxylated yielding 6-hydroxynicotine (CI) (58, 124) a reaction catalysed by an inducible nicotine oxidase. 6-Hydroxynicotine (CI) undergoes a dehydrogenation mediated by an oxidase and leading to the formation of 6-hydroxy-N-methylmyosmine (CII), which is in equilibrium with 6-hydroxypseudooxynicotine (CIII) (112, 125). An inducible ketone oxidase (109) catalyses the conversion of (CIII) to 2,6-dihydroxypseudooxynicotine (CIV) (228, 229) which undergoes cleavage to γ-methylaminobutyric acid (CV) and 2,6-dihydroxypyridine (CVI) (104). The latter is oxidised yielding 2,3,6-trihydroxypyridine (CVII), which may either serve as the precursor of·the nicotine blue pigments or may undergo ring cleavage yielding ultimately aliphatic acids (104, 127).

A species of bacteria, probably belonging to the *Pseudomonas*, has the ability to degrade nicotine (284) *via* the alternative pathway: nicotine (284), pseudooxynicotine (C), 3-succinoylpyridine (CVIII), 3-succinoyl-6-hydroxypyridine (CIX), aliphatic compounds (287, 289) (Scheme 31). Other, unclassified, bacteria seem to utilize both pathways for the degradation of nicotine (284) (93, 95).

The minor tobacco alkaloids nornicotine (289) and anabasine (294) are metabolised to aliphatic compounds by bacteria belonging to the *Pseudomonas*. 6-Hydroxymyosmine (CX) and 3-succinoyl-6-hydroxypyridine (CIX) were encountered as intermediates in the degradation of nornicotine (289) (286, 287) whereas anabasine (294) is decomposed *via* 1'2'-dehydro-6-hydroxyanabasine (CXI), and 3-glutaroyl-6-hydroxypyridine (CXII) (286).

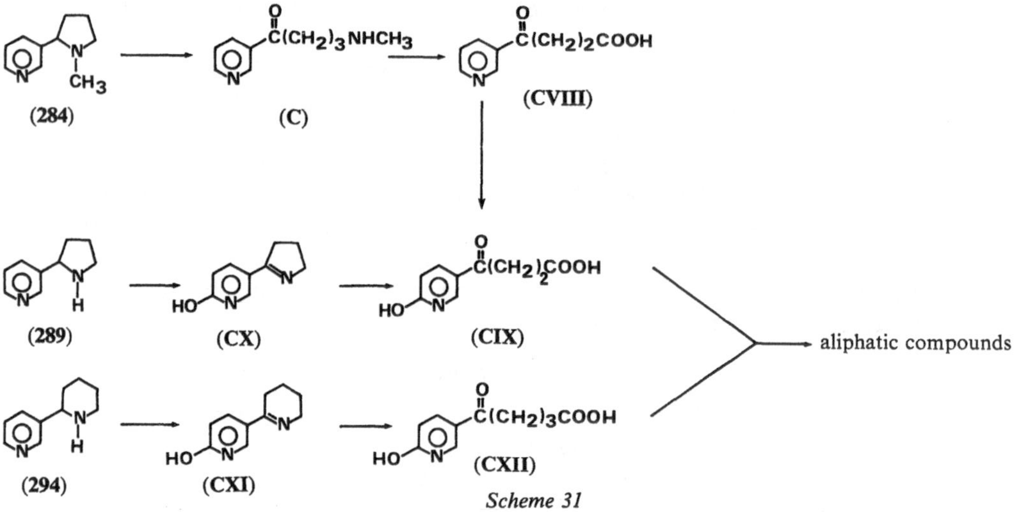

Scheme 31

2. Anabasine and Structurally Related Compounds

Eight compounds (**294, 308—314**) incorporating 3-pyridyl-piperidyl or 3-pyridyl-pyridyl moieties have been found in tobacco.

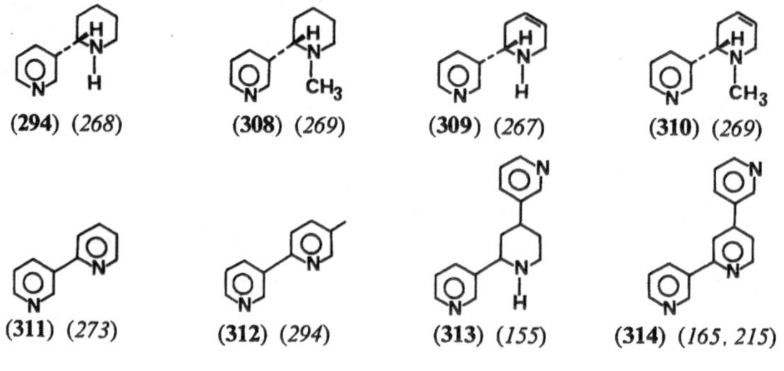

Of these, anabasine (**294**), first isolated from *Anabasis aphylla* (*209*), is the major alkaloid of *N. glauca* (*261*) and a minor component of *N. tabacum*. By analogy with the synthesis of myosmine (**288**) from ethyl nicotinoate (**LXXXII**) and *N*-benzoylpyrrolidone (**LXXXIV**) (*270*), anabasine (**294**) has been prepared (*271*) by condensation of ethyl nicotinoate (**LXXXII**) and *N*-benzoyl-2-piperidone (**CXIII**) and heating the product

(CXIV) with hydrochloric acid (Scheme 32). The anabasene thus formed, originally believed to have the double bond in the 2′,3′ position and later assigned the structure (CXV) on NMR evidence (73), is converted to anabasine (294) by catalytic hydrogenation.

Another synthetic procedure involves the acylation of 2-piperidone (315) with nicotinoyl chloride (CXVI). Pyrolysis of the product obtained (CXVII) yields anabasene (CXV), which is readily converted to anabasine (294) (199). This method has also been applied to the preparation of myosmine (288) by substituting 2-pyrrolidone (316) for 2-piperidone (315) (199, 200).

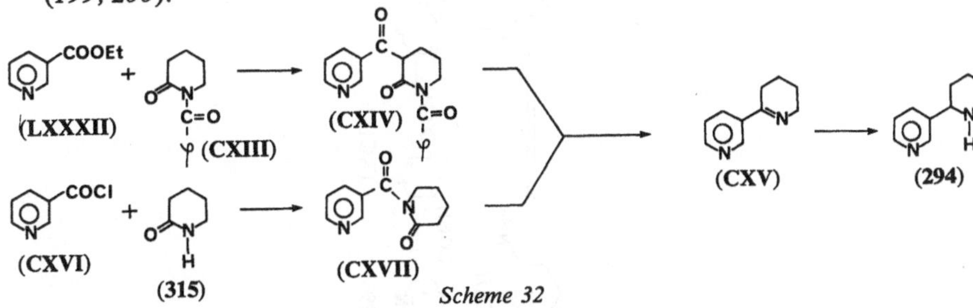

Scheme 32

N-methylanabasine (308), an alkaloid reported to be present in minute quantities in tobacco, is easily obtained from anabasine (294) by methylation using formaldehyde and formic acid or magnesium and methyl iodide. It has also been prepared in 11.6% yield (pH 10.6) by reacting glutaraldehyde (XCII), ammonia and N-methyl-Δ^1-piperideinium chloride in the presence of air (vide supra) (173).

The absolute configuration of the naturally occurring (−)-anabasine (294)* has been determined as S by catalytic hydrogenation of anabasine (294) to (−)2,3′-dipiperidyl (CXVIII) (226) and by oxidative degradation to L(−)-N-methyl-pipecolic acid (CXIX) (183). This result also establishes that (−)-N-methylanabasine (308), (−)-anatabine* (309) and (−)-N-methylanatabine (310) have the S-configuration, since these minor tobacco alkaloids are easily correlated with (−)-anabasine (294).

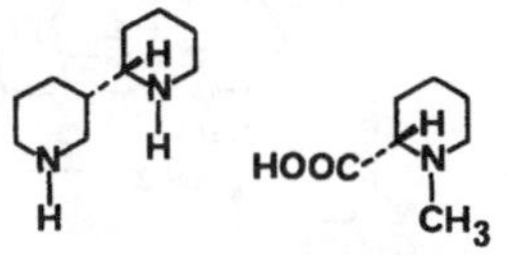

(CXVIII)   (CXIX)

Tracer experiments have shown that the piperidine and pyridine rings of (−)-anabasine (294) are derived from L-lysine (CXX) (105, 168) and nicotinic acid (293) (263) respectively. It has been suggested that the

---

* Anabasine (294) and anatabine (309) have been isolated in both (−) and (±) forms from different tobacco samples.

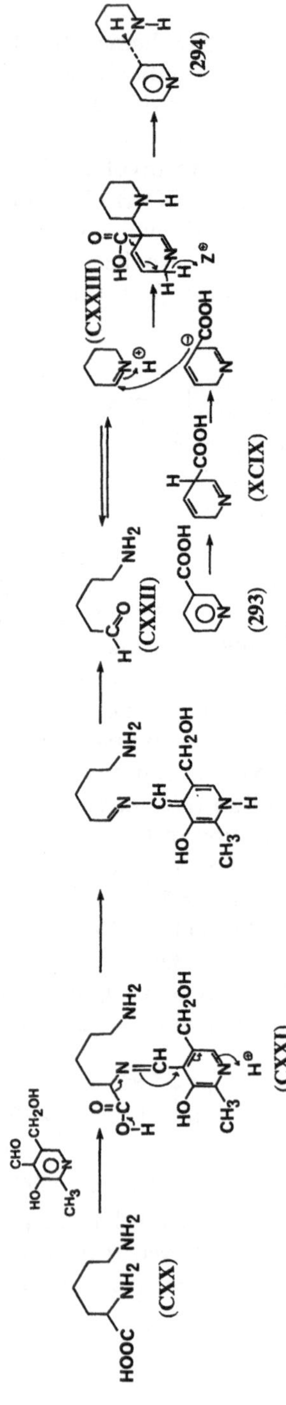

Scheme 33

biological conversion of these precursors to anabasine (294) proceeds as indicated in Scheme 33 (*174, 176*). Lysine (CXX) and pyridoxal form a Schiff base (CXXI). This undergoes decarboxylation and subsequent hydrolysis yielding pyridoxamine and 5-aminopentanal (CXXII), which cyclises to Δ^1-piperideine (CXXIII), the direct precursor of the piperidine ring (*171*). This series of transformations involves loss of the α-amino nitrogen (*179*) and retention of the C-2 hydrogen of lysine (CXX) (*115*) in agreement with the requirements indicated by feeding experiments. Δ^1-Piperideine (CXXIII) condenses with 3,6-dihydronicotinic acid (XCIX) yielding, after decarboxylation, (−)-anabasine (294).

The biosynthesis of *N*-methylanabasine (308) has not as yet been elucidated. However, it has been established by feeding experiments on *N. tabacum* and *N. glauca* that, in contrast to the case of anabasine (294), the piperidine ring of *N*-methylanabasine (308) is not derived from lysine (CXX) (*176*). Administration of $[2'-^{14}C]$-*N*-methyl-Δ^1-piperideinium chloride to these species, on the other hand, afforded fair amounts of (−)-$[2'-^{14}C]$-*N*-methylanabasine having virtually the same specific activity as the labelled precursor. However, since *N*-methylanabasine (308) was not present at a detectable level in plants which had not been fed the piperideinium salt, it was concluded that the formation of *N*-methylanabasine (308) in the feeding experiments represented an aberrant biosynthesis from an unnatural precursor (*176*).

A total synthesis of anatabine (309) and *N*-methylanatabine (310), which confirms their previous formulations as 1,2,3,6-tetrahydro-2,3'-dipyridine derivatives (*267, 269*), has been reported (*221*) (Scheme 34). It involves the condensation of the *bis*-carbamate (CXXIV), obtained from 3-formylpyridine (292) and ethyl carbamate, with 1,3-butadiene (CXXV) using a boron trifluoride di-(acetic acid) complex as a catalyst. The product obtained (CXXVI) was reduced with lithium aluminium hydride giving *N*-methylanatabine (310) in 28% yield. Hydrolysis of (CXXVI) afforded anatabine (309).

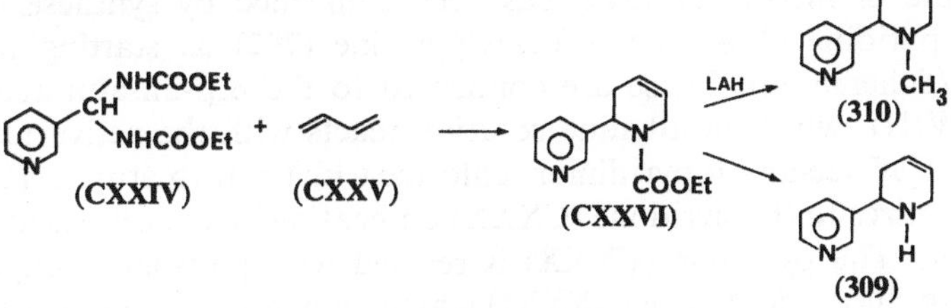

Scheme 34

Despite their small structural differences, tracer experiments have demonstrated that (−)-anatabine (309) and (−)-anabasine (294) are formed *via* different biosynthetic routes. Thus, administration to *N. glu-*

tinosa plants of [2-^{13}C]-lysine, which is efficiently incorporated in anabasine (294) (*168*), or of [2-^{14}C]-4-hydroxylysine fails to produce radioactive anatabine (309) (*155*). Moreover, whereas [6-^{14}C]-anabasine is
isolated from plants fed with [6-^{14}C]-nicotinic acid (*174, 263*), the activity
of the isolated anatabine (309) is equally divided between C-6 and C-6′,
i. e. both heterocyclic rings of anatabine (309) are derived from nicotinic
acid (293) (*174*). On the basis of this finding it has been suggested that the
formation of anatabine (309) may take place as shown in Scheme 35 (*174*).
Nicotinic acid (293) is initially converted to the 3,6-dihydroderivative
(XCIX), which is decarboxylated yielding 2,5-dihydropyridine (CXXVII).
This undergoes self-condensation and subsequent dehydrogenation to
afford (−)-anatabine (309).

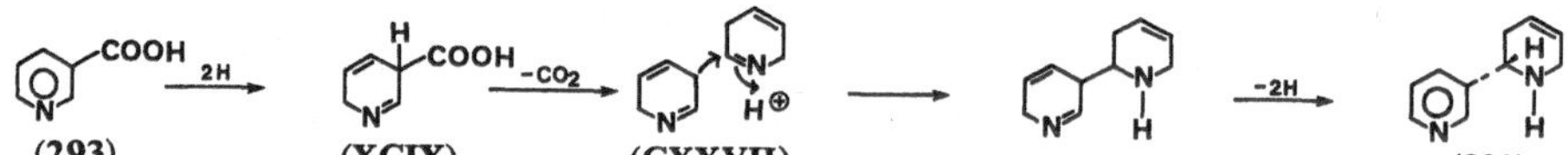

Scheme 35

2,3′-Dipyridyl (311), presumably identical with isonicoteine (*205*),
and 5-methyl-2,3′-dipyridyl (312) are minor components found in various
tobaccos. However, the quantity of 2,3′-dipyridyl (311) in a cigar tobacco
was reported to increase drastically on fermentation. It was assumed that
this compound is formed in the fermenting leaf by dehydrogenation of
anabasine (294) (*92*).

Nicotelline (314) was first discovered in tobacco in 1901 and later
recognised as 2,4-di(3-pyridyl)pyridine, whereas anatalline (313), a structurally related tobacco constituent, was isolated relatively recently. The
structure of nicotelline (314) has been confirmed by synthesis using 3-
acetyl-pyridine (296) and 3-formylpyridine (292) as starting materials
(*284*) (Scheme 36). These are condensed to the α,β-unsaturated ketone
(CXXVIII), which in alkaline solution reacts with the active methylene
group of N-acetamidopyridinium chloride with the formation of (CXXIX).
This cyclizes to the pyridone (CXXX) on heating in an acetic acid/ethanol
solution. The pyridone (CXXX) is reacted with phosphorusoxychloride
yielding chloronicotelline (CXXXI) which on hydrogenation using Raney-
nickel in the presence of sodium ethoxide is converted to nicotelline (314).

Although tracer experiments have not been carried out, it has been
suggested that anatalline (314) and nicotelline (315) are formed in the
tobacco plant by trimerization of dihydropyridines (*174*).

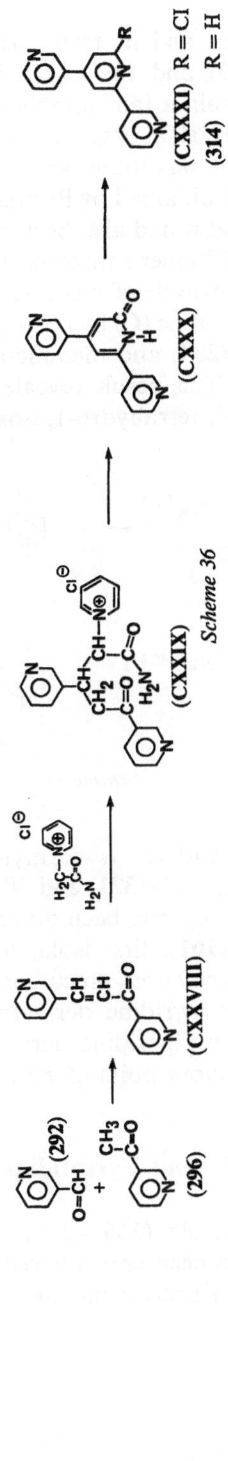

Scheme 36

3. Others

In addition to nicotine and its transformation products (**284—285, 287—290, 293, 295—299**) and the anabasine group (**294, 308—314**), tobacco is known to contain a fair number of pyridine derivatives (**291, 292, 300—307, 317—331, 100, 101**). Of these, metanicotine (**317**) had been known long before it was found as a constituent of a nicotine-free tobacco (1953). It was first obtained by PINNER on attempted benzoylation of nicotine (**284**) (*217*), and it had also been prepared in the course of the structure determination of Pinner's nicotone (**CXXXII**) (*223*) (Scheme 37). The latter, obtained by pyrolysis of nicotine N'-oxide (**299**) *in vacuo*, was converted to pseudooxynicotine (**C**) by heating with hydrochloric acid and to a mixture of nicotine (**284**) and metanicotine (**317**) by reduction and subsequent dehydration. This result revealed that nicotone is identical with 2-methyl-6-(3-pyridyl)-tetrahydro-1,2-oxazine (**CXXXII**).

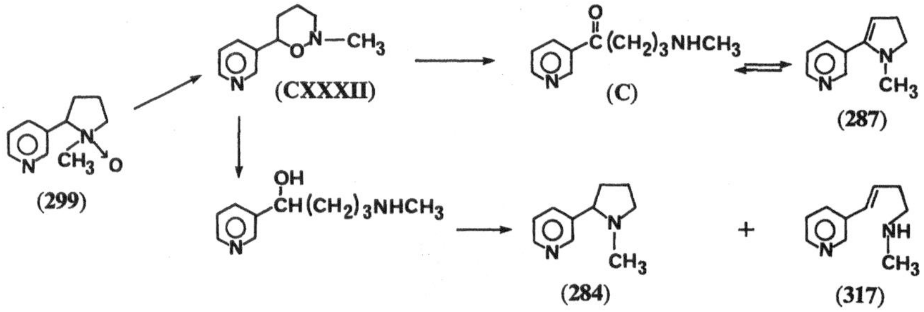

Scheme 37

The nornicotine derivatives N'-formyl- (**318**), N'-acetyl- (**319**), N'-hexanoyl- (**320**), N'-octanoyl- (**321**) and N'-nitrosonornicotine (**322**), a potential carcinogen, have recently been obtained from various tobaccos. The alkaloids (**100**) and (**101**), first isolated from Burley tobacco, are probably formed from a carotenoid precursor (*cf.* Scheme 7).

Several of the tobacco pyridine derivatives are structurally simple compounds. These and the piperidine derivatives (**315, 332—335**) may be of importance from a flavour point of view.

VII. Pyrrole and Pyrrolidine Derivatives

A fair number of pyrrole (**336—350**) and pyrrolidine (**286, 316, 351—359**) derivatives have been encountered in tobacco, some of which are formed during tobacco processing. Thus, it is evident from the gas

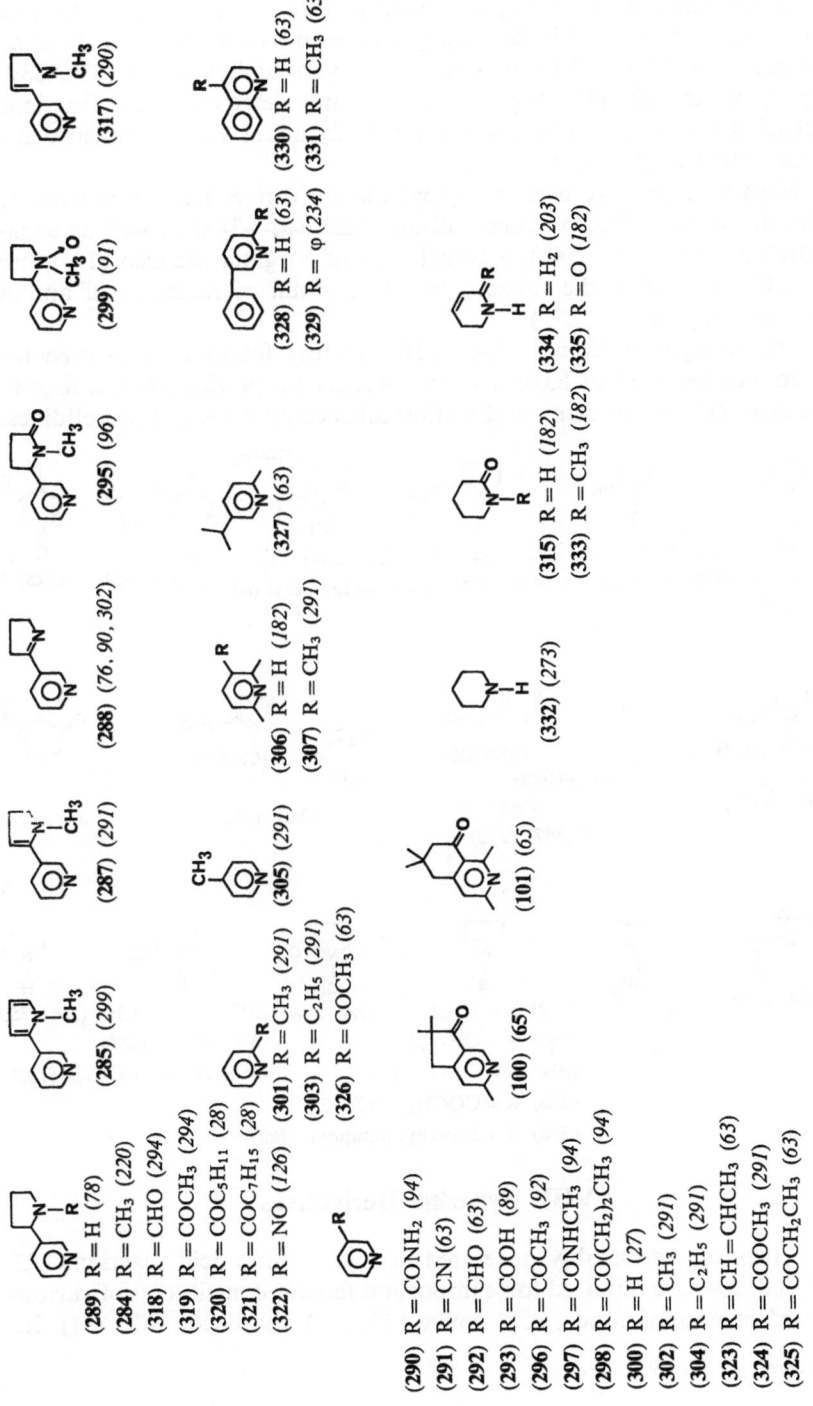

(317) (290)

(330) R = H (63)
(331) R = CH₃ (63)

(299) (91)

(328) R = H (63)
(329) R = φ (234)

(295) (96)

(327) (63)

(334) R = H₂ (203)
(335) R = O (182)

(288) (76, 90, 302)

(306) R = H (182)
(307) R = CH₃ (291)

(315) R = H (182)
(333) R = CH₃ (182)

(332) (273)

(287) (291)

(305) (291)

(101) (65)

(285) (299)

(301) R = CH₃ (291)
(303) R = C₂H₅ (291)
(326) R = COCH₃ (63)

(100) (65)

(289) R = H (78)
(284) R = CH₃ (220)
(318) R = CHO (294)
(319) R = COCH₃ (294)
(320) R = COC₅H₁₁ (28)
(321) R = COC₇H₁₅ (28)
(322) R = NO (126)

(290) R = CONH₂ (94)
(291) R = CN (63)
(292) R = CHO (63)
(293) R = COOH (89)
(296) R = COCH₃ (92)
(297) R = CONHCH₃ (94)
(298) R = CO(CH₂)₂CH₃ (94)
(300) R = H (27)
(302) R = CH₃ (291)
(304) R = C₂H₅ (291)
(323) R = CH=CH−CHCH₃ (63)
(324) R = COOCH₃ (291)
(325) R = COCH₂CH₃ (63)

chromatograms shown in Fig. 1 that the concentration of 2-acetylpyrrole (**337**; Fig. 1, peak 105) is drastically increased as a result of flue-curing and aging. Similarly, 2-formyl-and 2-formyl-5-methylpyrrole (**336, 339**; Fig. 1, peaks 109, 113; Fig. 2, peak 25) not detected in the fractions derived from the green leaves, are fairly abundant volatile constituents of the processed leaves (*291*).

Most recently, five new formylpyrrole derivatives have been isolated from flue-cured tobacco. These compounds (**344—348**) as well as tetra-hydroindolizin-8-one (**349**), a constituent of Virginia tobacco (*291*), are probably formed in the tobacco leaf as a result of reactions of amino acids with sugars (*182, 257*).

The *N*-acylpyrrolidines (**354—356**) recently found in flue-cured tobacco, are reported to have desirable flavour properties (*182*), a feature also characteristic of many of the other tobacco pyrroles and pyrrolidines.

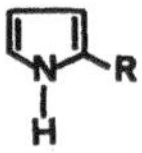

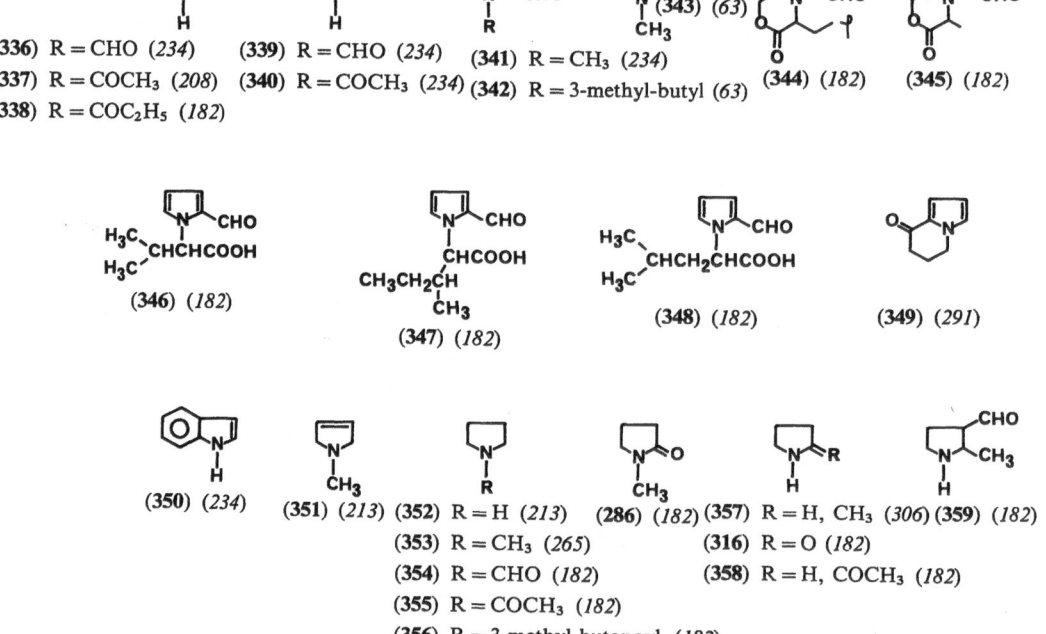

(**336**) R = CHO (*234*)   (**339**) R = CHO (*234*)   (**341**) R = CH₃ (*234*)

(**337**) R = COCH₃ (*208*)   (**340**) R = COCH₃ (*234*) (**342**) R = 3-methyl-butyl (*63*)   (**344**) (*182*) 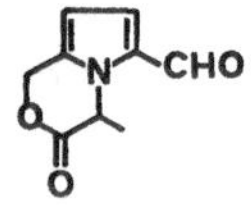   (**345**) (*182*)

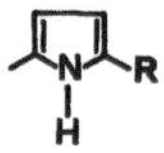

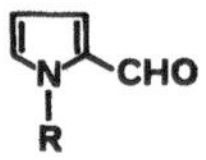

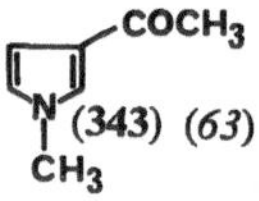

(**338**) R = COC₂H₅ (*182*)

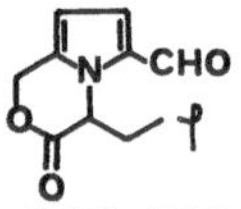

(**346**) (*182*)   (**347**) (*182*)   (**348**) (*182*)   (**349**) (*291*)

(**350**) (*234*)   (**351**) (*213*) (**352**) R = H (*213*)   (**286**) (*182*) (**357**) R = H, CH₃ (*306*) (**359**) (*182*)

(**353**) R = CH₃ (*265*)   (**316**) R = O (*182*)

(**354**) R = CHO (*182*)   (**358**) R = H, COCH₃ (*182*)

(**355**) R = COCH₃ (*182*)

(**356**) R = 3-methyl-butanoyl (*182*)

VIII. Pyrazine Derivatives

Pyrazines, presumably generated in sugar-amino acid reactions (*52, 161, 162, 204*) are reported to be important flavour constituents of various roasted or cooked foods, *e. g.* coffee (*30, 110*) and cocoa (*87, 191*). Investigations carried out in recent years have disclosed that a number of pyrazine derivatives (360—377) are also present in tobacco and that these are likely to contribute to the aroma of the leaf. Although several pyrazines are found in the green leaves of Virginia tobacco, curing and aging lead to an accumulation of a number of derivatives, e. g. 2,5-dimethyl, trimethyl and tetramethylpyrazine (370, 372, 376) (291).

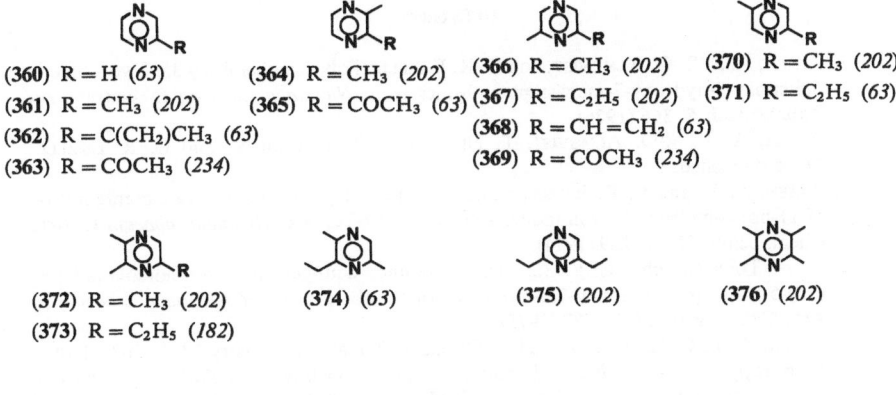

(360) R = H (63)
(361) R = CH₃ (202)
(362) R = C(CH₂)CH₃ (63)
(363) R = COCH₃ (234)

(364) R = CH₃ (202)
(365) R = COCH₃ (63)

(366) R = CH₃ (202)
(367) R = C₂H₅ (202)
(368) R = CH=CH₂ (63)
(369) R = COCH₃ (234)

(370) R = CH₃ (202)
(371) R = C₂H₅ (63)

(372) R = CH₃ (202)
(373) R = C₂H₅ (182)

(374) (63)

(375) (202)

(376) (202)

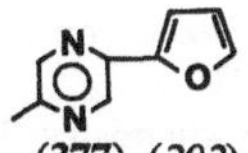

(377) (202)

IX. Miscellaneous Nitrogen-Containing Constituents

Apart from the compounds mentioned above a number of nitrogen-containing constituents have been found in tobacco. These are exemplified by a few nitrogen heterocyclics such as harmane (378), norharmane (379), benzothiazole (380), acetylimidazole (381), a series of succinimides (386—391) and maleimides (392—395) as well as a number of amines (203, 273, 306) and a few form- (63) and acetamides (63, 182). A study of Latakia tobacco, a fire-cured tobacco, led to the identification of 27 anilines and aliphatic amines (133).

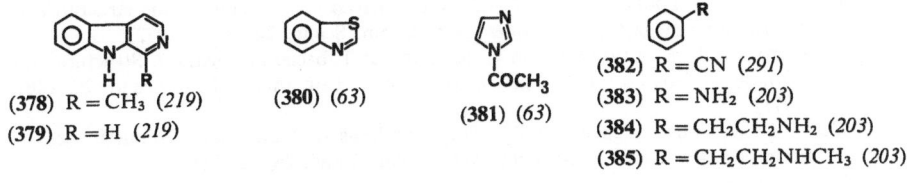

(378) R = CH₃ (219)
(379) R = H (219)

(380) (63)

(381) (63)

(382) R = CN (291)
(383) R = NH₂ (203)
(384) R = CH₂CH₂NH₂ (203)
(385) R = CH₂CH₂NHCH₃ (203)

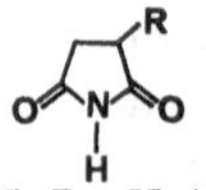

(**386**) R = H (*182*)
(**387**) R = CH_3 (*182*)

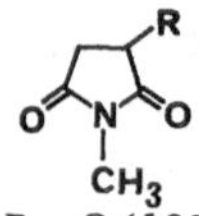

(**388**) R = Q (*182*)
(**389**) R = C_2H_5 (*182*)

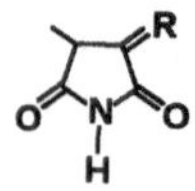

(**390**) R = H, C_2H_5 (*64*)
(**391**) R = $CHCH_3$ (*182*)

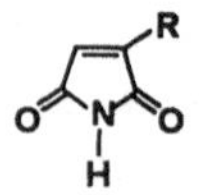

(**392**) R = H (*182*)
(**393**) R = C_3H_7 (*182*)

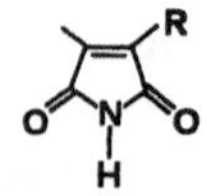

(**394**) R = CH_3 (*182*)
(**395**) R = C_2H_5 (*63*)

References

1. Aasen, A. J., S.-O. Almqvist, and C. R. Enzell: Tobacco chemistry 35. Two isomeric 5,6-epoxy-3-hydroxy-7-megastigmen-9-ones from *Nicotiana tabacum* L. Beiträge zur Tabakforsch. **8**, 366 (1976).

2. Aasen, A. J., S.-O. Almqvist, T. Nishida, J. R. Hlubucek, and C. R. Enzell: To be published.

3. Aasen, A. J., and C. R. Enzell: Tobacco chemistry 30. The absolute configuration of 11-nor-8-hydroxy-9-drimanone, a constituent of Greek *Nicotiana tabacum* L. Acta Chem. Scand. **B28**, 1239 (1974).

4. — — Tobacco chemistry 32. The absolute configuration of norsolanadione, (5*S*)-5-isopropyl-3*E*-nonen-2,8-dione, a nor-thunberganoid of *Nicotiana tabacum* L. Acta Chem. Scand. **B29**, 528 (1975).

5. Aasen, A. J., C. R. Enzell, and T. Chuman: Tobacco chemistry 33. (6*S*)-3-Methyl-6-isopropyl-9-oxo-2*E*,4*E*-decadienoic acid from Turkish *Nicotiana tabacum* L. Assignment of absolute configuration. Agr. Biol. Chem. **39**, 2085 (1975).

6. Aasen, A. J., J. R. Hlubucek, S.-O. Almqvist, B. Kimland, and C. R. Enzell: Tobacco chemistry 20. Structure and synthesis of three new tobacco constituents of probable isoprenoid origin. Acta Chem. Scand. **27**, 2405 (1973).

7. Aasen, A. J., J. R. Hlubucek, and C. R. Enzell: Tobacco chemistry 24. (9*R*)-9-Hydroxy-4-megastigmen-3-one, a new tobacco constituent. Acta Chem. Scand. **B28**, 285 (1974).

8. — — — Tobacco chemistry 27. The structures of four stereoisomeric 8,12ξ-epoxylabd-14-en-13ξ-ols isolated from Greek *Nicotiana tabacum* L. Acta Chem. Scand. **B29**, 589 (1975).

9. — — — Tobacco chemistry 29. (7*S*)-10-oxo-4ξ-methyl-7-isopropyl-5*E*-undecen-4-olide, a new thunbergane-type nor-isoprenoid isolated from Greek *Nicotiana tabacum* L. Acta Chem. Scand. **B29**, 677 (1975).

10. Aasen, A. J., N. Junker, C. R. Enzell, J.-E. Berg, and A. M. Pilotti: Tobacco chemistry 36. Absolute configuration of tobacco thunberganoids. Tetrahedron Letters 2607 (1975).

11. Aasen, A. J., B. Kimland, S.-O. Almqvist, and C. R. Enzell: Tobacco chemistry 9. 5-Methoxy-6,7-dimethylbenzofuran, a new tobacco constituent. Acta Chem. Scand. **25**, 3182 (1971).

12. — — — — Tobacco chemistry 13. 8,13-Epoxylabd-14-en-12-one and 8,13 β-epoxylabd-14-en-12-one, two new diterpenoids from tobacco. Acta Chem. Scand. **26**, 832 (1972).

13. — — — — Tobacco chemistry 15. New tobacco constituents — the structures of five isomeric megastigmatrienones. Acta Chem. Scand. **26**, 2573 (1972).

14. Aasen, A. J., B. Kimland, and C. R. Enzell: Tobacco chemistry 7. Structure and synthesis of 3-oxo-α-ionol, a new tobacco constituent. Acta Chem. Scand. **25**, 1481 (1971).

15. — — — Tobacco chemistry 11. Total synthesis of 5-methoxy-6,7-dimethylbenzofuran, a new tobacco constituent. Acta Chem. Scand. **25**, 3537 (1971).

16. — — — Tobacco chemistry 18. Absolute configuration of (9R)-9-hydroxy-4,7E-megastigmadien-3-one (3-oxo-α-ionol). Acta Chem. Scand. 27, 2107 (1973).
17. AASEN, A. J., B. KIMLAND, J. R. HLUBUCEK, and C. R. ENZELL: Unpublished result.
18. AASEN, A. J., T. NISHIDA, C. R. ENZELL, and M. DEVREUX: Tobacco chemistry 37. The absolute configuration of prenylsolanone, (9S)-6,12-dimethyl-9-isopropyltrideca-5E,10E.12-trien-2-one, a nor-thunberganoid of Nicotiana tabacum L. Acta Chem. Scand. 30, 178 (1976).
19. AASEN, A. J., Å. PILOTTI, C. R. ENZELL, J.-E. BERG, and A.-M. PILOTTI: To be published.
20. AASEN, A. J., C. H. G. VOGT, and C. R. ENZELL: Tobacco chemistry 28. Structure and synthesis of drim-8-en-7-one, a new tobacco constituent. Acta Chem. Scand. B29, 51 (1975).
21. ALMQVIST, S.-O., A. J. AASEN, J. R. HLUBUCEK, B. KIMLAND, and C. R. ENZELL: Tobacco chemistry 23. Structures and syntheses of four new nor-isoprenoid furans from tobacco. Acta Chem. Scand. B28, 528 (1974).
22. AYERS, J. E., M. J. FISHWICK, D. G. LAND, and T. SWAIN: Off-flavour of dehydrated carrot stored in oxygen. Nature 203, 81 (1964).
23. BAILEY, J. A., R. S. BURDEN, and G. G. VINCENT: Capsidiol: an antifungal compound produced in Nicotiana tabacum and Nicotiana clevelandii following infection with tobacco necrosis virus. Phytochem. 14, 597 (1975).
24. BALDWIN, J. E., H. H. BASSON, and H. KRAUSS JR.: The cleavage of aromatic nuclei with singlet oxygen: Significance in biosynthetic processes. Chem. Commun. 984 (1968).
25. BAYLEY, W. C., A. K. BOOSE, R. M. IKEDA, R. H. NEWMAN, H. V. SECOR, and C. VARSEL: The isolation from tobacco of 2-hydroxy-2,6,6-trimethylcyclohexylidene acetic acid-γ-lactone and its synthesis. J. Organ. Chem. 33, 2819 (1968).
26. BECKET, A. H., P. JENNER, and J. W. GORROD: Characterization of the diastereo-isomers of nicotine-1 -N-oxide, a metabolite of nicotine, and other possible oxidation products by nuclear magnetic resonance spectroscopy. Xenobiotica 3, 557 (1973).
27. BODNÁR, I., and L. NAGY: Pyridine content of tobacco. Z. Untersuch. Lebensm. 74, 302 (1937).
28. BOLT, A. J. N.: 1'-Hexanoylnornicotine and 1'-octanoylnornicotine from tobacco. Phytochem. 11, 2341 (1972).
29. BOLT, A. J. N., and R. E. CLARKE: Cholesterol glucoside in tobacco. Phytochem. 9, 819 (1970).
30. BONDAROVICH, H. A., P. FRIEDEL, V. KRAMPL, J. A. RENNER, F. W. SHEPHARD, and M. GIANTURCO: Volatile constituents of coffee. Pyrazines and other compounds. J. Agric. Food Chem. 15, 1093 (1967).
31. BOOTH, J., and E. BOYLAND: The metabolism of nicotine into two optically-active stereoisomers of nicotine-1'-oxide by animal tissues in vitro and by cigarette smokers. Biochem. Pharmacol. 19, 733 (1970).
32. BRANDÄNGE, S., and L. LINDBLOM: N-vinyl as N–H protecting group — a convenient synthesis of myosmine. Acta Chem. Scand. B30, 93 (1976).
33. BREUER, E., and D. MELUMAD: A one-step synthesis of nicotine from cyclopropyl 3-pyridyl ketone. Tetrahedron Letters 3595 (1969).
34. BURDEN, R. S., J. A. BAILEY, and G. G. VINCENT: Glutinosone, a new antifungal sesquiterpene from Nicotiana glutinosa infected with tobacco mosaic virus. Phytochem. 14, 221 (1975).
35. BURDEN, R. S., and H. F. TAYLOR: The structure and chemical transformations of xanthoxin. Tetrahedron Letters 4071 (1970).
36. CAMBIE, R. C., K. N. JOBLIN, and A. F. PRESTON: Chemistry of the Podocarpaceae. XXX. Conversion of 8α,13-epoxylabd-14-ene into a compound with an ambergris-type odour. Austral. J. Chem. 24, 583 (1971).

37. Cambie, R. C., K. N. Joblin, and A. F. Preston: Chemistry of the *Podocarpaceae.* XXXIV. Some oxidation products of (13*R*)-labda-8(17),14-dien-13-ol (manool). Austral. J. Chem. **24**, 2365 (1971).

38. Carruthers, W., and J. R. Plimmer: Sterols in green tobacco leaf. Chem. and Ind. 48 (1959).

39. Chakraborty, M. K., and J. A. Weybrew: The chemistry of tobacco trichomes. Tob. Sci. **7**, 122 (1963).

40. Chan, H. W.-S.: Singlet oxygen analogs in biological systems. Coupled oxygenation of 1,3-diens by soybean lipoxidase. J. Amer. Chem. Soc. **93**, 2357 (1971).

41. Chirkova, M. A., A. K. Dzizenko, and V. A. Pentegova: Neutral substances of the resin of *Abies sibirica.* II. Structure of a diterpenic hydroxy ketone. Khim. Prir. Soedin. **3**, 86 (1967).

42. Chuman, T., H. Kaneko, T. Fukuzumi, and M. Noguchi: Isolation of two terpenoid acids, 4-isopropyl-7-methyl-5*E*,7-octadienoic acid and 3-isopropyl-6-methyl-4*E*,6-heptadienoic acid from Turkish tobacco. Agr. Biol. Chem. **38**, 2295 (1974).

43. — — — — Acidic aroma constituents of Turkish tobacco. Terpenoid acids related to tobacco thunberganoids. Agr. Biol. Chem. **40**, 587 (1976).

44. Chuman, T., and M. Noguchi: Isolation of a new terpenoid acid 2-methyl-5-isopropyl-1-cyclopentene-1-carboxylic acid from Turkish tobacco. Agr. Biol. Chem. **39**, 567 (1975).

45. — — Isolation of new terpenoid acids (−)-3-methyl-6-isopropyl-9-oxo-2*E*,4*E*-decadienoic acid and 3-isopropyl-6-oxo-2*E*-heptenoic acid from Turkish tobacco. Agr. Biol. Chem. In press.

46. Colledge, A., W. W. Reid, and R. Russell: The diterpenoids of *Nicotiana* species and their potential technological significance. Chem. and Ind. 570 (1975).

47. Comes, R. A., M. T. Core, M. D. Edmonds, W. B. Edwards, and R. W. Jenkins: Preparation of carbon-14 labelled tobacco constituents. II. The synthesis of dl-nicotine (2′-^{14}C). J. Label. Compounds **9**, 253 (1973).

48. Cornforth, J. W., B. V. Milborrow, and G. Ryback: Synthesis of (±)abscisin II. Nature **206**, 715 (1965).

49. Courtney, J. L., and S. McDonald: A new C-20 α,β-unsaturated aldehyde (3,7,13-trimethyl-10-isopropyl-2,6,11,13-tetradecatetraen-1-al) from tobacco. Tetrahedron Letters 459 (1967).

50. Creasy, P. J., and M. J. Saxby: Steam volatile acids of Latakia tobacco leaf. Phytochem. **8**, 2427 (1969).

51. Dauben, W. G., W. E. Thiessen, and P. R. Resnick: Cembrene, a fourteen-membered ring diterpene hydrocarbon. J. Amer. Chem. Soc. **84**, 2015 (1962).

52. Dawes, I. W., and R. A. Edwards: Methyl-substituted pyrazines as volatile reaction products of heated aqueous aldose, amino acid mixtures. Chem. and Ind. 2203 (1966).

53. Dawson, R. F.: On biogenesis of nornicotine and anabasine. J. Amer. Chem. Soc. **67**, 503 (1945).

54. — Advances in enzymology, New York **8**, 203 (1948).

55. — Alkaloid biogenesis. III. Specificity of the nicotine-nornicotine conversion. J. Amer. Chem. Soc. **73**, 4218 (1951).

56. Dawson, R. F., D. R. Christman, A. D'Adamo, M. L. Solt, and A. P. Wolf: The biosynthesis of nicotine from isotopically labelled nicotinic acids. J. Amer. Chem. Soc. **82**, 2628 (1960).

57. Dawson, R. F., D. R. Christman, and R. Ch. Anderson: Alkaloid biogenesis. IV. The non-availability of nicotinic acid-[carboxyl-C^{14}] and its ethyl ester for nicotine biosynthesis. J. Amer. Chem. Soc. **75**, 5114 (1953).

58. Decker. K., H. Eberwein, F. A. Gries, und M. Brühmüller: Über den Abbau

des Nikotins durch Bakterienenzyme. VI. *L*-6-Hydroxynicotin als erstes Zwischenprodukt. Biochem. Z. **334**, 227 (1961).

59. DECKER, K., und R. SAMMECK: Enzymchemische Untersuchungen zum Nikotinabbau in der Kaninchenleber. Biochem. Z. **340**, 326 (1964).

60. DEMOLE, E.: Private communication.

61. — Chemistry of Burley tobacco flavor (*Nicotiana tabacum* L.). Novel constituents and newer syntheses. VI. International Congress of Essentials Oils. San Francisco U.S.A. 1974.

62. DEMOLE, E., et D. BERTHET: Identification de la damascénone et de la β-damascone dans le tabac Burley. Helv. Chim. Acta **54**, 681 (1971).

63. — — A chemical study of Burley tobacco flavour (*Nicotiana tabacum* L.). I. Volatile to medium volatile constituents. Helv. Chim. Acta **55**, 1866 (1972).

64. — — A chemical study of Burley tobacco flavour (*Nicotiana tabacum* L.). II. Medium volatile, free acidic constituents. Helv. Chim. Acta **55**, 1898 (1972).

65. DEMOLE, E., and C. DEMOLE: A chemical study of Burley tobacco flavour (*Nicotiana tabacum* L.) V. Identification and synthesis of the novel terpenoid alkaloids 1,3,6,6-tetramethyl-5,6,7,8-tetrahydro-isoquinolin-8-one and 3,6,6-trimethyl-5,6-dihydro-7*H*-2-pyrindin-7-one. Helv. Chim. Acta **58**, 523 (1975).

66. — — A chemical study of Burley tobacco flavour (*Nicotiana tabacum* L.). VII. Identification and synthesis of twelve irregular terpenoids related to solanone including 7,8-dioxabicyclo[3.2.1]octane and 4,9-dioxabicyclo[3.3.1]nonane derivatives. Helv. Chim. Acta **58**, 1867 (1975).

67. DEMOLE, E., C. DEMOLE, and D. BERTHET: A chemical study of Burley tobacco flavour (*Nicotiana tabacum* L.). III. Structure determination and synthesis of 5-(4-methyl-2-furyl)-6-methylheptan-2-one ("Solanofuran") and of 3,4,7-trimethyl-1,6-dioxa-spiro[4.5]dec-3-en-2-one ("Spiroxabovolide"). Two new flavour components of Burley tobacco. Helv. Chim. Acta **56**, 265 (1973).

68. — — — A chemical study of Burley tobacco flavour (*Nicotiana tabacum* L.). IV. Identification of seven new solanone metabolites including 7,8-dioxabicyclo[3.2.1]octane and 4,9-dioxabicyclo[3.3.1]nonane derivatives. Helv. Chim. Acta **57**, 192 (1974).

69. DEMOLE, E., and P. ENGGIST: Novel synthesis of 3,5,5-trimethyl-4-(2-butenylidene)-cyclohex-2-en-1-one, a major constituent of Burley tobacco flavour. Helv. Chim. Acta **57**, 2087 (1974).

70. — — A chemical study of Burley tobacco flavour (*Nicotiana tabacum* L.). VI. Identification and synthesis of four irregular terpenoids related to solanone, including a prenylsolanone. Helv. Chim. Acta **58**, 1602 (1975).

71. DEWEY, L. J., R. U. BYERRUM, and C. D. BALL: Biosynthesis of the pyrrolidine ring of nicotine. Biochim. Biophys. Acta **18**, 141 (1955).

72. DREW, M. G. B., D. H. TEMPLETON, and A. ZALKIN: The crystal and molecular structure of cembrene. Acta Cryst. **B25**, 261 (1969).

73. DUFFIELD, A. M., H. BUDZIKIEWICZ, and C. DJERASSI: Mass spectrometry in structural and stereochemical problems. LXXII. A study of the fragmentation processes of some tobacco alkaloids. J. Amer. Chem. Soc. **87**, 2926 (1965).

74. DYMICKY, M., and R. L. STEDMAN: Composition studies on tobacco. IV. Ergosterol, γ-sitosterol and a partially characterized steroidal glycoside from flue-cured leaves. Tob. Sci. **3**, 4 (1959).

75. — — Composition studies on tobacco. IX. Campesterol from flue-cured leaves. Tob. Sci. **3**, 179 (1959).

76. EDDY, C. R., and A. EISNER: Infrared spectra of nicotine and some of its derivatives. Anal. Chem. **26**, 1428 (1954).

77. EGGER, K., A. G. DABBAGH, und H. NITSCHE: Überführung von Neoxanthin in Diadinochrom. Tetrahedron Letters 2995 (1969).

78. EHRENSTEIN, M.: Zur Kenntnis der Alkaloide des Tabaks. Chem. Ber. **64**, 627 (1931).

79. Enzell, C. R.: Mass spectrometric studies of carotenoids. J. Pure and Applied Chem. **20**, 497 (1969).

80. Enzell, C. R., B. Kimland, and L.-E. Gunnarsson: Tobacco Chemistry 5. Norsolanesene, a C_{44}-isoprenoid hydrocarbon from tobacco. Tetrahedron Letters 1983 (1971).

81. Eppenberger, U., L. Hirth, und G. Ourisson: Anaerobische Cyclisierung von Squalen-2,3-epoxyd zu Cycloartenol in Gewebekulturen von *Nicotiana tabacum* L. Eur. J. Biochem. **8**, 180 (1969).

82. Erickson, R. E., C. H. Schunk, N. R. Trenner, B. H. Arison, and K. Folkers: Coenzyme Q. XI. The structure of solanesol. J. Amer. Chem. Soc. **81**, 4999 (1959).

83. Etoh, H., K. Ina, and M. Iguchi: Photosensitized oxygenation of α-pyran derived from β-ionone. Agr. Biol. Chem. **37**, 2241 (1973).

84. Feeney, J., and F. W. Hemming: Nuclear magnetic resonance spectrometry of naturally occurring polyprenols. Analyt. Biochem. **20**, 1 (1967).

85. Fikensher, L. H.: Nicotine, an alkaloid in *Erythroxylum coca*. Pharm. Weekblad **93**, 932 (1958).

86. Firn, R. D., and J. Friend: Enzymic production of the plant growth inhibitor xanthoxin. Planta **103**, 263 (1972).

87. Flament, I., B. Willhalm, et M. Stoll: Recherches sur les arômes. Sur l'arôme du cacao. III. Helv. Chim. Acta **50**, 2233 (1967).

88. Foote, C. S., and M. Brenner: Chemistry of singlet oxygen. VIII. An unusual allenic oxygenation product. Tetrahedron Letters 6041 (1968).

89. Frankenburg, W. G., and A. M. Gottscho: Nicotinic acid in processed cigar tobacco. Arch. Biochem. Biophys. **21**, 247 (1949).

90. — — Myosmine in cigar tobacco. Arch. Biochem. Biophys. **23**, 333 (1949).

91. — — The chemistry of tobacco fermentation. I. Conversion of the alkaloids. B. The formation of oxynicotine. J. Amer. Chem. Soc. **77**, 5728 (1955).

92. Frankenburg, W. G., A. M. Gottscho, E. W. Mayaud, and T.-C. Tso: The chemistry of tobacco fermentation: I. Conversion of the alkaloids. A. The formation of 3-pyridyl methyl ketone and of 2,3'-dipyridyl. J. Amer. Chem. Soc. **74**, 4309 (1952).

93. Frankenburg, W. G., A. M. Gottscho, and A. A. Vaitekunas: Biochemical conversions of some tobacco alkaloids. Tob. Sci. **2**, 9 (1958).

94. Frankenburg, W. G., A. M. Gottscho, A. A. Vaitekunas, and R. M. Zacharius: The chemistry of tobacco fermentation. I. Conversion of the alkaloids. C. The formation of 3-pyridyl propyl ketone, nicotinamide and *N*-methylnicotinamide. J. Amer. Chem. Soc. **77**, 5730 (1955).

95. Frankenburg, W. G., and A. A. Vaitekunas: Chemical studies on nicotine degradation by microorganisms derived from the surface of tobacco seeds. Arch. Biochem. Biophys. **58**, 509 (1955).

96. — — The chemistry of tobacco fermentation. I. Conversion of the alkaloids. D. Identification of cotinine in fermented leaves. J. Amer. Chem. Soc. **79**, 149 (1957).

97. Fredrickson, J. D.: β-Amyrenyl esters of tobacco. 20th Tobacco Chemists' Research Conference, Winston-Salem, N. C., 1966.

98. Fujimori, T., R. Kasuga, H. Kaneko, and M. Noguchi: Isolation of 3-(4,8,12-trimethyl-tridecyl)-furan ("Phytofuran") from Burley tobacco. Agr. Biol. Chem. **38**, 2293 (1974).

99. — — — — A new acetylenic diol, 3-hydroxy-7,8-dehydro-β-ionol from Burley *Nicotiana tabacum* L. Phytochem. **14**, 2095 (1975).

100. Fujimori, T., R. Kasuga, M. Noguchi, and H. Kaneko: Isolation of *R*-(−)-3-hydroxy-β-ionone from Burley tobacco. Agr. Biol. Chem. **38**, 891 (1974).

101. Fukuzumi, T., H. Kaneko, and H. Takahara: Studies on the chemical constituents

of tobacco leaves. III. Isolation of (−)-2-isopropyl-5-oxo-hexanoic acid from Turkish tobacco leaves and absolute configuration of solanone. Agr. Biol. Chem. **31**, 607 (1967).

102. FUKUZUMI, T., H. TAKAHARA, H. KANEKO, and I. ONISHI: Isolation of hydroxy acids from Turkish tobacco leaves. Agr. Biol. Chem. **29**, 967 (1965).

103. — — — — Isolation of 2-isopropylmalic acid from Turkish tobacco. Nippon Nogeikagaku Kaisha **39**, 204 (1965).

104. GHERNA, R. L., S. H. RICHARDSON, and S. C. RITTENBERG: The bacterial oxidation of nicotine. VI. The metabolism of 2,6-dihydroxypseudooxynicotine. J. Biol. Chem. **240**, 3669 (1965).

105. GILBERTSON, T. J.: Biosynthesis of the piperidine nucleus: metabolism of D- and L-lysine-2-^{14}C by Nicotiana glauca. Phytochem. **11**, 1737 (1972).

106. GILES, J. A., and J. N. SCHUMACHER: Turkish tobacco — I. Isolation and characterization of α- and β-levantenolide. Tetrahedron **14**, 246 (1961).

107. GILES, J. A., J. N. SCHUMACHER, S. S. MIMS, and E. BERNASEK: Turkish tobacco — II. Isolation and characterization of 12α-hydroxy-13-epimanoyloxide. Tetrahedron **18**, 169 (1962).

108. GILES, J. A., J. N. SCHUMACHER, and G. W. YOUNG: Turkish tobacco — III. Isolation and characterization of α$_2$-levantanolide. Tetrahedron **19**, 107 (1963).

109. GLOGER, M., und K. DECKER: Zum Mechanismus der Induktion nikotinabbauender Enzyme in Arthrobacter oxydans. Zeitschrift Naturforsch. **24B**, 1016 (1969).

110. GOLDMAN, I. M., J. SEIBL, I. FLAMENT, F. GAUTSCHI, M. WINTER, B. WILLHALM, et M. STOLL: Recherches sur les aromes. Sur l'arome de café. II. Pyrazines et pyridines. Helv. Chim. Acta **50**, 694 (1967).

111. GRAY, J. G., S. D. KUNG, S. G. WILDMAN, and S. J. SHEEN: Origin of Nicotiana tabacum L., detected by polypeptide composition of fraction I protein. Nature **252**, 226 (1974).

112. GRIES, F. A., K. DECKER, und M. BRÜHMÜLLER: Über den Abbau des Nikotins durch Bakterienenzyme. V. Abbau des L-6-Hydroxy-Nicotins zu [γ-Methylaminopropyl]-[6-hydroxy-pyridyl-(3)]-keton. Hoppe Seylers Z. Physiol. Chem. **325**, 229 (1961).

113. GRIFFITHS, W. T., D. R. THRELFALL, and T. W. GOODWIN: Observations on the nature and biosynthesis of terpenoid quinones and related compounds in tobacco shoots. Eur. J. Biochem. **5**, 124 (1968).

114. GRUNWALD, C.: Phytosterols in tobacco leaves at various stages of physiological maturity. Phytochem. **14**, 79 (1975).

115. GUPTA, R. N., and I. D. SPENSER: Biosynthesis of the piperidine nucleus: the occurrence of two pathways from lysine. Phytochem. **9**, 2329 (1970).

116. GUTCHO, S.: Tobacco flavoring substances and methods 1972. Noyes Data Corp. U.S.A., pp. 65, 66.

117. HAINES, P. G., and A. EISNER: Identification of pseudooxynicotine and its conversion to N-methylmyosmine. J. Amer. Chem. Soc. **72**, 1719 (1950).

118. HEWLINS, M. J. E., J. D. EHRHARDT, L. HIRTH, and G. OURISSON: Conversion of [^{14}C]-cycloartenol and [^{14}C]-lanosterol into phytosterols by cultures of Nicotiana tabacum. Eur. J. Biochem. **8**, 184 (1969).

119. HLUBUCEK, J. R., A. J. AASEN, S.-O. ALMQVIST, and C. R. ENZELL: Tobacco chemistry 21. Three new volatile tobacco constituents of probable isoprenoid origin. Acta Chem. Scand. **27**, 2232 (1973).

120. — — — — Tobacco chemistry 22. Structures and synthesis of a nor- and a seco-terpenoid of the drimane series isolated from tobacco. Acta Chem. Scand. **B28**, 18 (1974).

121. — — — — Tobacco chemistry 26. Synthesis of 14,15-bisnor-8-hydroxylabd-11E-en-13-one, a new tobacco constituent. Acta Chem. Scand. **B28**, 131 (1974).

122. Hlubucek, J. R., A. J. Aasen, S.-O. Almqvist, and C. R. Enzell: Tobacco chemistry 25. Two new drimane sesquiterpene alcohols from Greek *Nicotiana tabacum* L. Acta Chem. Scand. **B28,** 289 (1974).

123. Hlubucek, J. R., A. J. Aasen, B. Kimland, and C. R. Enzell: New volatile constituents of Greek *Nicotiana tabacum.* Phytochem. **12,** 2555 (1973).

124. Hochstein, L. I., and S. C. Rittenberg: Bacterial oxidation of nicotine. II. Isolation of the first oxidative product and its identification as (*l*)-6-hydroxynicotine. J. Biol. Chem. **234,** 156 (1959).

125. — — Bacterial oxidation of nicotine. III. Isolation and identification of 6-hydroxy-pseudooxynicotine. J. Biol. Chem. **235,** 795 (1960).

126. Hoffman, D., S. S. Hecht, R. M. Ornaf, and E. L. Wynder: *N'*-nitrosonornicotine in tobacco. Science **186,** 265 (1974).

127. Holmes, P. E., S. C. Rittenberg, and H. J. Knackmuss: The bacterial oxidation of nicotine. Synthesis of 2,3,6-trihydroxypyridine and accumulation and partial characterization of the product of 2,6-dihydroxypyridine oxidation. J. Biol. Chem. **247,** 7628 (1972).

128. Hu, M. W., W. E. Bodinell, and D. Hoffmann: Chemical studies on tobacco smoke. XXIII. Synthesis of carbon-14 labelled myosmine, nornicotine and *N'*-nitrosonornicotine. J. Label. Compounds **10,** 79 (1974).

129. Huber, C.: Vorläufige Notiz über einige Derivate des Nikotins. Liebigs Ann. Chem. **141,** 271 (1867).

130. — Vorläufige Mitteilung. Chem. Ber. **3,** 849 (1870).

131. Hudson, C. S., and A. Neuberger: The stereochemical formulas of the hydroxy-proline and allohydroxyproline enantiomorphs and some related substances. J. Organ. Chem. **15,** 24 (1950).

132. Il'in, G. S.: The interrelationship among the chief tobacco alkaloids. Biokhimiya (USSR) **13,** 193 (1948).

133. Irvine, W. J., and M. J. Saxby: Steam volatile amines of Latakia tobacco leaf. Phytochem. **8,** 473 (1969).

134. Isoe, S., S. B. Hyeon, S. Katsumura, and T. Sakan: Photo-oxygenation of caro-tenoids. II. The absolute configuration of loliolide and dihydroactinidiolide. Tetrahedron Letters 2517 (1972).

135. Isoe, S., S. B. Hyeon, and T. Sakan: Photo-oxygenation of carotenoids. I. The formation of dihydroactinidiolide and β-ionone from β-carotene. Tetrahedron Letters 279 (1969).

136. Isoe, S., S. Katsumura, S. B. Hyeon, and T. Sakan: Biogenetic type synthesis of grasshopper ketone and loliolide and a possible biogenesis of allenic carotenoids. Tetrahedron Letters 1089 (1971).

137. Isoe, S., S. Katsumura, and T. Sakan: The synthesis of damascenone and β-damascone and the possible mechanism of their formation from carotenoids. Helv. Chim. Acta **56,** 1514 (1973).

138. Johnson, R. R., and J. A. Nicholson: The structure, chemistry and synthesis of solanone — anomalous terpenoid ketone from tobacco. J. Organ. Chem. **30,** 2918 (1965).

139. Kallianos, A. G., and R. E. Means: Isoprenoid ketones in tobacco. CORESTA/ TCRC joint conference, Williamsburg, Va., Oct. 22—28, 1972. Abstr. 13, Coresta Information Bulletin 1972.

140. Kallianos, A. G., F. A. Shelburne, R. E. Means, R. K. Stevens, R. E. Lax, and J. D. Mold: Identification of the D-glucosides of stigmasterol, sitosterol and campesterol in tobacco and cigarette smoke. Biochem. J. **87,** 596 (1963).

141. Kaneko, H.: The aroma of cigar tobacco. Part II. Isolation of norambreinolide from cigar tobacco. Agr. Biol. Chem. **35,** 1461 (1971).

142. Kaneko, H., and M. Harada: 4-Hydroxy-β-damascone and 4-hydroxy-dihydro-β-damascone from cigar tobacco. Agr. Biol. Chem. **36,** 168 (1972).

143. — — Aroma of cigar tobacco. III. Isolation and synthesis of *R*-(+)-3-isopropyl-5-hydroxypentanoic acid lactone. Agr. Biol. Chem. **36**, 658 (1972).

144. KANEKO, H., and K. HOSHIMO: Isolation from cigar tobacco leaves of tetrahydro-actinidiolide (2-hydroxy-2,6,6-trimethylcyclohexyl acetic acid γ-lactone). Agr. Biol. Chem. **33**, 969 (1969).

145. KANEKO, H., and K. IJICHI: The aroma of cigar tobacco. Part I. Isolation of 2-hydroxy-2,6,6-trimethylcyclohexylidene-1-acetic acid lactone (dihydroactinidiolide) from ether extract of cigar leaves. Agr. Biol. Chem. **32**, 1337 (1968).

146. KARRER, P., und R. WIDMER: Konfiguration des Nikotins. Optisch aktive Hygrinsäure. Helv. Chim. Acta **8**, 364 (1925).

147. KATO, T., M. TANEMURA, T. SUZUKI, and Y. KITAHARA: Biogenetic-type synthesis of α- and β-levantenolides. Chem. Commun. 28 (1970).

148. KEARNS, D. R.: Physical and chemical properties of singlet molecular oxygen. Chem. Rev. **71**, 395 (1971).

149. KIMLAND, B., A. J. AASEN, S.-O. ALMQVIST, P. ARPINO, and C. R. ENZELL: Tobacco chemistry 16. Volatile acids of sun-cured Greek tobacco. Phytochem. **12**, 835 (1973).

150. KIMLAND, B., A. J. AASEN, and C. R. ENZELL: Tobacco chemistry 12. Volatile neutral constituents of Greek tobacco. Acta Chem. Scand. **26**, 1281 (1972).

151. — — — Tobacco chemistry 10. Volatile neutral constituents of Greek tobacco. Acta Chem. Scand. **26**, 2177 (1972).

152. KIMLAND, B., R. A. APPLETON, A. J. AASEN, J. ROERAADE, and C. R. ENZELL: Tobacco Chemistry 6. Neutral oxygen-containing volatile constituents of Greek tobacco. Phytochem. **11**, 309 ((1972).

153. KINZER, G. W., T. F. PAGE, and R. R. JOHNSON: Structure of two solanone precursors from tobacco. J. Organ. Chem. **31**, 1797 (1966).

154. KISAKI, T., S. MIZUSAKI, and E. TAMAKI: γ-Methylaminobutyraldehyde, a new intermediate in nicotine biosynthesis. Arch. Biochem. Biophys. **117**, 677 (1966).

155. — — — Phytochemical studies on tobacco alkaloids. XI. A new alkaloid in *Nicotiana tabacum* roots. Phytochem. **7**, 323 (1968).

156. KISAKI, T., and E. TAMAKI: Phytochemical studies on the tobacco alkaloids. I. Optical rotatory power of nicotine. Arch. Biochem. Biophys. **92**, 351 (1961).

157. — — Phytochemical studies of the tobacco alkaloids. III. Observations on the interconversion of *dl*-nicotine and *dl*-nornicotine in excised tobacco leaves. Arch. Biochem. Biophys. **94**, 252 (1961).

158. — — Phytochemical studies on the tobacco alkaloids. X. Degradation of the tobacco alkaloids and their optical rotatory changes in tobacco plants. Phytochem. **5**, 293 (1966).

159. KOBAYASHI, H., and S. AKIYOSHI: Thunbergene, a macrocyclic diterpene. Bull. Chem. Soc. Japan **35**, 1044 (1962).

160. — — Terpenoids. VI. Structure of thunbergene. Bull. Chem. Soc. Japan **36**, 823 (1963).

161. KOEHLER, P. E., M. E. MASON, and J. A. NEWELL: Formation of pyrazine compounds in sugar-amino acid model systems. J. Agric. Food Chem. **17**, 393 (1969).

162. KOEHLER, P. E., and G. V. ODELL: Factors affecting the formation of pyrazine compounds in sugar-amine reactions. J. Agric. Food Chem. **18**, 895 (1970).

163. KOFLER, M., A. LANGEMANN, R. RÜEGG, U. GLOOR, U. SCHWIETER, J. WÜRSCH, O. WISS, und O. ISLER: Struktur und Partialsynthese des pflanzlichen Chinons mit isoprenoider Seitenkette. Helv. Chim. Acta **42**, 2252 (1959).

164. KORTE, F., und H.-J. SCHULZE-STEINEN: Acyl-lacton-Umlagerung XXII. Umlagerung von α-Aroyl-pyrrolidonen in konz. Salzsäure zu Pyrrollinderivaten. Chem. Ber. **95**, 2444 (1962).

165. KUFFNER, F., and N. FADERL: Constitution of nicotelline. Monatsh. Chem. **87**, 71 (1956).

166. KURATA, S., Y. INOUYE, and H. KAKISAWA: Synthesis of dihydroactinidiolide and a trimethyloctalenedione. Tetrahedron Letters 5153 (1973).
167. LEETE, E.: The biogenesis of nicotine. Chem. and Ind. 537 (1955).
168. — The biogenesis of nicotine and anabasine. J. Amer. Chem. Soc. **78**, 3520 (1956).
169. — The biogenesis of nicotine. V. New precursors of the pyrrolidine ring. J. Amer. Chem. Soc. **80**, 2162 (1958).
170. — Biosynthesis of the *Nicotiana* alkaloids. XI. Investigation of tautomerism in *N*-methyl-Δ^1-pyrrolinium chloride and its incorporation into nicotine. J. Amer. Chem. Soc. **89**, 7081 (1967).
171. — Biosynthesis of the *Nicotiana* alkaloids. XIV. The incorporation of Δ^1-piperideine-6-^{14}C into the piperidine ring of anabasine. J. Amer. Chem. Soc. **91**, 1697 (1969).
172. — Biomimetic synthesis of nicotine. J. Chem. Soc. Chem. Commun. 1091 (1972).
173. — Biosynthesis and metabolism of the tobacco alkaloids. Proceedings of the First Philip Morris Science Symposium, 1973.
174. — Biosynthesis of anatabine and anabasine in *Nicotiana glutinosa*. J. Chem. Soc. Chem. Commun. 9 (1975).
175. LEETE, E., and V. M. BELL: The biogenesis of the *Nicotiana* alkaloids. VIII. The metabolism of nicotine in *N. tabacum*. J. Amer. Chem. Soc. **81**, 4358 (1959).
176. LEETE, E., and M. R. CHEDEKEL: The aberrant formation of (−)-*N*-methylanabasine from *N*-methyl-Δ^1-piperideinium chloride in *Nicotiana tabacum* and *N. glauca*. Phytochem. **11**, 2751 (1972).
177. — — Metabolism of nicotine in *Nicotiana glauca*. Phytochem. **13**, 1853 (1974).
178. LEETE, E., M. R. CHEDEKEL, and G. B. BODEM: Synthesis of myosmine and nornicotine using an acyl carbanion equivalent as an intermediate. J. Organ. Chem. **37**, 4465 (1972).
179. LEETE, E., E. G. GROS, and T. J. GILBERTSON: The biosynthesis of anabasine. Origin of the nitrogen of the piperidine ring. J. Amer. Chem. Soc. **86**, 3907 (1964).
180. — — — Biosynthesis of the pyrrolidine ring of nicotine — feeding experiments with ^{15}N-labelled ornithine-2-^{14}C. Tetrahedron Letters 587 (1964).
181. LEETE, E., and Y.-Y. LIU: Metabolism of [2-^{3}H]- and [6-^{3}H]-nicotinic acid in intact *Nicotiana tabacum* plants. Phytochem. **12**, 593 (1973).
182. LLOYD, R. A., C. W. MILLER, D. L. ROBERTS, J. A. GILES, J. P. DICKERSON, N. H. NELSON, C. E. RIX, and P. H. AYERS: Flue-cured tobacco flavor. I. Essence and essential oil components. Tob. Sci. **20**, 43 (1976).
183. LUKĒS, R., A. A. AROJAN, J. KOVÁŘ, und K. BLÁHA: Zur Konfiguration stickstoffhaltiger Verbindungen XV. Bestimmung der absoluten Konfiguration von Anabasin und Anatabin. Coll. Czech. Chem. Commun. **27**, 751 (1962).
184. MANSKE, R. H. F., and L. MARION: The alkaloids of *Lycopodium* species. I. *Lycopodium complanatum* L. Can. J. Res. B. **20**, 87 (1942).
185. — — Alkaloids of *Lycopodium* species. VII. *Lycopodium lucidulum*. Can. J. Res. B. **24**, 57 (1946).
186. MARION, L.: The occurrence of *L*-nicotine in *Asclepias syriaca*. Can. J. Res. B. **17**, 21 (1939).
187. — The alkaloids of *Sedum acre*. Can. J. Res. B. **23**, 165 (1945).
188. MARION, L., and R. H. F. MANSKE: Alkaloids of *Lycopodium* species. IV. *Lycopodium tristachyum* Pursh. Can. J. Res. B. **22**, 1 (1944).
189. — — Alkaloids of *Lycopodium* species. VI. *Lycopodium clavatum*. Can. J. Res. B **22**, 137 (1944).
190. — — Alkaloids of *Lycopodium* species. VIII. *Lycopodium sabinaefolium*. Can. J. Res. B **24**, 63 (1946).
191. MARION, J. P., F. MÜGGLER-CHAVAN, R. VIANI, J. BRICOUT, D. REYMOND, et R. H. EGLI: Sur la composition de l'arôme de cacao. Helv. Chim. Acta **50**, 1509 (1967).

192. McKennis, H., E. R. Bowman, L. D. Quin, and R. C. Denney: Structure of dibromoticonine, a bromination product of nicotine. J. Chem. Soc. Perkin I, 2046 (1973).

193. Milborrow, B. V.: Biosynthesis of abscisic acid by a cell-free system. Phytochem. 13, 131 (1974).

194. — The stereochemistry of cyclization in abscisic acid. Phytochem. 14, 123 (1975).

195. Mizusaki, S., Y. Tanabe, M. Noguchi, and E. Tamaki: Phytochemical studies on tobacco alkaloids. XIV. The occurrence and properties of putrescine N-methyltransferase in tobacco roots. Plant Cell Physiol. 12, 633 (1971).

196. — — — — N-Methylputrescine oxidase from tobacco roots. Phytochem. 11, 2757 (1972).

197. Mori, K.: Synthesis of the optically active dehydrovomifoliol. A synthetic proof of the absolute configuration of (+)-abscisic acid. Tetrahedron Letters 2635 (1973).

198. Mousseron-Canet, M., J.-P. Dalle, et J.-C. Mani: Photooxydation sensibilisée de composes apparentes aux carotenoides. Analogie avec la structure des carotenoides oxygenes, alleniques et autres. Tetrahedron Letters 6037 (1968).

199. Mundy, B. P., and B. R. Larsen: A new approach to pyrrolidine and piperidine alkaloids. Synthetic Commun. 2, 197 (1972).

200. Mundy, B. P., B. R. Larsen, L. F. McKenzie, and G. Braden: A convenient synthesis of myosmine. J. Organ. Chem. 37, 1635 (1972).

201. Nathan, R. A., and A. H. Adelman: Photosensitized generation of singlet molecular oxygen with near-infrared radiation. Chem. Commun. 674 (1974).

202. Neurath, G., und M. Dünger: Isolierung schwach basischer Heteroaromaten aus dem Tabakrauch. Beitr. Tabakforsch. 5, 1 (1969).

203. Neurath, G., A. Krull, B. Pirmann, und K. Wandrey: Untersuchung der flüchtigen Basen des Tabaks, II. Beitr. Tabakforsch. 3, 571 (1966).

204. Newell, J. A., M. E. Mason, and R. S. Matlock: Precursors of typical and atypical roasted peanut flavour. J. Agric. Food Chem. 15, 767 (1967).

205. Noga, E.: Über die Alkaloide im Tabakextrakt. Fachl. Mitt. Österr. Tabak-Regie I (1914).

206. Ohloff, G.: Classification and genesis of food flavours. Flavour Industry 501 (1972).

207. Ohloff, G., D. Rautenstrauch, und K. H. Schulte-Elte: Modellreaktionen zur Biosynthese von Verbindungen der Damascon-Reihe und ihre präparative Anwendung. Helv. Chim. Acta 56, 1503 (1973).

208. Onishi, I., H. Tomita, and T. Fukuzumi: Essential oils of tobacco leaves. IV. Neutral fraction. Bull. Agr. Chem. Soc. Japan 20, 61 (1956).

209. Oreykov, A., und G. Menshikov: Über die Alkaloide von Anabasis aphylla L. (I. Mitteilung.) Chem. Ber. 64, 266 (1931).

210. Pack, A. B.: The curing and quality of flue-cured tobacco: the effect of certain cultural and curing practices on the plastid pigment and carbohydrate contents. N. C. State College, Ph. D. Thesis 1950 (unpublished), from Weybrew, J. A., Tob. Sci. 1, 1 (1957).

211. Paris, R. R., et P. Frigot: Étude par chromatographie et par electrophorèse des alcaloïdes de diverse Crassulacceés indigènes; caractérisation de la nicotine chez le Sempervivum aracnoideum. C. R. Acad. Sci. 248, 1849 (1959).

212. Pelletier, S. W., S. Lajšić, Y. Ohtsuka, and Z. Djarmati: Naturally occurring terpenes. Synthesis of (+)- and (±)-14,15-bisnor-8α-hydroxylabd-11(E)-en-13-one, (+)-drimane-8,11-diol, and (−)-drimenol. J. Organ. Chem. 40, 1607 (1975).

213. Pictet, A., und G. Court: Über einige neue Pflanzenalkaloide. Chem. Ber. 40, 3771 (1907).

214. Pictet, A., und P. Crépieux: Über Phenyl- und Pyridylpyrrole und die Constitution des Nicotins. Chem. Ber. 28, 1904 (1895).

215. PICTET, A., und A. ROTSCHY: Über neue Alkaloide des Tabaks. Chem. Ber. 34, 696 (1901).
216. PINNER, A: Über Nicotin. (I. Mitteilung.) Arch. Pharmaz. 231, 378 (1893).
217. — Über Nikotin (Metanicotin). VII. Mitteilung. Chem. Ber. 27, 1053 (1894).
218. PINNER, A., und RÖWER: Über Nicotin. Die Konstitution des Alkaloids. Chem. Ber. 26, 292 (1893).
219. POINDEXTER, E. H., and R. D. CARPENTER: The isolation of harmane and norharmane from tobacco and cigarette smoke. Phytochem. 1, 215 (1962).
220. POSSELT, W., und L. REIMANN: Chemische Untersuchung des Tabaks und Darstellung eines eigentümlichen wirksamen Prinzips dieser Pflanze. Geigers Mag. Pharmac. 24, 138 (1828).
221. QUAN, P. M., T. K. B. KARNS, and L. D. QUIN: Total synthesis of dl-anatabine. Chem. and Ind. 1553 (1964).
222. RAMAMURTHY, V., and R. S. H. LIU: Photochemistry of dehydro-β-ionone and related compounds. Tetrahedron Letters 441 (1973).
223. RAYBURN, C. H., W. R. HARLAN, and H. R. HANMER: Rearrangement of nicotine oxide. J. Amer. Chem. Soc. 72, 1721 (1950).
224. REID, W. W.: The action of inhibitors on the incorporation of [2-¹⁴C]-mevalonate into the triterpenes and sterols of Nicotiana tabacum. Biochem. J. 100, 13P (1966).
225. — The phytochemistry of the genus Nicotiana. Ann. du Tabac S.E.I.T.A. 2, 145 (1974).
226. RIBAS-MARQUÉS, I., y Y. A. NODAR BLANCO: Alcaloides de las Papilionaceas XXXIX. Configuración absoluta de adenocarpina, santiaquina, ammodendrina, anabasina, anatabina y de sus metil-derivados. Configuración absoluta parcial del α,β'-dipiperidilo. Soc. Espan. Fis. Quim. 57, 781 (1961).
227. RICHARDSON, B., J. R. BAUR, R. S. HALLIWELL, and R. LANGSTON: Sterol metabolism in normal and tobacco mosaic virus infected tobacco plants. Steroids 11, 231 (1968).
228. RICHARDSON, S. H., and S. C. RITTENBERG: Bacterial oxidation of nicotine. IV. Isolation and identification of 2,6-dihydroxy-N-methylmyosmine. J. Biol. Chem. 236, 959 (1961).
229. — — Bacterial oxidation of nicotine. V. Identification of 2,6-dihydroxypseudooxynicotine as the third oxidative product. J. Biol. Chem. 236, 964 (1961).
230. ROBERTS, D. L.: Tobacco. U. S. Patent 3,217,717; November 16, 1965.
231. — The structure of a new sesquiterpene isolated from tobacco. Phytochem. 11, 2077 (1972).
232. ROBERTS, D. L., R. A. HECKMAN, B. P. HEGE, and S. A. BELLIN: Synthesis of (R,S)-abscisic acid. J. Organ. Chem. 33, 3566 (1968).
233. ROBERTS, D. L., C. W. MILLER, and R. A. LLOYD, JR.: Tobacco carotenoids. 27th Tobacco Chemist's Research Conference, Winston-Salem, N. C., October 3–5, 1973.
234. ROBERTS, D. L., and W. A. ROHDE: Isolation and identification of flavour components of Burley tobacco. Tob. Sci. 16, 107 (1972).
235. ROBERTS, D. L., and R. L. ROWLAND: Macrocyclic diterpenes. α- and β-4,8,13-duvatriene-1,3-diols from tobacco. J. Organ. Chem. 27, 3989 (1962).
236. ROBERTS, D. L., and J. N. SCHUMACHER: Tobacco. U. S. Patent No. 3,217,716. November 16, 1965.
237. ROWLAND, R. L.: Flue-cured tobacco. II. Neophytadiene. J. Amer. Chem. Soc. 79, 5007 (1957).
238. — Flue-cured tobacco. III. Solanachromene and α-tocopherol. J. Amer. Chem. Soc. 80, 6130 (1958).
239. ROWLAND, R. L., and J. A. GILES: Flue-cured tobacco. V. Polyisoprenoid compounds. Tob. Sci. 4, 29 (1960).

240. ROWLAND, R. L., and P. H. LATIMER: Flue-cured tobacco. IV. Isolation of solanesyl esters. Tob. Sci. **3**, 1 (1959).

241. ROWLAND, R. L., P. H. LATIMER, and J. A. GILES: Flue-cured tobacco. I. Isolation of solanesol, an unsaturated alcohol. J. Amer. Chem. Soc. **78**, 4680 (1956).

242. ROWLAND, R. L., and D. L. ROBERTS: Macrocyclic diterpenes isolated from tobacco, α- and β-4,8,13-Duvatriene-1,5-diols. J. Organ. Chem. **28**, 1165 (1963).

243. ROWLAND, R. L., A. RODGMAN, J. N. SCHUMACHER, D. L. ROBERTS, L. C. COOK, and W. E. WALKER, JR.: Macrocyclic diterpene hydroxyethers from tobacco and cigarette smoke. J. Organ. Chem. **29**, 16 (1964).

244. RÜEGG, R., U. GLOOR, A. LANGEMANN, M. KOFLER, C. VON PLANTA, G. RYSER, and O. ISLER: Total synthesis of solanesol. Helv. Chim. Acta **43**, 1745 (1960).

245. SABETAY, S., L. TRABAUD, and H. F. EMMANUEL: Constituents of the concrete oil of tobacco leaves *(Nicotiana tabacum)*. Comp. rend. **213**, 321 (1941).

246. SANDERSON, G. W., H. CO, and J. G. GONZALEZ: Biochemistry of tea fermentation: the role of carotenes in black tea aroma formation. J. Food Sci. **36**, 231 (1971).

247. SCHENK, H. R., H. GUTMANN, O. JEGER, und L. RUZICKA: Zur Kenntnis der Diterpene, XLII. Über eine neue, ergiebige Partialsynthese des Ambreinolides. Helv. Chim. Acta **35**, 817 (1952).

248. SCHMELTZ, I., R. L. MILLER, and R. L. STEDMAN: Gas chromatographic study of the steam-volatile fatty acids of various tobaccos. J. Gas Chromatog. **1**, 27 (1963).

249. SCHMELTZ, I., R. L. STEDMAN, and R. L. MILLER: Composition studies on tobacco. XVI. Steam-volatile acids. J. Assoc. Offic. Agr. Chemists **46**, 779 (1963).

250. SCHULTE-ELTE, K. H., M. GADOLA, und G. OHLOFF: Oxydative Umsetzungen in der Damascon-Reihe. $^1\Delta gO_2$- und SeO_2-Oxydation von β-Damascenon. Helv. Chim. Acta **56**, 2028 (1973).

251. SCHULTE-ELTE, K. H., B. L. MÜLLER, und G. OHLOFF: Die farbstoffsensibilisierte Photo-Oxygenierung von β-Damascol. Ein einfaches Verfahren zur Darstellung von β-Damascenon. Helv. Chim. Acta **54**, 1899 (1971).

252. SCHUMACHER, J. N., and R. A. HECKMAN: On the natural occurrence and relative configurations of the tetrahydroactinidiolide isomers. Phytochem. **10**, 421 (1971).

253. SCHUMACHER, J. N., and L. VESTAL: Isolation and identification of some components of Turkish tobacco. Tob. Sci. **18**, 43 (1974).

254. SCHÜTTE, H. R., W. MAIER, and K. MOTHES: Methylputrescine as possible precursor of nicotine in *Nicotiana rustica*. Acta Biochem. Polon. **13**, 401 (1966).

255. SCHWIETER, U., W. ARNOLD, W. E. OBERHÄNSLI, N. RIGASSI, und W. VETTER: Synthese in der Carotinoid-Reihe. Ein Beitrag zur Chromsäure-Oxydation von Polyenen. Helv. Chim. Acta **54**, 2447 (1971).

256. SCOTT, T. A., and J. P. GLYNN: The incorporation of [2,3,7-^{14}C]-nicotinic acid into nicotine by *Nicotiana tabacum*. Phytochem. **6**, 505 (1967).

257. SHIGEMATSU, H., S. SHIBATA, T. KURATA, H. KATO, and M. FUJIMAKI: 5-Acetyl-2,3-dihydro-1H-pyrrolizines and 5,6,7,8-tetrahydroindolizin-8-ones, odor constituents formed on heating L-proline with D-glucose. J. Agric. Food Chem. **23**, 233 (1975).

258. SKORIANETZ, W., und G. OHLOFF: Säurekatalysierte Umlagerung von *trans*-5,6-Dihydroxy-5,6-dihydro-β-jonon. Helv. Chim. Acta **56**, 2025 (1973).

259. — — Über vinyloge *English-Zimmerman*-Spaltungen in der Iononreihe. Helv. Chim. Acta **58**, 771 (1975).

260. SKRAUP, ZD. H., und A. COBENZL: Über α- und β-Naphthochinolin. Monatsh. Chem. **4**, 459 (1883).

261. SMITH, C. R.: Occurrence of anabasine in *Nicotiana glauca* R. Grah. *(Solanaceae)*. J. Amer. Chem. Soc. **57**, 959 (1935).

262. SMITH, H. H., and C. R. SMITH: Alkaloids in certain species and interspecific hybrids of *Nicotiana*. J. Agr. Res. **65**, 347 (1942).

263. Solt, M. L., R. F. Dawson, and D. R. Christman: Biosynthesis of anabasine and of nicotine by excised root cultures of *Nicotiana glauca*. Plant. Physiol. **35**, 887 (1960).

264. Springer, J. P., J. Clardy, R. H. Cox, H. G. Cutler, and R. J. Cole: The structure of a new type of plant growth inhibitor extracted from immature tobacco leaves. Tetrahedron Letters 2737 (1975).

265. Späth, E., und S. Biniecki: Tobacco Alkaloids. XVI. *N*-Methylpyrrolidin, ein neues Tabak-Alkaloid und zur Konstitution des Iso-Nicoteins. Chem. Ber. **72**, 1809 (1939).

266. Späth, E., und H. Bretschneider: Eine neue Synthese des Nicotins und einige Bemerkungen zu den Arbeiten Nagais über Ephedrine. Chem. Ber. **61**, 327 (1928).

267. Späth, E., und F. Kesztler: L-Anatabin, ein neues Tabakalkaloid. (XI. Mitteilung über Tabakbasen.) Chem. Ber. **70**, 239 (1937).

268. — — Über das Vorkommen von *d,l*-Nornicotin, *d,l*-Anatabin und *l*-Anabasin im Tabak. (XII. Mitteilung über Tabakalkaloide.) Chem. Ber. **70**, 704 (1937).

269. — — Über neue Basen des Tabaks. (XIII. Mitteilung über Tabakalkaloide.) Chem. Ber. **70**, 2450 (1937).

270. Späth, E., und L. Mamoli: Synthese des Myosmins (VI. Mitteilung über Tabakbasen) und Bemerkungen zu einer Notiz von T. Reynolds und R. Robinson. Chem. Ber. **69**, 757 (1936).

271. — — Eine neue Synthese des *d,l*-Anabasins. (VII. Mitteilung über Tabakalkaloide.) Chem. Ber. **69**, 1082 (1936).

272. Späth, E., J. P. Wibaut, und F. Kesztler: Über das *N*-Methylmyosmin. Chem. Ber. **71**, 100 (1938).

273. Späth, E., und E. Zajic: Über neue Tabakalkaloide (VIII. Mitteilung über Tabakbasen) und Bemerkungen zur Kenntnis des Rhoeadins, des *l*-Peganins und des Ammoresinols. Chem. Ber. **69**, 2448 (1936).

274. Steadman, J. R., and L. Sequeira: Abscisic acid in tobacco plants. Plant Physiol. **45**, 691 (1970).

275. Stevens, K. L., R. Lundin, and D. L. Davis: Acid catalyzed rearrangement of β-ionone epoxide. Tetrahedron **31**, 2749 (1975).

276. Stevens, M. A.: Relationship between polyene-carotene content and volatile compound composition of tomatoes. J. Amer. Soc. Horticult. Sci. **95**, 461 (1970).

277. Stevenson, J., F. W. Hemming, and R. A. Morton: The intracellular distribution of solanesol and plastoquinone in green leaves of the tobacco plant. Biochem. J. **88**, 52 (1963).

278. Stoll, M., et M. Hinder: Odeur et constitution VIII. Recherches sur quelques produits de dégradation du sclaréol. Helv. Chim. Acta **36**, 1984 (1953).

279. Strain, H. H.: Leaf xanthophylls. Carnagie Inst. Wash., Publ. No. **490** (1938).

280. — Leaf xanthophylls. J. Amer. Chem. Soc. **70**, 1672 (1948).

281. van Tamelen, E. E., and R. M. Coates: Biogenetic-type synthesis of (±)-farnesiferol A and (±)-farnesiferol C. Chem. Commun. 413 (1966).

282. van Tamelen, E. E., A. Storni, E. J. Hessler, and M. Schwartz: The biogenetically patterned *in vitro* oxidation-cyclization of farnesyl acetate. J. Amer. Chem. Soc. **85**, 3295 (1963).

283. Tanemura, M., T. Suzuki, T. Kato, and Y. Kitahara: Synthesis of levantenolides from acyclic progenitor. Tetrahedron Letters 1463 (1970).

284. Thesing, J., und A. Müller: Synthese des Nicotellins. Angew. Chem. **68**, 577 (1956).

285. Tso, T. C., and R. N. Jeffrey: Biochemical studies on tobacco alkaloids. I. The fate of labeled tobacco alkaloids supplied to *Nicotiana* plants. Arch. Biochem. Biophys. **80**, 46 (1959).

286. Wada, E.: Microbial degradation of nornicotine. Arch. Biochem. Biophys. **64**, 244 (1956).

287. — Microbial degradation of the tobacco alkaloids and some related compounds. Arch. Biochem. Biophys. **72**, 145 (1957).

288. WADA, E., T. KISAKI, and K. SAITO: Autooxidation of nicotine. Arch. Biochem. Biophys. **79**, 124 (1959).

289. WADA, E., and K. YAMASAKI: Degradation of nicotine by soil bacteria. J. Amer. Chem. Soc. **76**, 155 (1954).

290. WAHL, R.: Über das Vorkommen von Metanikotin in natürlich nikotinfreien Tabaksorten. Tabak-Forsch. Sonderheft **36** (1953).

291. WAHLBERG, I., K. KARLSSON, D. J. AUSTIN, N. JUNKER, J. ROERAADE, C. R. ENZELL, and W. H. JOHNSON: To be published.

292. WAHLBERG, I., K. KARLSSON, and C. R. ENZELL: To be published.

293. WAHLBERG, I., K. KARLSSON, T. NISHIDA, K.-P. CHENG, and C. R. ENZELL: To be published.

294. WARFIELD, A. H., W. D. GALLOWAY, and A. G. KALLIANOS: Some new alkaloids from Burley tobacco. Phytochem. **11**, 3371 (1972).

295. VON WARTBURG, B. R., H. R. WOLF, und O. JEGER: Photochemische Reaktionen. Neuartige Photoreaktionen des (±)-*trans*-β-Ionon-Epoxids. Helv. Chim. Acta **56**, 1948 (1973).

296. — — — Photochemische Reaktionen. Notiz zur Photoreaktivität ausgewählter ungesättigter Ketoalkohole der Iononreihe. Helv. Chim. Acta **56**, 1956 (1973).

297. WEIL, L., and J. MAHER: Photodynamic action of methylene blue on nicotine and its derivatives. Arch. Biochem. **29**, 241, (1950).

298. WEISS, G., M. KOREEDA, and K. NAKANISHI: Stereochemistry of theaspirone and the blumenols. Chem. Commun. 565 (1973).

299. WENUSCH, A.: Über das Auftreten von Nicotyrin im Tabak. Biochem. Z. **275**, 361 (1935).

300. WENUSCH, A., und R. SCHÖLLER: Beitrag zur Kenntnis der Zusammensetzung des Tabakrauches. Fachl. Mitt. Österr. Tabak-Regie **2** (1933).

301. WHITFIELD, D. M., and K. S. ROWAN: Changes in the chlorophylls and carotenoids of leaves of *Nicotiana tabacum* during senescence. Phytochem. **13**, 77 (1974).

302. WITKOP, B.: Infrared diagnosis of the hydrochlorides of organic bases. II. Structure of myosmine. J. Amer. Chem. Soc. **76**, 5597 (1954).

303. WOOLLEN, B. H., W. J. IRVINE, P. W. BROWN, and D. H. JONES: A thin-layer chromatographic method for tobacco lipid analysis. Tob. Sci. **16**, 101 (1972).

304. WRIGHT, JR., H. E., W. W. BURTON, and R. C. BERRY JR.: Carotenoids and related colorless polyenes of aged Burley tobacco. Arch. Biochem. Biophys. **82**, 107 (1959).

305. YANG, K. S., R. K. GHOLSON, and G. R. WALLER: Studies on nicotine biosynthesis. J. Amer. Chem. Soc. **87**, 4184 (1965).

306. YASUMATSU, N., and S. AKAIKE: Studies on the volatile bases in tobacco. Part 1. Volatile amines in cured leaves and tobacco plant. Nippon Nogeikagaku Kaishi **39**, 347 (1965).

307. ZANE, A.: 4,8,13-Duvatriene-1-ol-3-one and 11-isopropyl-4,8-dimethyl-3,7,12-pentadecatriene-2,14-dione isomers from *Nicotiana tabacum*. Phytochem. **12**, 731 (1973).

(Received February 19, 1976)

The Chemistry of the Eremophilane
and Related Sesquiterpenes

By A. R. PINDER, Department of Chemistry, Clemson University,
Clemson, South Carolina, U.S.A.

Contents

I. Introduction

The terpenes constitute the largest family of substances of natural occurrence, being widely distributed in the plant kingdom and to a lesser extent in the animal kingdom. They are divisible according to the number of carbon atoms present into several groups, one of which is labeled sesquiterpenes and is comprised of molecules containing fifteen carbon atoms. The sesquiterpenes are the largest group of terpenes, and are themselves subdivisible according to the arrangement of the carbon atoms in the molecular skeleton. In this chapter we shall be concerned with a subgroup having its carbon atoms arranged, in the main, in a bicyclic framework corresponding to a parent hydrocarbon formulated, for the moment without stereochemical implication, as (1) and named eremophilane. These sesquiterpenes have been reviewed earlier (*1, 2, 2a*). We shall include with them a small group of closely-related sesquiterpenes having a carbon skeleton in which ring B in (1) has been contracted to a five-membered carbocycle, as in (2)

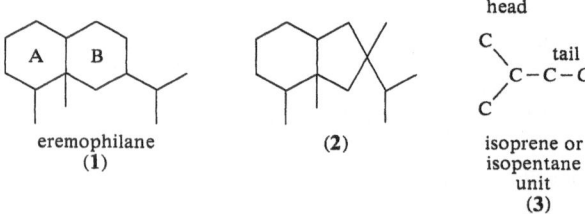

eremophilane   (2)   isoprene or
(1)   isopentane
            unit
            (3)

The most striking feature of the eremophilanoid sesquiterpenes is that they do not conform structurally to the well-known Isoprene Rule. This rule was first propounded by WALLACH in 1887 and has been prominent as an aid to the solution of structural problems in the terpene area generally. It was observed at that time that the carbon skeletons of terpenes of known structure were capable of being divided into isoprene (or isopentane) units (3) or, conversely, that the carbon frameworks of terpenes could be arrived at by an assembly of these units. A further point noted was that these units were arranged in the molecules in a head-to-tail manner. It was suggested as a consequence that terpene molecules were synthesized by nature from these or closely-related units. When the first members of the sesquiterpene group under discussion were discovered in 1932 (3) their formulation as non-isoprenoid structures was by no means immediate; in fact not until some seven years later (4) was it finally accepted that eremophilone, one of a trio of original members of the family, must be correctly represented by the non-isoprenoid gross structure (4).

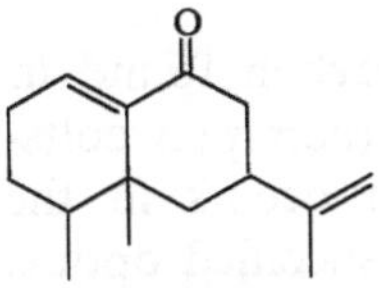

eremophilone
(4)

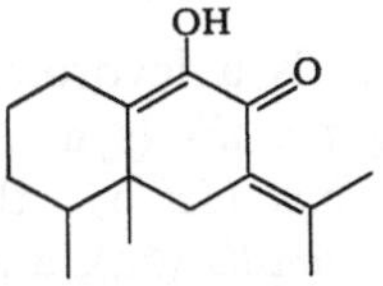

hydroxyeremophilone
(5)

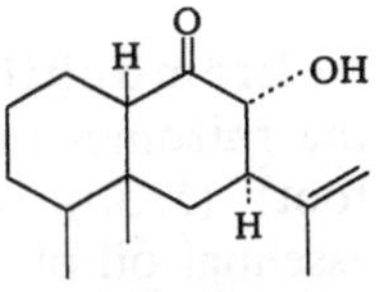

hydroxydihydro-
eremophilone
(6)

Eremophilone thus became the first naturally-occurring sesquiterpene whose structure violated the Isoprene Rule. For about thirty years the three compounds eremophilone (4), hydroxyeremophilone (5), and hydroxydihydroeremophilone (6), all occurring in the Australian shrub *Eremophila mitchelli*, remained the sole members of the new group of sesquiterpenes, but in more recent times over two hundred compounds belonging to the group have been isolated from diverse plant and even animal sources. Consequently these compounds now constitute a major subgroup of the sesquiterpenes.

For discussion it is convenient to arrange these sesquiterpenes into seven categories, mainly according to the carbon framework present and its stereochemistry.

II. Bicyclic Eremophilanes

The members of this subgroup are so called because they have the carbon skeleton found in the parent hydrocarbon eremophilane, which

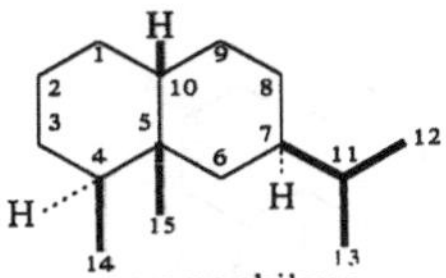

eremophilane
(7)

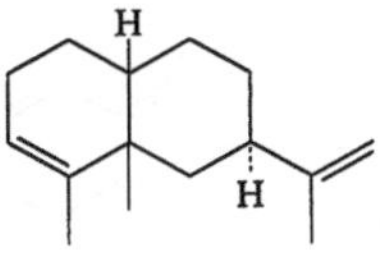

(8)

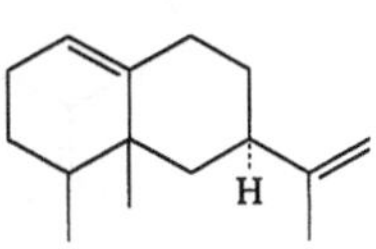

eremophilene
(9)

is represented in structure and stereochemistry by (7), but is not of natural occurrence. It is to be noted that in this stereostructure the three alkyl substituents have an all-*cis* orientation, the compound being 7β-iso-propyl-4β,5β-dimethyl-*cis*-decalin. It must also be mentioned that some of the generally accepted trivial names of these and other sesquiterpenes of the main family are generically incorrect. No attempt will be made to rectify this situation here, but attention will be drawn to such cases from time to time.

1. Hydrocarbons

Eremophilene, $C_{15}H_{24}$, is a levorotatory hydrocarbon found in the rhizomes of *Petasites officinalis* (syn. *P. hybridus*) (common colts-foot) (*5, 5a*), and of *P. albus* (*5a, 6*); the (+)-form occurs in the essential oil of *Valeriana officinalis* (*6a*), and one of unspecified optical activity in rhizomes of *P. japonicus* (*6b*). Quantitative hydrogenation revealed the presence of two double bonds, which pointed to a bicyclic structure; ultraviolet absorption measurement indicated the bonds were unconjugated. Quantitative ozonolysis showed that one of the olefinic linkages was present as part of a $C=CH_2$ group, confirmation being provided by a strong infrared band at 885 cm^{-1} in the spectrum of eremophilene. Total hydrogenation afforded eremophilane (7), indicating that eremophilene was an eremophiladiene (5). The ozonolysis referred to above furnished in addition to formaldehyde a methyl ketone (positive iodoform reaction), revealing the presence of an isopropenyl group at position 7. Analyses of the n.m.r. spectra of eremophilene and of

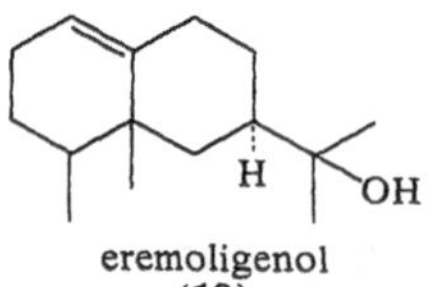

eremoligenol
(**10**)   (**11**)

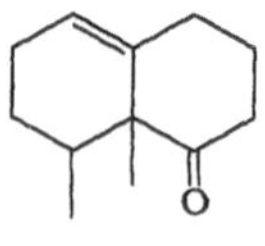

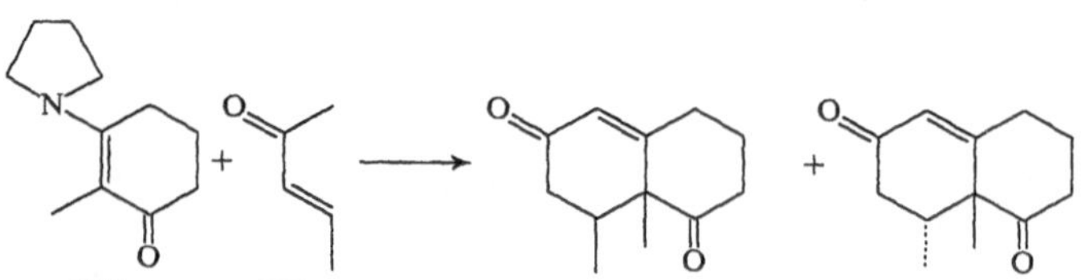

(**12**)   (**13**)   (**14a**)   (**14b**)

its partial hydrogenation product dihydroeremophilene (containing no $C=CH_2$ group) led to the formulation of eremophilene as (8) (7), but this was refuted by PIERS and KEZIERE (8, 9) who synthesized the racemic variety of hydrocarbon (8) and found it to be different from eremophilene. The structure of eremophilene was consequently reformulated as (9), largely on re-examination of n.m.r. spectral evidence (10), which indicated the presence of two upfield methyl groups and only one allylic methyl group. This structure was confirmed by the formation of eremophilene by dehydration of eremoligenol (10), of known structure

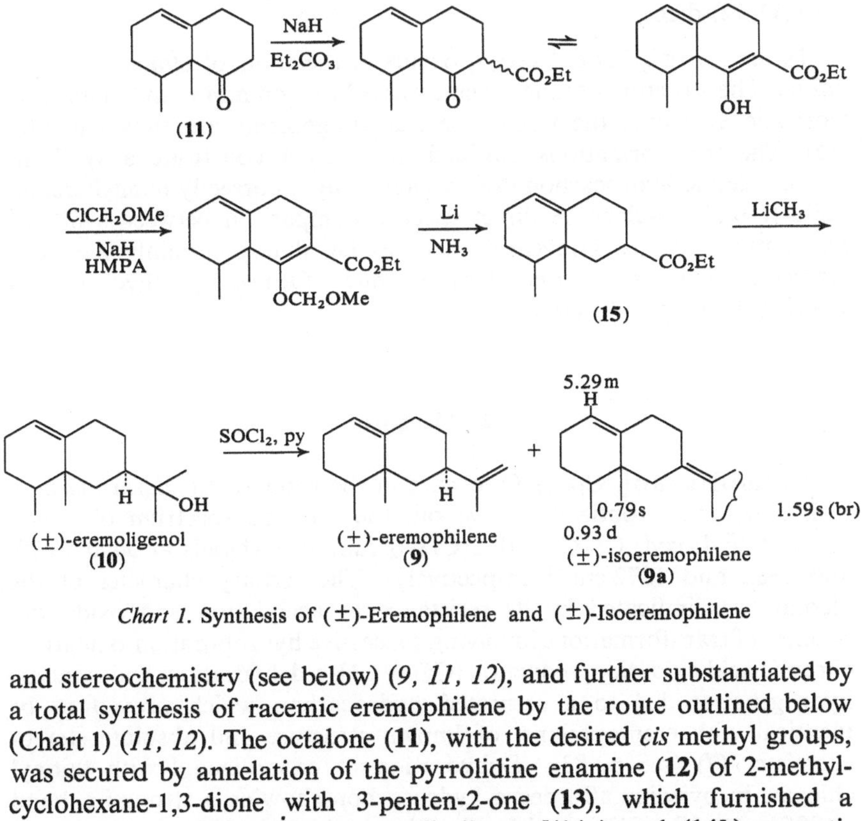

Chart 1. Synthesis of (±)-Eremophilene and (±)-Isoeremophilene

and stereochemistry (see below) (9, 11, 12), and further substantiated by a total synthesis of racemic eremophilene by the route outlined below (Chart 1) (11, 12). The octalone (11), with the desired cis methyl groups, was secured by annelation of the pyrrolidine enamine (12) of 2-methylcyclohexane-1,3-dione with 3-penten-2-one (13), which furnished a mixture of the cis and trans octalindiones [(14a) and (14b) respectively], the latter in the majority. Treatment of the mixture with ethanedithiol effected selective dithioketalization of the α,β-unsaturated carbonyl group, thus leading to a mixture of dithioketals which was desulfurized with Raney nickel to yield octalone (11) along with the corresponding trans epimeride. These two octalones were separated by fractional distillation, and configurations assigned on n.m.r. spectral evi-

dence (*11, 12, 13, 14, 15*). The final products (+)-eremophilene (**9**)
and its double bond isomer (**9a**), were separated by preparative g.l.c.
(*12*). The synthesis establishes that (**9**) represents the relative stereochemistry of eremophilene, since ester (**15**) was shown to have an
axial (β) carbethoxyl group by equilibration studies and by n.m.r. spectral
analysis (*12*). Consequently in eremophilene (**9**) the two methyl groups
and the isopropenyl group are all *cis*. Further, since natural (−)-eremo-
philene affords (+)-eremophilane (**7**) of known absolute stereochemistry
(**7**) (see page 83) it follows that (**9**) depicts the absolute stereochemistry
of (−)-eremophilene, which is more correctly named eremophila-
1(10),11(12)-diene.

Isoeremophilene, $C_{15}H_{24}$, occurs in the roots of *Othonna coronopifolia*. The structure (**9a**) has been assigned to it on n.m.r. and mass spectroscopic evidence; the n.m.r. spectral assignments are shown in (**9a**)
(*16*). The transformations outlined in Chart 1 constitute a synthesis
of the racemic hydrocarbon (**9a**), which is more correctly named eremo-
phila-1(10),7(11)-diene. A direct spectral comparison between synthetic
and natural products revealed them to be almost certainly identical.
Isoeremophilene is also one of the products formed by dehydration of
valerianol (see p. 148) (*16a*).

2. Alcohols

Eremoligenol, $C_{15}H_{26}O$, occurs in the roots of *Ligularia Fischeri*.
It is a colorless, optically active oil, the infrared spectrum of which
revealed hydroxyl, C=C, and $C(CH_3)_2$ functions (bands at 3400, 1672,
and 1382 and 1372 cm^{-1} respectively). The tertiary character of the
alcohol was indicated by its resistance to acetylation and oxidation.
A series of transformations involving successive hydroboration-oxidation,
chromic acid oxidation, reduction (CO→CH$_2$), dehydration, and catalytic
hydrogenation led to a saturated hydrocarbon which proved to be
identical with eremophilane, of known structure and absolute stereochemistry (**7**) (see p. 83). Dehydration of eremoligenol with thionyl
chloride in pyridine afforded a hydrocarbon showing a strong i.r. band
at 885 cm^{-1} (C=CH$_2$), and lacking the doublet at 1382 and 1372 cm^{-1}
(see above). The hydroxyl group is consequently part of a dimethyl-
carbinol grouping at $C_{(7)}$; thus partial structure (**16**), in which a
double bond has to be located, may be written for eremoligenol. The
dehydration product of eremoligenol referred to above proved not to be
a conjugated diene on u.v. spectral evidence; this precludes a 6,7 or
7,8 double bond in (**16**). The n.m.r. spectrum of eremoligenol shows

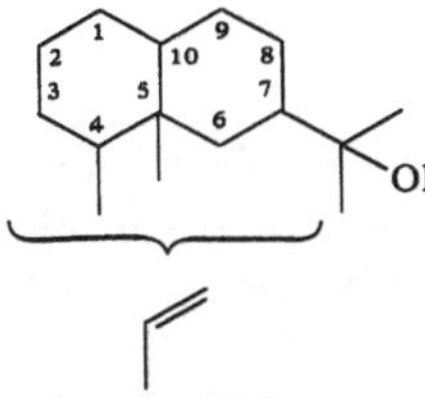

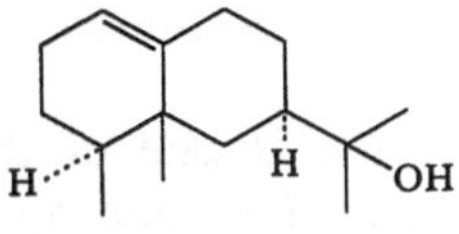

eremoligenol 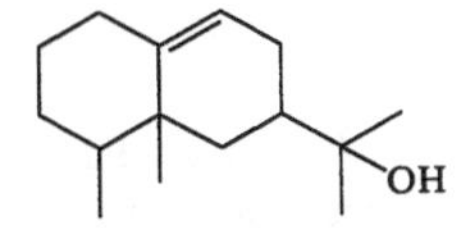   (18)

(16)   (17)

signals at δ 0.91 (s, angular CH₃), 0.87 (3 H, d, J = 6 Hz, $\rangle$CHCH₃),
and 5.31 ppm (1 H, m, =CH), which, collectively, indicate that the
double bond in eremoligenol cannot be 2,3, 3,4, or 8,9. The only two structural possibilities are consequently (17) and (18). Of these, the latter
may be ruled out on n.m.r. spectral evidence and because of the behavior
of the hydroboration–oxidation product of eremoligenol. On chromic
acid oxidation it yielded a hydroxyketone, (19) or (20), which on
dehydration gave a mixture of two enones, neither being conjugated,
nor were they converted into conjugated enones by acid or base
treatment. It would have been expected, were the enones represented

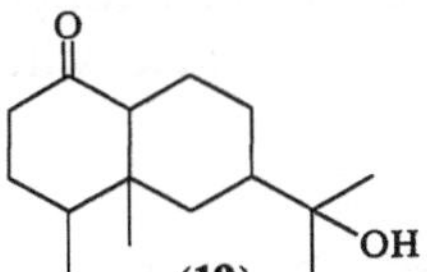

(19) 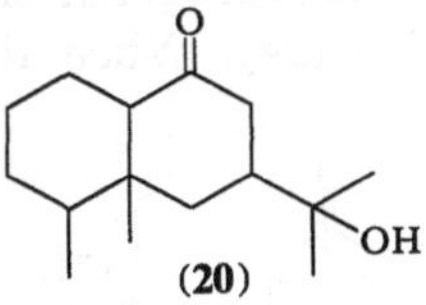   (20) 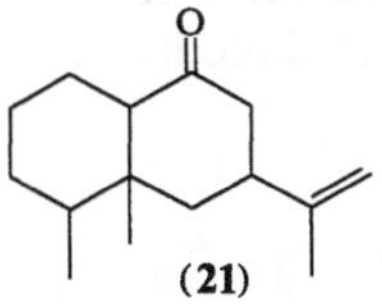   (21)

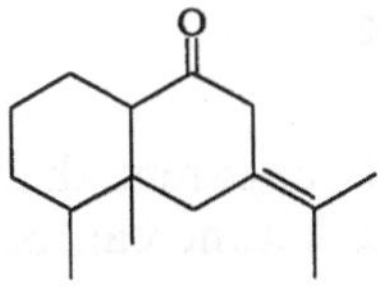

(22) 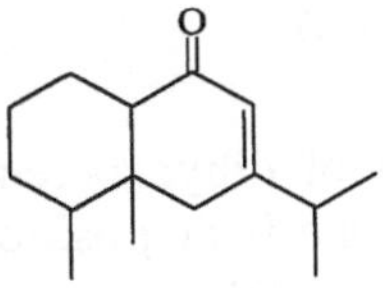   (23) 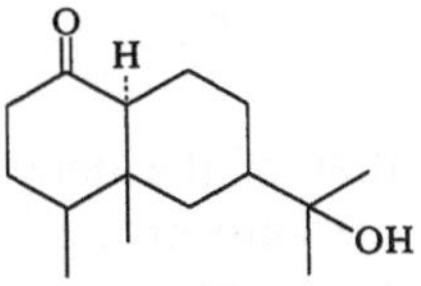   (24)

by (21) and (22), that some of the conjugated (23) would have resulted
from acid or base treatment. The hydroxyketone is therefore (24),
and the eremoligenol double bond is 1,10, as in (17). Because of the
degradation of (−)-eremoligenol to (+)-eremophilane, of known absolute
configuration (7), it follows that natural (−)-eremoligenol is (17) (17).
This structure has been confirmed by total synthesis of (±)-eremoligenol,
as outlined in Chart 1 (11, 12). Eremoligenol is thus 1,(10)-eremophilen-
11-ol.

3. Ketones

Eremophilone, $C_{15}H_{22}O$, is one of three closely related sesqui-terpenoid ketones occurring in the wood oil of *Eremophila mitchelli,* an Australian shrub known locally as "buddah" tree. Prolonged investi-gations by PENFOLD, SIMONSEN, and their collaborators have resulted in the elucidation of its structure (*3, 18, 19, 20, 21, 22*). Eremophilone is levorotatory and is an α,β-unsaturated ketone (λ_{max}^{EtOH} 243 nm, ε 7,950) (*22, 23*). On reduction with sodium and ethanol it afforded dihydroeremophilol, $C_{15}H_{26}O$, which on dehydrogenation with selenium gave eudalene (7-isopropyl-1-methylnaphthalene). This latter reaction indicated that eremophilone was a reduced naphthalenoid bicyclic sesquiterpene; during the dehydrogenation a carbon atom has been lost, which suggested that eremophilone contained a tertiary, probably angular, methyl group. Two double bonds were present, since eremophi-lone formed a saturated tetrahydro derivative on catalytic hydrogenation. Since dihydroeremophilol on ozonolysis afforded formaldehyde and a methyl ketone (positive iodoform reaction), one of the double bonds must be part of an isopropenyl group. The position of the keto-group was settled by reaction of tetrahydroeremophilone with methylmagnesium iodide, followed by selenium dehydrogenation, which yielded 7-isopropyl-1,5-dimethylnaphthalene (**25**) (*24*). When this result is compared with

that of the dehydrogenation of dihydroeremophilol it is apparent that the keto group in eremophilone is at position 9, and the ketone can be represented by partial structure (**26**).

Eremophilone formed a hydroxymethylene compound by reaction with ethyl formate; this means that the carbonyl group must be flanked by a methylene group, and therefore the second double bond cannot be located in the 7,8 position. This position can further be ruled out because the maximal ultraviolet absorption of eremophilone (see above) is characteristic of a conjugated enone and not of a dienone. The structure of eremophilone may therefore be expanded to (**27**).

Finally, there remained to be settled the location of the tertiary methyl group, which must be positions 4, 5, or 7 in (**27**). Alkaline hydrogen peroxide yielded with eremophilone a crystalline epoxide, which on

References, pp. 175—186

heating with acetic anhydride and hydrolysis gave hydroxyeremophilone (see p. 101), which accompanies eremophilone in *Eremophila mitchelli* oil; it is evident that eremophilone and hydroxyeremophilone have the same carbon skeleton. Ozonolysis of hydroxyeremophilone furnished several products including a monocyclic keto acid, $C_{10}H_{16}O_3$. Reduction of the keto group (Clemmensen) of this acid afforded a saturated mono-cyclic acid, $C_{10}H_{18}O_2$, which was identified by synthesis as 1,2-dimethylcyclohexylacetic acid (28) (21), though whether *cis* or *trans* was

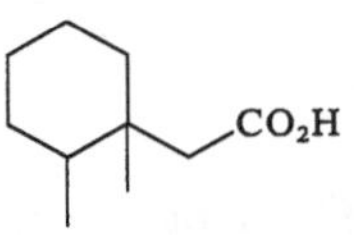

| (28) | eremophilone (29) | hydroxydihydro-eremophilone (30) |

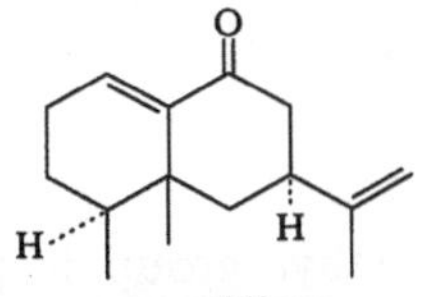

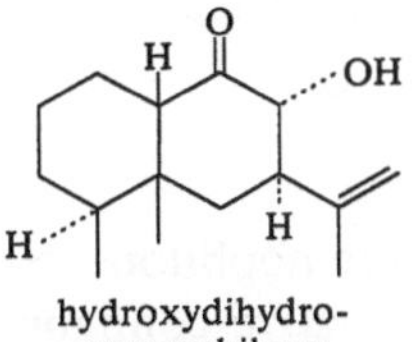

not established. Eremophilone must therefore be represented by (29), a gross structure which does not conform to the isoprene rule.

The stereochemistry of eremophilone is bound up with that of (+)-hydroxydihydroeremophilone (see p. 103), the third member of the trio of closely related sesquiterpenes of *Eremophila mitchelli* oil. The relative configuration of the latter compound has been shown to be (30) by X-ray diffraction analysis (25, 26). Eremophilone results from the pyrolysis of hydroxydihydroeremophilone acetate (31) (27).

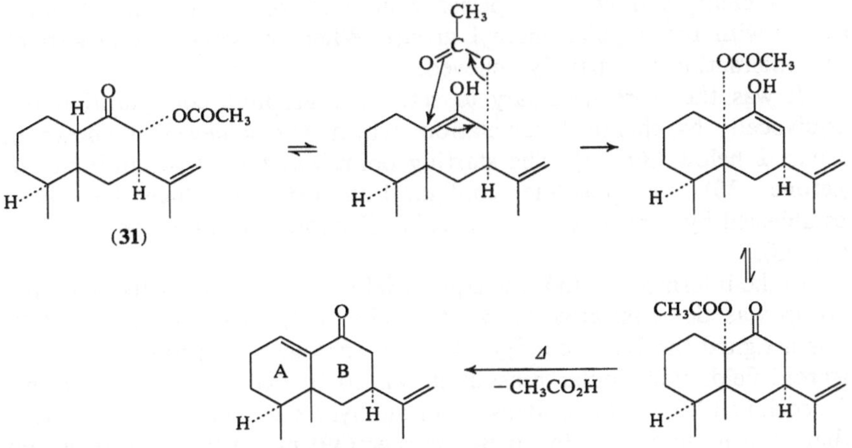

eremophilone
(32)

It follows that (32) must depict the relative configuration of eremophilone. The absolute configuration of the compound has been the subject
of some controversy. KLYNE (28) deduced, from a comparison of molecular
rotation differences, that eremophilone was of the same stereochemical
type as steroid 5-en-4-ones of known absolute configuration (33), so that in

(33)   (34)

eremophilone the angular methyl group is β. Since (32) depicts the
relative configuration of the ketone, it must also represent the absolute
configuration of natural (−)-eremophilone. However, a comparison of
the optical rotatory dispersion curves of eremophilone, 5-cholesten-4-one
(33), and 4-cholesten-6-one (34) revealed that the eremophilone curve
resembled that of (34), and bore a mirror image relationship to that of
(33). This required that the absolute configuration of eremophilone
be represented as the mirror image of (32) (29).

Some time later (30) it became apparent that optical rotatory
dispersion comparisons between eremophilone and steroids were invalid
because the bicyclic environment of the keto group in the three cases
is not identical, there being evidence that in eremophilone (32) ring B
does not adopt the chair conformation, but is a twist-boat. One reason for
this peculiarity is that the 7β-isopropenyl group would be axial if ring B
were a chair, and would experience an unfavorable 1,3-diaxial interaction with the angular methyl group. When the ring is a twist-boat
this interaction is partially relieved.

It was therefore necessary to settle the absolute configuration unequivocally by chemical correlation, which was achieved as shown in
Chart 2 below (30, 31). The starting point was the (+)-enantiomer of
ketone (35), the absolute configuration (35) of which had been
established by correlation with steroids of known absolute configuration
(32, 33).

In the intermediate (36) the equatorial (β) orientation of the 4-methyl
group was demonstrated by o.r.d. studies (31); likewise epoxide (37)
was assigned the 7α,8α configuration by analogy with precedents in the
steroid field. For similar reasons it was to be expected that the trans-
2-decalone (38) would condense with diethyl oxalate at position 7 rather
than 9, leading eventually to β-ketoester (39) in which the ester group
would be α (equatorial). The absolute stereochemistry of the final

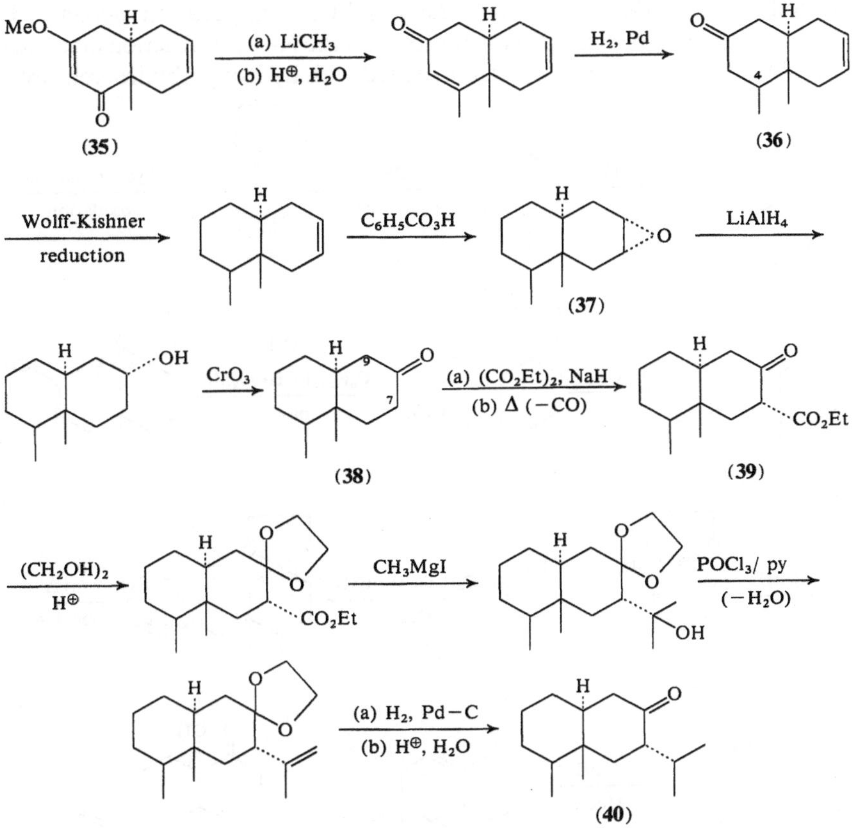
MeO
H
O
(35)
(a) LiCH3
(b) H⊕, H2O
H
O
H2, Pd
O
H
4
(36)
Wolff-Kishner
reduction
H
C6H5CO3H
H
O
(37)
LiAlH4
H
OH
CrO3
H
9
O
7
(38)
(a) (CO2Et)2, NaH
(b) Δ (−CO)
H
O
CO2Et
(39)
(CH2OH)2
H⊕
H
O
O
CO2Et
CH3MgI
H
O
O
OH
POCl3/ py
(−H2O)
H
O
O
(a) H2, Pd−C
(b) H⊕, H2O
H
O
(40)

Chart 2

decalone (**40**) is thus established. In Chart 3 is shown in outline how hydroxyeremophilone (**5**) was converted into the same decalone (**40**), (*30, 31*). Natural (+)-hydroxydihydroeremophilone therefore has absolute configuration (**5**). Next, hydroxydihydroeremophilone (**30**) is converted into hydroxyeremophilone (**5**) on oxidation with bismuth oxide (*34*).

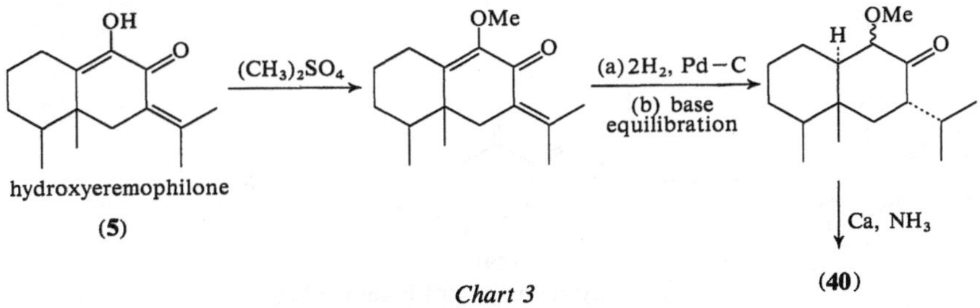
OH
O
hydroxyeremophilone
(5)
(CH3)2SO4
OMe
O
(a) 2H2, Pd−C
(b) base
equilibration
H
OMe
O
Ca, NH3
(40)

Chart 3

It follows that the former has the absolute configuration depicted in
(**30**), on the basis of this transformation and the X-ray diffraction evidence.
Finally, eremophilone has been correlated with hydroxydihydroeremophi-

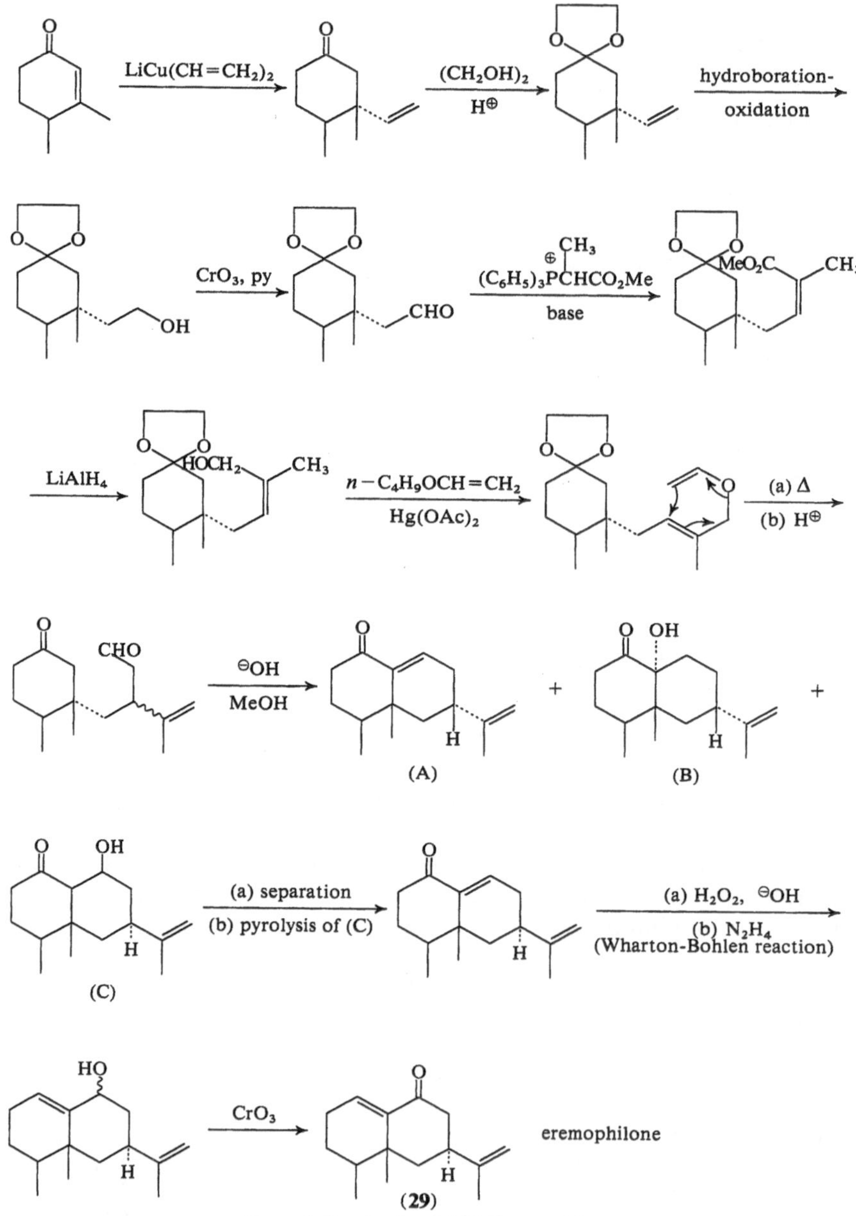

Chart 4. Synthesis of (±)-Eremophilone

lone acetate (27) as described on p. 89, and natural (−)-eremophilone must be allotted the absolute configuration (32). Confirmation is provided by other correlations in this area.

The structure and relative stereochemistry of eremophilone have been confirmed by two total syntheses of the racemic ketone. The first of these is outlined in Chart 4 (35).

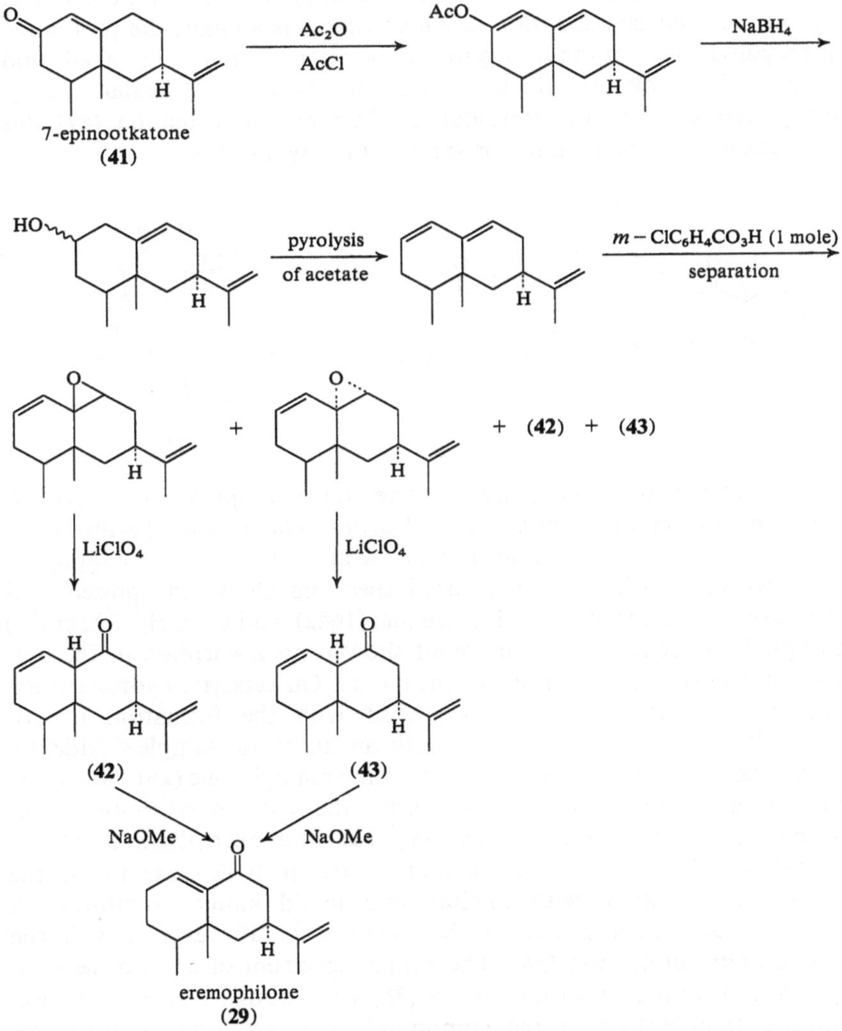

7-epinootkatone
(41)

Ac₂O / AcCl →

NaBH₄ →

HO⟩ pyrolysis of acetate →

m − ClC₆H₄CO₃H (1 mole) separation →

+

+ (42) + (43)

LiClO₄

LiClO₄

(42)

(43)

NaOMe →   ← NaOMe

eremophilone
(29)

Chart 5. Synthesis of (±)-Eremophilone

The second synthesis (*36*), which is stereoselective, is summarized in Chart 5. The starting material 7-epinootkatone (**41**) was secured by synthesis from β-pinene (*37*) (see p. 97). Ketone (**42**), eremophila-1, 11(12)-dien-9-one, occurs along with eremophilone in *E. mitchelli* oil (*38*) (see below).

An interesting point concerning 1,10-dihydroeremophilone (**44**) and 1,10,11,12-tetrahydroeremophilone (**45**) is that these 1-decalones are more stable when the ring junction is *cis* rather than *trans*, contrary to the usual stability relationship. The explanation lies in the fact that in the *trans* isomers (**46**), in which ring B is a chair, the β-oriented isopropenyl or isopropyl group at position 7 must be axial and suffer unfavorable 1,3-diaxial interaction with the angular methyl group. However, in the steroidal all-chair *cis* conformation (**47**) this interaction is absent, making for greater stability (*34, 39*).

(**44**)   (**45**)   (**46**) R
                                                        R= or
                                                        (**47**)

Eremophila-1,11(12)-dien-9-one (isoeremophilone), $C_{15}H_{22}O$, a congener of eremophilone in *Eremophila mitchelli* oil, was separated from other components by column and preparative gas chromatography. It is a crystalline solid whose infrared spectrum shows the presence of saturated ketone carbonyl (1710), olefinic (1642), and $C=CH_2$ (892 cm^{-1}) groups. In agreement the compound showed no absorption in the α,β-unsaturated ketone region of the ultraviolet. On catalytic hydrogenation two mols of hydrogen were absorbed with the formation of *cis*-tetrahydroeremophilone, identical with an authentic sample. Evidently the new compound differs structurally from eremophilone (**29**) only in the position of the cyclic double bond. Earlier the structure of another compound from *E. mitchelli*, 8α-hydroxy-7α(H)-eremophila-1,11(12)-dien-9-one (**48**), had been elucidated (*40*) (see p. 104). Reduction of the acetate of this ketol with calcium and liquid ammonia afforded a product (**49**), which proved to be identical in all respects with the ketone under discussion (*38*). The n.m.r. spectrum of the ketone is in agreement with the assigned structure (*38*), as is the observation that base-induced equilibration of the compound generates eremophilone (**29**) (*36, 40, 41*). It follows that (**49**) represents the absolute configuration of

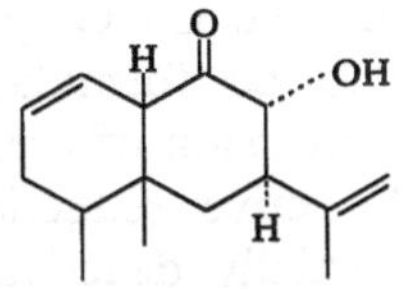

(48) 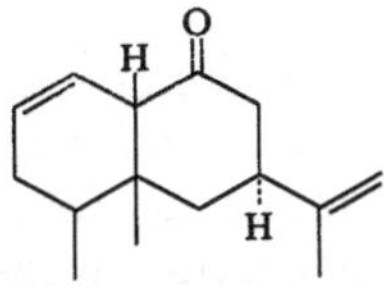 | isoeremophilone
(49)

the natural (+)-ketone. The structure has been confirmed by synthesis (Chart 5) (36).

Alloeremophilone, $C_{15}H_{22}O$, also occurs in E. mitchelli oil. It has not been investigated chemically, but was observed to be formed, along with eremophilone, by pyrolysis of hydroxydihydroeremophilone acetate (31) (see p. 89) (27) and is therefore formulated as (50).

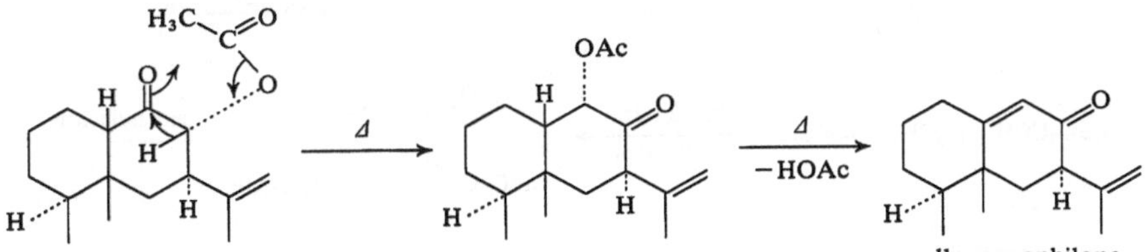

(31) | △ | △ −HOAc | alloeremophilone (50)

Eremofukinone, $C_{15}H_{24}O$, occurs in rhizomes of Petasites japonicus Maxim. Its chemical and spectral properties reveal that it is a saturated ketone (ν_{CO} 1705), and also contains a $C=CH_2$ group (890 cm^{-1}). On catalytic hydrogenation it yields a saturated ketone $C_{15}H_{26}O$, which must be bicyclic; on reduction of the carbonyl group via its ethylenedithioketal eremophilane (7) resulted. The double bond of eremofukinone was proved by n.m.r. spectroscopy (allylic CH_3 signal at δ 1.73 ppm) to be part of an isopropenyl group; it showed no tendency, when the ketone was exposed to acid or base, to migrate into conjugation with the carbonyl group. Treatment of eremofukinone with sodium carbonate dissolved in D_2O-methanol afforded, on mass spectrometric evidence, a trideuteroketone. Coupled with the chemical shift of the 4-methyl group (0.77 ppm) and the observed negative Cotton

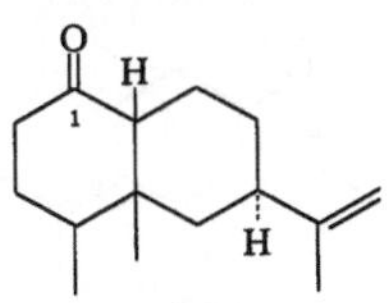

eremofukinone (51) | (52) 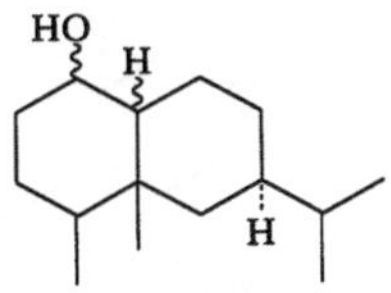 | (53) 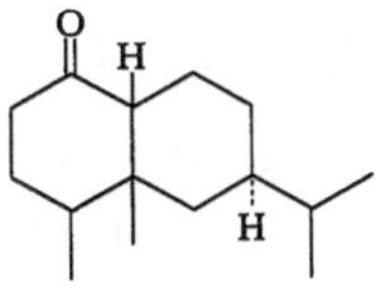 | (54)

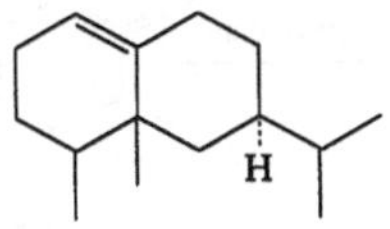

effect in the o.r.d. curve, this enables the C=O group to be placed at position 1, and eremofukinone to be formulated as (51). This formulation has been confirmed by conversion of eremophilene (9) into dihydroeremofukinone as follows. Selective reduction of the hydro-carbon by hydrogenation using a partially deactivated Raney nickel catalyst yielded 11,12-dihydroeremophilene (52), which on hydro-boration-oxidation yielded alcohol (53). Jones oxidation of the latter furnished ketone (54), which proved to be identical with dihydroeremo-fukinone (42).

7-Epinootkatone, $C_{15}H_{22}O$, is believed to occur in Haiti and Réunion vetiver oils, on gas chromatographic evidence (43). Its racemic

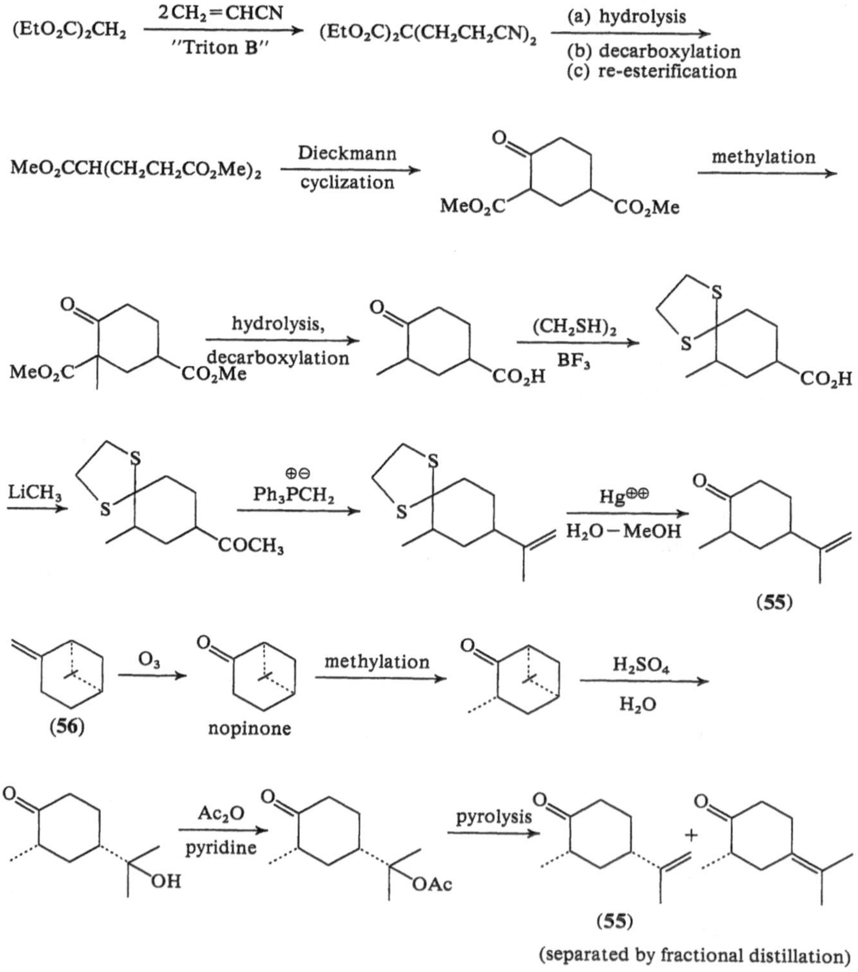

$(EtO_2C)_2CH_2$ $\xrightarrow[\text{"Triton B"}]{2\,CH_2=CHCN}$ $(EtO_2C)_2C(CH_2CH_2CN)_2$ $\xrightarrow[\substack{\text{(b) decarboxylation} \\ \text{(c) re-esterification}}]{\text{(a) hydrolysis}}$

$MeO_2CCH(CH_2CH_2CO_2Me)_2$ $\xrightarrow[\text{cyclization}]{\text{Dieckmann}}$ $\xrightarrow{\text{methylation}}$

$\xrightarrow[\text{decarboxylation}]{\text{hydrolysis,}}$ $\xrightarrow[\text{BF}_3]{(CH_2SH)_2}$

$\xrightarrow{\text{LiCH}_3}$ $\xrightarrow{Ph_3PCH_2^{\oplus\ominus}}$ $\xrightarrow[H_2O-MeOH]{Hg^{\oplus\oplus}}$

(55)

$\xrightarrow{O_3}$ $\xrightarrow{\text{methylation}}$ $\xrightarrow[H_2O]{H_2SO_4}$

(56)   nopinone

$\xrightarrow[\text{pyridine}]{Ac_2O}$ $\xrightarrow{\text{pyrolysis}}$ +

(55)

(separated by fractional distillation)

variety has been synthesized by two routes. The starting material in both cases was 4-isopropenyl-2-methylcyclohexanone (predominantly *cis* form) (55) obtained either by stepwise synthesis from diethyl malonate as follows (44, 45) or from β-pinene (56) (36, 37).

Annelation of (55) was effected with *trans*-3-penten-2-one, in the presence of sodium hydride (44, 45) or sodamide (36, 37) to yield almost exclusively 7-epinootkatone (57).

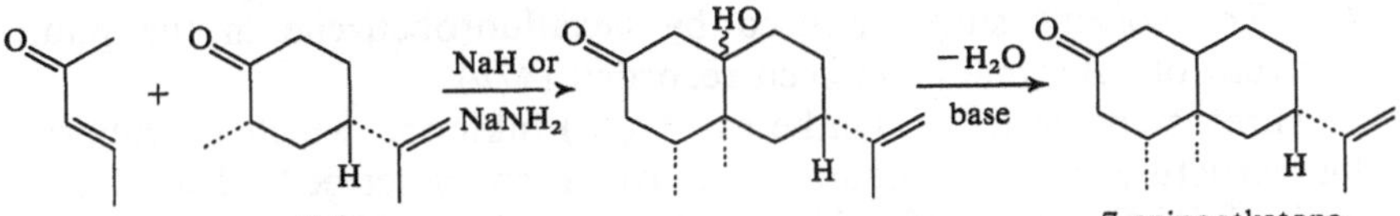

(55)   7-epinootkatone
                                                      (57)

Using sodium hydride (44, 45) as base gave a product containing a few percent of nootkatone (p. 149), and on one occasion (44) it was possible to isolate the crystalline racemic ketone. With sodamide a small amount of racemic α-vetivone (p. 153) was also obtained (37).

Fukinone, $C_{15}H_{24}O$, is a sesquiterpene ketone occurring in *Petasites japonicus* ("fuki" in Japanese). Its infrared spectrum showed it to be an α,β-unsaturated ketone (v_{CO} 1685, $v_{C=C}$ 1625 cm^{-1}), confirmed by its u.v. absorption (λ_{max} 251 nm, ε 6800). The n.m.r. spectrum of fukinone showed signals at δ 1.78 and 1.90 (allylic CH$_3$), 0.95 (singlet, angular CH$_3$), and 0.84 (doublet, $\rangle$CHCH$_3$). The suspicion that an isopropylidene group was present was confirmed by the formation of acetone on ozonolysis; acetone was also formed by alkaline hydrolysis of fukinone (retroaldol cleavage), indicating conjugation of the isopropylidene group with the keto group [cf. the behavior of the monoterpene pulegone (46)]. Catalytic hydrogenation of fukinone yielded dihydrofukinone which contained an isopropyl group [methyl signals (doublets) at 0.83 and 0.85 ppm]; no allylic methyl signals were observed. Simultaneously the infrared carbonyl band became characteristic of a saturated ketone (1710 cm^{-1}). Dihydrofukinone afforded an ethylenedithioketal which on desulfurization with Raney nickel yielded (+)-eremophilane (7). Optical rotatory dispersion studies on fukinone, dihydrofukinone, and

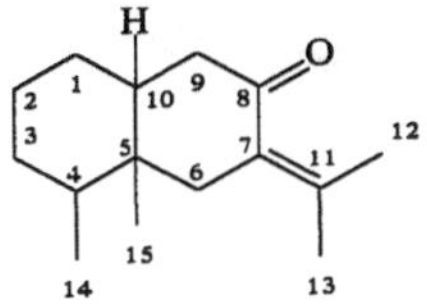

fukinone   isopetasol
(58)   (59)

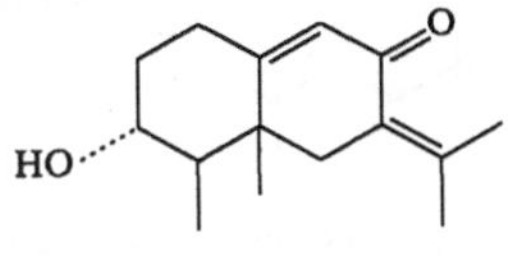

desisopropylidenefukinone (the retroaldol cleavage product) showed
that the curves were very similar to those of A/B *cis*-fused 3-keto steroids,
and it thus became possible to formulate fukinone as (**58**). Confirmation
was provided by a correlation with isopetasol, of known structure and
absolute configuration (**59**) (see p. 109). The latter was reduced to
tetrahydroisopetasol, shown to be a *cis*-2-decalone by its o.r.d. curve.
The tosylate of this compound was reduced with lithium aluminum hydride
to give an alcohol which on Jones oxidation yielded dihydrofukinone
(*47*). The solvent shifts induced by hexafluorobenzene in the p.m.r.
spectrum of fukinone have been recorded (*47a*).

Three stereoselective syntheses of (±)-fukinone which corroborate
this structure and configuration have been described. The first is
outlined in Chart 6 (*48*). The starting octalone (**60**) was synthesized
from 2,3-dimethylcyclohexanone and the *cis* relationship of the two
methyl groups established beyond doubt (*49, 50, 51*). The *cis*-2-decalone
stereochemistry of the intermediate (**61**) was proved by catalytic

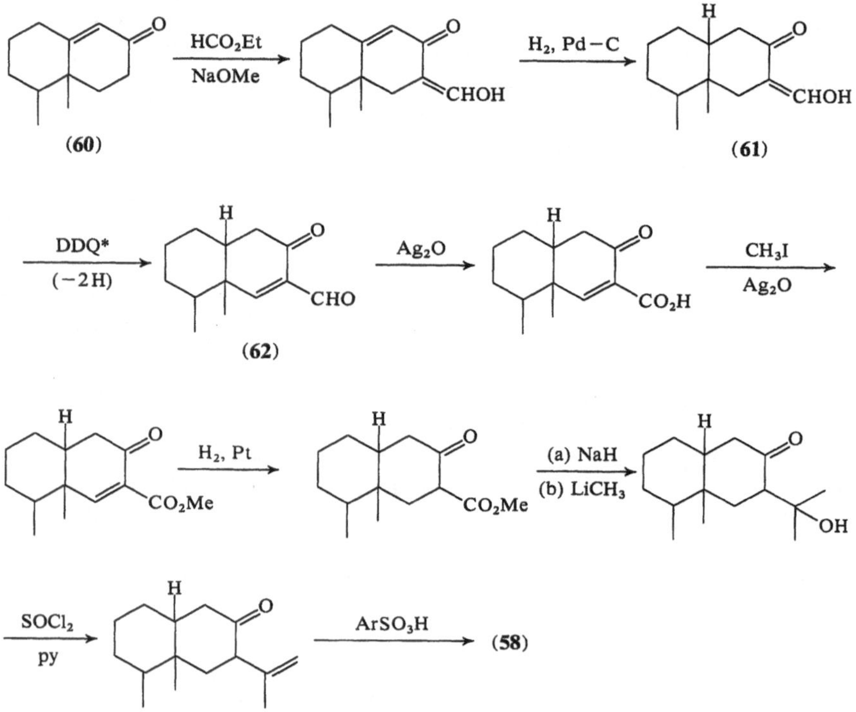

(* DDQ = 2,3-dichloro-5,6-dicyanobenzoquinone)

Chart 6. Synthesis of (±)-Fukinone

hydrogenation of (60) under basic conditions to the corresponding *cis*-fused 2-decalone and by reduction of the same compound with lithium in liquid ammonia, known to yield the corresponding *trans*-fused

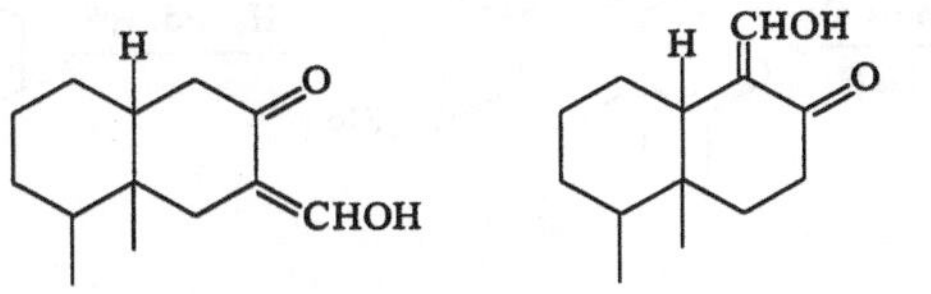

(63)   (64)

2-decalone; the two decalones were shown to be different. Ths *cis*-fused 2-decalone on treatment with ethyl formate and sodium methoxide, afforded a mixture of hydroxymethylene ketones (63) and (64); dehydrogenation of this mixture with DDQ gave a product from which was separated some of the keto aldehyde (62), proving that (61) has the desired *cis* ring junction (*49*).

The second synthesis (*52, 53*) (Chart 7) began with the known octalol (65) (*54*) secured by an improved route from *o*-anisaldehyde.

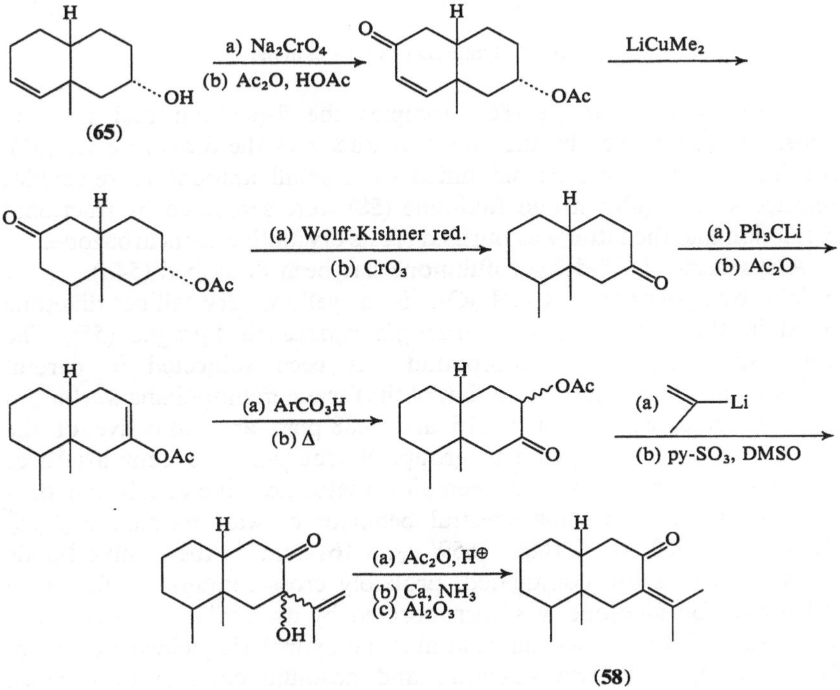

Chart 7. Synthesis of (±)-Fukinone

7*

Finally, a third synthesis (*50, 51*) used as starting material the same cis-4,5-dimethyloctalone (**59**) as used in the first; it is outlined in Chart 8.

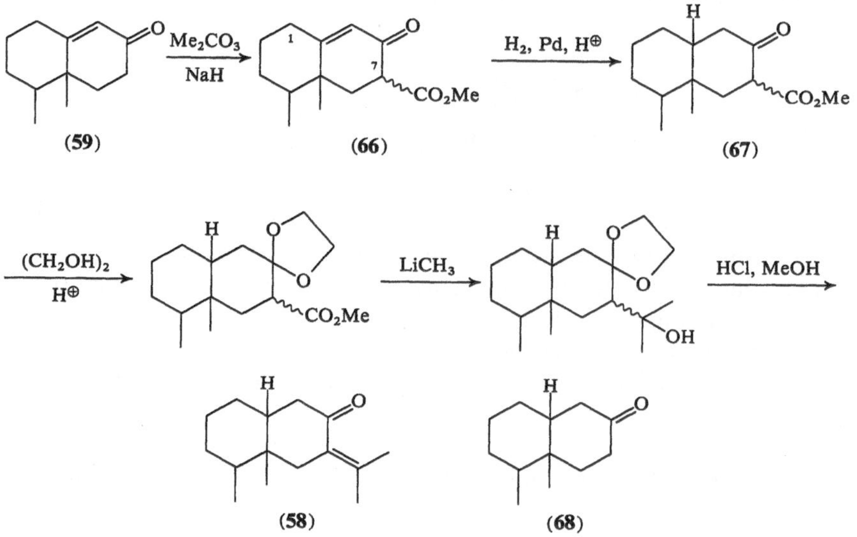

Chart 8. Synthesis of (±)-Fukinone

That the ester group in (**66**) occupies the 7-position and not the 1-position was proved by the enolic character of the dihydro ester (**67**). The final product was accompanied by a small amount of retroaldol cleavage ketone (**68**); it and fukinone (**58**) were separated by fractional distillation and the latter was purified *via* its crystalline semicarbazone.

A synthesis of Δ^9-dehydrofukinone has been described (*54a*).

Warburgiadione, $C_{15}H_{18}O_2$, is a yellow, crystalline diketone found in the heartwood of *Warburgia ugandensis* Sprague (*55*). The n.m.r. spectrum of the compound has been subjected to careful analysis, and its resemblance to that of the furanoeremophilane warburgin (p. 134) noted. Signals at δ 2.15 and 1.88 ppm are indicative of the presence of an isopropylidene group. Warburgiadione contains three double bonds, yielding on hydrogenation a saturated diketone, hexahydrowarburgiadione. From the spectral behavior of warburgiadione (λ_{max}^{EtOH} 292 nm, ε 21,500; $\nu_{C=O}$ 1686, 1659, $\nu_{C=C}$ 1614 cm^{-1}) the double bonds and keto groups are conjugated, probably cross-conjugated. Structure (**69**) for warburgiadione was corroborated by its partial synthesis from isopetasin (p. 108), which on mild alkaline hydrolysis yielded isopetasol (**70**) (p. 109), of known structure and absolute configuration. Since isopetasol has itself been totally synthesized as its racemic form recently (58), this sequence constitutes a total synthesis of warburgia-dione (56, 57). A more recent analysis of the n.m.r. spectrum of warburgia-dione using spin decoupling is in agreement with formulation (69) for natural (+)-warburgiadione (57).

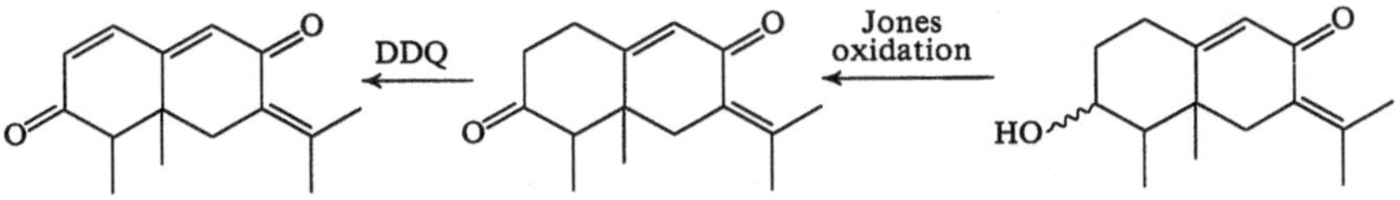

warburgiadione
(69)

isopetasone

isopetasol
(70)

4. Keto Alcohols

Hydroxyeremophilone (HE), $C_{15}H_{22}O_2$, is the second of the original trio of non-isoprenoid substances occurring in the wood oil of *E. mitchelli*. It is an enolic α-diketone, forming both carbonyl functional derivatives, and esters and ethers; its structure (5) has been elucidated, as with eremophilone (p. 88), largely due to the efforts of J. L. SIMONSEN and co-workers (18, 19, 20, 21, 22). Ozonolysis of its benzoate afforded acetone, suggestive of the presence of an isopropylidene group.

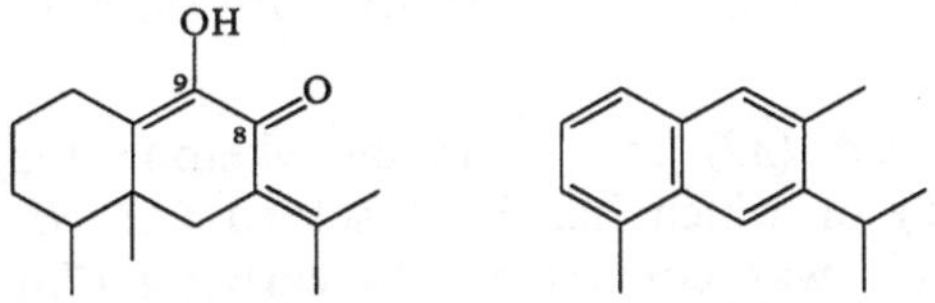

hydroxyeremophilone
(5)

(71)

Methylation yielded a methyl ether which on catalytic hydrogenation gave a saturated tetrahydro ketone. The latter reacted with methylmagnesium iodide to give a methyl carbinol, which on dehydrogenation gave 7-iso-propyl-1,6-dimethyl-naphthalene (71); the keto group in HE is therefore located at position 8 and the hydroxyl function at $C_{(9)}$. Consideration of this and other transformations of HE already discussed earlier (see p. 89) in connection with eremophilone permitted gross structure (5) to be advanced for the ketol. A correlation between hydroxyeremophilone and hydroxydihydroeremophilone (HDE) (outlined on p. 89), of known absolute configuration, established that natural (+)-hydroxyeremophilone has the absolute configuration (5).

This structure has been substantiated by a total synthesis of the natural compound (51), outlined in Chart 9. The starting point was the (+)-camphorsulfonate (72) [R = (+)-camphor-10], already synthesized

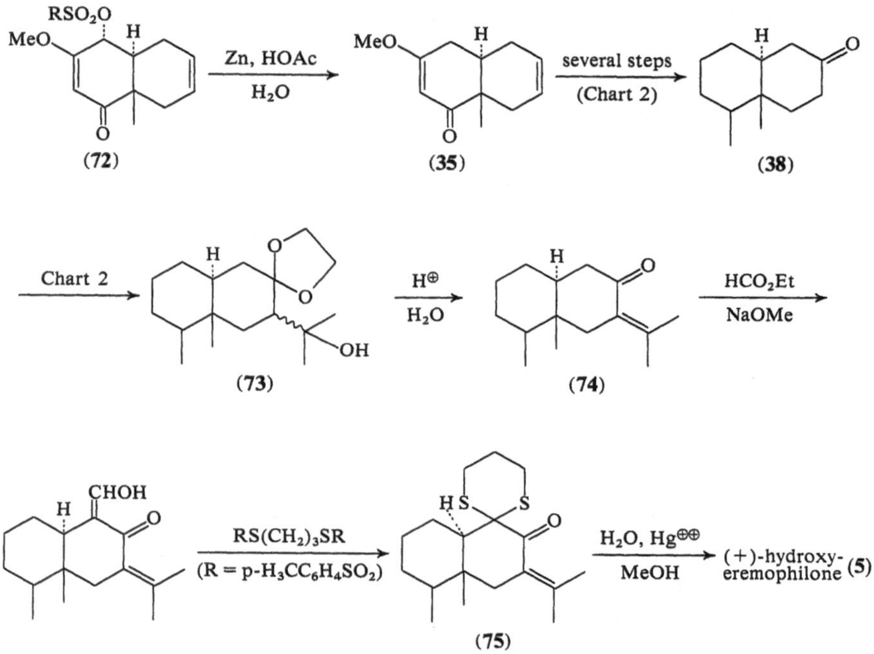

Chart 9. Synthesis of (+)-Hydroxyeremophilone

and hydrogenolyzed (32) to (+)-methoxyketone (35), the absolute configuration (35) of which has been settled (32, 33) by correlation with steroids. This was converted into carbinol (73) as outlined in Chart 2 (p. 91). and thence to *trans*-fukinone (74). The hydroxymethylene derivative of the latter was reacted with propane-1,3-dithiol ditosylate to give the 1,3-dithian (75), which on hydrolysis furnished (+)-hydroxyeremophilone (5).

The ozonolysis of hydroxyeremophilone benzoate referred to above gave along with acetone a compound $C_{15}H_{22}O_4$, formulated as the lactonic ketone (76) on the basis of its spectral and chemical behavior. This degradation product, which is also formed when hydroxyeremophilone and its methyl ether are oxidized, presumably arises *via* ring cleavage to acid (77)

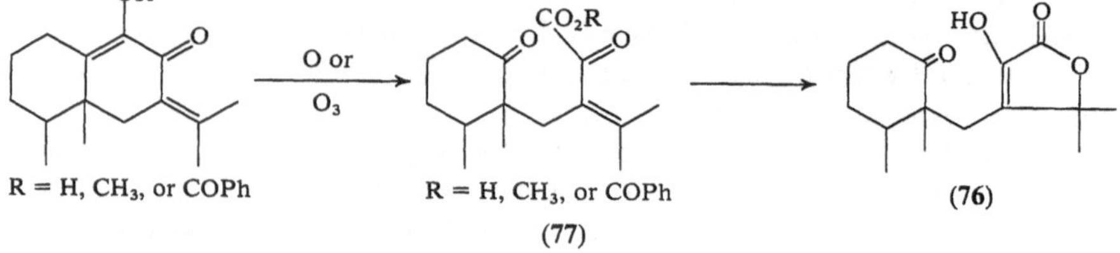

R = H, CH₃, or COPh

R = H, CH₃, or COPh   (76)

(77)

(or corresponding methyl ester or mixed benzoic anhydride from the methyl ether or benzoate respectively, which on hydrolysis would yield the same acid), which suffers lactonization to (76) (59).

Hydroxydihydroeremophilone (HDE), $C_{15}H_{24}O_2$, the third member of the trio of non-isoprenoid sesquiterpenes from *E. mitchelli*, has already been covered in some detail in the section on eremophilone (p. 88). Its structure has emerged, as with its two congeners, largely as a consequence of the researches of J. L. SIMONSEN and collaborators (18, 19, 20, 21, 22). The compound is a ketol which on catalytic hydrogenation takes up one mole of hydrogen to yield a saturated product, and on ozonolysis formaldehyde resulted; one double bond of the type $C=CH_2$ is therefore present. Sodium and ethanol reduction of HDE gave a 1,2-glycol which was oxidizable by lead tetra-acetate: HDE is therefore an α-ketol. The hydroxyl function in the above dihydro derivative could be hydrogenolyzed by sodium amalgam to give tetrahydroeremophilone, which behavior establishes the carbon skeleton of the ketol. HDE may consequently be formulated as (30), the relative configuration of the hydroxyl group being settled by X-ray diffraction analysis (25, 26), which shows that the group is equatorially bound, that the two cyclohexane rings have a chair conformation and are united

hydroxydihydroeremophilone
(30)

(78)

in the *cis* steroidal manner, with all non-angular substituents equatorially disposed. The more detailed picture of HDE is (78); that this also represents the absolute configuration of the natural (+)-ketol has been concluded from molecular rotation differences comparison between eremophilone and steroidal 5-en-4-ones (see p. 13) (28), and from the correlations outlined in Charts 2 and 3 (p. 91).

Petasitolone, $C_{15}H_{24}O_2$, is another constituent of *P. japonicus*, the infrared spectrum of which reveals the presence of alcohol and α,β-unsaturated ketone functions (bands at 3480 and 1665 cm^{-1}). Maximal u.v. absorption occurs at 237.5 nm. The n.m.r. spectrum is consistent with the presence of one olefinic proton (δ, 6.6 ppm, singlet), a hydroxyl proton (3.5), an isopropanol group (1.30, 6H), an angular methyl group (1.11), and a CH_3CH group (0.93, doublet, $J = 7 Hz$). Petasitolone formed a 2,4-dinitrophenylhydrazone in the usual manner, accompanied by another such derivative formed as a consequence of prior dehydration to anhydropetasitolone. The latter, $C_{15}H_{22}O$, is also formed by mild dehydration of petasitolone with phosphoryl chloride-pyridine; it has λ_{max}^{EtOH} 245 nm, indicating extension of the conjugation during elimination of water, and its n.m.r. spectrum shows additional olefinic hydrogen and allylic methyl signals. Evidently the hydroxyl group in petasitolone is allylic; it is also tertiary because the compound does not form an acetate and is not oxidizable by Jones reagent. Its allylic nature was confirmed by the observation that catalytic hydrogenation of petasitolone yielded a normal dihydro derivative (ν_{CO} 1690) as well as deoxydihydropetasitolone (no OH band in i.r. spectrum), the result of hydrogenolysis of the hydroxyl function. The latter compound proved to be identical with dihydrofukinone (**79**) (see p. 97), of known structure and stereochemistry. Petasitolone is therefore formulated as (**80**). Further, deoxydihydropetasitolone, on desulfurization of its ethylenedithioketal, afforded eremophilane (**7**), and dihydropetasitolone (**81**) underwent retroaldol cleavage to desisopropylidenefukinone (**68**). Finally, dehydration of dihydropetasitolone yielded fukinone (**58**) (*60*). A synthetic corroboration of structure (**80**) is provided by the observation that fukinone (**58**), on irradiation and exposure to oxygen in the presence of rose bengal followed by reduction with sodium sulfite, gave petasitolone in 45% yield (*60*).

8α-Hydroxy-7α(H)-eremophila-1,11(12)-dien-9-one, $C_{15}H_{22}O_2$, is another constituent of *E. mitchelli* wood oil. Its spectral properties show that it contains a saturated keto group, a hydroxyl group, a $C=CH_2$ group, and a *cis* CH=CH group (i.r. bands at 1705, 3480, 900 and 1645 cm^{-1}). The n.m.r. spectrum is consistent with the presence of an angular CH_3 group (1.06 ppm), a CH_3CH group (0.78, doublet), an allylic CH_3 group (1.67), a CHOH group (3.8, doublet), and four vinylic hydrogens (4.78, 5.7). On catalytic hydrogenation the compound

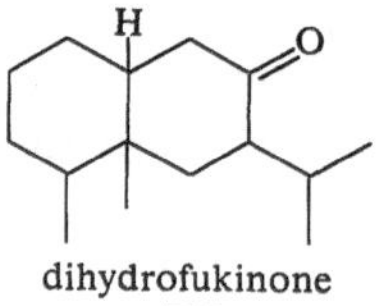

dihydrofukinone   petasitolone   (**81**)
(**79**)   (**80**)

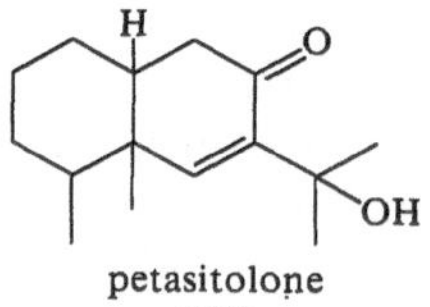

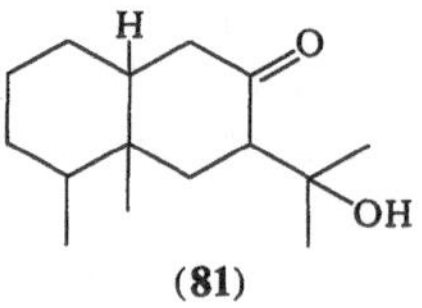

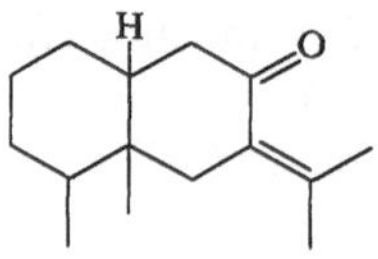

(**58**) 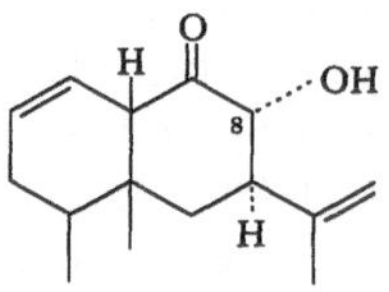   (**82**) 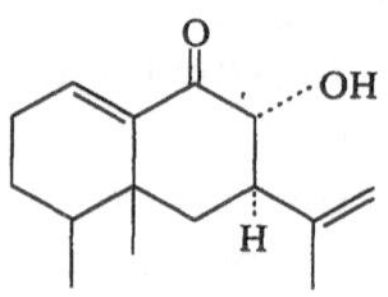   (**83**)

absorbed two moles of hydrogen to yield a saturated tetrahydro-
derivative. This, on acetylation and reduction with calcium and liquid
ammonia furnished isoeremophilone (**49**) (p. 94), which on base-
catalyzed equilibration gave eremophilone (**32**). The natural product may
therefore be represented by (**82**), in which the hydroxyl group is placed
at position 8 (α to the $C=O$ group) because of hydrogenolysis of the
acetate described above; the α (equatorial) orientation of this group
is deduced from the n.m.r. signal of the $C_{(8)}$-hydrogen atom (δ 3.8 ppm,
$J = 11.4$ Hz). The large value of the coupling constant indicates a
180° dihedral angle between the $C_{(8)}$-H and $C_{(9)}$-H bonds (*40*).

8α-Hydroxy-7α(H)-eremophila-1(10),11(12)-dien-9-one,
$C_{15}H_{22}O_2$, also from *E. mitchelli*, was observed to be very closely
related to the foregoing product. It is an α,β-unsaturated ketone
(λ_{max}^{EtOH} 246 nm, ε 7900), the n.m.r. spectrum of which reveals only
three vinylic hydrogens, one of which resonates at low field (δ 7.08).
Hydrogenation of the compound gave the same dihydroderivative as
the foregoing ketol. Acetylation of the natural product followed by
calcium-liquid ammonia hydrogenolysis afforded eremophilone (**32**).
The compound must therefore be formulated as (**83**) (*40*).

5. Acids

Flourensic acid, $C_{15}H_{22}O_3$, occurs in the flowering heads of
Flourensia cernua DC., a native Mexican shrub. It contains a carboxyl
group and, on n.m.r. spectral evidence, a CH_3CH group (δ 0.82, doublet),
a tertiary methyl group (1.02, singlet), and a $C=CH_2$ group (5.60,
6.32 ppm). A keto group is present, and the carboxyl group is α,β-
unsaturated (λ_{max} 203 nm, ε 8450). Catalytic hydrogenation resulted in
uptake of one mole of hydrogen to yield a saturated keto acid; the i.r.
spectrum of the methyl ester of the latter had $\nu_{C=O}^{CCl_4}$ 1736 and 1706 cm^{-1}.
Lithium aluminum hydride reduction of this ester, followed by
dehydrogenation, yielded an unidentified naphthalene. The ester formed
an ethyleneketal, which on similar hydride reduction gave a primary
alcohol, convertible to a mesylate. Treatment of the latter with lithium
aluminum hydride, followed by hydrolysis, afforded a ketone which
exchanged three hydrogens for deuterium on exposure to D_2O and base
and which appeared to be identical with *cis*-tetrahydroeremophilone
(**47**) (see p. 88) (o.r.d. and 2,4-dinitrophenylhydrazone comparison).
Epimerization at the $C_{(10)}$ ring junction was ruled out because the
o.r.d. curve of flourensic acid is similar to that of (**47**). Flourensic
acid is therefore formulated as (**84**); this is supported by mass spectral
studies on the dihydro ester and its ethylene-ketal (*61*).

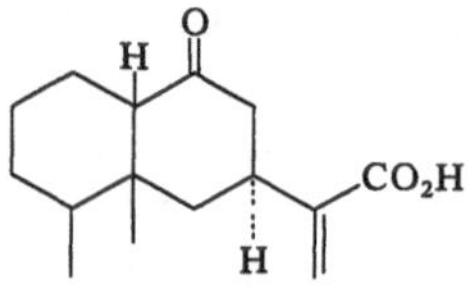

flourensic acid
(84)

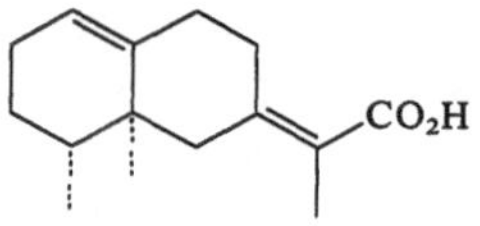

isovalencenic acid
(85)

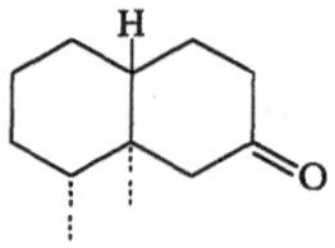

(86)

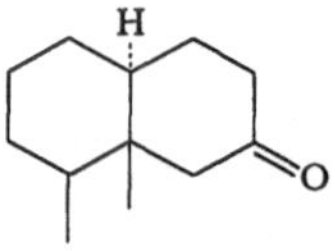

(87)

Isovalencenic acid, $C_{15}H_{22}O_2$, has been isolated from vetiver oil, of which it is a minor acidic constituent. Its structure (85) has been settled as follows (62). It shows u.v. maximal absorption at 225 nm (ε 8000) (α,β-unsaturated acid), and infrared bands at 1670, 1622, and 844 cm^{-1} are characteristic of carboxylic α,β C=O, C=C and C=CH$_2$ groups respectively. Its n.m.r. spectrum reveals a tertiary methyl (δ 0.86 ppm), a CH$_3$CH-residue (0.93, doublet), an allylic CH$_3$ (1.93), and a vinylic hydrogen (5.3, multiplet). The methyl ester was similarly examined and shows comparable spectral properties; catalytic hydrogenation yielded first a dihydro ester $C_{16}H_{26}O_2$ and ultimately an epimeric mixture of tetrahydro esters $C_{16}H_{28}O_2$. The mass spectrum of the latter shows a peak at m/e 88 (base peak), corresponding to the fragment ion $[CH_3CH{=}C{\diagup}^{OH}_{\diagdown OMe}]^+$ and suggesting the presence of a $-CH{\diagup}^{CO_2Me}_{\diagdown CH_3}$ grouping in the esters. The dihydro ester still absorbed strongly in the u.v. region, so that the partial hydrogenation was a consequence of attack upon an isolated double bond. Ozonolysis of this ester afforded a ketone in which the carbonyl group is flanked by two methylene groups (four hydrogens exchangeable by deuterium). The mass spectrum of this ketone, with peaks at m/e 122, 109, and 82, suggested an eremophilanoid structure for isovalencenic acid.

The structure (86) for the ketone was proved by its gross identity with and antipodal relation to ketone (87), synthesized unambiguously earlier (30) (i.r. and o.r.d. curve comparisons). Lithium aluminum hydride reduction of methyl isovalencenate furnished a primary alcohol, which with thionyl chloride gave the corresponding chloride. This was hydrogenolyzed by the same hydride to an unsaturated hydrocarbon $C_{15}H_{24}$, which proved to be identical with one of the dehydration products (**88**) of valerianol (*63*) (see p. 148).

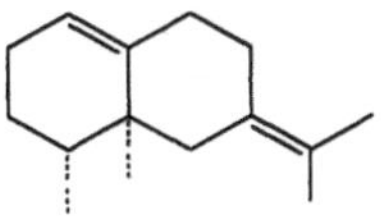

(**88**) 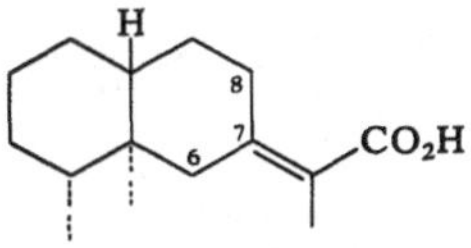   (**89**) 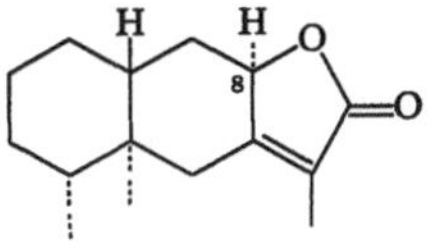   (**90**)

The steric structure of the carboxyl group in the natural acid was settled by oxidation of the dihydro acid (**89**) with selenium dioxide, which converted it into a γ-lactone (**90**). The n.m.r. spectrum of this product was examined; in particular the coupling pattern of the $\diagdown$CH—O proton was observed (δ 5.07 ppm, broad multiplet). From this it was apparent that lactonization had occurred toward position 8 in (**89**) rather than toward position 6. The carboxyl group in (**89**), and therefore in isovalencenic acid (**85**), is consequently *cis* to $C_{(8)}$ (*62*). The compound belongs to the antipodal eremophilane series, and occupies a borderline position between the eremophilane and nootkatane (see p. 149) types.

6. Esters

9β-Acetoxy-1,10-didehydrofukinone, $C_{17}H_{24}O_3$, occurs in the roots of *Senecio elegans* L. Its i.r. spectrum reveals both ester and α,β-unsaturated ketone carbonyl bands (1714, 1640 cm^{-1}), its mass spectrum suggested that an acetoxy group was present, and its n.m.r. spectrum pointed to its being an eremophilane derivative. The signals, based on structure (**91**), together with the effect of introducing the Eu(fod)$_3$ shift reagent, are given in the adjoining table.

δ	H at position		$+Eu(fod)^3$, $\Delta\delta$
5.71	1	ddd	+0.20
2.05	2	m	
1.5	3,4	m	
2.56, 2.34	6	dd (J = 14 Hz)	0.10
5.87	9	ddd	0.54
2.05	12	t	0.09
1.84	13	s (broad)	0.06
1.09	14	s	0.08
0.97	15	d (J = 6.5 Hz)	0.06

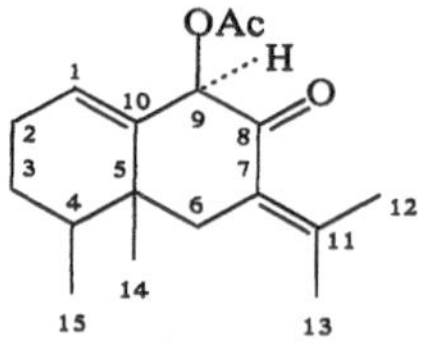

9β-acetoxy-1,10-
didehydrofukinone
(91)

Reduction of the natural product with sodium borohydride yielded
a mixture of epimeric alcohols, which on reduction with lithium
aluminum hydride afforded a diol, oxidizable to a dialdehyde by periodate. This sequence settled the position of the acetoxy group, and
enabled the product to be formulated as (91) (64).•

Petasin, $C_{20}H_{28}O_3$, is found in the roots of *Petasites hybridus* and
P. officinalis and is in part responsible for the spasmolytic action of
extracts thereof (65, 66, 67, 68). On hydrolysis it yielded angelic [(Z)-
α-methylcrotonic] acid and an alcohol isopetasol; however, isopetasol is
not the alcohol moiety present in the ester petasin, because the hydrolysis
was accompanied by a double bond shift in the true hydrolysis product
petasol, which has not been isolated. Petasin is a keto ester containing
two double bonds, only one of which is conjugated with the keto
group (λ_{max}^{EtOH} 232 nm, ε 22,600), but in isopetasol the second ethylenic
bond has migrated into conjugation (λ_{max}^{EtOH} 249 nm, ε 12,700). Isopetasin
(λ_{max}^{EtOH} 243 nm, ε 14,700), obtained by alumina chromatography of the
mother liquors of petasin (47), may be an artefact, although it has been
isolated from *P. kablikianus* (68a); it is derived from petasin by a similar
double bond migration.

The bicycle sesquiterpene framework of petasin was established by
its dehydrogenation to eudalene, suggesting either a eudesmane or
eremophilane framework. Isopetasol suffered retroaldol cleavage on
base treatment to acetone and a C_{12}-ketol, desisopropenylpetasol (or
desisopropylideneisopetasol); the o.r.d. curve of the latter proved to be
of the Δ^4-3-ketosteroid type. The angular methyl group in this ketol
(92) consequently has the β-orientation. The position of the hydroxyl
group was settled by two observations: (i) the compound does not
experience base-catalyzed dehydration to a conjugated dienone, so

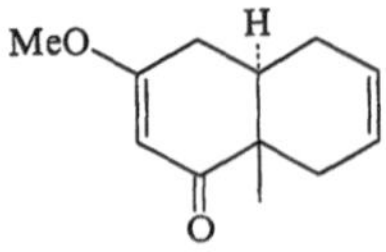

(35) 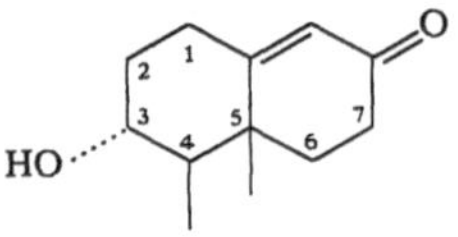   (92) 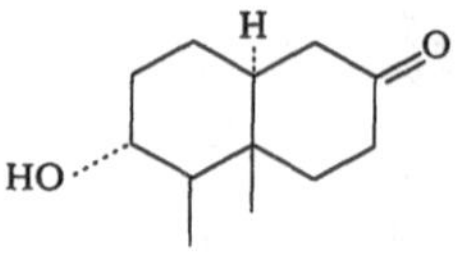   (93) 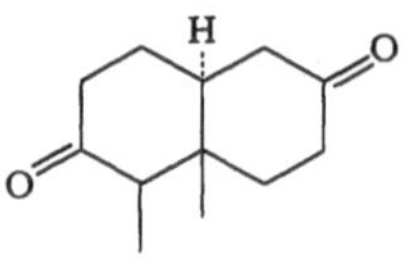   (94)

that the hydroxyl group cannot be at $C_{(2)}$, and (ii) the hydroxyl group, which is secondary, can be oxidized to yield a diketone, which shows both α,β and saturated $C=O$ bands in its i.r. spectrum; the OH group cannot, then, be at position 1. Similar arguments rule out positions 6 and 7, leaving only position 3. Lithium-ammonia reduction of (92) gave the *trans*-2-decalone (93) oxidizable to diketone (94). A diketone identical with (94) was synthesized from methoxyketone (35), of known absolute configuration (see p. 91). The α-orientation of the hydroxyl group in (92) was established by a study of the ease of hydrolysis of petasin, and of the rates of oxidation and acetylation of the same group in (92). Petasol, isopetasol, and petasin may therefore be formulated as (95), (96), and (97) respectively, except for the orientation of the iso-propenyl group in (95) and (97) (*69, 70*). The latter aspect of these

petasol	isopetasol	petasin
(95)	(96)	(97)

structures does not appear to have been settled, but it seems likely that the group is β-oriented.

Several other esters of petasol have been isolated from the same source, the acids involved being tiglic [(E)-α-methylcrotonic] and β-methylthioacrylic acids [(E) and (Z) forms] (*66*). The total synthesis of ($\pm$)-isopetasol starting from 2,3-dimethylcyclohexane-1,4-dione has been described (*58*).

Petasitin, $C_{20}H_{28}O_4$, found in the flower stalks of *P. japonicus*, is another ester, which on hydrolysis yields angelic acid (accompanied by a small amount of tiglic acid, formed by isomerization of angelic acid). The other product was a ketodiol, $C_{15}H_{22}O_3$, in which the keto group is α,β-unsaturated (λ_{max}^{EtOH} 244 nm, ε 11,600). One of the hydroxyl groups in the diol is tertiary since mild acetylation conditions led readily to a monoacetate, a diacetate resulting only under vigorous conditions. Catalytic hydrogenation caused uptake of three mols. of hydrogen with saturation of two double bonds and hydrogenolysis of the tertiary hydroxyl group, which must also be allylic. The product of this reaction was (98), since it proved to be identical with tetrahydroisopetasol, obtained by catalytic hydrogenation of isopetasol (96). The *cis* ring junction in (98) was assigned because of the mode of formation of the compound and because of the observed negative Cotton effect in its o.r.d. curve, very similar to that of A/B *cis* fused 3-ketosteroids. Dehydration of the monoacetate (see above) gave an anhydro compound showing spectral evidence (δ 5.11, 1.97 ppm) of $C=CH_2$ and allylic CH_3 groups respectively,

(**98**)

petasitin

(**99**)

(**100**)

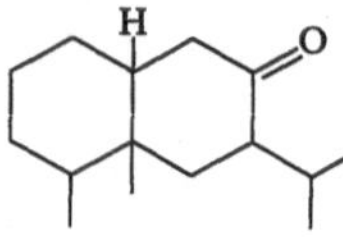

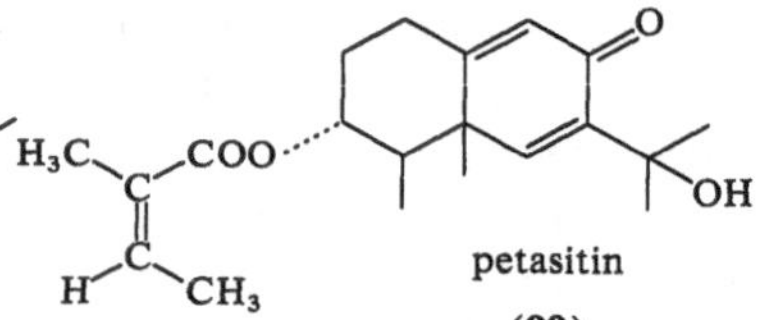

both signals being absent in the spectrum of the starting monoacetate. On hydrolysis, hydrogenation, and equilibration with base this product too yielded (**98**). Petasitin may consequently be formulated as (**99**). Confirmation was provided by reduction of the monoacetate to the tetrahydro compound, dehydration of which gave a mixture of isomers (**100**), from which the isopropylidene isomer was separable. Dehydrogenation of this with 2,3-dichloro-5,6-dicyanobenzoquinone afforded isopetasol acetate (**96**, HO replaced by AcO) (*71*).

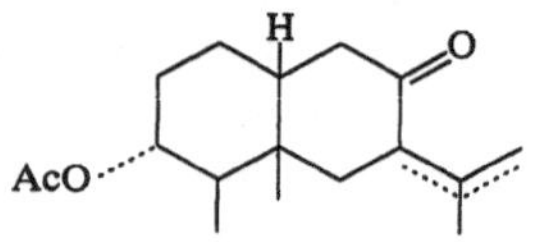

S-japonin, $C_{19}H_{28}O_3S$, has been isolated from the leaves of a cultivated variety of *P. japonicus* Maxim. It is an ester of (Z)-β-methylthioacrylic acid and a ketol, $C_{15}H_{24}O_2$, into which the compound is split on alkaline hydrolysis. The ketol is an α,β-unsaturated ketone (λ_{max} 250 nm, ε 9500, ν_{CO} 1680 and 1625 cm^{-1}), the spectra of which are closely similar to those of fukinone (**58**, p. 97). Successive dehydration and catalytic hydrogenation of the ketol yielded a saturated ketone identical with dihydrofukinone. The position of the hydroxyl function in the ketol (and therefore of the ester function in S-japonin) was established as follows. In the n.m.r. spectrum of S-japonin the signal ascribed to the hydrogen on the carbon carrying the ester group (δ 5.04 ppm) appeared as a septet, characteristic of an axial methine proton coupled with two flanking proton pairs. Consequently the ester function must be at position 2 and have the β (equatorial) configuration, and S-japonin must be represented by (**101**), the ketol hydrolysis product being the corresponding alcohol (*42*).

S-japonin

(**101**)

(**102**)

PR-toxin

(**103**)

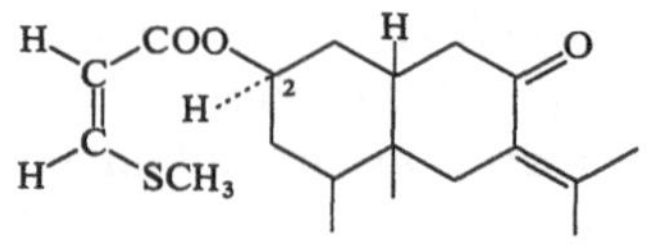

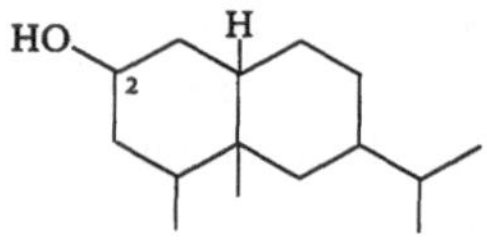

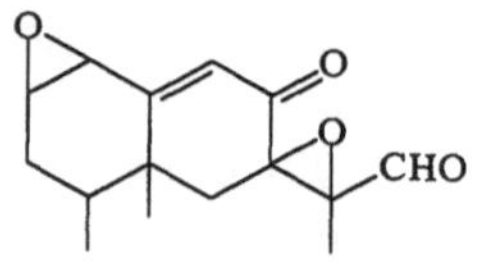

Confirmation was provided by hydrogenation of the ketol to a dihydroketol, which was converted into its ethylenedithioketal and thence, by treatment with Raney nickel, to an alcohol (102). The latter on chromic acid oxidation gave the corresponding ketone which was observed to exchange four deuterium atoms with NaOD-MeOD; consequently, the keto group must be at $C_{(2)}$ and the ester group of S-japonin at the same position (42).

PR Toxin, $C_{17}H_{20}O_6$, has been isolated from cultures of *Penicillium roqueforti* (72). It is a crystalline solid the structure of which has been elucidated mainly by spectroscopic study. Its functional groups proved to be an acetoxy group (hydrolysis yielding an alcohol and acetic acid, and acetylation of the alcohol regenerating the parent compound), an aldehyde function (v_{CO} 1720, and δ 9.75 ppm, singlet, 1 H), and an α,β-unsaturated ketone (v_{CO} 1680 cm^{-1}, δ 6.43 ppm, 1 H, for a proton on the double bond, and λ_{max}^{EtOH} 249 nm). Borohydride reduction furnished a primary-secondary diol. The two oxygens unaccounted for in the molecular formula were suspected of being epoxides because of absence of OH absorption in the i.r. spectrum, because of absence of C=O absorption in the i.r. spectrum of the product of successive borohydride reduction and hydrolysis (no lactone system present), because the toxin gave a positive result in a qualitative test for the epoxide group (73), and finally because the diol referred to above furnished on reduction with lithium aluminum hydride a pentahydric alcohol. The latter yielded a tetraacetate, suggesting that one of the five hydroxyl groups was tertiary. Two epoxide functions are apparently present in the toxin, the n.m.r. spectrum of which reveals two oxirane protons, coupled to each other and part of the same oxirane group. A disubstituted and a tetrasubstituted epoxide were consequently

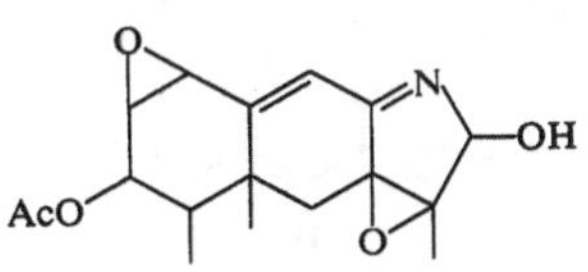

(104)

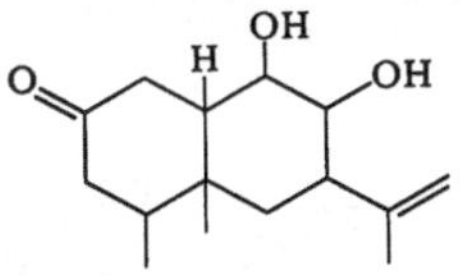

laterifior-2-one

(105)

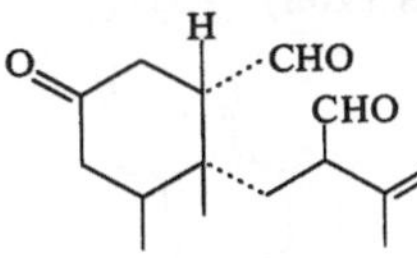

(106)

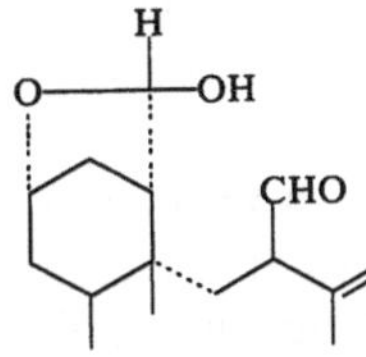

(107)

suspected. A detailed analysis of the spectrum, with spin decoupling, and mass spectral analysis of the compound and its derivatives, pointed to an eremophilane skeleton for the toxin, structure (**103**) being advanced without implied stereochemistry. The relative position and orientation of the aldehyde group were settled by the formation of a Schiff base (**104**) when the toxin was exposed to methanolic ammonia; this compound showed no aldehyde proton signal in its n.m.r. spectrum, and the aldehyde and α,β-unsaturated ketone bands at 1720 and 1680 cm^{-1} were replaced by a conjugated imine band at 1631 cm^{-1}. The carbon-13 n.m.r. spectrum of PR-toxin has been analyzed in detail and is in full accord with the proposed structure (*74*), the stereochemistry of which is yet to be elucidated. Other metabolites of *P. roqueforti* have been studied (*74a*).

10βH-Lateriflor-2-one 8,9-ditiglate, $C_{25}H_{36}O_5$, occurs in the roots of *Euryops virgineus* of S. African origin, which also contains several similarly-constituted sesquiterpenes. Separation was achieved by careful column chromatography. On hydrolysis it afforded a dihydroxy-ketone (lateriflor-2-one, **105**) and tiglic acid (two mols.). The former compound was already known as a breakdown product of several furano-eremophilanes encountered in the same or closely-related sources (see p. 113); periodate oxidation afforded a dialdehyde (**106**), indicating the vicinal relationship of the hydroxyl groups. Lithium aluminum hydride reduction of the natural product yielded a triol which also was oxidized by periodate to a monoaldehyde (**107**), the second aldehyde group being involved in hemiketal formation, with the third hydroxyl group arising from reduction of the keto-group. The structures (**105**), (**106**), and (**107**) were arrived at chiefly from a detailed analysis of their n.m.r. spectra and enabled the natural compound to be formulated as (**108**) (*75*).

(**108**)   (**109**)

A mixture of di-esters has been isolated from *Senecio purpureus* L. (*64*). They conform to the structure (**109**), hydrolysis yielding a mixture of angelic, tiglic, and senecioic acids. Successive borohydride and lithium aluminium hydride reduction afforded a triol of partial structure (**110**)

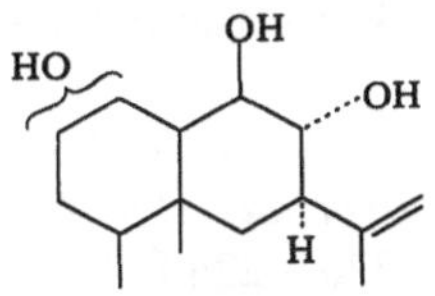

(110)

which on periodate oxidation afforded a dialdehyde in much the same manner as the foregoing products. A detailed interpretation of the n.m.r. spectra of the diesters and these breakdown products (64) allowed the esters to be assigned partial structure (109).

III. Furanoeremophilanes

Until very recently only a few natural furanoeremophilanes were known, but the investigations of BOHLMANN and his co-workers (16, 64, 75) during recent years on certain South African plants have succeeded in the isolation of many more so that this subgroup is now the largest amongst eremophilanes.

The basic framework of these compounds is one in which a furan or modified furan ring is fused onto the bicyclic eremophilane system at positions 7 and 8, the three-carbon substituent at $C_{(7)}$ providing two carbon atoms for the furan ring as in (111). In a few members the heterocyclic ring is that of γ-butenolactone as in (112).

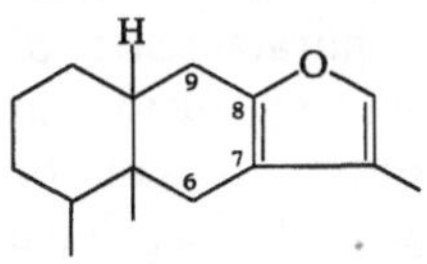

(111)

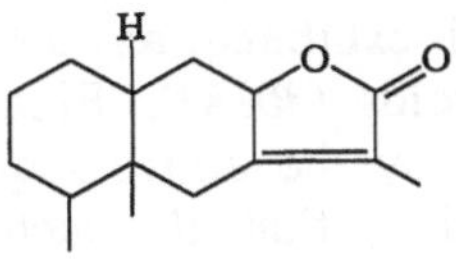

(112)

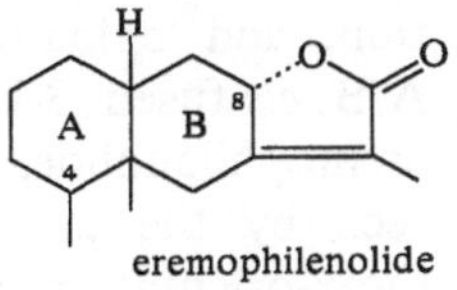

eremophilenolide

(113)

It is convenient to divide the furanoeremophilanes into two groups for discussion: butenolactones and furans. In the former the hetero-ring is a five-membered α,β-unsaturated lactone, and in the latter, much the larger sub-group, it is a furan ring. Earlier work in this area has been reviewed (2, 2a, 68). A connecting link between the two subgroups is the fact that mild oxidation of a furanoeremophilane leads to the corresponding butenolactone; this transformation has been useful in structure determination (75a, 75b, 75c).

1. Butenolactones

Eremophilenolide, $C_{15}H_{22}O_2$, occurs in the rhizomes of *Petasites officinalis* (*76, 77*) and *P. hybridus* (*5a, 77*). It manifests lactonic properties and on u.v. (λ_{max} 220–224 nm, log ε 4.16) and i.r. evidence (ν_{CO} 1760 cm^{-1}) is formulated as a conjugated butenolactone. On catalytic hydrogenation (uptake of one mol of H_2) a saturated lactone resulted (ν_{CO} 1780 cm^{-1}), which was reduced by lithium aluminum hydride to a diol. The ditosylate of the latter yielded eremophilane (**7**) on similar reduction. These observations are in harmony with structure (**113**) for eremophilenolide, the geometry of the decalin ring fusion remaining to be determined.

This aspect of the structure was settled by the following sequence. Dihydroeremophilenolide was reduced with lithium aluminum hydride under controlled conditions to aldehyde (**114**), which on Wolff-Kishner reduction yielded alcohol (**115**). Jones oxidation of the latter afforded

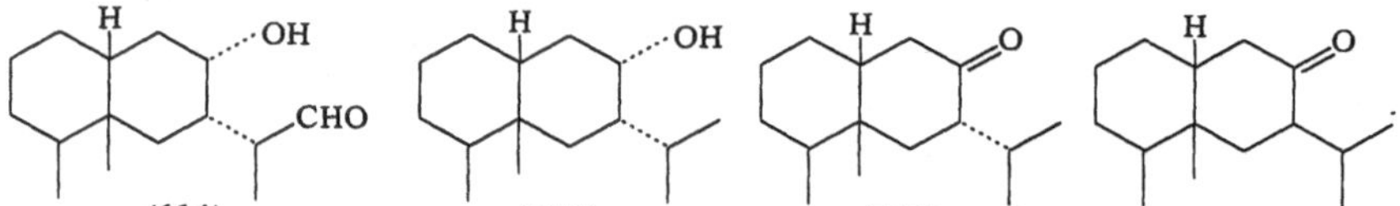

(**114**)   (**115**)   (**116**)   (**117**)

decalone (**116**), which was isomerized by base to ketone (**117**). An infrared band at 1430 cm^{-1} in the spectrum of the latter was indicative of a CH_2 group adjacent to the C=O function, in harmony with structure (**117**). The latter ketone proved to be identical with a ketone obtained from hydroxyeremophilone (**5**), of known absolute configuration, by sequential hydrogenation, base equilibration, calcium-ammonia reduction, and oxidation, and exhibited an o.r.d. curve characteristic of A/B *cis*-fused 3-keto-steroids (*30, 31*). Finally the *cis* ring fusion in eremophilenolide, proven by the above sequence, requires, as can be seen by the use of models, that the lactonic oxygen at $C_{(8)}$ have the α-orientation, assuming that the *cis*-decalin ring junction is steroidal in conformation. This seems certain because in that spatial arrangement the methyl group at $C_{(4)}$ can be equatorial. In the steroidal arrangement, a β-oxygen at $C_{(8)}$ would require ring B in (**113**) to adopt an unfavorable

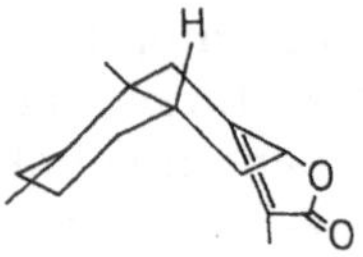

eremophilenolide
(**118**)

boat conformation. Eremophilenolide is thus represented by (113), which is also its absolute configuration, or more precisely by (118) (76).

This formulation and relative configuration have been corroborated by an X-ray diffraction analysis of the compound (78) and by total synthesis of the racemic form of eremophilenolide (79, 80) as outlined in Chart 10. The starting point was the cis-octalone (59), already

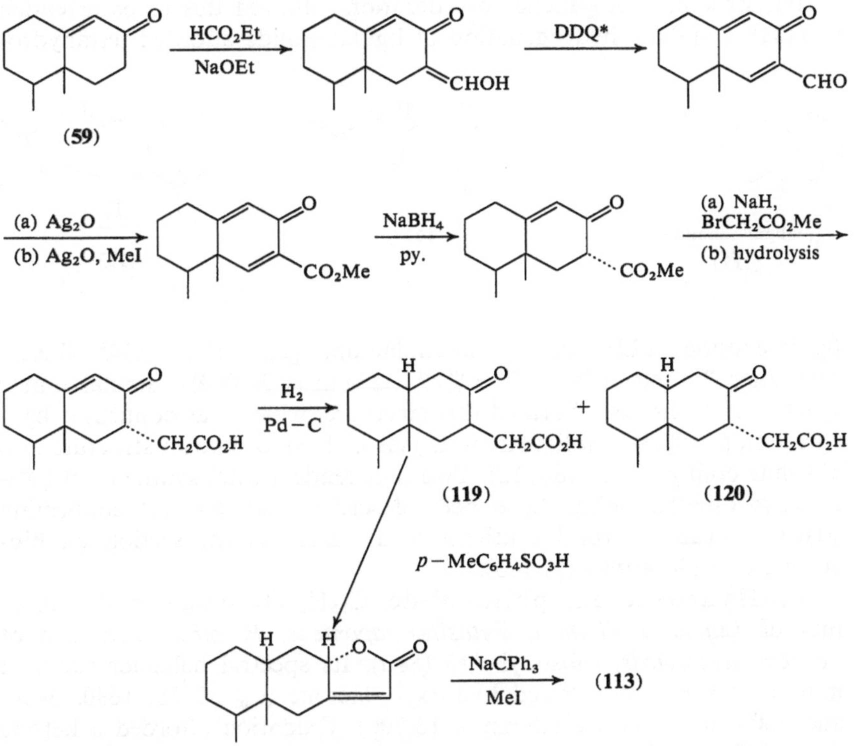

HCO₂Et / NaOEt → DDQ* →

(59)

(a) Ag₂O / (b) Ag₂O, MeI → NaBH₄ / py. → (a) NaH, BrCH₂CO₂Me / (b) hydrolysis →

H₂ / Pd—C → (119) + (120)

p—MeC₆H₄SO₃H

NaCPh₃ / MeI → (113)

* DDQ = 2,3-dichloro-5,6-dicyanobenzoquinone

Chart 10. Synthesis of (±)-Eremophilenolide

encountered in the synthesis of (±)-fukinone (Charts 6 and 8). The acids (119) and (120) were separated by fractional crystallization, and configurations assigned by n.m.r. spectral analysis and by independent, unambiguous syntheses of the trans acid (120). A second synthesis has also been reported (80a).

Ligularenolide, C₁₅H₁₈O₂, occurs in the roots of Ligularia sibirica, a Chinese drug known as "San-shion". It is a crystalline lactone the spectral characteristics of which indicate it to be an α,β-unsaturated butenolactone (λ_{max} 241 nm, ε 3,560; ν_{CO} 1764, 1648, and 1621 cm^{-1}). A u.v. absorption maximum of much greater intensity (ε 20,700) was observed at 331 nm, indicating extended conjugation. The n.m.r. spectrum showed signals corresponding to tertiary, secondary, and allylic methyl groups, and to two olefinic protons. A pair of allylic protons showed up as an AB quartet. A detailed analysis of the spectrum using double and triple resonance techniques led to advancement of a partial structure; biogenetic considerations allowed this to be extended to (121). Catalytic hydrogenation of ligularenolide afforded tetrahydro-

ligularenolide   (122)
(121)   (123)

ligularenolide (122), also a butenolactone [ν_{max} 1765, 1745 (Fermi resonance ?), and 1678 cm^{-1}, λ_{max}^{MeOH}, 222 nm (ε 24,000)]. Similar n.m.r. spectral analysis corroborated this structure, which was confirmed by a correlation with furanoligularenone (see p. 123), of known structure and absolute configuration (81, 82). Two independent total syntheses of ($\pm$)-tetrahydroligularenolide have been described (80, 83, 84) confirming structure (122). A third synthesis is discussed in the section on biogenetic considerations (p. 175).

6β-Hydroxyeremophilenolide, $C_{15}H_{22}O_3$, occurs in the rhizomes of *Ligularia Fischeri, Petasites japonicus, P. albus*, and also of *Senecio nemorensis*, subsp. *fuchsii* (84a). Its spectral behavior revealed it to be an α,β-unsaturated hydroxy-γ-lactone (ν_{max} 1755, 1690, 3450, and 3600 cm^{-1}; λ_{max} 218.5 nm, ε 13,700). Oxidation afforded a ketone (λ_{max} 240 nm), indicative of the secondary alcoholic nature of the hydroxyl group and of the presence of the $O=C-C=C-C=O$ chromophore in the ketone. The lactone under discussion is formed when the furanoeremophilane petasalbin (see below) is subjected to mild peracid oxidation. Since the structure and absolute configuration of the latter are known it follows that the natural lactone is (123), (6b, 17, 85). This structure and stereochemistry have been confirmed by an X-ray diffraction analysis of its *p*-bromobenzoate (78), and by analysis of its n.m.r. spectrum (85). The absolute configuration (123) has been deduced from molecular rotation differences and application of the Hudson-Klyne rule concerning the relation between lactone configuration and sign of optical rotation (85).

References, pp. 175—186

The petasitolides are four substances found in the rhizomes of *Petasites officinalis*, being esters distinguishable by their behavior on alkaline hydrolysis. All are derived from the hydroxylactone (124), 3α-hydroxyeremophilenolide, formed on hydrolysis of the compounds along with one of four acids, as summarized in the adjoining table.

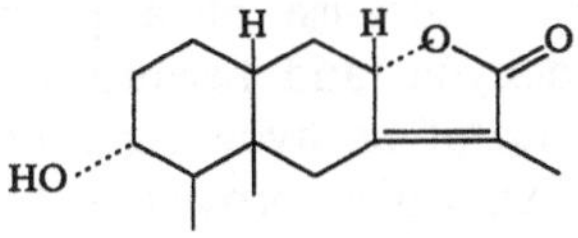

(124) 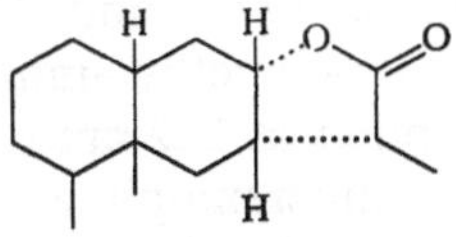 (125)

Compound	Acid hydrolysis product
Petasitolide-A	Angelic acid
Petasitolide-B	Tiglic acid
S-Petasitolide-A	*cis*-β-Methylthioacrylic acid
S-Petasitolide-B	*trans*-β-Methylthioacrylic acid

The B esters could possibly be artefacts, the acids of *trans* configuration being formed by the "natural" *cis* acids by isomerization during the hydrolysis (compare esters of petasol, p. 108) (*5a, 77, 86*). The determination of the constitution and stereochemistry of the petasitolides is thus reduced to an investigation of the alcohol (124). Spectral studies revealed that the compound, $C_{15}H_{22}O_3$, was an α,β-unsaturated γ-lactone containing a hydroxyl group. Hydrogenation yielded a saturated dihydro-derivative oxidisable to a keto-lactone $C_{15}H_{22}O_3$. Desulfurization of the ethylenethioketal of the latter afforded lactone (125) encountered earlier (dihydroeremophilenolide) (*76*). The alcohol portion of the petasitolides is thus a hydroxyeremophilenolide. The location of the hydroxyl group was deduced from a study of the n.m.r. spectrum of the ketone obtained by oxidation of the alcohol; in particular the CHCH$_3$ protons resonate at δ 0.89 and 1.0 ppm (doublet). The good resolution observed permits the placing of the keto group in the α-position to the CHCH$_3$ group, that is in position 3 (*87*). Consequently the hydroxyl group in (124) is also at $C_{(3)}$ and is presumed to have the α (equatorial) configuration. The petasitolides are therefore represented by (124) in which the 3α-hydroxyl group has been esterified by the appropriate acid (*75a*).

2. Furanoeremophilanes

These compounds, the nomenclature of which leaves much to be desired, are alcohols, ethers, epoxides, esters, and ketones derived from the parent substance furanoeremophilane (sometimes termed furoeremophilane), $C_{15}H_{22}O$, which itself occurs in the rhizomes of *Petasites officinalis* Moench. (*5, 5a*), of *P. albus* L. (*6, 88*), of *P. spurius* L. (*89*), of *P. japonicus* Maxim. (*90*), of *P. hybridus* (*91*), of *P. paradoxus* (*92*), and in the root of *Ligularia fischeri* Turcz. (*17*). A furan ring was diagnosed in the compound because of infrared absorption bands at 1576, 1660, 1776, and 1810 cm^{-1}, because of a positive color reaction with *p*-dimethylaminobenzaldehyde, and because of u.v. maximal absorption at 222 nm (ε 6,000). Catalytic hydrogenation afforded tetrahydrofuranoeremophilane, $C_{15}H_{26}O$, which proved to be identical with the product of hydrogenolysis of petasalbin of proven structure (see below), and eremophilane (7), a consequence of hydrogenolysis. Furanoeremophilane may consequently be formulated as (111) (*5*), this assignment being corroborated by a total synthesis of the racemic structure from ($\pm$)-fukinone (58), which was converted by sulfuric acid-acetic anhydride to the cyclic sulfonic ester (126). Pyrolysis of the latter yielded (111) (*93*). Another synthesis has been described (*80a*).

furanoeremophilane   (7)   (58)   (126)
(111)

a) Monohydric Alcohols and Their Ethers and Esters

Petasalbin, $C_{15}H_{22}O_2$, has been found in the rhizomes or roots of several *Petasites* and *Ligularia* spp. (*5a, 6b, 17, 85, 88, 89, 90, 91, 94*); it was known earlier as ligularol. Petasalbin shows in its i.r. spectrum bands at 1565 (furan ring) and 3485 and 3615 cm^{-1} (hydroxyl group). U.v. absorption at 220 nm was ascribed to the presence of a furan group and the n.m.r. spectrum was in harmony with an eremophilane system [δ 0.99 s (3H, angular CH_3); 0.85 d, CH_3; 2.07 s, 3H, aromatic CH_3; 4.73 s, aromatic H]. Catalytic hydrogenation afforded several products, including eremophilane (7) and tetrahydrofuranoeremophilane. The easy hydrogenolysis of the hydroxyl group attested to its benzylic character. Oxidation of petasalbin yielded the corresponding ketone, furanoeremophilane, the i.r. carbonyl absorption (1675 cm^{-1}) of which indicated the group to be conjugated with the furan ring (λ_{max} 269 nm). Since tetrahydrofuranoeremophilane is of known structure (*76*) it follows that the keto group in the furanoeremophilone, and therefore the hydroxyl group in petasalbin, must be placed at positions 6 or 9 in (111).

Autoxidation of petasalbin gave 6β-hydroxyeremophilenolide (123) (see p. 116), oxidation of which afforded a ketoeremophilenolide (127),

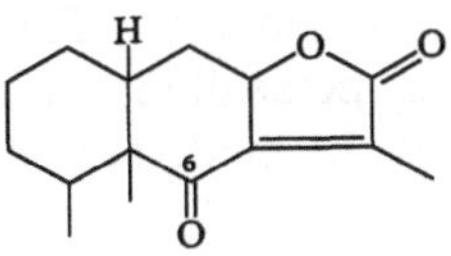

(127)

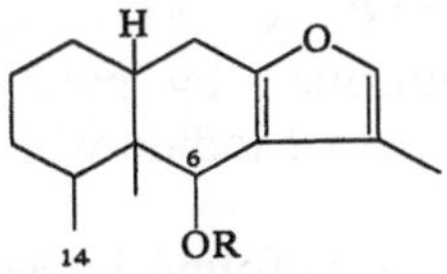

petasalbin (R = H)
(128)

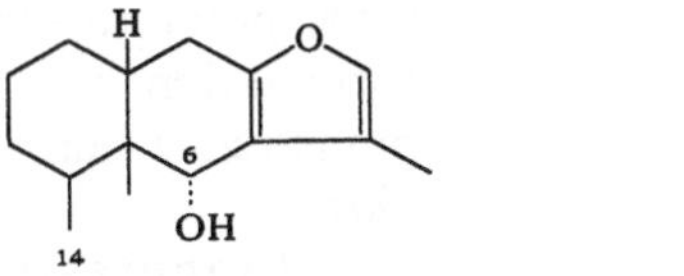

(129)

the ketonic C=O being conjugated with the double bond of the lactone ring (λ_{max} 240 nm, ε 10,750). Thus the carbonyl group in (127) must be located at position 6, and the hydroxyl group in petasalbin, to be formulated as (128) (R=H), in the same position (85).

Oxidation of petasalbin afforded a ketone 6-oxofuranoeremophilane, itself a natural product; reduction of this ketone yields an epimeride of petasalbin (128) (R = H), 6-epipetasalbin (129). The n.m.r. spectra of these epimerides show $C_{(4)}$–CH$_3$ signals (doublets) at δ 0.88 and 1.04 ppm respectively. The lower field signals in the latter case indicate that in this epimeride the methyl group at $C_{(4)}$ and the hydroxyl at $C_{(6)}$ have a 1,3-diaxial relationship. Next, on acetylation of petasalbin the $C_{(4)}$-methyl group signal is unchanged, and finally 6-epipetasalbin is not acetylated using conditions under which petasalbin is esterified. This behavior can be rationalized because in the epi-compound, rewritten as (130), there is severe steric hindrance of the hydroxyl group by two methyl groups. This situation is absent in petasalbin (131), which must consequently be assigned

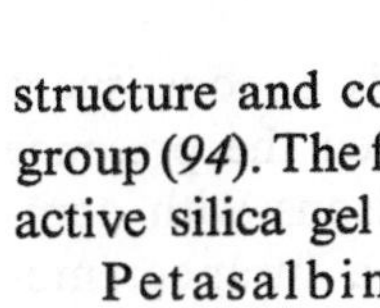

(130)

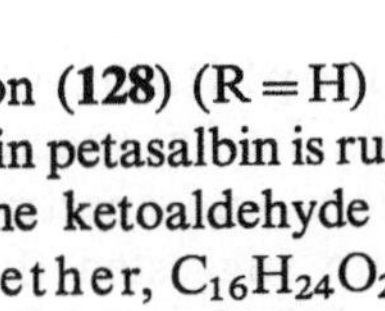

(131)

petasalbin

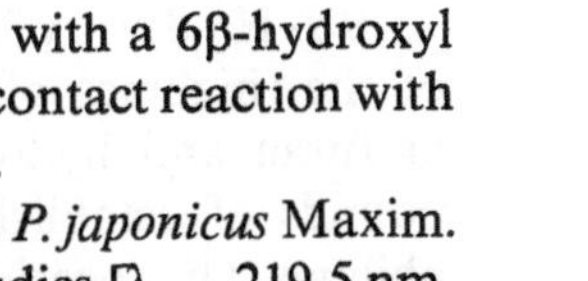

(132)

structure and configuration (128) (R=H) and (131), with a 6β-hydroxyl group (94). The furan ring in petasalbin is ruptured by contact reaction with active silica gel to give the ketoaldehyde (132) (95).

Petasalbin methyl ether, $C_{16}H_{24}O_2$, occurs in *P. japonicus* Maxim. rhizomes. Its structure has been settled by spectral studies [λ_{max} 219.5 nm, v_{max} 1638, 1565 cm^{-1} (furan ring); δ 2.03 d ($J = 1$) (furan methyl group, 7.0 q (furan H), 6.08 s ($C_{(6)}$-H), 3.35 s (OCH$_3$), 1.05 s (angular CH$_3$), and 0.75 d $(J = 6)$ (CHCH₃)]. In its mass spectrum an intense peak at m/e 138 appears, corresponding to the fragment (133), characteristic of a furan ring with an attendant allylic methoxyl group. These observations point to structure (128) (R = Me) for the compound (90). This formulation has been confirmed by methylation of petasalbin with potassium t-butoxide and methyl iodide which leads to the methyl ether (90).

Albopetasin, $C_{20}H_{28}O_3$, is found in several *Petasites* spp. (*5a, 6b, 85, 86, 88, 89, 90, 91*). It shows i.r. absorption (1572 cm⁻¹) typical of a furan ring and of an α,β-unsaturated ester (1718 and 1653 cm⁻¹; λ_{max} 224 nm, ε 11,900). On hydrolysis it yielded petasalbin and tiglic acid [(E)-α-methylcrotonic acid] with a trace of angelic acid [the (Z)-isomer], but pyrolysis of the ester gave only the latter acidic product.

Albopetasin is consequently petasalbin angelate (128)   (R = $\overset{OC}{\underset{H_3C}{>}}C = C\overset{CH_3}{\underset{H}{<}}$) (*85*).

(133)

9-oxofuranoeremophilane

(134)

(135)

Petasalbin acetate (ligularyl acetate) (128) (R = Ac), not itself a natural product, has been investigated by variable-temperature p.m.r. spectroscopy, and by circular dichroism, and it has been concluded that in solution this compound exists as an equilibrium mixture of steroidal and non-steroidal *cis*-decalin structures (*95a*).

Petasalbin senecioate, $C_{20}H_{28}O_3$, has been found in the roots of *Farfugium japonicum* (syn. *Ligularia tussilaginea*). Lithium aluminum hydride reduction afforded petasalbin (128) (R = H). Its spectral properties are consistent with the presence of a furan ring and an α,β-unsaturated ester grouping, and its mass spectrum indicated that it was an ester of senecioic acid. The compound is therefore represented by (128) (R = COCH=CMe₂) (*96*).

9α-Hydroxyfuranoeremophilane, $C_{15}H_{22}O_2$, is a constituent of the rhizomes of *P. hybridus*. Its i.r. spectrum revealed the presence of furan and hydroxyl functions and spectral comparisons with compounds of established structure suggested that it was a furanoeremophilane. The location of the hydroxyl group α to the furan ring followed from color reactions and from the n.m.r. spectrum. In particular the CHOH signal is a doublet at δ 4.49 ppm $(J = 4.5$ Hz); it follows that the hydroxyl group must be at $C_{(9)}$. The allylic nature of the alcohol group was inferred from its easy oxidation to 9-oxofuranoeremophilane (134), a compound obtainable from furanopetasol (p. 127) and kablicin (p. 131). The exposure of this ketone to base led to the *trans* isomer (135), revealing that the immediate oxidation product is a *cis*-1-decalone, and that the ring junction in the alcohol is *cis*. The n.m.r. spectrum of the alcohol does not enable a decision to be reached about the configuration of the hydroxyl group, because of the unknown conformation of ring A, but the $J_{9,10}$ value of 4.5 Hz suggests that the hydrogens at $C_{(9)}$ and $C_{(10)}$ are *cis* and thus, that the hydroxyl group is 9α, as in (136) (97).

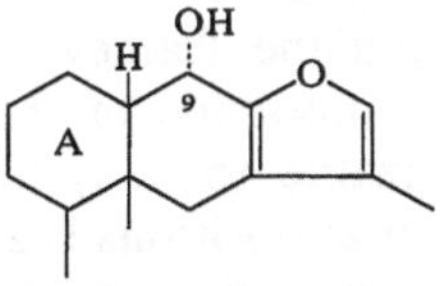

(136)

Tetradymol, $C_{15}H_{22}O_2$, is one of the toxic components of *Tetradymia glabrata*, a plant responsible for numerous cases of sheep poisoning. Its u.v. absorption (λ_{max} 222 nm) and color reactions indicated that a furan ring was present. This was confirmed by a comparison of its n.m.r. spectrum with the spectra of known furanoid compounds. A hydroxyl group (ν 3400 cm^{-1}) was revealed by the i.r. spectrum. Quantitative hydrogenation studies showed that the only unsaturation present was in the furan ring, the major product being a tetrahydrotetradymol. Careful analysis of the n.m.r. spectrum of tetradymol led to the conclusion that a furanoeremophilane system was present and that the hydroxyl group was to be located at a tertiary position, since no signals assignable to a > CH(OH) proton were present in the spectrum. Further, the hydroxyl group cannot be located at $C_{(4)}$, since the methyl group at that position is seen as a doublet. These observations leave structure (137) to be contemplated for tetradymol. A study of solvent-induced n.m.r. chemical shifts (pyridine-d$_5$, CDCl$_3$) demonstrated a Δ-value of 0.18 ppm for the tertiary CH$_3$ group and only 0.03 ppm for the secondary CH$_3$. This and other similar observations pointed to

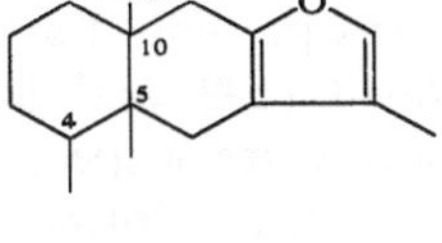

tetradymol

(137)

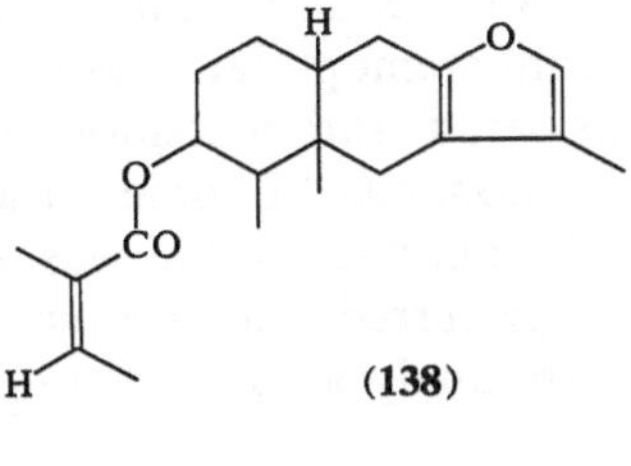

(138)

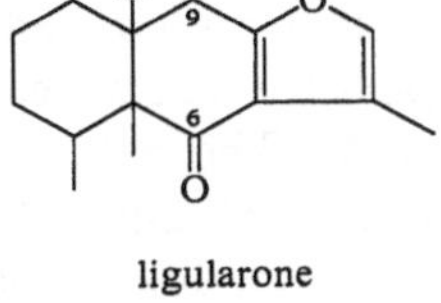

ligularone

(139)

structure and configuration (137) for tetradymol, the OH group at position 10 and the methyl groups at positions 4. (equatorial) and 5 having an all-*cis* arrangement. This was confirmed by an X-ray diffraction analysis of tetradymol 2-chloromercury derivative (98); tetradymol is thus 10β-hydroxyfuranoeremophilane (137).

3β-Angelyloxyfuranoeremophilane, $C_{20}H_{28}O_3$, is another constituent of *Farfugium japonicum*. Its positive qualitative furan test, its n.m.r. spectrum, and its lithium aluminum hydride reduction to 3β-hydroxyfuranoeremophilane established its structure as (138) (96).

Furanojaponin, $C_{20}H_{28}O_3$, occurs in the rhizomes of *Petasites japonicus* Maxim. At an early stage it was confidently anticipated that it possessed a furanoeremophilane framework because of its co-occurrence with several other compounds of this type. Hydrolysis afforded a secondary alcohol oxidizable to a cyclohexanone (v_{CO} 1707 cm^{-1}), proving that the OH group is at positions 1, 2, or 3. N.m.r. studies allowed position 3 to be eliminated and mass spectral investigation of the deuterated ketone showed that in an exchange with NaOMe—MeOD four D atoms had been incorporated. This observation placed the keto group, and therefore the hydroxyl group in the hydrolysis product and the ester group in the natural product, at $C_{(2)}$, and allowed furanojaponin to be formulated as (140), the angelic acid residue being diagnosed

furanojaponin
(140)

by n.m.r. study, and its β (equatorial) orientation inferred from the half-height width (*ca.* 15 Hz) of the $C_{(2)}$-H signal in the n.m.r. spectrum of the natural product and from its ease of saponification (90).

b) Monoketones

Ligularone, $C_{15}H_{20}O_2$, occurs in the roots of *Ligularia sibirica* Cass. (14), in *Petasites japonicus* rhizomes (90), and in the roots of *L. fischeri* (17). The usual color reactions and i.r. and n.m.r. spectral behavior indicated that a furanoeremophilane system was present. Two possible structures were considered: furanoeremophilone-6 (139) and -9 (134). The calculated dipole moments of these structures are respectively 2.35 and 3.66 D. As the observed value was 2.55 D, the former structure for ligularone was preferred and was supported by its u.v. absorption (λ_{max} 269 nm). An analysis of the n.m.r. spectrum is in agreement (94). The ketone, on n.m.r. spectral and circular dichroism evidence, exists in solution as a *cis* steroidal-nonsteroidal A/B equilibrating system (95a).

9-Oxofuranoeremophilane (134), $C_{15}H_{20}O_2$, is found in *P. officinalis* rhizomes (77) and in those of *P. hybridus* (91). Its structure (134) was proposed on spectral evidence, coupled with the facts that the ketone was different from ligularone (77) and was formed by manganese dioxide oxidation of 9α-hydroxyfuranoeremophilane (136) (97).

Furanoligularenone, $C_{15}H_{18}O_2$, is found in the roots of *Ligularia fischeri* (99, 100, 101), *L. stenocephala* (101), and *L. sibirica* Cass. (81). Qualitative tests established the presence of a furan ring and its spectral properties were diagnostic of an α,β-unsaturated ketone and a furan nucleus (λ max 226 nm, ε 22,000). Catalytic hydrogenation afforded a ketofuranoeremophilane (uptake one mol of H_2). An analysis of the n.m.r. spectrum of the natural product was consistent with two possible structures, (141) and (142). The latter can be excluded because the

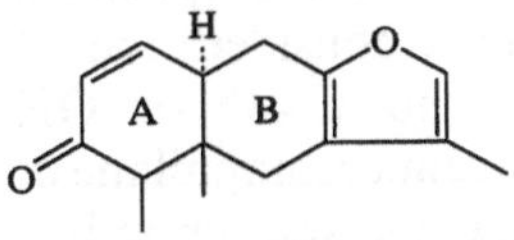

furanoligularenone
(141)

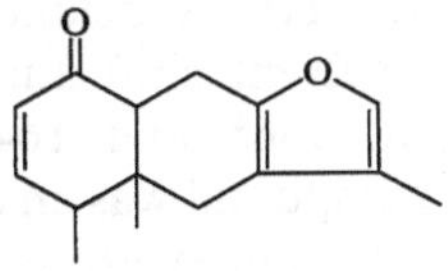

(142)

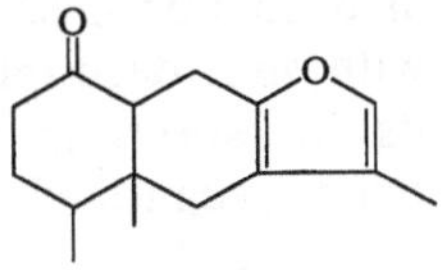

(143)

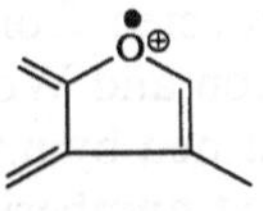

(144)

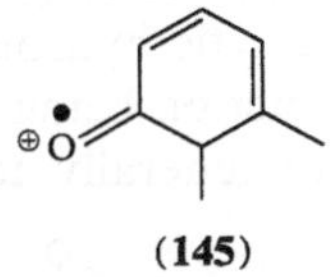

(145)

corresponding dihydro-derivative would be (143); treatment of the dihydro derivative with D_2O and NaOD in dioxan and observation of the change in the CH₃CH n.m.r. spectral signal showed the latter to be transformed from a doublet into a broad singlet. This behavior is inconsistent with (143), but would be in agreement with the dihydro derivative related to (141). Thus furanoligularenone may be formulated as (141). This structure is in harmony with the mass spectrum and accounts particularly for peaks at *m/e* 108 and 122 [fragments (144) and (145) respectively] (99). The stereochemistry of the A/B ring junction was settled as *trans* by o.r.d. studies, coupled with n.m.r. spectral study (100).

10β-H-Ligularenone, $C_{15}H_{18}O_2$, also found in *L. stenocephala* roots, resembles the foregoing ketone closely and is presumed to be its

10-epimeride **(146)** *(101)*. The nomenclature of these two ketones is inconsistent.

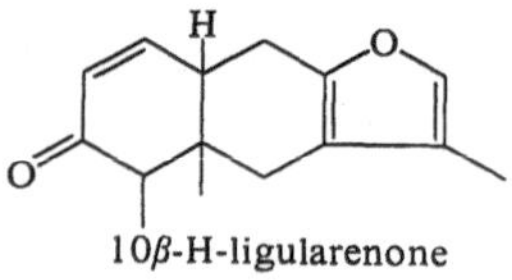

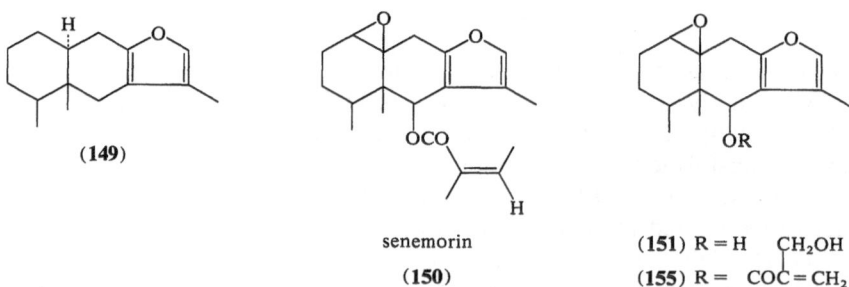

10β-H-ligularenone
(146)   **(147)** 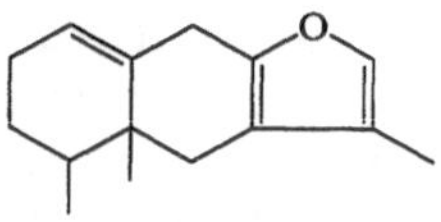   **(148)**

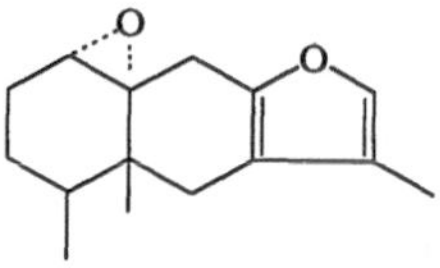

c) Epoxides

8,8a-Epoxyfuranoligularan, $C_{15}H_{20}O_2$, is present in *Senecio silvaticus* L. Its mass spectrum contains a base peak at *m/e* 108, characteristic of the furanoid fragment **(144)**, and its n.m.r. spectrum can be reconciled with a furanoeremophilane unit containing a trisubstituted epoxide, in which the CH group is flanked by a CH_2 group (triplet at δ 3.05 ppm, $J = 4.5$ Hz, 1 H). A detailed analysis of the spectrum with spin decoupling and solvent shift studies and a consideration of its i.r. spectrum (bands at 1565 and 1640 cm^{-1}, no C=O or OH bands), suggested that the compound was an epoxyfuranoeremophilane of structure **(147)**. Removal of the epoxy group with zinc and sodium iodide in buffered acetic acid yielded an olefin, to be formulated as **(148)**. Catalytic hydrogenation of the latter furnished a product which proved not to be identical with furanoeremophilane **(111**, p. 118), but was found to be identical with furanoligularane **(149)**, formed from furanoligularenone **(141)** by successive catalytic hydrogenation and Wolff-Kishner reduction *(100)*, the *trans* ring junction being confirmed by n.m.r. spectral analysis. According to the more generally accepted numbering of the ring system

(149)

senemorin

(150)

(151) R = H CH_2OH

(155) R = $COC=CH_2$

the compound is more correctly named 1,10-epoxyfuranoligularane *(102)*.

Senemorin, $C_{20}H_{26}O_4$, occurs in the rhizomes of *Senecio nemorensis,* subsp. *fuchsii.* Its u.v., i.r., and mass spectra and a detailed analysis of its n.m.r. spectrum revealed it to be 6β-angeloxy-1β,10β-epoxyfuranoeremophilane (150). Lithium aluminum hydride reduction of senemorin furnished two substances (151) and (152), respectively a consequence of partial and total reduction, the structures of both compounds being

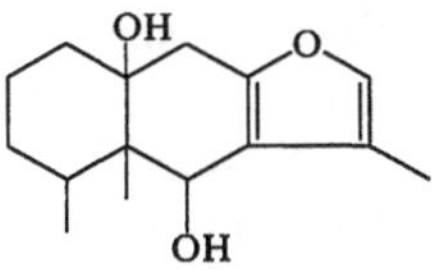

(152) 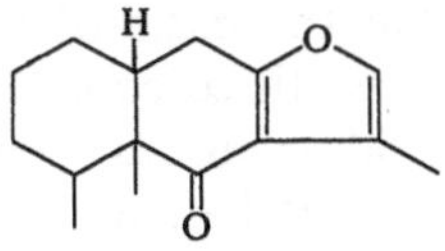   (153) 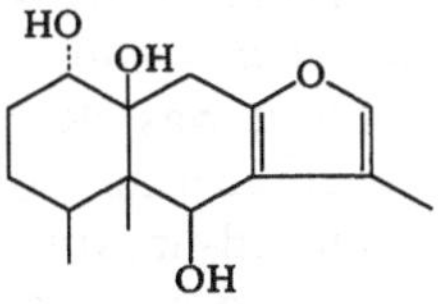   euryopsol
                                                                (154)

arrived at by detailed n.m.r. spectral analysis. Diol (152) was converted by sequential oxidation, dehydration, and hydrogenation to (153), which proved to be identical with petasalbone (ligularone) (139), the oxidation product of petasalbin (128) (p. 119). Further, alkaline hydrolysis of senemorin yielded a triol (154) identified with euryopsol (see p. 130). A detailed consideration of the coupling constants in the high resolution n.m.r. spectrum of senemorin pointed to the relative stereochemistry depicted in (150) for the compound (84a).

1β,10β-Epoxyfuranoeremophilan-6β-ol, $C_{15}H_{20}O_3$, already referred to above (151) as a partial reduction product of senemorin, occurs in the roots of *Ligularia fischeri* Turcz., along with its 2-hydroxymethyl-2-propenoate ester (155). Color reactions and u.v. spectral absorption indicated the presence of a furan ring in the alcohol (λ_{max} 215.5 nm, ε 7000); the furan ring was also consistent with i.r. bands at 1645 and 1570 cm^{-1}. A band at 3450 cm^{-1} was ascribed to a hydroxyl group. The n.m.r. spectrum was subjected to detailed analysis and its resemblance to that of furanoeremophilane-6β,10β-diol (p. 126) noted. Indeed, lithium aluminum hydride reduction of the compound yielded a diol which was shown to be identical with this compound (156); a

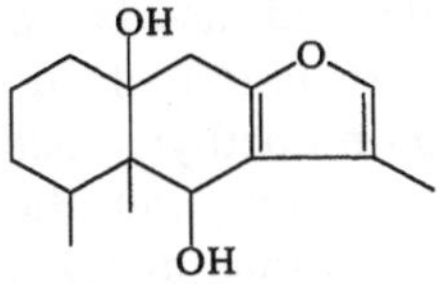

(156) 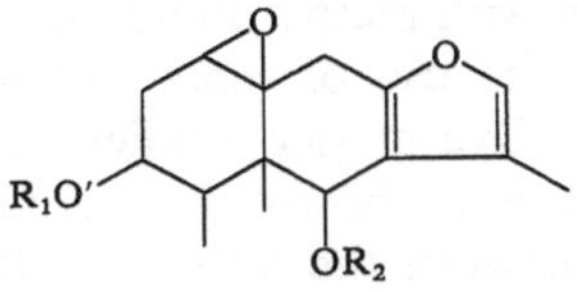   nemosenins   (158)
                                (157)

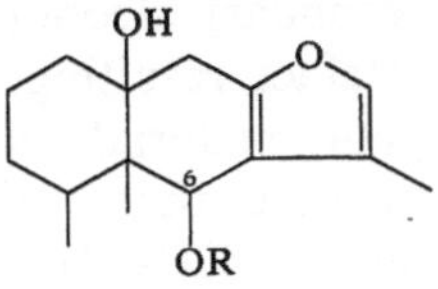

furanoeremophilane system was thus shown to be present. The lack of C=O absorption in the i.r. spectrum suggested that the third oxygen atom was ethereal and in view of the formation of diol (156) on complex hydride reduction the presence of an epoxide group was apparent;

in particular the epoxy group was 1β,10β. Detailed stereochemical features were elucidated by shift reagent studies, culminating in structure (151) for the compound (103, 104).

Nemosenin-A, -B, -C, and -D are four esters present in *Senecio nemorensis* subsp. *fuchsii*. On the basis of u.v., i.r., and mass spectral study it was clear that the first three compounds contained a hydroxyl group, an ester group, an epoxide group, and a furan nucleus. Nemosenin-A is an α,β-unsaturated ester, -B and -C are saturated esters, and -D is a saturated diester (no OH band in its i.r. spectrum). A detailed analysis of their n.m.r. spectra revealed that the compounds were respectively (157), (R_1 = H, R_2 = angelyl), (R_1 = H, R_2 = dihydroangelyl), (R_1 = H, R_2 = isobutyryl), and (R_1 = acetyl, R_2 = isobutyryl). Nemosenin-D yielded nemosenin-C on partial saponification (84a).

d) Di- and Trihydric Alcohols and Their Esters and Ethers

Furanoeremophilane-6β,10β-diol, $C_{15}H_{22}O_3$, occurs in the roots of *Ligularia japonica* Less., and has a parent ion peak in its mass spectrum at m/e 250. A positive Ehrlich test was indicative of the presence of a furan ring, confirmed by i.r. bands at 1645, 1565, and 1080 cm^{-1}. U.v. maximal absorption occurred at 215 nm (ε 8000). A detailed analysis of its n.m.r. spectrum revealed secondary and tertiary methyl groups, two allylic protons (AB quartet), a β-CH$_3$ on a furan ring, and a furan α-H proton. Addition of D_2O modified the spectrum so as to reveal the presence of two hydroxyl groups, one secondary and one tertiary. Taxonomic and biogenetic considerations suggested that a furanoeremophilane framework was present, and structure (158, R = H) was advanced.

A second compound, 10β-hydroxy-6β-methoxyfuranoeremophilane (158, R = CH$_3$) was found in the same source; it was readily formed from the foregoing diol by selective methylation of the secondary alcohol group with methyl iodide-silver oxide, and may be an artefact resulting from the use of methanol during extraction of the plant material. Nuclear Overhauser studies of the monomethyl ether revealed that the effect was shown by the CHOCH$_3$ atom and the tertiary CH$_3$, but not between the latter and the two protons at $C_{(9)}$; these observations are consistent with the presence of the tertiary (angular) CH$_3$ at $C_{(5)}$, and its equatorial disposition. Returning to the diol, intramolecular hydrogen bonding was observed in the i.r. spectrum between the two OH groups, but the diol did not react with periodate. The hydroxyl groups must consequently be in a 1,3-diaxial relationship to each other, requiring the tertiary OH group to be placed in the angular 10-position. A solvent shift n.m.r. spectral study (deuterochloroform to pyridine) confirmed this situation [$C_{(5)}$-CH$_3$ signal experienced a 0.23 ppm downfield shift].

A third substance isolated from the same source is formulated as the $C_{(6)}$-ester of the diol and 2-methylbutanoic acid (**158**,

$$\overset{\displaystyle CH_3}{\underset{\displaystyle |}{}}$$

R = COCHCH$_2$CH$_3$) (*105, 106*). Compound (**158**, R = H) is readily transformed into a mixture of three benzofurans on dehydration with phosphoryl chloride; a mechanism for these skeletal rearrangements has been advanced (*107*).

Furanopetasin C$_{20}$H$_{28}$O$_4$, occurs in several *Petasites* species (*5a, 77, 86, 91, 92, 108*). Spectral properties are consistent with the presence of an α,β-unsaturated ester group, a furan ring, and a hydroxyl group (*77*), the furan ring being confirmed by color reactions. On mild hydrolysis furanopetasol, C$_{15}$H$_{22}$O$_3$, was formed, together with angelic acid; furanopetasin is consequently an angelate ester of furanopetasol. The latter showed spectral absorption of a diol containing a furan ring. Furanopetasin was converted by catalytic hydrogenation into hexahydrofuranopetasin, which on tosylation and then lithium aluminum hydride reduction yielded the monotosylate of tetrahydrofuranopetasol. Oxidation then gave the corresponding ketone, the ethylenedithioketal of which with Raney nickel generated a tosylate which on detosylation with lithium aluminum hydride gave a product identified as tetrahydrofuranoeremophilane, already known as a reduction product of furanoeremophilane (see p. 118). Another informative degradation was performed by heating the above monotosylate with molybdenum disulfide at 300–320° in hydrogen at 150 ats., which led to eremophilane (**7**). It was thus evident that furanopetasin was derived from furanoeremophilane.

Of the two hydroxyl groups in furanopetasol, one is allylic to the furan ring, since it is selectively oxidized by activated manganese dioxide to yield a hydroxyketone (v_{CO} 1678 cm^{-1}; λ_{max} 280 nm, ε 27,000), the relatively long wavelength u.v. maximum signifying that the keto group is at position 9. This hydroxyketone was further

(**159**)

(**160**) R = H (furanopetasol)

(**161**) R = COÇ = C$\overset{\displaystyle CH_3}{\underset{\displaystyle H}{}}$
         |
        CH$_3$

(furanopetasin)

(**162**)

oxidisable to a diketone with chromic acid; this product had λ_{max} 283 nm, and consequently the new keto group is not to be placed at position 6. Treatment of the diketone with D_2O, followed by mass spectrometry, showed that five D atoms had been exchanged, and it was thus possible to formulate the compound as (159), and furanopetasol as (160). The orientation of the OH groups in the latter was settled by the observation that tetrahydropetasol formed a cyclic sulfite ester with thionyl chloride; inspection of models shows that this bridging is possible only if petasol is furanoeremophilane-2α,9α-diol.

That the ester group in furanopetasin is at position 2 was settled as follows: the monotosylate of tetrahydrofuranopetasol referred to above was oxidized to the corresponding ketone, then converted *via* its dithioketal to the corresponding deoxo-derivative, hydrolysis of which followed by vigorous oxidation (CrO_3) afforded a lactonecarboxylic acid. A formally analogous lactone formation was observed when the above ketone, without prior dithioketalization, was subjected to the same oxidation, leading in this case to a ketolactonecarboxylic acid. This means that the tosylated OH group in tetrahydrofuranopetasol mono-tosylate must be that at position 9, and consequently in furanopetasin (161) the angelyloxy group must be at $C_{(2)}$ (*108*).

Furanofukinol, $C_{15}H_{22}O_3$, is a diol of *Petasites japonicus*, which on oxidation yielded a diketone, $C_{15}H_{18}O_3$; this could be converted selectively into a monodithioketal which on desulfurization with Raney nickel afforded the known ligularone [6-oxofuranoeremophilane (139)]. One of the hydroxyl groups of furanofukinol is consequently at position 6. Co-occurring with furanofukinol is 6-angelylfuranofukinol, $C_{20}H_{28}O_4$, which on hydrolysis yielded furanofukinol, and which formed a monotosylate. The latter on reduction ($LiAlH_4$) gave petasalbin (128, R = H), of established structure and configuration. The angelyloxy group in the ester is therefore 6β. The diketone product mentioned above showed in its n.m.r. spectrum a quartet ($J = 6.5$ Hz, 1 H) at 2.8 ppm, clearly due to a $\underline{C}HCH_3$ group, this feature suggesting that a carbonyl group is at position 3. Further, it proved possible to effect a selective oxidation of furanofukinol to a ketol, with λ_{max} 269 nm, revealing the presence of the carbonyl group at position 6. Dehydration of this product afforded an enone, formulated as (162) on the basis of its n.m.r. spectral signals at 1.73 (3 H, d, $J = 1.2$ Hz, long range allylic coupling) and 5.43 ppm (1 H, m). Furanofukinol may consequently be formulated as (163), the 3α-OH stereochemistry being assigned because of the width at half height of the $C_{(3)}$-H signal (14 Hz), which indicates an axial C−H bond at this position. 6-Angelylfuranofukinol and 6-acetylfuranofukinol, which is also found in the same source, are thus respectively (163, R = angelyl, R' = H) and (163, R = Ac, furanofukinol (R = R' = H)

S-furanopetasitin (R = angelyl, R' = $\overset{OC}{\underset{H}{\diagdown}} C = C \overset{SMe}{\underset{H}{\diagup}}$)

(163)

R' = H) (*90*). 6-Acetyl-3-angelylfuranofukinol (**163**, R = Ac, R' = angelyl) and 3-angelylfuranofukinol (**163**, R = H, R' = angelyl) have been isolated from *Farfugium hiberniflorum* (syn. *Ligularia hiberniflora*). Reduction of both with lithium aluminum hydride gave furanofukinol (**163**, R = R' = H), and acetylation of the latter yielded the former. Spectral data of the latter were observed to be different from those of the known 6-angelylfuranofukinol (see above), so that the positions of the ester residues in the structures were revealed (*109*).

S-Furanopetasitin, $C_{24}H_{32}O_5S$, another *P. japonicus* constituent, yielded on alkaline partial hydrolysis *cis*-β-methylthioacrylic acid and 6-angelylfuranofukinol (**163**, R = angelyl, R' = H) (see above); a further step gave furanofukinol and angelic acid. S-Furanopetasitin is consequently to be formulated as (**163**, R = angelyl, R' = *cis*-β-methylthioacrylyl) (*90*).

Japonicin (Albopetasol), $C_{15}H_{22}O_3$, occurs in the roots of *Petasites albus*, and its mono- and di-angelyl esters have been isolated from *P. japonicus* (*2a, 5a, 88, 91, 92*). The usual chemical and spectral studies established that japonicin was a furanoeremophilane containing two hydroxyl groups. It seems probable from a consideration of spectral and chemical properties that japonicin is 3β,6β-dihydroxy-furanoeremophilane.

Two esters derived from 3β,9β-dihydrofuranoeremophilane (**164**, $R_1 = R_2 = H$) have been found in *Farfugium japonicum*. They are respectively the 3-angelate-9-senecioate (**164**; $R_1 = COCMe \doteq CHMe$, $R_2 = COCH = CMe_2$) and the 3-angelate (**164**; $R_1 = COCMe \doteq CHMe$, $R_2 = H$) of the diol. Their structures have been elucidated by chemical and spectroscopic study. Both esters, for example, afforded on lithium aluminum hydride reduction a diol, also formed by their alkaline hydrolysis. Oxidation of the diol with activated manganese dioxide afforded a diketone, which proved to be identical with 3,9-dioxofurano-eremophilane (**165**), the oxidant having attacked, unusually, the non-allylic hydroxyl group as well as the allylic. Diketone (**165**) was already known as a transformation product of kablicin (see p. 132), and its *cis* ring junction established. The foregoing diol is consequently 3,9-dihydroxyfuranoeremophilane; dehydration of it with tosyl chloride-pyridine gave an alcohol (dehydration *via* the allylic OH) and a diene. The former on catalytic hydrogenation yielded two $C_{(10)}$-epimeric alcohols, one of which was found to be identical with furanoliguranol (**166**), derived from furanoligularenone (**141**, see p. 123), of established structure and configuration. Thus the above-mentioned diol has a 3β-hydroxyl group. The monoester under discussion shows an allylic C$\underline{H}$OH signal in its n.m.r. spectrum at δ 4.30 ppm, whilst the corresponding proton in the diester resonates at δ 5.43 ppm. Consequently in the mono-ester the free hydroxyl group is in the allylic 9-position and the angelic acid residue is at the 3β-position. Further, in the diester the senecioic acid residue is at $C_{(9)}$, and is β in configuration, partial hydrolysis of the diester affording the monoester by selective elimination of the senecioyl group. The two esters may therefore be formulated as indicated. Finally, a non-steroidal *cis* ring fusion in (**164**) has been revealed by detailed n.m.r. spectral studies on the esters (*96*).

1 0 β - H y d r o x y - 6 β - s e n e c i o y l o x y f u r a n o e r e m o p h i l a n e, $C_{20}H_{28}O_4$, has also been isolated from the same source. On lithium aluminum hydride reduction it yielded a diol identical with furanoeremo-philane-6β,10β-diol (**158**, R = H; see p. 126), of known structure and con-figuration. A senecioyl residue was detected by n.m.r. spectral study, and placed in the 6β position. The compound may consequently be formulated as (**158**, R = OCCH=CMe$_2$) (*96*).

E u r y o p s o l, $C_{15}H_{22}O_4$, has been isolated from the resin of *Euryops floribundus* and *E. tenuissimus*. Its u.v. (λ max 220 nm, ε 6800) and i.r. (ν max 1560, 885 cm^{-1}) spectral absorption, and its positive color reaction towards Ehrlich's reagent (*p*-dimethylaminobenzaldehyde-HCl), pointed to the presence of a furan ring. Three hydroxyl groups were detected by deuterium exchange followed by mass spectrometry; a detailed examination of the n.m.r. spectrum of euryopsol revealed that two of these were secondary and one tertiary. The assumption that a furanoeremophilane skeleton was present was substantiated by n.m.r. spectral analysis. Thus the tertiary OH group must be at position 10 (the CH$_3$ signal at position 4 being a doublet) and one of the secondary OH groups must be at positions 6 or 9, because the spectrum shows that the hydrogen atom of the C$\underline{H}$OH group is not coupled to adjacent H atoms. Chromic acid oxidation of euryopsol afforded a diketone, both secondary alcohol groups having been attacked. This ketone had λ_{max} 270 nm, which is typical of 6- as opposed to 9-oxofuranoeremophilanes (λ_{max} 280—282 nm); accordingly, one of the secondary hydroxyl groups in euryopsol must be at position 6. Further, the diketone exchanged four D atoms (after back-exchange of the tertiary OD with water), indicating that one of the keto groups must be at position 1 (two α and two allylic

H atoms exchangeable) and allowing the diketone to be formulated as (167). Mass spectral fragmentation of the diketone is in agreement. Euryopsol may therefore by formulated as (154), the presence of a

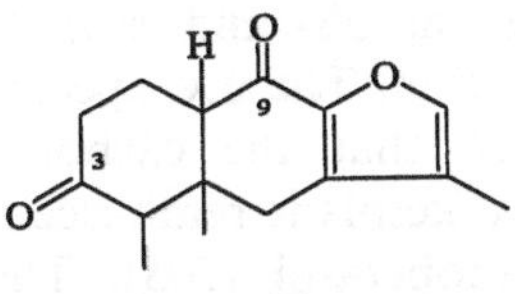

(165) 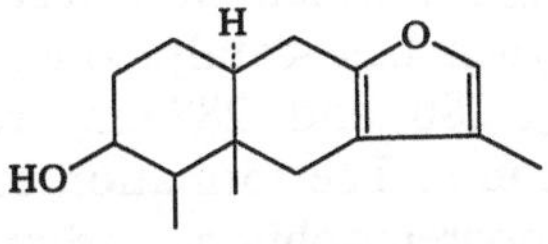   (166) 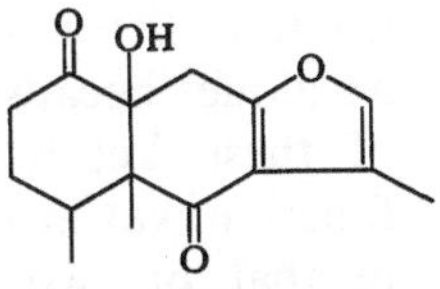   (167)

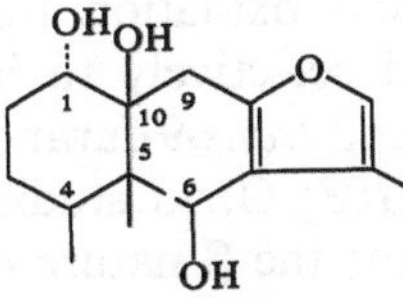

euryopsol
(154)

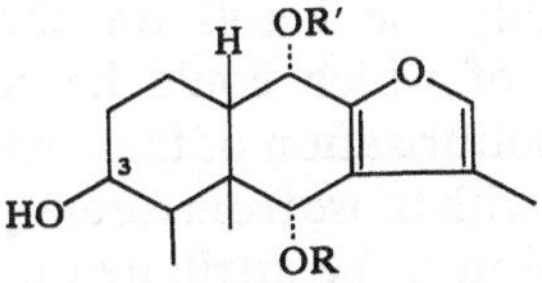

kablicin
(R = angelyl, R' = senecioyl)
(168)

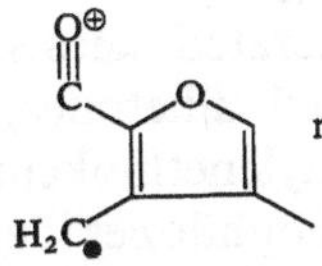

(169)

m/e 122

1,2-glycol unit being substantiated by the observed oxidation of euryopsol by periodate.

The stereochemistry of the hydroxyl groups emerged as follows. Firstly, a *cis* steroidal ring junction was assumed on the grounds that the great majority of furanoeremophilanes are so constructed. This view is supported by n.m.r. spectral considerations, especially solvent shift studies, which also indicate that the OH at position 1 is equatorial and therefore α. Thus in pyridine-d_5-D_2O the H at this position appears as a quartet, but in other solvents the signal is unresolved, due to virtual coupling. This latter effect is minimized in pyridine-d_5 because the protons at $C_{(2)}$ resonate at lower field (by *ca.* 0.2 ppm). The 1-proton forms the X part of an ABX system, with $J_{AX} + J_{BX} = 16$ Hz, and must thus be axial, and the OH there equatorial. Next the OH at position 6 must be β because, among other reasons, an α-configuration would be in a *quasi*-1,3-diaxial relationship with the methyl group at $C_{(4)}$; also an α-arrangement can be rejected by comparison with the chemical shifts of this methyl group in 6α- and 6β-hydroxyfuranoeremophilane (*94*). Thus euryopsol can be represented in detail by (154). An interesting reaction of euryopsol is its regioselective conversion to its 6-methyl ether by methanolic HCl (*110, 111*).

Kablicin, $C_{25}H_{34}O_6$, has been found in petroleum extracts of *Petasites kablikianus* and of *P. paradoxus* (*91, 92*). Its furanoeremophilanoid character was quickly discerned from its spectral properties.

The compound is a hydroxy-diester which on hydrolysis afforded senecioic, angelic, and tiglic (formed by isomerization) acids. Lithium aluminum hydride reduction of kablicin afforded a triol, $C_{15}H_{22}O_4$. Hydrolysis under mild, alkaline conditions gave a monoester and two isomeric ketols $C_{15}H_{20}O_3$, which were recognized as cis- and trans-1-decalones because of their base-catalyzed equilibration. The u.v. spectra of these ketols (λ_{max} 280 and 282 nm) revealed that the carbonyl function was at position 9. The formation of these ketols is reminiscent of that of two furanoeremophilones from furanopetasol (108). The position of the hydroxyl function in the $C_{15}H_{20}O_3$ ketols was found to be $C_{(3)}$ by n.m.r. spectral studies and spin decoupling experiments on the deuterated ketols. Finally the ketols on chromic acid oxidation furnished diketones, each of which could be converted selectively to its 3-ethylenethioketal; desulfurization of these gave cis- and trans-9-furanoeremophilones identical with those from furanopetasol (108). O.r.d. studies enabled a cis ring junction to be attributed to kablicin; the β nature of the hydroxyl function at $C_{(3)}$ was established with the help of n.m.r. spectral analysis. Kablicin may therefore be formulated as (168) (112).

e) Keto Alcohols and Keto Esters

Euryopsonol, $C_{15}H_{20}O_3$, occurs in the unsaponifiable fraction of the resin of Euryops floribundus (110, 111, 113). It was found to be doubly unsaturated and to contain a keto group and a secondary alcoholic group. A furan ring was detected by i.r. and n.m.r. spectroscopy in the usual way, though the compound gave no color with p-dimethyl-aminobenzaldehyde-HCl. Two double bonds were present, evidenced by the uptake of two moles of hydrogen on catalytic hydrogenation. Chromic acid oxidation afforded a diketone, λ_{max} 280 nm, ν_{CO} 1670 and 1720 cm^{-1}, compared with the single ν_{CO} of 1670 cm^{-1} in euryopsonol. If a furano-eremophilane framework can be assumed, the keyngroup of euryopsonol must be at $C_{(9)}$ and the new keto group in the diketone cannot be at $C_{(6)}$; it must therefore be in ring A, i.e. at positions 1,2, or 3. The diketone formed a monofurfurylidene derivative on reaction with furfural, proving that the new keto group must be flanked by a single methylene group and must consequently be at position 1 or 3. The diketone also formed a monothioketal with ethanedithiol, which on desulfurization gave trans-9-oxofuranoeremophilane (135) (see p. 120). This identity confirmed that the keto group of euryopsonol is at $C_{(9)}$, as did the appearance of a base peak at m/e 122 [characteristic of the fragment (169)] in the mass spectrum of the natural product. Next the diketone is not β-diketonic in its chemical properties; consequently the second keto group cannot be at $C_{(1)}$. Finally the compound proved to be different from 2,9-dioxofuranoeremophilane (**159,** see p. 127) (*108*); the diketone must therefore be formulated as 3,9-dioxofuranoeremophilane and euryopsonol as 3-hydroxy-9-oxo-furanoeremophilane (**170**).

The α-configuration of the hydroxyl group was demonstrated by the reduction of the above diketone with LiAlH(OMe)₃, a reagent known

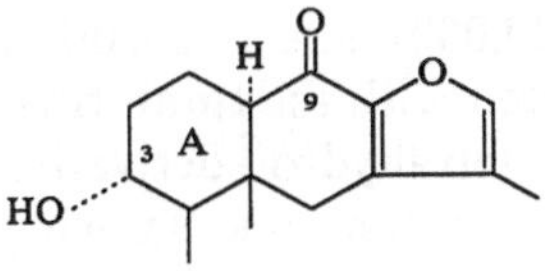

<table>
<tr><td align="center">euryopsonol
(**170**)</td><td align="center">warburgin
(**171**)</td><td align="center">furanoligularanone
(**172**)</td></tr>
</table>

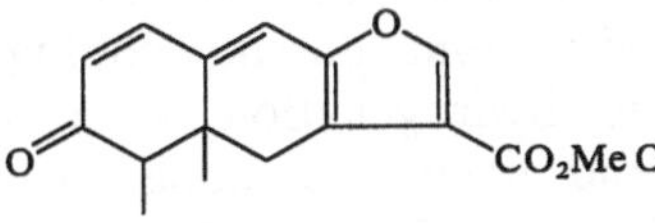

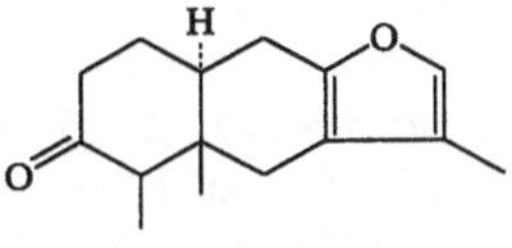

to yield stereospecifically the less stable alcohol in ketone reductions. The diol so realized was regiospecifically oxidized to the 9-ketone with manganese dioxide, to give *epi*-euryopsonol. Since the absolute configuration of furanoeremophilane is known (see p. 118) the *epi*-alcohol must be axial (3β), and euryopsonol has a 3α-hydroxyl group. Chemical and spectroscopic confirmations of this assignment have been presented: for example, euryopsonol is smoothly acetylated by pyridine-acetic anhydride but the *epi*-compound is not, owing to 1,3-diaxial interaction between the 3β-hydroxyl and 5-methyl groups (*111*).

The stereochemical nature of the ring fusion in euryopsonol has been settled as follows*. The stepwise formation of *trans*-9-oxofurano-eremophilane (**135**), discussed above, indicates that the junction should be *trans,* and a correlation with kablicin (**168**) (*2a*) confirms this and other features of the proposed structure. The latter correlation involved hydrolysis of kablicin to a *trans* keto-alcohol (**172a**) (one of three

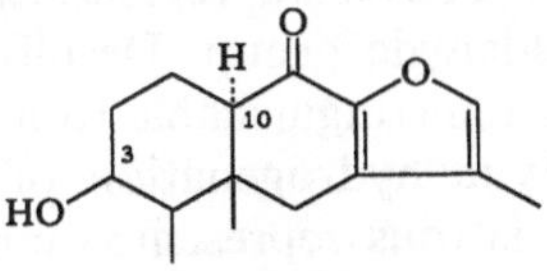

<table>
<tr><td align="center">(**172a**)</td><td align="center">(**172b**)</td></tr>
</table>

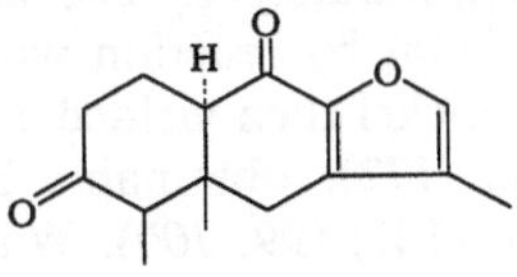

products; see p. 132). This on chromic acid oxidation gave a diketone (**172b**) which proved to be identical with the chromic acid oxidation product of euryopsonol (**170**). Euryopsonol and (**172a**) proved not to be identical, despite the identity of their melting points. The difference is one of configuration at C₍₃₎ since it has been proved (*111*) that euryopsonol has a 3α-hydroxyl group.

---

* The author wishes to thank Dr. D. E. A. RIVETT, Rhodes University, Grahamstown, S. Africa, for his help in the clarification of the literature relating to this aspect.

Warburgin, $C_{16}H_{16}O_4$, is a yellow keto ester occurring in the heartwood of *Warburgia ugandensis*. I.r. measurements revealed the presence of α,β-unsaturated ketone (1679 cm^{-1}) and ester (1734 cm^{-1}) functions. The latter was seen to be a methyl ester because hydrolysis followed by diazomethane methylation regenerated warburgin. The remaining oxygen was diagnosed as furanoid (i.r. bands at 3156 and 1566 cm^{-1}). The long wavelength (370 nm, ε 20,000) u.v. absorption suggested a conjugated dienone, possibly conjugated with the furan ring. Quantitative catalytic hydrogenation gave a tetrahydro derivative $C_{16}H_{20}O_4$, still furanoid, having no enone u.v. absorption, and showing ν_{CO} 1721 (ketone) and 1730 cm^{-1} (ester). The n.m.r. spectrum of warburgin confirmed its dienone nature; in tetrahydrowarburgin the only low field signal is at δ 7.93 ppm (1 H) due to a furan α-hydrogen, and the u.v. spectrum showed λ_{max}255 nm (ε 2560), typical of a furan-3-carboxylic ester; the presence of this fragment was confirmed by mass spectrometry (base peak m/e 152, arising from retro-Diels-Alder cleavage of the type

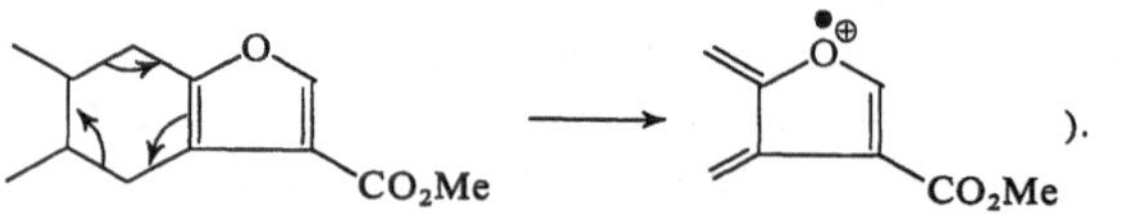

These and other considerations pointed to a furanoeremophilane framework for warburgin, and favored specifically structure (171), exclusive of stereochemistry. The correctness of this formulation was substantiated by a simple correlation with furanoligularenone [(141), see p. 123], of known structure and stereochemistry (*99, 100*). Lithium aluminum hydride reduction of tetrahydrowarburgin gave a diol oxidized by chromic acid to a keto-aldehyde. The latter was converted regioselectively to a monothioketal by reaction with the aldehyde group. Desulfurization with Raney nickel then yielded a product indistinguishable from furanoligularanone (172), obtainable by catalytic hydrogenation of furanoligularenone (141) (*99, 100*). Warburgin is thus represented completely by structure (171). O.r.d. measurements confirm the 4β,5β stereo-

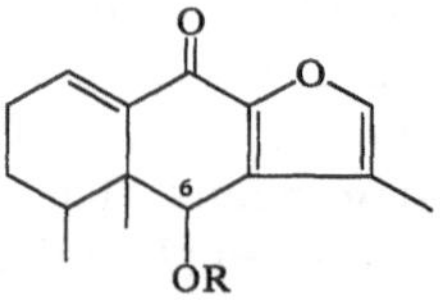

adenostylone (173) (R = COCHMe$_2$)

neoadenostylone (174) (R = COCMe $\overset{t}{=}$ CHMe

decompostin (175) (R = Ac)

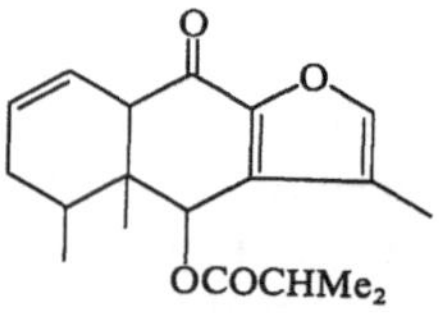

isoadenostylone

(176)

chemistry of the methyl groups, and point to (171) as the absolute stereo-chemistry of the compound (55, 57).

Four esters derived from 6β-hydroxy-9-furanoeremophilone are found in nature. They are adenostylone (173), neoadenostylone (174), decompostin (175), and isoadenostylone (176). Decompostin, $C_{17}H_{20}O_5$, occurs in *Cacalia decomposita* roots, contains a highly con-jugated system (λ_{max} 254, 302 nm, ε 4,600, 16,200), and is an α,β-un-saturated ketone. Spectral properties also indicate that a furan ring is present. The n.m.r. spectrum points to secondary and tertiary methyl groups, an acetoxy group on an allylic carbon atom, and a

$$CH_2-CH=C\begin{matrix} \diagup C \\ \diagdown C \end{matrix}$$

unit. Catalytic hydrogenation resulted in uptake of one mole of H_2, but the C=O i.r. absorption remained characteristic of a conjugated enone (1677 cm^{-1}). Reduction of decompostin with zinc-pyridine afforded a ketone which proved to be identical in all respects with *trans*-9-oxofuranoeremophilane (135) (see p. 120). Decom-postin is therefore a furanoeremophilane with an allylic acetoxy group; structure (175) accommodates the above and other spectral and chemical properties (*114*). Structural interrelationships between decompostin and the other three members of this group, coupled with n.m.r. spectral and o.r.d. measurements reveal that the acetoxy group is 6β (*115*).

The chemistry of adenostylone (173) and neoadenostylone, both found in *Adenostyles alliariae*, is closely parallel to that of decompostin; in the former, $C_{19}H_{24}O_4$, an isobutyryl ester residue was detected by n.m.r. spectroscopy and the compound is represented by (173) (*116*). The latter, $C_{20}H_{24}O_4$, is the corresponding 6β-angelate ester (174) (*115, 116, 117*). Isoadenostylone, $C_{19}H_{24}O_4$, also occurring in the same source, has an n.m.r. spectrum containing all the signals of adenostylone except that it shows two vinyl protons, and therefore contains a disubstituted double bond (broad singlet, 2H, at δ 5.8 ppm). Spin decoupling and solvent shift studies indicate that the double bond is 1,2, and iso-adenostylone is thus (176) (*115, 116, 117*).

2α-Angelyloxy-9-oxo-10αH-furanoeremophilane, $C_{20}H_{26}O_4$, has been found in *Petasites hybridus* (*68a*). On hydrolysis it yielded angelic

(176a)

acid and a ketol $C_{15}H_{20}O_3$, identical with 2α-hydroxy-9-oxo-10αH-furanoeremophilane described previously (*116*). The product must consequently be formulated as (**176a**) (*68a*).

f) Miscellaneous Furanoeremophilanes

8,12-Dimethoxydihydrofuranoeremophilane, $C_{17}H_{28}O_3$, is a crystalline constituent of *Petasites hybridus* rhizomes. Zeisel determination showed the presence of two methoxyl groups and the tetranitromethane test for unsaturation was positive. Quantitative catalytic hydrogenation (uptake 2.2 moles of hydrogen) resulted in the formation of tetrahydrofuranoeremophilane, identified earlier as a hydrogenolysis product of petasalbin (**128**) (*77*) (see p. 118), and also eremophilenolide (**113**) (see p. 113). The natural product is consequently derived from furanoeremophilane (**111**). The hydrogenolysis of the methoxyl groups indicated that they are of a labile character, possibly α,α'-furanoid, and allylic in nature. The structure (**177**) was contemplated, and corroborated by an n.m.r. spectral comparison with eremophilenolide (**113**).

(**177**)   (**178**)   (**179**)

A scheme to explain the formation of eremophilenolide has been proposed involving protonation of the angular methoxyl group (the reduction was effected in acid medium), followed by deprotonation and methanol elimination. Attack on $C_{(12)}$ by $^{\ominus}OH$ (or more likely by H_2O) then generated the hemiketal (**178**), hydrolysis of which would lead to eremophilenolide (**113**) (*91, 118*).

Furanoeremophilane-14β,6α-olide, $C_{15}H_{18}O_3$, occurs in the roots of *Ligularia hodgsoni* Hook. Chemical tests established readily the presence of furan and γ-lactone rings (v_{CO} 1767 cm^{-1}). Lithium aluminum hydride reduction afforded a diol $C_{15}H_{22}O_3$ readily cyclized to an internal ether $C_{15}H_{20}O_2$. Hydrogenolysis of the lactone with palladized carbon and hydrogen gave a carboxylic acid, convertible by lithium aluminum hydride into a primary alcohol. Sequential tosylation and lithium aluminum hydride reduction gave a product recognized as furanoeremophilane (**111**) (see p. 113). The lactone ring must therefore span positions 4 and 6, and the natural product may be formulated as (**179**). The α-orientation of the lactone oxygen at position 6 was inferred from the fact that there is an observed nuclear Overhauser effect between the hydrogen at $C_{(6)}$ and the neighboring angular methyl group at $C_{(5)}$; this requires these units to be *cis* to each other. Other features of structure (**179**) are confirmed by similar studies; for example an N.O.E. observed between 6-H and the β-CH$_3$ group indicates that $C_{(6)}$ is attached to the other β-position of the furan ring (*119*).

g) Furanoeremophilanes from South African Plants

A systematic investigation by BOHLMANN and his co-workers of South African plants of the genera *Euryops*, *Othonna*, and *Senecio* has revealed that many of the sesquiterpenes found therein are furanoeremophilanes. About one hundred new members of this sub-group have been so discovered. The species involved are *Euryops othonnoides*, *E. speciosissimum* (*120*), *Othonna quinquedentata* Thunb. (*121*), *E. hebecarpus*, *E. chrysanthemoides*, *E. tenuissimus*, *E. abrotani-folius*, *E. virgineus*, *E. linifolius*, *E. lateriflorus*, *E. spathaceus* (*75*), *Senecio elegans*, *S. glastifolius*, *S. rigidus*, *S. pterophorus*, *S. purpureus* (*64*), *Othonna amplexicaulis* Thunb., *O. quercifolia*, *O. dentata*, *O. arbo-rescens*, *O. coronopifolia*, *O. barkerae*, *O. filicaulis*, *O. bulbosa*, and possibly *O. multicaulis* (*16*). The natural products were extracted from the roots and other parts with ether or petroleum, and the components separated by careful column chromatography. In nearly all cases the amounts of individual constituents obtained were too small to allow much in the way of chemical investigation, so that structural assign-ments rest largely on u.v., i.r., n.m.r., and mass spectral interpretations. N.m.r. studies have involved spin decoupling and shift reagent techni-ques, and mass spectra were analyzed carefully and compared with the cracking patterns of furanoeremophilanes of known structure and stereochemistry. A typical example is the hydroxy ester (**180**) (see Chart 13), with spectroscopic data summarized (*75*).

$C_{20}H_{26}O_4$

(**180**)

i.r. $\begin{cases} C=C \ 1650 \ cm^{-1} \ (conj.) \\ OH \quad 3550 \ cm^{-1} \\ C=O \ in \ CO_2R \ 1705 \ cm^{-1} \ (br) \end{cases}$ $\left.\begin{array}{c} \\ \end{array}\right\}$ intramolecular H-bonding

u.v. λ_{max} 212 nm (α,β-unsaturated ester)

m.s. $M^{\oplus}$ 330.184 ($C_{20}H_{26}O_4$ = 330.183)

N.m.r. (in CCl₄)

H atom	δ, ppm	Eu(fod)₃ Δ, ppm	H atom	δ, ppm	Eu(fod)₃ Δ, ppm
1-H	5.57 m	0.31	13-H	1.75 d	0.05
2-H	2.1 m	0.5	14-H	1.23 s	0.72
3-H	1.2–2.0 m	0.3	15-H	1.11 s	0.97
6-H	6.39 s (br)	2.8	18-H	6.05 qq	0.15
9β-H	3.43 d (br)	0.35	19-H	2.02 dq	0.33
9α-H	2.96 d	0.24	20-H	1.91 dq	0.21
12-H	6.92 s (br)	0.01			

M.s.

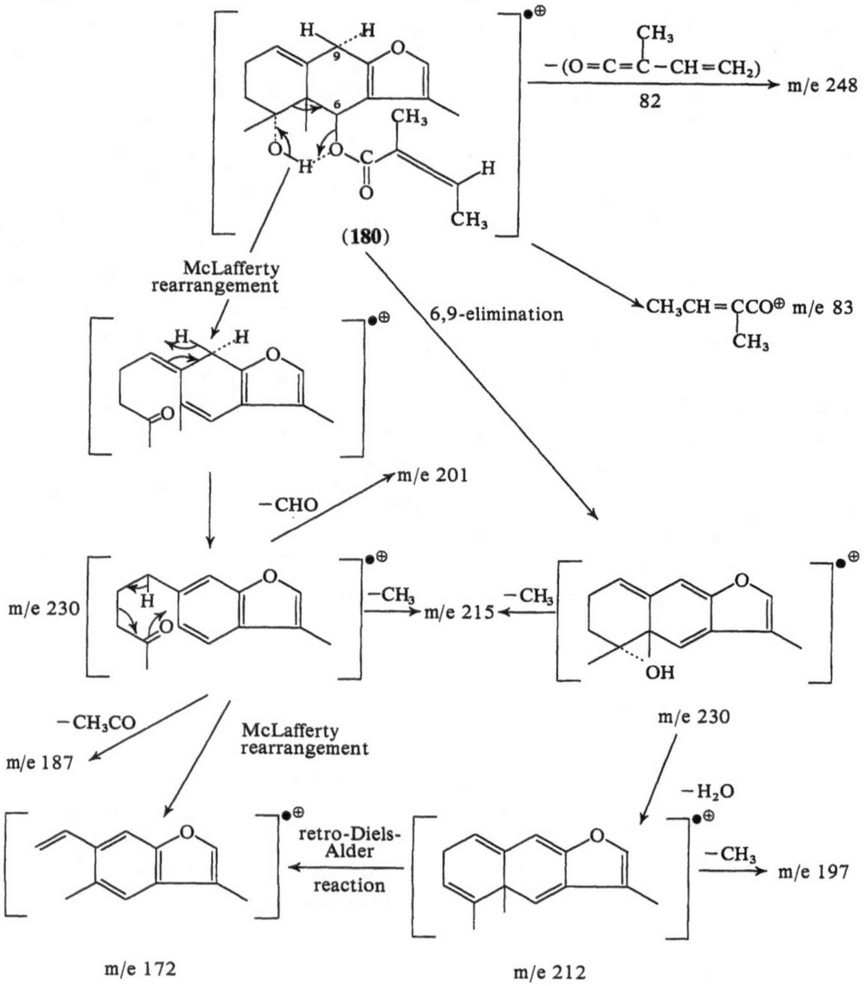

New compounds isolated are listed in Charts 11, 12, 13, 14, and 15; note that therein

angelyl = Me and tiglyl = H •

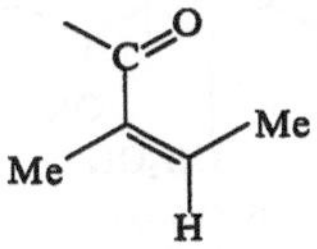

O angelyl

OR

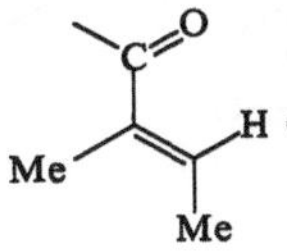

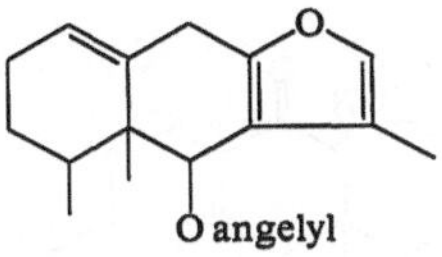

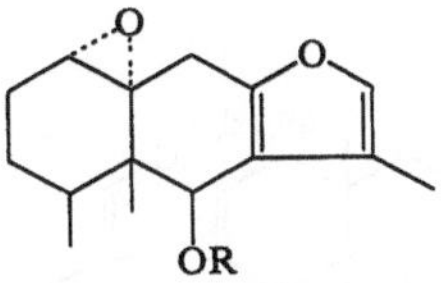

R = angelyl
R = Ac

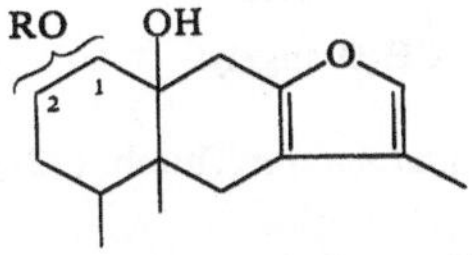

R = angelyl [at $C_{(1)}$ or $C_{(2)}$] } positions not established
R = Ac [at $C_{(1)}$ or $C_{(2)}$] } unequivocally

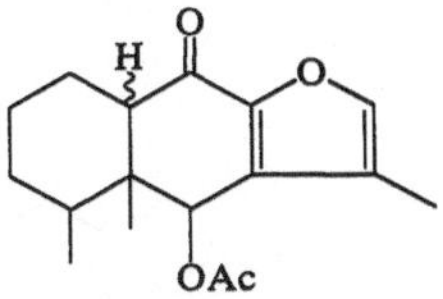

OAc

dihydrodecompostin
(see p. 135)

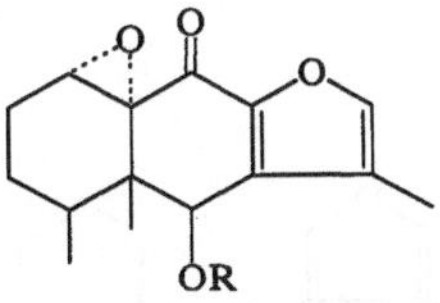

OR

R = Ac   R = α-methylacrylyl (75)
R = angelyl
R = isovaleryl
R = isobutyryl

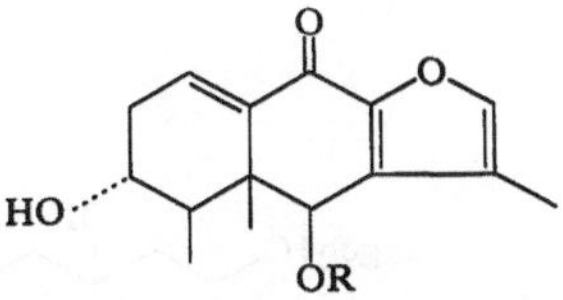

OR

R = isovaleryl
R = isobutyryl

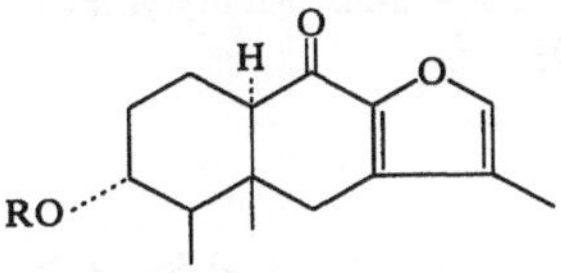

R = isovaleryl
R = isobutyryl

$$R = \underset{H}{\overset{H_3C}{C}}=C\underset{CH_2OH}{\overset{CO}{}}$$

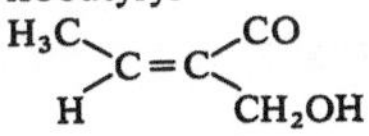

(euryopsonol 2′-hy-
droxymethylcrotonate)

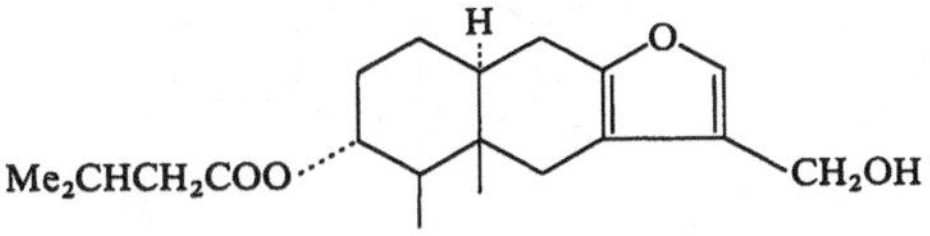

Me$_2$CHCH$_2$COO   CH$_2$OH

Chart 11. Compounds from *E. othonnoides* and *E. speciosissimum* (*120*)

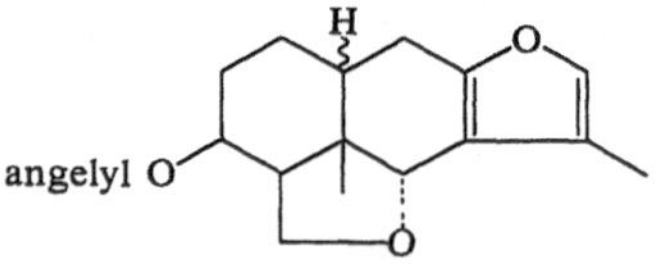

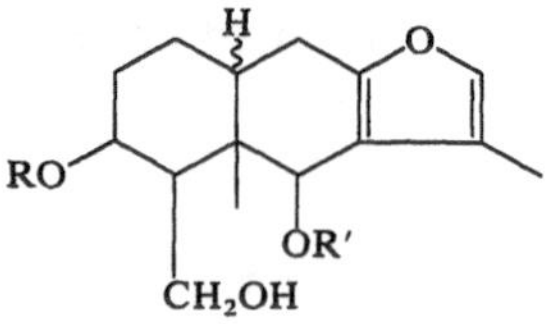

R = angelyl, R' = Ac ⎱ inseparable binary
R = Ac, R' = angelyl ⎰ mixture

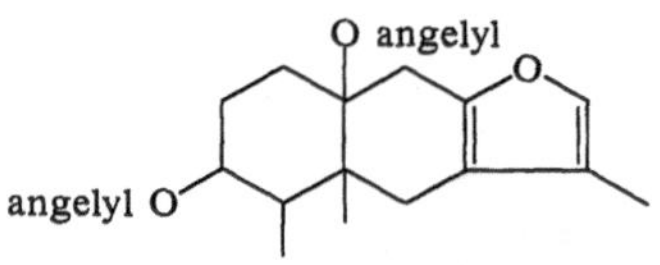

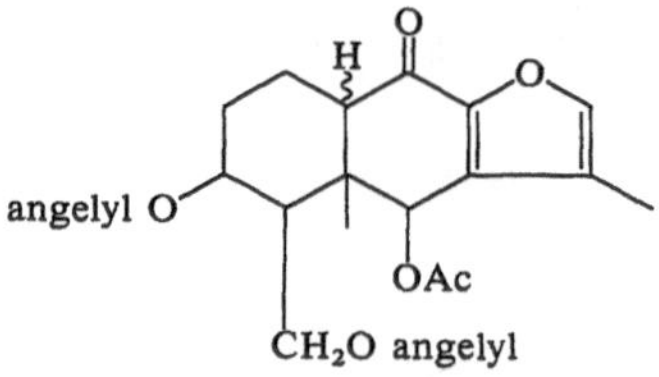

Chart 12. Compounds from O. quinquedentata Thunb. (121)

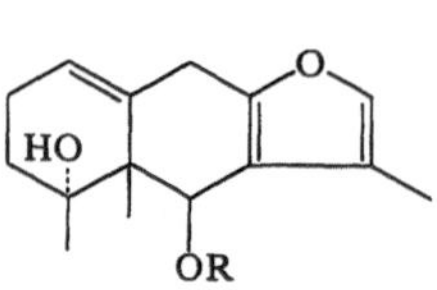

R = angelyl, R = tiglyl
R = isobutyryl, R = a-methylbutyryl (64)
R = senecioyl (64)

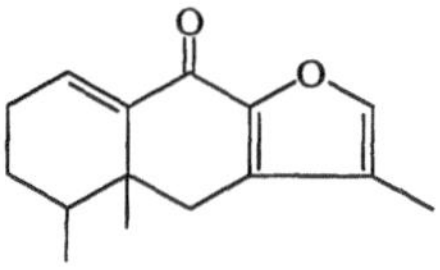

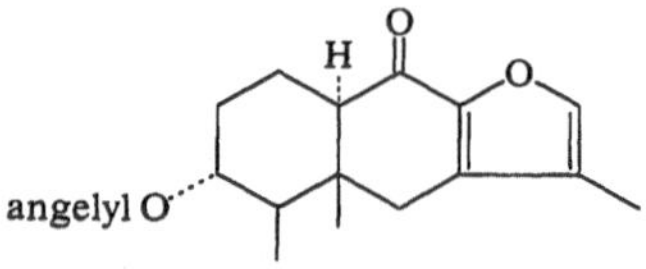

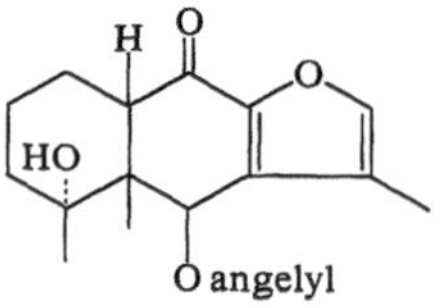

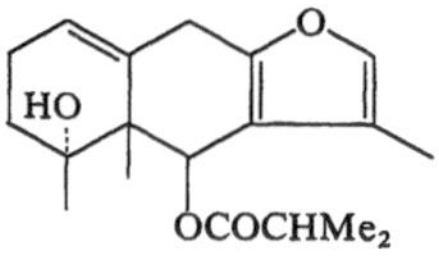

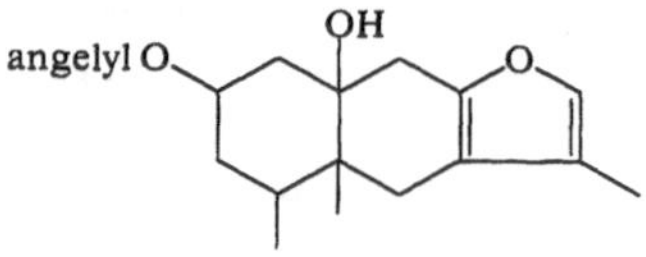

(see Chart 11)

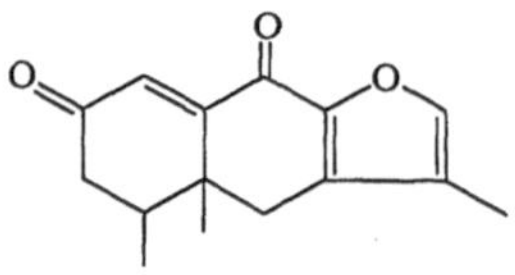

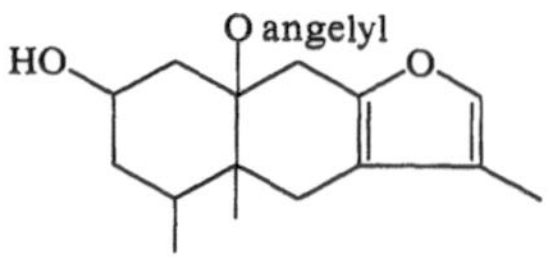

Chart 13. Further Compounds from Euryops Species (75)

References, pp. 175—186

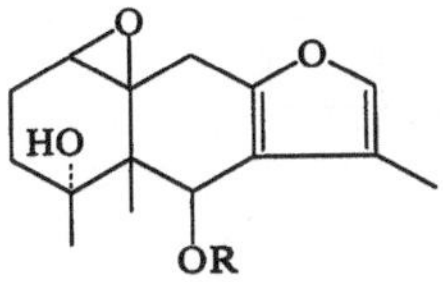

R = angelyl
R = isovaleryl
R = isobutyryl

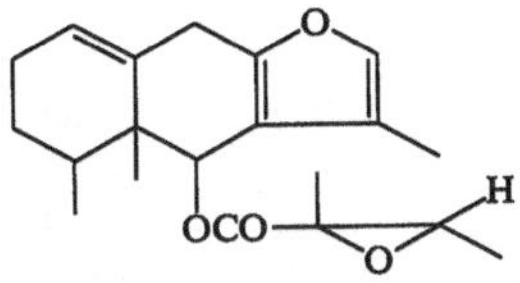

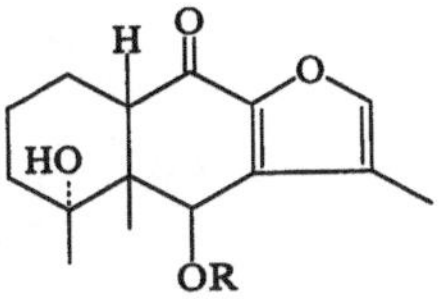

R = α-methylacrylyl
R = senecioyl
R = isovaleryl
R = tiglyl

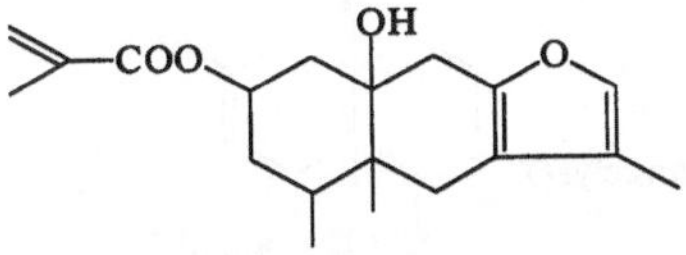

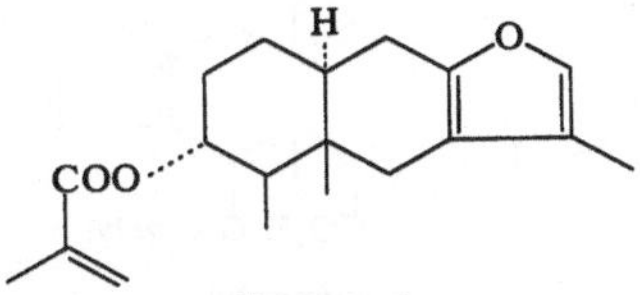

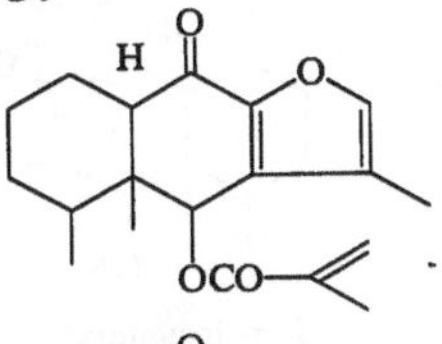

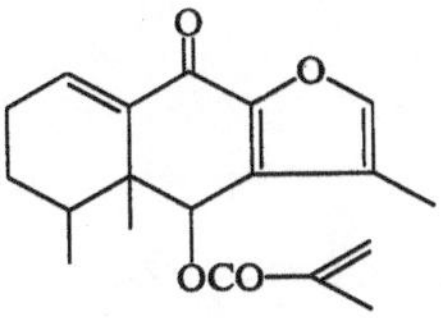

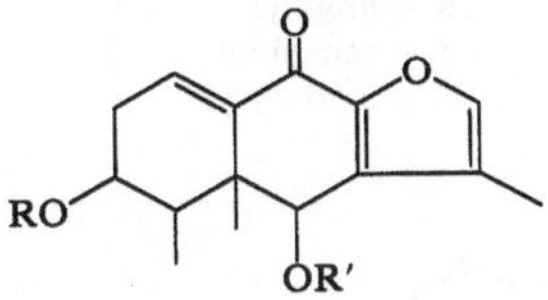

R = R' = α-methylacrylyl
R = R' = angelyl
R = α-methylacrylyl, R' = angelyl
R = H, R' = angelyl
R = H, R' = α-methylacrylyl

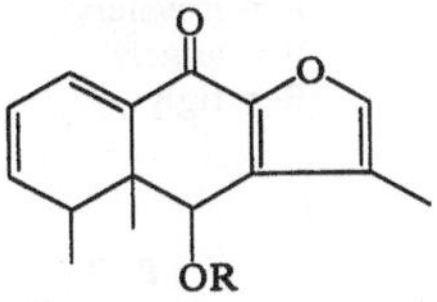

R = α-methylacrylyl
R = angelyl

Chart 13. Further Compounds from *Euryops* Species (*75*) (continued)

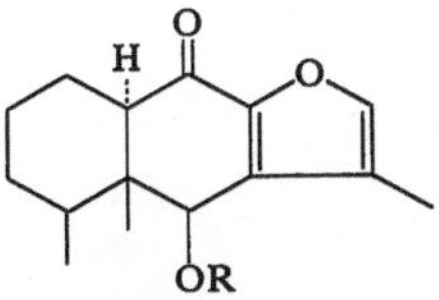

R = isobutyryl
R = angelyl
R = isovaleryl

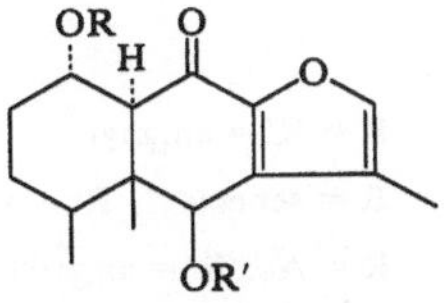

R = H, R' = isobutyryl
R = Ac, R' = isobutyryl
R = H, R' = tiglyl
R = Ac, R' = angelyl
R = angelyl, R' = Ac
R = H. R' = angelyl

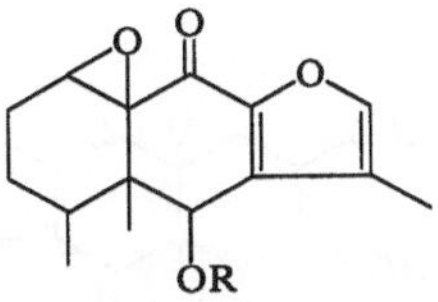

R = senecioyl
R = tiglyl

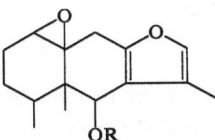

R = α-methylacrylyl
R = senecioyl

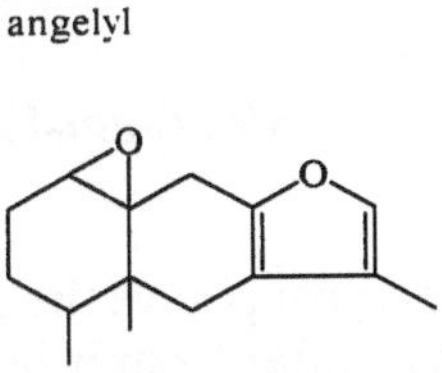

Chart 14. Compounds from *Senecio* Species (*64*)

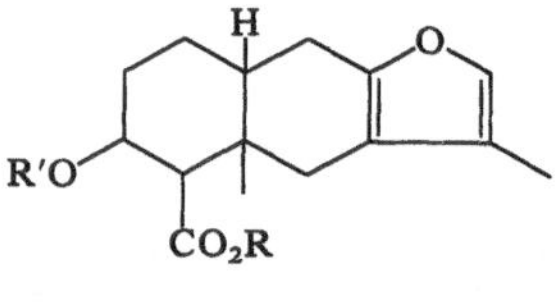

R = H, R' = angelyl
R = H, R' = a-methylacrylyl

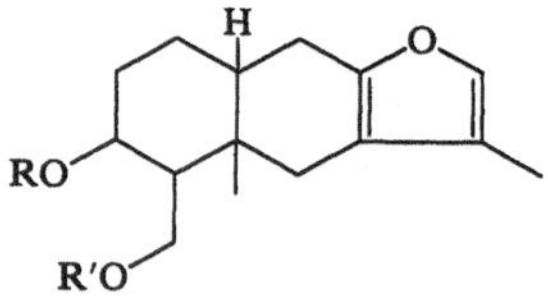

R = R' = angelyl
R = angelyl } or *vice versa*
R' = isovaleryl }

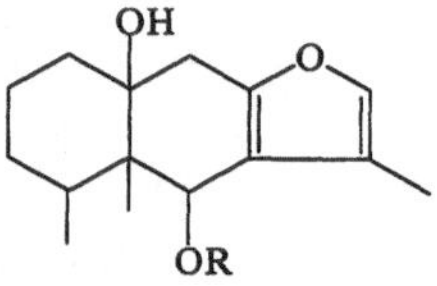

R = isovaleryl
R = angelyl
R = tiglyl

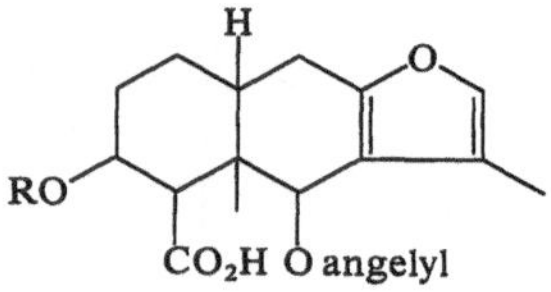

R = angelyl
R = senecioyl
R = isobutyryl

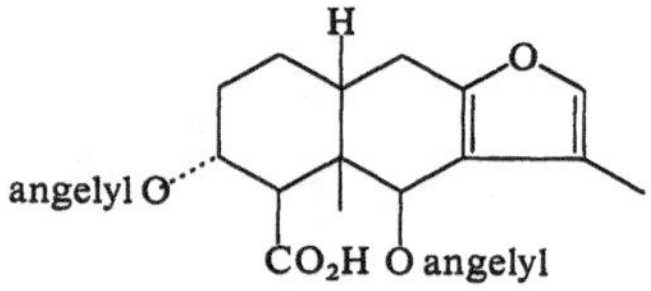

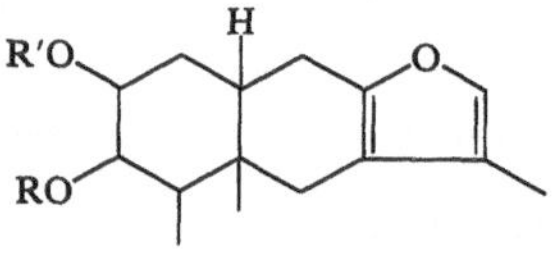

R = angelyl, R' = Ac
R = R' = angelyl

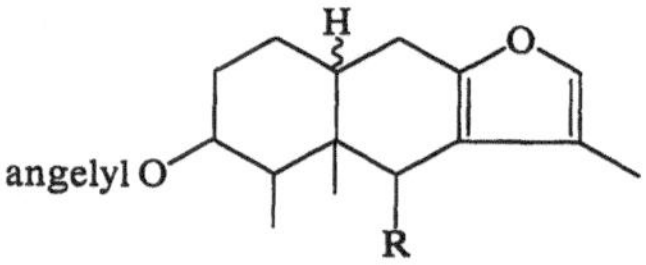

R = H
R = OAc
R = OH

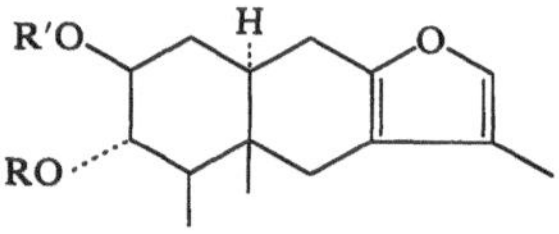

R = R' = angelyl

R = senecioyl, R' = angelyl

R = Ac, R' = angelyl

R = Ac, R' = senecioyl

Chart 15. Further Compounds from *Othonna* Species (*16*)

IV. 4-*epi*-Eremophilanes

Members of this group have the same carbon skeleton as eremophilane
(**7**) (see p. 83), but differ from it in that the methyl group at $C_{(4)}$ has the
α-configuration.

Capsidiol, $C_{15}H_{20}O_2$, is, to date, one of the two known members
of this family. It is an antifungal compound produced when sweet peppers

(*Capsicum frutescens* L.) (*122*) and tobacco plants (*123*) are inoculated with various fungi; it inhibits spore germination and mycelial growth. Capsidiol is crystalline and optically active. Two secondary alcoholic hydroxyl groups are present, as evidenced by the formation of an oily diacetate the CHOAc protons of which show characteristic downfield shifts in the n.m.r. spectrum in comparison with capsidiol itself. An isopropenyl group is responsible for a 3H singlet at δ 1.74, two 1H multiplets at δ 4.70 and 4.73 ppm, and a characteristic $C=CH_2$ i.r. band at 888 cm^{-1}. Catalytic hydrogenation afforded dihydrocapsidiol, $C_{15}H_{22}O_2$, which still contained a double bond; the n.m.r. spectrum revealed it to be trisubstituted. This bond is not conjugated with the isopropenyl one in capsidiol, since the compound is transparent to u.v. light above 215 nm.

A careful analysis of the n.m.r. spectrum of capsidiol, using spin decoupling, suggested that the two CHOH protons are coupled to the same CH_2 group: that is, the hydroxyl groups enjoy a 1,3-relationship. Oxidation of capsidiol (Jones reagent) gave a mixture of two keto-alcohols, one unconjugated and the other, named capsenone, conjugated (λ_{max} 238 nm, ϵ 6300). Capsenone is also formed when pepper fruit is inoculated with certain fungi; its u.v. spectrum is reminiscent of that of 3β-hydroxy-4-cholesten-6-one. On deuterium exchange capsenone captured two deuterium atoms, indicating the $C=O$ group is flanked by one CH_2. On mild treatment with acid capsenone was easily dehydrated to isocapsidienone, a cross-conjugated ketone, confirming that capsenone is a β-hydroxyketone and that capsidiol is a 1,3-diol.

The unconjugated keto alcohol (v_{CO} 1718 cm^{-1}) lost water extremely readily to give capsidienone, λ_{max} 290 nm (ϵ 12,200) (cf. 3,5-cholesta-dienone, λ_{max} 290 nm, ϵ 12,600). The same dienone resulted from somewhat more vigorous oxidation of capsidiol. Unexpectedly, the allylic hydroxyl group of capsidiol is here hardly attacked, which behavior suggests that this group is severely hindered sterically. This, view was corroborated by the formation of only a monomesylate on treatment of capsidiol with excess methansulfonyl chloride in pyridine; n.m.r. spectral study demonstrated that only the non-allylic OH group had

| capsidiol | nootkatane | aristolochene |
| (181) | (182) | (183) |

been mesylated. These studies and biogenetic considerations pointed to structure **(181)** for capsidiol *(124)*, excluding the stereochemistry at $C_{(7)}$. A detailed ^{13}C n.m.r. investigation confirmed that the vicinal methyl groups are *trans* (in contrast to all previously described eremophilanes). Further corroboration was provided by an X-ray diffraction analysis of capsidiol which also revealed that the isopropenyl group is equatorial and 7β *(125)*. For an account of experiments on the biosynthesis of capsidiol see p. 174.

V. Bicyclic 7-*epi*-Eremophilanes (Nootkatanes)

Members of the nootkatane family also have an eremophilane carbon skeleton, but differ from the eremophilane group in that the 3-carbon substituent at position 7 is *trans* to the methyl groups at positions 4 and 5. Further, almost all the structures are antipodal in stereochemistry at the latter positions, being derived from nootkatane **(182)**.

1. Hydrocarbons

Aristolochene, $C_{15}H_{24}$, has been isolated from the roots of *Aristolochia indica*. It shows i.r. absorption at 1648 and $810\,cm^{-1}$ (trisubstituted double bond), and at $886\,cm^{-1}$ ($C=CH_2$); apparently two double bonds are present, but these are not conjugated because the hydrocarbon shows only u.v. end absorption. Eudalene was formed on selenium dehydrogenation, pointing to a eudesmane or eremophilane carbon framework. The n.m.r. spectrum showed a doublet at δ 0.83 (3H, CHC$\underline{H}_3$), a singlet at 0.95 (3H, tertiary CH_3), a singlet at 1.70 (broad, 3H, allylic CH_3), and a singlet (broad, 2H, $=CH_2$) at 4.67 ppm. A broad doublet at 5.25 ppm (1H) is ascribed to a vinylic hydrogen of the

$$\text{type } CH_2-C\underline{H}=C\begin{smallmatrix}\diagup\ C \\[2pt] \diagdown\ C\end{smallmatrix}\quad ; \text{ the width at half height (8 Hz) of this signal}$$

pointed to an eremophilane type skeleton for the molecule. Selective catalytic hydrogenation afforded dihydroaristolochene, which showed no signals at 1.70 and 4.67 ppm (see above) in its n.m.r. spectrum, but the signal at 5.25 ppm persisted. Total hydrogenation afforded a mixture of two saturated hydrocarbons, $C_{15}H_{28}$, separable by preparative g.l.c. One of these (the *trans*-decalin) proved to be identical with a hydrocarbon (*trans*-nootkatane) obtainable from valencene, an isomeric hydrocarbon of known structure and configuration (see below). Aristolochene must consequently be represented by expression (183). The placing of the cyclic double bond 9,10 was a consequence of the result of sequential hydroboration-oxidation and oxidation of dihydro-aristolochene (see above); this led to a *trans*-1-decalone (184) with a positive Cotton effect which enabled it to be distinguished from the isomeric (185) (*126*).

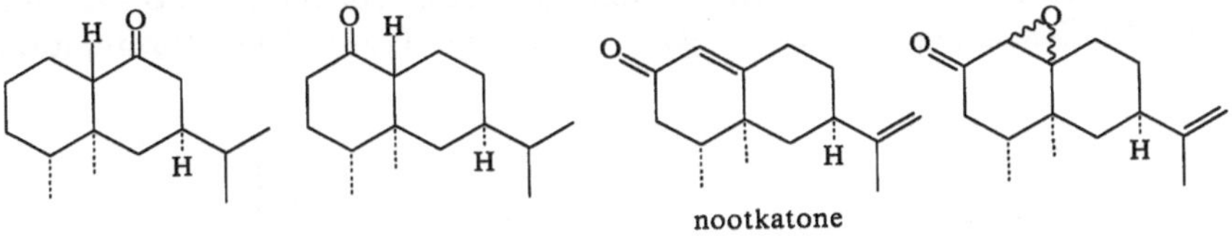

nootkatone

(184)   (185)   (186)   (187)

Nootkatene, $C_{15}H_{22}$, is a hydrocarbon occurring in the heartwood of Alaska yellow cedar *(Chamaecyparis nootkatensis)* (*127*) and in vetiver oil (*127a*). On total catalytic hydrogenation it afforded a saturated hydrocarbon $C_{15}H_{28}$ and must therefore contain three double bonds and be bicyclic. Selenium dehydrogenation gave eudalene. Nootkatene has two of its olefinic bonds conjugated (λ_{max} 235 nm, ε 13,500), and one is present as a $C=CH_2$ group (v 890 cm^{-1}); the latter deduction was confirmed by the formation of formaldehyde on ozonolysis (*127*). Because of the relatively short wavelength u.v. maximum the conjugated pair of double bonds is heteroannular. No further structural studies on the compound have been reported, but nootkatone of structure and absolute configuration (186) (see p. 149), has been converted into nootkatene in three simple steps. Treatment with alkaline hydrogen peroxide afforded 1,10-epoxy-1,10-dihydronootkatone (187), which with hydrazine (Wharton-Bohlen reaction) afforded the allylic alcohol (188). This when dehydrated with phosphoryl chloride-pyridine gave a triene identical in all respects with nootkatene, which must therefore be formulated as (189) (*128*). This formulation has been confirmed by synthesis from valencene (see below).

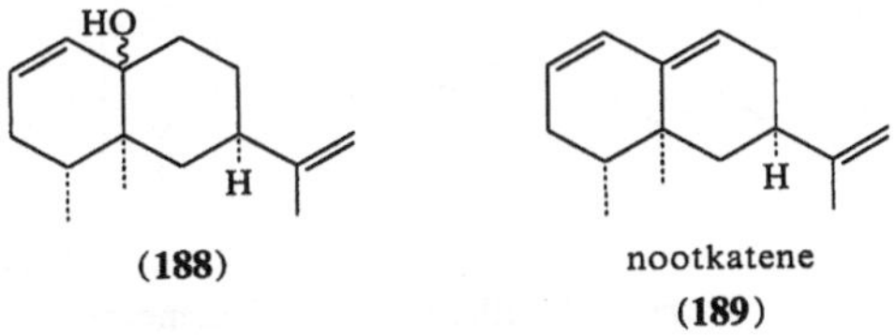

(188)   nootkatene
(189)

Valencene, $C_{15}H_{24}$, has been isolated from several sources. It is present in orange juice and peel oil (*129, 130*), in citrus flowers (*131*), in cold-pressed citrus oils (*132*), in camphor oil (*133*), and in vetiver oil (*127a*). Its spectral properties have been recorded (*130, 133*), and have revealed units such as tertiary, allylic, and secondary methyl groups, but the most valuable structural information has been the observation that on oxidation with *t*-butyl chromate valencene yielded nootkatone (**186**), of established structure and configuration (see p. 149) (*130, 133*). Valencene must thus be represented by (**190**). Conversely, nootkatone on Wolff-Kishner reduction gave valencene (*128*). This formulation

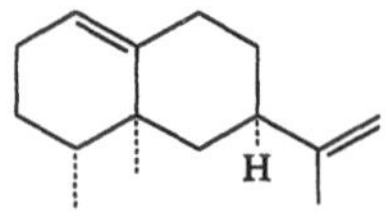

valencene	valerianol	(**192**)
(**190**)	(**191**)	

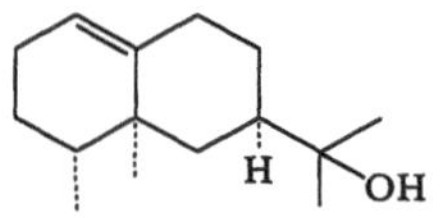

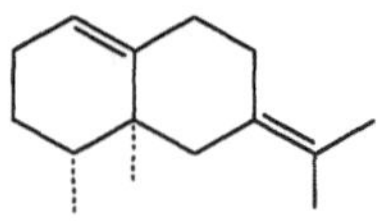

has been confirmed by the total synthesis of (±)-valencene by de-hydration of (±)-valerianol (**191**) (see below) of known structure and configuration with thionyl chloride in pyridine, which led to a mixture of valencene (**190**) and the isomeric diene (**192**) in 3:1 ratio.

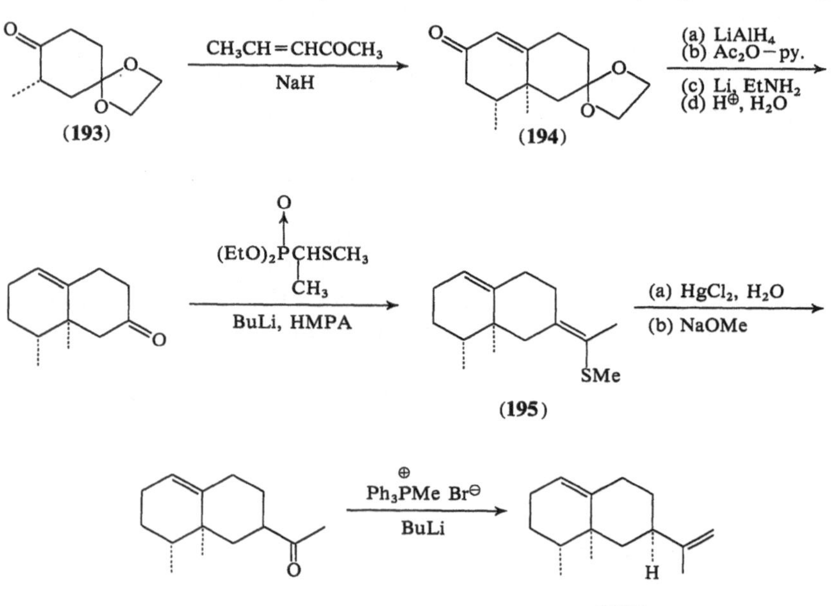

Chart 16. Synthesis of (±)-Valencene

These were separated by preparative g.l.c. (*12*). The total synthesis of vale-rianol is outlined on p. 148). Another synthesis of racemic valencene is outlined in Chart 16 (*134*). The starting keto-ketal (**193**) was available from furfural (*135, 136*). The *cis* disposition of the methyl groups in the annelation product (**194**) was established by a comparison of the effect of a shift reagent on its n.m.r. spectrum with that on the spectrum of nootkatone (**186**) of known stereochemistry. Compound (**195**) was secured by a Wadsworth-Emmons reaction (*137*) with diethyl 1-methyl-thioethylphosphonate (*138*).

Some oxidation studies on valencene have been described (*139, 140*); these have led to a synthesis of nootkatene (**189**) (see p. 145). The microbial oxidation of valencene has been found to yield dihydroagaro-furan, involving a methyl migration to generate a eudesmane skeleton (*141*).

β-vetivenene
(**196**)

γ-vetivenene
(**197**)

valerianol
(**191**)

β-Vetivenene, $C_{15}H_{22}$, is the major sesquiterpene hydrocarbon component of vetiver oil. It contains three double bonds, and its nootkatane skeleton was settled by hydrogenation studies. Its u.v. absorption [λ max 228.5 (ε 19,450), 263.5 (19,600), and 245 nm (inflexion, 13,300)] indicates that two of the double bonds are conjugated and heteroannular. An analysis of its n.m.r. spectrum revealed the locations of the double bonds and enabled structure (**196**) to be advanced for the hydrocarbon. Of interest is the low-field (δ 2.87 ppm) signal due to the doubly allylic CH_2 protons at position 8 (*127a*). The absolute configuration depicted is supported by the negative trend of its o.r.d. curve, as with nootkatene.

γ-Vetivenene, $C_{15}H_{22}$, also found in the same source, is another nootkatriene, which has two of its bonds conjugated and not homo-annular (λ max 234.5 nm, ε 21,400). The n.m.r. spectrum shows signals corresponding to angular (δ 1.04), secondary (0.92 d), and allylic (1.92) methyl groups, and also a $C=CH_2$ (4.91, 5.03) group. The spectrum best fits either of the two structures represented by (**197**), the 9,10-location for the third double bond being preferred on o.r.d. evidence and the nootkatane skeleton being established by catalytic hydrogenation. However, the protons at position 8 resonate at δ 2.32 ppm, which is at unexpectedly high field for doubly allylic protons (contrast β-vetivenene) (*127a*).

2. Alcohols

Valerianol, $C_{15}H_{26}O$, is found in the roots of *Valeriana officinalis* L.; it is optically active and forms a crystalline 3,5-dinitrobenzoate. On dehydrogenation it yields eudalene. Its i.r. spectrum shows bands at 3440, 3580 (OH), 1670 (C=C), 1375, and 1385 cm^{-1} (CMe$_2$). The n.m.r. spectrum revealed an olefinic H (δ 5.30, broad), geminal CH$_3$ groups (1.15), a hydroxyl group (2.42, confirmed by D$_2$O exchange), a tertiary CH$_3$ (0.92), and a secondary CH$_3$ group (0.90 ppm, m). The hydroxyl group is tertiary in its chemical behavior, and not allylic because it is not susceptible to hydrogenolysis. One double bond was detected by quantitative catalytic hydrogenation; it is not part of a C=CH$_2$ group on i.r. spectral evidence. However, when valerianol was dehydrated with phosphoryl or thionyl chloride a mixture of two separable dienes was obtained (3 : 1). Neither product was a conjugated diene, and the major one had a strong band at 886 cm^{-1} in its i.r. spectrum (C=CH$_2$). The minor diene had no such band and analysis of their n.m.r. spectra left no doubt that these hydrocarbons were isopropenyl-isopropylidene isomers. The alcohol function in valerianol is therefore attached to the tertiary position of an isopropyl group. The major diene proved to be identical with valencene (**190**) (see p. 146) in consonance with the view that valerianol has an eremophilanoid carbon framework, and the alcohol may consequently be formulated as (**191**) (*16a, 63*).

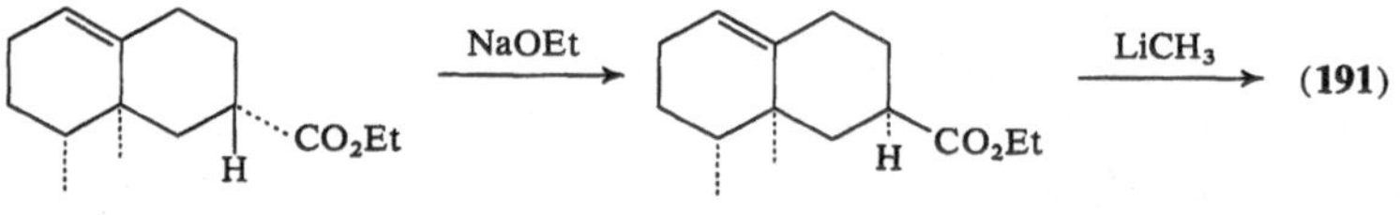

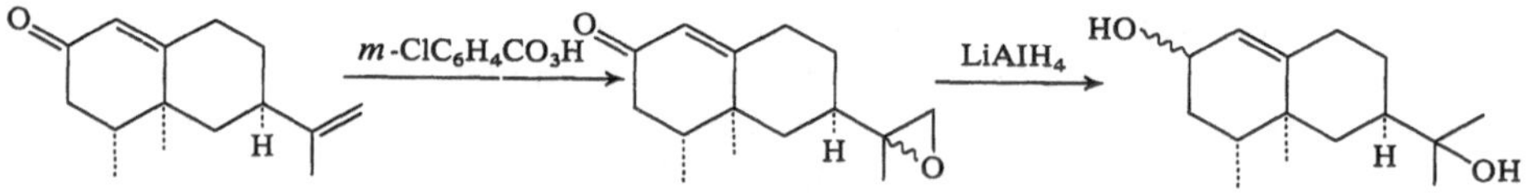

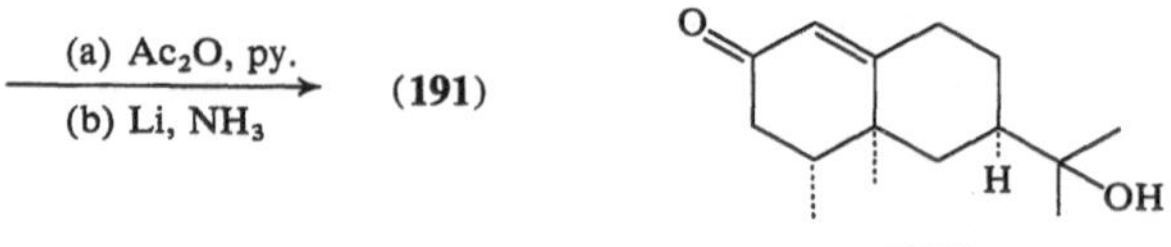

A total synthesis of ($\pm$)-valerianol, outlined below, corroborates this structure (*12*). The starting material, ester (**15**), was secured as described in Chart 1 (p. 85); treatment with base transformed it into the more stable (**198**) (CO$_2$Et equatorial). Reaction of this ester with methyllithium gave ($\pm$)-valerianol (191) in 84% yield (12). A partial synthesis of natural (+)-valerianol from natural (+)-nootkatone (186) proceeds as follows (45)

A compound called kusunol, isolated from camphor blue oil, proved to be identical with valerianol. Oxidation with t-butyl chromate gave ketol (199); this on dehydration with phosphoryl chloride-pyridine gave a mixture of nootkatone (186) (see below) and α-vetivone (p. 153), proving that (191) represents the absolute configuration of (+)-valerianol (142).

Bicyclovetivenol, $C_{15}H_{24}O$, is a further constituent of vetiver oil. It is an alcohol (v_{OH} 3400 cm^{-1}) containing two double bonds, which is oxidized smoothly by activated manganese dioxide to an α,β-unsaturated aldehyde (λ_{max} 256 nm). It is consequently an allylic primary alcohol. One of the double bonds is trisubstituted and the other tetrasubstituted on n.m.r. spectral evidence; further, the appearance of the =CH proton signal in the aldehyde at δ 5.40 ppm signifies that the C=CH group is not conjugated with aldehyde group, which must consequently be conjugated with the tetrasubstituted olefinic bond. The aldehyde with ethanedithiol afforded a dithioacetal, which on desulfurization with nickel yielded a diene (200). This same diene was also formed when α-vetivone (isonootkatone, see p. 153), of known constitution

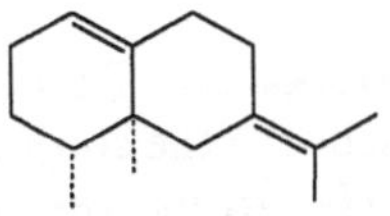

(200)

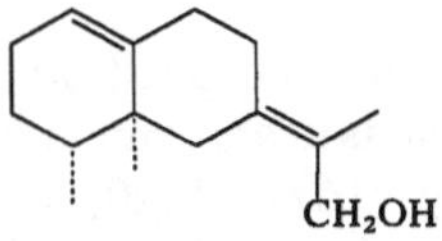

bicyclovetivenol
(201)

and stereochemistry, was similarly converted to its dithioketal and desulfurized. The structure of (200) is thus secure. Bicyclovetivenol must be represented by (201), the geometry of the exocyclic double bond remaining to be settled (143).

3. Ketones

Nootkatone, $C_{15}H_{22}O$, is a key substance in the nootkatane sesquiterpene group. It was first isolated from the heartwood of Alaska yellow cedar (Chamaecyparis nootkatensis) (144) (cf. valencene), and later from the peel oil and juice of grapefruit (Citrus paradisi); it also occurs in small amount in oils of bergamot, lemon, lime, orange, and tangerine (132). It is, along with valencene (see p. 146), responsible for the delicate flavor of citrus and is of commercial importance as a flavor additive in the soft drink industry. Nootkatone may be purified *via* its crystalline semicarbazone, and when pure is a low-melting, optically active, crystalline solid. Its u.v. absorption (λ_{max} 238 nm, ε 15,000) is that of an α,β-unsaturated ketone. An additional, unconjugated double bond is present, since on catalytic reduction tetrahydronootkatone, $C_{15}H_{26}O$, was formed. This could be reduced further to tetrahydronootkatol, $C_{15}H_{28}O$, a crystalline secondary alcohol. Ozonolysis of nootkatone afforded formaldehyde. Borohydride reduction gave nootkatol, the corresponding secondary alcohol, which on selenium dehydrogenation afforded eudalene.

The position of the keto group in the eudesmane or eremophilane bicyclic system was established by reaction of nootkatone with methylmagnesium iodide, which gave a methyl carbinol yielding on similar dehydrogenation 7-isopropyl-1,3-dimethylnaphthalene (**202**). The keto group of

nootkatone

(**202**)   (**186**)   (**203**)

nootkatone is thus at position 2 in the nootkatane framework (*144*). The i.r. spectrum is in agreement with the presence of the conjugated ketone functionality (ν 1672, 1620) and a $C=CH_2$ group (895 cm^{-1}). The n.m.r. spectrum is consistent with the presence of secondary (δ 0.9 d), tertiary (1.1), and allylic (1.7 ppm) methyl groups, with a $C=CH_2$

group (4.7 d), and with a $\overset{\displaystyle C}{\underset{\displaystyle C}{\diagdown\diagup}}C=CHCO$ (5.6) unit. The striking resem-

blance between the o.r.d. curves of nootkatone and tetrahydronootkatone and cholest-4-en-3-one and cholestanone respectively, pointed to an eremophilane rather than a eudesmane carbon skeleton for the ketone. Further, the addition of a drop of concentrated acid to a methanol solution of tetrahydronootkatone caused a marked reduction in the amplitude of the Cotton effect, so that the $C_{(4)}$-methyl group must be equatorial and α. Structure (**186**) was therefore advanced for the ketone, the 7β isopropenyl orientation being revealed by conversion of tetrahydronootkatone by reduction into hydrocarbon (**203**) [(+)-nootkatane]. The latter proved to be identical with but antipodal to the hydrocarbon

(204), obtainable in a similar fashion from ketone (205), derived from hydroxyeremophilone (p. 101) and of known absolute configuration (30, 31, 39).

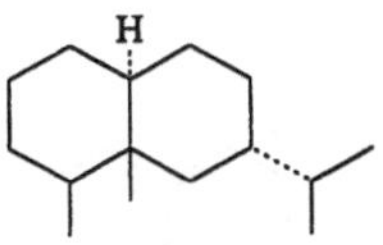

(204) 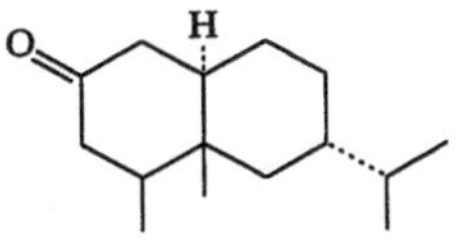   (205)

Several syntheses of (±)-nootkatone which corroborate the structure and relative configuration depicted in (186) have been described. The first, due to Pesaro, Bozzato, and Schudel (145), is outlined in Chart 17.

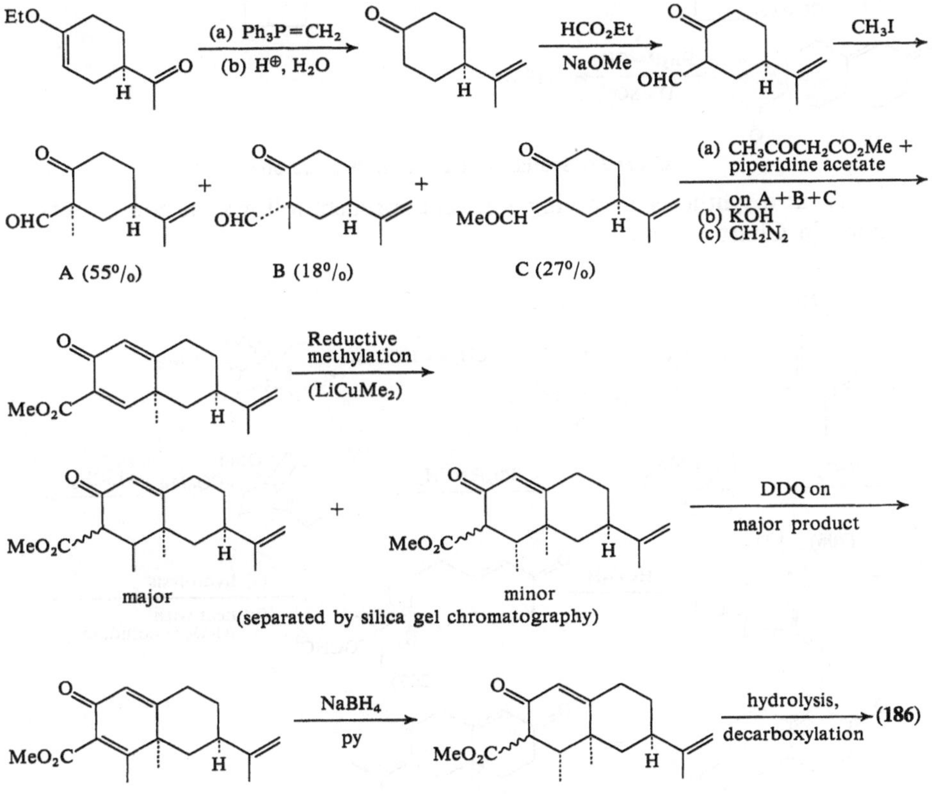

Chart 17. Synthesis of Racemic Nootkatone

Another synthesis (146, 147) is summarized in Chart 18.

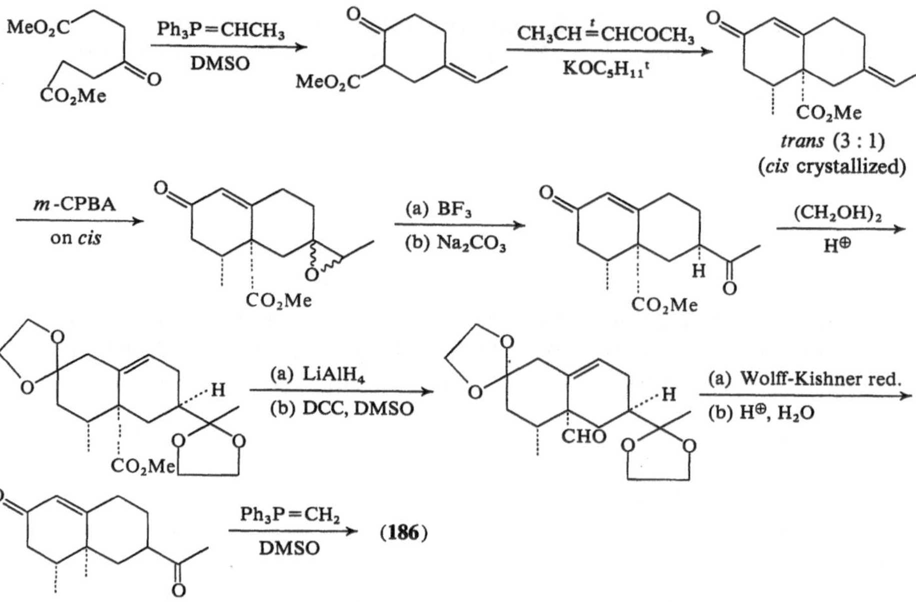

Chart 18. Synthesis of Racemic Nootkatone

A third synthesis, involving a novel approach, follows the course out-lined in Chart 19 (148, 149).

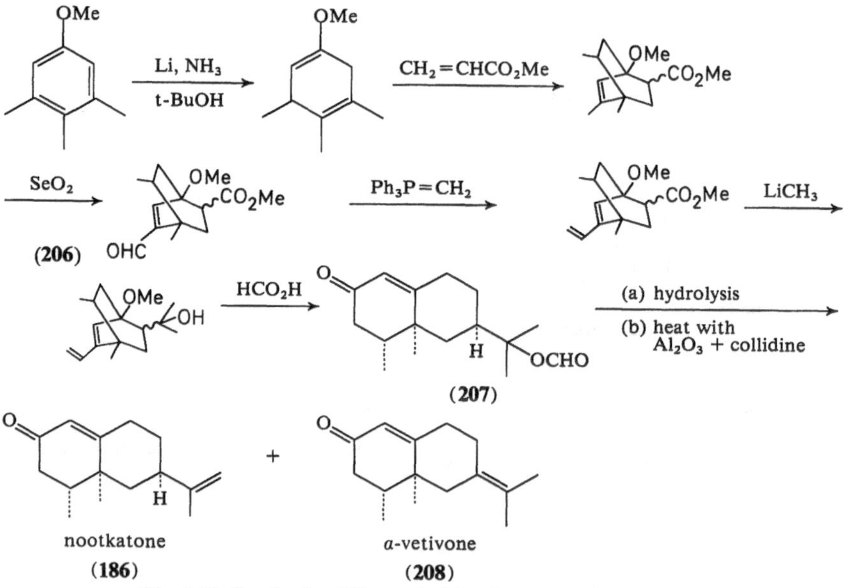

Chart 19. Synthesis of Racemic Nootkatone and α-Vetivone

The stereospecific transformation of (206) to (207) is rationalized as follows

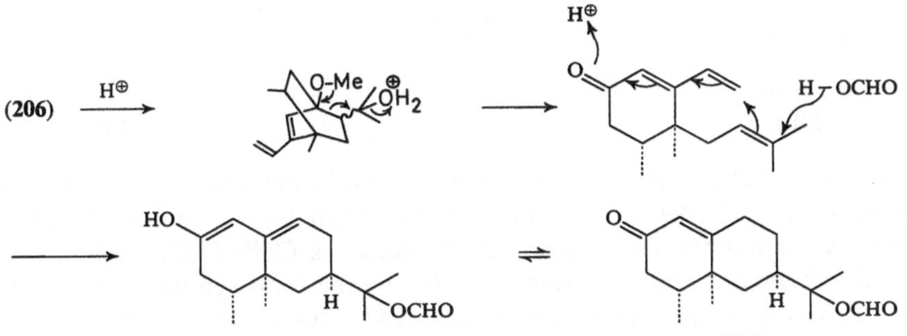

The final product consisted of a mixture of nootkatone (186) and α-vetivone (208) (see below), separable by preparative g.l.c.

Additional syntheses have been recorded (37, 134), and a partial synthesis of (−)-nootkatone from (+)-sabinene has been described (149a).

A conformational analysis of nootkatone has been made by n.m.r. spectral analysis with recourse to a shift reagent. As a result the molecule can be represented in greater detail as (209) (150). The photolysis of nootkatone in methanol has been investigated; it gives rise to several products including 1,10-dihydronootkatone and "photonootkatone" (210) (151, 152, 153).

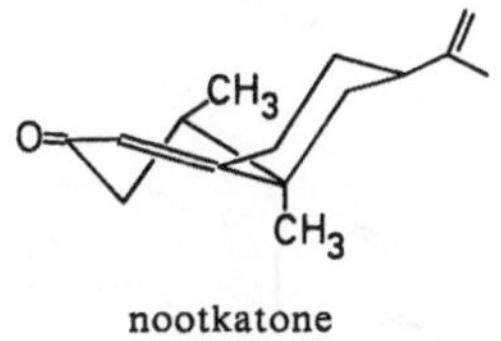

nootkatone
(209)

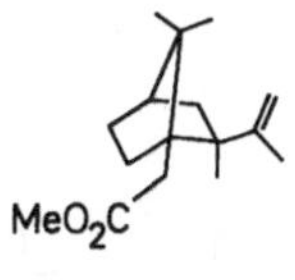

"photonootkatone"
(210)

α-Vetivone (Isonootkatone), $C_{15}H_{22}O$, (208), known for many years as a constituent of vetiver oil (154), is of particular interest because until recently it was generally believed to have a hydroazulenoid structure (155, 156). However n.m.r. spectral measurements proved not to be in agreement with such a formulation (157, 158, 159) and the total reduction product, vetivanol, prepared from carefully purified α-vetivone, did not yield any vetivazulene on dehydrogenation. The following sequence of reactions was carried out, starting with eremophilone, of known structure and absolute configuration (p. 88). Simultaneously,

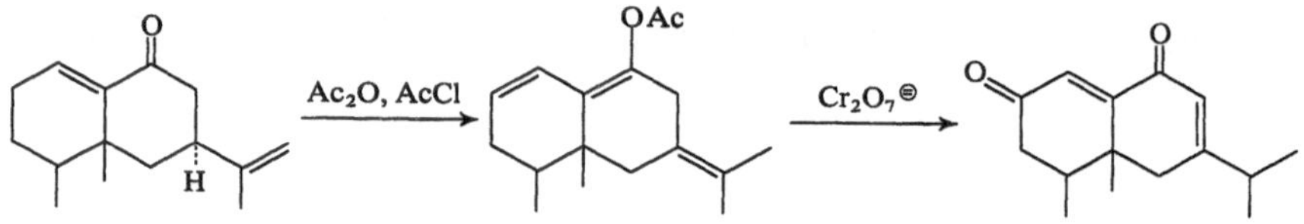

eremophilone
(29)

(211)

α-vetivone (208) was subjected to aerial oxidation under alkaline conditions and afforded a diketone identical with but antipodal to (211). α-Vetivone has therefore been reformulated as (208) (157, 158).

Synthetic confirmation was provided by the total synthesis of (±)-α-vetivone summarized in Chart 20 (160, 161); see also Chart 19.

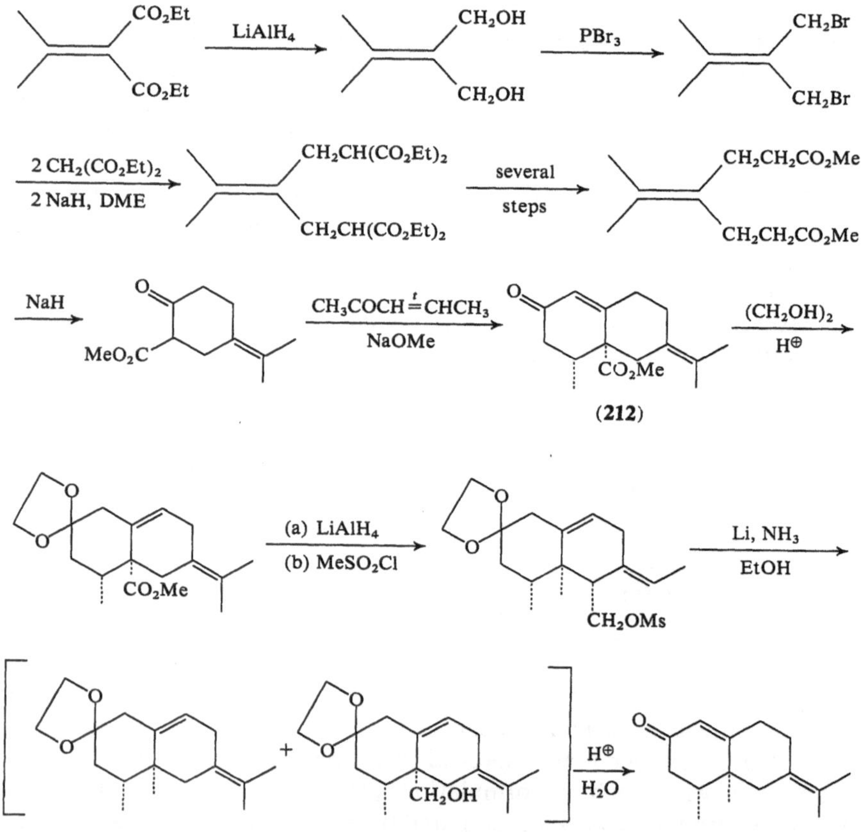

(212)

a-vetivone
(208)

Chart 20. Synthesis of (±)-α-Vetivone

The annelation step proceeded stereospecifically to give the single *cis* isomer (212), and the final product was separated by silica gel chromatography.

A partial synthesis of α-vetivone by acid treatment of nootkatone (186) has been described (*162*).

Nardostachone, $C_{15}H_{22}O$, is a ketone which occurs in spikenard (*Nardostachys jatamansi*) oil. Its the structure has not yet been settled unequivocally. Catalytic hydrogenation yielded tetrahydronardostachone identical with tetrahydronootkatone. Its spectral behavior has been interpreted (*163, 164*) in terms of formulation (213) but its u.v. maximal

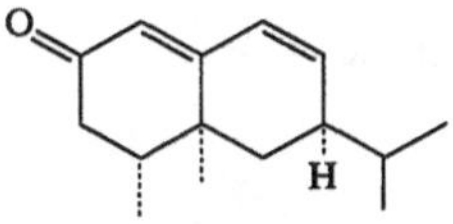

(213) 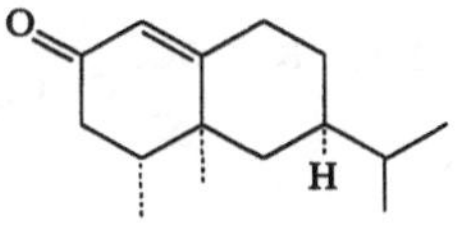   (214) 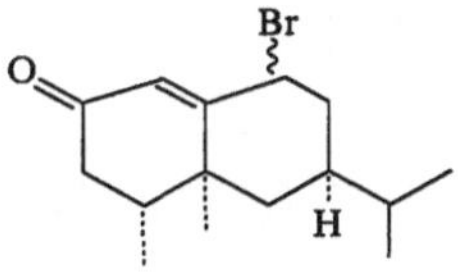   (215) 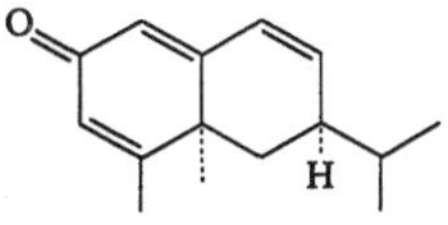   (216)

absorption (235 and 298 nm, ε 7478 and 15,200 respectively) is not in agreement with this structure (λ_{max}^{calc}; 280 nm) nor is the principal maximum for the semicarbazone (317 nm, ε 25,050) (*163*) consistent with (213) for the ketone.

A synthesis of compound (213) has been effected from nootkatone (186) by sequential regiospecific reduction to 11,12-dihydronootkatone (214), allylic bromination to (215) and dehydrobromination to (213). The latter (λ_{max} 283 nm, ε 18,500) differed noticeably in its spectral properties from nardostachone; further, the synthetic ketone semicarbazone was different from that of nardostachone (*45, 165*). The same ketone (213) was also prepared from nootkatone (186) by reduction followed by chloranil dehydrogenation, and by acid-catalyzed isomerization of nootkatone (*166*). Structure (216) was proposed as an alternative (*45, 165*); this, too has been synthesized (*166*) and found to have the desired u.v. maximum, but to differ from nardostachone in its n.m.r. spectrum. A third proposal (217) (*166*) remains to be evaluated.

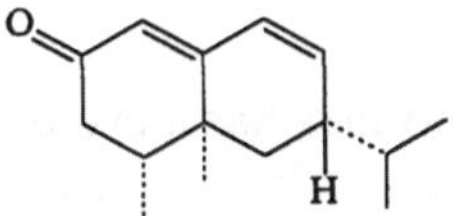

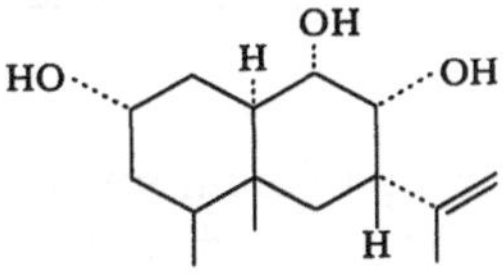

lateriflorol

(217)   (218)

4. Esters

Three esters with a nootkatane skeleton have been isolated from *Euryops lateriflorus* by BOHLMANN and his collaborators (75). All are derived from the alcohol (218), termed lateriflorol, and related to nootkatone but antipodal to it. One of these, $C_{21}H_{34}O_4$, (219) yielded

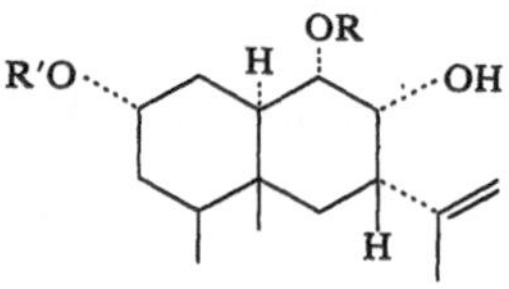

(219) R = CO$\diagdown$C=C$\diagup$Et, R' = H
              H   CH$_3$

(220) R = COCH=CMe$_2$, R' = H

(221) R = OC$\diagdown$C=C$\diagup$CH$_3$, R' = Ac
              H$_3$C   H

on hydrolysis lateriflorol (218) and 3-methyl-2-pentenoic acid as a mixture of (E)- and (Z)-isomers. Lateriflorol also resulted from lithium aluminum hydride reduction of the ester. On periodate oxidation lateriflorol reacted rapidly to yield a dialdehyde, indicating the presence of a *cis* 1,2-glycol unit of the type $-CH(OH)-CH(OH)-$. A combination of i.r., n.m.r., and mass spectral analysis allowed structure (218) to be assigned to lateriflorol and (219) to the ester. The second ester, $C_{20}H_{32}O_4$, also yielded lateriflorol on lithium aluminum hydride reduction; spectroscopic behavior indicated a senecioic acid residue was present and structure (220) was proposed. Finally, the third ester also afforded lateriflorol. Spectral study revealed it to be a diester, the acid moieties being acetic and angelic. Hydrolysis removed the acetyl group yielding the angelate mono-ester, and also the isomeric 2-acetate-8-angelate, formed by base-catalyzed $O_{(9)} \rightarrow O_{(8)}$ migration of the angelyl group. Again spectroscopy played a vital role in enabling structure (221) to be proposed (75).

VI. Tricyclic 7-*epi*-Eremophilanes

A small group of eremophilanoid compounds has a cyclopropane ring fused across positions 6 and 7. Stereochemically the members are all related to nootkatane or its enantiomer and have the carbon framework shown in (222).

References, pp. 175—186

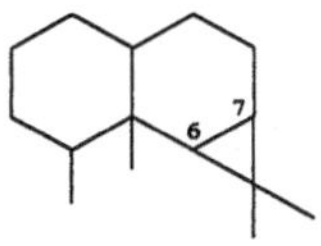

calarene

(222) 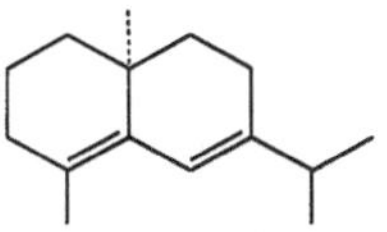 (223) (224) (225)

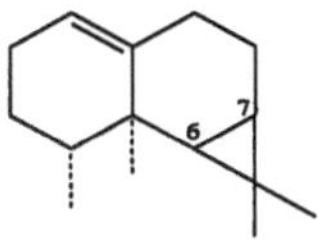

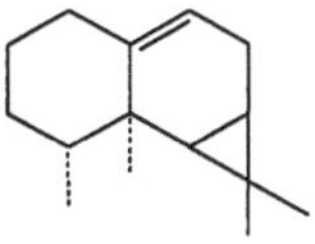

1. Hydrocarbons

Calarene, $C_{15}H_{24}$, is fairly widely distributed in nature. It was first encountered as long ago as 1914 when SEMMLER and JAKUBOWICZ isolated it (under the name β-gurjunene) from gurjun balsam (*Diptero-carpus* spp.) (*167*). More recently it has been found in oil of sweet flag (*Acorus calamus* L.) (*168*), in Chinese spikenard oil (*Nardostachys jata-mansi*) (*169*), and in Indian valerian root oil (*Valeriana wallichi*) (*170*). Catalytic hydrogenation afforded calarane, $C_{15}H_{26}$, a saturated hydro-carbon; calarene must consequently be tricyclic. That one of the rings present is cyclopropanoid was revealed by a multiplet (2H) in the n.m.r. spectrum in the high field region (δ 0.60 ppm) characteristic of cyclopropane protons. A multiplet at 5.25 ppm (1H) indicated a single vinylic proton and singlets at 0.98, 1.02, and 1.08 (each 3H) were assignable to tertiary methyl groups. Finally a doublet at 0.97 (3H) was due to a CH_3CH group. Treatment of calarene with formic acid at 100° for a prolonged period furnished a conjugated dienic hydro-carbon found to be identical with the dehydration product of maaliol, and formulated as (223). These results, coupled with biogenetic con-siderations, point to two possible double bond isomeric structures for calarene, (224) and (225).

The correctness of the former was demonstrated as follows. The angular methyl group must be α, to accommodate the formation of (223). On the assumption that the cyclopropane ring is 6β,7β hydroboration-oxidation of calarene would be expected to yield (226) (stereospecific α-attack) if the double bond is 1,10, or (227) if it is 9,10. Models of calarenes of these types show that only one two-chair conformation is possible (instead of steroidal and nonsteroidal possibilities with most

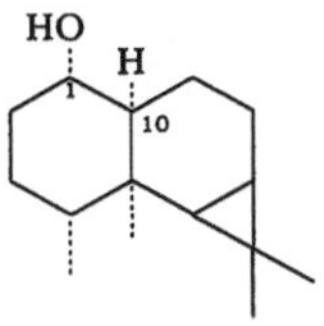

(226) 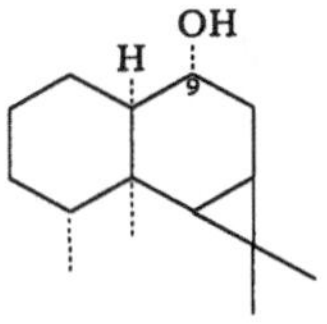 (227)

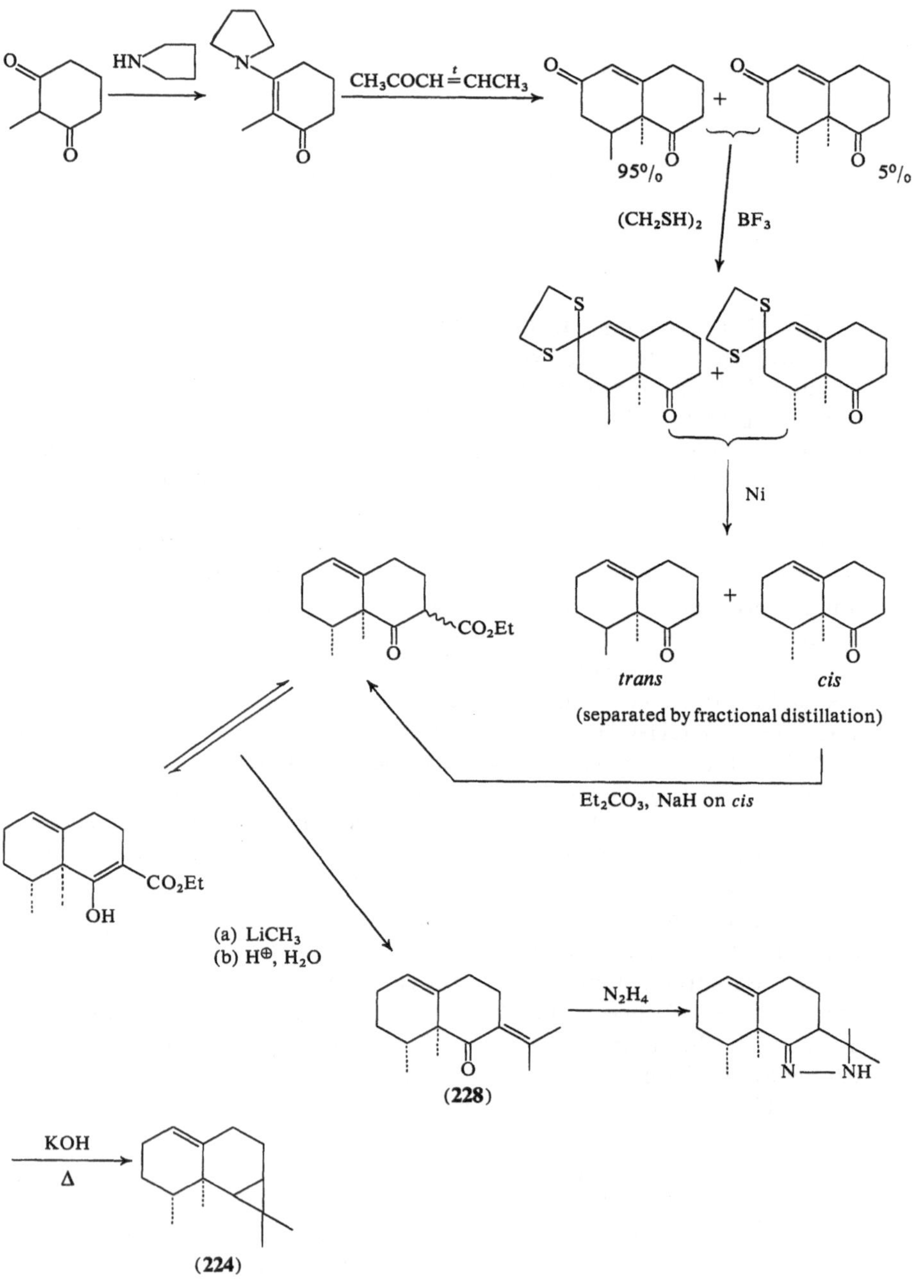

Chart 21. Synthesis of (±)-Calarene

cis-decalins). These models further reveal that in (226) the OH group is axial while in (227) it is equatorial. In fact i.r. measurements on the alcohol and its derived acetate show that the group is axial. The alcohol must consequently be (226) and calarene (224). The hitherto assumed 6β,7β configuration for the cyclopropane ring was proved correct by a link-up with epimaaliol of known structure and configuration.

This structural assignment is supported by detailed n.m.r. analysis, by mass spectrometric studies, by biogenetic considerations (*169, 171, 172, 173*), and also by a total synthesis of (±)-calarene (*14, 15*), summarized in Chart 21.

In the formation of (228) the ester group is regiospecifically attacked by the methyllithium, the first mole of reagent serving to convert the β-keto ester precursor to its enolate anion with protection of the keto-group against attack.

α-Ferulene, $C_{15}H_{24}$, occurs in the neutral fraction of the latex of *Ferula communis* L. Dehydrogenation yielded eudalene and catalytic hydrogenation gave a saturated dihydro-derivative, $C_{15}H_{26}$. α-Ferulene is thus tricyclic. Its n.m.r. spectrum showed a broad olefinic H signal at δ 4.93 (1H), and signals at 2.05 (broad, 4H, 2 allylic CH_2 groups), 1.0 (12H, broad, 4 methyl groups), and 0.61 ppm (2H, multiplet). The last signal is diagnostic of the presence of a cyclopropane ring. α-Ferulene was easily converted by aerial oxidation to a ketone $C_{15}H_{22}O$ which on Wolff-Kishner reduction regenerated α-ferulene. The i.r. spectrum of this ketone was exactly superimposable on that of aristolone (229), of known structure and configuration (see p. 161).

aristolone	α-ferulene [(+)-9-aristolene]	1,10-aristolene	(−)-9-aristolene
(229)	(230)	(231)	(232)

The two ketones proved to have identical physical properties but equal and opposite optical rotations and therefore have an antipodal relationship. Further, α-ferulene and its dihydro-derivative are similarly related to 9-aristolene (see below) and its dihydro-derivative (aristolane or calarane) respectively. Therefore structure (230) may be written for α-ferulene, which belongs to the antipodal nootkatane series (*174*).

1,10-Aristolene, $C_{15}H_{24}$, has been isolated from the sesquiterpene hydrocarbon fraction of the gorgonian *Pseudopterogorgia americana*, along with the isomeric (+)-9-aristolene. Comparisons were made be-

tween them and the known calarene and α-ferulene respectively. 1,10-
Aristolene proved to be identical with but antipodal to calarene (224),
and must therefore be formulated as (231), while (+)-9-aristolene was
found to be identical with α-ferulene (230) (*175*). (−)-9-Aristolene
(232) occurs in the roots of *Aristolochia debilis,* along with calarene
and (+)-1,10-aristolene·(*176*), and in *Nardostachys jatamansi* oil (*176a*).

2. Alcohols

"Calarenol", $C_{15}H_{24}O$, has been isolated from the roots of
Nardostachys jatamansi (black brown variety); it proved to be an
inseparable mixture of two isomers. A hydroxyl function was responsible
for an i.r. spectral band at 3336 cm⁻¹, with trisubstituted C=C bands
at 1665 and 847 cm⁻¹. The n.m.r. spectrum revealed a multiplet at
δ 5.55 (1H, C=CH), a singlet at 1.08 (CH₃C−OH), and signals at
1.0 and 0.96 ppm (9H) for 3 tertiary methyl groups. Two cyclopropanoid protons resonated at 0.51 ppm. Quantitative catalytic hydrogenation resulted in uptake of one mole of H_2 and formation of
saturated dihydrocalarenol, $C_{15}H_{26}O$, which proved to be homo-
geneous and which must be tricyclic. The hydroxyl group in calarenol
is, from its chemical behavior, tertiary. Selenium dehydrogenation yields
eudalene.

A key to the structures was the observation that when dihydro-
calarenyl acetate was pyrolyzed two alkenes were obtained; catalytic
hydrogenation of this mixture afforded a saturated hydrocarbon identified with calarane, the known reduction product of calarene (224)
(see p. 157). Dihydrocalarenol may therefore be formulated as (233).
The pyrolysis of calarenyl acetate afforded a mixture of four unsaturated

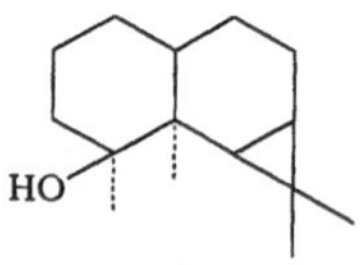

(233) 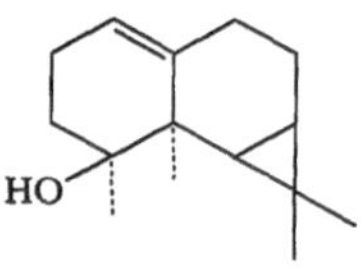   (234) 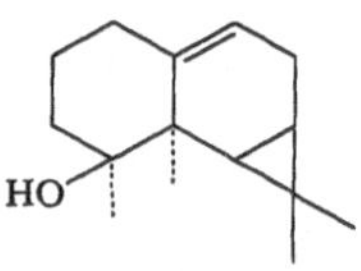   (235) 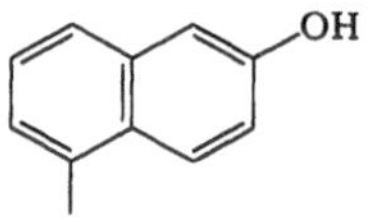   (236)

hydrocarbons, some containing a C=CH₂ group on i.r. spectral
evidence. A partial catalytic reduction of this mixture resulted in a
preferential attack on this group and the product was found to contain
some calarene (224). Evidently one of the components of "calarenol"
is the alcohol (234). On the basis of the evidence outlined the other
component is probably the double bond isomer (235) (*163, 177*).

3. Ketones

Aristolone, $C_{15}H_{22}O$, is a crystalline levorotatory constituent of *Aristolochia debilis* roots. It formed the familiar ketone functional derivatives and the carbonyl band in its i.r. spectrum (1667 cm^{-1}) indicated the compound was an α,β-unsaturated ketone (*178*). Catalytic hydrogenation afforded dihydroaristolone, a saturated ketone, so that aristolone must be tricyclic. The u.v. spectrum of aristolone suggests that the keto group is cross-conjugated with a double bond and a cyclopropane ring; this is confirmed by similar studies on dihydroaristolone and aristolol (the secondary alcohol corresponding to aristolone). Wolff-Kishner reduction of aristolone afforded the hydrocarbon deoxyaristolone. Dehydrogenation (Se) of aristolone gave 5-methyl-2-naphthol (**236**), and of deoxyaristolone and aristolol gave 1-methylnaphthalene (*179*). Permanganate oxidation of aristolone and deoxyaristolone yielded aristoic acid, $C_{15}H_{20}O_5$, which was transformed by acid into crotonic acid and 5-hydroxy-3-isopropyl-2-methyl-benzoic acid (**237**) (*180, 181*). Aristoic acid is formulated as (**238**), its acid-catalyzed cleavage being

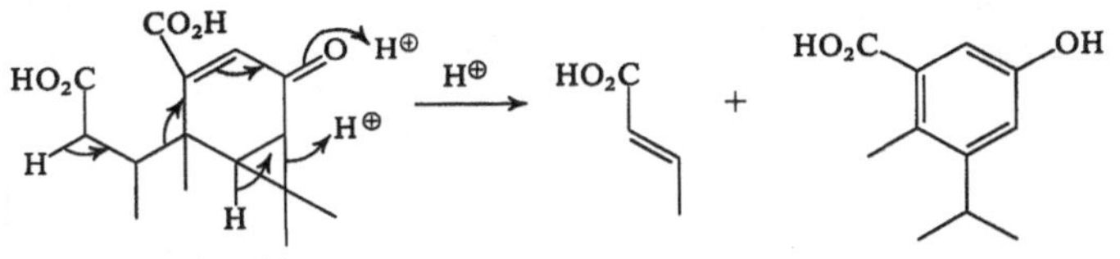

aristoic acid
(**238**)

(**237**)

rationalized as shown. Aristolone, from which aristoic acid is formed by an unanticipated oxidation of the cyclohexane ring, is assigned structure (**239**) (*181*). The stereochemistry depicted in this formula has emerged

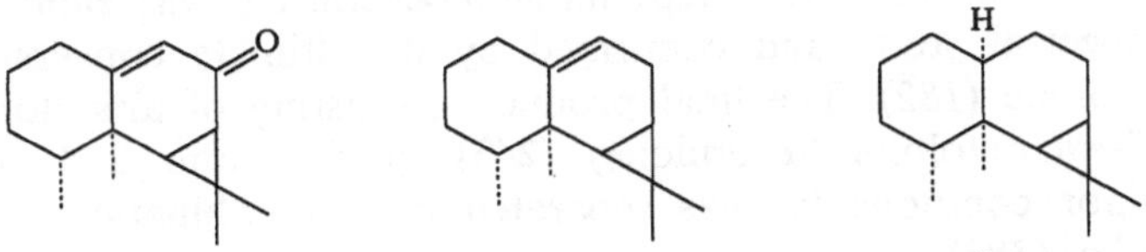

aristolone
(**239**)

(**240**)

calarane
(**241**)

from a correlation with calarene (**224**) (see p. 157). Raney nickel desulfurization of aristolone ethylenedithioketal afforded deoxyaristolone (**240**), different from calarene. However, on catalytic hydrogenation this hydrocarbon gave a product identical with calarane (**241**), the catalytic reduction product of calarene, of known absolute stereochemistry (*169*).

The gross structure (239) and the relative stereochemistry for aristolone have been confirmed by two total syntheses of the racemic form, outlined in Charts 22 (182) and 23 (183).

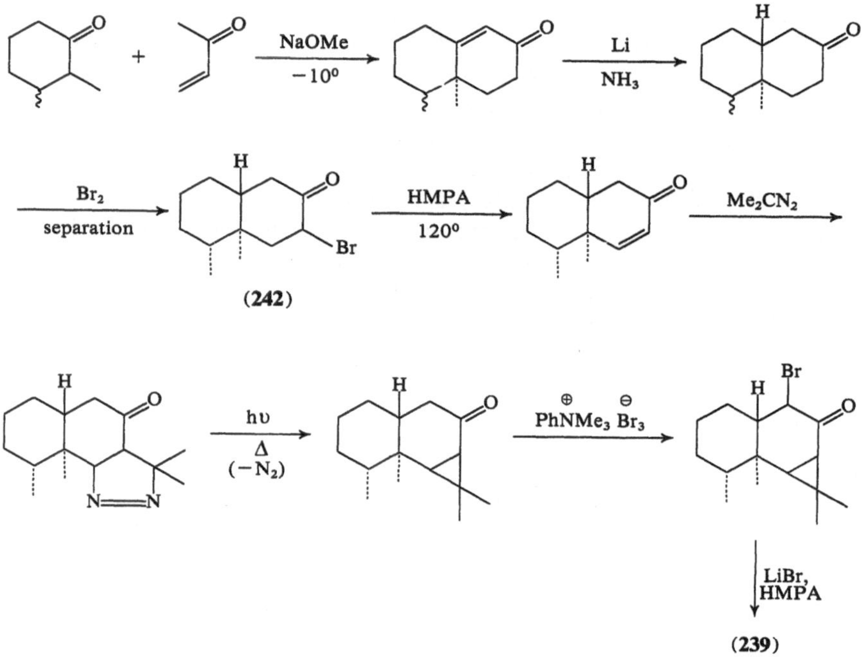

NaOMe −10°

Li NH₃

Br₂ separation

(242)

HMPA 120°

Me₂CN₂

hʋ Δ (−N₂)

⊕ ⊖ PhNMe₃ Br₃

LiBr, HMPA

(239)

Chart 22. Synthesis of (±)-Aristolone

Bromoketone (242) crystallized from the mixture of epimerides resulting from the bromination step; its stereochemistry was supported by n.m.r. spectral study and confirmed by its ultimate conversion into (±)-aristolone (182). The final product, consisting of aristolone (239) and 6,7-*epi*aristolone (α-ferulone) (243) in the ratio 2:1, together with minor components, was separated by a combination of p.t.l.c. and p.g.l.c. (183).

The u.v. irradiation of aristolone results in decarbonylation, ring-contraction, and cleavage of the cyclopropane ring to give the hydrindene (244) (184).

$\Delta^{1,(10)}$-Aristolen-2-one, $C_{15}H_{22}O$ and 1,8,9,10-tetradehydro-aristolan-2-one, $C_{15}H_{20}O$, are two substances co-occurring in the underground parts of *Nardostachys chinensis* Batalin; the former also occurs in *Aristolochia debilis* (185). They are closely related in that the

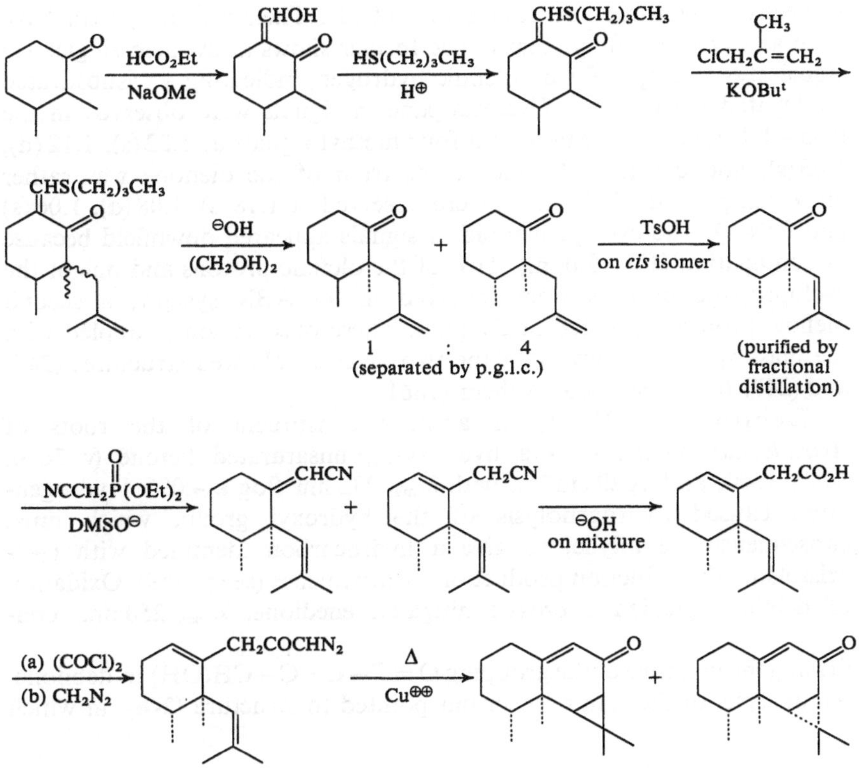

2 : 1

(239) **(243)**

Chart 23. Synthesis of (±)-Aristolone

latter on mild reduction yields the former, with uptake of one mole of H_2. Their respective u.v. spectra leave no doubt that the former is a conjugated enone (λ_{max} 243 nm, ε 5330) and the latter a corresponding conjugated dienone (λ_{max} 313 nm, ε 12,850, and 226 nm, ε 5680). Both on vigorous reduction give calarane (**245**) (see p. 157) with uptake

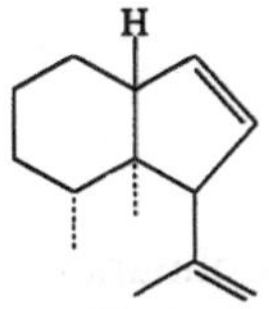

(244)

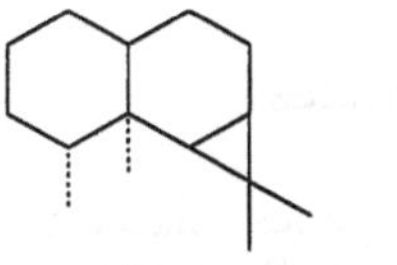

calarane **(245)**

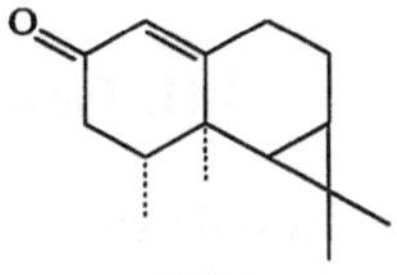

(246)

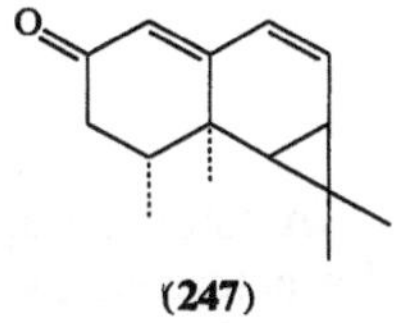

(247)

respectively of three and four moles of H_2. Their n.m.r. spectra have been analyzed carefully; that of the former shows a low-field singlet 1H signal at δ 5.78 ppm for an olefinic hydrogen, indicating a trisubstituted conjugated C=C bond. Cyclopropane H signals were observed in the 0.55—1.1 ppm (2H) region, and four methyl signals at 1.22 (s), 1.12 (d), 1.03 (s), and 0.98 (s). The n.m.r. spectrum of the dienone was rather more complex: methyl signals were observed at 1.18 (s), 1.08 (d), 1.06 (s) and 0.84 (s). The cyclopropanoid H signals appeared downfield because of conjugation (1.5—2.0, m). Two of the olefinic protons and one of the cyclopropane protons were involved in an ABX system; a second olefinic proton resonated at 5.6 ppm. These observations, coupled with the mass spectral behavior of the two ketones, allowed structures (246) and (247) to be assigned to them (186).

　Debilone, $C_{15}H_{22}O_2$, is another constituent of the roots of *Aristolochia debilis*. It is a hydroxy-α,β-unsaturated ketone (ν 3610, 3489, 1660, and 1620 cm^{-1}), with λ$_{max}$ 232 nm (log ε 4.09). Hydrogenation caused hydrogenolysis of the hydroxyl group, which must consequently be allylic, to give a hydrocarbon identified with (−)-aristolane, the reduction product of the aristolenes (see p. 159). Oxidation of debilone yielded a cross-conjugated enedione, λ$_{max}$ 250 nm, con-

firming the presence of the grouping $O=C-C=C-CH(OH)$ in debilone. An analysis of the n.m.r. spectrum pointed to structure (248), in which

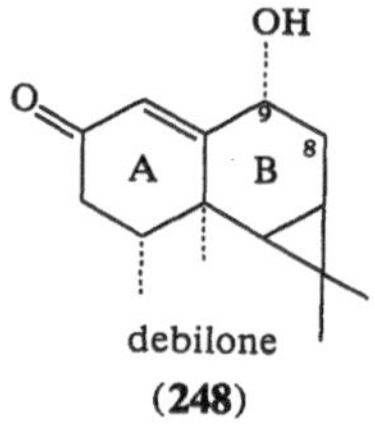

debilone
(248)　　　　　　　　(249) 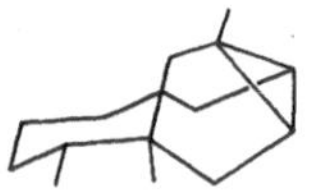　　　　　　　　(250)

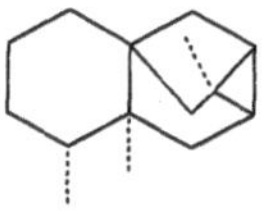

the OH at $C_{(9)}$ is assigned the α (axial) configuration because the coupling constant for both $H_{(8)}$ protons and $H_{(9)}$ is small, suggesting $H_{(9)}$ is equatorial. Ring B may be distorted into a twist-boat or boat conformation as a consequence (176).

VII. Ishwaranes

　A small group of tetracyclic sesquiterpenes related to eremophilane is characterized by the carbon framework (249), depicted in greater detail as (250). All are related stereochemically to nootkatane.

Ishwarane, $C_{15}H_{24}$, was first isolated from the essential oil of the roots of *Aristolochia indica* L. (*187*), named "ishwarene" incorrectly, and has since been found in the oil of *Cymbopetalum penduliflorum* Dunal (*188*). Assignment of structure is based in part on n.m.r. spectral study; the compound shows signals at 0.51 (m) (2 H, cyclopropane

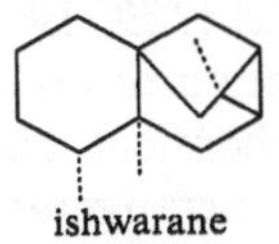

ishwarane
(**251**)

protons), 0.73 (d) (C$\underline{H}_3$CH), 0.78 (s) (tert. CH$_3$), and 1.12 (s) (cyclo-propyl CH$_3$). No olefinic hydrogen signal was present, so that ishwarane

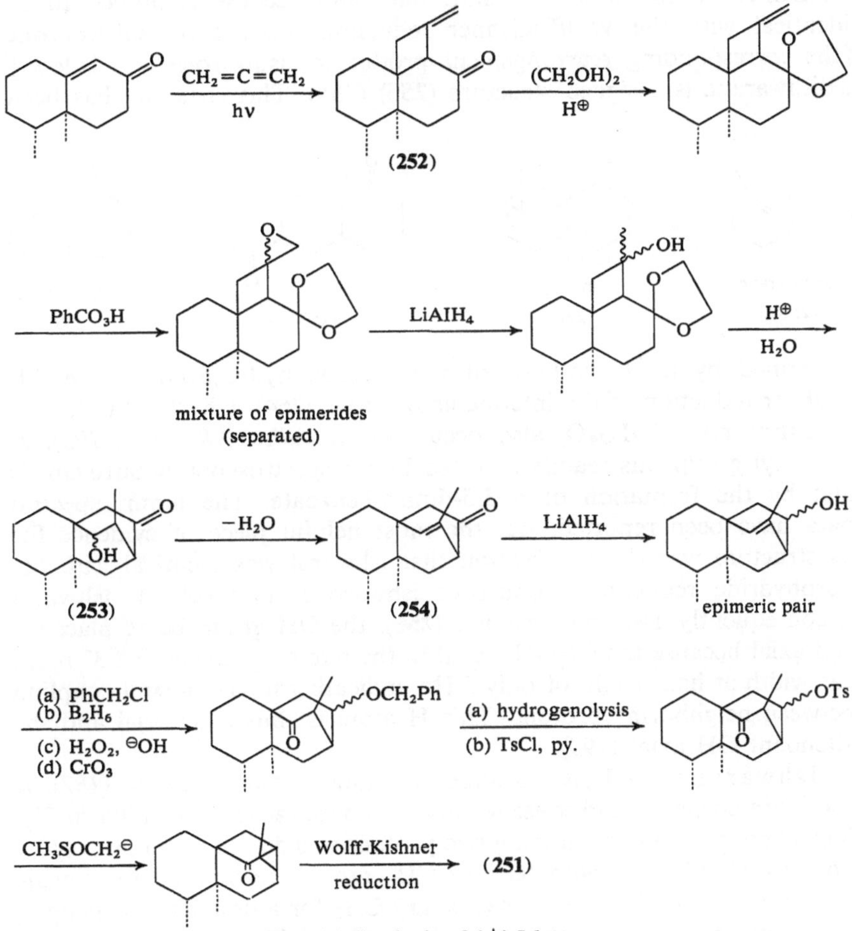

Chart 24. Synthesis of (±)-Ishwarane

must be tetracyclic. But the most important observation was that ishwarane was identical with the hydrocarbon obtained by Wolff-Kishner reduction of ishwarone, a congeneric ketone of known structure and stereochemistry (see below). Ishwarane may consequently be formulated as (251) (126, 188). This structure and relative stereochemistry have been substantiated by a total synthesis (189, 190) of the racemic hydrocarbon as outlined in Chart 24. The configuration of (252) is based on the known stereospecificity of the photochemical allene addition, and there are literature precedents for the rearrangement leading to the formation of (253) (189, 190).

When ishwarane is treated with cupric acetate-acetic acid it is isomerized to isoishwarane, which contains one C=C bond, despite its name; on the basis of n.m.r. data and because it proved to be identical with the Wolff-Kishner reduction product of isoishwarone [the corresponding rearrangement product of ishwarone (see below)] isoishwarane is assigned structure (255) (188). This structure has been

isoishwarane (255)   ishwarol (256)   (257)   isoishwarone (258)

confirmed by total synthesis of the racemic hydrocarbon by Wolff-Kishner reduction of the intermediate ketone (254) (Chart 24) (191).

Ishwarol, $C_{15}H_{24}O$, also occurs in A. indica (187, 192, 193). A hydroxyl group was readily detected by i.r. spectroscopy (v 3610 cm^{-1}) and by the formation of a 3,5-dinitrobenzoate. The n.m.r. spectral data have been reported, but the most helpful piece of evidence for its structure was the observation that ishwarol was identical with the borohydride reduction product of ishwarone (see below). Ishwarol is consequently assigned structure (256), the OH group being placed α and axial because the CHOH signal in the n.m.r. spectrum (δ 3.37 ppm) has width at half height of only 5 Hz, indicating no axial-axial coupling between neighboring protons. This H atom is thus equatorial and the attendant OH axial (193).

Ishwarone, $C_{15}H_{22}O$, is another component of A. indica (187). It is a saturated ketone and consequently must be tetracyclic (v_{CO} 1706 cm^{-1}). A cyclopropane ring containing two protons was detected by n.m.r. spectroscopy (δ 0.55, m), with tertiary CH$_3$ groups accounting for signals at 0.75 (s) and 1.15 (s), and a secondary CH$_3$ for a doublet at 0.85 ppm. It is isomerized to isoishwarone by sequential treatment with HCl gas and pyridine; the iso-ketone contains one olefinic hydrogen and one allylic methyl group, and its ozonolysis afforded a diketone (257). The latter was synthesized from valerianol (p. 148) by sequential hydroboration-oxidation, Jones oxidation, dehydration (SOCl₂), and ozonolysis. These observations mean that isoishwarone may be formulated as (258), and ishwarone as (259), although in the latter case the possibility of the fusion of the cyclopropane at positions 6 and 7 has not been eliminated (*194*).

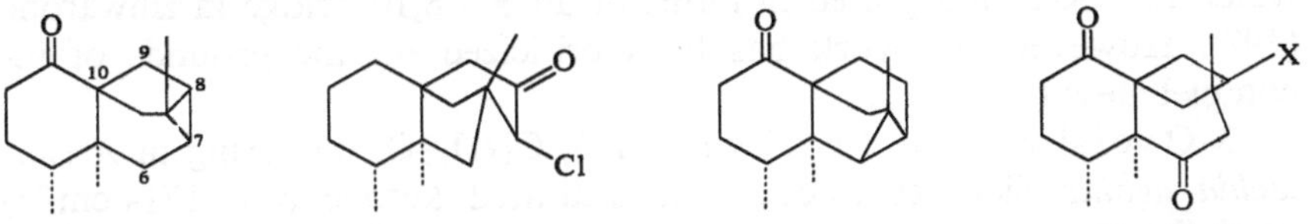

| ishwarone | | | X = Cl, Br |
| (259) | (260) | (261) | (262) |

That the union is 6, 7 was demonstrated unequivocally as follows. Ishwarone on ozonolysis afforded oxoishwarone, $C_{15}H_{20}O_2$, by oxidation of a CH_2 group α to the cyclopropane ring. This compound with conc. hydrochloric acid gave a "hydrochloride", $C_{15}H_{21}ClO_2$, a consequence of cleavage of the same ring. The n.m.r. spectrum of this product showed on octet signal at 3.93 ppm assignable to the CHCl proton. This multiplicity rules out (260) as the structure of oxoishwarone "hydrochloride", which would give rise to a singlet CHCl signal, and which would be the "hydrochloride" expected if (261) were ishwarone. The "hydrobromide" likewise shows a CHBr multiplet at 4.05 ppm. On the other hand the "hydrohalides" derived from (259), *viz.* (262), would be expected to have a multiplet CHX signal (*195*). The other possible structure for the "hydrohalides" (derived from an oxo-

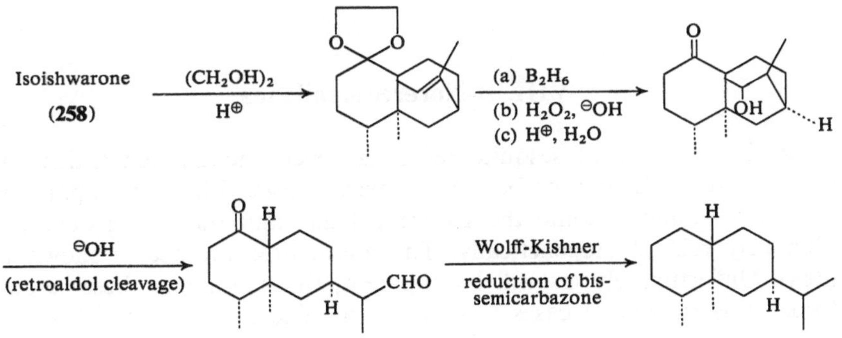

Chart 25. Absolute Configuration of Ishwarone

ishwarone with a $C=O$ group at position 9 rather than 6) is ruled out by n.m.r. spectral study.

The absolute stereochemistry of ishwarone was settled by the sequence outlined in Chart 25.

The formation of (+)-nootkatane settles the absolute configuration of isoishwarone as (258) and of ishwarone (259) (195, 196). These assignments are at variance with the result of an X-ray diffraction analysis on bromooxoishwarone (oxoishwarone "hydrobromide", see above). which has been interpreted in terms of an α 7,8,10 bridge in ishwarone (197). However this work has been criticized on the grounds of incompleteness (195).

3-Oxoishwarane (3-ishwarone), $C_{15}H_{22}O$, occurring in *Aristolochia debilis* Sieb. et Zucc., is a saturated ketone (ν_{CO} 1714 cm^{-1}) and is thus tetracyclic. In its n.m.r. spectrum it shows signals characteristic of two tertiary methyl groups (δ 0.73, 1.16), a secondary methyl (0.89, d), an AB quartet at 1.32 and 2.31 (isolated CH_2 group), and a multiplet signal for one (?) cyclopropyl H at 0.6 ppm. On exchange with D_2O in the presence of base a trideuteroketone resulted, indicating a $-CHCOCH_2-$ environment for the carbonyl group. An n.m.r. spectral analysis of the deuterated product and the fact that the natural ketone on Wolff-Kishner reduction afforded ishwarane (255) point to structure (263) for the compound (185).

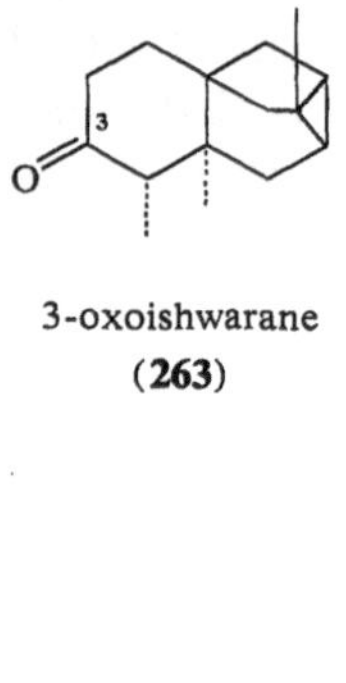

3-oxoishwarane
(263)   (264) 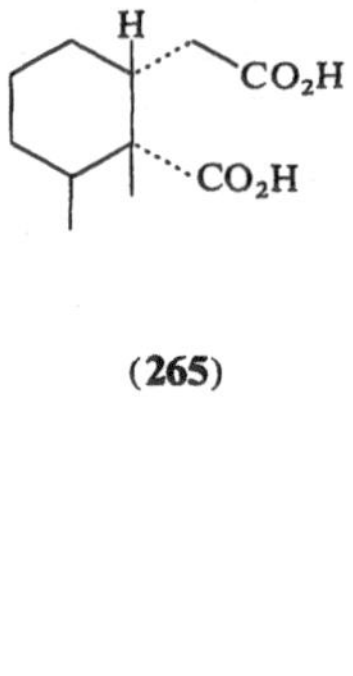   (265)

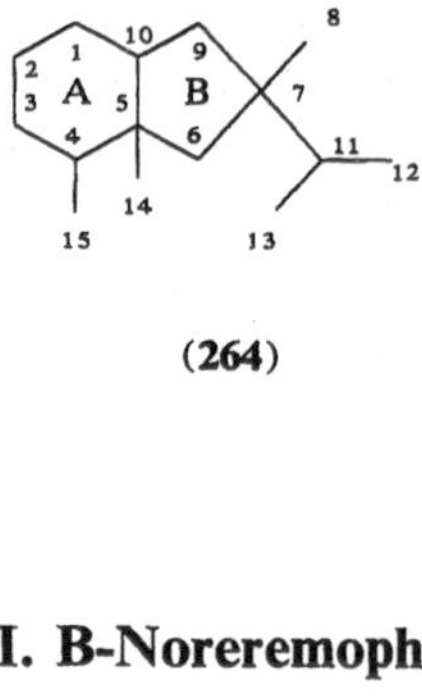

VIII. B-Noreremophilanes

A small family of sesquiterpenes has been encountered which has a B-noreremophilane carbon framework (264). Like eremophilanes these compounds violate the Isoprene Rule, and the two groups are obviously related biogenetically. The numbering adopted is shown in (264). Unfortunately, as will be apparent, the naming of these sesquiterpenes is in several cases unsatisfactory. The hydrocarbon (264) for example has been named "fukinane", which suggests it is related structurally to the normal eremophilanoid ketone fukinone (58) (see

p. 97), from which it has been synthesized (*198*). Earlier the hydro-carbon had been called "bakkane" (*199*) to point out its relationship to the bakkenolides, a subgroup of this family. All the B-noreremophilanes so far isolated occur in the buds or rhizomes of *Petasites japonicus* and its subspecies *giganteus*.

1. The Bakkenolides

Five substances, known as bakkenolide-A, -B, -C, -D, and -E, have been characterized.

Bakkenolide-A, $C_{15}H_{22}O_2$, also found in *P. albus* (*68a*), shows only weak end-absorption in the u.v. region, but has i.r. bands at 1767 (γ-lactone C=O), 1668 (C=C), and 899 cm^{-1} (C=CH$_2$); the C=CH$_2$ group is not conjugated with the C=O group. Lactonic properties were revealed in the behavior towards aqueous alkali. The n.m.r. spectrum was consistent with the presence of secondary and tertiary methyl groups, and a detailed analysis pointed to the occurrence in the structure of the unit $-\mathrm{CO-O-CH_2-C=CH_2}$. Catalytic hydrogena-tion afforded a dihydro compound as a mixture of two stereoisomers (two secondary and one tertiary CH$_3$ signals in n.m.r. spectrum), and lithium aluminum hydride treatment gave a diol $C_{15}H_{26}O_2$. Oxidation with alkaline permanganate of both bakkenolide-A and the diol gave a dicarboxylic acid $C_{11}H_{18}O_4$, in which one carboxyl group was shown to be hindered (tertiary) by deuterium exchange experiments on its dimethyl ester (only one CH$_3$ converted into CD$_3$). The acid afforded easily a cyclic anhydride $C_{11}H_{16}O_3$, with a six- or higher-membered anhydride ring on the basis of its i.r. spectrum. The acid is assigned structure (**265**) on n.m.r. spectral evidence; its formulation points to ex-pression (**266**) for bakkenolide-A. The stereochemistry depicted therein

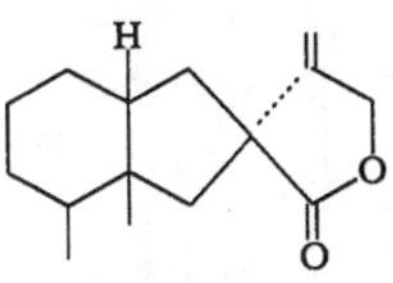

bakkenolide-A
(**266**) (**267**) 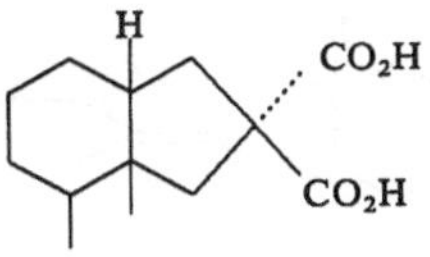 (**268**) 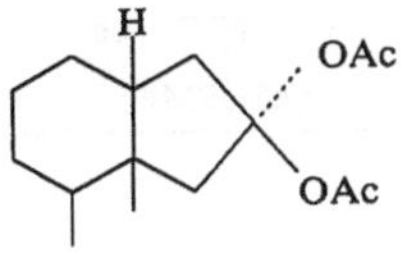 (**269**)

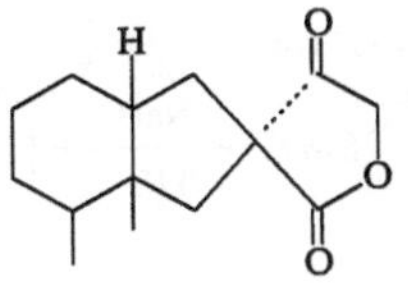

was settled by transformation of the compound by osmium tetroxide-periodate oxidation into ketolactone (**267**), then hypobromite oxidation to dicarboxylic acid (**268**), lead tetraacetate oxidation to (**269**), and finally hydrolysis to ketone (**270**). The final product was identical with a ketone prepared by the stepwise degradation of fukinone (**58**), of known

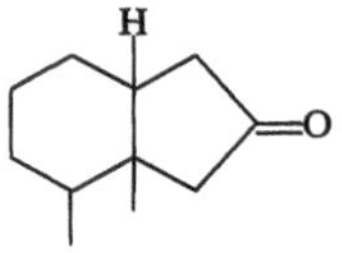

(270)

structure and absolute configuration (see p. 97) (*200, 201*); it has also been totally synthesized as its racemic form (see Chart 26). Similar structural conclusions have been reached relating to a compound named fukinanolide, which is clearly identical with bakkenolide-A (*202*). It is suggested that the latter name be retained.

This formulation and stereochemistry have been corroborated by a total synthesis of (±)-bakkenolide-A as in Chart 26 (*203*). The starting material had already been synthesized (*183*) (see Chart 23). The step resulting in the formation of (**271**) represents a stereospecific [2,3]-sigmatropic rearrangement.

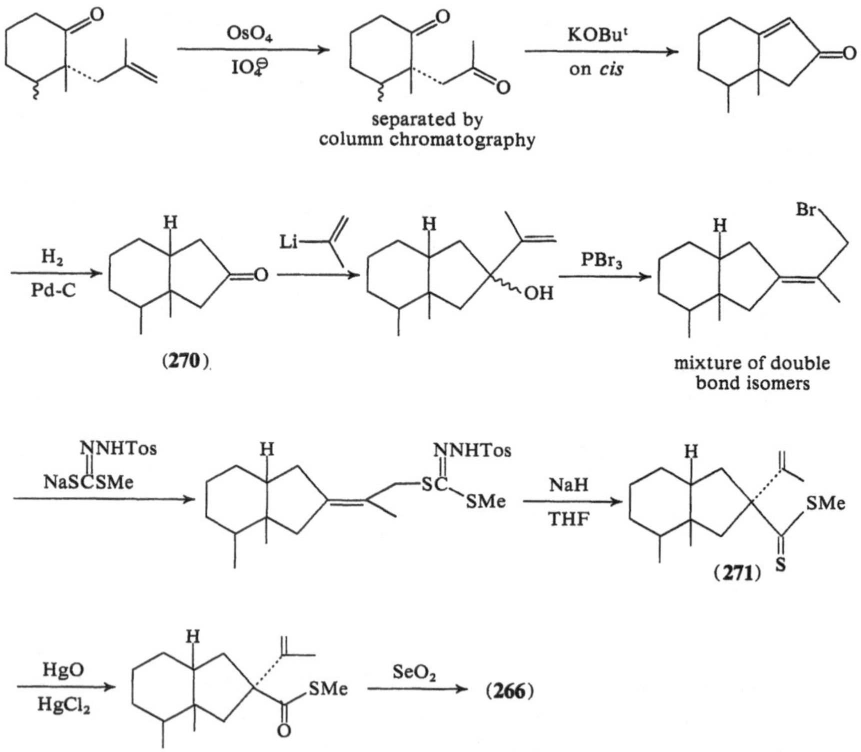

Chart 26. Synthesis of (±)-Bakkenolide-A

Bakkenolide-B, $C_{22}H_{30}O_6$, molecular wt. 390 (mass spectrometry), showed in its i.r. spectrum bands at 1775 (γ-lactone CO), 1736 (ester CO), 1706 (α,β-unsaturated ester CO), 1669 (isolated C=C), 1646 (conjugated C=C), 902 (C=CH$_2$), and 848 cm^{-1} (*trans* CH=CH). Hydrolysis generated angelic acid and a diol $C_{15}H_{22}O_4$, ν_{OH} 3490, ν_{CO} 1764, $\nu_{C=C}$ 1669, and $\nu_{C=CH_2}$ 895 cm^{-1}. Hydrolysis under somewhat milder conditions afforded a hydroxyangelate ester $C_{20}H_{28}O_5$, which proved to be identical with the third member of the group, bakkenolide-C, and which on acetylation regenerated bakkenolide-B. It was apparent that the B-compound was an acetate-angelate diester of the diol, and that the C-compound was one of the two possible angelate monoesters. Likewise, bakkenolide-D, $C_{21}H_{28}O_6S$, on hydrolysis gave the same diol and *cis*-β-methylthioacrylic acid. An extensive and detailed n.m.r. spectral study of the three compounds and the common diol, using the nuclear Overhauser effect, allowed structures (272), (273), and (274) to be advanced for bakkenolide-B, -C, and -D respectively (204). Structures (272) and (274) have also been advanced independently for two compounds fukinolide and S-fukinolide (202), which appear to be identical with bakkenolide-B and -C.

(272) bakkenolide-B (R = angelyl, R' = Ac)

(273) bakkenolide-C (R = angelyl, R' = H)

(274) bakkenolide-D (R = *cis*-β-methylthioacrylyl, R' = Ac)

(275) bakkenolide-E (R = tiglyl, R' = Ac)

Bakkenolide-E, $C_{22}H_{30}O_6$, is a γ-lactone (ν_{CO} 1792), a diester (ν_{CO} 1740, 1700, saturated and α,β-unsaturated), and contains an unconjugated C=CH$_2$ group (ν 1650, 910 cm^{-1}). The ester moieties were identified as acetate and tiglate, and the hydrolysis product proved to be the same diol obtained from bakkenolides-B, -C, and -D. Discernible similarities between the n.m.r. spectra of the -E and -B compounds led to the formulation of bakkenolide-E as (275) (199).

The structures advanced for the five bakkenolides are supported by the results of a detailed investigation of their high resolution mass spectral behavior and that of their derivatives (205). The structure and absolute configuration of the common diol $C_{15}H_{22}O_4$ have been confirmed by an X-ray diffraction analysis of its bisbromoacetate (206).

9-Acetoxybakkenolide-A, $C_{17}H_{24}O_4$, is a γ-lactone (ν_{CO} 1782) and a saturated ester (ν_{CO} 1730). It contains an unconjugated C=CH$_2$ group (ν 1670 and 900), and a band at 1225 cm^{-1} is assigned to an acetoxy group. Its n.m.r. spectrum was closely similar to those of known bakkane sesquiterpenes; it showed signals at 0.98 d ($\underline{CH_3}$CH),

1.05 s (tert. CH₃), 2.04 s (OCOCH₃), 4.63 t (OCH₂–C=CH₂), 5.15 d (CH₂=C), and 5.30 d ppm (HC–O). A link-up with the previously described diol $C_{15}H_{22}O_4$ (see above) was realized as follows. Selective Jones oxidation of the diol afforded a monoketone (276) the acetate of which was converted into its ethylenedithioketal. This on treatment with Raney nickel gave (277), which was identical with the dihydro-

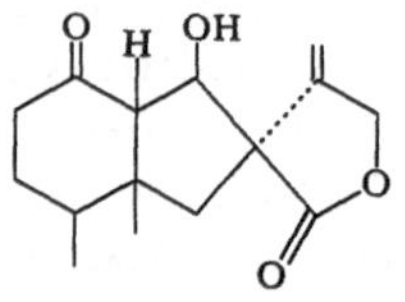

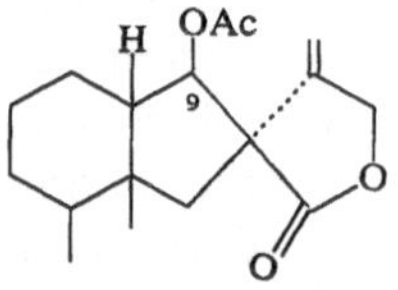

9-acetoxy-
bakkenolide-A

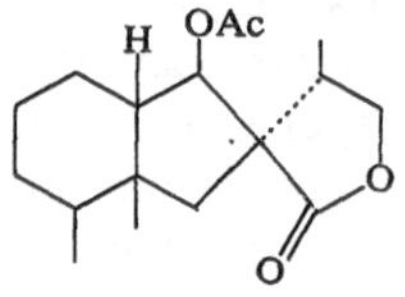

(276)　　　　　　　　　(277)　　　　　　　　　(278)

derivative of the natural product; the latter must consequently be formulated as (278) (42).

2. Other Noreremophilanes

A keto-acid, $C_{18}H_{24}O_5$, has been isolated from the roots of *Othonna quercifolia* DC. Treatment of it with diazomethane yielded the corresponding methyl ester $C_{19}H_{26}O_5$, the n.m.r. spectrum of which revealed an angelyl residue and a methyl ketone group. A detailed analysis of the spectrum, using double resonance, and its mass spectrum, suggested structure (278a), that of a *bis*-noreremophilanoid sesquiterpene, for the ester and consequently (278b) for the acid. A possible biogenetic route to the latter has been suggested (16).

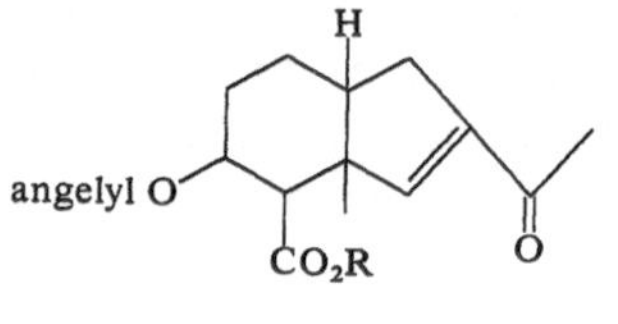

angelyl O

ĊO₂R

(278a) (R = Me)
(278b) (R = H)

IX. Biogenetic Considerations

Few investigations on the biogenetic aspect of eremophilanes have been described. In 1939, ROBINSON, to explain the failure of the eremophilane skeleton to conform to the Isoprene Rule, suggested that the biogenesis of these compounds involved a eudesmanoid intermediate (obeying the Rule), which experienced migration of a methyl group from one angular position to the other (*4, 206a*).

This proposal has led to suggestions that structures such as (**279**) and (**280**) could be precursors of eremophilone by the mechanisms indicated

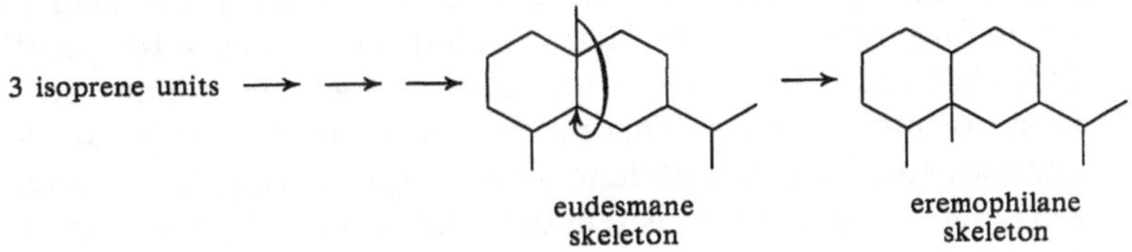

3 isoprene units → → →

eudesmane
skeleton

eremophilane
skeleton

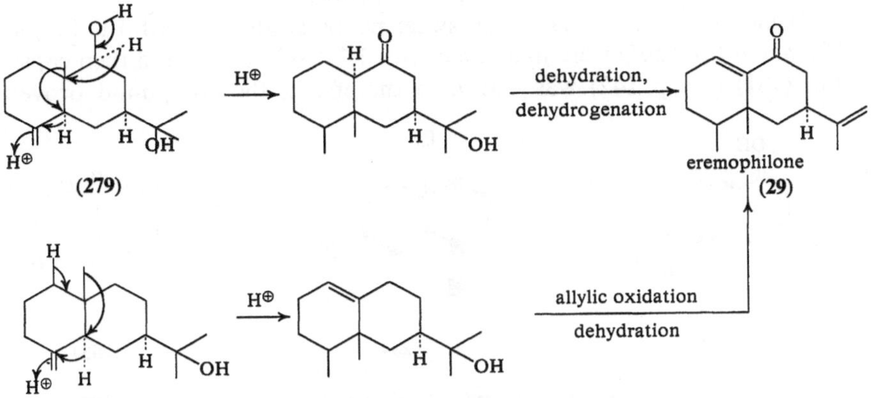

(**279**)

dehydration,
dehydrogenation

eremophilone
(**29**)

(**280**)

allylic oxidation
dehydration

(*30, 207*). It is to be noted that the stereochemistry of these precursors is favorable for Wagner-Meerwein rearrangements to yield the desired configuration in the final product. It was pointed out by ERDTMAN and NORIN (*208*) that similar schemes culminating in nootkatanes (see p. 144) would require as starting point sesquiterpenes antipodal in] configuration to normal eudesmanes. At that time these were not known, but have since been found to be natural products; one of them, intermedeol (**281**), has even been found in grapefruit oil alongside nootkatone (**186**) and valencene (**190**) (*209, 210*), and another has been

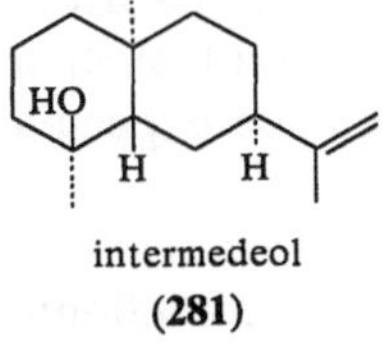

intermedeol
(**281**)

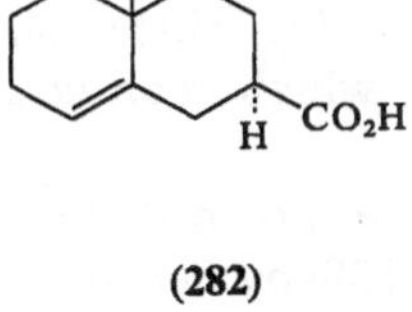

(**282**)

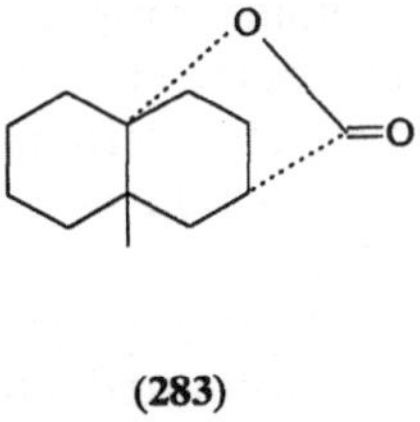

(**283**)

found in vetiver oil, where valerianol (191) also occurs (211). These discoveries provide powerful support for the methyl migration biogenetic hypothesis.

Several attempts have been made to simulate *in vitro* postulated type of methyl migration. In one of these (212) the acid (282) was successfully rearranged to lactone (283), but it was subsequently shown (213) that the methyl migration observed was not direct but occurred *via* spiro intermediates. This revelation led to the proposal of an alternative scheme for eremophilane biogenesis involving similar intermediates (213), but this suggestion is rendered unnecessary by the known existence of antipodal eudesmane types (see above), and unlikely as a consequence of recent biogenetic studies described below.

It has been found that when sweet peppers infected with the fungus *Monolinia fructicola* were incubated with 1,2-^{13}C$_2$-acetate and then capsidiol (181) (see p. 142) was extracted therefrom, the compound proved

capsidiol
(181)

(284)

(285)

to be doubly labelled to an extent of *ca.* 5%, on mass spectrometric evidence. A careful carbon-13 n.m.r. spectral analysis of the enriched capsidiol enabled coupling constants between carbon atoms 1 and 2, 4 and 14, 5 and 6, 7 and 8, and 11 and 12 to be evaluated. The relative intensities of these signals gave no evidence of gross differences in the degree of carbon-13 enrichment at these positions. Further, *no coupling was observed between carbons 5 and 15*. The conclusion to be drawn from these observations is that in capsidiol carbon atoms 5 and 15 are derived from separate acetate units, in agreement with migration of a methyl group from position 10 to position 5 in (284), the eudesmanoid precursor. The pattern of double-labelling in the latter to be expected from the farnesyl pyrophosphate biogenetic route is as shown, six intact C$_2$ units being present. Finally the absence of C$_{(5)}$–C$_{(15)}$ coupling also rules out the possibility of the transformation of (284) to (181) *via* a pathway involving spiro intermediates (213), since the end product of such a route would be a molecule in which such coupling would be expected (214).

Other biogenetic studies have shown that (±)-2-^{14}C-mevalonolactone is incorporated into petasin (p. 108) by *Petasites hybridus* (215). Mevalonic acid is likewise incorporated into petasin esters in the same plant in a manner consistent with the postulated pathway involving *trans,trans*-farnesyl pyrophosphate (*215a*).

The only example of what appears to be a direct 10 → 5 methyl migration *in vitro* is the case of dihydroalantolactone $5\alpha,6\alpha$-epoxide (**285**), which on treatment with formic acid gave three products, two of which were found to be eremophilanoid in structure. One proved to be the alcohol (**286**), and the other its formate. A rationalization depicted in expression (**287**) has been proposed to explain the rearrangement (*216*). Compound (**286**) was subsequently converted into tetrahydroligularenolide (**122**), to provide the first *in vitro* correlation between eudesmanoid and eremophilanoid types (*217, 218, 219*).

(286)   (287)

References

1. PINDER, A. R.: Recent Developments in the Eremophilane Sesquiterpene Group. I. New Bicyclic Structures. Perf. Ess. Oil Rec. **59**, 280 (1968).

2. — Recent Developments in the Eremophilane Sesquiterpene Group. II. Tricyclic Structures. Perf. Ess. Oil Rec. **59**, 645 (1968).

2a. SORM, F.: Advances in Terpene Chemistry. Pure and Applied Chem. **21**, 263 (1970).

3. BRADFIELD, A. E., A. R. PENFOLD, and J. L. SIMONSEN: The Constitution of Eremophilone and of two Related Hydroxyketones from the Wood Oil of *Eremophila mitchelli*. J. Chem. Soc. (London) **1932**, 2744.

4. ROBINSON, R., quoted by PENFOLD, A. R., and J. L. SIMONSEN: The Constitutions of Eremophilone, Hydroxyeremophilone, and Hydroxydihydroeremophilone. Part III. J. Chem. Soc. (London) **1939**, 87.

5. HOCHMANNOVA, J., L. NOVOTNY, and V. HEROUT: Sesquiterpenic Hydrocarbons from Coltsfoot Rhizomes. Collect. Czech. Chem. Comm. **27**, 1870 (1962).

5a. NOVOTNY, L., V. HEROUT, and F. SORM: Substances from *Petasites officinalis* and *P. albus*. Tetrahedron Letters **1961**, 697.

6. HOCHMANNOVA, J., L. NOVOTNY, and V. HEROUT: Hydrocarbons from *Petasites albus* Rhizomes. Collect. Czech. Chem. Comm. **27**, 2711 (1962).

6a. WITEK, S., and J. KREPINSKY: The Composition of Valerian Oil. Collect. Czech. Chem. Comm. **31**, 1113 (1966).

6b. NOVOTNY, L., and V. HEROUT: Constituents of Rhizomes of *Petasites japonicus*. Collect. Czech. Chem. Comm. **30**, 3579 (1965).

7. HOCHMANNOVA, J., and V. HEROUT: Structure of Eremophilene. Collect. Czech. Chem. Comm. **29**, 2369 (1964).

8. PIERS, E., and R. J. KEZIERE: Stereoselective Synthesis of (±)-Eremophil-3,11-diene and Related Compounds. Comments on the Proposed Structure of Eremophilene. Tetrahedron Letters **1968**, 583.

9. — — Stereoselective Synthesis of (±)-Eremophil-3,11-diene and Related Compounds. Concerning the Structure of Eremophilene. Canad. J. Chem. **47**, 137 (1969).

10. KREPINSKY, J., O. MOTL, L. DOLEJS, L. NOVOTNY, V. HEROUT, and R. B. BATES: The Structure of Eremophilene, the Sesquiterpenic Hydrocarbon from *Petasites* genus. Tetrahedron Letters **1968**, 3315.

11. COATES, R. M., and J. E. SHAW: The Total Synthesis of (±)-Eremophilene and (±)-Eremoligenol. Tetrahedron Letters **1968**, 5405.

12. — — Stereoselective Total Synthesis of (±)-Eremoligenol, (±)-Eremophilene, (±)-Valerianol, and (±)-Valencene. J. Organ. Chem. (USA) **35**, 2597 (1970).

13. — — Stereoselectivity in the Synthesis of *cis*- and *trans*-4,4a-Dimethyl-2-octalone Derivatives. Chem. Commun. **1968**, 47.

14. — — Total Synthesis of (±)-Calarene. J. Amer. Chem. Soc. **92**, 5657 (1970).

15. — — The Total Synthesis of (±)-$\Delta^{1(10)}$-Aristolene (Calarene). Chem. Commun. **1968**, 515.

16. BOHLMANN, F., C. ZDERO, and M. GRENZ: New Sesquiterpenes of the Genus *Othonna*. Chem. Ber. **107**, 3928 (1974).

16a. JOMMI, G., J. KREPINSKY, V. HEROUT, and F. SORM: The Structure of Valerianol, a Sesquiterpenic Alcohol of Eremophilane Type from Valerian Oil. Collect. Czech. Chem. Commun. **34**, 593 (1969).

17. ISHII, H., T. TOZYO, and H. MINATO: Components of the Roots of *Ligularia Fischeri*. J. Chem. Soc. (London) **1966**, 1545.

18. BRADFIELD, A. E., A. R. PENFOLD, and J. L. SIMONSEN: The Essential Oil from the Wood of *Eremophila mitchelli*. Proc. Roy. Soc. New South Wales **66**, 420 (1933).

19. BRADFIELD, A. E., N. HELLSTRÖM, A. R. PENFOLD, and J. L. SIMONSEN: The Constitutions of Eremophilone, Hydroxyeremophilone, and Hydroxydihydroeremophilone. Part II. J. Chem. Soc. (London) **1938**, 767.

20. PENFOLD, A. R., and J. L. SIMONSEN: The Constituents of Eremophilone, Hydroxyeremophilone, and Hydroxydihydroeremophilone. Part III. J. Chem. Soc. (London) **1939**, 87.

21. COPP, F. C., and J. L. SIMONSEN: Experiments on the Synthesis of 1,2-Dimethylcyclohexylacetic Acid. J. Chem. Soc. (London) **1940**, 415.

22. GILLAM, A. E., J. I. LYNAS-GRAY, A. R. PENFOLD, and J. L. SIMONSEN: The Constitutions of Eremophilone, Hydroxyeremophilone, and Hydroxydihydroeremophilone. Part IV. J. Chem. Soc. (London) **1941**, 60.

23. GILLAM, A. E.: The Absorption Spectrum of α-Cyperone and its Indications as to the Probable Structure of the Compound. J. Chem. Soc. (London) **1936**, 676.

24. BRADFIELD, A. E., R. R. PRITCHARD, and J. L. SIMONSEN: The Constitution of α-Cyperone. J. Chem. Soc. (London) **1937**, 760.

25. GRANT, D. F., and D. ROGERS: The Structure of Hydroxydihydroeremophilone. Chem. and Ind. **1956**, 278.

26. GRANT, D. F.: The Crystal Structure of Hydroxydihydroeremophilone. Acta Crystallogr. **10**, 498 (1957).

27. BATES, R. B., and S. K. PAKINIKAR: Eremophilone and Alloeremophilone from Hydroxydihydroeremophilone. Chem. and Ind. **1966**, 2170.

28. KLYNE, W.: The Molecular Rotations of Polycyclic Compounds. Part II. Diterpenoids and Sesquiterpenoids. J. Chem. Soc. (London) **1953**, 3072.

29. DJERASSI, C., R. RINIKER, and B. RINIKER: Optical Rotatory Dispersion Studies. VII. Application to Problems of Absolute Configurations. J. Amer. Chem. Soc. **78**, 6362 (1956).

30. ZALKOW, L. H., F. X. MARKLEY, and C. DJERASSI: Terpenoids. XLVIII. The Absolute Configuration of Eremophilone and Related Compounds. J. Amer. Chem. Soc. **82,** 6354 (1960).

31. — — — Terpenoids. XL. Absolute Configuration of Eremophilone. J. Amer. Chem. Soc. **81,** 2914 (1959).

32. SPEZIALE, A. J., J. A. STEPHENS, and Q. E. THOMPSON: Studies in Steroid Total Synthesis. I. Resolution of a Bicyclic Intermediate. J. Amer. Chem. Soc. **76,** 5011 (1954).

33. BARKLEY, L. B., M. W. FARRAR, W. S. KNOWLES, H. RAFFELSON, and Q. E. THOMPSON: Studies in Steroid Total Synthesis. II. Correlation of Optically Active Bicyclic Intermediates with Natural Steroids. J. Amer. Chem. Soc. **76,** 5014 (1954).

34. DJERASSI, C., R. MAULI, and L. H. ZALKOW: Terpenoids. XXXVIII. Interconversion of Eremophilone, Hydroxyeremophilone, and Hydroxydihydroeremophilone. The Relative Stereochemistry of Eremophilone and its Reduction Products. J. Amer. Chem. Soc. **81,** 3424 (1959).

35. ZIEGLER, F. A., and P. A. WENDER: The Stereospecific Generation of the *cis* Vicinal Methyls in Eremophilane and Valencane Sesquiterpenes. The Total Synthesis of (±)-Eremophilone and (±)-7-Epieremophilone. Tetrahedron Letters **1974,** 449.

36. McMURRY, J. E., J. H. MUSSER, M. S. AHMAD, and L. C. BLASZCZAK: Total Synthesis of Eremophilone. J. Organ. Chem. (USA) **40,** 1829 (1975).

37. VAN DER GEN, A., L. M. VAN DER LINDE, J. G. WITTEVEEN, and H. BOELENS: Stereoselective Synthesis of Eremophilane Sesquiterpenes from β-Pinene. Rec. trav. chim. Pays-Bas **90,** 1034 (1971).

38. CHETTY, G. L., L. H. ZALKOW, and R. A. MASSY-WESTROPP: 7α(H)-Eremophila-1,11-dien-9-one. A New Sesquiterpene of the Eremophilane Type. Tetrahedron Letters **1969,** 307.

39. ZALKOW, L. H., A. M. SHALIGRAM, S.-E. HU, and C. DJERASSI: Studies in the Chemistry of the Eremophilane Sesquiterpenes. Tetrahedron **22,** 337 (1966).

40. MASSY-WESTROPP, R. A., and G. D. REYNOLDS: Eremophilane Sesquiterpenes from *Eremophila mitchelli.* Austral. J. Chem. **19,** 303 (1966).

41. ZALKOW, L. H., and G. L. CHETTY: Interconversion of Eremophilone and Isoeremophilone and Related Reactions. J. Organ. Chem. (USA) **40,** 1833 (1975).

42. NAYA, K., M. KAWAI, M. NAITO, and T. KASAI: The Structures of Eremofukinone, 9-Acetoxyfukinanolide, and S-Japonin from *Petasites japonicus* Maxim. Chemistry Letters **1972,** 241.

43. ANDERSEN, N. H.: Biogenetic Implications of the Antipodal Sesquiterpenes of Vetiver Oil. Phytochem. **9,** 145 (1970).

44. ODOM, H. C., and A. R. PINDER: Total Synthesis of (±)-Nootkatone. Chem. Commun. **1969,** 26.

45. — — Synthetic Experiments in the Eremophilane Sesquiterpene Group. Synthesis of 7-Epinootkatone and Partial Synthesis of Valerianol. The Structure of Nardostachone. J. Chem. Soc. (London) Perkin Trans. I **1972,** 2193.

46. RODD, E. H. (ed.): The Chemistry of Carbon Compounds, Vol. IIB, p. 526 (by D. H. R. BARTON). Amsterdam: Elsevier. 1953.

47. NAYA, K., I. TAKAGI, Y. KAWAGUCHI, Y. ASADA, Y. HIROSE, and H. SHINODA: The Structure of Fukinone, a Constituent of *Petasites japonicus* Maxim. Tetrahedron **24,** 5871 (1968).

47a. HORIBE, I., H. SHIGEMATO, and K. TORI: ¹H NMR Shifts Induced by Hexafluorobenzene in Germacrones, Ten-membered Ring Sesquiterpenes. Tetrahedron Letters **1975,** 2849.

48. PIERS, E., and R. D. SMILLIE: Stereoselective Total Synthesis of (±)-Fukinone. J. Organ. Chem. (USA) **35,** 3997 (1970).

49. PIERS, E., R. S. BRITTON, and W. DE WAAL: Annelation of 2,3-Dimethylcyclo-hexanone. Synthetic Proof for the Stereochemistry of the Sesquiterpene Aristolone. Canad. J. Chem. **47**, 4307 (1969).

50. TORRENCE, A. K., and A. R. PINDER: Total Synthesis of (±)-Fukinone. Tetrahedron Letters **1971**, 745.

51. PINDER, A. R., and A. K. TORRENCE: Total Synthesis of Racemic Fukinone and of Natural (+)-Hydroxyeremophilone. J. Chem. Soc. (London) (C) **1971**, 3410.

52. MARSHALL, J. A., and G. M. COHEN: The Stereoselective Total Synthesis of Racemic Fukinone. Tetrahedron Letters **1970**, 3865.

53. — — The Stereoselective Total Synthesis of Racemic Fukinone. J. Organ. Chem. (USA) **36**, 877 (1971).

54. JOHNSON, W. S., and K. E. HARDING: Olefinic Cyclizations. IX. Further Observations on the Butenylcyclohexanol System. J. Organ. Chem. (USA) **32**, 478 (1967).

54a. OHASHI, M.: A New Isoxazole Annelation; Application to the Synthesis of Dehydrofukinone. Chem. Commun. **1969**, 893.

55. BROOKS, C. J. W., and G. H. DRAFFAN: Warburgin, a New Sesquiterpenoid of the Eremophilane Group. Chem. Commun. **1966**, 393.

56. — — The Constitution of Warburgiadione. Chem. Commun. **1966**, 701.

57. — — Sesquiterpenes of *Warburgia* species. I. Warburgin and Warburgiadione. Tetrahedron **25**, 2865 (1969).

58. YAMAKAWA, K., I. IZUTA, H. OKA, and R. SAKAGUCHI: Total Synthesis of (±)-Isopetasol, (±)-3-Epi-isopetasol, and (±)-Warburgiadione. Tetrahedron Letters **1974**, 2187.

59. GEISSMAN, T. A.: On the Structure of Eremophilone. J. Amer. Chem. Soc. **75**, 4008 (1953).

60. NAYA, K., F. YOSHIMURA, and I. TAKAGI: The Structure of Petasitolone, a New Constituent of *Petasites japonicus* Maxim. Bull. Chem. Soc. Japan **44**, 3165 (1971).

61. KINGSTON, D. G. I., M. M. RAO, and T. D. SPITTLER: Isolation and Structure Determination of Fluorensic Acid, a New Sesquiterpene of the Eremophilane Type. Tetrahedron Letters **1971**, 1613.

61a. KINGSTON, D. G. I., M. M. RAO, T. D. SPITTLER, R. C. PETTERSEN, and D. L. CULLEN: Sesquiterpenes from *Flourensia cernua*. Phytochem. **14**, 2033 (1975).

62. HANAYAMA, N., F. KIDO, R. SAKUMA, H. UDA, and A. YOSHIKOSHI: Minor Acidic Constituents of Vetiver Oil. Tetrahedron Letters **1968**, 6099.

63. JOMMI, G., J. KREPINSKY, V. HEROUT, and F. SORM: The Structure of Valerianol, a Sesquiterpenoic Alcohol of Eremophilane Type from Valeriana Oil. Tetrahedron Letters **1967**, 677.

64. BOHLMANN, F., and C. ZDERO: Naturally Occurring Terpene Derivatives. XLI. On New Sesquiterpenes from the Genus *Senecio*. Chem. Ber. **107**, 2912 (1974).

65. AEBI, A., J. BÜCHI, T. WAALER, E. EICHENBERGER, and J. SCHMUTZ: Inhaltsstoffe von *Petasites hydridus*(L.). Fl. Wett. Pharmac. Acta Helv. **30**, 277 (1955).

66. STOLL, A., R. MORF, A. RHEINER, and J. RENZ: Über Inhaltsstoffe aus *Petasites officinalis* Moench. I. Petasin und die Petasolester B und C. Experientia **12**, 360 (1956).

67. AEBI, A., T. WAALER, and J. BÜCHI: Petasin und S-Petasin, die spasmolytisch wirk-samen Inhaltsstoffe von *Petasites officinalis* (L.). Fl. Wett. Pharm. Weekblad **93**, 397 (1958).

68. AEBI, A., and T. WAALER: Über die Inhaltsstoffe von *Petasites hydridus*. Basel: Helbing und Lichtenhahn. 1958.

68a. NOVOTNY, L., K. KOTVA, J. TOMAN, and V. HEROUT: Sesquiterpenes from *Petasites*. Phytochem. **11**, 2795 (1972).

69. AEBI, A., and C. DJERASSI: Die absolute Konfiguration des Sesquiterpenoids Petasin. Helv. Chim. Acta **42**, 1785 (1959).

70. HERBST, D., and C. DJERASSI: Terpenoids. XLIV. Synthetic Confirmation of the Structure and Absolute Configuration of Petasin. J. Amer. Chem. Soc. 82, 4337 (1960).

71. NAYA, K., and I. TAKAGI: The Structure of Petasitin. A New Sesquiterpene from Petasites japonicus Maxim. Tetrahedron Letters 1968, 629.

72. WEI, R., P. E. STILL, E. B. SMALLEY, H. K. SCHNOES, and F. M. STRONG: Isolation and Partial Characterization of a Mycotoxin from Penicillium roqueforti. Appl. Microbiol. 25, 111 (1973).

73. ROSS, W. C. J.: The Reactions of Certain Epoxides in Aqueous Solutions. J. Chem. Soc. (London) 1950, 2257.

74. WEI, R., H. K. SCHNOES, P. A. HART, and F. M. STRONG: The Structure of PR-Toxin, a Mycotoxin from Penicillium roqueforti. Tetrahedron 31, 109 (1975).

74a. MOREAU, S., A. GAUDEMER, A. LABLACHE-COMBIER, et J. BIGUET: Metabolites de Penicillium roqueforti: PR Toxine et Metabolites Associés. Tetrahedron Letters 1976, 833.

75. BOHLMANN, F., C. ZDERO, and M. GRENZ: Über die Inhaltsstoffe der Gattung Euryops. Chem. Ber. 107, 2730 (1974).

75a. NOVOTNY, L., V. HEROUT, and F. SORM: Constitution of Petasitolides and S-Petasitolides. Collect. Czech. Chem. Comm. 29, 2182 (1964).

75b. HIKINO, H., Y. HIKINO, and I. YOSIKA: Structure and Autoxidation of Atractylon. Chem. Pharm. Bull. (Japan) 10, 641 (1962).

75c. NOVOTNY, L., and F. SORM: Beiträge zur Biochemie und Physiologie von Naturstoffen, p. 327. Jena: Fischer Verlag. 1965.

76. NOVOTNY, L., J. JIZBA, V. HEROUT, F. SORM, L. H. ZALKOW, S. HU, and C. DJERASSI: Constitution and Absolute Configuration of Eremophilenolide. Tetrahedron 19, 1101 (1963).

77. NOVOTNY, L., J. JIZBA, V. HEROUT, and F. SORM: Plant Substances. XVI. The Constituents of Coltsfoot Rhizomes (Petasites officinalis Moench). Collect. Czech. Chem. Comm. 27, 1393 (1962).

78. KABUTO, C., N. TAKADA, S. MAEDA, and Y. KITAHARA: X-ray Structure Determination of Eremophilenolide. Chemistry Letters 1973, 371.

79. PIERS, E., M. B. GERAGHTY, and R. D. SMILLIE: Total Synthesis of Eremophilenolide. Chem. Commun. 1971, 614.

80. PIERS, E., and M. B. GERAGHTY: Total Synthesis of Eremophilane-type Sesquiterpenoids: (±)-Eremophilenolide, (±)-Tetrahydroligularenolide, and (±)-Aristolochene. Canad. J. Chem. 51, 2166 (1973).

80a. NAGAKURA, I., S. MAEDA, M. UENO, M. FUNAMIZU, and Y. KITAHARA: Eremophilanes. Part I. Total Syntheses of (±)-Eremophilenolide and (±)-Furanoeremophilane. Chemistry Letters 1975, 1143.

81. TANAHASHI, Y., Y. ISHIZAKI, T. TAKAHASHI, and K. TORI: Ligularenolide. A New Sesquiterpene Lactone of Eremophilane Type. Tetrahedron Letters 1968, 3739.

82. ISHIZAKI, Y., Y. TANAHASHI, T. TAKAHASHI, and K. TORI: The Structure of Ligularenolide. A New Sesquiterpene Lactone of Eremophilane Type. Tetrahedron 26, 5387 (1970).

83. TATEE, T., and T. TAKAHASHI: Synthesis of (±)-Tetrahydroligularenolide. Chemistry Letters 1973, 929.

84. — —Total Synthesis of (±)-Tetrahydroligularenolide. Bull. Chem. Soc. Japan 48, 281 (1975).

84a. NOVOTNY, L., M. KROJIDLO, Z. SAMEK, J. KOHOUTOVA, and F. SORM: The Structure of Nemosenins-A, -B, -C, -D, and Senemorin, New Furoemophilane Derivatives from Senecio nemorensis, subsp. fuchsii. Collect. Czech. Chem. Comm. 38, 739 (1973).

85. NOVOTNY, L., V. HEROUT, and F. SORM: Constitution of Petasalbin, Albopetasine,

and Hydroxyeremophilenolide, the Components of *Petasites albus* Rhizomes. Collect. Czech. Chem. Comm. **29,** 2189 (1964).

86. HOERHAMMER, L., H. WAGNER, and B. LAY: New Methods in Pharmacognosy. V. Chromatography of *Tussilago farfara* and *Petasites* leaves. Deut. Apotheker-Zeit. **103,** 429 (1963) [Chem. Abstr. **59,** 10469e (1963)].

87. ANET, F. A. L.: Some Aspects of the Nuclear Magnetic Resonance Spectra of Compounds Containing *C*-Methyl Groups. Canad. J. Chem. **39,** 2262 (1961).

88. NOVOTNY, L., V. HEROUT, and F. SORM: Plant Substances. XVII. Constituents of *Petasites albus* L. Rhizomes. Collect. Czech. Chem. Comm. **27,** 1400 (1962).

89. NOVOTNY, L., and V. HEROUT: Plant Substances XIX. Constituents of *Petasites spurius* L. Rhizomes. Collect. Czech. Chem. Comm. **27,** 2462 (1962).

90. NAYA, K., M. NAKAGAWA, M. HAYASHI, K. TSUJI, and M. NAITO: The Constituents of *Petasites japonicus* Maxim. Rhizomes. Tetrahedron Letters **1971,** 2961.

91. NOVOTNY, L., J. TOMAN, F. STARY, A. D. MARQUEZ, V. HEROUT, and F. SORM: Contribution to the Taxonomy of Some European *Petasites* Species. Phytochem. **5,** 1281 (1966).

92. NOVOTNY, L., J. TOMAN, and V. HEROUT: Terpenoids of *P. paradoxus* and *P. kablikianus* in Relation to their Phylogeny. Phytochem. **7,** 1349 (1968).

93. BOHLMANN, F., and C. FISCHER: Notiz über die Synthese des Furanoeremophilans. Chem. Ber. **107,** 1767 (1974).

94. ISHII, H., T. TOZYO, and H. MINATO: Structure of Ligularol and Ligularone from *Ligularia sibirica* Cass. Tetrahedron **21,** 2605 (1965).

95. NOVOTNY, L., and K. KOTVA: Opening of the Furan Ring. Reaction of 6β-Hydroxy-furoeremophilane with Activated Silica Gel. Collect. Czech. Chem. Comm. **39,** 2949 (1974).

95a. TADA, M., and T. TAKAHASHI: Conformational Isomers of Ligularol Acetate and Ligularone, *cis*-Decalin Derivatives. Tetrahedron Letters **1973,** 5169.

96. NAGANO, H., Y. TANAHASHI, Y. MORIYAMA, and T. TAKAHASHI: New Furanoeremo-philane Derivatives from *Farfugium japonicum* Kitamura. Bull. Chem. Soc. Japan **46,** 2840 (1973).

97. NOVOTNY, L., Z. SAMEK, J. HARMATHA, and F. SORM: 9-Hydroxyfuranoeremophilane, a Further Dominating Component from Light Petroleum Extract of *Petasites hybridus* L. Rhizomes. Collect. Czech. Chem. Comm. **34,** 336 (1969).

98. JENNINGS, P. W., S. K. REEDER, J. C. HURLEY, C. N. CAUGHLAN, and G. D. SMITH: Isolation and Structure of One of the Toxic Constituents from *Tetradymia glabrata*. J. Organ. Chem. (USA) **39,** 3392 (1974).

99. PATIL, F., J.-M. LEHN, G. OURISSON, Y. TANAHASHI, and T. TAKAHASHI: La Furannoligularenone. Bull. soc. chim. France **1965,** 3085.

100. PATIL, F., G. OURISSON, Y. TANAHASHI, M. WADA, and T. TAKAHASHI: La Furannoligularénone. Bull. soc. chim. France **1968,** 1047.

101. TAKAHASHI, T.: Furanoligularenones from *Ligularia* Species. Japanese Pat. 10,924 (1967) [Chem. Abstr. **67,** 99986u (1967)].

102. SCHILD, W.: 8,8a-Epoxyfuranoligularan, ein neues Sesquiterpen aus *Senecio silvaticus* L. Tetrahedron **27,** 5735 (1971).

103. MORIYAMA, Y., T. SATO, H. NAGANO, Y. TANAHASHI, and T. TAKAHASHI: 1β,10β-Epoxyfuranoeremophilan-6β-ol, a New Furanosesquiterpene from *Ligularia fischeri* Turcz. Chemistry Letters **1972,** 637.

104. SATO, Y., Y. MORIYAMA, H. NAGANO, Y. TANAHASHI, and T. TAKAHASHI: New Furanosesquiterpenes from *Ligularia fischeri* Turcz. 1β,10β-Epoxyfuranoeremophilan-6β-ol and its Derivatives. Bull. Chem. Soc. Japan **48,** 112 (1975).

105. TADA, M., Y. MORIYAMA, Y. TANAHASHI, and T. TAKAHASHI: New Furanosesqui-terpenes from *Ligularia japonica* Less. Tetrahedron Letters **1971,** 4007.

106. — — — — Furanoeremophilane-6β,10β-diol and its Derivatives. Bull. Chem. Soc. Japan 47, 1999 (1974).
107. TADA, M., Y. TANAHASHI, Y. MORIYAMA, and T. TAKAHASHI: Skeletal Rearrangement of Furanoeremophilane-6β,10β-diol into Farfugin-A and Farfugin-B. Tetrahedron Letters 1972, 5255.
108. NOVOTNY, L., C. TABACIKOVA-WLOTZKA, V. HEROUT, and F. SORM: Constitution of Furanopetasin, the Main Constituent of Petasites officinalis Moench. Rhizomes. Collect. Czech. Chem. Comm. 29, 1922 (1964).
109. NAGANO, H., and T. TAKAHASHI: 6β-Acetoxyfuranoeremophilane-3α-yl Angelate and 6β-Hydroxyfuranoeremophilan-3α-yl Angelate. New Furanoeremophilane Derivatives from Farfugium hibernifllorum Kitamura. Bull. Chem. Soc. Japan 45, 1935 (1972).
110. EAGLE, G. A., D. E. A. RIVETT, D. H. WILLIAMS, and R. G. WILSON: The Structure of Euryopsol, a Furanoeremophilane from Euryops Species. Tetrahedron 25, 5227 (1969).
111. RIVETT, D. E. A., and G. R. WOOLARD: The Structure of Euryopsonol. Tetrahedron 23, 2431 (1967).
112. NOVOTNY, L., Z. SAMEK, V. HEROUT, and F. SORM: The Structure of Kablicin, the Main Component of the Light Petroleum Extracts of Petasites kablikianus and of P. paradoxus Rhizomes. Tetrahedron Letters 1968, 1401 (see erratum facing p. 2281).
113. HORN, D. H. S., J. R. NUNN, and D. E. A. RIVETT: Some Constituents of the Resin of Euryops floribundus. J. S. African Chem. Inst. 7, 22 (1954).
114. RODRIGUEZ-HAHN, A. GUZMAN, and J. ROMO: The Constituents of Cacalia decomposita A. Gray. IV. Structure of Decompostin. Tetrahedron 24, 477 (1968).
115. SAMEK, Z., J. HARMATHA, L. NOVOTNY, and F. SORM: Absolute Configuration of Adenostylone, Neoadenostylone, and Isoadenostylone from Adenostyles alliariae, and of Decompostin from Cacalia decomposita. Collect. Czech. Chem. Comm. 34, 2792 (1969).
116. HARMATHA, J., Z. SAMEK, L. NOVOTNY, V. HEROUT, and F. SORM: The Structure of Adenostylone and Isoadenostylone, two Furanoeremophilanes from Adenostyles alliariae. Tetrahedron Letters 1968, 1409.
117. — — — — — The Structure of Adenostylone, Isoadenostylone and Neoadenostylone-Components of the Rhizomes of Adenostyles alliariae. Collect. Czech. Chem. Commun. 34, 1739 (1969).
118. NOVOTNY, L., Z. SAMEK, and F. SORM: Isolation and Structure of Dimethoxydihydrofuranoeremophilane. Collect. Czech. Chem. Comm. 30, 371 (1965).
119. ISHIZAKI, Y., Y. TANAHASHI, T. TAKAHASHI, and K. TORI: Furanoeremophilan-14β-6α-olide, a New Furanosesquiterpene Lactone from Ligularia hodgsoni Hook. The Structure and Nuclear Overhauser Effects. Chem. Commun. 1969, 551.
120. BOHLMANN, F., C. ZDERO, and N. RAO: Neue Furanosesquiterpene aus Euryops-Arten. Chem. Ber. 105, 3523 (1972).
121. BOHLMANN, F., and N. RAO: Neue Furanoeremophilane aus Othonna quinquedentata Thunb. Tetrahedron Letters 1973, 613.
122. STOESSL, A., C. H. UNWIN, and E. W. B. WARD: Capsidiol, an Antifungal Compound from Capsicum frutescens. Phytopath. Zeitschr. 74, 141 (1972).
123. BAILEY, J. A., R. S. BURDEN, and G. G. VINCENT: Capsidiol: An Antifungal Compound Produced in Nicotiana tabacum and N. clevelandii following Infection with Tobacco Necrosis Virus. Phytochem. 14, 597 (1975).
124. GORDON, M., A. STOESSL, and J. B. STOTHERS: The Structure of Capsidiol, an Antifungal Sesquiterpene from Sweet Peppers. Canad. J. Chem. 51, 748 (1973).
125. BIRNBAUM, G. I., A. STOESSL, S. H. GROVER, and J. B. STOTHERS: The Complete Stereostructure of Capsidiol. X-ray Analysis and ^{13}C Nuclear Magnetic Resonance

of Eremophilane Derivatives Having *trans* Vicinal Methyl Groups. Canad. J. Chem. **52**, 993 (1974).

126. Govindachari, T. R., P. A. Mohamed, and P. C. Parthasarathy: Ishwarane and Aristolochene, Two New Sesquiterpene Hydrocarbons from *Aristolochia indica*. Tetrahedron **26**, 615 (1970).

127. Erdtman, H., and J. G. Topliss: Nootkatene, a New Sesquiterpene Type Hydrocarbon from the Heartwood of *Chamaecyparis nootkatensis*. Acta Chem. Scand. **11**, 1157 (1957).

127a. Andersen, N. H., M. S. Falcone, and D. D. Syrdal: Structures of Vetivenenes and Vetispirenes. Tetrahedron Letters **1970**, 1759.

128. MacLeod, W. D.: The Constitution of Nootkatone, Nootkatene, and Valencene. Tetrahedron Letters **1965**, 4779.

129. Hunter, G. L. K., and W. B. Brogden: Terpenes and Sesquiterpenes in Coldpressed Orange Oil. J. Food Sci. **30**, 1 (1965).

130. — — Conversion of Valencene to Nootkatone. J. Food Sci. **30**, 876 (1965).

131. Attaway, J. A., A. P. Pieringer, and L. J. Barabas: The Origin of Citrus Flavor Components. II. Identification of Volatile Components from Citrus Blossoms. Phytochem. **5**, 1273 (1966).

132. MacLeod, W. D., and Buigues, N.: Sesquiterpenes. I. Nootkatone, a New Flavor Constituent. J. Food Sci. **29**, 565 (1964).

133. Ishida, T., M. Nishimura, S. Hayashi, T. Matsuura, and M. Araki: Identification of Valencene as a Sesquiterpene of Camphor Oil. Chem. and Ind. **1970**, 312.

134. McGuire, H. M., H. C. Odom, and A. R. Pinder: A New Total Synthesis of (±)-Valencene and (±)-Nootkatone. J. Chem. Soc. (London) Perkin Trans. I **1974**, 1879.

135. Lukes, R. M., G. I. Poos, and L. H. Sarett: Approaches to Total Synthesis of Adrenal Steroids. III. 5-Carbomethoxy-5-methylcyclohexene-1,4-dione as a Dienophile. J. Amer. Chem. Soc. **74**, 1401 (1952).

136. Narang, S. A., and P. C. Dutta: Synthetical Studies of Terpenoids. V. A New Synthesis of (±)-6β-Acetoxy-5,5,9β-trimethyl-*trans*-2-decalone. J. Chem. Soc. (London) **1960**, 2842.

137. Wadsworth, W. S., and W. D. Emmons: The Utility of Phosphonate Carbanions in Olefin Synthesis. J. Amer. Chem. Soc. **83**, 1733 (1961).

138. Corey, E. J., and J. Shulman: The Application of Lithium Reagents from (1-Methylthio)alkylphosphonate Esters to the Synthesis of Ketones. J. Organ. Chem. (USA) **35**, 777 (1970).

139. Shaffer, G. W., E. H. Eschinasi, K. L. Purzycki, and A. B. Doerr: Oxidations of Valencene. J. Organ. Chem. (USA) **40**, 2181 (1975).

140. Ohloff, G.: Ger. Patent Applicn. 2,065,461 (Aug. 30, 73).

141. Paknikar, S. K., and C. G. Naik: Stereochemistry of Dihydroagarofurans and Evidence in Support of the Structure of 4,11-Epoxy-*cis*-eudesmane. Tetrahedron Letters **1975**, 1293.

142. Hikino, H., N. Suzuki, and T. Takemoto: Structure and Absolute Configuration of Kusunol. Chem. Pharm. Bull. (Japan) **16**, 832 (1968).

143. Takahashi, S.: A Constituent of Vetiver Oil: Bicyclic Primary Alcohol. Chem. Pharm. Bull. (Japan) **16**, 2447 (1968).

144. Erdtman, H., and Y. Hirose: The Chemistry of the Natural Order Cupressales. 46. The Structure of Nootkatone. Acta Chem. Scand. **16**, 1311 (1962).

145. Pesaro, M., G. Bozzato, and P. Schudel: Total Synthesis of Racemic Nootkatone. Chem. Commun. **1968**, 1152.

146. Marshall, J. A., and R. A. Ruden: The Stereoselective Total Synthesis of Racemic Nootkatone. Tetrahedron Letters **1970**, 1239.

147. — — The Stereoselective Total Synthesis of Racemic Nootkatone. J. Organ. Chem. (USA) **36**, 594 (1971).

148. DASTUR, K. P.: A Stereoselective Approach to Eremophilane Sesquiterpenes. A Synthesis of (±)-Nootkatone. J. Amer. Chem. Soc. **95**, 6509 (1973).

149. — A Stereoselective Approach to Eremophilane Sesquiterpenes. A Synthesis of (±)-Nootkatone and (±)-α-Vetivone. J. Amer. Chem. Soc. **96**, 2605 (1974).

149a. VAN DER GEN, A., L. M. VAN DER LINDE, J. G. WITTEVEEN, and H. BOELENS: Synthesis of Optically Active Eremophilane Sesquiterpenoids from Sabinene. Rec. trav. Chim. Pays-Bas **90**, 1045 (1971).

150. JONES, R. A.: Application of Lanthanide Shift Reagents to the Analysis of NMR Spectra of Flavour Constituents — A Configurational and Conformational Analysis of Nootkatone. Flavour Industry **5**, 125 (1974).

151. STEVENS, K. L., and J. R. SCHERER: Photolysis of Nootkatone. J. Agric. Food Chem. **16**, 673 (1968).

152. STEVENS, K. L.: Photo-induced Reduction of Nootkatone. J. Food Sci. **34**, 484 (1969).

153. — Photolysis of Nootkatone — an Unusual Intramolecular Photocyclisation. J. Sci. Food Agric. **21**, 593 (1970).

154. ST. PFAU, A., and P. A. PLATTNER: Sur les Vétivones, Constituants Odorants des Essences de Vétiver. Helv. Chim. Acta **22**, 640 (1939).

155. — — Sur la Constitution de la β-Vétivone. Helv. Chim. Acta **23**, 768 (1940).

156. NAVES, Y.-R., and E. PERROTTET: Sur les α- et β-Vétivones. Helv. Chim. Acta **24**, 3 (1941).

157. ENDO, K., and P. DE MAYO: α-Vetivone. Chem. Commun. **1967**, 89.

158. — — α-Vetivone. Chem. Pharm. Bull. (Japan) **17**, 1324 (1969).

159. MARSHALL, J. A., and N. H. ANDERSEN: The Structure of α-Vetivone (Isonootkatone). Tetrahedron Letters **1967**, 1611.

160. MARSHALL, J. A., H. FAUBL, and T. M. WARNE: The Total Synthesis of Racemic Isonootkatone (α-Vetivone). Chem. Commun. **1967**, 753.

161. MARSHALL, J. A., and T. M. WARNE: The Total Synthesis of (±)-Isonootkatone. Stereochemical Studies of the Robinson Annelation Reaction with 3-Penten-2-one. J. Organ. Chem. (USA) **36**, 178 (1971).

162. STEVENS, K. L., D. G. GUADAGNI, and D. J. STERN: Odour Character and Threshold Values of Nootkatone and Related Compounds. J. Sci. Food Agric. **21**, 590 (1970).

163. SASTRY, S. D., M. L. MAHESWARI, K. K. CHAKRAVARTI, and S. C. BHATTACHARYYA: Chemical Constituents of *Nardostachys jatamansi*. Perf. Essent. Oil Rec. **58**, 154 (1967).

164. — — — — The Structure of Nardostachone. Tetrahedron **23**, 2491 (1967).

165. PINDER, A. R.: The Structure of Nardostachone. Tetrahedron Letters **1970**, 413.

166. KÖNST, W. M., L. M. VAN DER LINDE, and J. G. WITTEVEEN: Recent Developments in the Chemistry of Eremophilanes. Internat. Flavours and Food Add. **6**, 121 (1975).

167. SEMMLER, F. W., and W. JAKUBOWICZ: Zur Kenntnis der Bestandteile ätherischer Öle [Trennung und Eigenschaften der im Ostindischen Copaivabalsam-Öl vorkommenden Sesquiterpene (Gurjunene); Derivate dieser Sesquiterpene]. Ber. dtsch. Chem. Ges. **47**, 1141 (1914).

168. SORM, F., M. HOLUB, V. SYKORA, J. MLEZIVA, M. STREIBL, J. PLIVA, B. SCHNEIDER, and V. HEROUT: Sesquiterpenic Hydrocarbons from Oil of Sweet Flag. Collect. Czech. Chem. Comm. **18**, 512 (1953).

169. BÜCHI, G., F. GREUTER, and T. TOKOROYAMA: Structure of Calarene and Stereochemistry of Aristolone. Tetrahedron Letters **1962**, 827.

170. NARAYANAN, C. S., K. S. KULKARNI, A. S. VAIDYA, S. KANTHAMANI, G. L. KUMARI, B. V. BAPAT, S. K. PAKNIKAR, S. N. KULKARNI, G. R. KELKAR, and S. C. BHATTACHARYYA: Components of Indian Valerian Root Oil. Tetrahedron **20**, 963 (1964).

171. VRKOC, J., J. KREPINSKY, V. HEROUT, and F. SORM: The Nature of Sesquiterpenic

Hydrocarbon Calarene and Structure of β-Gurjunene. Tetrahedron Letters **1963**, 225.

172. STREITH, J., P. PESNELLE, and G. OURISSON: Le β-Gurjunène. Identification avec le Calarène. Bull. soc. chim. France **1963**, 518.

173. VRKOC, J., J. KREPINSKY, V. HEROUT, and F. SORM: The Identity of β-Gurjunene with Calarene. Collect. Czech. Chem. Comm. **29**, 795 (1964).

174. CARBONI, S., A. DA SETTIMO, V. MALAGUZZI, A. MARSILI, and P. L. PACINI: Structure of α-Ferulene, a Sesquiterpene Having a (+)-Aristolane Skeleton. Tetrahedron Letters **1965**, 3017.

175. WEINHEIMER, A. L., P. H. WASHECHECK, D. VAN DER HELM, and M. B. HOSSAIN: The Sesquiterpene Hydrocarbons of the Gorgonian *Pseudopterogorgia americana;* the Non-isoprenoid β-Gorgonene. Chem. Commun. **1968**, 1070.

176. KREPINSKI, J., G. JOMMI, Z. SAMEK, and F. SORM: The Sesquiterpenenic Constituents of *Aristolochia debilis.* The Structure of Debilone. Collect. Czech. Chem. Comm. **35**, 745 (1970).

176a. PESNELLE, P., and G. OURISSON: Isolement de l'Aristolène dans l'Essence de Nard. Bull. soc. chim. France **1963**, 912.

177. SASTRY, S. D., M. L. MAHESWARI, K. K. CHAKRAVARTI, and S. C. BHATTACHARYYA: The Structure of Calarenol. Tetrahedron **23**, 1997 (1967).

178. KARIYONE, T., and S. NAITO: Components of *Aristolochia debilis.* I. J. Pharm. Soc. Japan **75**, 1511 (1955) [Chem. Abstr. **50**, 10684a (1956)].

179. FURUKAWA, S., and N. SOMA: Structure of Aristolone, a Constituent of *Aristolochia debilis.* I. J. Pharm. Soc. Japan **81**, 559 (1961). [Chem. Abstr. **55**, 21160d (1961)].

180. FURUKAWA, S., K. OYAMADA, and N. SOMA: Structure of Aristolone, a Constituent of *Aristolochia debilis.* II. J. Pharm. Soc. Japan **81**, 565 (1961) [Chem. Abstr. **55**, 21160i (1961)].

181. FURUKAWA, S.: Structure of Aristolone, a Constituent of *Aristolochia debilis.* III. J. Pharm. Soc. Japan **81**, 570 (1961) [Chem. Abstr. **55**, 21161g (1961)].

182. BERGER, C., M. FRANCK-NEUMANN, and G. OURISSON: Synthèse Totale de l'Aristolone. Tetrahedron Letters **1968**, 3451.

183. PIERS, E., R. W. BRITTON, and W. DE WAAL: Total Synthesis of (±)-Aristolone. Canad. J. Chem. **47**, 831 (1969).

184. BERGER, C., M. FRANCK-NEUMANN, and G. OURISSON: Photodécarbonylation de l'Aristolone et d'Autres Bicyclo[4.1.0]hepténones. Tetrahedron Letters **1970**, 3531.

185. NISHIDA, R., and Z. KUMAZAWA: 3-Oxoishwarane, a New Sesquiterpene Ketone from *Aristolochia debilis* Sieb. et. Zucc. Agric. and Biol. Chem. (Japan) **37**, 341 (1973).

186. RÜCKER, G.: $\Delta^{1,(10)}$-Aristolenon-(2) und 1,8,9,10-Tetradehydroaristolanon-(2), neue Sesquiterpen-Ketone vom Aristolan-Typ. Liebigs Ann. Chem. **717**, 221 (1968).

187. RAO, U. S. K., B. L. MANJUNATH, and K. N. MENON: Chemical Examination of the Roots of *Aristolochia indica* L. II. The Essential Oil. J. Indian Chem. Soc. **12**, 494 (1935).

188. TENG, L. C., and J. F. DEBARDELEBEN: A Novel Tricyclic Sesquiterpene from the Oil of Orejuela of *Cymbopetalum penduliflorum* Dunal. Experientia **27**, 14 (1971).

189. KELLY, R. B., J. ZAMECNIK, and B. A. BECKETT: Total Synthesis of the Sesquiterpenoid Ishwarane: Structure of Ishwarone. Chem. Commun. **1971**, 479.

190. — — — Total Synthesis of Ishwarane, a Tetracyclic Sesquiterpene. Canad. J. Chem. **50**, 3455 (1972).

191. KELLY, R. B., and J. ZAMECNIK: The Total Synthesis of Isoishwarone: Structures of the Sesquiterpenoids Ishwarane and Ishwarone. Chem. Commun. **1970**, 1102.

192. DESAI, H. K., D. H. GAWAD, T. R. GOVINDACHARI, B. S. JOSHI, V. N. KAMAT, J. D. MODI, P. A. MOHAMED, P. C. PARTHASARATHY, S. J. PATNAKAR, A. R. SIDHAYE, and N. VISWANATHAN: Chemical Investigations of Some Indian Plants. V. Indian J. Chem. **8**, 851 (1970).

193. GOVINDACHARI, T. R., and P. C. PARTHASARATHY: Ishwarol, a New Tetracyclic Sesquiterpene Alcohol from *Aristolochia indica* L. Indian J. Chem. **9**, 1310 (1971).

194. GANGULY, A. R., K. W. GOPINATH, T. R. GOVINDACHARI, K. NAGARAJAN, B. R. PAI, and P. C. PARTHASARATHY: Ishwarone, a Novel Tetracyclic Sesquiterpene. Tetrahedron Letters **1969**, 133.

195. FUHRER, H., A. K. GANGULY, K. W. GOPINATH, T. R. GOVINDACHARI, K. NAGARAJAN, B. R. PAI, and P. C. PARTHASARATHY: Ishwarone. Tetrahedron **26**, 2371 (1970).

196. GOVINDACHARI, T. R., K. NAGARAJAN, and P. C. PARTHASARATHY: Absolute Stereochemistry of Ishwarone. Chem. Commun. **1969**, 823.

197. SWAMINATHAN, S., and G. S. MURTHY: The Structure of Bromo-oxoishwarone. Current Sci. (India) **38**, 135 (1969).

198. NAYA, K., and M. KOBAYASHI: The Preparation of Fukinane, a New Skeletal Sesquiterpenic Hydrocarbon. Bull. Chem. Soc. Japan **44**, 258 (1971).

199. SHIRAHATA, K., N. ABE, T. KATO, and Y. KITAHARA: Bakkenolide-E, a Minor Component of the Bud of *Petasites japonicus* subsp. *giganteus* Kitam. Bull. Chem. Soc. Japan **41**, 1732 (1968).

200. ABE, N., R. ONODA, K. SHIRAHATA, T. KATO, M. C. WOODS, and Y. KITAHARA: The Structure of Bakkenolide-A. Tetrahedron Letters **1968**, 369.

201. SHIRAHATA, K., T. KATO, Y. KITAHARA, and N. ABE: Constituents of the Genus *Petasites*. IV. Bakkenolide-A, a Sesquiterpene of Novel Carbon Skeleton. Tetrahedron **25**, 3179 (1969).

202. NAYA, K., I. TAKAGI, M. HAYASHI, S. NAKAMURA, M. KOBAYASHI, and S. KATSUMURA: Structure of New Skeletal Sesquiterpenoids from *Petasites japonicus* Maxim. Chem. and Ind. **1968**, 318.

203. EVANS, D. A., and C. L. SIMS: The Total Synthesis of (+)-Bakkenolide-A. Tetrahedron Letters **1973**, 4691.

204. ABE, N., R. ONODA, K. SHIRAHATA, T. KATO, M. C. WOODS, Y. KITAHARA, K. RO, and T. KURIHARA: The Structure of Bakkenolides-B, -C, and -D as Determined by the Use of an N.O.E. Tetrahedron Letters **1968**, 1993.

205. SHIRAHATA, K., T. KATO, Y. KITAHARA, and N. ABE: Mass Spectra of Bakkenolides and Their Derivatives. Tetrahedron **25**, 4671 (1969).

206. KATAYAMA, C., A. FURUSAKI, I. NITTA, M. HAYASHI, and K. NAYA: Crystal Structure of Fukinolidol Di-bromoacetate. Bull. Chem. Soc. Japan **43**, 1976 (1970).

206a. ROBINSON, R.: The Structural Relations of Natural Products, p. 12. London: Oxford University Press. 1955.

207. HENDRICKSON, J. B.: Stereochemical Implications in Sesquiterpene Biogenesis. Tetrahedron **7**, 82 (1959).

208. ERDTMAN, H., and T. NORIN: The Chemistry of the Order Cupressales: Fortschr. Chem. Org. Naturstoffe **24**, 206 (1966) (see particularly p. 245).

209. SULSER, H., J. R. SCHERER, and K. L. STEVENS: The Structure of Paradisiol, a New Sesquiterpene Alcohol from Grapefruit Oil. J. Organ. Chem. (USA) **36**, 2422 (1971).

210. HUFFMAN, J. W., and L. H. ZALKOW: The Structure of Paradisiol and its Identity to Intermedeol. Tetrahedron Letters **1973**, 751.

211. HOMMA, A., M. KATO, M.-D. WU, and A. YOSHIKOSHI: Minor Sesquiterpene Alcohols of Vetiver Oil. Tetrahedron Letters **1970**, 231.

212. HEATHCOCK, C. H., and T. R. KELLY: Sesquiterpenoids. IV. Acid-catalyzed Methyl Migration in 9-Methyldecalins. Tetrahedron **24**, 3753 (1968).

213. DUNHAM, D. J., and R. G. LAWTON: Spiro Intermediates in Sesquiterpenes Rearrangements and Synthesis. J. Amer. Chem. Soc. **93**, 2075 (1971).

214. Baker, F. C., C. J. W. Brooks, and S. A. Hutchinson: Biosynthesis of Capsidiol in Sweet Peppers *(Capsicum frutescens)* Infected with Fungi: Evidence for Methyl Group Migration from ^{13}C Nuclear Magnetic Resonance Spectroscopy. Chem. Commun. **1975**, 293.

215. Zabkiewicz, J. A., R. A. B. Keates, and C. J. W. Brooks: Incorporation of Mevalonolactone into *Petasites hybridus:* Effect of Synthetic Inhibitors on Sesquiterpenoid and Sterol Production. Phytochem. **8**, 2087 (1969).

215a. Brooks, C. J. W., and R. A. B. Keates: Biosynthesis of Petasin in *Petasites hybridus.* Phytochem. **11**, 3235 (1972).

216. Kitagawa, I., Y. Yamazoe, R. Takeda, and I. Yosioka: Conversion of Dihydroalantolactone to Eremophilane-type Derivatives: a Biogenetic-type Transformation. Tetrahedron Letters **1972**, 4843.

217. Kitagawa, I., H. Shibuya, Y. Yamazoe, H. Takeno, and I. Yosioka: Conversion of Dihydroantolactone to Tetrahydroligularenolide. A Biogenetic-Type Transformation of Eudesmanolide to Eremophilanolide. Tetrahedron Letters **1974**, 111.

218. Kitagawa, I., Y. Yamazoe, H. Shibuya, R. Takeda, H. Takeno, and I. Yosioka: Biogenetically Patterned Transformation of Eudesmanolide to Eremophilanolide. I. Angular Methyl Migration of 5α,6α-Epoxydihydroalantolactone. Chem. Pharm. Bull. (Japan) **22**, 2662 (1974).

219. Kitagawa, I., H. Shibuya, H. Takeno, T. Nishino, and I. Yosioka: Biogenetically Patterned Transformation of Eudesmanolide to Eremophilanolide. II. Structures of Minor Products Obtained by Acid Treatment of 5α,6α-Epoxyeudesman-8β,12-olide. Chem. Pharm. Bull. (Japan) **24**, 56 (1976).

(Received April 5, 1976)

Phytoalexine und verwandte Pflanzenstoffe

Von D. Gross, Institut für Biochemie der Pflanzen, Forschungszentrum für Molekularbiologie und Medizin der Akademie der Wissenschaften der DDR, Halle (Saale), Deutsche Demokratische Republik

Mit 4 Abbildungen

Inhalt

I. Einführung

Die intensive Bearbeitung des Resistenzphänomens im Pflanzenreich hat ergeben, daß höhere Pflanzen vielfach Abwehrstoffe enthalten, die einen Befall und eine Erkrankung des Pflanzengewebes durch pilzliche und bakterielle Krankheitserreger sowie durch Viren hemmen oder verhindern. Dieser stoffliche Abwehrmechanismus sowie spezielle organspezifische Strukturbarrieren und andere Resistenzfaktoren stellen allgemein einen wirksamen Schutz gegen pathogene Mikroorganismen dar. Verschiedene Pflanzenarten, vor allem einige Sorten unserer hochgezüchteten Kulturpflanzen, sind jedoch oft anfällig gegen bestimmte Erreger. Nach KRANZ (173) beruhen Pflanzenkrankheiten zu über 90% auf Infektion. Dabei bilden Mykosen den Hauptanteil. Gegenüber dem ansteigenden Auftreten von Virosen sind Bakteriosen bisher nur vereinzelt von wirtschaftlicher Bedeutung.

Die vorliegende Übersicht befaßt sich mit fungitoxisch wirksamen endogenen Pflanzenstoffen, wobei vor allem chemische und biochemische Probleme strukturbekannter Phytoalexine und phytoalexinartiger Verbindungen behandelt werden. Methodische Einzelheiten der Isolierung, Strukturaufklärung und Biotestung können nur angedeutet werden. Bei der überaus großen Fülle an phytopathologisch orientierten Arbeiten muß in den meisten Fällen auf die zitierte Originalliteratur verwiesen werden.

II. Präinfektionelle Abwehrstoffe

Fungitoxische Pflanzeninhaltsstoffe können in Anlehnung an STOESSL (277) in präinfektionelle und postinfektionelle Abwehrstoffe unterteilt werden, obgleich eine derartige Aufgliederung problematisch und eine eindeutige Zuordnung nicht immer möglich ist. Die Gruppe der präinfektionellen Verbindungen umfaßt fungitoxische und fungistatische Substanzen, die bereits in der gesunden, d. h. nicht infizierten Pflanze in ausreichender Konzentration vorliegen und die mit dazu beitragen, das Keimen von Sporen bzw. das Eindringen oder Ausbreiten pathogener Mikroorganismen zu unterdrücken oder zu verhindern. Zu diesen prä-infektionellen oder präformierten Inhibitoren gehören zahlreiche Pflanzen-inhaltsstoffe, die auf Grund ihrer chemischen Struktur unterschiedlichen Naturstoffklassen angehören: hydroxylierte alipathische oder olefinische Carbonsäuren, phenolische Verbindungen wie Catechol und Protocatechu-säure, Aminosäuren, Chinone, Cumarine, Tannine, Flavonoide, Tropo-lone, Stilbene, Alkaloide und weitere Sekundärstoffe. Als ein Beispiel charakteristischer präformierter Abwehrstoffe seien die Tuliposide an-geführt, die 1967 als fungitoxische Substanzen in der Narbe, aber auch in Stengel, Blatt und Zwiebel von *Tulipa gesneriana* L. nachgewiesen worden sind (*256*). Nach Isolierung von α-Methylen-γ-butyrolacton (**3**) (*31*) sowie einer zu diesem Lacton und D-Glucose hydrolysierbaren glucosidischen Verbindung (*44*) wurden 1968/1969 die Strukturen von Tuliposid A (**1**) und Tuliposid B (**4**) veröffentlicht (*290, 291*). Tuliposid A konnte als Acyl-β-D-glucopyranosid der γ-Hydroxy-α-methylenbutter-säure (**2**) identifiziert werden, während Tuliposid B als Glucoseester der β,γ-Dihydroxy-α-methylenbuttersäure (**5**) charakterisiert wurde.

HOCH₂–CH₂–C–COOR
$$\text{HOCH}_2-\text{CH}_2-\underset{\underset{\text{CH}_2}{\|}}{\text{C}}-\text{COOR}$$

(**1**) R = glucosyl
(**2**) R = H

$$\text{HOCH}_2-\text{CHOH}-\underset{\underset{\text{CH}_2}{\|}}{\text{C}}-\text{COOR}$$

(**4**) R = glucosyl
(**5**) R = H

(**3**)

(**6**)

Bei saurer Hydrolyse entstehen aus Tuliposid A und B D-Glucose und α-Methylen-γ-butyrolacton (Tulipalin A) (**3**) bzw. β-Hydroxy-α-methylen-γ-butyrolacton (Tulipalin B) (**6**). Alkalische Hydrolyse mit Barium-hydroxid führt unter Abspaltung von Glucose zu den entsprechenden Hydroxy-α-methylenbuttersäuren (**2**) und (**5**).

Die Tuliposide A und B besitzen antibakterielle und fungitoxische Aktivität, während die beiden nach Glucoseabspaltung entstehenden Hydroxy-α-methylenbuttersäuren (**2**) und (**5**) keine nennenswerte biolo-gische Aktivität aufweisen. Dagegen sind die Lactone (**3**) und (**6**) anti-bakteriell und stark fungitoxisch wirksam. In diesem Zusammenhang erscheint es interessant, daß eine Anzahl wachstumsregulatorisch wirk-samer Pflanzeninhaltsstoffe als charakteristisches Strukturelement eben-falls eine α-Methylen-γ-butyrolactongruppierung enthält (vgl. *105*).

Nach neueren Untersuchungen sind die Tuliposide bei Pflanzen der Ordnung *Liliiflorae* weit verbreitet (*266*). Tuliposid A wurde in allen untersuchten Arten von *Erythronium, Tulipa, Gagea, Bomarea* und *Alstroemeria* nachgewiesen, während Tuliposid B vor allem in *Erythronium*- und *Tulipa*-Arten vorkommt.

Die Tatsache, daß Tulpen von *Botrytis tulipae,* nicht aber von *B. cinerea* befallen werden, hat zu interessanten Untersuchungsergebnissen geführt (*23, 257, 258*). Es wurde nachgewiesen, daß die Tuliposide A und B von *B. tulipae* zu den fungitoxisch unwirksamen Hydroxy-α-methylenbuttersäuren (**2**) und (**5**) umgewandelt und dadurch entgiftet werden. Bei *B. cinerea* entstehen dagegen die fungitoxisch stärker wirksamen Lactone (**3**) und (**6**), die eine Erkrankung des Gewebes durch *B. cinerea* verhindern. Weitere präinfektionelle Abwehrstoffe sind in Übersichten von STOESSL (*277*) sowie von FAWCETT und SPENCER (*86*) umfassend behandelt.

III. Postinfektionelle Abwehrstoffe

Zu dieser Gruppe zählen fungitoxische Pflanzenstoffe, deren Bildung erst nach einer Infektion des Pflanzengewebes durch einen Erreger und dessen Stoffwechseltätigkeit induziert wird. Infolge einer Wechselwirkung zwischen Wirtspflanze und Erreger wird im infizierten Gewebe eine Umsteuerung bestimmter Stoffwechselwege induziert, was zu einer Neubildung oder starken Vermehrung fungistatisch oder fungitoxisch wirksamer Substanzen führt. Als postinfektionelle Abwehrstoffe sind vor allem bestimmte phenolische Substanzen und aromatische Hydroxysäuren, aber auch verschiedene Chinone, Flavonoide, Anthocyane, Steroide und weitere sekundäre Pflanzenstoffe zu nennen.

Zu den postinfektionellen Inhibitoren gehört die wichtige Gruppe der Phytoalexine und phytoalexinähnlichen Pflanzenstoffe, auf die im nachfolgenden näher eingegangen wird. Der Begriff Phytoalexine (griechisch *phytos* Pflanze, *alekein* abwehren) geht auf Arbeiten des Phytopathologen K. O. MÜLLER zurück, der auf Grund umfangreicher Untersuchungen über die durch *Phytophthora infestans* (Mont.) de Bary bewirkte Kraut- und Knollenfäule der Kartoffel 1940 gemeinsam mit BÖRGER die Phytoalexintheorie veröffentlicht hat (*210*). Es wurde postuliert, daß durch den Kontakt eines Parasiten mit pflanzlichen Wirtszellen eine an die lebende Zelle gebundene Abwehrreaktion induziert wird, wobei das hemmende Prinzip stofflicher Natur ist. Diese in späteren Jahren erweiterte und modifizierte Phytoalexintheorie (vgl. *171, 172, 208*) hat die Bearbeitung stofflicher Abwehrmechanismen sehr stimuliert und zur Isolierung zahlreicher Phytoalexine und phytoalexinähnlicher Substanzen geführt. Die

Abgrenzung der Phytoalexine von anderen antimikrobiell wirkenden Pflanzeninhaltsstoffen ist oft nicht eindeutig, und vielfach entsprechen die heute in der Literatur als Phytoalexine bezeichneten Substanzen nicht streng der MÜLLERschen Definition.

In vorliegender Arbeit wird der Begriff Phytoalexin weit gefaßt. Es werden darunter niedermolekulare fungitoxische Verbindungen verstanden, die in der gesunden Pflanze nicht oder in nur geringen Konzentrationen vorkommen und die nach Infektion oder Inokulation mit Pilzen, Bakterien oder Viren in wirksamer Menge entweder *de novo* synthetisiert oder verstärkt akkumuliert werden.

Über Phytoalexine existieren bereits einige kurze Einführungen und Abhandlungen sowie umfassendere Übersichten (*58, 59, 76, 77, 145, 174, 180, 198a, 200, 201, 245, 277, 281b, 301, 331*). In vorliegender Arbeit erfolgt die Einteilung der Phytoalexine nicht nach ihrem Vorkommen in bestimmten Pflanzen, sondern rein nach strukturchemischen und biochemischen Gesichtspunkten.

IV. Phytoalexine und phytoalexinähnliche Verbindungen

1. Polyacetylene

a) Safinol und Dehydrosafinol

Verschiedene der überaus zahlreichen Polyacetylene der Compositen besitzen fungitoxische Aktivität. Als Phytoalexin bzw. phytoalexinähnliche Substanz dieser Stoffklasse ist das Safinol, $C_{13}H_{12}O_2$, zu nennen, das aus *Carthamus tinctorius* L. (*Compositae*) nach Infektion mit *Phytophthora drechsleri* Tucker isoliert (*285, 286*) und als trans,trans-3,11-Tridecadien-5,7,9-triin-1,2-diol (**7**) charakterisiert worden ist (*8, 9, 286*). Dieses Polyacetylen war als Inhaltsstoff von *C. tinctorius* bereits bekannt (*39*).

$$CH_3CH=CH-(C\equiv C)_3-CH=CH-CHOH-CH_2OH$$

(**7**)

Infizierte Hypokotyle von *C. tinctorius* enthalten als weitere fungitoxische Substanz 3,4-Dehydrosafinol, eine Verbindung der Summenformel $C_{13}H_{10}O_2$, deren Struktur als trans-11-Tridecen-3,5,7,9-tetrain-1,2-diol (**8**) ermittelt wurde (*9*).

$$CH_3CH=CH-(C\equiv C)_4-CHOH-CH_2OH$$

(**8**)

Safinol und Dehydrosafinol hemmen das Mycelwachstum von *P. drechsleri*, wobei die ED_{50} für Safinol 12 µg/ml (*286*) und für Dehydrosafinol 1,7 µg/ml (*9*) beträgt.

Während Dehydrosafinol bisher nur in infizierten Pflanzen nachgewiesen werden konnte (*9*), findet sich Safinol auch in gesundem Gewebe, allerdings erhöht sich der Safinolgehalt nach Infektion mit *P. drechsleri* um mehr als das 5fache (55 µg Safinol/100 g Frischgewicht) (*10, 286, 287*).

b) Wyeron

Eine der bereits Anfang der sechziger Jahre in *Vicia faba* L. (*Papilionaceae*) beobachteten fungitoxischen Verbindungen (*84, 315*) konnte 1965 in ihrer Struktur aufgeklärt werden (*87, 89*). Es handelt sich um einen acetylenischen furanoiden Ketoester der Summenformel $C_{15}H_{14}O_4$, der als Wyeron (**10**) bezeichnet wurde und auch synthetisch zugänglich ist. Nach *Botrytis*infektion lassen sich in *Vicia faba* neben Wyeron und der nachfolgend zu besprechenden Wyeronsäure weitere 5 fungitoxische Verbindungen nachweisen (*116a*), von denen eine kürzlich als 4,5-Wyeronepoxid charakterisiert werden konnte (*116c, 116d*). Dieses Epoxid besitzt gegenüber *Botrytis cinerea* und *B. fabae* eine ED_{50} von 6,4 bzw. 16 µg/ml und ist damit fungitoxisch wirksamer als Wyeron. Konidien oder Mycel von *B. cinerea* und *B. fabae* wandeln Wyeron (**10**) und sein Epoxid *in vitro* zu Wyerol (**10a**) um (*116b, 116c*). Letzteres wird von *B. fabae* zu 4,5-Dihydro-4,5-dihydroxywyerol metabolisiert (*116c*).

	R = H	R₁ = O
(9)	R = H	R_1 = O
(10)	R = CH_3	R_1 = O
(10a)	R = CH_3	R_1 = H, OH

Wyeron ist gegen zahlreiche Phytopathogene und Dermatophyte fungitoxisch wirksam (*87, 88*). Die ED_{50} beträgt gegenüber *Botrytis cinerea* Pers. 100 µg/ml, gegenüber *Alternaria brassicicola* und *Glomeralla cingulata* 10 µg/ml. Für die fungitoxische Aktivität wird vor allem der $R-CO-C \equiv C$-Teil des Wyeronmoleküls verantwortlich gemacht.

Der Wyerongehalt steigt in Blättern von *Vicia faba* nach Infektion mit *Botrytis fabae* Sardina innerhalb 4 Tagen von < 0,1 µg/g auf 45 µg/g Frischgewicht an und erhöht sich damit gegenüber dem gesunden Blattgewebe um mehr als das 500fache (*85*). Das spätere Absinken der Wyeron-konzentration beruht möglicherweise auf einer Umwandlung zur fungitoxisch wirksamen Wyeronsäure. Inokulation von *Vicia faba* L. mit *Phytophthora megasperma var. sojae* A. A. Hildb. bewirkt ebenfalls eine Akkumulation von Wyeron (*158*). Nach 6tägiger Inkubation mit *Botrytis fabae* bzw. *B. cinerea* beträgt der Gehalt an Wyeronepoxid im Kotyledonengewebe von *Vicia faba* 179 bzw. 93 µg/g Frischgewicht (*116c*).

c) Wyeronsäure

Intensive Suche nach weiteren fungitoxischen Substanzen in *Vicia faba* nach Infektion mit *Botrytis spp.*, vor allem *B. cinerea* Pers. und *B. fabae* Sardina, hat zur Isolierung eines wasserlöslichen Inhibitors geführt (*75, 79, 246*), der 1970 als Wyeronsäure (*9*) identifiziert worden ist (*189*) und sich mit Diazomethan zu vorstehend beschriebenem Wyeron verestern läßt.

Die quantitativen Veränderungen im Gehalt an Wyeronsäure nach Infektion mit *B. cinerea* und *B. fabae* sind eingehend untersucht worden (*77, 116a*), ebenso die fungitoxische Aktivität von Wyeronsäure (*77, 78, 189, 193, 246*). Infizierte Bohnenblätter enthalten 40—50 µg/g Frischgewicht, während der Wyeronsäuregehalt gesunder Blätter unter 0,1 µg/g Frischgewicht liegt. Die unterschiedliche Pathogenität von *B. fabae* und *B. cinerea* beruht offensichtlich darauf, daß Wyeronsäure von *B. fabae* zu einem Detoxikationsprodukt abgebaut wird und daß dadurch eine Akkumulation der fungitoxisch wirksameren Wyeronsäure verhindert wird. Als Metabolit des Wyeronsäurestoffwechsels tritt bei *B. fabae* eine Verbindung auf, die als Hexahydroderivat der Wyeronsäure (*11*) identifiziert worden ist (*195, 196*). Dieses Reduktionsprodukt ist gegenüber *B. fabae* und *B. cinerea* weit weniger toxisch als Wyeronsäure. Dieser Befund unterstreicht die Bedeutung der acetylenischen Ketofunktion für die biologische Wirksamkeit. *B, cinerea* kann dagegen diese Reduktion der Wyeronsäure nicht durchführen, wodurch sich die unterschiedliche Pathogenität von *B. fabae* und *B. cinerea* erklären läßt.

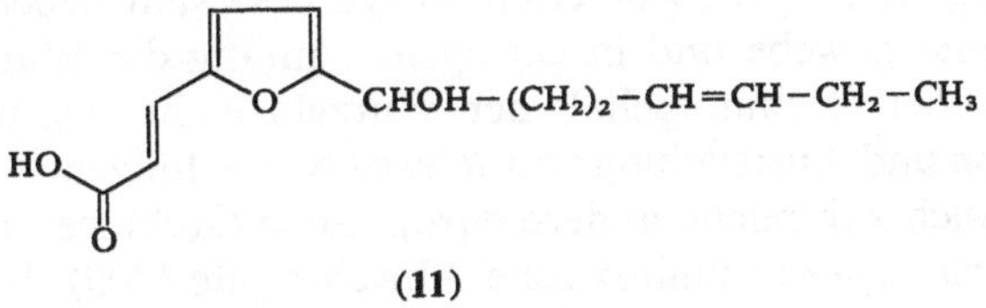

(11)

2. Dihydrophenanthrene

a) Orchinol

Es ist schon seit den klassischen Untersuchungen von NOEL BERNARD bekannt, daß bei manchen Orchideen die Wurzeln, nicht aber die Knollen von einer Infektionskrankheit befallen werden, deren Erreger bestimmte Mykorrhizapilze wie *Rhizoctonia repens* Bern. sind. Die Krankheit greift jedoch nicht auf Knollen und Stengel über und kommt nach einiger Zeit durch chemische Abwehrreaktionen zum Stillstand. Dieser Befund hat zu umfangreichen Arbeiten über fungitoxische Abwehrstoffe bei Orchideen geführt.

1957 konnte aus Knollen von *Orchis militaris* L. nach Infektion mit *R. repens* eine schwach phenolische Substanz der Summenformel $C_{16}H_{16}O_2$ isoliert werden, die als Orchinol bezeichnet wurde (*40*). Eine intensive Bearbeitung, insbesondere durch GÄUMANN und Mitarbeiter (*93, 94, 96, 97, 98, 99*), hat in den nachfolgenden Jahren ergeben, daß es sich um einen postinfektionellen Abwehrstoff handelt, der nur in infizierten, nicht aber in gesunden Knollen nachzuweisen ist.

Orchinol konnte 1963 als 2-Hydroxy-5,7-dimethoxy-9,10-dihydrophenanthren (**12**) charakterisiert werden (*114, 115, 116*). Dehydroorchinol (*211*) und Orchinol (*272, 279*) sind seit kurzem auch synthetisch zugänglich. Die experimentell noch nicht untersuchte Biosynthese des Orchinols soll nach BIRCH über Stilbene verlaufen (*36*).

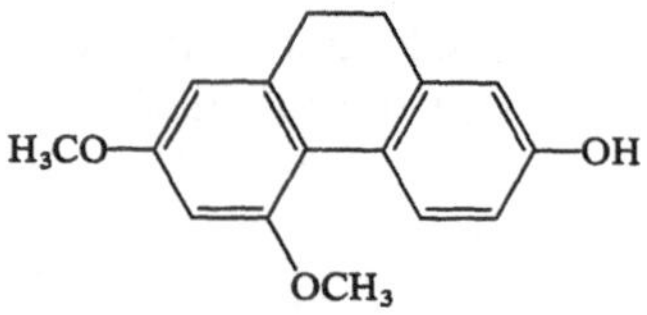

(12)

Durch physiologische Experimente mit infizierten Gewebestücken von *Orchis militaris* ließ sich zeigen, daß die Orchinolbildung etwa 36 h nach Infektion einsetzt und nach etwa 8 Tagen ihr Maximum erreicht. Der Orchinolgehalt künstlich infizierter Knollen lag zwischen 130—500 mg/kg Frischgewicht (*96*). Zur Orchinolsynthese sind neben der Knolle auch das Stengelgewebe und in geringem Umfang die Wurzel befähigt. Jedoch reicht der Orchinolgehalt der Wurzeln nicht aus, um diese vor einer Infektion und Ansiedelung von *R. repens* zu schützen. Außer *O. militaris* bilden auch zahlreiche andere europäische Orchideen bei Infektion mit *Rhizoctonia repens* fungitoxische Abwehrstoffe (*100*). Bei 16 von 24 getesteten Orchideenarten wurde Orchinol nachgewiesen. Neben *R. repens* sind weitere Mykorrhizapilze zur Induktion der Orchinolbildung bei *O. militaris* fähig.

Orchinol besitzt eine starke fungitoxische Wirkung gegen *R. repens* und weist gegenüber anderen Mykorrhizapilzen ein breites Wirkungsspektrum auf (*90, 98*). Durch die mehrmonatige Verweilzeit von Orchinol

(*90*) und seine geringe Translokalisation in den Knollen (*93, 96*) sind diese langfristig gegen eine Infektion geschützt.

b) Hircinol

Weitere Untersuchungen über chemische Abwehrstoffe der Orchideen haben zur Auffindung einer zweiten fungitoxisch wirksamen Substanz geführt, die neben geringen Mengen Orchinol aus *Loroglossum hircinum* (L.) Rich. isoliert und als Hircinol bezeichnet wurde (*95, 296*). Hircinol, $C_{15}H_{14}O_3$, wurde in seiner Struktur als 4,7-Dihydroxy-5-methoxy-9,10-dihydro-phenanthren (13) aufgeklärt und besitzt damit eine dem Orchinol verwandte Dihydrophenanthrenstruktur (*90, 296*).

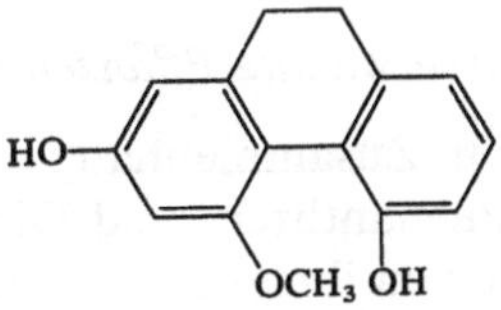

(13)

Sterile Knollengewebe von *Loroglossum hircinum* enthalten kein Hircinol; seine Bildung wird erst durch bestimmte Mykorrhizapilze, meistens Ascomyceten, seltener Basidiomyceten, induziert. Infizierte Knollen enthalten 250—400 mg Hircinol pro kg Frischgewicht. Hircinol ließ sich auch in Spuren in *O. militaris* nachweisen. Beide Orchideenarten besitzen somit die Fähigkeit, Orchinol und Hircinol als Abwehrstoffe zu synthetisieren, allerdings bildet *O. militaris* 100mal mehr Orchinol als Hircinol, während die Verhältnisse bei *L. hircinum* entgegengesetzt liegen. Die fungitoxische Aktivität des Hircinols entspricht in seiner Spezifität und wirksamen Dosis etwa dem Orchinol.

c) Loroglossol

Als weiteres Dihydrophenanthrenderivat wurde aus infizierten Knollen von *Loroglossum hircinum* Loroglossol isoliert (*115, 116, 296*) und als 4-Hydroxy-5,7-dimethoxy-9,10-dihydro-phenanthren (14) identifiziert (90).

Nach älteren Literaturangaben soll Loroglossol keine fungitoxische Aktivität besitzen (*116, 296*) und im Gegensatz zu Orchinol und Hircinol

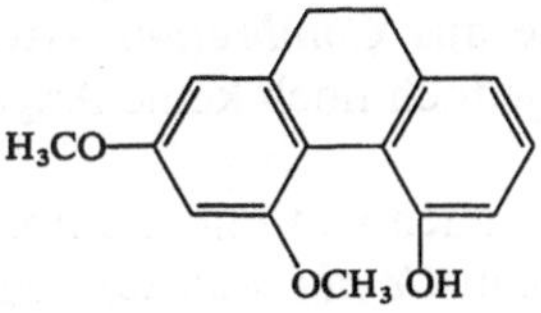

(14)

13*

auch nicht das Wachstum von *Candida lipolytica* BY 17 beeinflussen (*90*). Dagegen ist in neueren Untersuchungen eindeutig gezeigt worden, daß Loroglossol ähnlich fungitoxisch wirksam ist wie Orchinol und Hircinol (*322*). Darüber hinaus haben eingehende Arbeiten über die fungitoxische Wirksamkeit synthetisch dargestellter Dihydrophenanthren- und Phenanthrenabkömmlinge sowie verschiedener Stilbene und Dihydrostilbene ergeben, daß sowohl Dehydroorchinol und Dehydroloroglossol als auch Orchinol und Loroglossol fungitoxische Aktivität besitzen und die Sporenkeimung und das Wachstum verschiedener Pilze unterdrücken (*321*). Im Sporenkeimungstest mit *Phytophthora infestans* wurde für Orchinol und Loroglossol eine ED_{50} von $5 \cdot 10^{-5}$ m ermittelt. Dehydroorchinol erwies sich sogar noch wirksamer (ED_{50} $5 \cdot 10^{-6}$ m).

d) Strukturverwandte Pflanzeninhaltsstoffe

Es erscheint in diesem Zusammenhang interessant, einige weitere natürlich vorkommende Phenanthren- und Dihydrophenanthrenderivate aufzuführen, da sich einige von ihnen durch hohe wachstumsinhibierende Wirkung auszeichnen. Das trifft vor allem auf das kürzlich aufgefundene Batatasin I zu, ein endogener Wachstumsregulator aus *Dioscorea batatas*, der als 3-Hydroxy-2,5,7-trimethoxy-phenanthren ($C_{17}H_{16}O_4$) (**15**) identifiziert worden ist (*119, 120, 184*). *Dioscorea prazeri* enthält zwei 9,10-Dihydrophenanthrene, die als 3,4-Dihydroxy-5,6,8-trimethoxy- bzw. 3,4-Dihydroxy-5,7-dimethoxy-9,10-dihydrophenanthren (**15a** und **15b**) charakterisiert worden sind (*247a*).

		R_1 = H
(15a)	= OCH_3	R_1 = H
(15b)	H	R_1 = OCH_3

Enge strukturelle Verwandtschaft zeigen auch einige tetrasubstituierte Phenanthrene aus *Tamus communis* L. (*186, 250*), das aus Opium isolierte 5-Hydroxy-3,7-dimethoxy-phenanthren (**16**) (*101*) sowie eine Anzahl unterschiedlich hydroxylierter und methoxylierter Phenanthrene und 9,10-Dihydro-phenanthrene aus *Combretum*-Arten (*185, 187, 188*). Für diese Verbindungen liegen jedoch noch keine Angaben zur Fungitoxizität vor.

Das gemeinsame Vorkommen von Phenanthrenen und Stilbenen bzw. ihrer Dihydroderivate, z. B. in *Combretum spp.* oder in *Dioscorea batatas* [neben Batatasin I wurde das als Batatasin III bezeichnete 3,3'-Dihydroxy-

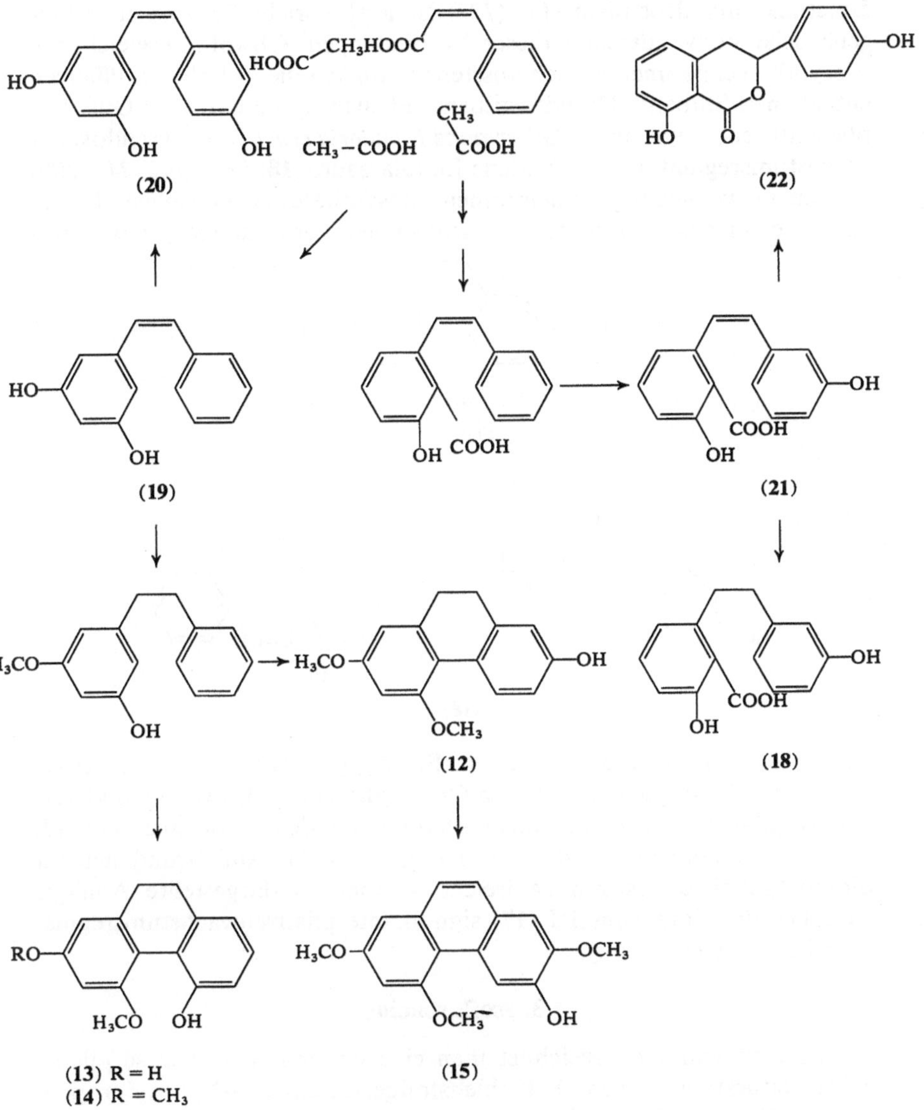

Abb. 1. Mögliche Biosynthesewege pflanzlicher Stilbene und Phenanthrene

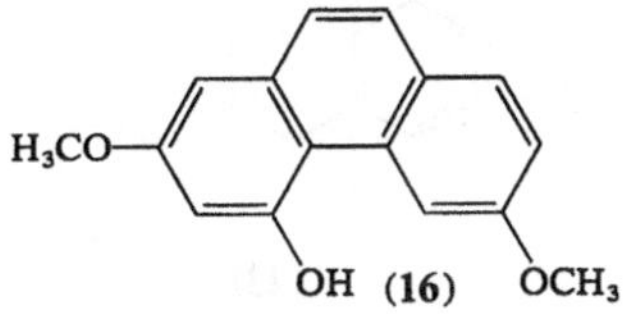

(16)

5-methoxy-dihydrostilben (**17**) (*121*) isoliert], spricht für eine enge biogenetische Verwandtschaft dieser Verbindungen. Obwohl vergleichende
Biosyntheseexperimente noch ausstehen, dürften die in höheren Pflanzen
enthaltenen Stilbene, Dihydrostilbene, Phenanthrene und 9,10-Dihydro-
phenanthrene sowie die im Lebermoos *Lunularia cruciata* L. als endogener
Wachstumsregulator aufgefundene Lunularsäure (**18**) (*241, 242, 243, 300*)
auf einem weitgehend gemeinsamen Biosyntheseweg entstehen. Dieser
könnte einerseits zu fungitoxisch wirksamen Verbindungen, z. B. Pino-

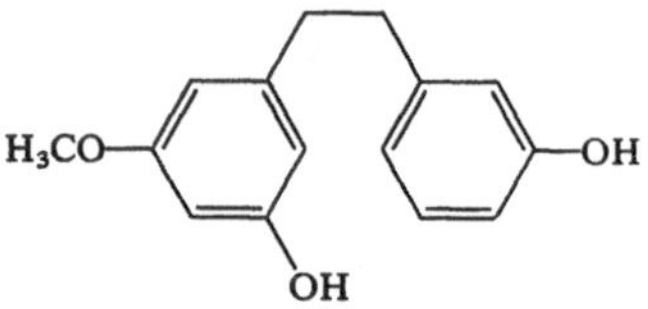

(**17**)

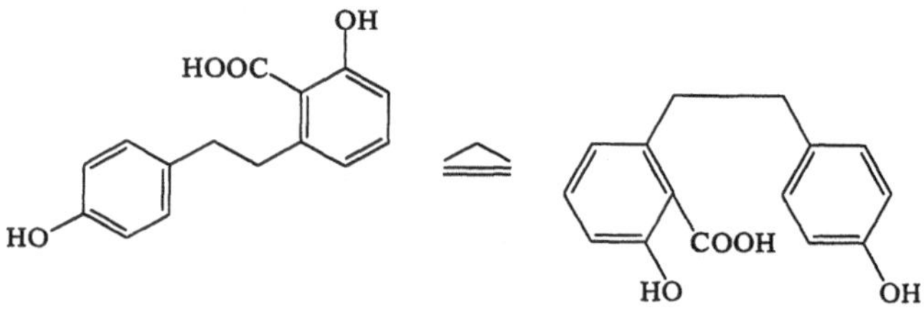

(**18**)

sylvin (**19**) und anderen Stilbenen [z. B. (**20**)] [vgl. (*118*) und dort zitierte
Literatur] oder Phytoalexinen vom Orchinoltyp (**12, 13, 14**), und andererseits zu pflanzlichen Wachstumsinhibitoren wie den Batatasinen (**15, 17**)
oder zur Lunularsäure (**18**) führen (vgl. Abb. 1). Auf Grund neuerer
Biotestergebnisse besitzen zahlreiche synthetisch dargestellte Analoga
vom Typ des Batatasins III (**17**) signifikante pflanzenwachstumsregula-
torische Aktivität (*250 a*).

3. Isoflavonoide

Als Isoflavonoide bezeichnet man eine umfangreiche Klasse pflanzlicher Naturstoffe, denen als Kohlenstoffgerüst ein 1,2-Diphenylpropan-
skelett (**23**) zugrunde liegt (vgl. *330*). Nachfolgend werden zu dieser Gruppe
gehörende Phytoalexine beschrieben.

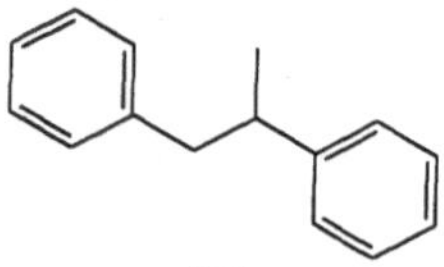

(**23**)

Literaturverzeichnis: SS. 229—247

a) Isoflavane

Sativan, Vestitol und Isosativan. — Die Leguminosen *Medicago sativa* L. und *Lotus corniculatus* L. enthalten nach Infektion mit einer Sporensuspension von *Helminthosporium turcicum* Pass. ein als Sativan bezeichnetes fungitoxisches Isoflavanderivat der Summenformel $C_{11}H_{18}O_4$, das als (−)-7-Hydroxy-2′,4′-dimethoxy-isoflavan (25) identifiziert wurde (*41, 146*). Als Begleitsubstanz wurde in beiden Pflanzen nach Infektion mit *H. turcicum* bzw. *Stemphylium botryosum* das ebenfalls fungitoxisch wirksame (−)-Vestitol (24) aufgefunden, das sich vom Sativan durch die fehlende 2′-O-Methyl-Gruppe unterscheidet. Sativan und Vestitol finden sich auch in einigen *Helminthosporium carbonum*-infizierten *Trifolium*-Arten, z. B. *T. hybridum* (*145a*). Begleitsubstanzen sind Medicarpin (33), Maackiain (40), 4-Methoxymaackiain sowie das neu aufgefundene Isosativan (25a), dessen ED_{50} 16 µg/ml beträgt (*145a*).

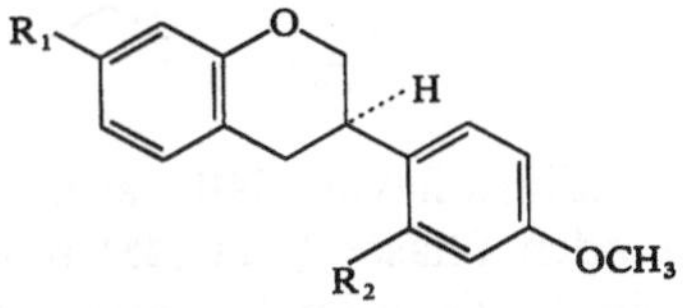

(24) $R_1 = R_2 = OH$
(25) $R_1 = OH$ $R_2 = OCH_3$
(25a) $R_1 = OCH_3$ $R_2 = OH$

Vestitol und Sativan hemmen das Mycelwachstum von *H. turcicum* und *S. botryosum* in hohem Maße. Für eine 50%ige Hemmung sind pro ml Nährlösung 15 µg Sativan bzw. 35 µg Vestitol notwendig.

Als biogenetische Vorstufe des Sativans wird das gleichfalls in *Medicago sativa* L. vorkommende (−)-Medicarpin (33) postuliert, wobei dessen Dihydrofuranring enzymatisch gespalten und die 2′-Hydroxyl-Gruppe selektiv methyliert werden soll (vgl. Abb. 2). Diese Annahme wird durch die Tatsache untermauert, daß für *Stemphylium botryosum in vitro* ein Abbau von Medicarpin (33) zu Vestitol (24) nachgewiesen ist (*273*). Aus diesen biogenetischen Überlegungen heraus wird die 6aR,11aR-Konfiguration des (−)-Medicarpins (33) auf (−)-Vestitol (24) und (−)-Sativan (25) übertragen und die angegebene 3R-Konfiguration für diese beiden Isoflavane angenommen, was in Übereinstimmung zu der ermittelten Absolutkonfiguration von (+)-3S-Vestitol steht (*181*).

2′-O-Methyl-phaseollidinisoflavan. — Kürzlich wurde aus der Leguminose *Vigna unguiculata* (L.) Walp. nach Inokulation mit *Colletotrichum lindemuthianum* ein weiteres isoflavonoides Phytoalexin der Summenformel $C_{20}H_{24}O_4$ isoliert, das durch spektroskopische Unter-

suchungen sowie durch biogenetische Überlegungen bezüglich der
Stellung der O-Substituenten als 2′-O-Methyl-phaseollidinisoflavan (**26**)
charakterisiert werden konnte (*239*). Dieses Isoflavanderivat unterscheidet
sich von dem vorstehend beschriebenen Sativan durch die 3,3-Dimethyl-
allylgruppe in Position 3′ und das Fehlen der 4′-O-Methyl-Gruppe. Über
die Absolutkonfiguration am C-3 liegen noch keine Angaben vor. In
Analogie zu Sativan und Vestitol ist möglicherweise 3R-Anordnung
zu diskutieren.

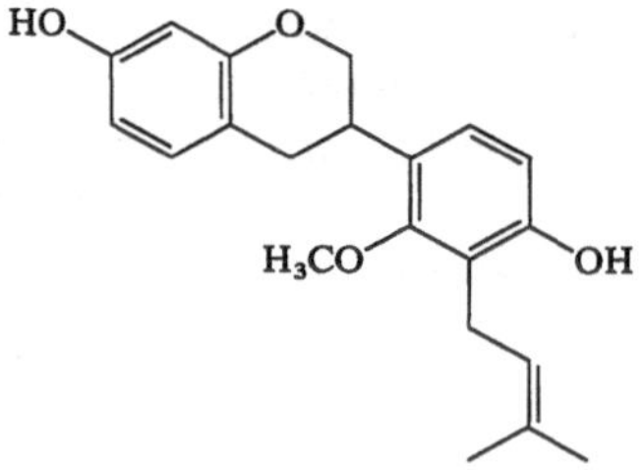

(**26**)

2 -O-Methyl-phaseollidinisoflavan läßt sich dünnschichtchromatographisch vom begleitenden Phaseollidin (**45**) nur schwierig abtrennen,
beide Verbindungen zeigen jedoch im UV-Spektrum signifikante Unterschiede.

In Konzentrationen von 10—15 ppm wird die Konidiensporenkei-
mung von *C. lindemuthianum* I 47 und I 57 vollständig gehemmt. Für
Phaseollidin (**45**) betragen die entsprechenden Hemmkonzentrationen
20—25 ppm.

Phaseollinisoflavan und 2′-O-Methyl-phaseollinisoflavan. — *Phaseolus vulgaris* L. akkumuliert nach Infektion mit *Colletotrichum lindemuthianum* (*16, 17*), *Rhizoctonia solani* (*270*) sowie nach Infektion mit Tabak-
nekrosevirus (TNV) (*17, 47*) neben den beiden Pterocarpanderivaten
Phaseollin (**46**) und Phaseollidin (**45**) sowie dem Isoflavon Kieviton (**29**)
eine fungitoxische Substanz der Summenformel $C_{20}H_{20}O_4$, die als Phaseollinisoflavan (**27**) charakterisiert worden ist (*47*).

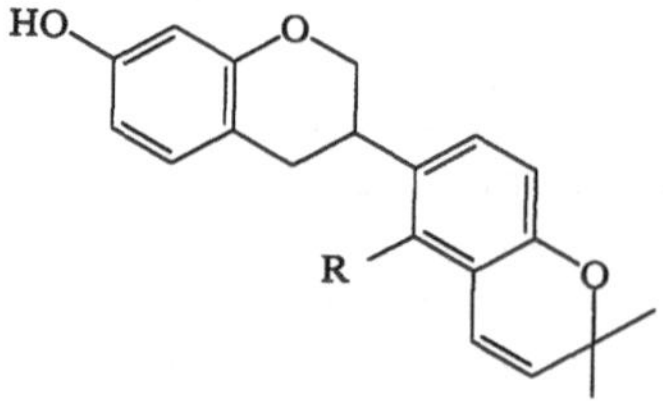

(**27**) R = OH
(**28**) R = OCH$_3$

Phaseollinisoflavan wurde auch in nicht-inokulierten Wurzelexudaten von *Phaseolus vulgaris* L. nachgewiesen (*50*). Nach Infektion von *Phaseolus vulgaris* mit *Rhizoctonia solani* enthält das pflanzliche Gewebe bis zu 50 μg Phaseollinisoflavan (*270*). Die ED_{50} wird im Wachstumstest mit *R. solani* mit 27 μg/ml angegeben. Es ist noch offen, ob dieses Isoflavan durch eine spezifische Induktion seiner Synthese im Pflanzengewebe gebildet wird oder ob es aus dem Stoffwechsel anderer fungitoxischer Isoflavonoide wie Phaseollin (**46**), Phaseollidin (**45**) oder Kieviton (**29**) hervorgeht. Für letztere Annahme sprechen Versuche, nach denen Phaseollin (**46**) von *Stemphylium botryosum* zu Phaseollinisoflavan (**27**) umgewandelt wird (*139*).

Das 2'-O-Methylderivat (**28**) des Phaseollinisoflavans wurde aus *P. vulgaris* L. nach Infektion mit *Fusarium solani* (Mart.) Sacc. *f. sp. phaseoli* (Burk.) Snyd. and Hans. isoliert (*305*). Eine strukturelle Verwandtschaft zu dem vorstehend besprochenen 2'-O-Methylphaseollidinisoflavan (**26**) ergibt sich formal durch die Annahme der Ausbildung eines sauerstoffhaltigen Sechsringes (2,2-Dimethylchromenring) unter Einbeziehung der 3,3-Dimethylallylgruppe und der 4'-ständigen Hydroxylgruppe des Phaseollidinisoflavans. 12 μg 2'-O-Methylphaseollinisoflavan pro ml bewirken eine 50%ige Hemmung des nichtpathogenen *F. solani f. sp. cucurbitae*, während der pathogene Pilz *F. solani f. sp. phaseoli* selbst bei doppelter Konzentration im Wachstum nicht hemmend beeinflußt wird (*305*).

b) Isoflavanone

Kieviton. — Hypokotyle von *Phaseolus vulgaris* L. enthalten nach Infektion mit *Rhizoctonia solani* Kühn (*268, 269*) oder TNV (*47*) sowie nach chemischem Stress durch $CuCl_2$-Behandlung (*269*) eine fungitoxische Substanz der Summenformel $C_{20}H_{20}O_6$. Bei dieser erstmals aus der Sorte Kievitsboon koekoek isolierten und daher als Kieviton bezeichneten Verbindung handelt es sich um 2',4',5,7-Tetrahydroxy-8-isopentenylisoflavanon (**29**) (*47, 269, 271*). Im Gegensatz zum bereits beschriebenen 2'-O-Methylphaseollidinisoflavan (**26**) findet sich beim Kieviton (**29**) die 3,3-Dimethylallylgruppe in Position 8. Die Absolutkonfiguration ist noch nicht zugeordnet. Das gemeinsame Vorkommen mit anderen Isoflavonoiden wie Phaseollin (**46**), Phaseollidin (**45**) und Phaseollinisoflavan (**27**) (*15, 50, 159, 212*) spricht für eine biogenetische Verwandtschaft und daher für eine 3R-Konfiguration.

Das Vorkommen von Kieviton in *Phaseolus vulgaris* nach Pilzinfektion ist auch von anderen Autoren bestätigt worden, wobei *R. solani* (*270*), *Colletotrichum lindemuthianum* (*16, 17*) sowie Viren wie TNV (*17*) als induzierende Erreger eingesetzt worden sind. In der Leguminose *Vigna sinensis* Endl. läßt sich nach Virus- oder Pilzinfektion ebenfalls Kieviton

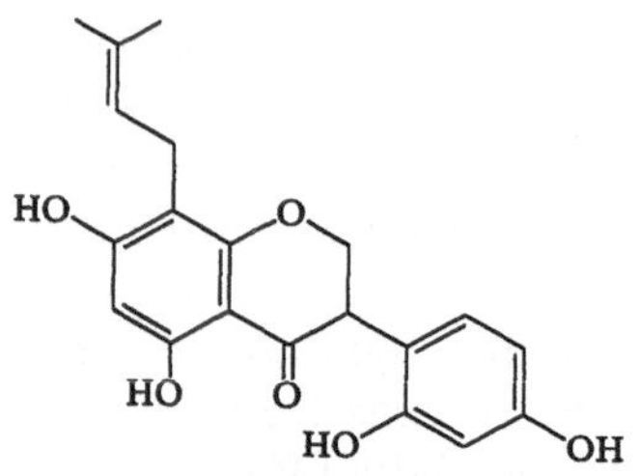

(29)

(29) nachweisen, das von Phaseollin (46) und Phaseollidin (45) begleitet wird (15, 159, 212).

Der Kievitongehalt *Rhizoctonia solani*-infizierter Hypokotyle von *Phaseolus vulgaris* beträgt 50 µg/g Frischgewicht; die ED_{50} beträgt im Mycelwachstumstest mit *R. solani* 60 µg/ml (269). In neueren Untersuchungen ist gezeigt worden, daß der Kievitongehalt auf 640 µg/g Frischgewicht ansteigen kann und daß Kieviton offenbar eine wichtige Funktion als chemischer Abwehrstoff besitzt (270). Andere Autoren haben nach Einwirkung von $CuCl_2$ oder Antibiotika wie Actinomycin D oder Cycloheximid eine erhöhte Kievitonakkumulation in Hypokotylen von *Vigna sinensis* gefunden und gezeigt, daß die Bildung von Kieviton durch derartige abiotische Faktoren induziert werden kann (212).

2'-Hydroxy-5-methoxy-6,7-methylendioxyisoflavon. — Mit *Cercospora beticola* infizierte Blätter von *Beta vulgaris* L. enthalten ein fungitoxisch wirksames Isoflavon (30) der Summenformel $C_{17}H_{12}O_6$, das als

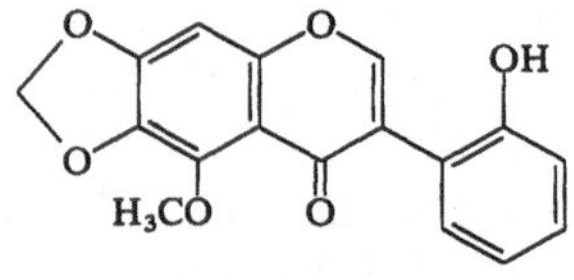

(30)

2'-Hydroxy-5-methoxy-6,7-methylendioxyisoflavon identifiziert wurde und sich gegenüber *C. beticola* und *Monilinia fructicola* durch hohe Fungitoxizität auszeichnet (102, 149a). Das Vorkommen eines isoflavonoiden Phytoalexins in einer Chenopodiacee erscheint besonders interessant, da alle anderen Phytoalexine mit Isoflavonoidstruktur bisher aus Leguminosen isoliert worden sind. Als Begleitsubstanz findet sich in *Beta vulgaris* eine zweite phytoalexinartige Verbindung mit der Summenformel $C_{18}H_{16}O_6$, die als 2',5-Dimethoxy-6,7-methylendioxyflavanon (31) charakterisiert wurde und die — obwohl kein Isoflavon — an dieser Stelle mit aufgeführt werden soll. Dieses Flavanonderivat (31) ist weit weniger fungitoxisch als das vorstehend besprochene Isoflavon (30).

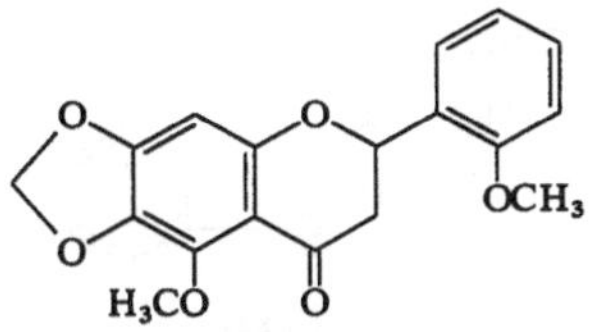

(31)

c) Pterocarpane

Zur umfangreichen Naturstoffklasse der Isoflavonoide gehören aus biogenetischen Gründen die Pterocarpane. Diese leiten sich vom Ringgerüst des Cumaranochromans (32) ab, das systematisch als 6a,11a-Dihydro-6H-benzofuro-[3,2-c] [I] benzopyran bezeichnet wird. Man kennt heute über 25 natürlich vorkommende Pterocarpane sowie einige Pterocarpanglucoside, wobei für verschiedene (−)-Pterocarpane die im vorliegenden Text allgemein durchgängig angegebene 6aR,11aR-Konfiguration abgeleitet worden ist (vgl. 330).

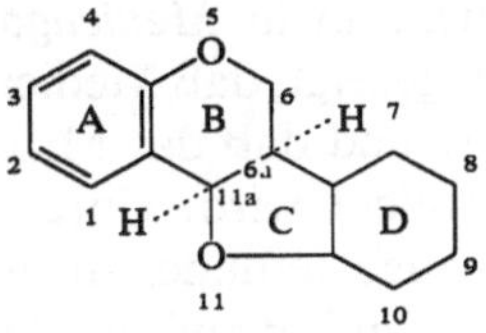

(32)

Zur Gruppe der Pterocarpane gehören einige charakteristische Phytoalexine der Leguminosen wie Medicarpin (33), Pisatin (39), Phaseollin (46), 6-Hydroxyphaseollin (50) und Phaseollidin (45), die nachfolgend besprochen werden.

Medicarpin. — In Blättern von Medicago sativa L. kann durch Infektion mit Sporensuspensionen bestimmter pathogener und nicht-pathogener Pilze wie Colletotrichum phomoides (Sacc.) Chester, Helminthosporium turcicum Pass. und Phoma herbarum sowie durch Behandlung mit sporenfreiem Kulturfiltrat von Stemphylium loti und S. botryosum die Bildung einer fungitoxischen phytoalexinartigen Substanz induziert werden (132, 134, 135). Die Synthese dieser Substanz kann auch durch chemischen Stress, z. B. Behandlung des Pflanzenmaterials mit $3 \cdot 10^{-3}$ m CuCl$_2$-Lösung, ausgelöst werden (80, 81). Auf Grund der durchgeführten Strukturaufklärung handelt es sich um (−)-3-Hydroxy-9-methoxypterocarpan (33) (137, 267), eine Verbindung, die in ihrer (−)-, (+)- und (±)-Form als Inhaltsstoff verschiedener tropischer Bäume schon bekannt war. Der (−)-Form ist 6aR,11aR-Konfiguration zugeordnet worden (148).

(33)

Dieses als Medicarpin bezeichnete Demethylhomopterocarpin ist auch in einigen anderen pilzinfizierten Leguminosen aufgefunden worden, z. B. in *Canavalia ensiformis* (L.) DC. (*158*) und *Vigna unguiculata* nach Beimpfen mit *Colletotrichum lindemuthianum* oder TNV (*159, 182*), in *Trifolium pratense* nach Infektion mit *H. turcicum* (*138*) sowie in *Cicer arietinum* L. (*159*). Darüber hinaus ist Medicarpin in *Trifolium hybridum* und einigen anderen *Trifolium*-Arten nach Infektion mit *Helmintho-sporium carbonum* (*145 a*) sowie neben Wyeron (**10**), Wyeronepoxid und Wyeronsäure (**9**) in *Botrytis cinerea*-infizierter *Vicia faba* nachgewiesen worden (*116 d*). Quantitative Untersuchungen über den Einfluß einiger äußerer Faktoren wie Temperatur oder Licht auf die Bildung und/oder Akkumulation von Medicarpin in *Medicago sativa* nach Beimpfen mit *Monilia fructicola* haben gezeigt, daß Medicarpin bereits 6 Stunden nach Infektion nachzuweisen ist und daß das Maximum im Medicarpingehalt nach etwa 30 Stunden erreicht wird. Unter Langtagsbedingungen wird eine höhere Medicarpinmenge gefunden als unter Kurztagsbedingungen; die Lichtintensität ist ohne Einfluß auf die Medicarpinkonzentration (*70*).

Die ED_{50} beträgt im Mycelwachstumstest mit *Rhizoctonia solani* 0,15 mM (*304*). Umfangreiche Arbeiten zur Struktur-Wirkungsbeziehung der Pterocarpane und strukturverwandter Verbindungen haben interessante Aspekte bezüglich der fungitoxischen Aktivität gegenüber *Monilinia fructicola* (*233*) sowie *Fusarium solani* und *Aphanomyces eureiches* ergeben (*306 a*).

Die Biosynthese des Medicarpins, dessen Bildung in Keimlingen durch Pilzinfektion oder durch chemischen Stress wie $CuCl_2$-Behandlung induziert werden kann, ist kürzlich mit verschiedenen in Frage kommenden Pterocarpanprecursoren untersucht worden (*80, 81*). Die durchgeführten Fütterungsexperimente haben ergeben, daß die *de novo*-Synthese des Medicarpins über Phenylalanin verläuft und daß 2',4',4-Trihydroxychalcon-^{14}CO (**34**) und 7-Hydroxy-4'-methoxyisoflavon-$^{14}CH_3$ (Formononetin) (**35**) gut in Medicarpin eingebaut werden. Dagegen zeigen das entsprechende 4-Methoxychalcon sowie 7,4'-Dihydroxyisoflavon (Daidzein) geringe Inkorporationsraten. 2',7-Dihydroxy-4'-methoxyisoflavon (**36**) und das entsprechende Isoflavanon (**37**) erwiesen sich ebenfalls als effektive Medicarpinvorstufen, so daß der in Abb. 2 gezeigte Biosyntheseweg des Medicarpins (**33**) zu postulieren ist.

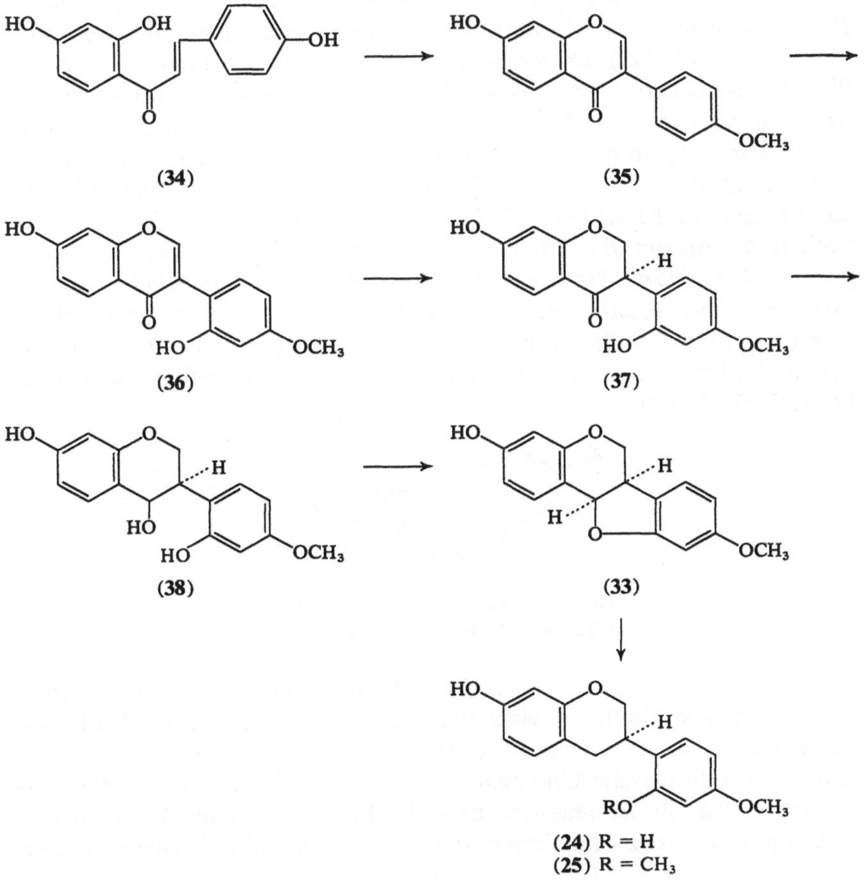

(34)   (35)

(36)   (37)

(38)   (33)

(24) R = H
(25) R = CH₃

Abb. 2. Biosynthese von Medicarpin

Zum Abbau des Medicarpins liegen ebenfalls eingehendere Untersuchungen vor (*123, 132, 135, 136*). Das kürzlich in *Trifolium pratense*
nach Infektion mit *Botrytis cinerea* oder *Sclerotinia trifoliorum* aufgefundene 6a-Hydroxymedicarpin stellt offensichtlich ein Abbauprodukt des
Medicarpins dar (*35a*). Von *Stemphylium botryosum* wird Medicarpin
zu zwei Abbauprodukten umgewandelt, von denen eines mit dem
bereits besprochenen Isoflavanderivat Vestitol (**24**) identisch ist (vgl.
Abb. 2) (*273*). Der Abbau von Medicarpin (**33**) zum entsprechenden
2′-Hydroxyisoflavan kann durch Maackiain (**40**) induziert werden (*133*).
Der bei *Stemphylium loti* und *C. phomoides* langsamer verlaufende Abbau-
weg führt wahrscheinlich zu anderen Produkten als bei *S. botryosum*. Von
Helminthosporium turcicum wird Medicarpin nicht metabolisiert.

Pisatin. — Sporensuspensionen von *Ascochyta pisi* (Lib.) können bei *Pisum sativum* L. die Bildung einer fungitoxisch wirksamen Substanz induzieren (*293, 295*). Dieser Befund führte 1960 zur Isolierung einer als Pisatin bezeichneten fungitoxischen Substanz ($C_{17}H_{14}O_6$) aus *Monilinia fructicola*-infiziertem endokarpem Hülsengewebe von *Pisum sativum* (*61, 62*). Pisatin liegt ein pterocarpanoides Ringgerüst mit einer alkoholischen Hydroxylgruppe, einer Methoxygruppe und einer 8,9-ständigen Methylen-dioxygruppe zugrunde (**39**) (*37, 230, 231, 234*). *Trifolium pratense* enthält nach Infektion mit *Botrytis cinerea* oder *Sclerotinia trifoliorum* Homo-pisatin (**39a**) (*35a*). Im Gegensatz zu den meisten natürlichen Ptero-carpanen sind Pisatin und Homopisatin rechtsdrehend und sollten somit 6aS,11aS-Konfiguration besitzen (*148*). Pisatin und einige struktur-verwandte Pterocarpane sind heute partial- und totalsynthetisch zugäng-lich (*32, 37, 91, 92*).

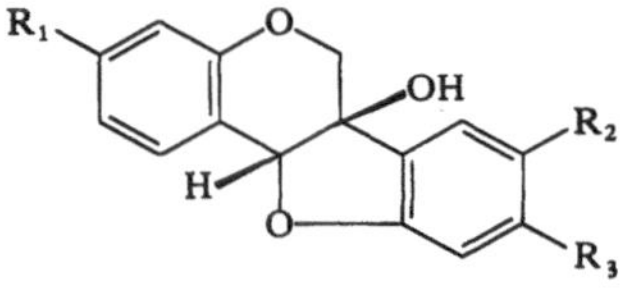

(**39**) $R_1 = OCH_3$, $R_2 = R_3 = OCH_2O$
(**39a**) $R_1 = OCH_3$, $R_2 = H$, $R_3 = OCH_3$

Die ED_{50} von Pisatin gegenüber *M. fructicola* beträgt 10^{-4} m (*61*). Das schwach antibiotisch wirkende Pisatin weist ein breites Wirkungs-spektrum auf und zeigt gegenüber Pathogenen und Nichtpathogenen der Erbse charakteristische Unterschiede in der Fungitoxizität (*57, 306*). Als strukturverwandte Begleitsubstanzen des Pisatins sind aus *M. fructicola*-oder *Fusarium solani*-infizierten Erbsen (−)-Inermin (Maackiain) (**40**)

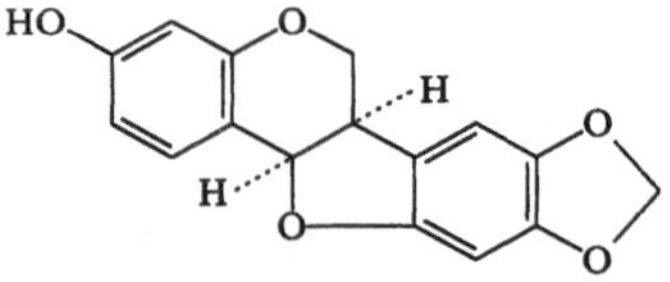

(**40**)

(*278*) sowie 3-Hydroxy-2,9-dimethoxypterocarpan ($C_{17}H_{16}O_5$) (**41**), 2,3,9-Trimethoxypterocarpan ($C_{18}H_{18}O_5$) (**42**) und 4-Hydroxy-2,3,9-trimethoxy-pterocarpan ($C_{18}H_{18}O_6$) (**43**) (*244*) isoliert worden. Die Linksdrehung dieser ebenfalls fungitoxischen Verbindungen spricht für eine möglicher-weise übereinstimmende 6aR,11aR-Absolutkonfiguration.

Eingehendere Untersuchungen liegen zur Bildung bzw. Akkumulation von Pisatin nach *Monilinia fructicola*-Infektion in verschiedenen Arten und Linien von *Pisum sativum* und *P. arvense* (*66*), in Embryonen nach

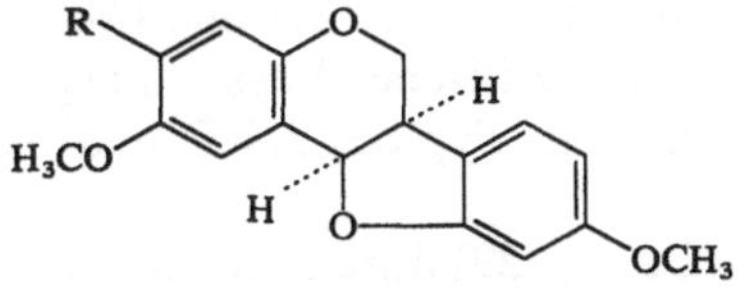

(41) R = OH
(42) R = OCH₃

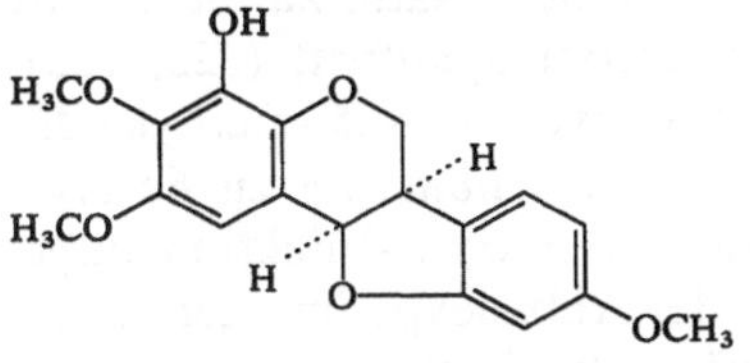

(43)

Infektion mit *Alternaria alternata* (*236*), *Erysiphe pisi* und *E. graminis hordei* (*221a, b*), sowie zum Einfluß äußerer physiologischer Faktoren wie UV-Licht, O_2-Druck und Temperaturänderungen auf die Pisatinbildung vor (*64, 65, 67, 112*). Darüber hinaus sind speziellere phytopathologische Aspekte näher bearbeitet worden (*12, 54, 57, 263, 306*). Für das System *Pisum sativum — Ascochyta pisi* wurde gezeigt, daß sich die einzelnen Sorten von *P. sativum* im Bildungsvermögen der Phytoalexine unterscheiden und daß die einzelnen *A. pisi*-Stämme gegenüber den *Pisum*-Phytoalexinen unterschiedlich empfindlich sind (*117*). Als Induktoren der Pisatinsynthese erwiesen sich nicht nur die bereits genannten Pilze, wie z. B. *Fusarium solani* (*54*), sondern auch *Penicillium expansum* (*12*), *Aureobasidium pullulans, Cladosporium herbarum, Epicoccum nigrum* und *Botrytis fabae* (*194*), außerdem Viren (*15*) sowie Kokosmilch (*13*) und verschiedene basische Polypeptide und Polyamine (*110, 111*). Wie bei einigen anderen Phytoalexinen führt auch chemischer Stress durch Behandlung des Pflanzenmaterials mit Schwermetallionen (Ag, Hg, Cu) oder Enzyminhibitoren wie Natrium-jodacetat, -fluorid, -azid, -cyanid, p-Chlormercuribenzoat usw. (*64, 194, 232, 260*) oder mit zahlreichen Acridinabkömmlingen (*108*) zu einer erhöhten Pisatinsynthese. Hemmstoffe der Proteinsynthese wie Cycloheximid oder Actinomycin D unterdrücken die Pisatinbildung (*261*).

Untersuchungen zur Biosynthese haben ergeben, daß die vom pilzlichen Parasit produzierten Induktoren nicht mit in die Pisatinsynthese eingehen (*303*). Die weiteren Experimente mit isotop-markierten Verbindungen zeigten, daß der Ring A des Pisatins aus Acetateinheiten aufgebaut wird und daß die restlichen C-Atome des Pterocarpangerüstes (C-6,6a,11a und Ring D) aus einer C_6-C_3-Verbindung vom Zimtsäuretyp stammen (*106, 107*). Möglicherweise stellt (−)-Inermin (**40**) eine direkte

Vorstufe dar, die zu Pisatin hydroxyliert und O-methyliert wird. Andererseits ist auch eine stereospezifische Wasseranlagerung an ein hypothetisches Zwischenprodukt vom Typ des 3,9-Dihydroxypterocarp-6a-en (47) denkbar (vgl. Abb. 3).

Zum Abbau des Pisatins (39) durch verschiedene pilzliche Organismen, z. B. *Fusarium oxysporum, F. pinades, F. solani* oder *Ascochyta pisi*, liegen ebenfalls Untersuchungen vor (*53, 326, 327, 328, 329*). So ist *Stemphylium botryosum in vitro* imstande, Pisatin zu metabolisieren, wobei ein fungitoxisches Stoffwechselprodukt auftritt (*122, 123*). Maackiain wird von *S. botryosum* zu 7,2′-Dihydroxy-4′,5′-methylendioxyisoflavan umgesetzt (*133*) und von *Sclerotinia trifoliorum* zu 6a-Hydroxymaackiain, einem in *Trifolium pratense* nach *Botrytis*-Infektion aufgefundenen Pterocarpan, hydroxyliert (*35 a*). Bei Abbauexperimenten an *Fusarium solani f. sp. pisi* wurde ein schon früher beobachteter Metabolit (*53*) kürzlich als 3,6a-Dihydroxy-8,9-methylendioxypterocarpan (44) identifiziert (*309*).

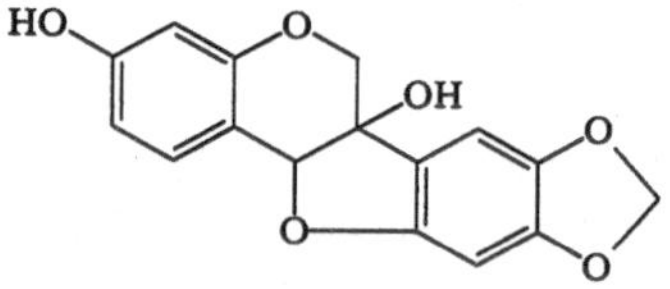

(44)

Phaseollidin. — *Phaseolus vulgaris* enthält nach Infektion mit *Monilinia fructicola* (Wint.) Honey neben dem nachfolgend abzuhandelnden Phaseollin eine weitere fungitoxische Substanz der Summenformel $C_{20}H_{20}O_4$, die als 10-substituiertes 3,9-Dihydroxypterocarpan (45) charakterisiert und als Phaseollidin bezeichnet wurde (*229, 235*). Die Struktur dieses lipophilen Diphenols wurde auch von anderen Autoren bestätigt, die Phaseollidin nach Infektion mit TNV aus *Phaseolus vulgaris* isoliert haben (*47*). Phaseollidin enthält im Unterschied zum Medicarpin (33) am C-9 eine freie Hydroxylgruppe und am C-Atom 10 eine 3,3-Dimethylallylgruppe, die beim später zu besprechenden Phaseollin (46) zu einem Chromenring cyclisiert ist. Phaseollidin weist wie das sterisch zuge-

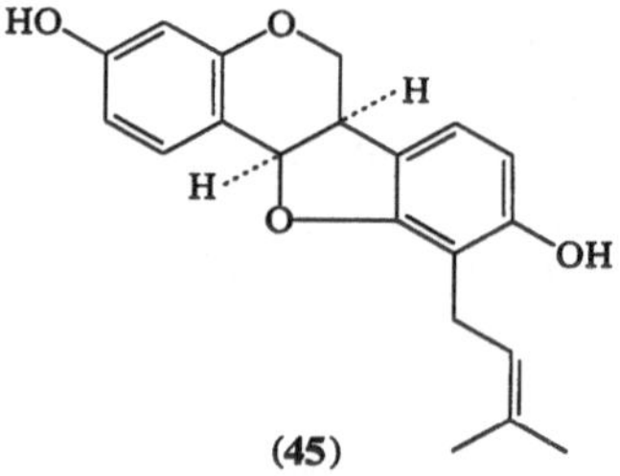

(45)

ordnete (−)-6aR,11aR-Phaseollin (**46**) (*148*) einen negativen Drehwert $[\alpha]_D$ auf, was möglicherweise für eine übereinstimmende Absolutkonfiguration beider Verbindungen und darüber hinaus für eine gemeinsame biogenetische Vorstufe sprechen könnte (*229, 235*).

Phaseollidin ist in *Phaseolus vulgaris* auch nach Infektion mit *Colletotrichum lindemuthianum* (*16, 17*), *Rhizoctonia solani* (*270*) oder TNV (*17*) sowie in der virus-infizierten Leguminose *Vigna sinensis* Endl. (*15, 212*) nachgewiesen worden. Strukturverwandte Begleitstoffe mit Phytoalexincharakter sind Phaseollin (**46**), Kieviton (**29**) und Phaseollinisoflavan (**27**), wobei Phaseollin meist als Hauptkomponente anzutreffen ist (*60*). Die genannten Isoflavonoide sind interessanterweise auch in Exudaten nichtinokulierter Wurzeln von *Phaseolus vulgaris* aufgefunden worden (*50*).

Phaseollidin zeigt gegenüber zahlreichen pathogenen und nichtpathogenen Pilzen starke Fungitoxizität, z. B. gegenüber *Colletotrichum lindemuthianum* (*16, 17, 47, 229*), *Rhizoctonia solani* (*270*), *Monilinia fructicola* (*60, 229*) und weiteren Pilzarten, wobei eine unterschiedliche Sensitivität gegenüber Phaseollidin nachgewiesen wurde (*229*).

Phaseollin. — Das Vorkommen einer fungitoxischen Substanz mit Phytoalexincharakter, die nach Inokulation von *Phaseolus vulgaris* L. mit *Sclerotinia fructicola* oder *Phytophthora infestans* beobachtet wurde (*149, 209*), konnte 1963 durch CRUICKSHANK und PERRIN bestätigt werden. Diesen Autoren gelang es, aus *Monilinia fructicola* (Wint.) Honey-infizierten *Phaseolus vulgaris*-Pflanzen eine als Phaseollin (**46**) bezeichnete Substanz der Summenformel $C_{20}H_{18}O_4$ zu isolieren und als phenolisches Homopterocarpinderivat zu identifizieren (*63*). Phaseollin liegt das heterocyclische Ringgerüst des Pterocarpans (**32**) zugrunde, das am C-Atom 3 eine phenolische Hydroxylgruppe sowie am Ring D einen aus dem Isoprenstoffwechsel hervorgehenden 2,2-Dimethylchromenring besitzt, der formal aus der 3,3-Dimethylallylgruppe des Phaseollidins (**45**) entstehen könnte (*227, 228*). Als (−)-Pterocarpanabkömmling ist Phaseollin 6aR,11aR-Konfiguration zugeordnet worden (*148*).

(**46**)

Vorkommen, Akkumulation und Isolierung von Phaseollin sind in der Folgezeit an verschiedenen *Phaseolus vulgaris*-Sorten und an anderen Arten der Gattung *Phaseolus* untersucht worden, wobei zur Infektion vor allem *Colletotrichum lindemuthianum* (Sacc. & Magn.) Bris Cav. (*16, 17,*

19, 247, 248), Monilinia fructicola (*69, 226*), Penicillium expansum (*249*), Rhizoctonia solani Kühn. (*238, 270, 307*) sowie weitere Pilzarten (*50 b, 69, 83 a, 237*) eingesetzt worden sind. Darüber hinaus sind Pseudomonas phaseolicola (Burk.) Dows. und P. morsprunorum (*191, 274*) und andere Bakterien (*69*) sowie TNV (*17, 20*) imstande, die Bildung von Phaseollin zu induzieren. Obwohl Phaseollin im allgemeinen nur im infizierten Gewebe bzw. in unmittelbarer Nähe des Infektionsherdes nachzuweisen ist (z. B. *69, 247*), gibt es Angaben über das Vorkommen von Phaseollin in Exudaten nichtinokulierter Wurzeln von Phaseolus vulgaris (*50*).

Vielfach werden neben Phaseollin einige weitere strukturverwandte Isoflavonoide mit Phytoalexincharakter aufgefunden wie Phaseollidin (**45**), Kieviton (**29**) oder Phaseollinisoflavan (**27**) (*16, 17, 50, 224 a, 270*). Das trifft auch für die Leguminose Vigna sinensis zu, die nach Virusinfektion neben Phaseollin die genannten Isoflavonoidabkömmlinge enthält (**15**). In Phaseolus vulgaris stellt Phaseollin aber im allgemeinen die Hauptkomponente dar (*60*). Wie bei einigen anderen Phytoalexinen führt auch bei Phaseolus vulgaris chemischer Stress, z. B. Behandlung des Pflanzengewebes mit Quecksilberchlorid (5×10^{-5} m) (*226*) oder 9-Aminoacridin (*126*), zur Induktion der Phaseollinbildung. Besonders interessant erscheint der Befund, daß nicht nur Konidiensuspensionen, sondern auch ein aus dem Mycel von Monilinia fructicola isoliertes wasserlösliches Polypeptid die Phaseollinsynthese zu induzieren vermögen (*68, 226*). Dieses als Monilicolin A bezeichnete schwefelhaltige Polypeptid besitzt ein Molekulargewicht von etwa 8000 und ist aus 65 Aminosäureresten aufgebaut. Das selbst nicht fungi- oder phytotoxische Monilicolin A ist in äußerst niedrigen Konzentrationen (oberhalb $2,5 \times 10^{-9}$ m) induktionsauslösend und besitzt hohe Wirkungsspezifität. Es ist beispielsweise nicht in der Lage, in Erbsen die Pisatinbildung hervorzurufen. Es wird diskutiert, daß Monilicolin A die Biosynthese des Phaseollins im inokulierten Gewebe reguliert und kontrolliert. Andererseits bewirkt das von dem Bakterium Pseudomonas phaseolicola gebildete Exotoxin Phaseotoxin bei Bohnen eine Erniedrigung im Gehalt an Phaseollin, Phaseollidin (**45**), Phaseollinisoflavan (**27**) und Kieviton (**29**) (*224 a*).

Die Fungitoxizität des Phaseollins ist ebenfalls intensiv untersucht worden. So wird beispielsweise das Mycelwachstum von Monilinia fructicola gehemmt, wobei die ED_{50} 3 µg/ml beträgt (*63*). Phaseollin ist auch gegenüber anderen Pilzen fungitoxisch, z. B. Colletotrichum lindemuthianum (ED_{50} 2—4 µg/ml) (*14, 16, 17*), Rhizoctonia solani Kühn (ED_{50} 18 µg/ml) (*238, 270, 308*), Fusarium-Arten (*306*) (ED_{50} F. solani f. sp. cucurbitae 0,05 mM) (*304*). Es wird diskutiert, daß Phaseollin über die Plasmamembran oder deren Funktion wirkt (*308*).

Die Hemmung des Wachstums von M. fructicola ist mit einer Vielzahl strukturverwandter Pterocarpanderivate untersucht worden (*233*), wobei

Phaseollin neben Maackiain (**40**) und Homopterocarpin mit die höchste Hemmwirkung zeigt. Eine vergleichende Betrachtung der Ergebnisse bezüglich der Struktur-Wirkungsbeziehung hat ergeben, daß für eine fungitoxische Aktivität der Pterocarpane die aromatischen Ringe A und D nicht in derselben Ebene liegen dürfen und daß sauerstoffhaltige Substituenten (OH, OCH₃), besonders am C-3 oder C-9, notwendig sind. Im Gegensatz dazu ist in neueren Untersuchungen gefunden worden, daß sowohl planar als auch aplanar gebaute Pterocarpane und Isoflavane fungitoxische Aktivität besitzen und daß die Fungitoxizität der getesteten Verbindungen offensichtlich nicht ausschließlich von der Raumstruktur abhängig ist (*306 a*).

Durch Biosyntheseversuche ist gezeigt worden, daß die Vorstufen des Phaseollins aus dem Stoffwechsel der Wirtspflanze *Phaseolus vulgaris* und nicht aus dem Metabolismus des Parasiten *Sclerotinia fructicola* stammen (*303*). Dieser Befund bedeutet, daß vom Pilz nur Induktoren und keine Precursoren geliefert werden, was in Übereinstimmung zur postulierten Phytoalexintheorie steht.

Der spezifische Einbau von Phenylalanin-U-¹⁴C, Zimtsäure-¹⁴COOH, Acetat-1-¹⁴C und Daidzein-U-³H (*127, 128*) weist darauf hin, daß Phaseollin *de novo* synthetisiert wird und daß die Biosynthese entsprechend dem allgemeinen Bildungsschema (vgl. Abb. 3) der Isoflavonoide abläuft [vgl. (*259, 330*) und dort zitierte Literatur]. Danach entsteht der Ring A durch Kopf-Schwanz-Kondensation dreier Acetat-Einheiten, während der aromatische Ring D sowie die C-Atome 6, 6a und 11a des Phaseollins aus einem C₆-C₃-Körper unter 6,6a-Arylwanderung hervorgehen. Die

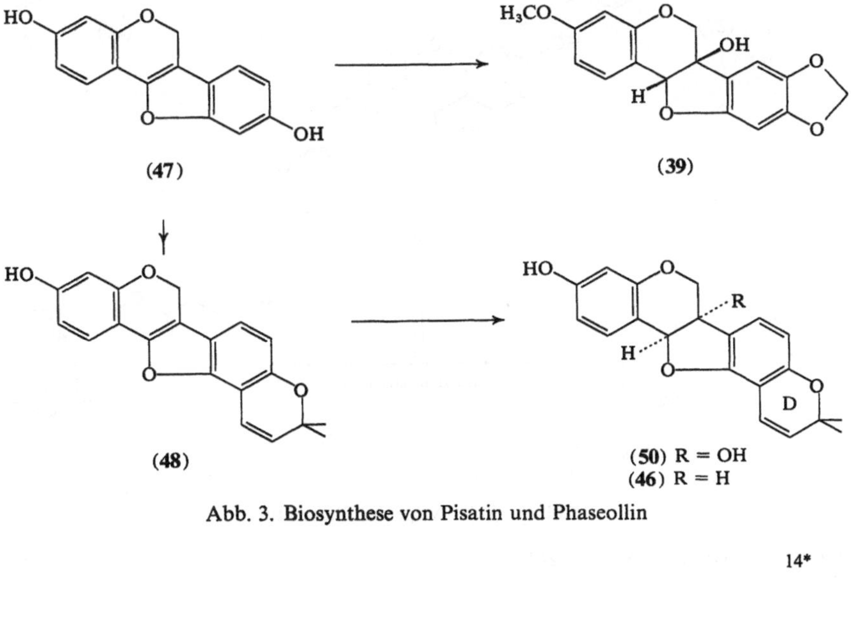

(47)   (39)

(48)   (50) R = OH
                                        (46) R = H

Abb. 3. Biosynthese von Pisatin und Phaseollin

Inkorporation von Daidzein zeigt, daß die den Chromenring bildende isoprenoide Seitenkette offensichtlich erst auf einer isoflavonoiden Zwischenstufe, wahrscheinlich einem bereits vorgebildeten Pterocarpan, eingeführt wird. Bisher konnte allerdings der Einbau von Mevalonat in Phaseollin nicht nachgewiesen werden (*127*).

Aus biogenetischen Überlegungen heraus wird ein 3,9-Dihydroxypterocarp-6a-en (**47**) als wichtiges Schlüsselprodukt der Biosynthesekette diskutiert, dessen Reduktion zum Phaseollintyp (**46**), (**50**), (**48**) führt, während Wasseranlagerung die Pisatinstruktur (**39**) ergibt (vgl. *140*) (Abb. 3).

Untersuchungen über die pilzinduzierte Bildung des Phaseollins bei gleichzeitiger Ermittlung der Aktivität der Phenylalanin-Ammoniak-Lyase haben ergeben, daß die durch Pilzinfektion oder chemischen Stress induzierte Phaseollinsynthese mit einer erhöhten PAL-Konzentration korreliert ist (*109, 126*). Actinomycin D, Cyclohemixid und 6-Methylpurin hemmen sowohl die Phaseollinbildung als auch die PAL-Synthese, was bedeutet, daß das induzierte Phaseollin und die PAL neu synthetisiertes Protein und RNA benötigen.

In Untersuchungen zum Phaseollinabbau (vgl. Abb. 4) wurde nachgewiesen, daß *Stemphylium botryosum* Wallr., ein luzernepathogener

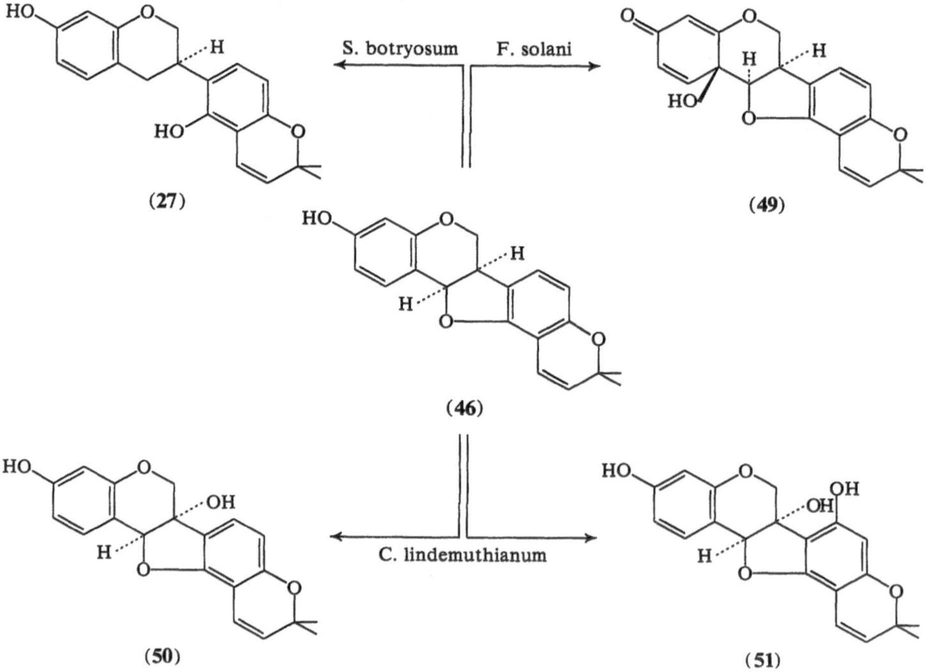
S. botryosum
F. solani
(27)
(49)
(46)
C. lindemuthianum
(50)
(51)

Abb. 4. Abbauwege des Phaseollins

Literaturverzeichnis: SS. 229—247

Pilz, Phaseollin *in vitro* zu einem Abbauprodukt umwandelt, dessen Fungitoxizität der des Phaseollins entspricht (*122*); es wurde kürzlich als Phaseollinisoflavan (**27**) identifiziert (*139*). *Colletotrichum lindemuthianum* baut Phaseollin zu zwei fungitoxischen Verbindungen, $C_{20}H_{18}O_5$ und $C_{20}H_{18}O_6$, ab, die als 6a-Hydroxyphaseollin (**50**), einer in Sojabohnen nach Induktion durch *P. megasperma* vorkommenden Verbindung (*264*), und 6a,7-Dihydroxyphaseollin (**51**) aufgeklärt worden sind (*48*). Durch *in vitro*-Versuche an *Fusarium solani f. sp. phaseoli* ist gezeigt worden, daß bei Inkubation mit Phaseollin ein Oxydationsprodukt auftritt, das weniger toxisch als Phaseollin ist (*129, 130*). Dieser Phaseollinmetabolit der Summenformel $C_{20}H_{18}O_5$ konnte kürzlich als 1a-Hydroxyphaseollon (**49**) charakterisiert werden (*131, 310*).

Diese Verbindung stellt ein von seiner Struktur her überraschendes Detoxikationsprodukt dar. Da keine phenolische Hydroxylgruppe mehr vorhanden ist, kann dieser Abbaumechanismus als ungewöhnliche Entgiftung von Phenolen angesehen werden. Auf Grund von NMR-Daten diskutieren die Autoren die in der Formel (**49a**) angegebene 6aS,11aS-Konfiguration (*131*). Andererseits wird für (—)-Pterocarpane, wie (—)-Phaseollin (**46**) 6aR,11aR-Anordnung angenommen, die bei der Hydroxylierung in 1a-Position erhalten bleiben sollte (vgl. Abb. 4).

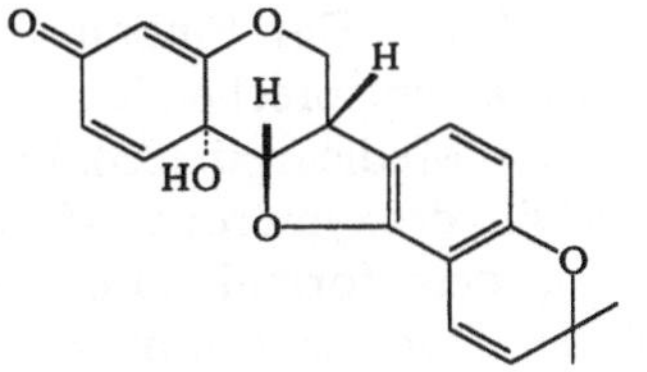

(49a)

6a-Hydroxyphaseollin — 1971/72 wurde aus Hypokotylen von Sojabohnen [*Glycine max* (L.) Merr.] nach Inokulieren mit *Phytophthora megasperma* Drechs. *var. sojae* A. A. Hildeb. eine fungitoxische Substanz der Summenformel $C_{20}H_{18}O_5$ isoliert und als 6a-Hydroxyphaseollin (**50**) aufgeklärt (*157, 163, 264*). Dieses Phytoalexin konnte bisher nur in infiziertem Gewebe nachgewiesen werden (*163, 164*) und ist sehr wahr-

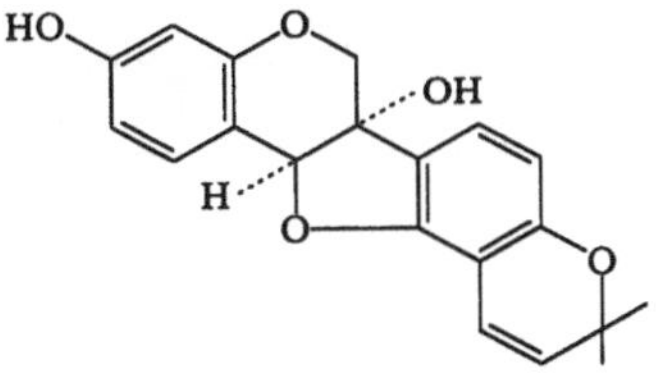

(50)

scheinlich mit einem der schon früher in pilzinfizierten oder stress-behandelten Sojabohnen aufgefundenen fungitoxischen Phenole mit Phytoalexincharakter identisch (*34, 35, 42, 52, 168, 170, 214, 225, 292, 294*). UV-Bestrahlung bewirkt ebenfalls einen erhöhten Hydroxyphaseollingehalt (*43*).

6a-Hydroxyphaseollin wird auch nach TNV-Infektion in *Glycine max* gebildet (*169*). Bakterien wie *Xanthomonas phaseoli var. sojae* (*214a*) und *Pseudomonas*-Arten wie *P. glycine* und *P. lachcrymans* induzieren ebenfalls die Akkumulation von Hydroxyphaseollin und einiger weiterer Isoflavonoide (*161*). Neben seiner Fungitoxizität (ED_{50} 25 µg/ml gegen *P. megasperma*) weist 6a-Hydroxyphaseollin auch antibakterielle Eigenschaften auf (*161*). *P. megasperma var. sojae*-resistente Sorten von *Glycine max* enthalten mehr 6a-Hydroxyphaseollin als unverträgliche Sorten (*104, 157, 162*). Dieser Befund wird durch neuere Untersuchungen gestützt, wonach die derepressive Bildung des Hydroxyphaseollins die Basis für die Resistenz zu sein scheint (*160*). Interessant ist die in dieser Arbeit beschriebene Isolierung eines spezifischen Auslösers, der in resistenten Sorten die erhöhte Hydroxyphaseollinproduktion bewirkt.

Die bisher zur Biosynthese des 6a-Hydroxyphaseollins vorliegenden Experimente haben gezeigt, daß eine *de novo*-Synthese abläuft und daß 6a-Hydroxyphaseollin nicht aus präformierten Glykosiden oder anderen Konjugaten gebildet wird (*164*). Der Einbau von Isoliquiritigenin (**34**) steht in Übereinstimmung zu entsprechenden Untersuchungen am Phaseollin (**46**) und vor allem am Medicarpin (**33**). Es wird postuliert, daß der Biosyntheseweg über 3,9-Dihydroxypterocarp-6a-en (**47**) und 6a-Dehydrophaseollin (**48**) führt, aus dem formal durch Wasseranlagerung unter Ausbildung der angularen Hydroxygruppe 6a-Hydroxyphaseollin (**50**) entsteht (*259*) (vgl. Abb. 3).

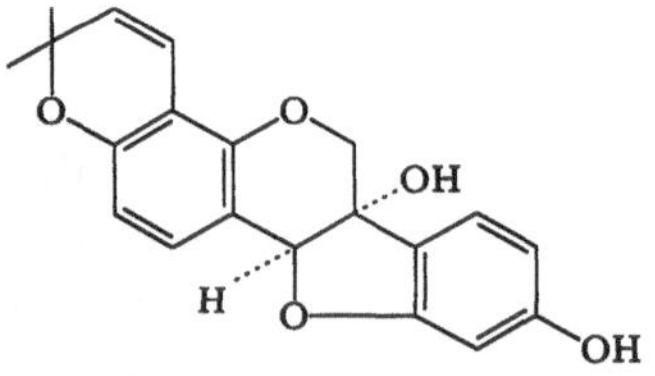

(52)

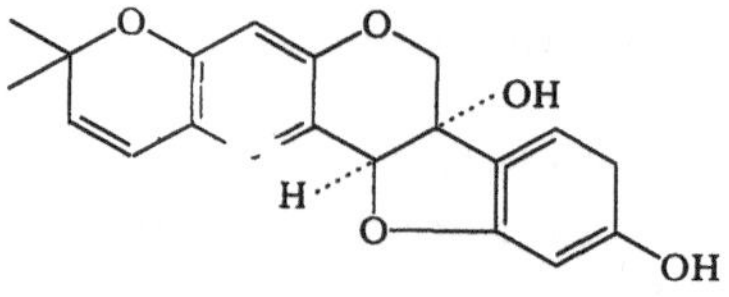

(52a)

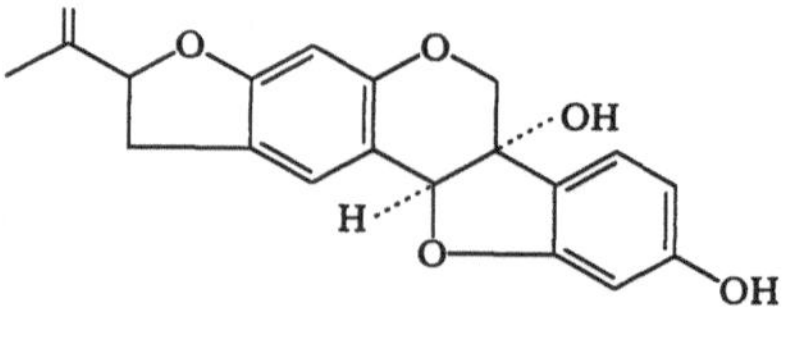

(52b)

Weitere Pterocarpane. — Neuere analytische Untersuchungen haben zur Auffindung eines weiteren Pterocarpanabkömmlings (52) geführt, der aus CuCl₂-behandelten Kotyledonen von *Glycine max* isoliert wurde (46). Diese gegen *Cladosporium cucumerinum* hoch wirksame fungitoxische Substanz wurde dünnschichtchromatographisch vom begleitenden Hydroxyphaseollin abgetrennt. Eingehende Strukturuntersuchungen ergaben, daß sich beide isomeren Verbindungen in der Stellung des 2,2-Dimethylchromenringes unterscheiden, der sich bei der neu aufgefundenen Substanz am Ring A des Pterocarpangerüstes befindet. ORD-Daten sprechen für eine 6aR, 11aR-Konfiguration (*189 a*). Kürzlich wurden aus *Glycine max* nach CuCl₂-Behandlung die beiden isomeren Pterocarpane (52 a) und (52 b) isoliert und strukturell aufgeklärt (*189 a*).

4. Sesquiterpene

a) Rishitin

1968 wurde aus Kartoffelknollen (*255, 289*) und aus Tomaten (*254*), die mit der Kraut- und Knollenfäule [Erreger *Phytophthora infestans* (Mont.) de Bary] befallen waren, eine fungitoxische Substanz isoliert, die wegen der als Versuchsobjekt dienenden Kartoffelsorte Rishiri als Rishitin bezeichnet wurde. Es handelt sich um einen bicyclisch gebauten norsesquiterpenoiden Alkohol (53), $C_{14}H_{22}O_2$, dessen Struktur und relative Konfiguration bereits 1968 aufgeklärt werden konnte (*156*). Rishitin besitzt ein dihydroxyliertes Eudesmangerüst mit einer beiden Ringen gemeinsamen Doppelbindung. 1969 erfolgte die Zuordnung der Absolutkonfiguration (*45, 113*). Das totalsynthetisch zugängliche Sesquiterpen Santonin kann auf chemischem Wege in Rishitin umgewandelt werden (*213*). Biosyntheseversuche haben ergeben, daß ^{14}C-markiertes Acetat und Mevalonat-2-^{14}C als spezifische Precursoren des Rishitins fungieren (*262*) und daß der Bildungsweg wahrscheinlich über Isolubimin (60 b), Lubimin (58) und 4-Hydroxylubimin (59) verläuft (*150 b*). Durch Biosyntheseversuche mit ^{13}C-markiertem Acetat an *Monilinia fructicola*-infizierten Kartoffeln und Auswertung der ^{13}C-NMR-Spektren des isolierten Rishitins ließ sich zeigen, daß das C-Atom 15 des Lubimins (−CHO) bei der Umwandlung zum Rishitin abgespalten wird (*281 c*).

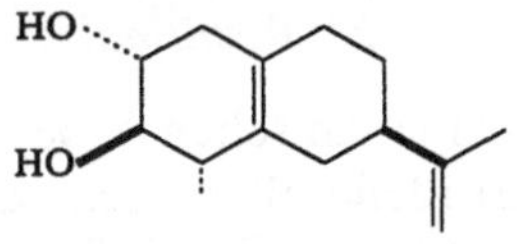

(53)

Erkrankte Kartoffeln enthalten zwei Tage nach Infektion mit *P. infestans* etwa 120 µg Rishitin pro g Frischgewicht, während Rishitin weder in gesundem Kartoffelgewebe noch im Kulturfiltrat oder Mycelextrakt von *P. infestans* nachgewiesen werden konnte (*289*). Ähnliche Befunde wurden 1974 auch für Tomaten erhalten, wonach mit *Verticillium alboatrum* inokulierte Stammsegmente von Tomatenpflanzen Rishitin akkumulieren (4,4 µg/g Frischgewicht), während in nichtinfiziertem Gewebe kein Rishitin nachzuweisen war (*288*).

Untersuchungen zur Fungitoxizität des Rishitins haben gezeigt, daß die Keimung der Zoosporen und das Wachstum des Keimschlauches von *P. infestans* durch 10^{-3} m Rishitin vollständig unterdrückt werden (*147,289*). Die ED_{50} beträgt $2,1 \times 10^{-4}$ m. Rishitin hemmt auch *Alternaria kikuchiana* Tanaka (ED_{50} $1,3 \times 10^{-4}$ m) (*289*) und *Fusarium solani* (ED_{50} 10^{-4} m) (*244, 289, 314*).

Durch eingehendere Versuche zur Struktur-Wirkungsbeziehung von Rishitin und 15 Strukturanaloga wurde nachgewiesen, daß die Hydroxylgruppe am C-Atom 3 für die fungitoxische Wirkung notwendig ist und daß eine Sättigung der Doppelbindung im Ring und/oder der Isopropylengruppe zu keiner wesentlichen Aktivitätsveränderung führt, während zusätzliche Sauerstoffunktionen in der Seitenkette eine Aktivitätserniedrigung bewirken (*147*). Interessanterweise konnte in diesen Arbeiten gleichzeitig auch eine wachstumsretardierende Wirkung des Rishitins nachgewiesen werden. Auf Grund dieses Befundes diskutieren die Autoren eine mögliche Doppelfunktion des Rishitins, wonach es einerseits steuernd in Wachstums- und Entwicklungsprozesse der Wirtspflanze eingreifen und zum anderen als fungitoxischer Abwehrstoff gegenüber Parasiten fungieren soll.

Zur Rishitinproblematik sind zahlreiche weitere Arbeiten erschienen, die sich insbesondere mit dem Vorkommen bzw. der Akkumulation des fungitoxisch wirksamen Rishitins in Kartoffelknollen oder Tomaten nach Infektion mit *P. infestans* und anderen Erregern befassen (*51, 70a, 82, 190, 190a, 191a, 199, 202, 203, 205, 224, 252, 253, 311, 312, 313, 314*). Es ist zu erwähnen, daß die Rishitinbildung nicht nur durch Pilzinfektion, sondern auch durch synthetische Hemmstoffe wie Jodacetat, p-Chlormercuribenzoat und Natriumfluorid sowie durch einige Fungizide wie Zineb, Kupferchlorid oder Tetramethylthiuramdisulfid induziert werden kann (*199*).

b) Rishitinol

1971—72 gelang es einer japanischen Arbeitsgruppe, aus *Phytophthora infestans*-befallenen Kartoffelknollen den dem Rishitin strukturverwandten Sesquiterpenalkohol Rishitinol (**54**), $C_{15}H_{22}O_2$, zu isolieren und die Struktur aus chemisch-physikalischen und spektroskopischen Daten abzuleiten (*153, 154*).

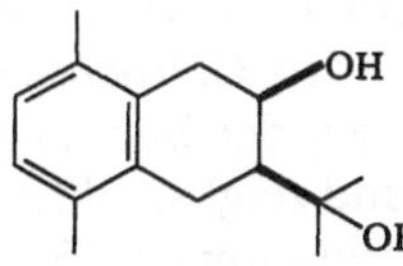

(54)

Das Racemat dieses Sesquiterpenes ist synthetisch zugänglich. In Analogie zum Rishitin ist für die Hydroxyisopropylgruppe des Rishitinols β-Orientierung angenommen worden. Über die Fungitoxizität des Rishitinols liegen bisher keine Angaben vor.

Darüber hinaus wurden als terpenoide fungitoxische Begleitstoffe des Rishitins aus *Phytophthora infestans*- und *Erwinia carotovora*-infizierten Kartoffelknollen die später zu besprechenden Sesquiterpene Phytuberin (**67**) und Lubimin (**58**) isoliert.

c) Glutinoson

Das vor kurzem aus TMV-infizierter *Nicotiana glutinosa* isolierte Glutinoson (**55**) gehört ebenfalls zur Gruppe der Sesquiterpene und weist eine enge strukturelle Verwandtschaft zu dem vorstehend genannten Rishitin (**53**) auf (*49*). Glutinoson ist ein α,β-ungesättigtes Keton mit einer bicyclischen Grundstruktur und einer charakteristischen Isopropylengruppe. Die Fungitoxizität (*20 a*) und die chemische Ähnlichkeit zu den bereits erwähnten Solanaceen-Phytoalexinen erscheinen phytochemisch besonders interessant. Bisher ist jedoch noch nicht nachgewiesen, ob Glutinoson auch in pilzinfiziertem Gewebe von *Nicotiana glutinosa* gebildet wird.

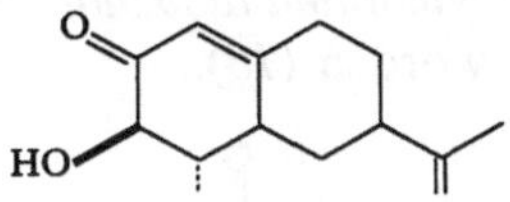

(55)

d) Capsidiol

Die von Müller (*209*) und Van den Ende (*302*) gemachte Beobachtung, daß in *Capsicum fructescens* L. nach Pilzinfektion phytoalexinähnliche Substanzen gebildet werden, konnte 1972 von Stoessl *et al.* (*282*) bestätigt werden. Diese Autoren isolierten aus Diffusaten von Paprikafrüchten, die mit verschiedenen Pilzen wie *Monilinia fructicola* (Wint.) Honey, *Phytophthora capsici* Leonian, *Botrytis cinerea* Pers. u. a. inokuliert worden waren, eine als Capsidiol (**56**) bezeichnete Verbindung, die sich auch in pilzinfizierten Blättern von *C. fructescens* finden läßt (*315 a*). Dieses bicyclisch gebaute Sesquiterpen der Summenformel $C_{15}H_{24}O_2$ leitet sich vom 4-epi-Eremophilan ab und zeigt strukturell große Ähnlichkeit zu den vorstehend aufgeführten Verbindungen (**53**) und (**55**) sowie zum nachfolgend beschriebenen Lubimin (**58**).

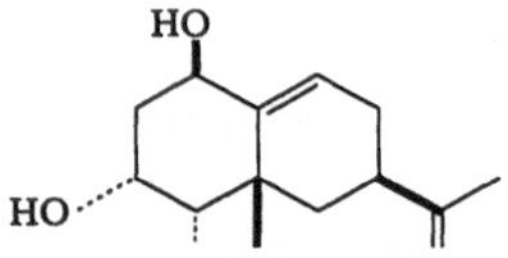

(**56**)

Nach Röntgenstruktur- und Kernresonanzuntersuchungen sind die beiden vicinalen Methylgruppen des Capsidiols im Gegensatz zu anderen natürlich vorkommenden Eremophilanabkömmlingen *trans*-ständig und die äquatorial orientierte Isopropylengruppe *cis*-ständig zur angularen Methylgruppe angeordnet (*38, 103*). Durch Biosyntheseexperimente mit ^{13}C-markiertem Acetat an *Capsicum fructescens*, inokuliert mit *Monilina fructicola*, ließ sich zeigen, daß die angulare Methylgruppe im Zuge einer Methylgruppenwanderung von C-10 nach C-5 entsteht (*20 b, 21*).

Capsidiol findet sich auch in geringen Spuren ($< 5 \times 10^{-7}$ m) in nicht inokulierten Früchten. Nach Inokulation mit verschiedenen pathogenen und apathogenen Pilzen steigt der Gehalt an Capsidiol je nach Pilz bis zu $7,5 \times 10^{-4}$ m an. Die Experimente haben eindeutig gezeigt, daß Capsidiol ein Produkt des pflanzlichen Stoffwechsels ist. In den Nährmedien oder Pilzextrakten ließ sich kein Capsidiol nachweisen.

Die Capsidiolbildung wird auch durch *Alternaria alternata* (*318*) sowie das Bakterium *Erwinia carotovora* (*319*) induziert. Kürzlich ist Capsidiol interessanterweise auch in *Datura stramonium* nach Pilzinfektion (*281, 317 a*) sowie in *Nicotiana tabacum* und *N. clevelandii* nach TNV-Infektion nachgewiesen worden (*18*).

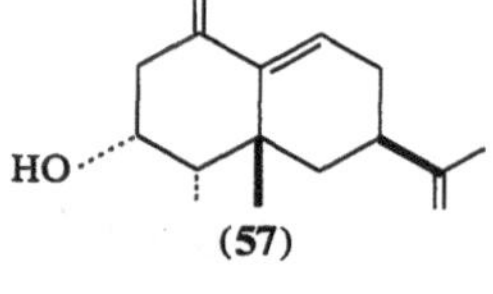

(**57**)

Capsidiol wird von verschiedenen Pilzen wie *Botrytis cinerea* oder *Fusarium oxysporum f. vasinfectum* oxydativ zu Capsenon (**57**) (*316, 318*) und weiter zu einem noch nicht näher identifizierten Abbauprodukt (*283*) umgewandelt. Capsenon ist weniger fungitoxisch als Capsidiol, so daß dieser oxydative Abbau als eine Art Entgiftung durch den Pilz angesehen werden kann.

1974 haben WARD *et al.* die Fungitoxizität von Capsidiol und mehr als 20 strukturverwandten Sesquiterpenen untersucht und eine Vielzahl pathogener und apathogener Pilze in ihre Testung einbezogen (*320*). Dabei hat sich ergeben, daß Capsidiol von den überprüften Verbindungen die höchste Fungitoxizität aufwies (ED_{50} 1×10^{-5} m gegen *Phytophthora infestans*). Trotz dieser umfangreichen Arbeiten konnten noch keine eindeutigen Ergebnisse bezüglich einer Struktur-Wirkungsbeziehung erhalten werden. Weiterführende Experimente haben dazu geführt, Capsidiol als potentielles Fungizid gegen Mehltau bei Tomaten zu überprüfen (*323*).

e) Lubimin, 4-Hydroxylubimin und Isolubimin

Das fungitoxisch wirksame Sesquiterpen Lubimin (**58**), Summenformel $C_{15}H_{24}O_2$, wurde in *P. infestans*-infizierten Kartoffelknollen aufgefunden und eingehend unter phytopathologischen Aspekten untersucht (*155, 199, 202, 203, 204, 205, 206*). Außerdem wurde Lubimin aus

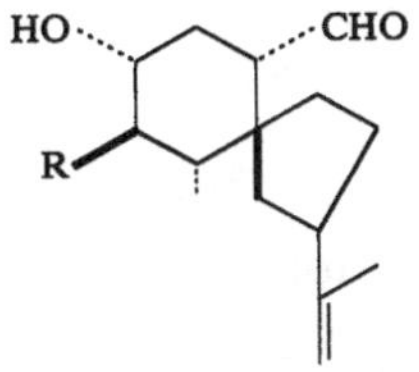

(**58**) R = H
(**59**) R = OH

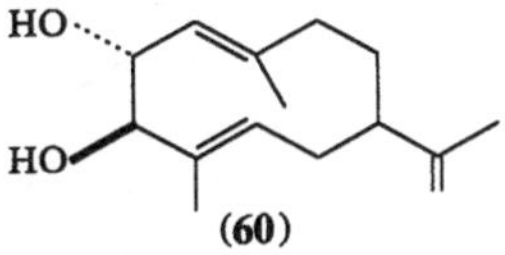

(**60**)

Solanum melongena und *Datura stramonium* L. nach Inokulation mit *Monilinia fructicola* und einigen anderen Pilzen isoliert (*37a, 280, 281, 317, 317a*). Die 1971 veröffentlichte Struktur des Lubimins (*204*) ist kürzlich auf Grund chemischer und spektroskopischer Untersuchungen sowie biogenetischer Überlegungen revidiert worden (*155, 280*). Diese unabhängig von zwei verschiedenen Laboratorien durchgeführte Korrektur hat übereinstimmend zu der hier angegebenen Strukturformel für Lubimin (**58**) geführt, dem somit ein Agarospirangerüst zugrunde liegt. Gleichzeitig ist neben Lubimin in *Phytophthora*- oder *Monilinia fructicola*-infizierten Kartoffeln (*155, 281c*) sowie in *Datura stramonium* nach Inokulation mit *Monilinia fructicola* oder anderen Pilzen (*37a, 281, 317a*) 4-Hydroxylubimin (**59**) aufgefunden worden. Strukturverwandt sind Solavetivon (**65**) (*56, 281c*) und Dihydrolubimin ($-CH_2OH$ anstelle von $-CHO$) (*155, 281a, 281c*).

Lubimin und das durch Röntgenstrukturanalyse in seiner Konformation aufgeklärte 4-Hydroxylubimin haben am C-Atom 1 übereinstimmende Absolutkonfiguration (*37a, 281, 281a*), so daß eine gemeinsame biogenetische Vorstufe beider Verbindungen zu diskutieren ist. Möglicherweise leiten sich Lubimin und 4-Hydroxylubimin sowie weitere strukturverwandte Verbindungen biogenetisch vom 2,3-Dihydroxygermacren (**60**) ab, einem in *Datura stramonium* neben Lubimin (**58**), 4-Hydroxylubimin (**59**) und Capsidiol (**56**) als Stressmetabolit aufgefundenen Sesquiterpen (*37a, 281, 317a*). Durch Biosyntheseexperimente mit ^{14}C-markiertem Acetat an *Datura stramonium* und an Kartoffeln ist der isoprenoide Bildungsweg für die Verbindungen (**58**), (**59**) und (**60**) gesichert worden (*37a, 281c*). In neueren Versuchen wurde nach Verfütterung von Spirovetiva-1(10),11-dien-2-on (**60a**) an *P. infestans*-infizierte Kartoffeln neben Lubimin und Rishitin (**53**) ein weiteres Vetispiran isoliert und als 14-Hydroxy-spiravetiva-11-en-2-on (**60b**) identifiziert (*150b*). Diese als Isolubimin bezeichnete Verbindung wird als Zwischenprodukt der Biosynthesefolge (**60a**) → Isolubimin (**60b**) → Lubimin (**58**) → 4-Hydroxylubimin (**59**) → Rishitin (**53**) diskutiert (*150b*).

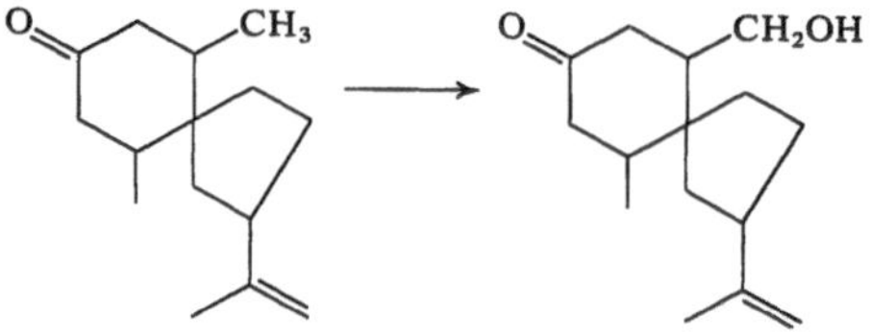

(60a)   (60b)

f) Strukturverwandte Sesquiterpene

Zu den vorstehend aufgeführten Verbindungen zeigen einige weitere phytoalexinartige Sesquiterpene eine enge strukturelle Ähnlichkeit.

Dazu gehören die kürzlich neben Lubimin (**58**) aus *Solanum melongena* isolierten Nerolidolabkömmlinge (**61**), (**62**) und (**63**) sowie das Eudesmanderivat (**64**), deren Bildung durch *Monilinia fructicola, Aspergillus fumigatus* Fres., *Penicillium frequetans* Westling, *Botrytis cinerea* Pers.

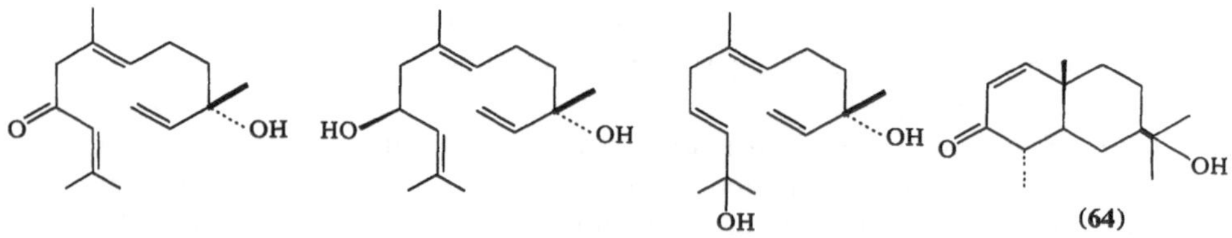

(61)   (62)   (63)   (64)

und *Fusarium oxysporum f. vasinfectum* (Atkinson) Snyder und Hanson induziert wird (*281a, 317*).

In diesem Zusammenhang sind auch zwei Vetispiranderivate der Summenformel $C_{15}H_{22}O$ bzw. $C_{15}H_{20}O$ zu nennen, die in *Solanum tuberosum* nach Infektion mit *Erwinia carotovora* und *Phytophthora infestans* aufgefunden und als Spirovetiva-1(10),11-dien-2-on (Solavetivon) (**65**) bzw. Spirovetiva-1(10),3,11-trien-2-on (**66**) identifiziert worden sind (*56*).

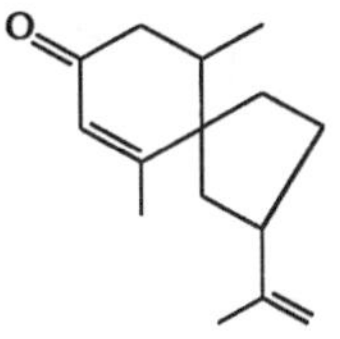

(**65**) 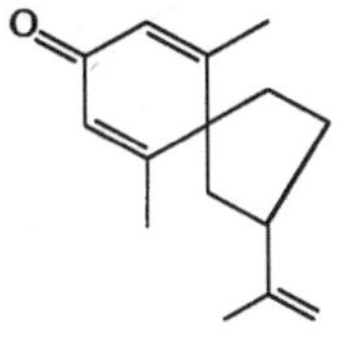   (**66**)

Kartoffelknollen enthalten nach Inokulieren mit *Phytophthora infestans, Monilinia fructicola, Erwinia carotovora var. astroseptica* oder *Glomerella cingulata* (*55, 190, 190a, 281c*) einen als Phytuberin bezeichneten Stressmetaboliten, dem auf Grund chemischer und spektroskopischer Daten (*55*) sowie Röntgenstrukturanalyse der Dihydroverbindung (*141a*) die Struktur eines tricyclischen Sesquiterpenacetats (**67**) zukommt. Die Annahme, daß das fungitoxisch wirksame Phytuberin, Summenformel $C_{17}H_{26}O_4$, mit Lubimin (**58**), Rishitin (**53**) und Solavetivon (**65**) biogenetisch verwandt ist, wird durch das gemeinsame Vorkommen der genannten Sesquiterpenoide in Kartoffeln sowie durch Biosyntheseversuche mit Acetat-1,2-^{13}C untermauert (*141a, 281c*).

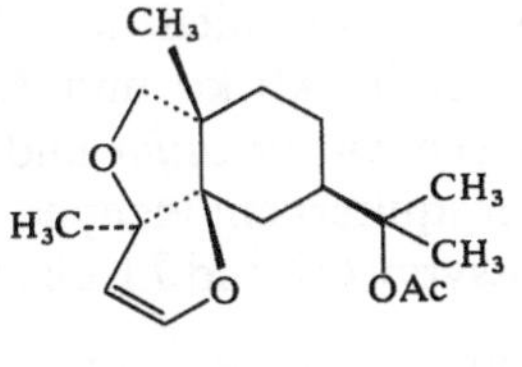

(**67**)

Als weitere strukturverwandte endogene Pflanzenstoffe seien Agarol (**68**) aus *Aquilaria agalocha* Roxb., Costol (**69**) aus *Sausurea lappa* Clarke sowie Occidiol (**70**) und Occidentalol (**70a**) aus *Thuja occidentalis* aufgeführt [vgl. (*145*) und dort zitierte Literatur], über deren Fungitoxizität allerdings noch keine Angaben vorliegen.

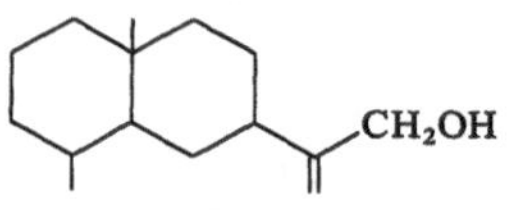

(68) 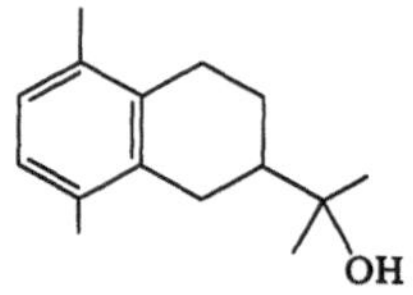   (70)

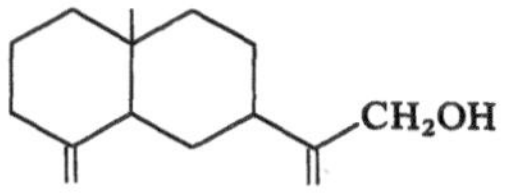

(69)   (70 a)

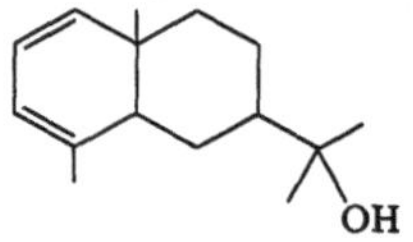

g) Gossypol

Das vor allem in Samen, aber auch in Wurzeln und subepidermalen Blattdrüsen von *Gossypium*-Arten vorkommende gelbe Pigment wird als Gossypol (**71**) bezeichnet und ist trotz seiner für Naturstoffe ungewöhnlichen 2,2'-Binaphthylstruktur als dimeres Sesquiterpen, $C_{30}H_{30}O_8$, aufzufassen. Die chemischen, stereochemischen und toxikologischen Aspekte des Gossypols werden in neueren Übersichten umfassend behandelt (*1, 30, 73*); Konformation (*332*) und Synthese (*83*) des Gossypols sind ebenfalls beschrieben. Es ist bekannt, daß die Biosynthese auf dem Isoprenweg erfolgt, wobei Neryl- und cis,cis-Farnesyl-2-^{14}C-pyrophosphat besonders hohe Inkorporationsraten zeigen (*124, 125*). Versuche über eine mikrobielle Detoxifikation haben ergeben, daß das gegenüber Säugetieren stark giftige Gossypol von einem bestimmten *Diplodia*-Stamm abgebaut werden kann (*22*).

Der symmetrische Bau des Gossypolmoleküls bedingt eine durch Atropisomerie hervorgerufene optische Aktivität. Neben dem Racemat findet sich die (+)-Form in verschiedenen *Gossypium*-Arten (*74*) sowie in *Thespesia populnea*, einer in Afrika und Asien beheimateten Malvacee (*33, 72, 167*). Aus *Gossypium hirsutum* und *G. barbadense* sind kürzlich zwei weitere dimere Sesquiterpenaldehyde isoliert und als 6-Methoxy- und 6,6'-Dimethoxygossypol (**72** und **73**) charakterisiert worden (*192, 276*).

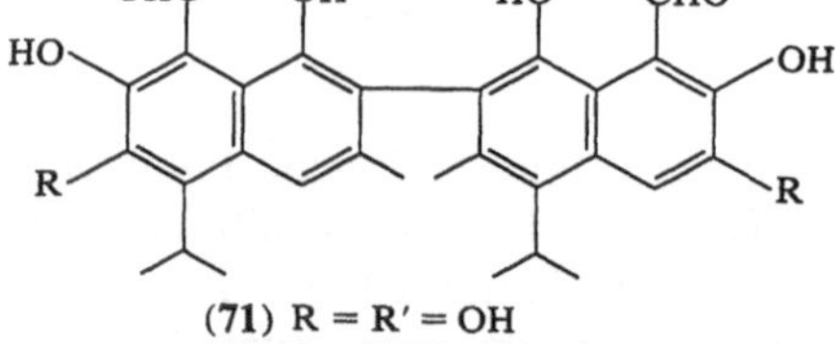

(**71**) R = R' = OH
(**72**) R = OCH₃, R' = OH
(**73**) R = R' = OCH₃

Beide Verbindungen zeigen fungitoxische Aktivität gegen verschiedene *Penicillium-* und *Cladosporium*-Arten.

Obwohl Gossypol ein „normaler" Inhaltsstoff bestimmter Teile der Baumwollpflanze ist, zeigt es andererseits ein gewisses phytoalexin-ähnliches Verhalten. Die Arbeitsgruppe um BELL hat bereits 1967 nachgewiesen, daß Gossypol im Stengelgewebe verschiedener *Gossypium*-Arten nach Infektion mit Konidien von *Verticillium albo-atrum* oder *Rhizopus nigricans* Ehr. stark akkumuliert wird (*24*). Weiterführende Untersuchungen über den Einfluß verschiedener Faktoren auf die Bildung und Anreicherung von Gossypol und anderen phytoalexin-ähnlichen Substanzen in *Gossypium*-Gewebe (*11, 25, 27, 211a, b*) haben ergeben, daß im Stengelgewebe nach Infektion mit *Verticillium dahliae* neben Gossypol etwa 5 fungitoxische sesquiterpenoide Aldehyde nachzuweisen sind, die mit für die Resistenz der Pflanzen gegenüber pilzlichen Krankheitserregern verantwortlich gemacht werden (*26, 28, 192*).

h) Hemigossypol und strukturverwandte Verbindungen

Eine vertiefte chemische Bearbeitung hat zur Isolierung und Charakterisierung von drei sesquiterpenoiden Aldehyden geführt, die als Hemigossypol (**74**) (*29, 192, 276, 334, 335*), 6-Methoxyhemigossypol (**75**) (*29, 192, 276*) und 6-Desoxyhemigossypol (**76**) (*29*) identifiziert wurden.

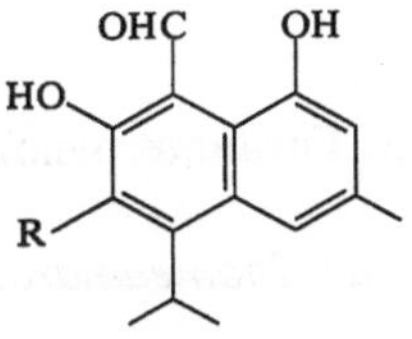

(**74**) R = OH
(**75**) R = OCH₃
(**76**) R = H

Das 1974 isolierte Isohemigossypol (*251*) ist auf Grund neuerer strukturchemischer Untersuchungen mit Hemigossypol (**74**) identisch (*314a*). Hemigossypol weist gegenüber *Verticillium albo-atrum* eine höhere Fungitoxizität auf als Gossypol (*334*).

Vergleichende Untersuchungen über die Verbreitung dieser drei C_{15}-Aldehyde innerhalb der Malvaceen haben ergeben, daß Hemigossypol (**74**) im *Verticillium*-infizierten Stengelgewebe von 21 überprüften *Gossypium*-Arten und verschiedenen Genera der Malvaceen nachweisbar ist (*29*). 6-Methoxyhemigossypol (**75**) zeigt eine ähnliche Verbreitung, während 6-Desoxyhemigossypol (**76**) als vermutliche Vorstufe von Hemigossypol nur in geringerer Konzentration vorkommt und nur in einer kleineren Anzahl der untersuchten *Gossypium*-Arten mit Sicherheit nachzuweisen war. Die Malvaceen *Hibiscus esculentus* L., *H. rosa-sinensis* L. und *H. syriacus* L. enthalten nach Inokulation mit *V. dahliae* keinen der genannten sesquiterpenoiden Aldehyde.

Kürzlich ist gezeigt worden, daß sich in *V. dahliae*-infiziertem Stengelgewebe von *G. barbadense* zwei nicht-aldehydische Verbindungen (**77**) und (**78**) anreichern, die autoxydativ in Hemigossypol (**74**) und sein 6-Methoxyderivat (**75**) übergehen und somit deren natürliche Vorstufe darstellen dürften (*275*). Beide Substanzen sind als 2H-Naphtho(1,8-b-c)-furane identifiziert worden. Mit hoher Wahrscheinlichkeit ist die Methoxyverbindung (**78**) mit dem früher von anderen Autoren isolierten Vergosin (*334, 335*) identisch. Abschließend sei das kürzlich aus *Gossypium hirsutum* isolierte und aus Hemigossypol auf chemischen Wege darstellbare p-Hemigossypolon (**79**) genannt (*104 a*). Dieses Chinonderivat hemmt das Wachstum von *Heliothis virescens*.

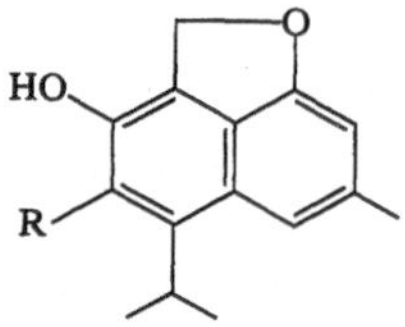

(**77**) R = OH
(**78**) R = OCH₃

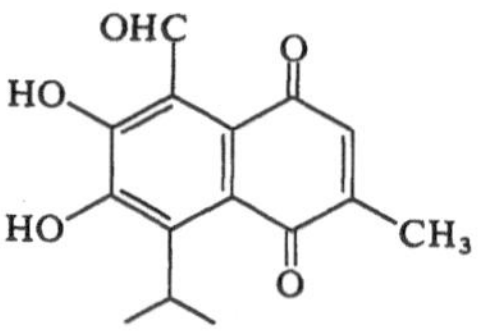

(**79**)

5. Furanoterpenoide

a) Ipomeamaron

Eine bereits 1943 von HIURA (*141*) in *Ipomoea batatas* (L.) Lam. (*Convolvulaceae*) nach Infektion mit *Ceratocystis fimbriata* (Ell. & Halst.) Elliot aufgefundene postinfektionelle Substanz wurde Anfang der fünfziger Jahre von verschiedenen japanischen Arbeitsgruppen erneut isoliert, strukturell aufgeklärt und chemisch intensiv bearbeitet (*2, 7, 175, 176, 178, 179, 197, 198, 221, 284, 324*). Diese als Ipomeamaron (**80**) bezeichnete äußerst bitter schmeckende Verbindung gehört zur Gruppe der Terpene und stellt ein furanosesquiterpenoides Keton dar, das auch auf synthetischem Wege zugänglich ist. Ipomeamaron hemmt das Mycelwachstum und die Sporulation von *C. fimbriata* (*297*) und ist außerdem hepatotoxisch wie das nachfolgend zu besprechende Ipomeamaronol (*325*). Ipomeamaron findet sich in gesundem Gewebe, wenn überhaupt, nur in äußerst niedrigen Konzentrationen. Bei Infektion mit verschiedenen pilzlichen Organismen

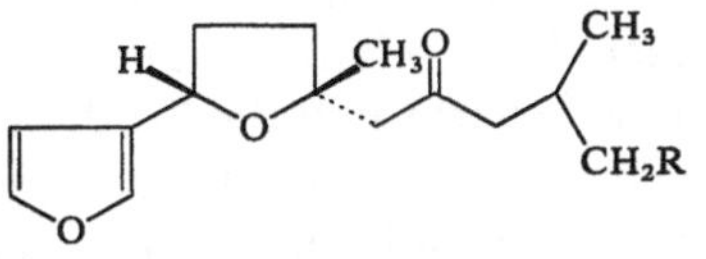

(**80**) R = H
(**81**) R = OH

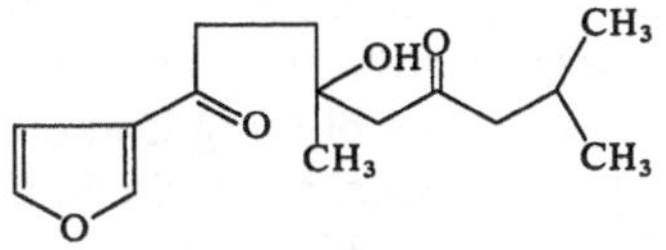

(**81b**)

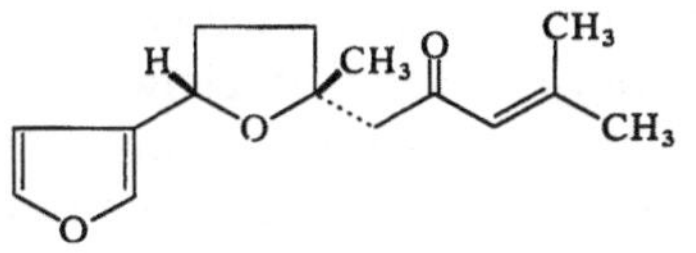

(**81a**)

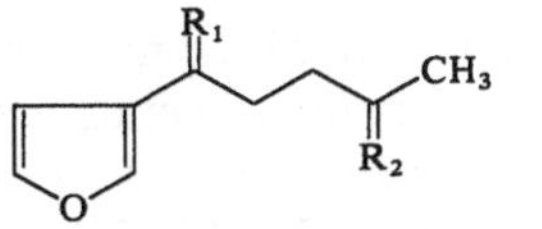

(**81c**) $R_1 = R_2 = O$
(**81d**) $R_1 = H, OH$ $R_2 = O$
(**81e**) $R_1 = O$ $R_2 = H, OH$
(**81f**) $R_1 = R_2 = H, OH$

wie *C. fimbriata, Fusarium javanicum* Koorders (*325*), *Helicobasidium mompa* Tanaka (vgl. *145*) u. a. sowie durch chemischen Stress, z. B. HgCl$_2$-Behandlung (*165, 299*) steigt am Infektionsort der Gehalt an Ipomeamaron stark an. Resistente Sorten enthalten nach 96—120 h 40 mg Ipomeamaron pro g Frischgewicht. Das Vorkommen von Ipomeamaron scheint auf die Art *Ipomoea* beschränkt zu sein. [Verteilung und quantitative Bestimmung von Ipomeamaron in Süßkartoffeln vgl. (*54 a, 330 a*)]. Die Bildung von Furanoterpenoiden hängt offenbar vom Ausmaß des Mycelwachstums und den damit verbundenen Stoffwechselveränderungen im Wirtsgewebe ab (*142*). Beispielsweise bewirkt *C. fimbriata*-Infektion einen Anstieg der 3-Hydroxy-3-methylglutaryl-coenzym A-Reduktase (*283 a*).

Interessante Ergebnisse bringt eine Arbeit über die Isolierung von aktiven Substanzen aus wäßrigen Mycel- und Konidienextrakten von *C. fimbriata*, die in pilzfreien Wurzelscheiben von *Ipomoea batatas* die Bildung von Ipomeamaron und strukturverwandten Terpenen induzieren (*166*). Die isolierten dialysierbaren Faktoren sind niedermolekulare hitzestabile Verbindungen; sie sind wasser- oder 0,02 m KCl-löslich und in organischen Lösungsmitteln unlöslich. Sie haben kein kationisches oder anionisches Verhalten und werden nicht in die Nährlösung ausgeschieden. Untersuchungen über den Einfluß verschiedener proteinogener Aminosäuren auf die Ipomeamaronbildung ergaben, daß Alanin, Valin, Cystein und Glycin die Terpenbildung signifikant beeinflussen (*165*).

Befall durch tierische Schädlinge wie *Cylas formicarius elegantulus* Summers führt ebenfalls zur erhöhten Ipomeamaronbildung im nekrotischen Gewebe (*6*). Aus Larvenhomogenaten von *Cylas formicarius* Fabricius und *Euscepes postfasciatus* Fairmaire ist kürzlich ein die Nekrose auslösender und die Furanoterpenoidbildung induzierender Faktor isoliert worden, der höchstwahrscheinlich von den Larven in das befallene

Gewebe ausgeschieden wird (298 a). Es handelt sich um eine hoch-molekulare Verbindung mit proteinartigen Eigenschaften, die im Ge-webe eine Äthylenbildung und eine Nekrose bewirkt und die Produktion von Furanoterpenoiden induziert. Diese Arbeiten zeigen, daß die Phyto-alexinbildung im Pflanzengewebe auch durch Schadinsekten bzw. deren Stoffwechselprodukte ausgelöst werden kann.

Die über den Terpenweg verlaufende Biosynthese des Ipomeamarons ist mit radioaktiv markierten Precursoren intensiv untersucht worden, wobei vor allem Acetat, Mevalonat, Leucin, Pyruvat, Citrat sowie Farnesol auf ihren spezifischen Einbau in Ipomeamaron überprüft wurden (3, 4, 5, 143, 144, 215, 217, 218 a, 218 c, 222, 223). Dabei zeigte Äthanol-2-^{14}C wesentlich bessere Inkorporationsraten als Acetat-2-^{14}C (218). C$_{10}$- und C$_{15}$-Terpenalkohole wie Farnesol, Nerodiol oder Geraniol unterdrücken zwar den Einbau von Acetat in Ipomeamaron, nicht aber die Ipomeamaron-bildung (216).

b) Ipomeamaronol

In Ceratocystis fimbriata-infizierten Wurzelscheiben von Ipomoea batatas findet sich neben Ipomeamaron und dessen biologischen Oxidationsprodukten wie Ipomeanin und Batatsäure (177) das biogene-tisch verwandtes Sesquiterpen Ipomeamaronol (81) (151, 152, 333). Diese phytoalexinartige Verbindung dürfte biosynthetisch aus Ipomeamaron durch Oxydation einer der beiden endständigen Methylgruppen hervor-gehen.

c) Dehydroipomeamaron und 4-Hydroxymyoporon

Schließlich ist Dehydroipomeamaron (81 a) zu nennen, eine kürzlich aus C. fimbriata-infizierten Wurzelscheiben von Ipomoea batatas (220) isolierte Verbindung, die als direkte Vorstufe des Ipomeamarons anzusehen ist (218 b, 219). Darüber hinaus enthält Ipomoea batatas nach Infektion mit F. solani oder C. fimbriata sowie nach Quecksilberchlorid-Behandlung das furanoide Sesquiterpen 4-Hydroxymyoporon (81 b) sowie einige Toxine (81 c—81 f), die biogenetisch als Abbauprodukte von (81 b) aufzufassen sind (50 a).

6. Verschiedenes

Abschließend sollen einige Phytoalexine vorgestellt werden, die zu den vorstehend genannten, nach chemischen Gesichtspunkten geordne-ten Verbindungen keine unmittelbare strukturmäßige Beziehung erkennen lassen.

Literaturverzeichnis: SS. 229—247

Mit Konidien von *Perenospora tabacina* infizierte Tabakblätter enthalten einen Hemmstoff, $C_{17}H_{26}O_3$, der als Quieson bezeichnet und als 5-Isobutyroxy-β-jonon (82) identifiziert worden ist (*183*). Das auch synthetisch zugängliche Quieson (*207*) zeigt enge strukturelle Verwandtschaft zu Vomifoliol und ähnlich gebauten pflanzlichen Jononabkömmlingen, zu denen auch im weiteren Sinn das Phytohormon Abscisinsäure zählt.

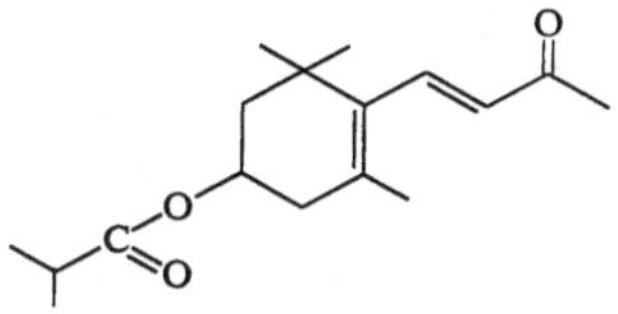

(82)

Die Leguminose *Vigna unguiculata* (L.) Walp. enthält nach Infektion mit *Colletotrichum lindemuthianum* eine fungitoxische Substanz (30 mg/kg Frischgewicht), die als substituiertes 2-Arylbenzofuran (83) identifiziert worden ist (*240*). Sie wurde als Vignafuran bezeichnet und konnte in uninfizierten Kontrollpflanzen nicht nachgewiesen werden. Unter den in *Vigna unguiculata* vorkommenden phytoalexin-ähnlichen Verbindungen stellt Vignafuran die wirksamste dar. Vignafuran bewirkt in einer Konzentration von 8 ppm bei zwei getesteten *C. lindemuthianum*-Stämmen eine vollständige Hemmung der Sporenkeimung.

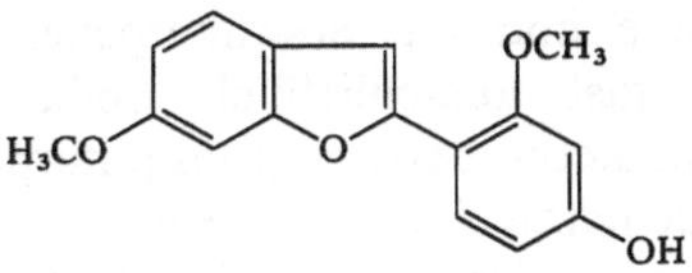

(83) 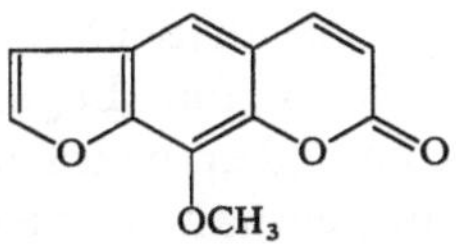 (84)

Weiterhin sei das Xanthotoxin (84) genannt, eine Verbindung, die aus *Pastinaca sativa*-Wurzeln nach Inokulieren mit *Ceratocystis fimbriata, Helminthosporium carbonum, Alternaria spp.* oder *Colletotrichum lindemuthianum* isoliert worden ist (*150*). 72 h nach Inokulieren ist 1 mg Xanthotoxin pro g Frischgewicht enthalten, während der Xanthotoxingehalt nicht infizierter Kontrollpflanzen bei 0,5 mg/g liegt. Die ED_{50} beträgt gegenüber *C. fimbriata* 1×10^{-4} m.

Zellfreie Extrakte junger Keimlinge von *Ricinus communis* L. bilden ein fungitoxisches Diterpen, wobei infizierte Keimlinge eine wesentlich höhere Syntheserate aufweisen (*265*). Dieses als Casben (85) bezeichnete Diterpen ist vor kurzem synthetisch dargestellt worden (*56 a*). Es besitzt einen aus 14 C-Atomen bestehenden Makrocyclus mit drei Methylgruppen, zwei Doppelbindungen und einen *cis*-ständigen dimethylierten Cyclopropanring. Das als Phytoalexin angesehene Casben zeigt neben fungitoxischer Aktivität signifikante wachstumshemmende Wirkung, indem es GA_3-stimuliertes Wachstum einer Zwergmaismutante inhibiert.

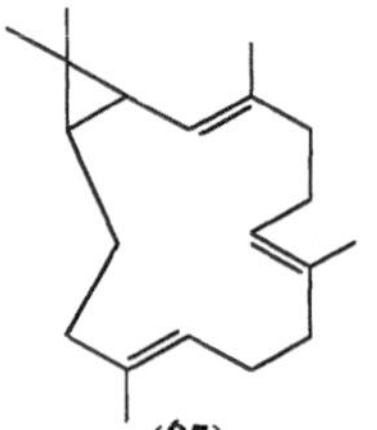

(85) 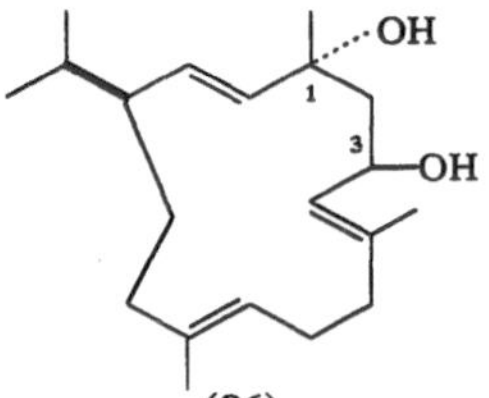   (86)

In diesem Zusammenhang erscheint ein strukturmäßig sehr ähnlich gebauter pflanzlicher Wachstumsregulator von Interesse, der aus Tabakblättern isoliert und als 4,8,13-Duvatrien-1,3-diol (86) charakterisiert worden ist (*71*). Dieser Inhibitor hemmt das Wachstum von Weizenkoleoptilen.

V. Schlußbetrachtung

Bei den derzeitig über 50 strukturbekannten Phytoalexinen handelt es sich im wesentlichen um Polyacetylene, Dihydrophenanthrene, Isoflavonoide einschließlich Pterocarpane und um Sesquiterpene. Es ist auffallend, daß die Leguminosen fast ausschließlich isoflavonoide Phytoalexine bilden, während die Solanaceen bevorzugt sesquiterpenoide Abwehrstoffe enthalten. Die verschiedenen zu einer dieser Substanzklassen gehörenden Phytoalexine zeigen untereinander keine großen strukturellen Unterschiede. Meist handelt es sich um biogenetisch sehr nahe stehende und durch Biosynthese oder Abbau ineinander überführbare Verbindungen, die vielfach vergesellschaftet vorkommen.

Es gilt heute als gesichert, daß die mit einer Wirtspflanze in Wechselwirkung tretenden pilzlichen oder bakteriellen Krankheitserreger oder Viren nur eine induzierende Wirkung haben, d. h. die pathogenen Mikroorganismen stellen dem pflanzlichen Stoffwechsel nur Induktoren und keine Precursoren für eine Phytoalexinbiosynthese zur Verfügung. In einigen wenigen Fällen hat man aus pilzlichen Mycelextrakten bereits spezifische Induktoren isolieren können. Es scheint sich dabei um spezifisch wirkende proteinartige Verbindungen zu handeln.

Trotz umfangreicher Untersuchungen sind Einzelheiten über den Mechanismus und die Regulation der Phytoalexinbildung jedoch noch weitgehend unbekannt.

Neben den vom Mikroorganismus produzierten Induktoren kann auch ein chemischer Stress, z. B. Behandlung des Pflanzenmaterials mit Schwermetallchloriden wie Kupfer- oder Quecksilberchlorid, die Phytoalexinbildung auslösen. Das ist beispielsweise für die experimentell eingehender untersuchten Phytoalexine Medicarpin, Pisatin und Phaseollin nachgewiesen. Aus diesem Grund wird von einigen Autoren auch die Bezeichnung ‚Stressmetabolite' verwendet. Mit wenigen Ausnahmen hat man bisher bevorzugt Kulturpflanzen wie Kartoffel und Körnerleguminosen auf ihren Phytoalexingehalt untersucht. Es ist aber zu erwarten, daß bei intensiver Suche in anderen höheren Pflanzen ebenfalls Phytoalexine oder phytoalexin-artige Substanzen aufgefunden werden.

Literatur

1. ADAMS, R., T. A. GEISSMAN, and J. D. EDWARDS: Gossypol, a Pigment of Cotton Seed. Chem. Reviews **60**, 555 (1960).

2. AKAZAWA, T.: Chromatographic Isolation of Pure Ipomeamarone and Reinvestigation on its Chemical Properties. Arch. Biochem. Biophysics **90**, 82 (1960).

3. — Biosynthesis of Ipomeamarone. II. Synthetic mechanism. Arch. Biochem. Biophysics **105**, 512 (1964).

4. AKAZAWA, T., and J. URITANI: Biosynthesis of Ipomeamarone. The Incorporation of Acetate-2-^{14}C into Ipomeamarone. Agric. Biol. Chem. **26**, 131 (1962).

5. AKAZAWA, T., I. URITANI, and Y. AKAZAWA: Biosynthesis of Ipomeamarone. I. The Incorporation of Acetate-2-^{14}C and Mevalonate-2-^{14}C into Ipomeamarone. Arch. Biochem. Biophysics **99**, 52 (1962).

6. AKAZAWA, T., I. URITANI, and H. KUBOTA: Isolation of Ipomeamarone and Two Coumarin Derivatives from Sweet Potato Roots Injured by the Weevil, *Cylas formicarius elegantulus.* Arch. Biochem. Biophysics **88**, 150 (1960).

7. AKAZAWA, T., and K. WADA: Analytical Study of Ipomeamarone and Chlorogenic Acid Alterations in Sweet Potato Roots Infected by *Ceratocystis fimbriata.* Plant Physiol. **36**, 139 (1961).

8. ALLEN, E. H., and C. A. THOMAS: trans-trans-3,11-Tridecadiene-5,7,9-triyne-1,2-diol, an antifungal Polyacetylene from Diseased Safflower (*Carthamus tinctorius*). Phytochemistry **10**, 1579 (1971).

9. — — A Second Antifungal Polyacetylene Compound from *Phytophthora*-infected Safflower. Phytopathology **61**, 1107 (1971).

10. — — Time Course of Safynol Accumulation in Resistant and Susceptible Safflower Infected with *Phytophthora drechsleri.* Physiol. Plant Pathol. **1**, 235 (1971).

11. AVAZKHODZHAEV, M. KH., and Z. M. MUSLIMOV: Physiological-biochemical Study of the Wilt Resistance of the Cotton Plant. Fiziol. Biokhim. Khlop. **1972**, 213. Edit. YULDASHEV, S. KH.,"Fan", Taschkent; Chem. Abstr. **80**, 118327s (1974).

12. BAILEY, J. A.: Phytoalexin Production by Leaves of *Pisum sativum* in Relation to Senescence. Ann. Appl. Biol. **64**, 315 (1969).

13. — Pisatin Production by Tissue Cultures of *Pisum sativum* L. J. gen. Microbiol. **61**, 409 (1970).

14. — Phytoalexins and the Ability of Leaf Tissues to Inhibit Fungal Growth. Ecol. Leaf Surface Micro-Organisms, Proc. Int. Symp. **1970** (Pub. 1971), edited by PREECE, T. F., Academic Press, p. 519.

15. BAILEY, J. A.: Production of Antifungal Compounds in Cowpea *(Vigna sinensis)* and Pea *(Pisum sativum)* after Virus Infection. J. gen. Microbiol. **75**, 119 (1973).
16. — Relation Between Symptom Expression and Phytoalexin Concentration in Hypocotyls of *Phaseolus vulgaris* Infected with *Colletotrichum lindemuthianum*. Physiol. Plant Pathol. **4**, 477 (1974).
17. BAILEY, J. A., and R. S. BURDEN: Biochemical Changes and Phytoalexin Accumulation in *Phaseouls vulgaris* Following Cellular Browning Caused by Tobacco Necrosis Virus. Physiol. Plant Pathol. **3**, 171 (1973).
18. BAILEY, J. A., R. S. BURDEN, and G. G. VINCENT: Capsidiol: An Antifungal Compound Produced in *Nicotiana tabacum* and *Nicotiana clevelandii* Following Infection with Tobacco Necrosis Virus. Phytochemistry **14**, 597 (1975).
19. BAILEY, J. A., and B. J. DEVERALL: Formation and Activity of Phaseollin in the Interaction Between *Phaseolus vulgaris* Hypocotyls and Physiological Races of *Colletotrichum lindemuthianum*. Physiol. Plant Pathol. **1**, 435—449 (1971).
20. BAILEY, J. A., and J. L. INGHAM: Phaseollin Accumulation in *Phaseolus vulgaris* in Response to Infection by Tobacco Necrosis Virus and the Rust *Uromyces appendiculatus*. Physiol. Plant Pathol. **1**, 451 (1971).
20 a. BAILEY, J. A., G. G. VINCENT, and R. S. BURDEN: The Antifungal Activity of Glutinosone and Capsidiol and Their Accumulation in Virus-Infected Tobacco Species. Physiol. Plant Pathol. **8** (1), 35 (1976).
20 b. BAKER, F. C., and C. J. W. BROOKS: Biosynthesis of the Sesquiterpenoid Capsidiol, in Sweet Pepper Fruits Inoculated with Fungal Spores. Phytochemistry **15**, 689 (1976).
21. BAKER, F. C., C. J. BROOKS, and S. A. HUTCHINSON: Biosynthesis of Capsidiol in Sweet Peppers *(Capsicum frutescens)* Infected with Fungi: Evidence for Methyl Group Migration from ^{13}C Nuclear Magnetic Resonance Spectroscopy. Chem. Commun. **1975**, 293.
22. BAUGHER, W. L., and T. C. CAMPBELL: Gossypol Detoxication by Fungi. Science **164**, 1526 (1969).
23. BEIJERSBERGEN, J. C. M., and C. B. G. LEMMERS: Enzymic and Non-enzymic Liberation of Tulipalin A (α-methylene butyrolactone) in Extracts of Tulip. Physiol. Plant Pathol. **2**, 265 (1972).
24. BELL, A. A.: Formation of Gossypol in Infected or Chemically Irritated Tissues of *Gossypium* species. Phytopathology **57**, 759 (1967).
25. — Phytoalexin Production and *Verticillium* Wilt Resistance in Cotton. Phytopathology **59**, 1119 (1969).
26. — In: Biological Control of Plant Insects and Diseases (MAXWELL, F. G., ed.), p. 403. Mississippi State: University Press. 1974.
27. BELL, A. A., and J. T. PRESLEY: Temperature Effects upon Resistance and Phytoalexin Synthesis in Cotton Inoculated with *Verticillium albo-atrum*. Phytopathology **59**, 1141 (1969).
28. BELL, A. A., and R. D. STIPANOVIC: Proc. Beltwide Cotton Prod. Res. Conf., p. 87, National Cotton Council, Memphis, Tenn. 1972.
29. BELL, A. A., R. D. STIPANOVIC, C. R. HOWELL, and P. A. FRYXELL: Antimicrobial Terpenoids of *Gossypium:* Hemigossypol, 6-Methoxyhemigossypol and 6-Deoxyhemigossypol. Phytochemistry **14**, 225 (1975).
30. BERARDI, L. C., and L. A. GOLDBLATT: In: Toxic Constituents of Plant Foodstuffs, p. 211. New York: Academic Press. 1969.
31. BERGMAN, B. H. H., J. C. M. BEIJERSBERGEN, J. C. OVEREEM, and A. K. SIJPESTEIJN: Isolation and Identification of α-Methylenebutyrolactone, a Fungitoxic Substance from Tulips. Recueil Trav. chim. Pays-Bas **86**, 709 (1967).
32. BEVAN, C. W. L., A. J. BIRCH, B. MOORE, and S. K. MUKERJEE: A Partial Synthesis of

(±)-Pisatin: Some Remarks on the Structure and Reactions of Pterocarpin. J. chem. Soc. (London) **1964**, 5991.

33. BHAKUNI, D. S., M. M. DHAR, and V. N. SHARMA: The Chemistry of Thespesin. Experientia **24**, 109 (1968).

34. BIEHN, W. L., J. KUC, and E. B. WILLIAMS: Accumulation of Phenols in Resistant Plant-fungi Interactions. Phytopathology **58**, 1255 (1968).

35. BIEHN, W. I., E. B. WILLIAMS, and J. KUC: Fungitoxicity of Phenols Accumulating in *Glycine max*-fungi Interactions. Phytopathology **58**, 1261 (1968).

35a. BILTON, J. N., J. R. DEBNAM, and I. M. SMITH: 6a-Hydroxypterocarpans from Red Clover. Phytochemistry **15**, 1411 (1976).

36. BIRCH, A. J., and F. R. S. PHIL: Some Natural Antifungal Agents. Chem. Ind. **1966**, 1173.

37. BIRCH, A. J., B. MOORE, S. K. MUKERJEE, and C. W. L. BEVAN: A Partial Synthesis of (±)-Pisatin from Pterocarpin. Tetrahedron Letters **1962**, 673.

37a. BIRNBAUM, G. I., C. P. HUBER, M. L. POST, J. B. STOTHERS, J. R. ROBINSON, A. STOESSL, and E. B. WARD: Sesquiterpenoid Stress Compounds of *Datura stramonium*: Biosynthesis of the Three Major Metabolites from (1,2-^{13}C) Acetate and the X-Ray Structure of 3-Hydroxylubimin. Chem. Commun. **1976**, 330.

38. BIRNBAUM, G. I., A. STOESSL, S. H. GROVER, and J. B. STOTHERS: The Complete Stereostructure of Capsidiol. X-ray Analysis and ^{13}C-Nuclear Magnetic Resonance of Eremophilane Derivatives Having trans-Vicinal Methyl Groups. Canad. J. Chem. **52**, 993 (1974).

39. BOHLMANN, F., S. KÖHN, und G. ARNDT: Polyacetylenverbindungen. CXIV. Die Polyine der Gattung Carthamus L. Chem. Ber. **99**, 3433 (1966); BOHLMANN, F., and C. ZDERO: Polyacetylenverbindungen. 182. Weitere Acetylenverbindungen aus Carthamus tinctorius L. Chem. Ber. **103**, 2853 (1970).

40. BOLLER, A., H. CORRODI, E. GÄUMANN, E. HARDEGGER, H. KERN, und N. WINTERHALTER-WILD: Welkstoffe und Antibiotika. Über induzierte Abwehrstoffe bei Orchideen. I. Helv. Chim. Acta **40**, 1062 (1957).

41. BONDE, M. R., R. L. MILLAR, and J. L. INGHAM: Induction and Identification of Sativan and Vestitol as Two Phytoalexins from *Lotus corniculatus*. Phytochemistry **12**, 2957 (1973).

42. BRIDGE, M., and W. L. KLARMAN: Ultraviolet Induction of an Antifuncal Chemical in Soybeans. Phytopathology **60**, 1013 (1970).

43. BRIDGE, M. A., and W. L. KLARMAN: Soybean Phytoalexin, Hydroxyphaseollin, Induced by UV-Irradiation. Phytopathology **63**, 606 (1973).

44. BRONGERSMA-OOSTERHOFF, U. W.: Structure Determination of the Allergenic Agent, Isolated from Tulip Bulbs. Recueil Trav. chim. Pays-Bas **86**, 705 (1967).

45. BUKHARI, S. T. K., and R. D. GUTHRIE: Structure of Rishitin. An Example of the Use of Cuprammonium Complexing in Structural Elucidation. J. Chem. Soc. (London) C **1969**, 1073.

46. BURDEN, R. S., and J. A. BAILEY: Structure of the Phytoalexin from Soybean. Phytochemistry **14**, 1389 (1975).

47. BURDEN, R. S., J. A. BAILEY, and G. W. DAWSON: Structures of 3 New Isoflavanoids from *Phaseolus vulgaris* Infected with Tobacco Necrosis Virus. Tetrahedron Letters **1972**, 4175.

48. BURDEN, R. S., J. A. BAILEY, and G. G. VINCENT: Metabolism of Phaseolin by *Colletotrichum lindemuthianum*. Phytochemistry **13**, 1789 (1974).

49. — — — Glutinosone, a New Antifungal Sesquiterpene from *Nicotiana glutinosa* Infected with Tobacco Mosaic Virus. Phytochemistry **14**, 221 (1975).

50. BURDEN, R. S., P. M. ROGERS, and R. L. WAIN: Fungicides. XVI. Natural Resistance of Plant Roots to Fungal Pathogens. Ann. Appl. Biol. **78**, 59 (1974).

50a. Burka, L. T., L. Kuhnert, B. J. Wilson, and T. M. Harris: 4-Hydroxymyoporone, a Key Intermediate in the Biosynthesis of Pulmonary Toxins Produced by *Fusarium solani* Infected Sweet Potatoes. Tetrahedron Letters **1974**, 4017.

50b. Cardoso, C. O. N., and M. O. Garraway: Production of Phenols and Phytoalexins in Hypocotyls of Beans Infected with *Fusarium solani f. phaseoli.* Summa Phytopathol. **1**, 92 (1975); Chem. Abstr. **84**, 14763x (1976).

51. Chalova, L. I., N. L. Vasynkova, O. L. Ozeretshovskaya, and L. V. Metlitskii: Chemical Identifikation of One of the Potato Phytoalexins. Prikl. Biochim. Microbiol. **7**, 55 (1971); C. A. **74**, 95634 (1971).

52. Chamberlain, D. W., and J. D. Paxton: Protection of Soybean Plants by Phytoalexin. Phytopathology **58**, 1349 (1968).

53. Christenson, J. A.: The Degradation of Pisatin by Pea Pathogens. Phytopathology **59**, 10 (1969).

54. Christenson, J. A., and L. A. Hadwiger: Induction of Pisatin Formation in the Pea Foot Region by Pathogenic and Nonpathogenic Clones of *Fusarium solani.* Phytopathology **63**, 784 (1973).

54a. Coxon, D. T., and R. F. Curtis: Ipomeamarone, a Toxic Furanoterpenoid in Sweet Potatoes *(Ipomea batatas)* in the United Kingdom. Food Cosmet.Toxicol. **13**, 87 (1975).

55. Coxon, D. T., R. F. Curtis, K. R. Price, and B. Howard: Phytuberin: a Novel Antifungal Terpenoid from Potato. Tetrahedron Letters **1974**, 2363.

56. Coxon, D. T., K. R. Price, B. Howard, S. F. Osman, E. B. Kalan, and R. M. Zacharius: Two New Vetispirane Derivatives: Stress Metabolites from Potato *(Solanum tuberosum)* Tubers. Tetrahedron Letters **1974**, 2921.

56a. Crombie, L., G. Kneen, and G. Pattenden: Synthesis of Casbene. Chem. Commun. **1976**, 66.

57. Cruickshank, I. A. M.: Studies on Phytoalexins IV. The Antimicrobial Spectrum of Pisatin. Austral. J. biol. Sci. **15**, 147 (1962).

58. — Phytoalexins. Annu. Rev. Phytopathol. **1**, 351 (1963).

59. — Phytoalexins in the *Leguminosae* with Special Reference to Their Selective Toxicity. Tag. Ber. dt. Akad. Landw. Wiss. Berlin **74**, 313 (1965).

60. Cruickshank, I. A. M., D. R. Biggs, D. R. Perrin, and C. P. Whittle: Phaseollin and Phaseollidin Relationships in Infection-Droplets on Endocarp of *Phaseolus vulgaris.* Physiol. Plant Pathol. **4**, 261 (1974).

61. Cruickshank, I. A. M., and D. R. Perrin: Isolation of a Phytoalexin from *Pisum sativum* L. Nature (London) **187**, 799 (1960).

62. — — Studies on Phytoalexins. III. The Isolation, Assay and General Properties of a Phytoalexin from *Pisum sativum.* Austral. J. biol. Sci. **14**, 336 (1961).

63. — — Phytoalexins of the *Leguminosae.* Phaseollin from *Phaseolus vulgaris* L. Life Sciences **2**, 680 (1963).

64. — — Studies on Phytoalexins. VI. Pisatin; the Effect of Some Factors on its Formation in *Pisum sativum* L. and the Significance of Pisatin in Disease Resistance. Austral. J. biol. Sci. **16**, 111 (1963).

65. — — Studies on Phytoalexins. VIII. The Effect of Some Further Factors on the Formation, Stability and Localization of Pisatin *in vivo.* Austral. J. biol. Sci. **18**, 817 (1965).

66. — — Studies on Phytoalexins. IX. Pisatin Formation by Cultivars of *Pisum sativum* L. and Other *Pisum species.* Austral. J. biol. Sci. **18**, 829 (1965).

67. — — Studies of Phytoalexins. X. Effect of Oxygen Tension on the Biosynthesis of Pisatin and Phaseollin. Phytopathol. Z. **60**, 335 (1967).

68. — — The Isolation and Partial Characterization of Monilicolin A, a Polypeptide with Phaseollin-inducing Activity from *Monilinia fructicola.* Life Science **7**, 449 (1968).

69. — — Studies on Phytoalexins. XI. The induction, Antimicrobial Spectrum and Chemical Assay of Phaseollin. Phytopathol. Z. **70**, 209 (1971).

70. CRUICKSHANK, I. A. M., J. VEERARAGHAVAN, and D. R. PERRIN: Physical Factors Affecting the Formation and/or Net Accumulation of Medicarpin in Infection Droplets on White Clover Leaflets. Austral. J. Plant Physiol. 1, 149 (1974).

70a. CURRIER, W. W., and J. KUC: Effect of Temperature on Rishitin and Steroid Glycoalkaloid Accumulation in Potato Tuber. Phytopathology 65, 1194 (1975).

71. CUTLER, H. G., and R. J. COLE: Properties of a Plant Growth Inhibitor Extracted from Immature Tobacco Leaves. Plant Cell Physiol. 15, 19 (1974).

72. DATTA, S. C., V. V. S. MURTI, and T. R. SESHADRI: Isolation und Study of (+)-Gossypol from Thespesia populnea. Indian J. Chem. 10, 263 (1972).

73. — — — Stereochemistry of Gossypol. Current Sci. 41, 545 (1972).

74. DECHARY, J. M., and P. PRADEL: Occurrence of (+)-Gossypol in Gossypium species. J. Amer. Oil Chem. Soc. 48, 563 (1971).

75. DEVERALL, B. J.: Biochemical Changes in Infection Droplets Containing Spores of Botrytis spp. Incubated in the Seed Carities of Pods of Bean (Vicia faba L.) Ann. Appl. Biol. 59, 375 (1967).

76. DEVERALL, B. J.: Phytoalexins. Phytochem. Ecol., Proc. Phytochem. Soc. Symp. 1971 (Pub. 1972), 217; Edit. J. B. HARBORNE. London: Academic Press.

77. — Phytoalexins and Disease Resistance. Proc. Roy. Soc. (London), Ser. B 181, 233 (1972).

78. DEVERALL, B. J., and P. M. ROGERS: The Effect of pH and Composition of Test Solutions on the Inhibitory Activity of Wyerone Acid Towards Germination of Fungal Spores. Ann. Appl. Biol. 72, 301 (1972).

79. DEVERALL, B. J., and J. C. VESSEY: Role of a Phytoalexin in Controlling Lesion Development in Leaves of Vicia faba after infection by Botrytis spp. Ann. Appl. Biol. 63, 449 (1969).

80. DEWICK, P. M.: Pterocarpan Biosynthesis: 2'-Hydroxyisoflavone and -isoflavanon Precursors of Demethylhomopterocarpin in Red Clover. Chem. Commun. 1975, 656.

81. — Pterocarpan Biosynthesis: Chalcone and Isoflavone Precursors of Demethylhomopterocarpin and Maackiain in Trifolium pratense. Phytochemistry 14, 979 (1975).

82. DYAKOV, YU. T., L. V. METLITSKII, O. L. OZERETSKOVSKAYA, and L. A. YURGANOVA: Inflectivity and Virulence of Phytophthora infestans in Relation to the Ability of the Fungus to Induce Rishitin Production. Mikol. Fitopatol. 7, 208 (1973); C. A. 80, 24741 v (1974).

83. EDWARDS, J. D., JR.: Synthesis of Gossypol and Gossypol Derivates. J. Amer. Oil Chem. Soc. 47, 441 (1970).

83a. ELUAGHY, M. A., and R. HEITEFUSS: Permeability Changes and Production of Antifungal Compounds in Phaseolus vulgaris infected with Uromyces phaseoli. II. Role of Phytoalexins. Physiol. Plant Pathol. 8, 269 (1976).

84. FAWCETT, C. H.: Antifungal Compounds in Plants; Some Recent Developments. Int. Pestic. Congr. 5, 18 (1963); Antifungal Compounds in Seedlings of Vicia faba. Chemical Results. Society of Chem. Industry Monographs 15, 119 (1961).

85. FAWCETT, C. H., R. D. FIRN, and D. M. SPENCER: Wyerone Increase in Leaves of Broad Bean (Vivia faba L.) after infection by Botrytis fabae. Physiol. Plant Pathol. 1, 163 (1971).

86. FAWCETT, C. H., and D. M. SPENCER: Plant Chemotherapy with Natural Products. Annu. Rev. Phytopathol. 8, 403 (1970).

87. FAWCETT, C. H., M. SPENCER, and R. L. WAIN: The Isolation and Properties of a Fungicidal Compound Present in Seedlings of Vicia faba. Neth. J. Plant Pathol. 75, 72 (1969).

88. FAWCETT, C. H., D. M. SPENCER, R. L. WAIN, A. G. FALLIS, E. R. H. SIR JONES, M. LE QUAN, C. B. PAGE, V. THALLER, D. C. SHUBROOK, and P. M. WHITHAM: Natural Acetylenes. XXVII. An Antifungal Acetylenic Furanoid Keto-ester (Wyerone) from Shoots of the Broad Bean (Vicia faba L.; Fam. Papilionaceae). J. chem. Soc. (C) 1968, 2455.

89. FAWCETT, C. H., D. M. SPENCER, R. L. WAIN, E. R. H. JONES, M. LE QUAN, C. B. PAGE, and V. THALLER: An Antifungal Acetylenic Keto-ester from a Plant of the *Papilionaceae* Family. Chem. Commun. **1965**, 422.

90. FISCH, M. H., B. H. FLICK, and J. ARDITTI: Structure and Antifungal Activity of Hircinol, Loroglossol and Orchinol. Phytochemistry **12**, 437 (1973).

91. FUKUI, K., and M. NAKAYAMA: Total Synthesis of (±)-Pterocarpin and (±)-Pisatin. Tetrahedron Letters **1966**, 1805.

92. FUKUI, K., M. NAKAYAMA, and T. HARANO: The Synthesis of 3-Hydroxy-8,9-dimethoxy-pterocarpan. Bull. chem. Soc. Japan **42**, 233 (1969).

93. GÄUMANN, E.: Nouvelles domées sur les réactions chimiques de défense chez les Orchidées. C. R. hebd. Séances Acad. Sci. **250**, 1944 (1960).

94. — Sur les réactions de défense chimique chez les Orchidées. C. R. hebd. Séances Acad. Sci. **257**, 2372 (1963).

95. — Weitere Untersuchungen über die chemische Infektabwehr der Orchideen. Phytopathol. Z. **49**, 211 (1963).

96. GÄUMANN, E., und H. R. HOHL: Weitere Untersuchungen über die chemischen Abwehrreaktionen der Orchideen. Phytopathol. Z. **38**, 93 (1960).

97. GÄUMANN, E., und H. KERN: Über die Isolierung und den chemischen Nachweis des Orchinols. Phytopathol. Z. **35**, 347 (1959).

98. — — Über chemische Abwehrreaktionen bei Orchideen. Phytopathol. Z. **36**, 1 (1959).

99. — — Sur les réactions de défense chimiques chez les orchidées. C. R. hebd. Séances Acad. Sci. **248**, 2542 (1959).

100. GÄUMANN, E., J. NUESCH, und R. H. RIMPAU: Weitere Untersuchungen über die chemische Abwehrreaktion der Orchideen. Phytopathol. Z. **38**, 274 (1960).

101. GOMBOS, M., K. SZENDREI, J. NOVAK, and J. REISCH: Nitrogenmentes fenantrenszarmazek izolalasa az opiumbol. Herb. Hungarica **13**, 63 (1974).

102. GEIGERT, J., F. R. STERMITZ, G. JOHNSON, D. MAAG, and D. K. JOHNSON: Two Phytoalexins from Sugar Beet *(Beta vulgaris)* Leaves. Tetrahedron **29**, 2703 (1973).

103. GORDON, M., A. STOESSL, and J. B. STOTHERS: NMR Studies XXV. Postinfectional Inhibitors from Plants. IV. Structure of Capsidiol, an Antifungal Sesquiterpene from Sweet Pappers. Canad. J. Chem. **51**, 748 (1973).

104. GRAY, G., W. L. KLARMAN, and M. BRIDGE: Relative Quantities of Antifungal Metabolites Produced in Resistant and Susceptible Soybean Plants Inoculated with *Phytophthora megasperma var. sojae* and Closely Related Non-pathogenic Fungi. Canad. J. Bot. **46**, 285 (1968).

104a. GRAY, J. R., T. J. MABRY, A. A. BELL, R. D. STIPANOVIC, and M. J. LUKEFAHR: para-Hemigossypolone: a Sequiterpenoid Aldehyde Quinone from *Gossypium hirsutum*. Chem. Commun. **1976**, 109.

105. GROSS, D.: Growth Regulating Substances of Plant Origin. Phytochemistry **14**, 2105 (1975).

106. HADWIGER, L. A.: The Biosynthesis of Pisatin. Phytochemistry **5**, 523 (1966).

107. — Changes in Host-Metabolism Associated with Phytoalexin Production. Phytopathology **57**, 813 (1967).

108. — Specificity of DNA Intercalating Compounds in the Control of Phenylalanine Ammonia Lyase and Pisatin Levels. Plant Physiol. **47**, 346 (1971).

109. HADWIGER, L. A., S. L. HESS, and S. VON BROEMBSEN: Stimulation of Phenylalanine Ammonialyase Activity and Phytoalexin Production. Phytopathology **60**, 332 (1970).

110. HADWIGER, L. A., A. JAFRI, S. VON BROEMBSEN, and E. ROBERT: Mode of Pisatin Induction. Increased Template Activity and Dye-Bindung Capacity of Chromatin Isolated and from Polypeptide Treated Pea Pods. Plant Physiol. **53**, 52 (1974).

111. HADWIGER, L. A., and M. E. SCHWOCHAU: Induction of Phenylalanine Ammonialyase and Pisatin in Pea Pods by Polylysine, Spermidine or Histone Fractions. Biochem. Biophys. Res. Commun. **38**, 683 (1970).

112. — — Ultraviolet Light-induced Formation of Pisatin and Phenylalanine Ammonia Lyase. Plant Physiol. **47**, 588 (1971).

113. HARADA, N., and K. NAKANISHI: A Method for Determining the Chiralities of Optically Active Glycols. J. Am. Chem. Soc. **91**, 3989 (1969).

114. HARDEGGER, E., H. R. BILAND, and H. CORRODI: Welkstoffe und Antibiotika. Synthese von 2,4-Dimethoxy-6-hydroxy-phenanthren und Konstitution des Orchinols. Helv. Chim. Acta **46**, 1354 (1963).

115. HARDEGGER, E., N. RIGASSI, J. SERES, C. EGLI, P. MÜLLER, and K. O. FITZI: Welkstoffe und Antibiotika. Synthese von 2,4-Dimethoxy-6-hydroxy-9,10-dihydrophenanthren. Helv. Chim. Acta **46**, 2543 (1963).

116. HARDEGGER, E., M. SCHELLENBAUM, und H. CORRODI: Welkstoffe und Antibiotika. Über induzierte Abwehrstoffe der Orchideen. II. Helv. Chim. Acta **46**, 1171 (1963).

116a. HARGREAVES, J. A., and J. W. MANSFIELD: Phytoalexin Production by *Vicia faba* in Response to Infection by *Botrytis*. Ann. Appl. Biol. **81**, 271 (1975).

116b. HARGREAVES, J. A., J. W. MANSFIELD, and D. T. COXON: Conversion of Wyerone to Weyerol by *Botrytis cinerea* and *B. fabae in vitro*. Phytochemistry **15**, 651 (1976).

116c. HARGREAVES, J. A., J. W. MANSFIELD, D. T. COXON, and K. R. PRICE: Wyerone Epoxide as a Phytoalexin in *Vicia faba* and its Metabolism by *Botrytis cinerea* and *B. fabae in vitro*. Phytochemistry **15**, 1119 (1976).

116d. HARGREAVES, J. A., J. W. MANSFIELD, and D. T. COXON: Identification of Medicarpin as a Phytoalexin in the Broad Bean Plant (*Vicia faba* L.). Nature **262**, 318 (1976).

117. HARROWER, K. M.: Differential Effects of Phytoalexin from *Pisum sativum* on Two Races of *Ascochyta pisi*. Trans. Brit. Mycol. Soc. **61**, 383 (1973).

118. HART, J. H., and W. E. HILLIS: Inhibition of Wood-rotting Fungi by Stilbenes and Other Polyphenols in *Eucalyptus sideroxylon*. Phytopathology **64**, 939 (1974).

119. HASEGAWA, K., and T. HASHIMOTO: Quantitative Changes of Batatasins and Abscisic Acid in Relation to the Development of Dormancy in Yam Bulbils. Plant Cell Physiol. **14**, 369 (1973); Gibberellin-induced Dormancy and Batatasin Content in Yam Bulbils. Plant Cell Physiol. **15**, 1 (1974).

120. HASHIMOTO, T., K. HASEGAWA, and A. KAWARADA: Batatasins: New Dormancy-inducing Substances of Yam Bulbils. Planta (Berlin) **108**, 369 (1972).

121. HASHIMOTO, T., K. HASEGAWA, H. YAMAGUCHI, M. SAITO, and S. ISHIMOTO: Structure and Synthesis of Batatasins, Dormancy-inducing Substances of Yam Bulbils. Phytochemistry **13**, 2849 (1974).

122. HEATH, M. C., and V. J. HIGGENS: Degradation of Phaseollin and Pisatin by *Stemphylium botryosum*. Phytopathology **62**, 763 (1972).

123. — — In vitro and in vivo Conversion of Phaseollin and Pisatin by an Alfalfa Pathogen *Stemphylium botryosum*. Physiol. Plant Pathol. **3**, 107 (1973).

124. HEINSTEIN, P. F., D. L. HERMAN, S. B. TOVE, and F. H. SMITH: Biosynthesis of Gossypol. Incorporation of Mevalonate-2-^{14}C and Isoprenyl Pyrophosphates. J. Biol. Chem. **245**, 4658 (1970).

125. HEINSTEIN, P. F., F. H. SMITH, and S. B. TOVE: Biosynthesis of ^{14}C-Labeled Gossypol. J. Biol. Chem. **237**, 2643 (1962).

126. HESS, S., and L. HADWIGER: The Induction of Phenylalanine Ammonia Lyase and Phaseollin by 9-Aminoacridine and Other Deoxyribonucleic Acid Intercalating Compounds. Plant Physiol. **48**, 197 (1971).

127. HESS, S. L., L. A. HADWIGER, and M. E. SCHWOCHAU: Studies on Biosynthesis of Phaseollin in Excised Pods of *Phaseolus vulgaris*. Phytopathology **61**, 79 (1971).

128. HESS, S. L., and M. E. SCHWOCHAU: Induction, Purification and Biosynthesis of Phaseollin in Excised Pods of *Phaseolus vulgaris*. Phytopathology **59**, 1030 (1969).

129. HEUVEL, J. VAN DEN, and H. D. VAN ETTEN: Alteration of Phaseollin by *Fusarium solani f. sp. phaseoli*. Phytopathology **62**, 794 (1972).

130. Heuvel, J. van den, and H. D. van Etten: Detoxification of Phaseollin by *Fusarium solani f. sp. phaseoli.* Physiol. Plant Pathol. **3,** 327 (1973).

131. Heuvel, J. van den, H. D. van Etten, J. W. Serum, D. L. Coffen, and T. H. Williams: Identification of 1 a-Hydroxyphaseollone, a Phaseollin Metabolite Produced by *Fusarium solani.* Phytochemistry **13,** 1129 (1974).

132. Higgins, V. J.: Role of the Phytoalexin Medicarpin in Three Leaf Spot Diseases of Alfalfa. Physiol. Plant Pathol. **2,** 289 (1972).

133. — Induced Conversion of the Phytoalexin Maackiain to Dihydromaackiain by the Alfalfa Pathogen *Stemphylium botryosum.* Physiol. Plant Pathol. **6,** 5 (1975).

134. Higgins, V. J., and R. L. Millar: Phytoalexin Production by Alfalfa in Response to Infection by *Colletrotrichum phomoides. Helminthosporium turcicum, Stemphylium loti,* and *S. botryosum.* Phytopathology **58,** 1377 (1968).

135. — — Comparative Abilities of *Stemphylium botryosum* and *Helminthosporium turcicum* to Induce and Degrade a Phytoalexin from Alfalfa. Phytopathology **59,** 1493 (1969); Degradation of Alfalfa Phytoalexin by *Stemphylium botryosum.* Phytopathology **59,** 1500 (1969).

136. — — Degradation of Alfalfa Phytoalexin by *Stemphylium loti* and *Colletotrium phomoides.* Phytopathology **60,** 269 (1970).

137. Higgins, V. J., R. L. Millar, D. G. Smith, and A. G. McInnes: Purification and Identification of Alfalfa Phytoalexin. Phytopathology **60,** 1295 (1970).

138. Higgins, V. J., and D. G. Smith: Separation and Identification of Two Pterocarpanoid Phytoalexins Produced by Red Clover Leaves. Phytopathology **62,** 235 (1972).

139. Higgins, V. J., A. Stoessl, and M. C. Heath: Conversion of Phaseollin to Phaseollinisoflavan by *Stemphylium botryosum.* Phytopathology **64,** 105 (1974).

140. Hijwegen, T.: Autonomous and Induced Pterocarpanoid Formation in the *Leguminosae.* Phytochemistry **12,** 375 (1973).

141. Hiura, M.: Studies on Storage and Rot of Sweet Potato. Rep. Gifu. Agric. Coll. **50,** 1 (1943).

141a. Hughes, D. L., and D. T. Coxon: Phytuberin; Revised Structure from the X-Ray Crystal Analysis of Dihydrophytuberin. Chem. Commun. **1974,** 822.

142. Hyodo, H., I. Uritani, and S. Akai: Production of Furanoterpenoids and Other Compounds in Sweet Potato Root Tissue in Response to Infection by Various Isolates of *Ceratocystis fimbriata.* Phytopathol. Z. **65,** 332 (1969).

143. Imaseki, H., S. Takei, and I. Uritani: Ipomeamarone Accumulation and Lipid Metabolism in Sweet Potato Infected by the Black Rot Fungus. I. Identification of Sterol and Changes in Lipid Metabolism During Infection Process. Plant Cell Physiol. **5,** 119 (1964).

144. Imaseki, H., and I. Uritani: Ipomeamarone Accumulation and Lipid Metabolism in Sweet Potato Infected by the Black Rot Fungus. II. Accumulation Mechanism of Ipomeamarone in the Infected Region with Special Regard to Contribution of the Non-infected Tissue. Plant Cell Physiol. **5,** 133 (1964).

145. Ingham, J. L.: Phytoalexins and Other Natural Products as Factors in Plant Disease Resistance. Bot. Rev. **38,** 343 (1972).

145a. Ingham, J. L.: Isosativan: an Isoflavan Phytoalexin from *Trifolium hybridum* and Other *Trifolium* species. Z. Naturforsch. **31c,** 331 (1976).

146. Ingham, J. L., and R. L. Millar: Sativin. Induced Isoflavan from the Leaves of *Medicago sativa.* Nature (London) **242,** 125 (1973).

147. Ishizaka, N., K. Tomiyama, N. Katsui, A. Murai, and T. Masamune: Biological Activities of Rishitin, an Antifungal Compound Isolated from Diseased Potato Tubers, and Its Derivatives. Plant Cell Physiol. **10,** 183 (1969).

148. Ito, S., Y. Fujise, and A. Mori: Absolute Configuration of Pterocarpinoids. Chem. Commun. **1965,** 595.

149. JEROME, S. M. R., and K. O. MÜLLER: Studies on Phytoalexins. II. Influence of Temperature on Resistance of *Phaseolus vulgaris* toward *Sclerotinia fructicola* with Reference to Phytoalexin Output. Austral. J. biol. Sci. **11**, 301 (1958).

149a. JOHNSON, G., D. D. MAAG, D. K. JOHNSON, and R. D. THOMAS: The Possible Role of Phytoalexins in the Resistance of Sugarbeet *(Beta vulgaris)* to *Cercospora beticola*. Physiol. Plant Pathol. **8**, 225 (1976).

150. JOHNSON, C., D. R. BRANNON, and J. KUC: Xanthotoxin: A Phytoalexin of *Pastinaca sativa* Root. Phytochemistry **12**, 2961 (1973).

150a. JONES, D. R., C. H. UNWIN, and E. W. B. WARD: The Significance of Capsidiol Induction in Pepper Fruit During an Incompatible Interaction with *Phytophthora infestans*. Phytopathology **65**, 1286 (1975).

150b. KALAN, E. B., and S. F. OSMAN: Isolubimin: a Possible Precursor of Lubimin in Infected Potato Slices. Phytochemistry **15**, 775 (1976).

151. KATO, N., H. IMASEKI, N. NAKASHIMA, and I. URITANI: Structure of a New Sesquiterpenoid, Ipomeamaronol, in Diseased Sweet Potato Root Tissue. Tetrahedron Letters **1971**, 843.

152. KATO, N., H. IMASEKI, N. NAKASHIMA, T. AKAZAWA, and I. URITANI: Isolation of a New Phytoalexin-like Compound, Ipomeamaronol, from Black-Rot Fungus Infected Sweet Potato Root Tissue, and Its Structural Elucidation. Pant Cell Physiol. **14**, 597 (1973).

153. KATSUI, N., A. MATSUNAGA, K. IMAIZUMI, T. MASAMUNE, and K. TOMIYAMA: The Structure and Synthesis of Rishitinol, a New Sesquiterpene Alcohol from Diseased Potato Tubers. VI. Tetrahedron Letters **1971**, 83.

154. — — — — — The Structure and Synthesis of Rishitinol. A Sesquiterpenes Alcohol from Diseased Potato Tubers. VII. Bull. Chem. Soc. Japan **45**, 2871 (1972).

155. KATSUI, N., A. MATSUNAGA, and T. MASAMUNE: Studies on the Phytoalexins XI. Structure of Oxylubimin, Antifungal Metabolites From Diseased Potato Tubers. Tetrahedron Letters **1974**, 4483.

156. KATSUI, N., A. MURAI, M. TAKASUGI, K. IMAIZUMI, T. MASAMUNE, and K. TOMIYAMA: The Structure of Rishitin, a New Antifungal Compound from Diseased Potato Tubers. Chem. Commun. **1968**, 43.

157. KEEN, N. T.: Hydroxyphaseollin Production by Soybean Resistant and Susceptible to *Phytophthora megasperma var. sojae*. Physiol. Plant Pathol. **1**, 265 (1971).

158. — Accumulation of Wyerone in Broadbean and Demethylhomopterocarpin in Jack Bean After Inoculation With *Phytophthora megasperma var. sojae* A. A. Hildb. Phytopathology **62**, 1365 (1972).

159. — The Isolation of Phytoalexins from Germinating Seeds of *Cicer arietinum, Vigna sinensis, Arachis hypogaea*, and Other Plants. Phytopathology **65**, 91 (1975).

160. — Specific Elicitors of Plant Phytoalexin Production. Determinants of Race Specificity in Pathogens. Science (Washington) **187**, 74 (1975).

161. KEEN, N. T., and B. W. KENNEDY: Hydroxyphaseollin and Related Isoflavanoids in the Hypersensitive Resistance Reaction of Soybeans to *Pseudomonas glycinea*. Physiol. Plant Pathol. **4**, 173 (1974).

162. KEEN, N. T., and J. D. PAXTON: Coordinate Production of Hydroxyphaseollin and the Yellow-fluorescent Compound PA_K in Soybeans Resistant to *Phytophthora megasperma var. sojae*. Phytopathology **65**, 635 (1975).

163. KEEN, N. T., J. J. SIMS, D. C. ERWIN, E. RICE, and J. E. PARTRIDGE: 6a-Hydroxyphaseollin: an Antifungal Chemical Induced in Soybean Hypocotyls by *Phytophthora megasperma var. sojae*. Phytopathology **61**, 1084 (1971).

164. KEEN, N. T., A. I. ZAKI, and J. J. SIMS: Biosynthesis of Hydroxyphaseollin and Related Isoflavanoids in Disease-Resistent Soy-bean Hypocotyls. Phytochemistry **11**, 1031 (1972).

165. Kim, W. K., I. Oguni, and I. Uritani: Phytopathological Chemistry of Sweet Potato with Black Rot and Injury. Phytoalexin Induction in Sweet Potato Roots by Amino Acids. Agric. Biol. Chem. **38**, 2567 (1974).

166. Kim, W. K., and I. Uritani: Fungal Extracts That Induce Phytoalexins in Sweet Potato Roots. Plant Cell Physiol. **15**, 1093 (1974).

167. King, T. J., and L. B. de Silva: Optically Active Gossypol from *Thespesia populnea.* Tetrahedron Letters **1968**, 261.

168. Klarman, W. L., and J. W. Gerdemann: Induced Susceptibility in Soybean Plants Genetically Resistant to *Phytophthora sojae.* Phytopathology **53**, 863 (1963); Resistance of Soybeans to Three Phytophthora Species Due to the Production of a Phytoalexin. Phytopathology **53**, 1317 (1963).

169. Klarman, W. L., and F. Hammerschlag: Production of the Phytoalexin, Hydroxyphaseolin, in Soybean Leaves Inoculated With Tobacco Necrosis Virus. Phytopathology **62**, 719 (1972).

170. Klarman, W. L., and J. B. Stanford: Isolation and Purification of an Antifungal Principle From Infected Soybeans. Life Science **7**, 1095 (1968).

171. Klinkowski, M.: Phytoalexine: Begriff und methodische Fragen. Ein Beitrag zur Phytoalexin-Theorie von K. O. Müller. Forsch. Fortschr. **40**, 321 (1966).

172. — Die Phytoalexin-Theorie von K. O. Müller. Abh. Sächs. Akad. Wiss. Leipzig, Math.-Naturw. Kl. **1966**, Kl. 49, No. 3, 1.

173. Kranz, J.: Pflanzenkrankheiten bedrohen unsere Ernte-Phytopathologie heute. Umschau **75**, 691 (1975).

174. Krzywanski, Z.: Phytoalexins. Wiad. Bot. **14**, 109 (1970).

175. Kubota, T., and T. Matsuura: Proc. Imp. Acad. (Tokyo) **28**, 44, 83, 198 (1952); J. Inst. Polytechn. Osaka City Univ., Ser. C. Chem. **2**, 94, 103 (1952); **4**, 104, 108, 248 (1953); J. chem. Soc. Japan **74**, 101, 197, 248, 666 (1953).

176. Kubota, T., T. Matsuura, and N. Ichikawa: Chemical Studies on the Black Rot Disease of Sweet Potato. VIII. The Reaction of Phenyl Magnesium Bromide on Ipomeamarone. J. chem. Soc. Japan **75**, 447 (1954).

177. Kubota, T., and K. Naya: On the Chemical Constitution of Batatic Acid. A New Furan Keto-acid from the Black Rotted Sweet Potato. Chem. Ind. **1954**, 1427.

178. Kubota, T., H. Yamaguchi, K. Naya, and T. Matsuura: Chemical Studies on the Black Rot Disease of Sweet Potato. I. On Volatile Substances of Black-Rotted Sweet Potato. J. Inst. Polytechn. Osaka City Univ. Ser. C. Chem. **2**, 82 (1952).

179. — — — — Chemical Studies on the Black Rot Disease of Sweet Potato. I. On the Volatile Constituents of Black Rotted Sweet Potato. J. chem. Soc. Japan **73**, 897 (1952); Chemical Studies on the Black Rot Disease of Sweet Potato. II. Some Properties of Ipomeamarone. J. chem. Soc. Japan **74**, 44 (1953).

180. Kuc, J.: Phytoalexins. Annu. Rev. Phytopathol. **10**, 207 (1972).

181. Kurosawa, K., W. D. Ollis, B. T. Redman, J. O. Sutherland, O. R. Gottlieb, and H. M. Alves: The Absolute Configurations of the Animal Metabolite, Equol, Three Naturally Occurring Isoflavans, and One Natural Isoflavan Quinone. Chem. Commun. **1968**, 1265.

182. Lampard, J. F.: Demethylhomopterocarpin: An Antifungal Compound in *Canavalia ensiformis* and *Vigna unguiculata* Following Infection. Phytochemistry **13**, 291 (1974).

183. Leppik, R. A., D. W. Hollomon, and W. Bottomley: Quiesone: An Inhibitor of the Germination of *Peronospora tabacina conidia.* Phytochemistry **11**, 2055 (1972).

184. Letcher, R. M.: Structure and Synthesis of the Growth Inhibitor Batatasin I from *Dioscorea batatas.* Phytochemistry **12**, 2789 (1973).

185. Letcher, R. M., and L. R. M. Nhamo: Chemical Constituents of the *Combretaceae.* I. Substituted Phenanthrenes and 9,10-Dihydrophenanthrenes from the Heartwood of *Combretum apiculatum.* J. chem. Soc. C **1971**, 3070.

186. — — A Revised Structure for the Tetra-substituted Phenanthrene from *Tamus communis.* Tetrahedron Letters **1972,** 4869.

187. — — Chemical Constituents of the *Combretaceae.* III. Substituted Phenanthrenes, 9,10-Dihydrophenanthrenes and Bibenzyls from the Heartwood of *Combretum psidioides.* J. chem. Soc. Perkin I **1972,** 2941.

188. LETCHER, R. M., L. R. M. NHAMO, and I. T. GUMIRO: Chemical Constituents of the *Combretaceae.* II. Substituted Phenanthrenes and 9,10-Dihydrophenanthrenes and a Substituted Bibenzyl from the Heartwood of *Combretum molle.* J. chem. Soc. Perkin I **1972,** 206.

189. LETCHER, R. M., D. A. WIDDOWSON, B. J. DEVERALL, and J. W. MANSFIELD: Identification and Activity of Wyerone Acid as a Phytoalexin in Broad Bean *(Vicia faba)* After Infection by *Botrytis.* Phytochemistry **9,** 249 (1970).

189a. LYNE, R. L., L. J. MULHEIRN, and D. P. LEWORTHY: New Pterocarpinoid Phytoalexins of Soybean. Chem. Commun. **1976,** 497.

190. LYON, G. D.: Occurrence of Rishitin and Phytuberin in Potato Tubers Inoculated with *Erwinia carotovora var. astroseptica.* Physiol. Plant Pathol. **2,** 411 (1972).

190a. LYON, G. D., B. M. LUND, C. E. BAYLISS, and G. M. WYATT: Resistance of Potato Tubers to *Erwinia carotovora* and Formation of Rishitin and Phytuberin in Infected Tissues. Physiol. Plant Pathol. **6,** 43 (1975).

191. LYON, F. M., and R. K. S. WOOD: Production of Phaseollin, Coumestrol, and Related Compounds in Bean Leaves Inoculated with *Pseudomonas* species. Physiol. Plant Pathol. **6,** 117 (1975).

191a. MCCANCE, D. J., and R. B. DRYSDALE: Production of Tomatine and Rishitin in Tomato Plants Inoculated with *Fusarium oxysporum f. sp. lycopersici.* Physiol. Plant Pathol. **7,** 221 (1975).

192. MACE, M. E., A. A. BELL, and R. D. STIPENOVIC: Histochemistry and Isolation of Gossypol and Related Terpenoids in Roots of Cotton Seedlings. Phytopathology **64,** 1297 (1974).

193. MANSFIELD, J. W., and B. J. DEVERALL: Mode of Action of Pollen in Breaking Resistance of *Vicia faba* to *Botrytis cinerea.* Nature (London) **232,** 339 (1971).

194. MANSFIELD, J. W., N. J. DIX, and A. M. PERKIN: Role of the Phytoalexin Pisatin in Controlling Saprophytic Fungal Growth on Pea Leaves. Trans. Br. Mycol. Soc. **64,** 507 (1975).

195. MANSFIELD, J. W., A. E. A. PORTER, and D. A. WIDDOWSON: Structure of a Fungal Metabolite of the Phytoalexin Wyerone Acid from *Vicia faba.* J. chem. Soc. Perkin I **1973,** 2557.

196. MANSFIELD, J. W., and D. A. WIDDOWSON: The Metabolism of Wyerone Acid (a Phytoalexin from *Vicia faba* L.) by *Botrytis fabae* and *B. cinerea.* Physiol. Plant Pathol. **3,** 393 (1973).

197. MATSUURA, T.: Chemische Untersuchungen über Schwarzflecke der Batate. XI. Mitt. Synthese des Ipomeamarons und seine damit zusammenhängenden Verbindungen II. Synthese des Phenylanalogs von Ipomeamaron. J. Inst. Polytechn. Osaka City Univ. Ser. C Chem. **5,** 42 (1956).

198. MATSUURA, T., K. NAYA, and T. KUBOTA: Chemical Studies on the Black Rot Disease of Sweet Potato XI. Synthesis of Phenylanalôg of Ipomeamarone. J. chem. Soc. Japan **77,** 248 (1956).

198a. METLITSKII, L. V.: Plants Protect Themselves. Khim. Zhizn. **1974,** 31; Chem. Abstr. **84,** 27952 (1976).

199. METLITSKII, L. V., YU. T. D'YAKOV, O. L. OZERETSKOVSKAYA, L. A. YURGANOVA, L. I. CHALOVA, and N. I. VASYUKOVA: Induction of Potato Phytoalexins. Izv. Akad. Nauk SSSR, Ser. Biol., **1971,** 399; Chem. Abstr. **75,** 45732e (1971).

200. METLITSKII, L. V., and O. L. OZERETSKOVSKAYA: Phytoncides and Phytoalexins and

Their Role in Plant Immunity. Mikol. Fitopatol. **4**, 146 (1970); Chem. Abstr. **73**, 95531t (1970).

201. METLITSKII, L. V., and O. L. OZERETSKOVSKAYA: Phytoalexins and Plant Self-defense. Priroda (Moscow) **1975**, 12; Chem. Abstr. **83**, 25007 (1975).

202. METLITSKII, L. V., O. L. OZERETSKOVSKAYA, O. N. SAVEL'EVA, YU. T. D'YAKOV, N. J. VASYUKVA, M. A. DAVYDOVA, L. I. CHALOVA, and G. I. CHALENKO: Isolation of Rishitin and Lubimin in an Infection Drop Placed on the Potato Tuber Surface. Prikl. Biokhim. Mikrobiol. **9**, 744 (1973); Chem. Abstr. **80**, 26033g (1974).

203. METLITSKII, L. V., O. L. OZERETSKOVSKAYA, N. J. VASYUKOVA, M. A. DAVYDOVA, O. N. SAVEL'EVA, and YU. T. D'YAKOV: Role of Phytoalexins in the Vertical Resistance of the Potato to *Phytophthora infestans*. Mikol. Fitopatol. **8**, 42 (1974); Chem. Abstr. **81**, 132973b (1974).

204. METLITSKII, L. V., O. L. OZERETSKOVSKAYA, N. S. VUL'FSON, and L. I. CHALOVA: Chemical Nature of Lubimin, a New Phytoalexin of Potatoes. Dokl. Akad. Nauk SSSR **200**, 1470 (1971); Chem. Abstr. **76**, 96945e (1972).

205. — — — — Effects of Lubimin on Potato Resistance to *Phytophthora infestans* and its Chemical Identification. Mikol. Fitopatol. **5**, 439 (1971); Chem. Abstr. **76**, 56808 (1972).

206. METLITSKII, L. V., O. L. OZERETSKOVSKAYA, N. S. VUL'FSON, N. I. VASYUKOVA, L. I. CHALOVA, and M. A. DAVYDOVA: Biological Activity and Chemical Characteristics of Lubimine, a New Phytoalexin of Potatoes. Immunitet Pokoi Rast. **1972**, 3, edit. by METLITSKII, L. V. ,,Nauka", Moskau; Chem. Abstr. **78**, 156736r (1973).

207. MORI, K.: Synthesis of dl-3-Isobutyroxy-β-ionone and dl-Dehydrovomifoliol. Agr. Biol. Chem. **37**, 2899 (1973).

208. MÜLLER, K. O.: Einige einfache Versuche zum Nachweis von Phytoalexinen. Phytopathol. Z. **27**, 237 (1956); Relationship between Phytoalexin Output and the Number of Infections Involved. Nature **182**, 167 (1958); The Phytoalexin Concept and its Methodological Significance. Recent Advan. Bot. **1**, 396 (1959); Die Phytoalexine, in Sicht einer allgemeinen Immunbiologie. Zentralbl. Bakteriol. Hyg. 2. Abt. **123**, 259 (1969).

209. — Studies on Phytoalexins I. The Formation and the Immunological Significance of Phytoalexin Produced by *Phaseolus vulgaris* in Response to Infection with *Sclerotinia fructicola* and *Phytophthora infestans*. Austral. J. Biol. Sci. **11**, 275 (1958).

210. MÜLLER, K. O., and H. BÖRGER: Experimentelle Untersuchungen über die *Phytophthora*-Resistenz der Kartoffel. Arb. Biol. Anst. (Reichsanst.) Berlin **23**, 189 (1941).

211. MÜLLER, P., J. SERES, K. STEINER, S. E. HELALI, und E. HARDEGGER: Welkstoffe und Antibiotika. Synthese von Dehydroorchinolmethyläther und Dehydroorchinol. Helv. Chim. Acta **57**, 790 (1974).

211a. MUKHAMEDOVA, R. A., and YA. KH. TURAKULOV: Phytoalexin Activity of Cotton Plants and their Wilt Resistance. Uzb. Biol. Zh. **18**, 3 (1974); Chem. Abstr. **84**, 102285y (1976).

211b. MUKHAMEDOVA, R. A., N. V. LYUBIMOVA, M. KH. AVAZKODZHAYEV, and L. U. METLITSKII: Phytoalexin Activity of Cotton Plants as a Wilt Resistance Factor. Mikhol. Fitopatol. **9**, 505 (1975); Chem. Abstr. **84**, 147871 (1976).

212. MUNN, C. B., and R. B. DRYSDALE: Kievitone Production and Phenylalanine Ammonia-lyase Activity in Cowpea. Phytochemistry **14**, 1303 (1975).

213. MURAI, A., K. NISHIZAKURA, N. KATSUI, and T. MASAMUNE: The Synthesis of Rishitin. Tetrahedron Letters (London) **1975**, 4399.

214. NONAKA, F., S. ISAYAMA, and H. FURUKAWA: On the Phytoalexin Produced by the Results of the Interaction Between Pods and Phytopathogens. Agric. Bull. Saga Univ. **22**, 51 (1966).

214a. NONAKA, F., and M. MATSUZAKI: Production of Hydroxyphaseollin in Soybean Leaves Infected with the Leaf Blight Bacterium *Xanthomonas phaseoli var. sojae* and its Antifungal Action. Saga Daigaku Nogaku Iho **40**, 1 (1976); Chem. Abstr. **85**, 17300 (1976).

215. OBA, K., H. SHIBATA, and I. URITANI: The Mechanism Supplying Acetyl-CoA for Terpene Biosynthesis in Sweet Potato with Black Rot: Incorporation of Acetate-2-^{14}C, Pyruvate-3-^{14}C and Citrate-2,4-^{14}C into Ipomeamarone. Plant Cell Physiol. **11**, 507 (1970).

216. OGUNI, I., K. OSHIMA, H. IMASEKI, and I. URITANI: Biochemical Studies on the Terpene Metabolism in Sweet Potato Root Tissue with Black Rot. Effect of C_{10}- and C_{15}-Terpenols on Acetate-2-^{14}C Incorporation into Ipomeamarone. Agr. Biol. Chem. **33**, 50 (1969).

217. OGUNI, I., and I. URITANI: The Incorporation of Farnesol-2-^{14}C into Ipomeamarone. Agr. Biol. Chem. **34**, 156 (1970).

218. — — Utilization of Ethanol-2-^{14}C for the Biosynthesis of Ipomeamarone by Sweet Potato Root Tissue Infected with *Ceratocystis fimbriata*. Agr. Biol. Chem. **35**, 357 (1971).

218a. OGUNI, I., and I. URITANI: Participation of Farnesol in the Biosynthesis of Ipomeamarone. Plant Cell Physiol. **12**, 507 (1971).

218b. OGUNI, I., and I. URITANI: Isolation of Dehydro-ipomeamarone, a New Sesqui-Terpenoid from the Black-Rot Fungus Infected Sweet Potato Root Tissue and its Relation to the Biosynthesis of Ipomeamarone. Agr. Biol. Chem. **37**, 2443 (1973).

218c. OGUNI, I., and I. URITANI: Effect of (−)-Hydroxycitrate on Ipomeamarone Biosynthesis from Pyruvate in Sweet Potato with Black Rot. Plant Cell Physiol. **15**, 179 (1974).

219. — — Dehydroipomeamarone as an Intermediate on the Biosynthesis of Ipomeamarone, a Phytoalexin from Sweet Potato Root Infected with *Ceratocystis fimbriata*. Plant Physiol. **53**, 649 (1974).

220. — — Dehydroipomeamarone from Infected *Ipomoea batatas* Root Tissue. Phytochemistry **13**, 521 (1974).

221. OHNO, T.: The Bitter Substance produced in Black Rotten Sweet Potato. II. On the Constitution of Ipomeamarone. Part I. Bull. Chem. Soc. Japan **25**, 222 (1952).

221a. OKU, H., E. OUCHI, T. SHIRAISHI, and T. BUBA: Pisatin Production in Powdery Mildewed Pea Seedlings. Phytopathology **65**, 1263 (1975).

221b. OKU, H., T. SHIRAISHI, and S. OUCHI: Role of Phytoalexin as the Inhibitor of Infection Establishment in Plant Disease. Naturwissenschaften **62**, 486 (1975).

222. OSHIMA-OBA, K., I. SUGIURA, and I. URITANI: The Incorporation of LEUCINE-U-^{14}C into Ipomeamarone. Agric. Biol. Chem. **33**, 586 (1969).

223. OSHIMA, K., and I. URITANI: Participation of Mevalonate in the Biosynthetic Pathway of Ipomeamarone. Agric. Biol. Chem. **32**, 1146 (1968).

224. OZERETSKOVSKAYA, O. L., N. I. VASYUKOVA, and L. V. METLITSKII: Potato Phytoalexins. Dokl. Acad. Nauk SSSR **189**, 1146 (1969); Chem. Abstr. **72**, 107943y (1970).

224a. PATIL, S. S., and S. S. GNANAMANICKAM: Suppression of Bacterially Induced Hypersensitive Reaction and Phytoalexin Accumulation in Bean by Phaseotoxin. Nature **259**, 486 (1976).

225. PAXTON, J. D., and D. W. CHAMBERLAIN: Phytoalexin Production and Disease Resistance in Soybean as Affected by Age. Phytopathology **59**, 775 (1969).

226. PAXTON, J. D., J. GOODCHILD, and I. A. M. CRUICKSHANK: Phaseollin Production by Live Bean Endocarp. Physiol. Plant Pathol. **4**, 167 (1974).

227. PERRIN, D. R.: The Structure of Phaseolin. Tetrahedron Letters **1964**, 29.

228. — Physicochemical Properties of Phaseollin. Phytopathol. Z. **70**, 227 (1971).

229. PERRIN, D. R., D. R. BIGGS, and J. A. M. CRUICKSHANK: Phytoalexins XV. Phaseollidin, a Phytoalexin from *Phaseolus vulgaris*. Isolation, Physicochemical Properties, and Antifungal Activity. Aust. J. Chem. **27**, 1607 (1974).

230. Perrin, D. R., and W. Bottomley: Pisatin: An Antifungal Substance from *Pisum sativum* L. Nature **191**, 76 (1961).
231. — — Studies in Phytoalexins V. The Structure of Pisatin from *Pisum sativum* L. J. Amer. Chem. Soc. **84**, 1919 (1962).
232. Perrin, D. R., and I. A. M. Cruickshank: Studies on Phytoalexins. VII. Chemical Stimulation of Pisatin Formation in *Pisum sativum*. Austral. J. Biol. Sci. **18**, 803 (1965).
233. — — The Antifungal Activity of Pterocarpans towards *Monilinia fructicola*. Phytochemistry **8**, 971 (1969).
234. Perrin, D. D., and D. R. Perrin: The N.m.r. Spectrum of Pisatin. J. Amer. Chem. Soc. **84**, 1922 (1962).
235. Perrin, D. R., C. P. Whittle, and T. J. Batterham: The Structure of Phaseollidin. Tetrahedron Letters **1972**, 1673.
236. Pfleger, F. L., and G. E. Harman: Inability of Storage Fungi to Invade Pea Embryos: Evidence Against Phytoalexin Involvement. Phytopathology **65**, 642 (1975).
237. Pierre, R. E.: Phytoalexin Induction in Beans Resistant or Susceptible to *Fusarium* and *Thielaviopsis*. Phytopathology **61**, 322 (1971).
238. Pierre, R. E., and D. F. Batemann: Induction and Distribution of Phytoalexins in *Rhizoctonia*-infected Bean Hypocotyls. Phytopathology **57**, 1154 (1967).
239. Preston, N. W.: 2'-O-Methylphaseollidinisoflavan from Infected Tissue of *Vigna unguiculata*. Phytochemistry **14**, 1131 (1975).
240. Preston, N. W., K. Chamberlain, and R. A. Skipp: A 2-Arylbenzofuran Phytoalexin from Cowpea *(Vigna unguiculata)*. Phytochemistry **14**, 1843 (1975).
241. Pryce, R. J.: Lunularic Acid, a Common Endogenous Growth Inhibitor of Liverworts. Planta (Berlin) **97**, 354 (1971).
242. — The Occurrence and Metabolism of Lunularic Acid, an Endogenous Stilbene Growth Inhibitor in Liverworts. Phytochemistry **11**, 872 (1972); Metabolism of Lunularic Acid to a New Plant Stilbene by *Lunularia cruciata*. Phytochemistry **11**, 1355 (1972); The Occurrence of Lunularic and Abscisic Acid in Plants. Phytochemistry **11**, 1759 (1972).
243. Pryce, R. J., and L. Linton: Lunularic Acid Decarboxylase from the Liverwort *Conocephalum conicum*. Phytochemistry **13**, 2497 (1974).
244. Pueppke, S. G., and H. D. van Etten: Identification of Three New Pterocarpans (6a,11a-dihydro-6H-benzofuro-[3,2-c] [1] benzopyrans) from *Pisum sativum* Infected with *Fusarium solani f. sp. pisi*. J. Chem. Soc. Perkin I **1975**, 946.
245. Purkayastha, R. P.: Phytoalexins. Plant Antigens and Disease Resistance. Sci. Cult. **39**, 528 (1973).
246. Purkayastha, R. P., and B. J. Deverall: The Detection of Antifungal Substances Before and After Infection of Beans *(Vicia faba* L.) by *Botrytis* spp. Ann. Appl. Biol. **56**, 269 (1965); The Growth of *Botrytis fabae* and *B. cinerea* Into Leaves of Bean *(Vicia faba* L.). Ann. Appl. Biol. **56**, 139 (1965).
247. Rahe, J. E.: Occurrence and Levels of the Phytoalexin Phaseollin in Relation to Delimitation at Sites of Infection of *Phaseolus vulgaris* by *Colletotrichum lendemuthianum*. Can. J. Bot. **51**, 2423 (1973).
247a. Rajaraman, K., and S. Rangaswami: Structures of Two New 9,10-Dihydrophenanthrenes from *Dioscorea prazeri*. Indian J. Chem. **13**, 1137 (1975).
248. Rathmell, W. G.: Phenolic Compounds and Phenylalanine Ammonia Lyase Activity in Relation to Phytoalexin Biosynthesis in Infected Hypocotyls of *Phaseolus vulgaris*. Physiol. Plant Pathol. **3**, 259 (1973).
249. Rathmell, W. G., and D. S. Bendall: Phenolic Compounds in Relation to Phytoalexin Biosynthesis in Hypocotyls of *Phaseolus vulgaris*. Physiol. Plant Pathol. **1**, 351 (1971).

250. Reisch, J., M. Bathory, K. Szendrei, I. Novak, and E. Minher: *Dioscoreaceae.* Weitere Phenanthrene aus dem Rhizom von *Tamus communis.* Phytochemistry **12**, 228 (1973).

250a. Saito, M., N. Kondo, H. Yamaguchi, and T. Hashimoto: Plant Growth-Regulating Activities of Batatasin III Analogs. Plant Cell Physiol. **17**, 411 (1976).

251. Sandykov, A. S., L. V. Melitskii, A. K. Karimdzhanov, A. I. Ismailov, R. A. Mukhamedova, M. Kh. Avazkhodzhaev, and F. G. Kamaev: Isohemigossypol as a Phytoalexin of a Cotton Plant. Dokl. Akad. Nauk SSSR **218**, 1472 (1974); Chem. Abstr. **82**, 82996j (1975).

252. Sato, N., K. Kitazawa, and K. Tomiyama: The Role of Rishitin in Localizing the Invading Hyphae of *Phytophthora infestans* in Infection Sites at the Cut Surfaces of Potato Tubers. Physiol. Plant Pathol. **1**, 289 (1971).

253. Sato, N., and K. Tomiyama: Localized Accumulation of Rishitin in the Potato-tuber Tissue Infected by an Incompatible Race of *Phythophthora infestans.* Ann. Phytopathol. Soc. Japan **35**, 202 (1969).

254. Sato, N., K. Tomiyama, N. Katsui, and T. Masamune: Isolation of Rishitin from Tomato Plants. Ann. Phytopathol. Soc. Japan **34**, 344 (1968).

255. — — — — Isolation of Rishitin from Tubers in Interspecific Potato Varieties Containing Different Late-blight Resistance Genes. Ann. Phytopathol. Soc. Japan **34**, 140 (1968).

256. Schönbeck, F.: Untersuchungen über Blüteninfektionen. V. Untersuchungen an Tulpen. Phytopathol. Z. **59**, 205 (1967).

257. Schönbeck, F., and C. Schröder: Role of Antimicrobial Substances (Tuliposides) in Tulips Att cked by *Botrytis spp.* Physiol. Plant Pathol. **2**, 91 (1972).

258. Schroeder, C.: Influence of γ-Hydroxylic Acids on the Hot-parasite Ratio of Tulips and *Botrytis* species. Phytopathol. Z. **74**, 175 (1972).

259. Schütte, H. R.: Flavonoid Biosynthesis and their Regulation. Fortschr. Bot. **36**, 108 (1974).

260. Schwochau, M. E., and L. A. Hadwiger: Stimulation of Pisatin Production in *Pisum sativum* by Actinomycin D and Other Compounds. Arch. Biochem. Biophysics **126**, 731 (1968).

261. — — Regulation of Gene Expression by Actinomycin D and Other Compounds Which Change the Conformation of DNA. Arch. Biochem. Biophysics **134**, 34 (1969).

262. Shih, M., and J. Kuc: Incorporation of ^{14}C from Acetate and Mevalonate into Rishitin and Steroid Glycoalkaloids by Potato Tuber Slices Inoculated with *Phytophthora infestans.* Phytopathology **63**, 826 (1973).

263. Shiraishi, T., H. Oku, M. Isono, and S. Ouchi: The Injurious Effect of Pisatin on the Plasma membrane of Pea. Plant Cell Physiol. **16**, 939 (1975).

264. Sims, J. J., N. T. Keen, and V. K. Honwad: Hydroxyphaseollin, an Induced Antifungal Compound from Soybeans. Phytochemistry **11**, 827 (1972).

265. Sitton, D., and C. A. West: Casbene: an Antifungal Diterpene Produced in Cell-free Extracts of *Ricinus communis* Seedlings. Phytochemistry **14**, 1921 (1975).

266. Slob, A., B. Jekel, B. de Jong, and E. Schlatmann: On the Occurrence of Tuliposides in the *Liliiflorae.* Phytochemistry **14**, 1997 (1975).

267. Smith, D. G., A. G. McInnes, V. J. Higgins, and R. L. Millar: Nature of the Phytoalexin Produced by Alfalfa in Response to Fungal Infection. Physiol. Plant Pathol. **1**, 41 (1971).

268. Smith, D. A., H. D. van Etten, and D. F. Bateman: Isolation of Substance II, an Antifungal Compound from *Rhizoctonia solani*-Infected Bean Tissue. Phytopathology **61**, 912 (1971).

269. — — — Kievitone. Principal Antifungal Component of "Substance II" Isolated from *Rhizoctonia* Infected Bean Tissues. Physiol. Plant Pathol. **3**, 179 (1973).

16*

270. SMITH, D. A., H. D. VAN ETTEN, and D. F. BATEMAN: Accumulation of Phytoalexins in *Phaseolus vulgaris* Hypocotyls Following Infection by *Rhizoctonia solani*. Physiol. Plant Pathol. **5**, 51 (1975).

271. SMITH, D. A., H. D. VAN ETTEN, J. W. SERUM, T. M. JONES, D. F. BATEMAN, T. H. WILLIAMS, and D. L. COFFEN: Confirmation of the Structure of Kievitone, an Antifungal Isoflavanone Isolated from *Rhizoctonia*-Infected Bean Tissues. Physiol. Plant Pathol. **3**, 293 (1973).

272. STEINER, K., C. EGLI, N. RIGASSI, S. E. HELALI, and E. HARDEGGER: Welkstoffe und Antibiotika. Zur Synthese des Orchinols. Helv. Chim. Acta **57**, 1137 (1974).

273. STEINER, P. W., and R. L. MILLAR: Degradation of Medicarpin and Sativan by *Stemphyllum botryosum*. Phytopathology **64**, 586 (1974).

274. STHOLASUTA, P., J. A. BAILEY, V. SEVERIN, and B. J. DEVERALL: Effect of Bacterial Inoculation of Bean and Pea Leaves on the Accumulation of Phaseollin and Pisatin. Physiol. Plant Pathol. **1**, 177 (1971).

275. STIPANOVIC, R. D., A. A. BELL, and C. R. HOWELL: Naphthofuran Precursors of Sesquiterpenoid Aldehydes in Diseased *Gossypium*. Phytochemistry **14**, 1809 (1975).

276. STIPANOVIC, R. D., A. A. BELL, M. E. MACE, and C. R. HOWELL: Antimicrobial Terpenoids of *Gossypium*: 6-Methoxygossypol and 6,6'-Dimethoxygossypol. Phytochemistry **14**, 1077 (1975).

277. STOESSL, A.: Antifungal Compounds Produced by Higher Plants. Rec. Adv. Phytochem. **3**, 143 (1972).

278. — Inermin Associated with Pisatin in Peas Inoculated With the Fungus *Monilinia fructicola*. Can. J. Biochem. **50**, 107 (1972).

279. STOESSL, A., G. L. ROCK, and M. H. FISCH: An Efficient Synthesis of Orchinol and Other Orchid Phenanthrenes. Chem. Ind. **1974**, 703.

280. STOESSL, A., J. B. STOTHERS, and E. W. B. WARD: Lubimin: A Phytoalexin of Several *Solanaceae*. Structure Revision and Biogenetic Relationship. Chem. Commun. **1974**, 709.

281. — — — A 2,3-Dihydroxygermacrene and Other Stress Metabolites of *Datura stramonium*. Chem. Commun. **1975**, 431.

281a. — — — The Structures of Some Stress Metabolites from *Solanum melongena*. Can. J. Chem. **53**, 3351 (1975).

281b. STOESSL, A., J. B. STOTHERS, and E. W. B. WARD: Sesquiterpenoid Stress Compounds of the *Solanaceae*. Phytochemistry **15**, 855 (1976).

281c. STOESSL, A., E. W. B. WARD, and J. B. STOTHERS: Incorporation of Doubly Labelled Sodium Acetate-$^{13}C_2$ into Phytuberin and Other Sesquiterpenes in Potatoes; Experimental Confirmation of Postulated C-C-Cleavages. Tetrahedron Letters **1976**, 3271.

282. STOESSL, A., C. H. UNWIN, and E. W. B. WARD: Postinfectional Inhibitors from Plants. I. Capsidiol, an Antifungal Compound from *Capsicum frutescens*. Phytopathol. Z. **74**, 141 (1972).

283. — — — Postinfectional Inhibitors From Plants: Fungal Oxidation of Capsidiol in Pepper Fruit. Phytopathology **63**, 1225 (1973).

283a. SUZUKI, H., K. OBA, and I. URITANI: The Occurrence and Some Properties of 3-Hydroxy-3-methylglutaryl coenzyme A reductase in Sweet Potato Roots Infected by *Ceratocystis fimbriata*. Physiol. Plant Pathol. **7**, 265 (1975).

284. TAIRA, T., and Y. FUKAGAWA: On the Bitter Substance Separated from Alcohol Destillation of Sweet Potato Mash. J. Agric. Chem. Soc. Japan **32**, 513 (1958).

285. THOMAS, C. A., and E. H. ALLEN: An Antifungal Polyacetylene Compound from *Phytophthora*-infected Safflower Hypocotyls. Phytopathology **59**, 1053 (1969).

286. — — An Antifungal Polyacetylene Compound from *Phytophthora*-Infected Safflower. Phytopathology **60**, 261 (1970).

287. — — Concentration of Safynol in *Phytophthora*-Infected Safflower. Phytopathology **60**, 1153 (1970).

288. TJAMOS, E. C., and I. M. SMITH: Role of Phytoalexins in the Resistance of Tomato to *Verticillium* wilt. Physiol. Plant Pathol. **4**, 249 (1974).
289. TOMIYAMA, K., T. SAKUMA, N. ISHIZAKA, N. SATO, N. KATSUI, M. TAKASUGI, and T. MASAMUNE: A New Antifungal Substance Isolated From Resistant Potato Tuber Tissue Infected by Pathogens. Phytopathology **58**, 115 (1968).
290. TSCHESCHE, R., F. J. KÄMMERER, and G. WULFF: Über die Struktur der antibiotisch aktiven Substanz der Tulpe *(Tulipa gesneriana* L.). Chem. Ber. **102**, 2057 (1969).
291. TSCHESCHE, R., F. J. KÄMMERER, G. WULFF, und T. SCHÖNBECK: Über die antibiotisch wirksamen Substanzen der Tulpe *(Tulipa gesneriana)*. Tetrahedron Letters **1968**, 701.
292. UEHARA, K.: On the Phytoalexin Production of the Soybean Pod in Reaction to *Fusarium spp.*, the Causal Fungus of Pod Blight. I. Ann. Phytopathol. Soc. Japan **23**, 225 (1958).
293. — On Some Propertied of Phytoalexins Produced as a Result of the Interaction Between Pea *(Pisum sativum* L.) and *Ascochyta pisi* Lib. I. Ann. Phytopathol. Soc. Japan **23**, 230 (1958).
294. — On the Phytoalexin Production of the Soybean Pod in Reaction to *Fusarium spp.*, the Causal Fungus of Pod Blight. II. Ann. Phytopathol. Soc. Japan **24**, 224 (1959).
295. — On Some Properties of Phytoalexin Produced as a Result of the Interaction Between Pea *(Pisum sativum* L.) and *Ascochyta pisi* Lib. II. Effect of Duration of Mounting the Spore Suspension on the Pea Pod and Pre-infectional Treatment of Pea Pods with Ether or Heat Upon Phytoalexin Production. Ann. Phytopathol. Soc. Japan **25**, 85 (1960).
296. URECH, J., B. FECHTIG, J. NÜESCH, und E. VISCHER: Hircinol, eine antifungisch wirksame Substanz aus Knollen von *Loroglossum hircinum* (L.) Rich. Helv. Chim. Acta **46**, 2758 (1963).
297. URITANI, I., and T. AKAZAWA: Antibiotic Effect on *Ceratostomella fimbriata* of Ipomeamarone, an Abnormal Metabolite in Black Rot of Sweet Potato. Science **121**, 216 (1955).
298. URITANI, I., and K. OSHIMA: Effects of Ipomeamarone on Respiratory Enzyme System in Mitochondria. Agric. Biol. Chem. **29**, 641 (1965).
298a. URITANI, I., T. SAITO, H. HONDA, and W. K. KIM: Induction of Furano-terpenoids in Sweet Potato Roots by the Larval Components of the Sweet Potato Weevils. Agric. Biol. Chem. **39**, 1857 (1975).
299. URITANI, I., M. URITANI, and H. YAMADA: Similar Metabolic Alterations Induced in Sweet Potato by Poisonous Chemicals and by *Ceratostomella fimbriata*. Phytopathology **50**, 30 (1960).
300. VALIO, I. F. M., and W. W. SCHWABE: Growth and Dormancy in *Lunularia cruciata* (L.) Dum. VII. The Isolation and Bioassay of Lunularic Acid. J. Exp. Bot. **21**, 138 (1970).
301. VAN DEN ENDE, G.: Neue Untersuchungen über die Phytoalexin-Bildung. Tag. Ber. Dtsch. Akad. Landw. Wiss. Berlin **74**, 283 (1965).
302. — Phytoalexin-Bildung bei der Wechselwirkung zwischen *Sclerotinia fructicola* und Wirtsgeweben. Phytopathol. Z. **64**, 68 (1969).
303. VAN DEN ENDE, G., und K. O. MÜLLER: Zur Kinetik der Phytoalexinbildung. Naturwissenschaften **51**, 317 (1964).
304. VAN ETTEN, H. D.: Antifungal and Hemolytic Activities of Four Pterocarpan Phytoalexins. Phytopathology **62**, 795 (1972).
305. — Identification of a Second Antifungal Isoflavan From Diseased *Phaseolus vulgaris* Tissue. Phytochemistry **12**, 1791 (1973).
306. — Differential Sensitivity of Fungi to Pisatin and to Phaseollin. Phytopathology **63**, 1477 (1973).
306a. VAN ETTEN, H. D.: Antifungal Activity of Pterocarpans and other Selected Isoflavonoids. Phytochemistry **15**, 655 (1976).

307. Van Etten, H. D., and D. F. Bateman: Isolation of Phaseollin from *Rhizoctonia*-Infected Bean Tissue. Phytopathology **60**, 385 (1970).
308. — — Studies on the Mode of Action of the Phytoalexin Phaseollin. Phytopathology **61**, 1363 (1971).
309. Van Etten, H. D., S. G. Pueppke, and T. C. Kelsey: 3,6a-Dihydroxy-8,9-methylenedioxypterocarpan as a Metabolite of Pisatin Produced by *Fusarium solani f. sp. Pisi*. Phytochemistry **14**, 1103 (1975).
310. Van Etten, H. D., and D. A. Smith: Accumulation of Antifungal Isoflavonoids and 1a-Hydroxyphaseollone, a Phaseollin Metabolite in Bean Tissue Infected With *Fusarium solani f. specialis phaseoli*. Physiol. Plant Pathol. **5**, 225 (1975).
311. Varns, J. L., W. W. Currier, and J. Kuc: Specificity of Rishitin and Phytuberin Accumulation by Potato. Phytopathology **61**, 968 (1971).
312. Varns, J. L., and J. Kuc: Suppression of Rishitin and Phytuberin Accumulation and Hypersensitive Response in Potato by Compatible Races of *Phytophthora infestans*. Phytopathology **61**, 178 (1971).
313. Varns, J. L., J. Kuc, and E. B. Williams: Terpenoid Accumulation as a Biochemical Response of the Potato Tuber to *Phytophthora infestans*. Phytopathology **61**, 174 (1971).
314. Vasyukova, N. I., O. L. Ozeretskovskaya, and L. V. Metlitskii: Phytoalexins of Potatoes. Prikl. Biokhim. Microbiol. **6**, 431 (1970); Chem. Abstr. **73**, 127885 (1970).
314a. Veech, J. A., R. D. Stipanovic, and A. A. Bell: Peroxidative Conversion of Hemigossypol to Gossypol. A Revised Structure for Isohemigossypol. Chem. Commun. **1976**, 144.
315. Wain, R. L., D. M. Spencer, and C. H. Fawcett: Antifungal Compounds in Seedlings of *Vicia faba*. Chem. Ind. **1961**, 343.
315a. Ward, E. W. B.: Capsidiol Production in Pepper Leaves in Incompatible Interactions with Fungi. Phytopathology **66**, 175 (1976).
316. Ward, E. W. B., and A. Stoessl: Postinfectional Inhibitors from Plants. III. Detoxification of Capsidiol, an Antifungal Compound from Peppers. Phytopathology **62**, 1186 (1972).
317. Ward, E. W. B., C. H. Unwin, J. Hill, and A. Stoessl: Sesquiterpenoid Phytoalexins from Fruits of Eggplants. Phytopathology **65**, 859 (1975).
317a. Ward, E. W. B., C. H. Unwin, G. L. Rock, and A. Stoessl: Postinfectional Inhibitors from Plants. Sesquiterpenoid Phytoalexins from Fruit Capsules of *Datura stramonium*. Can. J. Bot. **54**, 25 (1976).
318. Ward, E. W. B., C. H. Unwin, and A. Stoessl: Postinfectional Inhibitors from Plants. VII. Tolerance of Capsidiol by Fungal Pathogens of Pepper Fruit. Can. J. Bot. **51**, 2327 (1973).
319. — — — Postinfectional Inhibitors from Plants. VI. Capsidiol Production in Pepper Fruit Infected With Bacteria. Phytopathology **63**, 1537 (1973).
320. — — — Postinfectional Inhibitors from Plants. XIII. Fungitoxicity of the Phytoalexin, Capsidiol and Related Sesquiterpenes. Can. J. Bot. **52**, 2481 (1974).
321. — — — Postinfectional Inhibitors from Plants. XV. Antifungal Activity of the Phytoalexins Orchinol and Related Phenanthrenes and Stilbenes. Can. J. Bot. **53**, 964 (1975).
322. — — — Loroglossol: An Orchid Phytoalexin. Phytopathology **65**, 632 (1975).
323. — — — Experimental Control of Late Blight of Tomatoes with Capsidiol, the Phytoalexin from Peppers. Phytopathology **65**, 168 (1975).
324. Watanabe, H., and S. Nishiyama: Studies on the Black-Rotten Sweet Potato. Part 3. Chemical Properties of Ipomeamarone. J. Agric. Chem. Soc. Japan **26**, 200 (1952).
325. Wilson, B. J., D. T. C. Yang, and M. R. Boyd: Toxicity of Mould-Damaged Sweet Potatoes *(Ipomoea batatas)*. Nature (London) **227**, 521 (1970).
326. Wit-Elshove, A. de: Breakdown of Pisatin by Some Fungi Pathogenic to *Pisum sativum*. Netherlands J. Plant Pathol. **74**, 44 (1968).

327. — The Role of Pisatin in the Resistance of Pea Plants — Some Further Experiments on the Breakdown of Pisatin. Netherlands J. Plant Pathol. **75**, 164 (1969).

328. — Some Aspects of the Degradation of Pisatin by Fungi, Pathogenic to *Pisum sativum* L. Acta bot. neerl. **19**, 113 (1970).

329. WIT-ELSHOVE, A. DE, and A. FUCHS: The Influence of the Carbohydrate Source on Pisatin Breakdown by Fungi Pathogenic to Pea *(Pisum sativum)*. Physiol. Plant Pathol. **1**, 17 (1971).

330. WONG, E.: Structural and Biogenetic Relationships of Isoflavonoids. Fortschr. Chem. Org. Naturst. **28**, 1 (1970).

330a. Wood, G., and A. HUANG: Detection and Quantitative Determination of Ipomeamarone in Damaged Sweet Potatoes *(Ipomoea batatas)*. J. Agr. Food Chem. **23**, 239 (1975).

331. WOOD, H. K. S.: Hypersensitivity, Phytoalexins and Disease Resistance. Mitt. Biol. Bundesanst. Land-Forstwirtsch., Berlin-Dahlem No. **154**, 95 (1973).

332. WOOD, A. B., F. V. ROBINSON, and R. C. A. LAGO: Conformation and Hydrogen Bonding of Gossypol. Chem. Ind. **1969**, 1738.

333. YANG, D. T. C., B. J. WILSON, and T. M. HARRIS: The Structure of Ipomeamaronol: A New Toxic Furanosesquiterpene From Moldy Sweet Potatoes. Phytochemistry **10**, 1653 (1971).

334. ZAKI, A. I., N. T. KEEN, and D. C. ERWIN: Implication of Vergosin and Hemigossypol in the Resistance of Cotton to *Verticillium alboatrum*. Phytopathology **62**, 1402 (1972).

335. ZAKI, A. I., N. T. KEEN, J. J. SIMS, and D. C. ERWIN: Vergosin and Hemigossypol, Antifungal Compounds Produced in Cotton Plants Inoculated with *Verticillium albo-atrum*. Phytopathology **62**, 1398 (1972).

(Received April 21, 1976)

Studies in Secondary Metabolism
with Plant Tissue Cultures

By K. H. OVERTON and D. J. PICKEN, Department of Chemistry,
University of Glasgow, Scotland

Contents

I. Introduction

The notion of culturing plant cells *in vitro* goes back to the beginning of this century. However, successful experiments in the culturing of unorganized plant cells for prolonged periods were first reported in 1939 independently by GAUTHERET (*40*) and WHITE (*112*), who in essence established the technique still in use today.

A number of excellent monographs and reviews (*20, 41, 92, 93, 110, 113*) concerned with the initiation and maintenance of plant tissue cultures are available and nothing more will be attempted here than to make intelligible some of the terms used in this chapter and to provide a brief outline of the basic techniques that will be unfamiliar to many chemists.

It is convenient for present purposes to consider two types of plant culture that can be grown *in vitro*: (a) organ cultures, i.e. isolated roots, leaves, flowers etc. which retain the organization of the intact organ; (b) callus cultures (usually referred to as tissue cultures) which normally consist of a mass of mostly *undifferentiated* cells. Crown gall tumour cultures (bacterially induced) and wound virus tumour cultures (virally induced) may have potential use in biosynthetic studies of secondary metabolism but have apparently been little used until now. Most of the work discussed in this chapter utilizes callus cultures. It is worth noting that such cultures are species-specific, both morphologically and biochemically, but also that callus of a particular species may be derived from different parts of the plant. A further property of callus cultures of interest in biosynthetic studies is their ability, in response to certain stimuli, to regenerate intact plantlets which, unlike the callus, usually resemble the parent plant in their secondary metabolism. However, the factors controlling organogenesis in callus cultures are not at all well understood and its induction calls for empirical and not always successful experimentation. Nevertheless, this extension of plant tissue culture has potential in the solution of biosynthetic problems.

Tissue culturing techniques call for rigorously aseptic conditions since plant tissue cultures readily succumb to infection by bacteria, fungi and viruses. Nutrient media are usually sterilized by autoclaving and thermolabile substrates by filtration through bacterial filters.

The initiation and transfer of culture material requires a sterile area or laboratory and all the precautions employed in handling bacterial cultures. A common procedure for initiating a callus culture is to germinate a sterilized seed and to dissect aseptically from the resulting seedling a portion of stem, root or leaf tissue. This is placed on the surface of a suitable nutrient medium solidified with agar, which contains growth hormones (auxins and cytokinins) to promote the formation of undifferentiated callus tissue. On prolonged subculture, some callus tissues lose their requirement for exogenously supplied auxin and are then said to be "habituated" (French "anergie"). A mass of light-coloured spongy or friable tissue grows from the original inoculum and a portion must be periodically transferred to fresh nutrient medium at intervals of 2—3 weeks. Such cell-lines have been maintained for years, even decades, but genetic changes have been noted to occur on continued subculturing. The consequent biochemical changes are clearly undesirable in prolonged biosynthetic studies and hence the provision of long-term storage methods that will maintain stock cultures genetically intact is a matter of some urgency. Some recent progress in this area is therefore of particular interest (70a). Cultures grown on agar are referred to as static cultures. Most of the work described in this review utilizes suspension cultures which are readily obtained by inoculating callus grown on agar into liquid medium (usually of the same composition but lacking agar) and arranging for continuous agitation of the resulting suspension which consists microscopically of single cells and small cell clusters. Such suspension cultures are preferable to static cultures for biosynthetic experiments since both the administration of precursors and extraction of products are facilitated. Temperatures of 23°—28° C and a fairly narrow pH range of 5.3—6.5 are commonly found satisfactory. The effect of light intensity and wavelength on growth and metabolism requires careful attention in each particular case.

Numerous growth media have been described but it is striking that many of the studies included in this review feature a few basic types [e. g. WHITE's (113), MURASHIGE and SKOOG's (70)] with minor modifications. Although entirely synthetic media are sometimes effective, the inclusion of coconut milk (5—20%) (also casein hydrolysate or yeast extract) is a common practice. The constituents of coconut milk have been analyzed at length but its peculiar and potent effect on tissue culture growth remains undefined and this, in view of natural variation in its complex composition, makes its role in plant tissue culture rather unsatisfactory. The chemically defined growth regulators in common use are kinetin (6-N-furfurylamino-purine), α-naphthalene acetic acid (NAA), 3-indolyl-acetic acid (IAA), 2,4-dichlorophenoxyacetic acid (2,4-D), gibberellins and abscisic acid. These substances have profound effects on culture growth, the production of metabolites (23, 38) and differentiation (92) and are usually varied empirically towards a desired end.

II. The Chemical Potential of Plant Tissue Cultures

The cells of many plant tissue cultures are totipotent, that is to say they possess all the information necessary to the functioning and replication of the whole plant including its secondary metabolism. This is evident from the fact, already mentioned, that callus tissues are often capable of regenerating whole plants which are in many respects comparable with the parent plant. In view of the many difficulties of growing intact higher plants under controlled and reproducible conditions, chemists have been attracted by the potential of tissue culture in two distinct areas: (1) the production of medicinally active secondary plant metabolites like steroids and alkaloids either by *de novo* synthesis by the culture (22, 79) or through biotransformations of more advanced but accessible intermediates (89, 96); (2) the study of biosynthetic and also biodegradative pathways. This review is concerned entirely with work in the latter area. In both areas the stimulus has undoubtedly come from the highly successful exploitation of micro-organisms in the industrial synthesis of medicinals and in the exploration of primary and secondary metabolic pathways. To attempt to isolate and discipline the chemistry of higher plants in comparably simple and reproducible systems is understandably a powerful challenge.

A major obstacle in studying the biosynthesis of secondary metabolites in intact higher plants has been the frequently very low level or on occasion total failure of precursor incorporation. The basic cause must be a failure to transport precursors in their appropriate form to the site of biosynthesis and this could result from a variety of factors including problems of permeability, translocation and the segregation of metabolic pools. Tissue cultures possess an altogether simpler organization so that, compared with whole plants, these problems should be greatly simplified. They moreover can be grown reproducibly under standard conditions, have short growth cycles and are not subject to seasonal variations. It is a particular attraction in biosynthetic studies that the incorporation and turnover of labelled precursors can be studied over very short periods of time. The need for rigorous sterility has the automatic advantage of ensuring that metabolic processes under observation must be promoted by the plant tissue and not by micro organisms associated with it, a condition that cannot be met with certainty in working with intact plants. It has also been found that active cell-free systems and purified enzyme preparations can be prepared rather readily from tissue cultures, possibly because of the absence of substances, notably phenols and quinones which are responsible for the deactivation of enzymes in the course of their extraction from intact plants.

It must seem remarkable indeed that tissue cultures have not been used more widely in biosynthetic studies. There are in fact two good reasons for this. The first stems from the technical problems associated with initiating and maintaining sterile cultures and the long times, sometimes years, that may be needed to establish a system manifesting good growth and metabolic activity. But a more serious disincentive is that tissue cultures quite frequently do not synthesize the secondary metabolites characteristic of the parent plant at all or do so in only minute amounts. Too little is known at present about what factors inhibit secondary metabolite synthesis in tissue cultures or cause its re-emergence on redifferentiation to attack this major problem in a systematic way. However, a relatively large number of higher plants have been established in culture and in many cases biosynthetic pathways of major interest have been identified. By a judicious choice of problems, it has been possible to exploit the special advantages of tissue cultures for biosynthetic studies.

In this review we have selected from recently published work examples that illustrate the successful application of tissue and organ culture techniques to the solution of problems in secondary plant metabolism.

III. Biosynthesis of Polyisoprenoids

1. Plant Sterol Metabolism

In what are probably the most extended biosynthetic investigations with plant tissue cultures to date, OURISSON, BENVENISTE and their colleagues have studied the biosynthesis of plant sterols, using tissues of *Nicotiana tabacum* (tobacco) and *Rubus fruticosus* (bramble).

Whereas lanosterol (1) is the precursor of steroids in animals and fungi, it has never been detected in higher plants. On the other hand the ubiquitous occurrence of cycloartenol (2) and 24-methylenecycloartanol (3) and cycloeucalenol (4) in photosynthetic plant tissues has led to the suggestion that $9\beta,19\beta$-cyclopropylsterols may take the place of lanosterol in plant sterol biosynthesis (2, 12, 42). Sustained investigations over the past ten years by the group of OURISSON and BENVENISTE have amply supported this.

Their earliest studies (13, 14), exploiting one of the peculiar advantages of plant tissue cultures, set out to detect the incorporation of radioactive label into short-lived intermediates of the plant sterol biosynthetic pathway. By monitoring the labelled products formed from [1-^{14}C]acetate and [$^{14}CH_3$]methionine after incubation times as short as five minutes, two points emerged: (a) labelled lanosterol could not be detected and (b) cycloartenol was rapidly and heavily labelled but must be rapidly

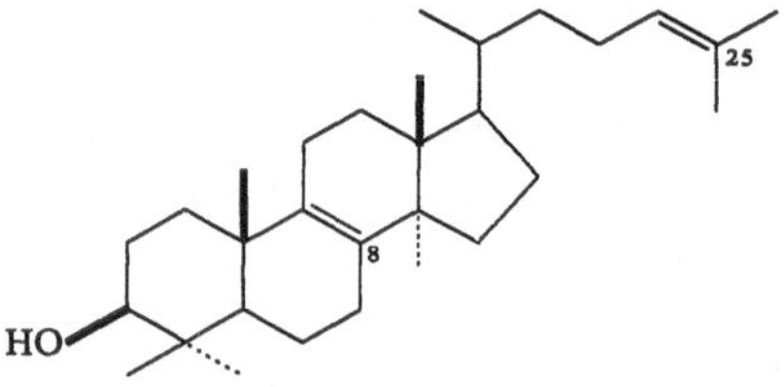

(1)

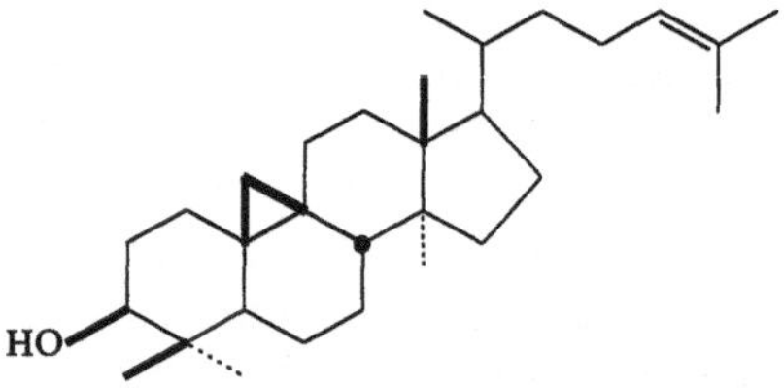

(2)

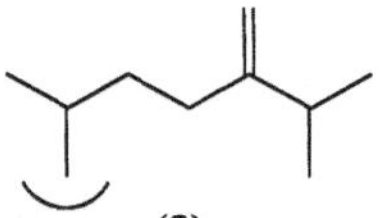

(3)

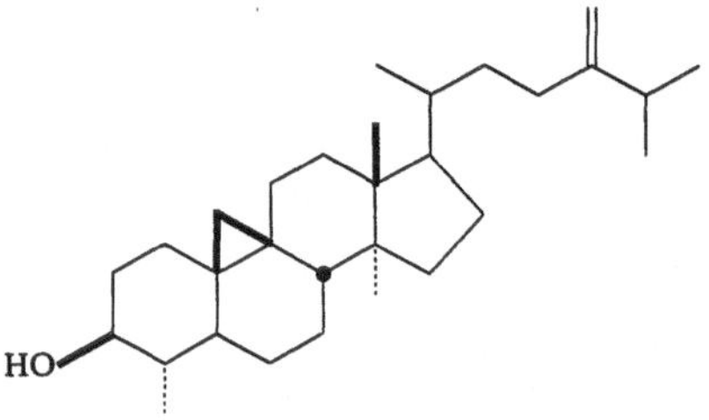

(4)

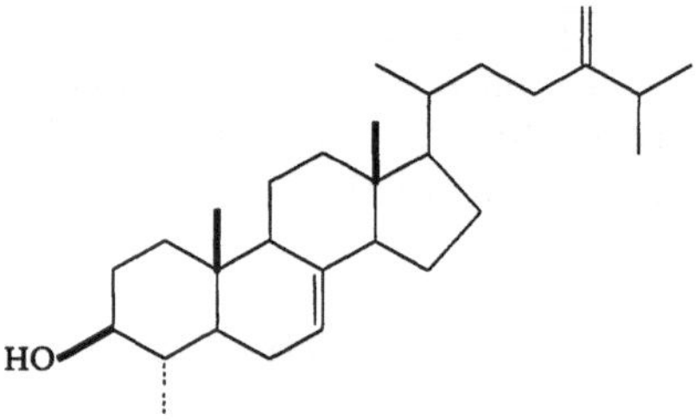

(5)

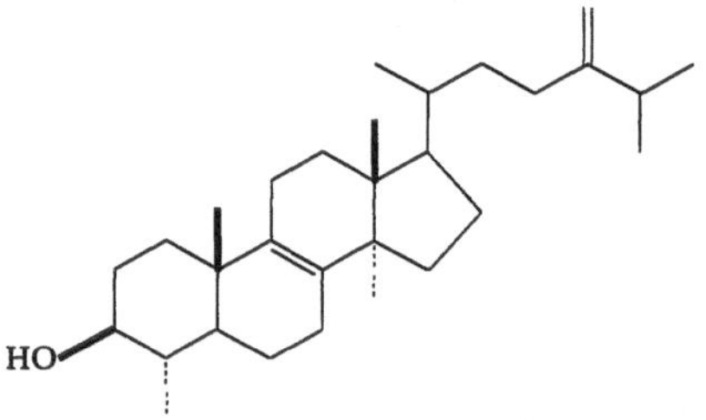

(6)

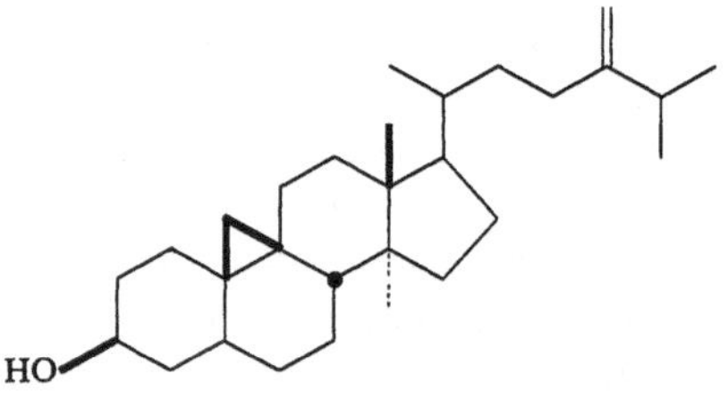

(7)

transformed into steroids further along the pathway, since it could be detected in only minute concentrations in the cultures. It was possible to identify in addition to phytosterols, squalene, cycloartenol (2), 24-methylenecycloartanol (3), 24-methylenelophenol (5) and its ethylidene homologue, and to show that the order of labelling was squalene, cyclo-artenol, 24-methylenelophenol and phytosterols. These early experiments provided compelling evidence that cycloartenol may take the place of lanosterol in plant sterol biosynthesis. In a later report (6) cycloeucalenol (4) and obtusifoliol (6) labelled from [1-^{14}C]acetate were identified as constituents of the 4-methyl steroid fraction from tobacco cultures, thus supporting their possible role as intermediates in phytosterol biosynthesis.

More direct evidence for the intermediacy of cycloartenol in plant sterol biosynthesis was forthcoming from experiments directed towards the kinetics of formation of sterols and their precursors (11) which again took advantage of short incorporation times. With [^{14}C]acetate all the sterols found in tobacco cultures incorporate label within five minutes. Thus the rate of incorporation of precursors could be followed in experiments of different duration. However since the various intermediates appear in widely differing concentrations, some very small indeed, a method of estimating specific activities had to be devised. This consisted of acetylating sterols labelled from [1-^{14}C]acetate with [^{3}H]acetic anhydride of known specific activity. By measuring the tritium : carbon ratios one then obtains the specific activity of each [^{14}C]-labelled steroid. In this way it is possible to establish the kinetics of formation and transformation of each intermediate thought to be involved in a biosynthetic pathway. On the basis of measured specific activities a tentative scheme was proposed for phytosterol biosynthesis in which cycloartenol appears to play a direct role. Its conversion into sterols does not apparently follow a unique route but can proceed by a number of alternative routes that constitute a grid akin to that recently proposed for yeast sterol biosynthesis (4).

In further attempts (52) to probe the possible involvement of lanosterol in phytosterol biosynthesis, [25-^{14}C]lanosterol and [25-^{14}C]cycloartenol were incubated with tobacco tissue cultures. Cycloartenol was converted into all the phytosterols previously found to be produced from acetate. Lanosterol was actually metabolized to 24-methylenelanost-8-enol and obtusifoliol (6), but this can probably be ascribed to a lack of specificity in the enzymes responsible for C-24 methylation and C-4 demethylation. Isotope dilution experiments failed to detect lanosterol. The inference must be that if lanosterol plays a biosynthetic role, then it must have a very high turnover rate or remain particle-bound.

The intermediacy of cycloartenol in phytosterol biosynthesis suggests that plants possess an enzyme capable of opening the 9β,19β-cyclopropane ring. The ubiquitous presence of cycloeucalenol (4) and obtusifoliol (6) further suggests that such an enzyme probably acts on cycloeucalenol.

The appropriate enzyme has now been obtained (9) in cell-free extracts from bramble and tobacco tissues and shown to be tightly bound to microsomal membranes. An analogous microsomal fraction from rabbit livers was inactive. 4,4-Dimethylsterols are very poor substrates, but cycloeucalenol (4) and 24-methylenepollinastanol (7), are good substrates, suggesting that the 4β-methyl group of 4,4-dimethylsterols inhibits the enzyme. The enzyme functions without addition of exogenous ATP or NADH and therefore effects the equivalent of acid-catalyzed cyclopropane cleavage. Neither cycloartenol nor 24-methylenecycloartanol are substrates and this probably eliminates lanosterol and 24-methylene lanosterol from the major phytosterol pathway. The results, taken together with previous relevant information, suggest a modified pathway for phytosterol biosynthesis as depicted in Scheme 1.

The interesting observation has been made that tissue cultures of *Helianthus tuberosus* (Jerusalem artichoke) incorporate [1-^{14}C]acetate into phytosterols and their precursors with reproducibly high specific activity, when metabolizing cultures are irradiated with far red light (*7*).

Two key intermediates in animal steroid biosynthesis have recently been investigated in plant tissue cultures. Presqualene pyrophosphate (**12**) the probable intermediate between farnesyl pyrophosphate and squalene, has been isolated (*10*) from bramble tissue cultures. When a cell-free extract was incubated with [2-^{14}C]mevalonate together with all the co-factors required for squalene formation, the squalene formed was heavily labelled, while in the absence of NADH squalene was only poorly labelled. The product of the second experiment after hydrolysis by alkaline phosphatase was identified as presqualene alcohol (**13**).

(**12**) R^2 = P$_2$O$_6$
(**13**) R^2 = H

The proven role of 2,3-oxidosqualene (**8**) in animal steroid biosynthesis has led to its investigation in plant tissue cultures also. BENVENISTE and MASSEY-WESTROPP (*15*) were able to demonstrate the synthesis of 2,3-oxidosqualene in tobacco tissues by trapping radio-activity from [1-^{14}C]acetate in added oxidosqualene. [^{14}C]2,3-Oxidosqualene was converted (*33*) under anaerobic conditions by tobacco tissue cultures into cycloartenol (**2**) and 24-methylenecycloartanol (**9**), but not into lanosterol (**1**) or phytosterols. In a complementary experiment (*8*) microsomes obtained from an acetone powder of bramble tissue cultures produced labelled cycloartenol and 24-methylenecycloartanol when tritiated 2,3-oxidosqualene and unlabelled adenosyl methionine were substrates.

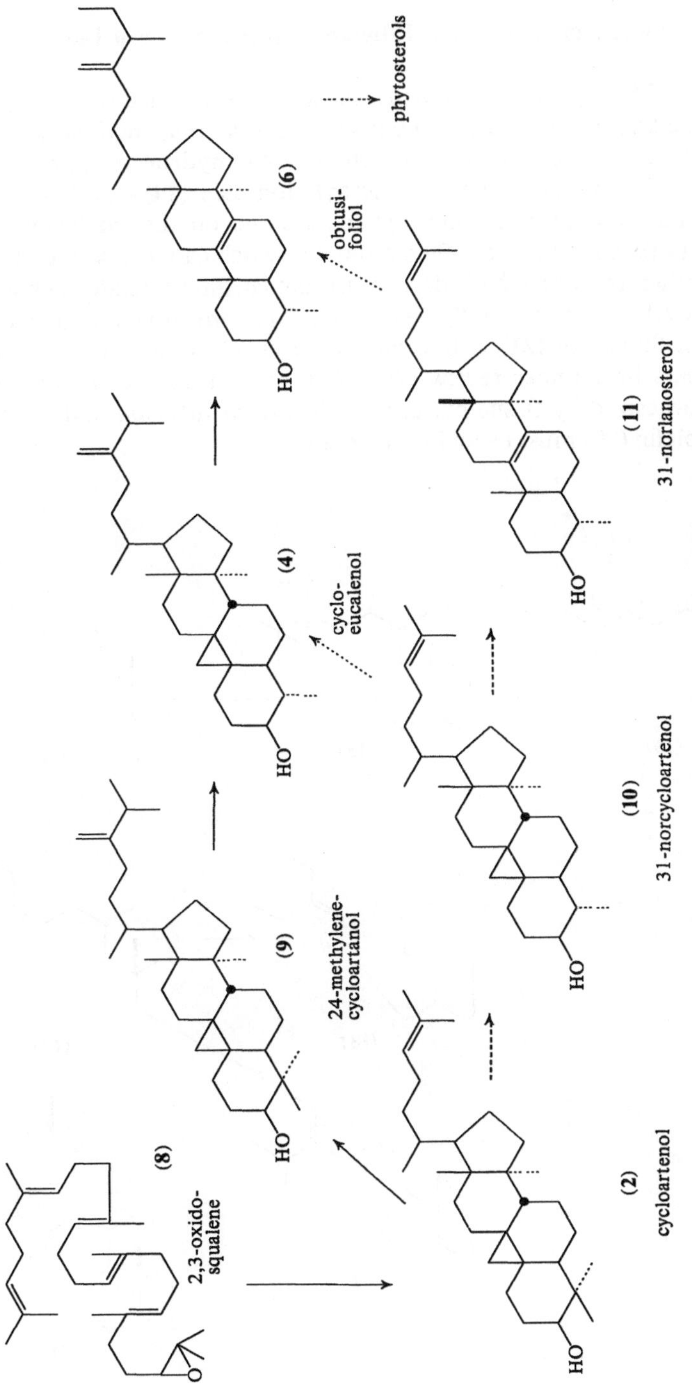

Scheme 1. Probable pathways in phytosterol biosynthesis

2. Mechanism of Side Chain Ethylation in Stigmasterol Biosynthesis

The C-24 ethyl and ethylidene substituents characteristic of phyto-sterols are known to be formed by two successive *trans*-methylations from adenosyl methionine. It has been shown that hydrogen migrates from C-24 to C-25 following the first methylation step (14)→(15). Attack of the second equivalent of adenosyl methionine on the methylene inter-mediate (16) generates the C-24 cation (17) which can lead to a C-24 ethyl group either by direct hydride capture or *via* intermediate olefins (18), (19), or (20). Compound (19) was shown (*51, 86a*) to be an intermediate when poriferasterol (21) is biosynthesized by *Ochromonas malhamensis*. The papers here under review (*106, 108*) show that, when stigmasterol is biosynthesized by tissue cultures of *Nicotiana tabacum* and *Dioscorea tokoro*, olefin (20) must be an intermediate.

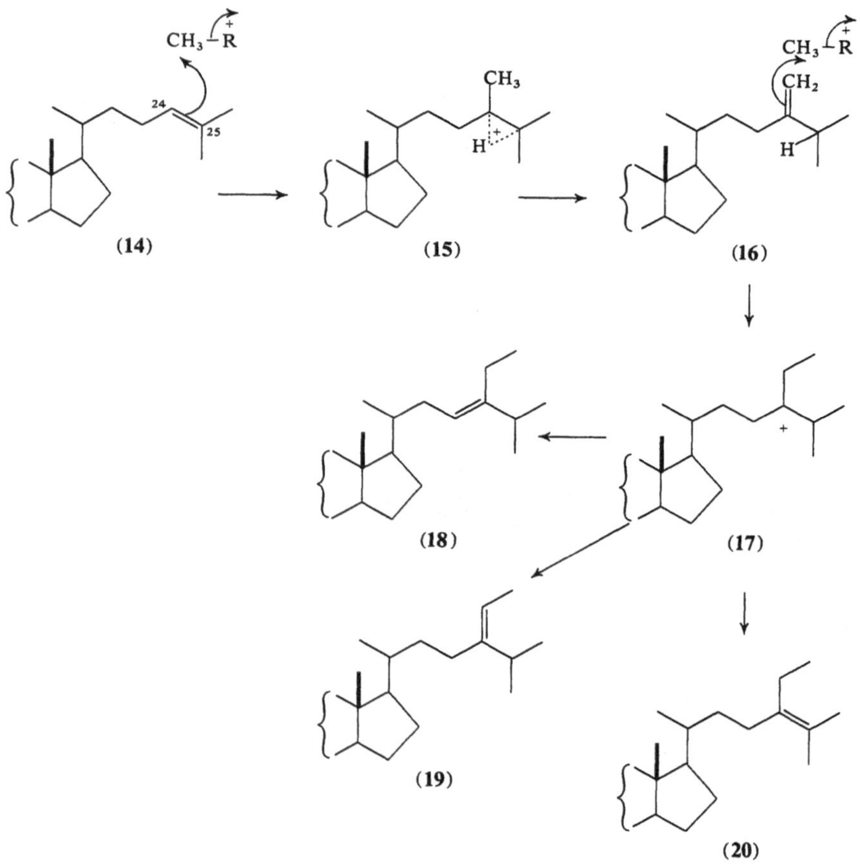

3R-[2-^{14}C,(4R)-4-^{3}H$_1$]Mevalonate was incubated with tobacco cultures and labelled stigmasterol (**22**), cycloartenol (**23**) and 24-methylene-cycloartanol (**24**) acetates were isolated after acetylation. Stigmasteryl acetate (**22**) had a tritium: carbon ratio of 2:5. Since cholesterol formed from the same precursor is known to retain tritium label at C-17, C-20 and C-24, this suggests loss of tritium from C-24 in the stigmasterol acetate. This was confirmed by ozonolysis, when the 2-ethyl-3-methyl-pentanal formed was essentially devoid of tritium. Both cycloartenyl and 24-methylenecycloartanyl acetates had tritium:carbon ratios of 6:6. After exchange with base of the ketone (**25**) obtained from ozonolysis of (**24**), the ratio fell to 5:6. This indicates that tritium migrates in the normal manner from C-24 to C-25 during the first transmethylation, but is then lost in the second transmethylation which therefore presumably proceeds *via* a Δ^{24,25}-intermediate (**20**). Analogous results were obtained with

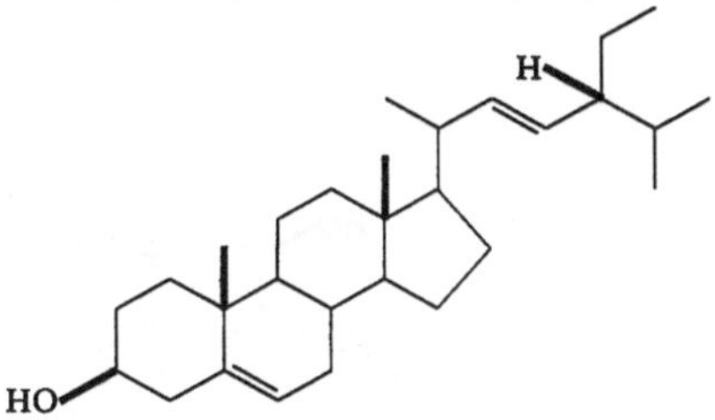

(**21**)

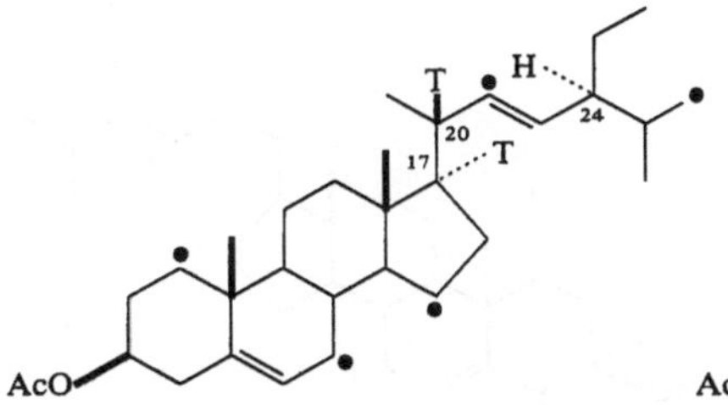

(**22**)

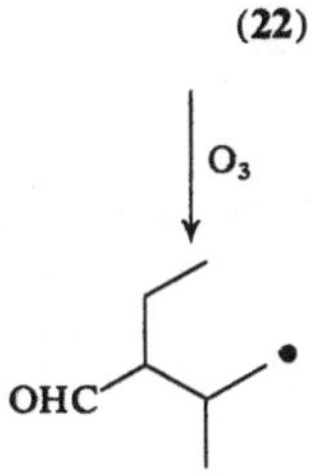

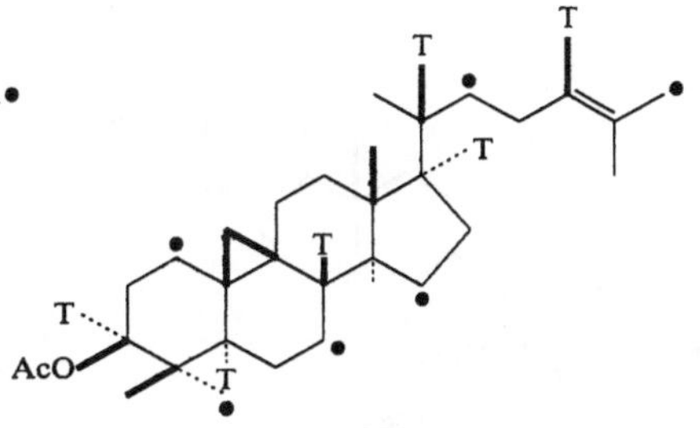

(**23**)

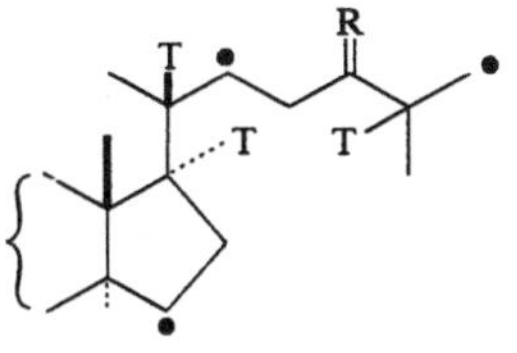

O$_3$

(**24**) R = CH$_2$
(**25**) R = O

17*

D. tokoro cultures and additionally it was shown that when [24-³H]cyclo-artenol was the precursor, the stigmasterol isolated had lost the tritium label.

3. Biosynthesis of Steroidal Sapogenins

Early experiments (*95*) with cultures of *Dioscorea deltoidea* had shown that [4-¹⁴C]- and [26-¹⁴C]cholesterol were incorporated to a comparable extent into diosgenin (**26**), suggesting that the cholesterol side chain is probably incorporated intact into diosgenin, a matter under dispute at that time.

[4-¹⁴C-22,23-³H₄]Sitosterol (**26a**) was subsequently incorporated (*97*) into diosgenin by the same culture with loss of half the tritium label, thus excluding a Δ²²,²³-intermediate to diosgenin. The sequence of oxygen

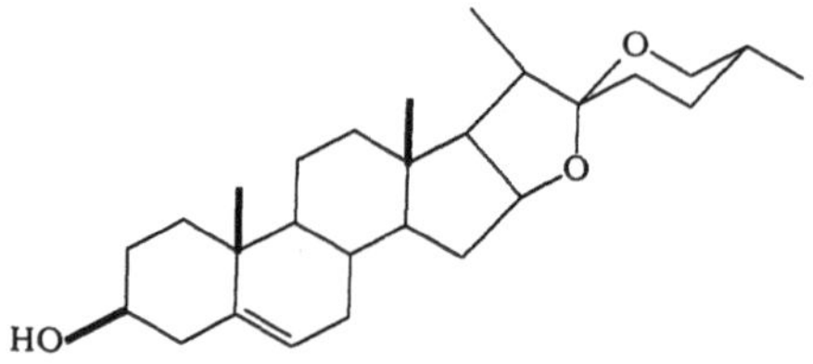

(**26**)

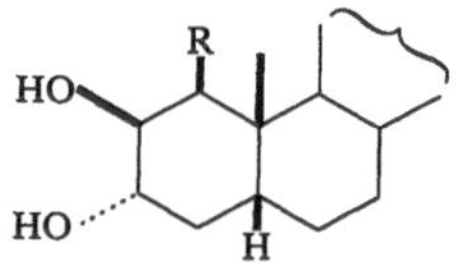

(**27**) R = H
(**28**) R = OH

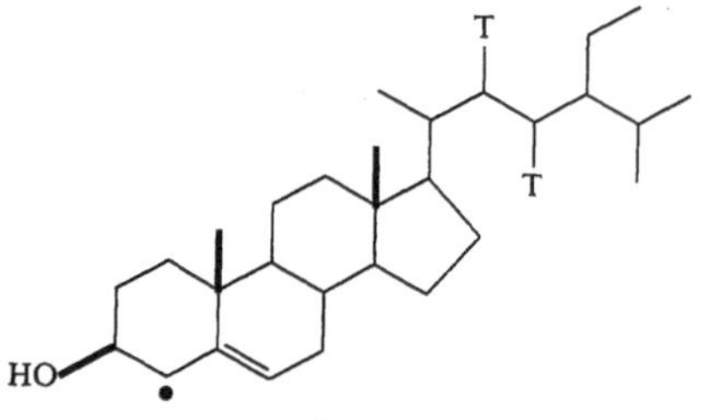

(**26a**)

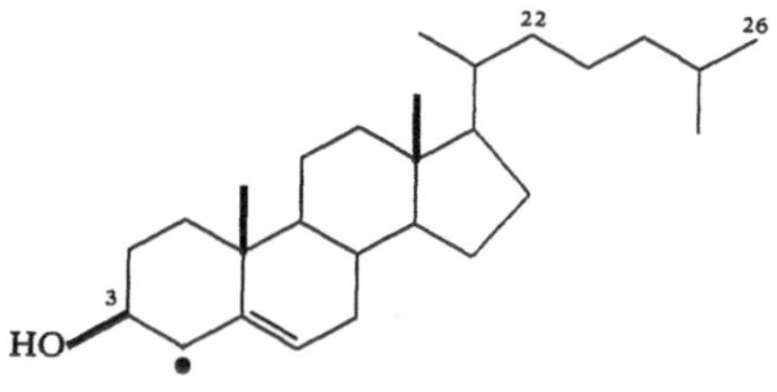

(**29**)

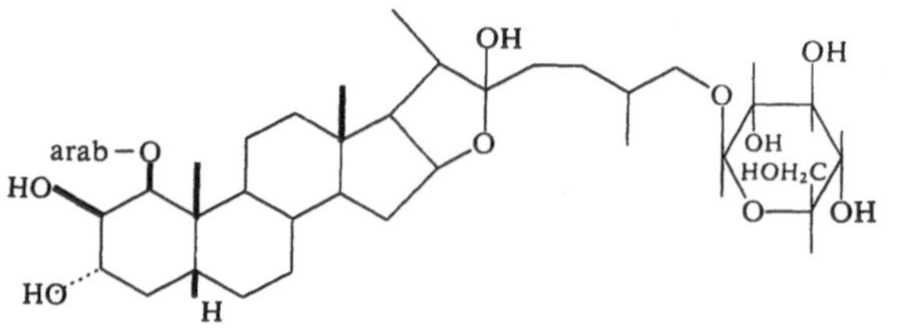

(**30**)

introduction at C-16, C-22 and C-26 remained problematical. The first step in the cyclisation of the steroid side chain may involve oxygenation at C-26 for which there is independent evidence. But direct hydroxylation at C-22 *via* a mixed-function oxidase is an alternative possibility.

Japanese work (*107*) with tissue cultures of *Dioscorea tokoro* has thrown further light on these questions. These cultures synthesize diosgenin (**26**), yonogenin (**27**) and tokorogenin (**28**), which are also produced by the intact plant. In these experiments the following labelled precursors were administered to the cultures and the total radioactivity incorporated from each precursor into diosgenin, yonogenin and tokorogenin determined: [24-^{3}H$_1$]cycloartenol, [4-^{14}C]cholesterol (**29**) [26-^{3}H$_1$]3,16,26-trihydroxycholest-5-ene, and [16-^{3}H$_1$,22-^{3}H$_1$]3,16,22,26-tetrahydroxycholest-5-ene. The results show that (a) cycloartenol is a key presursor of sapogenins; (b) hydroxylation of the side chain of cholesterol precedes ring-A hydroxylation in the biosynthesis of yonogenin and tokorogenin. The following biosynthetic sequence is suggested by these results: mevalonate, cycloartenol, cholesterol, 3,16,26-trihydroxycholest-5-ene, 3,16,22,26-tetrahydroxycholest-5-ene, diosgenin, [yonogenin], tokorogenin.

More recently TOMITA and UOMORI have isolated (*109*) from the same culture prototokoronin (**30**) and shown that labelled yonogenin and diosgenin were incorporated into it by the culture.

4. Biosynthesis of Triterpenoids

TOMITA and his colleagues have recently used tissue cultures of *Isodon japonicus* to investigate a number of points related to triterpenoid biosynthesis.

They were able to show (*102*) that while maslinic acid (**31**) formed from 3R-[2-^{14}C,(4R)-4-^{3}H$_1$]mevalonate retained all the tritium label, 3-epi-maslinic acid (**32**) lost one-sixth of the tritium label. Moreover, [^{14}C$_6$]-maslinic acid biosynthesized by the cultures from [2-^{14}C]mevalonate,

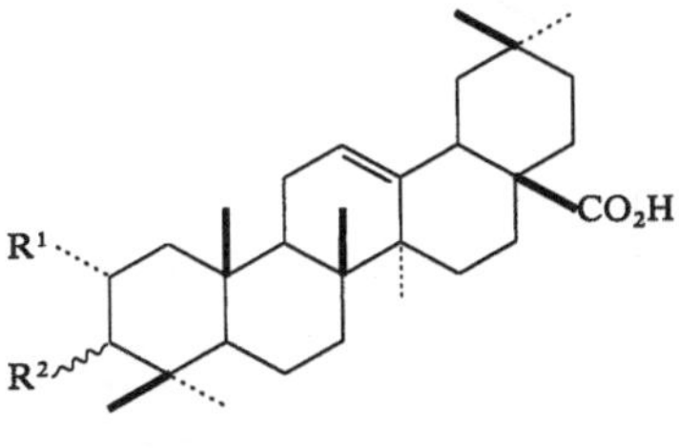

(**31**) R^1 = OH; R^2 = βOH
(**32**) R^1 = OH; R^2 = αOH
(**33**) R^1 = H; R^2 = βOH

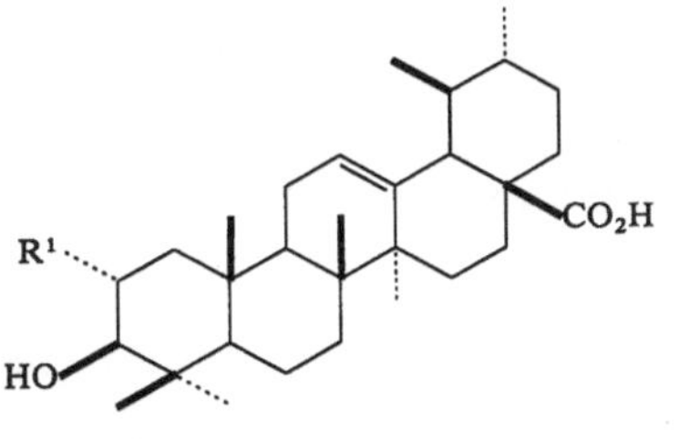

(**38**) R^1 = H
(**39**) R^1 = OH

was converted by them (0.35%) into 3-epi-maslinic acid, suggesting that
the 3-epi-acid is formed *via* the 3-ketone or its equivalent. Previous
suggestions (*24, 69*) that 3α-hydroxy triterpenoids may arise directly from
cyclisation of (3R)-2,3-oxidosqualene folded in a boat-chair-chair conformation are not borne out in this instance.

In further experiments (*104*) with the *Isodon* culture, Tomita and his
colleagues have demonstrated that the high incorporations of precursors
obtainable with tissue cultures make it possible to determine the labelling
patterns in secondary metabolites of higher plants by ^{13}C NMR spectroscopy. Incorporation of [4-^{13}C]mevalonate into oleanolic (**33**), maslinic
(**31**) and 3-epi-maslinic (**32**) acids showed in the ^{13}C NMR spectra of
the derived methyl esters enrichment at C-3, C-5, C-9, C-13, C-18, and
C-19, in accord with Ruzicka's proposal for the formation of β-amyrin
from squalene, as indicated in Scheme 2 [(**34**)→(**35**)→(**36**)→(**37**)].

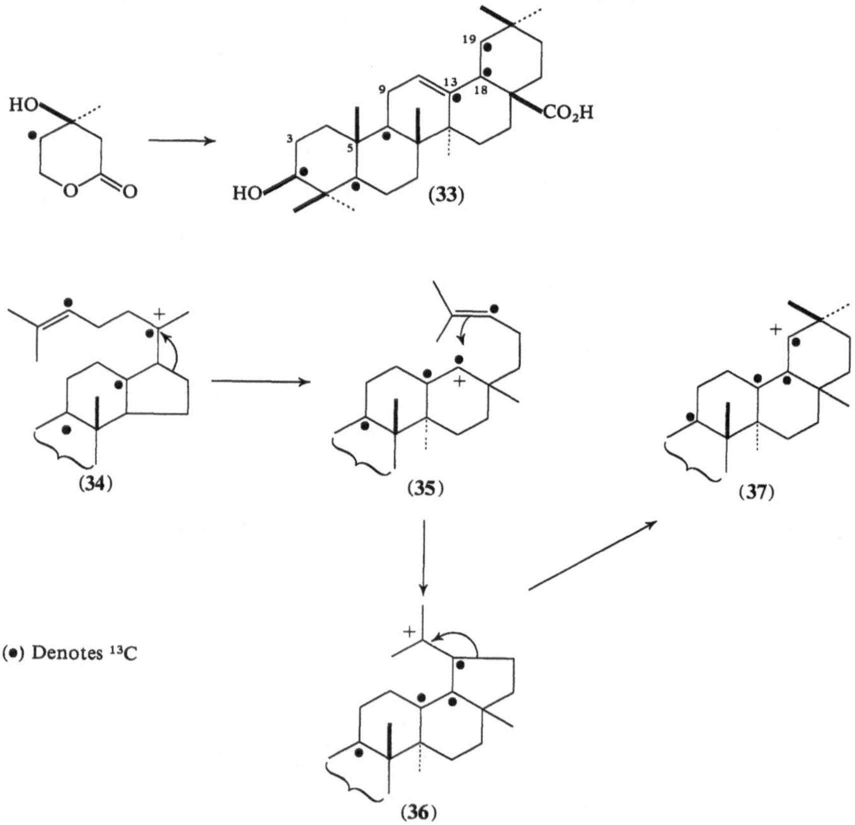

(•) Denotes ^{13}C

Scheme 2. Biosynthetic pathway to oleananes in *isodon japonicus* cultures

It was not possible from the ^{13}C NMR spectra of the methyl esters of ursolic (**38**) and 2α-hydroxy-ursolic (**39**) acids, produced from [4-^{13}C]-mevalonate in the same experiment, to distinguish between the alternative pathways (A) and (B) for formation of the ursane skeleton, since the critical signals for C-19 and C-20 in the methyl esters of (**38**) and (**39**) could not be assigned with certainty (Scheme 3).

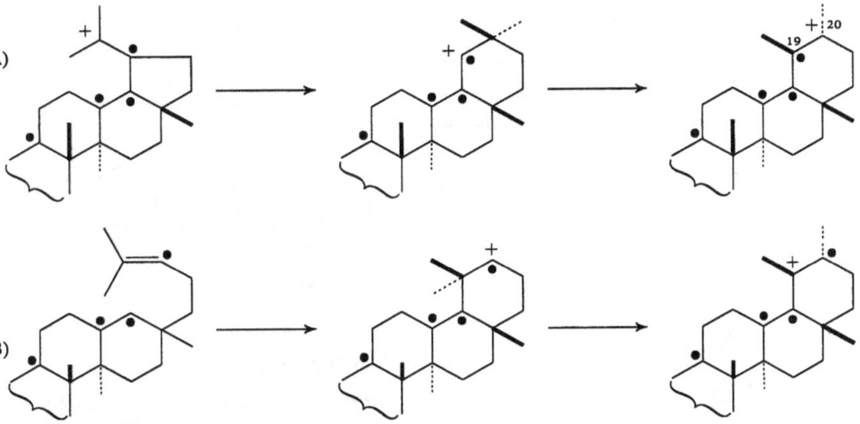

(•) Denotes [13]C

Scheme 3. Alternative biosynthetic pathways to ursanes

Returning to the problem (*103, 105*) of ursane biosynthesis, TOMITA and his colleagues subsequently employed [1,2-[13]C$_2$]acetate. (For a further example see the biosynthesis of γ-bisabolene below). The ^{13}C NMR spectra of the methyl esters of ursolic (**38**) and 2α-hydroxy-ursolic (**39**) acids secured in these experiments, showed eighteen doublets and twelve singlets. Singlet signals particularly for C-19, C-20, C-21, C-29 and C-30 are in accord with the biosynthetic pathway depicted in Scheme 4. An intermediate of type (**40**) is therefore excluded from ursane bio-synthesis which thus proceeds by route (A) and not route (B) (Scheme 3). Thus ursanes and oleananes are biosynthesized *via* a common inter-mediate (**37**) = (**41**). The ^{13}C spectra of methyl oleanolate, methyl mas-linate and methyl 3-epi-maslinate also obtained from the experiment with doubly labelled acetate support the conclusions previously reached for oleanane biosynthesis.

5. Isomerisation of trans,trans- and cis,trans-Farnesol and the Biosynthesis of γ-Bisabolene

Tissue cultures of *Andrographis paniculata* synthesize (*1*) three sesquiterpenoid lactones, paniculides A (**42**), B (**43**) and C (**44**). It is

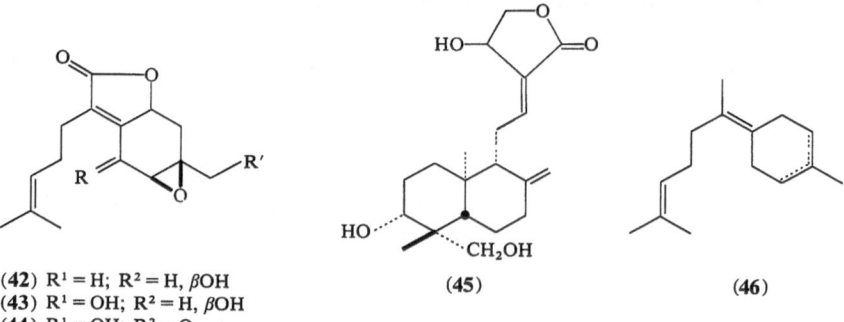

CH_3——CO_2H →

(41)

(40)

(•) Denotes uncoupled ^{13}C atom

(——) Denotes two coupled ^{13}C atoms

Scheme 4. Biosynthetic pathway to ursanes in *isodon japonicus* cultures

an unusual feature of this tissue that these or related substances are not found in the intact plant. Neither, on the other hand, does the tissue culture produce the diterpenoids characteristic of the intact plant and exemplified by andrographolide (**45**). On the reasonable assumption that the paniculides must be formed from *cis,trans*- or *cis, cis*-farnesol *via* γ-bisabolene (**46**), a search was made for these potential intermediates.

(**42**) $R^1 = H$; $R^2 = H$, βOH
(**43**) $R^1 = OH$; $R^2 = H$, βOH
(**44**) $R^1 = OH$; $R^2 = O$

(**45**)

(**46**)

While they could not be detected in the suspension cultures, a cell-free extract prepared from the callus incorporated [2-^{14}C]mevalonate into *trans,trans*- and *cis,trans*-farnesols and into γ-bisabolene.

The mechanism of the *trans,trans*- to *cis,trans*-farnesol isomerisation was studied first (*75, 76*). When [4,8,12-^{14}C$_3$,1-^{3}H$_2$]*trans,trans*-farnesol was isomerised to *cis,trans*-farnesol, 50% of the tritium label was lost; the same loss occurred on isomerisation in the opposite direction. These results would support an isomerisation mechanism *via* aldehydes. The stereochemistry of hydrogen loss from C-1 of farnesol during this iso-merisation was next investigated with [4,8,12-^{14}C$_3$]*trans,trans*- and *cis,trans*-farnesols (biosynthesized by the tissue from [2-^{14}C]mevalonate), each tritium-labelled at C-1 in either the *pro*-R or *pro*-S position. This produced the unexpected result that the pro-1S hydrogen is exchanged in the conversion of *trans,trans*- into *cis,trans*-farnesol but the pro-1R hydrogen in the conversion of *cis,trans*- to *trans,trans*-farnesol [(47) ⇌ (48)].

The *Andrographis* cultures were next used (*74*) to investigate the bio-synthesis of γ-bisabolene, a key intermediate in the formation of a variety of sesquiterpenoid skeletons found in Nature. The basic questions to be settled are: (a) whether the ring carbon atom derived from C-2 of mevalonate is anti (**49**) or syn (**50**) to the side chain and (b) whether 2-*cis*,6-*trans*- (**51**) or 2-*cis*,6-*cis*- (**52**) farnesyl pyrophosphate is the intermediate.

On the simplest basis, γ-bisabolene (**49**) should lead to paniculide B (**53**) in which C-9 is labelled from C-2 of mevalonic acid, whereas in paniculide B (**54**), derived from γ-bisabolene (**50**), C-11 should be so labelled. The ^{13}C NMR spectra of paniculide B enriched by administering [1,2-^{13}C$_2$]acetate to the cultures left no doubt that the labelling pattern corresponded to (**53**) and not (**54**), suggesting γ-bisabolene (**49**) rather than (**50**) as intermediate (see Scheme 5). In support of this conclusion, γ-bisabolene biosynthesized by the cell-free system from [2-^{14}C]meva-lonate corresponded on radio-glc comparison to authentic Z- (**49**) but not E- (**50**) γ-bisabolene. It was also found that [4,8,12-^{14}C$_3$]*cis,trans*-farnesol was incorporated (1.2%) into γ-bisabolene and its pyrophosphate (1.6%)

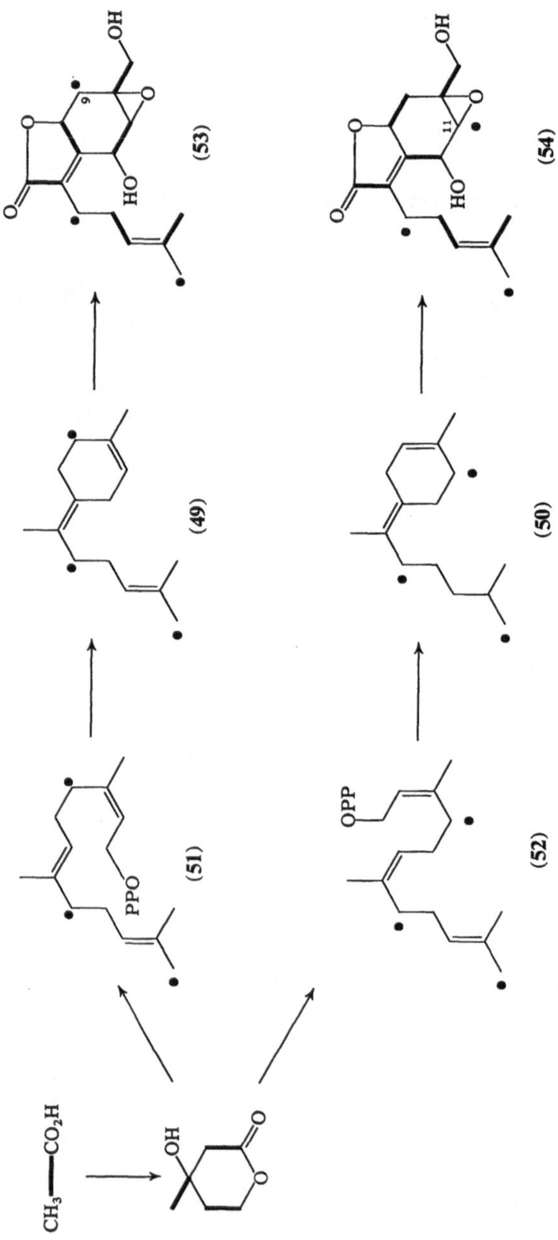

Scheme 5. Biosynthesis of γ-bisabolene and paniculides in *andrographis paniculata* cultures

(━━) Denotes two coupled ^{13}C atoms   (•) Denotes uncoupled ^{13}C atoms.

into paniculide B; [2-^{14}C]*cis,cis*-farnesol was not incorporated into γ-bisabolene. These experiments show that *Andrographis* tissue cultures convert mevalonate into Z-γ-bisabolene (**49**) *via cis,trans*-farnesyl pyrophosphate and that this is probably converted into the paniculides without isomerisation about the central double bond.

6. Biosynthesis of ent-Kaurene

ent-Kaurene (**55**) is a key intermediate in the biosynthesis of gibberellins. YAFIN and SCHECHTER have compared (*114*) its formation in germinating tomato seeds on the one hand and cell suspension cultures of tobacco and tomato *(Solanum lycopersicum)* on the other. It was their intention to compare *ent*-kaurene formation in a rapidly differentiating system (germinating seeds) in which gibberellin synthesis is maximal and in a non-differentiating system. Incubation of extracts derived from the two cell-cultures with [2-^{14}C]mevalonate or [1-^{14}C]isopentenyl pyrophosphate afforded labelled *trans*-farnesol (**56**) and *trans*-geranylgeraniol (**57**) after alkaline phosphatase hydrolysis. [^{14}C]Geranylgeranyl pyrophosphate (**57**) did not label *ent*-kaurene with these preparations but [^{14}C]copalyl pyrophosphate (**58**) did. Cell-free extracts derived from germinating tomato seeds, on the other hand, catalyzed formation of labelled *ent*-kaurene from [^{14}C]isopentenyl pyrophosphate, [^{14}C]mevalonate, [^{14}C]*trans*-geranylgeranyl pyrophosphate and [^{14}C]copalyl pyrophosphate. It appears therefore that *ent*-kaurene and therefore gibberellin biosynthesis is blocked at the stage of geranylgeranyl pyrophosphate to copalyl pyrophosphate cyclisation (see Scheme 6).

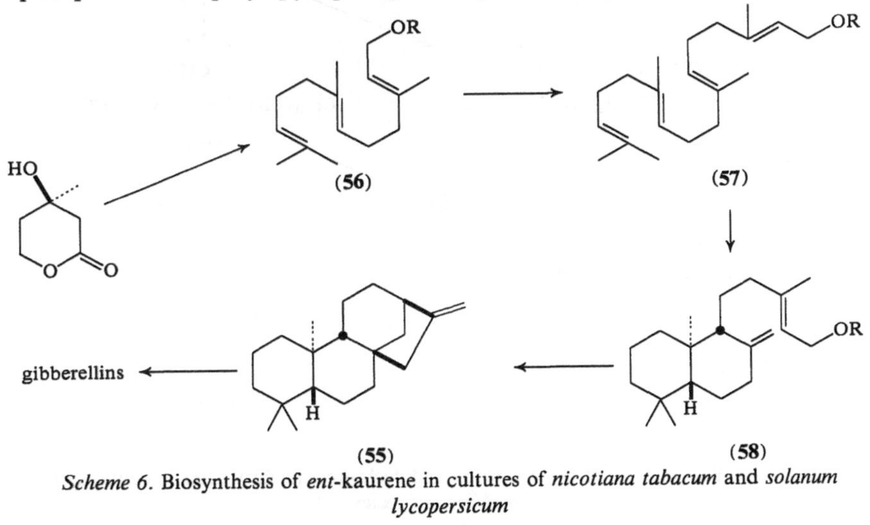

Scheme 6. Biosynthesis of *ent*-kaurene in cultures of *nicotiana tabacum* and *solanum lycopersicum*

IV. Biosynthesis of Polyketides

1. Biosynthesis of Cyclopropane and Cyclopropene Fatty Acids in Higher Plants

The biosynthesis of cyclopropane and cyclopropene fatty acids has been investigated (115) in immature seeds, leaves and callus tissue cultures of several species of *Malvaceae*. The ring methylene group was found to arise from methionine. With labelled L-[^{14}C]methionine and [2-^{14}C]acetate, the change of radio-activity with time in the precursors suggested that the biosynthetic pathway involved initial formation of dihydrosterculic acid (59) from oleic acid (60) with subsequent desaturation to sterculic acid (61) and α-oxidation to malvalic (62) and dihydromalvalic (63) acids. Direct evidence in favour of this pathway was provided by the conversion of [1-^{14}C]oleic acid into dihydrosterculic and sterculic acids and by desaturation of [1-^{14}C]dihydrosterculic acid to sterculic acid, the first time this process has been demonstrated in higher plants. Stearolic acid (64), which in principle might methylenate directly to sterculic acid, was not metabolized. The presence of an active fatty acid α-oxidation sytem was demonstrated in the callus (see Scheme 7).

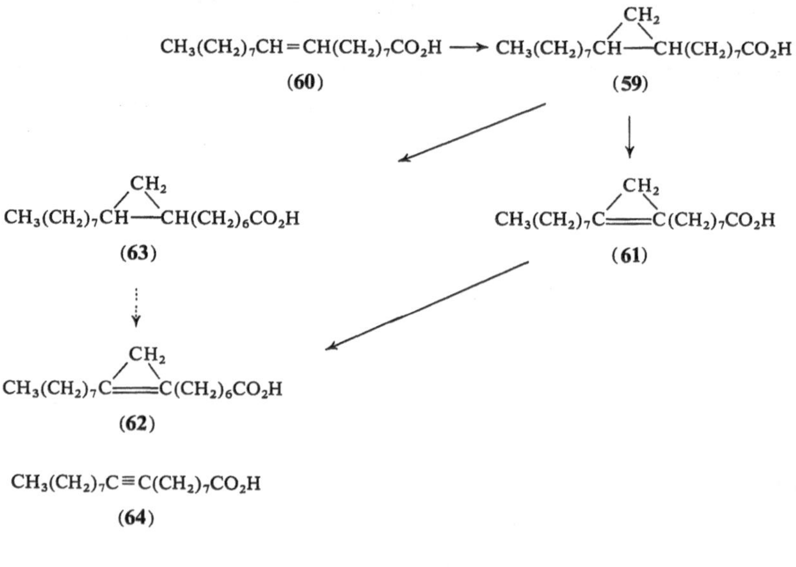

$$CH_3(CH_2)_7CH=CH(CH_2)_7CO_2H \longrightarrow CH_3(CH_2)_7\overset{\displaystyle CH_2}{\overset{\diagup\diagdown}{CH}}\!\!-\!\!CH(CH_2)_7CO_2H$$

(60)   (59)

$$CH_3(CH_2)_7\overset{\displaystyle CH_2}{\overset{\diagup\diagdown}{CH}}\!\!-\!\!CH(CH_2)_6CO_2H$$

(63)

$$CH_3(CH_2)_7\overset{\displaystyle CH_2}{\overset{\diagup\diagdown}{C}}\!\!=\!\!=\!\!C(CH_2)_7CO_2H$$

(61)

$$CH_3(CH_2)_7\overset{\displaystyle CH_2}{\overset{\diagup\diagdown}{C}}\!\!=\!\!=\!\!C(CH_2)_6CO_2H$$

(62)

$$CH_3(CH_2)_7C\equiv C(CH_2)_7CO_2H$$

(64)

Scheme 7. Biosynthesis of cyclopropane and cyclopropene fatty acids in cultures of *malvaceae* species

V. Metabolism of Aromatic Plant Constituents

Both the formation and catabolism of a range of aromatic plant constituents have recently been studied with plant tissue cultures. Much of this work is complemented by parallel work with intact plants or in some cases, micro-organisms.

The special advantages of plant tissue cultures for detailed bio-synthetic studies at the enzymic level are well illustrated by the work of GRISEBACH and HAHLBROCK on anthocyanin biosynthesis and of ZENK's group on the homogentisate pathway for tyrosine degradation. For catabolic studies in particular, tissue culture offers important advantages discussed below.

An informative review (29) dealing with the degradation of aromatic compounds in plants has appeared recently.

1. Flavonoid Biosynthesis

HAHLBROCK, GRISEBACH and their colleagues are engaged in a detailed study of the regulation of flavonoid biosynthesis in cell suspension cultures of *Petroselenium hortense* (parsley). This is a particularly apt system for a study of enzymic interdependence in the biosynthetic pathway to flavone glycosides, since only two flavonoids, apigenin (71) and graveobioside B (75) are produced and nine of the eleven postulated enzymes in the biosynthetic pathway have been identified. The enzymes responsible for steps a, b, c, e, g, h, i, and k (see Scheme 8) have been investigated (50). They can be divided into two groups on the basis of their response to light. The first group (mediating steps a, b and c) comprises three enzymes acting on the substrates of the phenylpropanoid type [phenylalanine (65), cinnamic acid (66) and p-coumaric acid (67)] which are not restricted to flavonoid biosynthesis. The second group (me-diating steps e, g, h, i and k) includes the enzymes involved in the for-mation of flavonoid compounds. It is tempting to infer from these results that in this system flavonoid biosynthesis is governed by two or more differently regulated sequences of enzymic steps.

Two of the enzymes of this pathway have been further characterized. *p*-Coumaric acid (67) was the most effective substrate for a preparation from parsley cell suspension cultures (46) that catalyzed formation of coenzyme A thiol esters. Its activity was markedly enhanced by light. An *O*-methyl transferase, presumably that involved in step k, has been isolated (44) from the parsley culture and purified 82-fold. Only o-di-hydric phenols are effective substrates for this enzyme and only the meta position of 1-substituted-3,4-dihydric phenols is methylated.

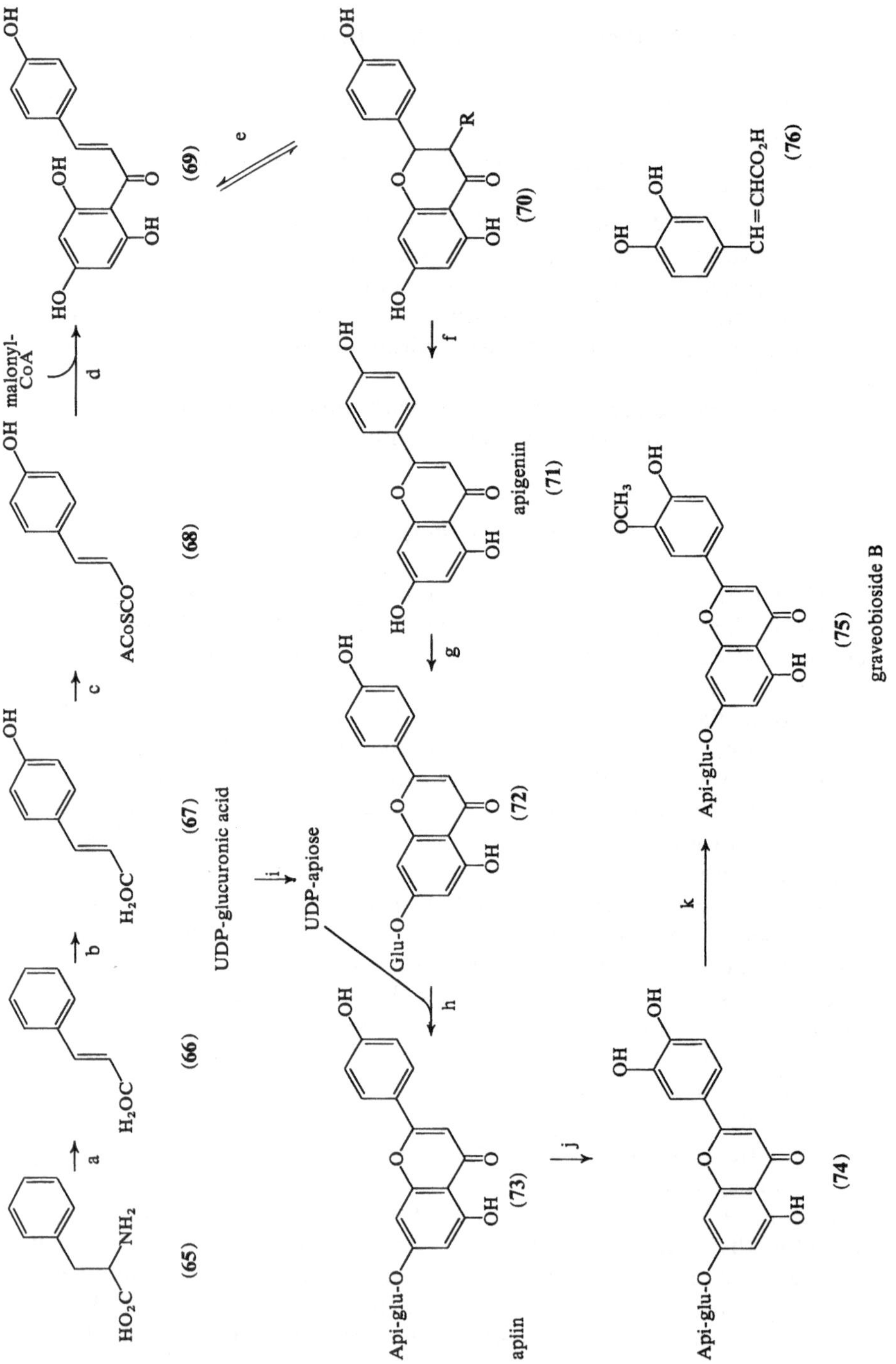

Scheme 8. Flavonoid biosynthesis in *petroselenium hortense* cultures

Luteolin (5,7,3′,4′-tetrahydroxyflavone) and its 7-*O*-glucoside were the best substratès. Again, the enzymic activity was markedly affected by light.

Attempts to establish (*48*) whether the B-ring oxygenation pattern of flavonoids is determined by hydroxylation at the stage of *p*-coumaric acid or at a later point, led to isolation of a phenolase preparation from parsley cultures that hydroxylated *p*-coumaric acid to caffeic acid (**76**) but acted only weakly on naringenin (**70**; R = H) and not at all on dihydrokaempferol (**70**; R = OH) or apigenin (**71**). Its activity was not affected by illumination and it would appear not to be specifically involved in flavonoid biosynthesis.

2. Anthocyanidin Biosynthesis

The anthocyanidins are polyhydroxyflavylium salts, exemplified by cyanidin (**77**). Their glycosides, the anthocyanins, are mainly responsible for the conspicuous blue, violet and red colours of flowers and fruits. Cyanidin biosynthesis has been studied in some detail by the group of Grisebach and Hahlbrock with cell suspension cultures of *Happlopappus gracilis*. Anthocyanins were produced efficiently (*47*) in blue light, dihydrokaempferol (**78**) being a more effective precursor than L-phenyl-

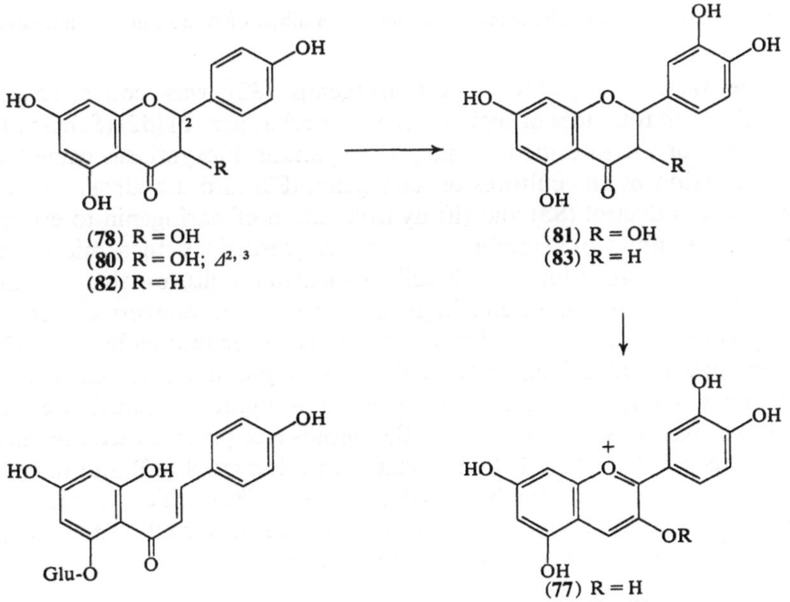

(**78**) R = OH
(**80**) R = OH; Δ²,³
(**82**) R = H

(**81**) R = OH
(**83**) R = H

(**79**)

(**77**) R = H
(**84**) R = Glu

Scheme 9. Anthocyanidin biosynthesis in *happlopappus gracilis* cultures

alanine, 4,2',4',6'-tetrahydroxychalcone-2'-glucoside (79) or kaempferol
(80). Incorporation times were notably shorter and incorporation rates
higher than in parallel experiments with *Happlopappus* seedlings.

Efficient conversion of dihydroflavanols into anthocyanins was
further demonstrated (45) by incorporation with a low dilution factor
of dihydroquercetin (81) into cyanidin (77). Alternative mechanisms have
been proposed for the conversion of dihydroflavanols into anthocyanidins
[e.g. (81) to (77)], that involve either [2,4]- or [3,4]hydride shifts, (a)
or (b).

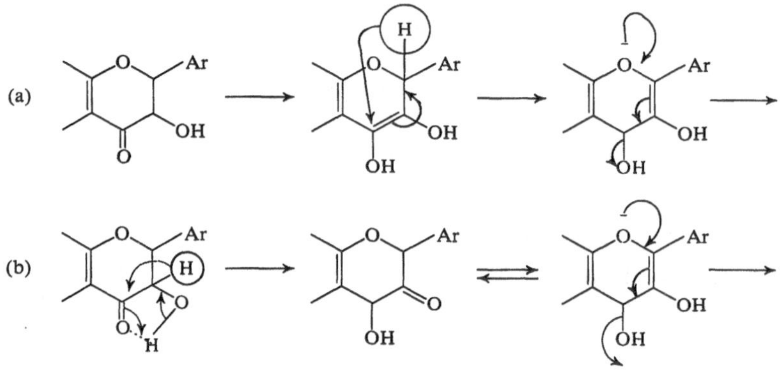

Scheme 10. Alternative mechanisms for conversion of dihydroflavanols into anthocyanidins

The fact that [2-^{14}C,2-^{3}H$_1$]naringenin (82) was converted into
cyanidin with total loss of tritium favours mechanism (b) (62) (Scheme 10).
Evidence for the intermediate stages in cyanidin biosynthesis came from
(i) conversion by the cultures of naringenin (82) into dihydrokaempferol
(78) and eriodyctiol (83) and (ii) hydroxylation of naringenin to eriodyctiol and of dihydrokaempferol to dihydroquercetin with a microsomal
preparation from illuminated cell suspension cultures that required
NADPH and oxygen. Interestingly this preparation converted cinnamic
into *p*-coumaric acid but did not convert the latter into caffeic acid (76).
Other enzymes identified in the culture were *p*-coumarate: CoA ligase,
flavanone synthetase and anthocyanidin-3-O-glucosyltransferase. The
proposed pathway follows that for flavonoids (see previous section) as far
as naringenin (82). This is hydroxylated to eriodyctiol (83) and dihydrokaempferol (78) and finally dihydroquercetin (81) which is converted
into cyanidin (77) and its 3-O-glucoside (84) (see Scheme 9). The sequence
may operate partly or completely at membranes.

3. Retrochalcone Biosynthesis

Three new substances have been isolated from *Glycyrrhiza* species: licochalcone A (**85**), licochalcone B (**86**) and echinatin (**87**), the latter also

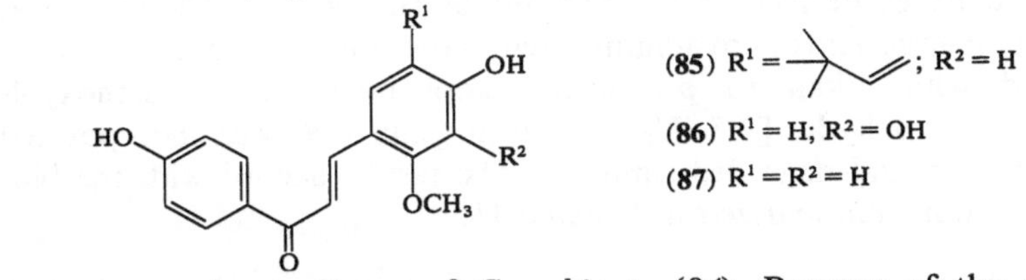

(**85**) $R^1 = $ —⟨ ; $R^2 = H$

(**86**) $R^1 = H$; $R^2 = OH$

(**87**) $R^1 = R^2 = H$

from suspension cultures of *G. echinata* (*84*). Because of the unusual oxygenation pattern compared with normal chalcones, it was suspected that these substances are formed by formal inversion of the enone system

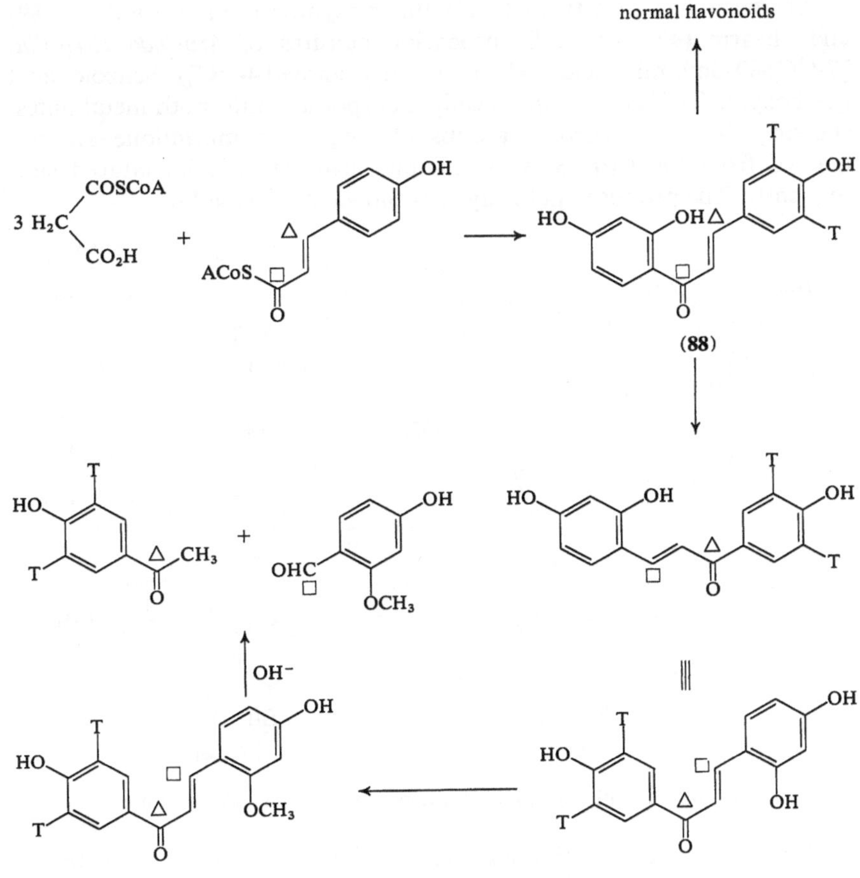

Scheme 11. Retrochalcone biosynthesis in *glycyrrhiza echinata* cultures

of normal chalcones, which co-occur in both the roots and tissue cultures. The members of the new group were therefore named "retrochalcones".

The biosynthesis of echinatin (87) was studied (85) with callus suspension cultures derived from seedlings of *G. echinata*. [3-^{14}C]- And [1-^{14}C]cinnamic acids were administered to the cultures and the echinatin degraded with alkali to *p*-hydroxyacetophenone and 2-methoxy-4-hydroxybenzaldehyde. [3,5-^{3}H$_2$]Isoliquiritigenin (88) was incorporated into echinatin and degraded similarly. The results accord with the biosynthetic route summarized in Scheme 11.

4. Anthraquinone Biosynthesis

LEISTNER has investigated (65) the biosynthesis of morindone (89) and alizarin (90) with cell suspension cultures of *Morinda citrifolia*. [7-^{14}C]-D-Shikimic acid (91) and *o*-(succinoyl-[4-^{14}C]) benzoic acid [carboxyl-^{14}C] (92) were specifically incorporated into both metabolites, showing that the hydroxyl groups of ring A in morindone are not derived from the hydroxyls of shikimic acid, but are introduced subsequently. The proposed pathway is as shown in Scheme 12.

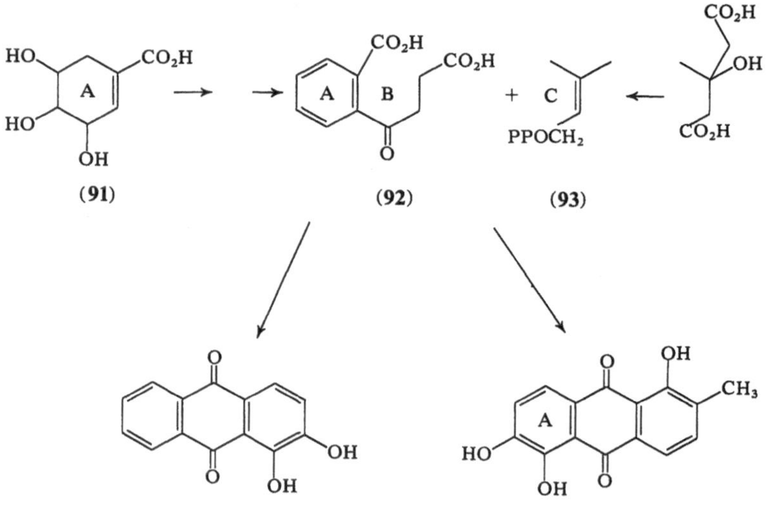

Scheme 12. Anthraquinone biosynthesis in *morinda citrifolia* cultures

Chorismate [from shikimate (91)] and α-ketoglutarate condense to afford o-succinoylbenzoic acid (92) which cyclises to a naphthalene of unknown structure (1,4-dihydroxy-2-naphthoic acid?). Condensation with γγ-dimethylallyl pyrophosphate (93) from mevalonate, decarboxylation and ring closure then lead, *via* anthraquinone, to alizarin (90) and morindone (89). The actual incorporation of labelled mevalonate with the tissues was poor and its involvement is based on parallel work (64) with roots of *Rubia tinctorum*.

FURUYA and his colleagues have isolated (36) from callus cultures of *Digitalis lanata* four 3-methylanthraquinones, pachybasin (94), 3-methylquinizarin (95), 3-methylpurpurin (96) and 3-methylalizarin (97), whose structures suggest an analogous involvement of mevalonate in their biosynthesis.

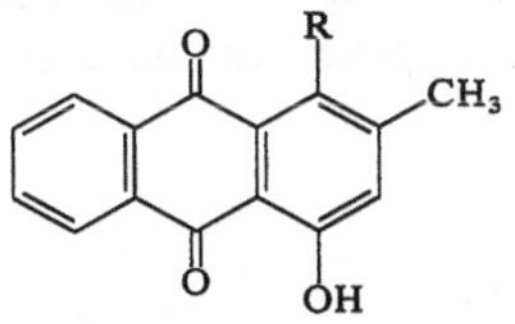

(94) R = H
(95) R = OH

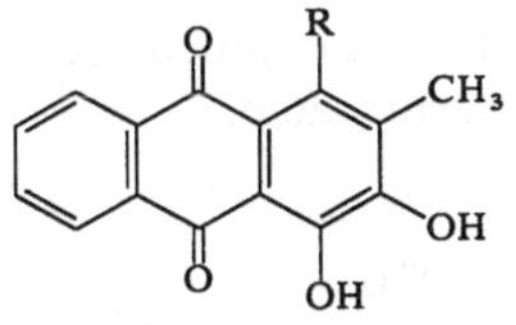

(96) R = OH
(97) R = H

The following additional anthraquinones have been detected in plant tissue cultures:

Rheum palamatum — chrysophanol (98) and emodin (99) (34); *Cassia tora* — chrysophanol, emodin and physcion (100) (98); *Cassia senna* — chrysophanol, physcion, emodin, rhein (101) and aloe emodin (102) (80).

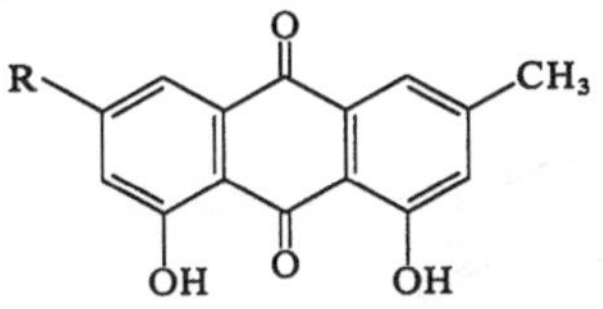

(98) R = H
(99) R = OH
(100) R = OCH₃

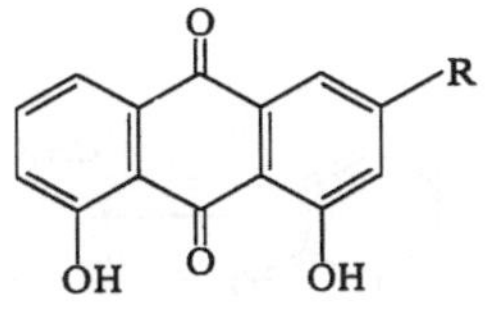

(101) R = CO₂H
(102) R = CH₂OH

5. Coumarin Biosynthesis

Scopoletin (103) and its glucoside scopolin (104) are found in large quantities in habituated tobacco tissue cultures, chiefly during the growing period. These substances appear to be connected with various metabolic disturbances in plants. FRITIG, HIRTH and OURISSON have found (73) good incorporation of U-[14C]phenylalanine into scopoletin, cinnamic (105), *p*-coumaric (106), caffeic (107) and ferulic (108) acids in tobacco cultures. The relative specific activities of scopoletin and scopolin and particularly their change with time suggested that the aglycone is probably synthesized first. L-Phenylalanine is a thousand times more effective as a precursor than L-tyrosine. Scopoletin and scopolin metabolism in tobacco cultures has been further investigated (67) by LOEWEN-BERG. He found that specific activity of the coumarins formed from U-[^{14}C]phenylalanine reached a maximum after 18 hours. Labelled scopolin was subsequently incorporated into proteins and cell-wall fractions.

It seems that the metabolic pathway to scopoletin and scopolin in tobacco tissues (Scheme 13) may differ in detail from that in tobacco leaves, where the glucoside scopolin appears to be formed first *via* glucosido-ferulic acid.

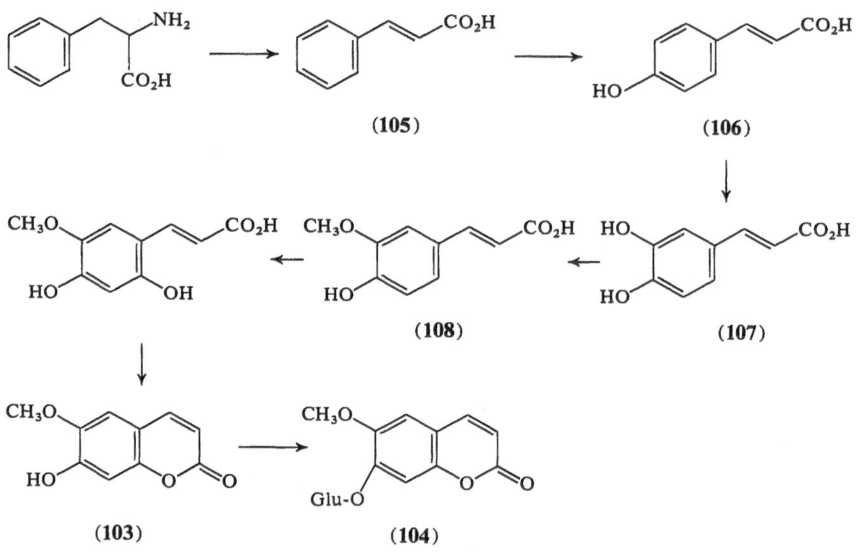

Scheme 13. Biosynthesis of scopoletin and scopolin in *nicotiana tabacum* cultures

Cell cultures of *Ruta graveolens* (garden rue) produce (87) a range of linear furanocoumarins including psoralen (109), xanthotoxin (110), bergapten (111) and isopimpinellin (112), also rutamarin (113) and a new related prenyl coumarin, rutacultin (114).

The biosynthesis of these substances by the cultures has been investigated (3) with a range of ^{14}C-labelled precursors. In contrast to [2-^{14}C]umbelliferone, [2-^{14}C]mevalonic acid was found to be only poorly

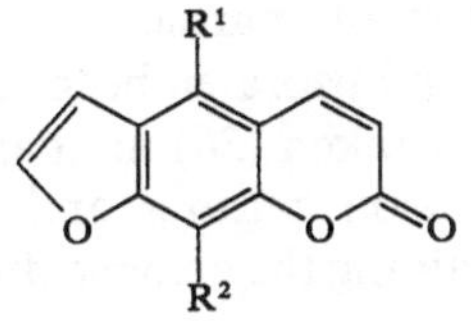

(109) R¹ = R² = H
(110) R¹ = H; R² = OCH₃
(111) R¹ = OCH₃; R² = H
(112) R¹ = R² = OCH₃

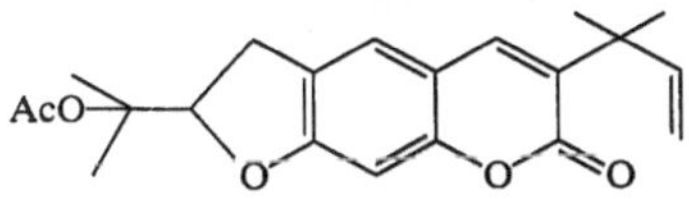

(113)

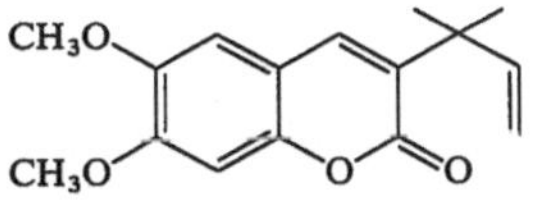

(114)

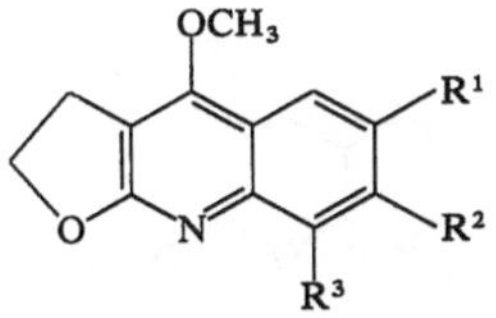

(118) R¹ = H; R² = R³ = OCH₃
(119) R¹ = R² = OCH₃; R³ = H
(120) R¹ = OCH₃; R² = R³ = H

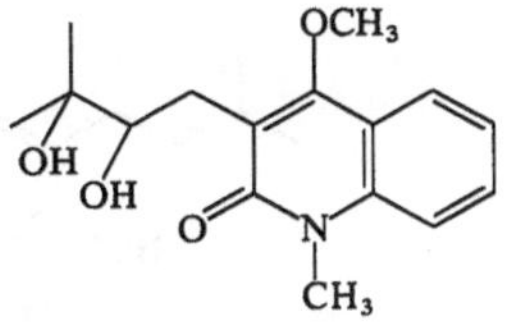

(121)

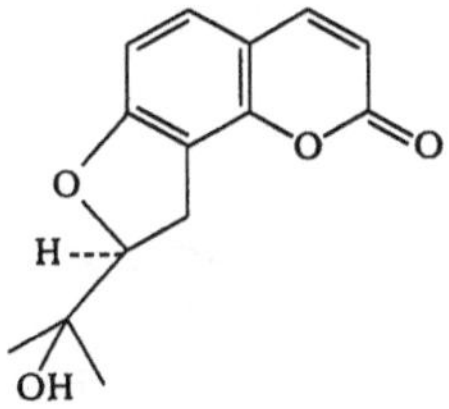

(122)

incorporated; [2-^{14}C]7-demethyl-suberosin (115) and [3-^{14}C]($\pm$)mar-mesin (116) were very good precursors, while [2-^{14}C]7-O-prenylumbelli-ferone (117) was also incorporated but possibly *via* prior ether cleavage. Psoralen was well incorporated into bergapten and xanthotoxin but not into dimethoxylated isoimpinellin. Interesting differences were noted in metabolic products between the intact plant and the derived cultures. Thus in addition to rutacultin, the cultures contained four alkaloids not found in the plant: skimmianine (118), kokusaginine (119), 6-me-thoxydictamnine (120) and edulinine (121). The results lend strong support to the pathway for linear furanocoumarin biosynthesis shown in Scheme 14. There are at present no suggestions for the loss of the elements of isopropanol in the conversion of marmesin into psoralen, nor have the details of oxygenation beyond psoralen been explored.

A further step has been taken (*30*) in characterizing this pathway by the isolation, from suspension cultures and leaves of *R. graveolens,* of a particulate fraction containing the enzyme dimethylallyl pyrophosphate:

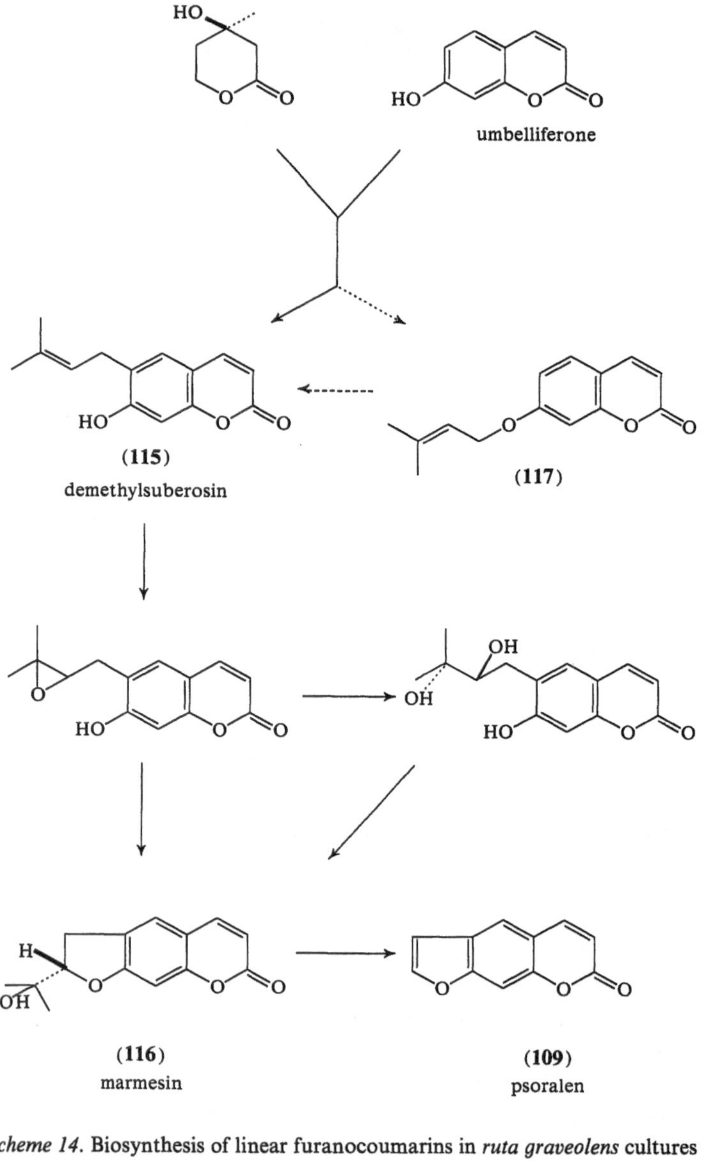

Scheme 14. Biosynthesis of linear furanocoumarins in *ruta graveolens* cultures

umbelliferone dimethylallyl transferase, which mediates the first step specifically directed towards furanocoumarin biosynthesis. In presence of dimethylallyl pyrophosphate and Mn^{2+}, umbelliferone is converted into 6-dimethylallyl umbelliferone (demethylsuberosin), an established intermediate to linear furanocoumarins, but not into 8-dimethylallyl-umbelliferone (osthenol), the intermediate to angular furanocoumarins such as columbianetin (122).

6. Catabolism of Aromatic Plant Constituents

6.1 General

It has become clear in recent years that aromatic plant constituents are not necessarily the end-products of metabolism but that they may be subject to catabolic transformations in the plant. The disadvantages of studying catabolism in intact plants are several. Most serious probably is the difficulty of working with aseptically grown and maintained plants in order to ensure that the observed catabolic pathway is mediated by the plant and not by epiphytic micro-organisms. Additional drawbacks are the low concentrations of catabolic products normally found in plants and frequently extensive binding of phenolic substrates or products to polymeric structures in differentiated plant tissues. Since plant tissue cultures demand sterile conditions for successful maintenance, this automatically removes the difficulty of correctly attributing an observed pathway. They offer in addition the possibility of working anaerobically, thus inhibiting possibly deleterious oxygen-dependent reactions. It is not surprising therefore that plant tissue cultures are making an increasing contribution in this area of plant metabolism.

ELLIS and TOWERS probably provided the first unequivocal demonstration (31) that plant tissues can degrade aromatic rings to carbon dioxide. Working with cell suspension cultures of *Ruta graveolens* and *Melitotus alba,* they found that labelled CO_2 was liberated when ring-labelled phenylalanine, cinnamic acid and tryptophan were incubated with the cultures. Shortly afterwards, BARZ and his colleagues studied (17) the degradation of various benzoic and cinnamic acids with cell suspension cultures of *Phaseolus aureus* (mung bean) and *Glycine max* (soy bean) and found that they rapidly degraded the aromatic rings of dihydroxyphenolic substrates, such as catechol, protocatechuic acid (123) and caffeic acid (124) to CO_2. ELLIS detected (27) similar catabolic activity with catechol in tissue cultures of eighteen additional plants. More recently BARZ's group have followed up (18) their earlier observation that the soy and mung bean cultures are able to decarboxylate substituted

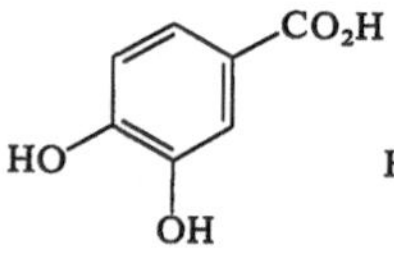

(123) 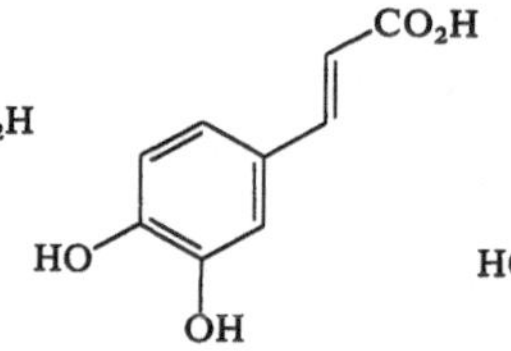　　　　　(124) 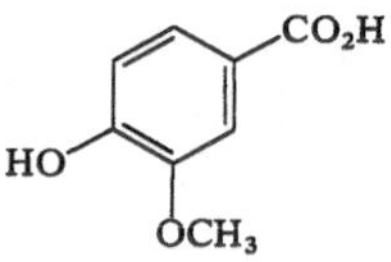　　　　　(125) 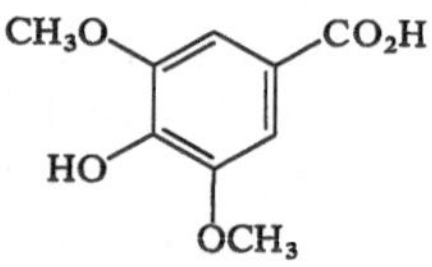　　　　　(126)

benzoic acids, specifically those bearing a *p*-hydroxyl group, such as vanillic (125) and syringic (126) acids. The expected phenolic products could not be detected with these systems and are probably subject to further degradation. These cultures were also able specifically to demethylate 4-methoxybenzoic acids. Surprisingly, however, the endogenously demethylated *p*-hydroxybenzoic acids (*p*-hydroxybenzoic and syringic acids) were not decarboxylated, while the same acids administered to the cultures were decarboxylated, implying that these functions are compartmented in the tissues.

6.2 Flavonoid Catabolism

The flavonols quercetin (127), kaempferol (128) and isorhamnetin (129) are extensively degraded (56) by cell suspension cultures of *Cicer arietinum* (55), *Phaseolus aureus*, *Glycine max* and *Petroselenium hortense*. The B-ring of the flavonol appears as the appropriate hydroxy-benzoic acid, but neither phloroglucinol nor phloroglucinol carboxylic acid, representing ring A, could be detected. All four cultures were able to

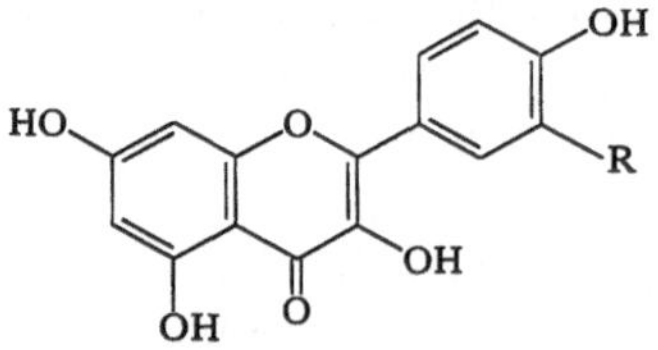

(127) R = OH
(128) R = H
(129) R = OCH$_3$

transform flavonols to 2,3-dihydroxyflavanones, and the diol (130) from datiscetin (131) was transformed by the cultures into salicylic acid. It would seem therefore that flavonols are catabolized *via* 2,3-dihydroflavanones to substituted benzoic acids representing the original ring B, as in the conversion of datiscetin to salicylic acid.

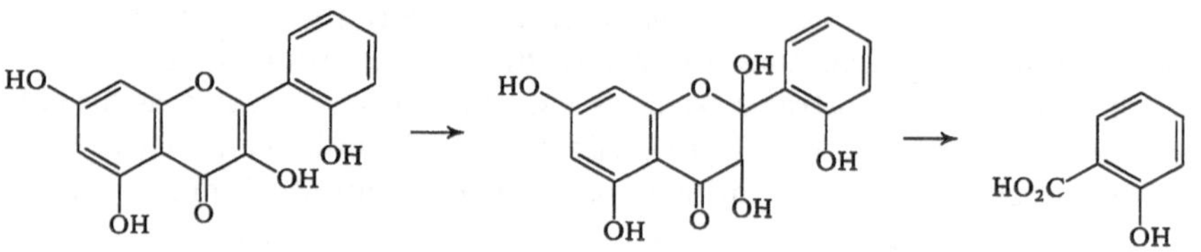

(131)　　　　　　　　　　　　(130)

BARZ and his colleagues have shown (5) that the catabolism of aurones resembles that of flavanols and chalcones in tissue cultures and in plants. Thus cell suspension cultures of *Glycine max, Phaseolus aureus, Cicer arietinum* and *Petroselenium hortense* catabolize hispidol-6-glucoside, ($[\alpha\text{-}^{14}C]$-4′,6′-dihydroxyaurone-6β-D-glucoside) (132) to $[^{14}C]p$-hydroxy-benzoic acid and labelled CO_2.

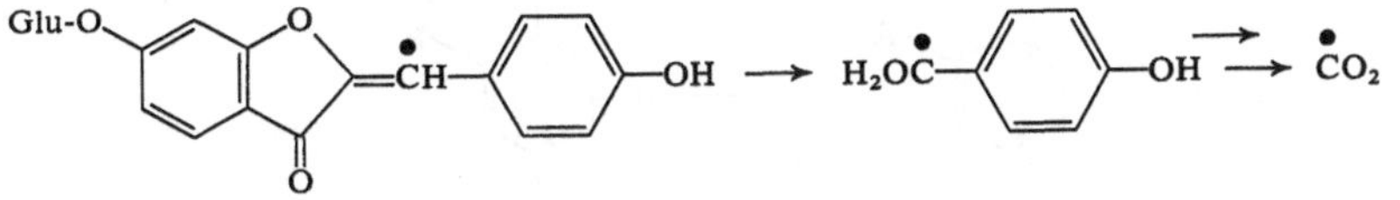

(132)

In recent studies of chalcone and isoflavone catabolism, using cultures of *Phaseolus aureus, Glycine max* and *Pisum sativum*, the group of BARZ has shown (16, 19, 60) that the A-rings of chalcones and isoflavones with either resorcinol or phloroglucinol substitution patterns are degraded to CO_2, and the B-rings of chalcones to p-coumaric and p-hydroxybenzoic acids (Scheme 15).

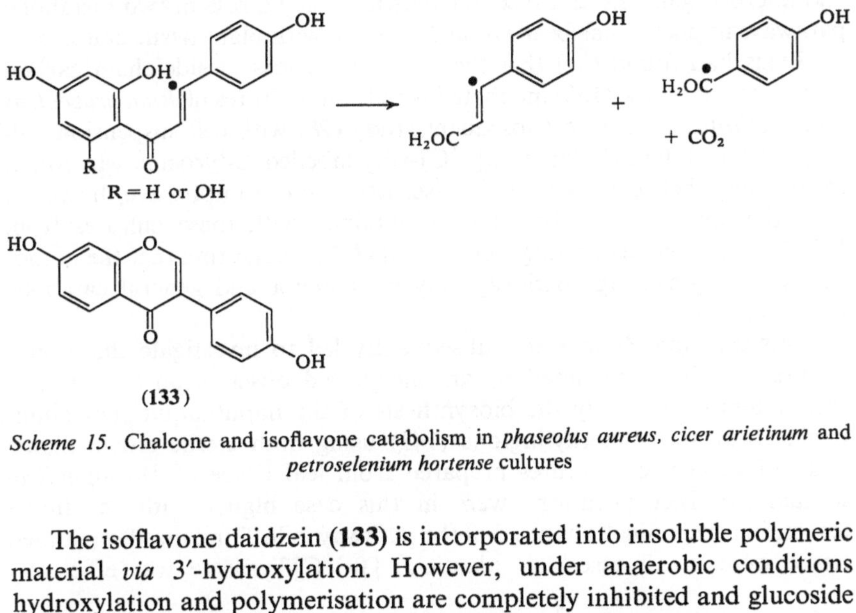

(133)

Scheme 15. Chalcone and isoflavone catabolism in *phaseolus aureus, cicer arietinum* and *petroselenium hortense* cultures

The isoflavone daidzein (133) is incorporated into insoluble polymeric material *via* 3′-hydroxylation. However, under anaerobic conditions hydroxylation and polymerisation are completely inhibited and glucoside formation occurs instead. Similarly methylation of the 4′-hydroxyl group of isoflavones inhibits phenolase-catalyzed 3′-hydroxylation and thus incorporation into polymers.

It is well documented that cinnamyl alcohols are the primary building units of lignins. Following up the observation that cell suspension cultures of *Glycine max* deposit a lignin-like substance during growth, Ebel and Grisebach have isolated (*43*) from the cultures a preparation that reduced the CoA ester of [2-^{14}C]ferulic acid (**134**) to coniferyl alcohol (**135**) and of *p*-coumaric acid (**136**) to the alcohol (**137**) in presence of NADPH and CoASH.

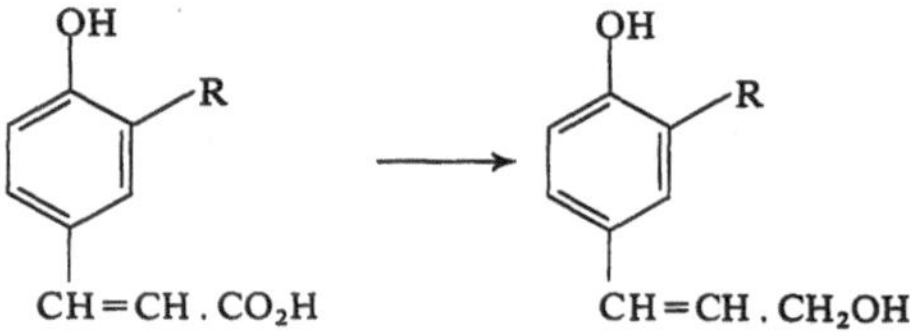

CH=CH.CO$_2$H   CH=CH.CH$_2$OH

(**134**) R = OCH$_3$ (**135**) R = OCH$_3$
(**136**) R = OH   (**137**) R = OH

6.3 Degradation of Tyrosine by the Homogentisate Pathway

A major pathway for tyrosine degradation, well-known in animals and micro-organisms and recently shown to play a role in two metabolic pathways in plants, has been studied in detail with plant tissue cultures.

Ellis had found (*31*) that there is a rapid loss of side-chain carbon atoms from tyrosine (**138**) incubated with tissue cultures of *Ruta graveolens* and *Melitotus alba*. In a subsequent study (*28*) with cell suspension cultures of ten different plants, U[^{14}C]-ring labelled L-tyrosine was found to liberate labelled CO$_2$ in each case. Much more impressive, however, were the amounts of labelled CO$_2$ obtained with these cultures from U[^{14}C]-ring labelled homogenetisic acid (**139**), suggesting that the homogentisate ring-cleavage pathway may be a major and general catabolic route in plants.

Durand and Zenk were subsequently led to investigate the homogentisate pathway in detail by an unexpected observation (*26*). It was their intention to study the biosynthesis of the naphthaquinones plumbagin (**140**) and 7-methyljuglone (**141**) using both sterile-grown plants and cell suspension cultures prepared from leaf tissue of *Drosophyllum lusitanicum*. Incorporations were in this case higher with the intact leaves, but the indications are that the pathway in leaf and undifferentiated tissue were in all essentials identical. [β-^{14}C]Tyrosine was efficiently

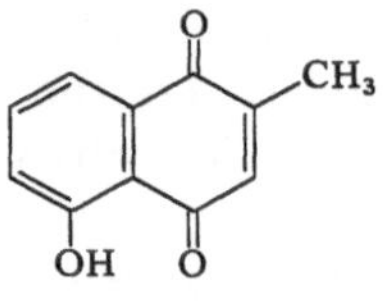

(**140**) 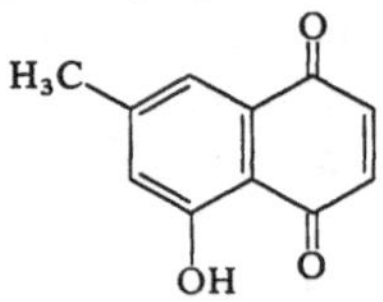   (**141**)

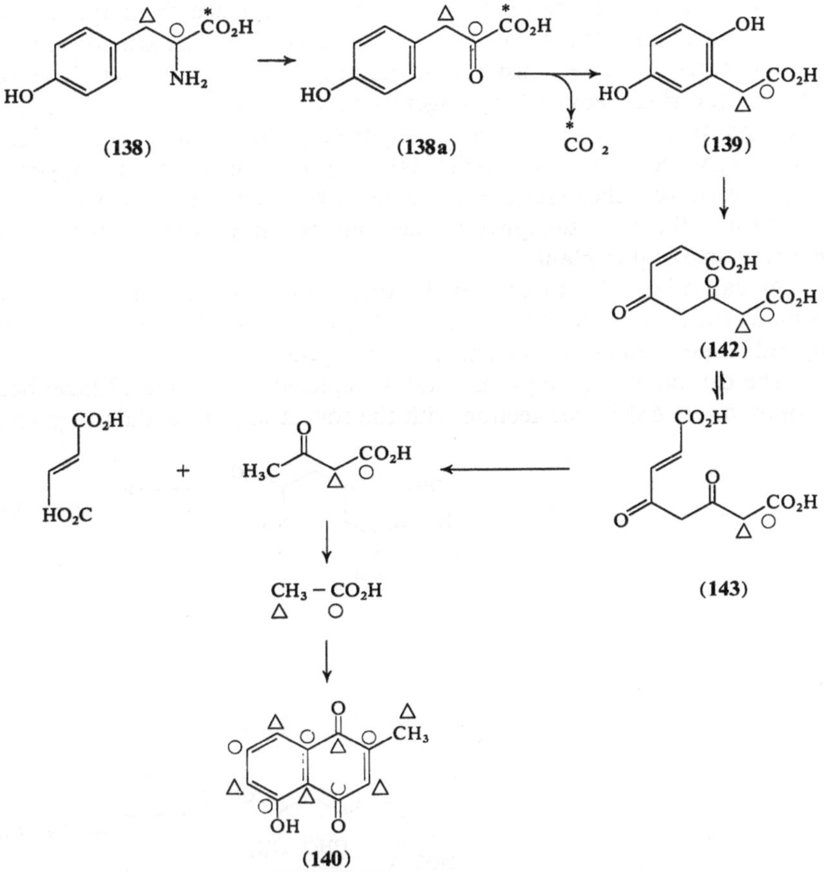

Scheme 16. Biosynthesis of plumbagin from tyrosine *via* the homogentisate pathway in *drosophyllum lusitanicum* cultures

incorporated into the naphthaquinones. However, degradation revealed that both are biosynthesized *via* the acetate-malonate pathway, since the label from tyrosine appeared at alternate carbon atoms throughout the carbon skeleton. Tyrosine must accordingly be degraded by the homogentisate pathway to acetate. Both the degradation of tyrosine to acetate and its incorporation into plumbagin occur with remarkable efficiency. As would be expected on this basis, [2-^{14}C]homogentisate, [2-^{14}C]4-maleyl- (**142**) and [2-^{14}C]4-fumaryl (**143**) acetoacetates were well incorporated into plumbagin and when αα′-bipyridyl (a homogentisate oxygenase inhibitor) was administered together with [β-^{14}C]tyrosine, labelled homogentisate accumulated. The results indicated the pathway set out in Scheme 16.

The presence of the complete catabolic sequence from tyrosine to acetoacetate and fumarate was subsequently demonstrated (25) in homogenates from cell suspension cultures of *D. lusitanicum*. The key enzyme for ring-cleavage, homogentisate oxygenase, was purified 190-fold and tyrosine transaminase, *p*-hydroxyphenylpyruvic acid oxidase and fumarylacetoacetic acid hydrolase assayed. Homogentisate oxygenase was isolated from the tissue cultures of ten other plants. This work probably constitutes the first unequivocal demonstration of the homogentisate pathway in a higher plant.

At least one further mechanism for tyrosine catabolism in tissue cultures is indicated (28) by the release of labelled CO_2 from ring-labelled DOPA by cell suspension cultures from ten plant species.

The catabolic pathways *via* DOPA depicted in Scheme 17 have been proposed (59, 68) in connection with the formation of betalain pigments.

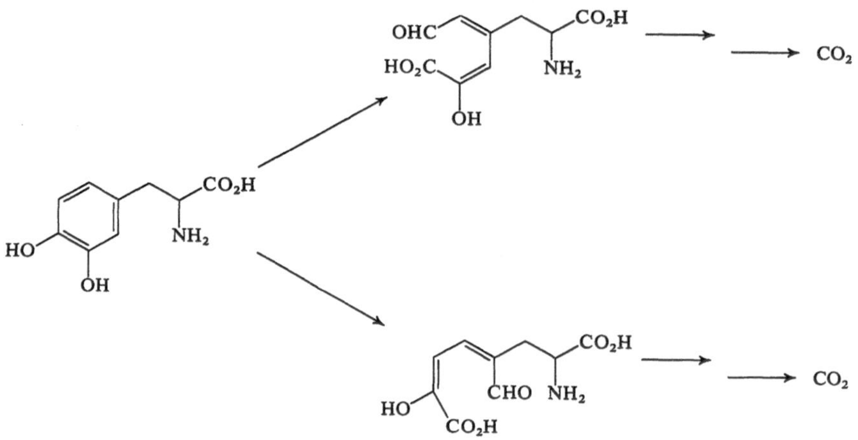

Scheme 17. Possible mechanisms for tyrosine catabolism *via* DOPA in plant tissue cultures

VI. Biosynthesis of Alkaloids

1. General

A substantial number of alkaloid-producing plants have been propagated in cell and organ culture, mainly with a view to the production of pharmacologically active alkaloids. The yields of alkaloids, where they are formed, have almost always been inferior to those from the whole plant (usually less than 10%) and critically dependent on added growth factors in the medium. In some genera, alkaloids are not produced by

References, pp. 293—298

the callus culture but re-appear upon organogenesis, usually root formation (e. g. *Datura* and *Nicotiana*). Undetected re-differentiation of this kind may well account for discrepancies in biosynthetic results that have been reported with supposedly comparable tissues.

2. Indole Alkaloids

Callus cultures of *Catharanthus roseus* have been most thoroughly investigated for the production of indole alkaloids, since the plant contains over sixty alkaloids of which four possess moderate anti-tumour activity.

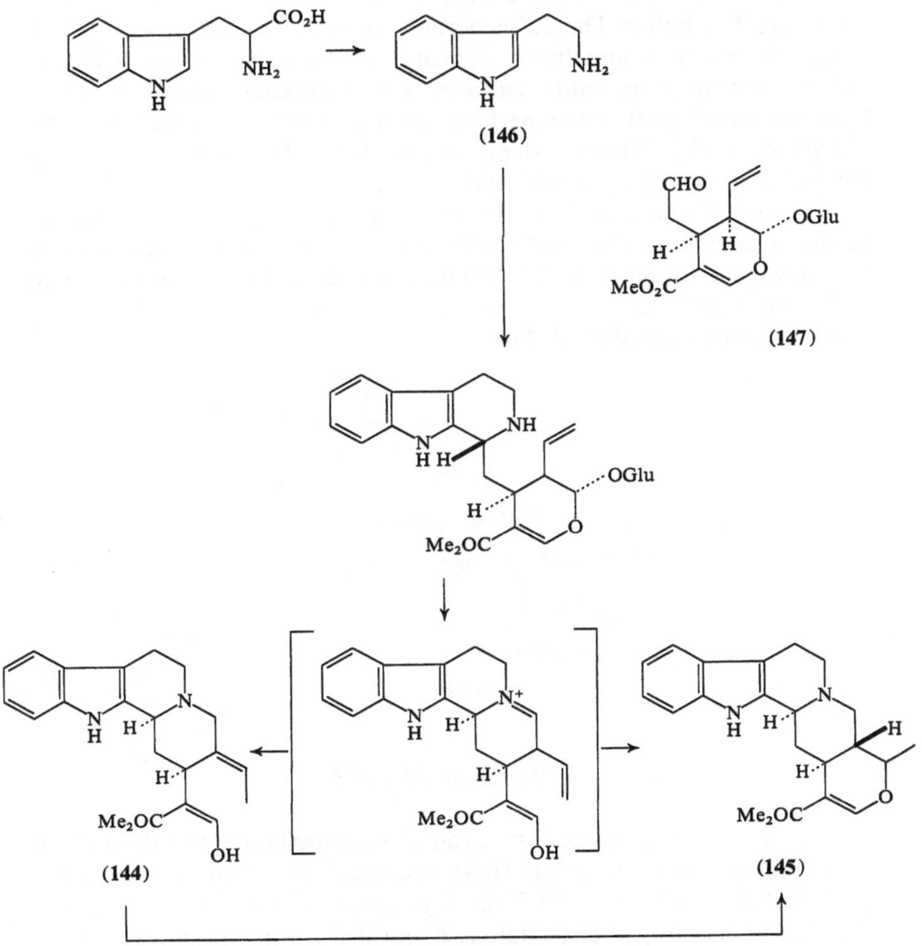

Scheme 18. Biosynthesis of geissoschizine and ajmalicine in *catharantus roseus* cultures

Patterson and Carew were able, in an extensive study (77), to detect ten indole alkaloids, none of which were of the dimeric type characteristic of the plant.

Studies of the detailed mechanism of indole alkaloid biosynthesis have been severely hampered by the very low levels of incorporation of labelled precursors with intact plants of the species *Vinca, Catharanthus* and *Aspidosperma*. Scott and Lee have recently studied (86) indole alkaloid biosynthesis with cell-free preparations from both seedlings and tissue cultures of *Catharanthus roseus*. These soluble systems clearly contained the range of enzymes necessary to synthesize the *Corynanthe* alkaloids geissoschizine (144) and ajmalicine (145) from [2-^{14}C]tryptamine (146) and [OC-^{3}H$_3$]secologanin (147) in presence of NADPH, thiols and Tris buffer. The callus-derived system also converted [Ar-^{3}H]-geissoschizine into ajmalicine (Scheme 18). Incorporations with the cell-free system from callus cultures were markedly superior to those from the seedlings (tryptamine into ajmalicine 18%; geissoschizine into ajmalicine 7.7%). These systems clearly have the potential for more detailed studies at the enzymic level.

Cultures of *Vinca minor* have been investigated recently (39). Contrary to previous reports (78, 78a) with cultures of this plant, Garnier and his colleagues found that [2-^{14}C]DL-tryptophan was not incorporated and were unable to detect vincamine. The major constituent detected was the lignan lirioresinol B (148).

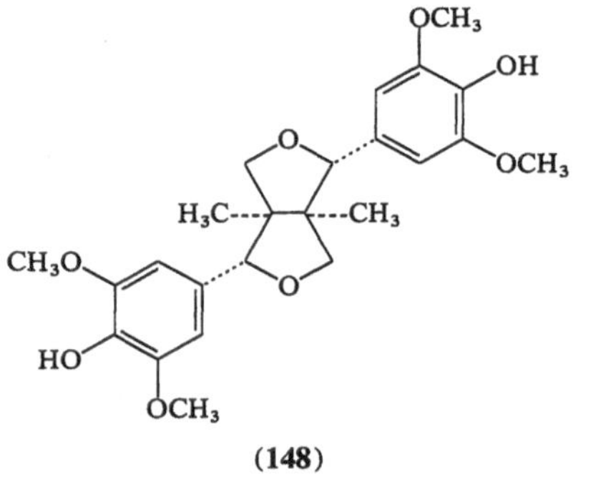

(148)

3. Berberine Alkaloids

An early keystep in the formation of protoberberine-based alkaloids is the conversion of reticuline (149) into scoulerine (150), a tetrahydro-protoberberine alkaloid, the N-methyl group of the former becoming the "berberine bridge" of the latter. A purified enzyme preparation from cell cultures of *Maclaya microcarpa* effected (81) this transformation, for which oxygen is required, with ($\pm$) N-[^{14}CH$_3$]reticulene. This is an early step in the formation of protopine (151), allocryptopine and sanguinarine (153), the major alkaloids in both the cultures and the parent plants.

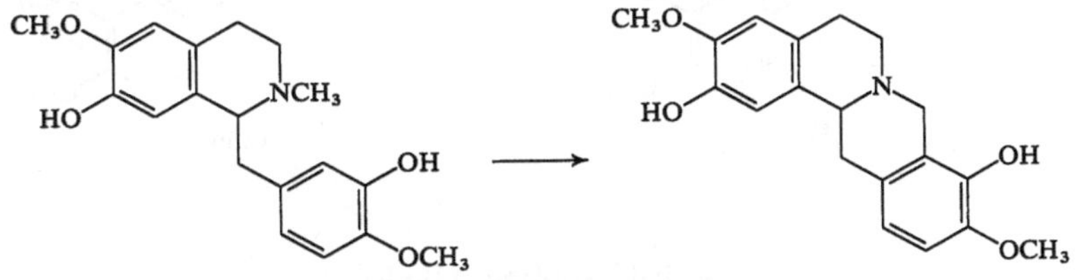

(149)   (150)

Norsanguinarine (154), dihydrosanguinarine (155), oxysanguinarine (156), protopine (151), and cryptopine (157) have been identified (35) in callus cultures of *Papaver somniferum*.

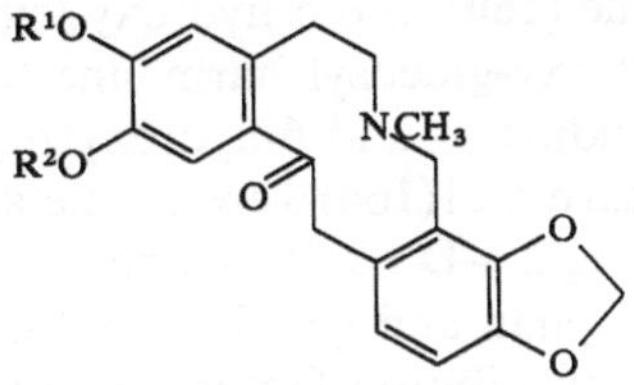

(151) R^1 + R^2 = CH$_2$
(157) R^1 = R^2 = CH$_3$

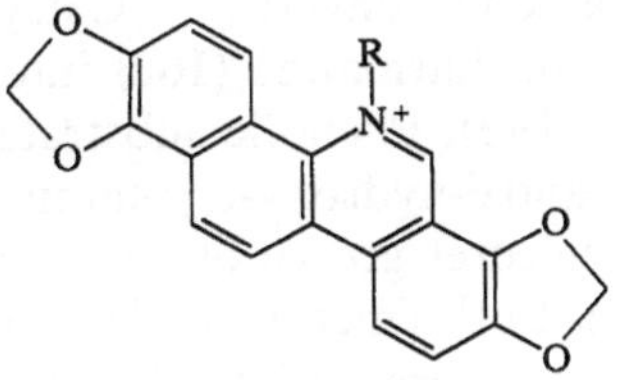

(153) R = CH$_3$
(154) R = H

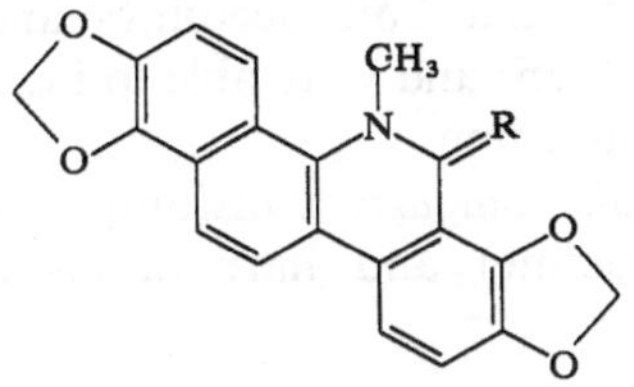

(155) R = H$_2$
(156) R = O

Callus cultures of *Coptis japonica* produce the alkaloids, mainly berberine (158) and jatrorrhizine (159), characteristic of the rhizome, but in smaller amounts (37, 58). Jatrorrhizine levels in plantlets regenerated from the cultures exceed those in the parent rhizome. FURUYA and his colleagues have compared (57) the alkaloid composition of callus tissues and plantlets regenerated from them in eleven representative species of the *Papaveraceae*. All the callus tissues produce benzophenanthridine, protopine and aporphine alkaloids which are simpler and more widely distributed than the morphinane type characteristic of the parent plants. Alkaloids in the regenerated plants resemble those of the parents.

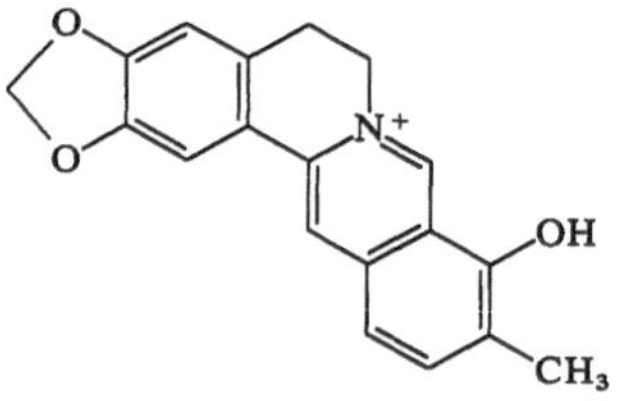

(158) 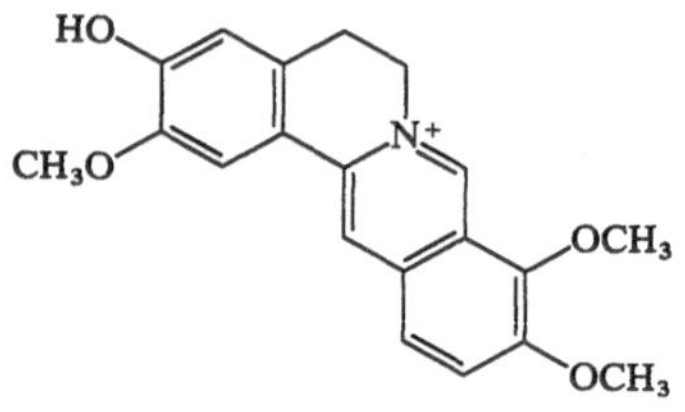   (159)

4. β-Carboline Alkaloids

NETTLESHIP and SLAYTOR were able to obtain (72) two biosynthetically distinct callus cultures from *Peganum harmala*, depending on the culture medium used. Cultures grown in a medium either with or without 2,4-D were able to convert [2-^{14}C]tryptamine (160) into 5-hydroxy tryptamine (161) and harmaline (162) into 8-hydroxy-glucosyl harmaline (dihydroruine) (163). Phenolic substrates, including 5- and 6-hydroxytryptophan and 5- and 6-hydroxy-tryptamine and harmalol (164) were not metabolized. Only cultures grown on medium lacking 2,4-D could synthesize harmine (162a). Both [methylene-^{14}C]-L-tryptophan and [methyl-^{14}C]harmaline were incorporated into the alkaloid with dilution factors of respectively 30,000 and 2. The probable biosynthesis and metabolism of alkaloids in *P. harmala* is therefore as depicted in Scheme 19.

It had previously been found (71) that both seedlings and callus cultures of *P. harmala* produced ruine (165) and were able to incorporate [CH$_3$-1-^{14}C]harmine hydrochloride into ruine.

Cell suspension cultures of *Phaseolus vulgaris* transform (111, 111a) tryptophan into *nor*-harman (β-carboline) and harman (1-methyl-β-carboline).

5. Furoquinoline Alkaloids

Cell suspension cultures of *Ruta graveolens* elaborated (88) administered 4-hydroxy-2-quinolone (166) into furoquinoline alkaloids and edulinine (167). 4-Hydroxy (168) and 4-methoxy-3-(3-methyl-2-butenyl)-2-quinolone (169), known from previous studies with whole plants to be precursors, were also transformed into furoquinolines. Dictamnine (170) was directly transformed into γ-fagarine (171) and skimmianine (172). Edulinine was formed by the cultures from 4-hydroxy-N-methyl-2-quinolone (173), the 3-prenylated derivative (174) and its methyl ether. The biosynthetic sequence shown in Scheme 20 is thus indicated.

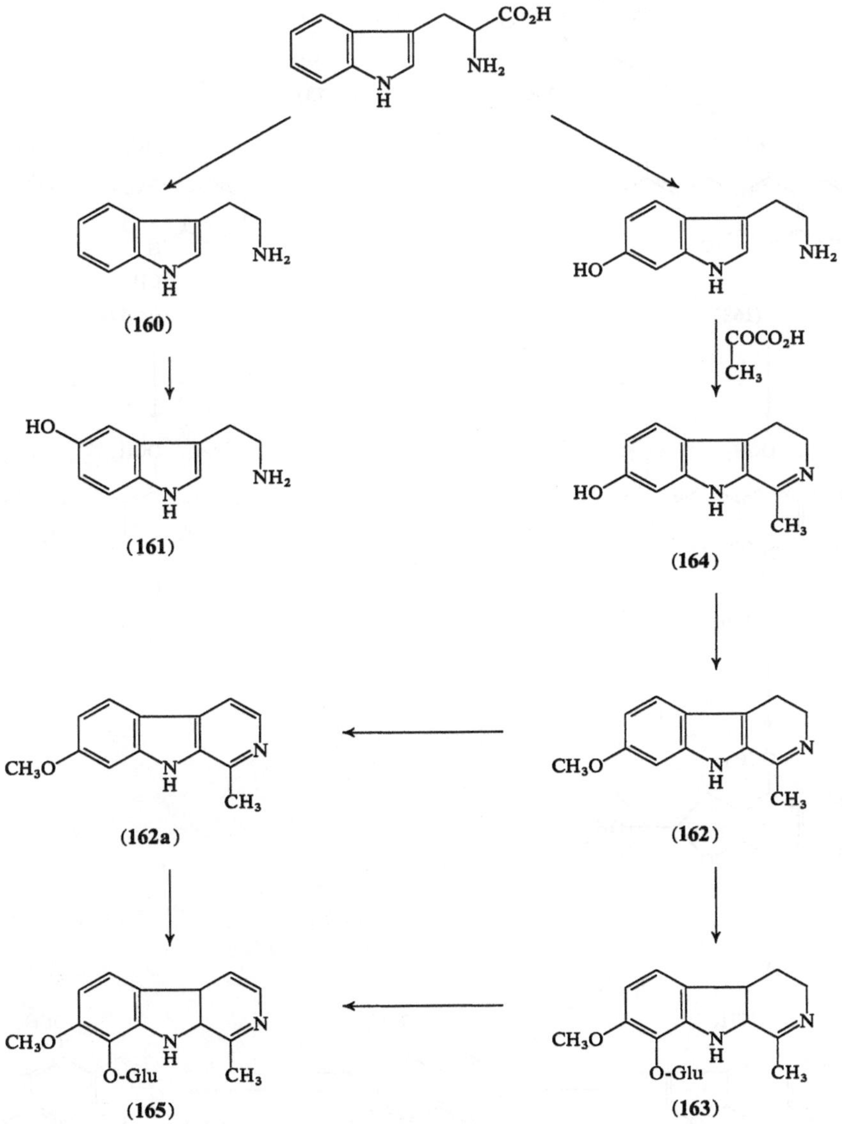

Scheme 19. Metabolism of β-carboline alkaloids in *peganum harmala* cultures

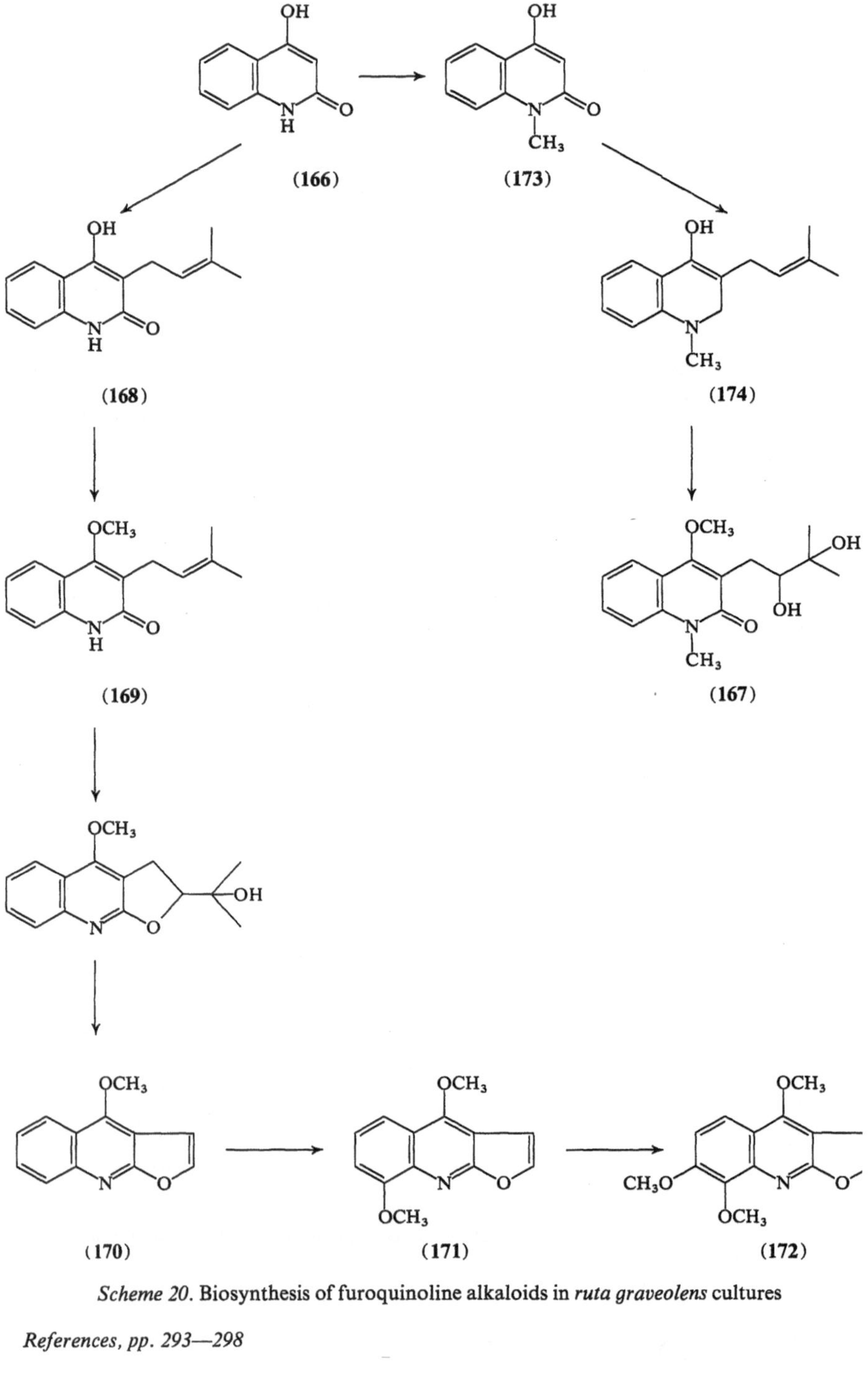

Scheme 20. Biosynthesis of furoquinoline alkaloids in *ruta graveolens* cultures

Changes in alkaloid yields were noted on prolonged subculture of the tissues and newly established cultures gave different proportions both of coumarins and alkaloids (see p. 276). There is however no reason to suppose that the biosynthetic pathway differs from that in whole plants.

In separate experiments (*90*) 4-hydroxy-2-quinolone was bio-transformed by *Ruta* cell cultures into dictamnine which accumulated progressively. However, when all the quinolone was used up, dictamnine was rapidly catabolized to γ-fagarine.

6. Tropane Alkaloids

A number of investigations into tropane alkaloid biosynthesis have been reported, using undifferentiated callus tissues, root tissues grown *in vitro* or root tissue regenerated from callus. The biosynthetic pathway to tropane alkaloids in plants is well established in outline. The work with tissue cultures and root organ cultures here reviewed is concerned with previously undefined details of biosynthesis and also with the factors that inhibit effective alkaloid synthesis in tissue cultures as compared with the intact plants. Formation of tropane alkaloids have been reported with callus tissues from several *Datura* species (*21, 63, 82, 91, 99, 100*). Interestingly callus tissues from seed, root, leaf and stem have been shown (*18*) to contain tropane alkaloids, despite the fact that in the intact plant these alkaloids are confined to the root. However, alkaloid levels are generally low and in some tissues they are entirely absent.

The biosynthetic deficiencies of *Datura* cell cultures with regard to tropane alkaloid biosynthesis have been investigated by ELZE and TEUSCHER (*32, 101*). When (±)-[2-^{14}C]ornithine (**175**) was incubated with cell suspension cultures of *Datura ferox*, proline or tropane alkaloids were not detected in labelled form. However, a number of pyrroline precursors including glutamate, arginine, aspartate and particularly α-keto-δ-aminovalerate were labelled; the latter is known to form 1-pyrroline-2-carboxylate spontaneously. It must therefore be concluded that the cultures lack the ability to condense pyrroline derivatives with active acetoacetate to form hygrine-α-carboxylate (**176**), the proposed intermediate to tropine. They are also apparently unable to synthesize tropic acid from phenylalanine. Thus in experiments (*54, 83*) with callus cultures of *D. innoxia,* when tropine was added to the medium there was no alkaloid production, but instead relatively large quantities of 3α-acetyltropine were formed from endogenous acetate. Addition of tropic acid, but not of phenylalanine, its established precursor in the intact plant, then finally stimulated formation of hyoscyamine (**177**) and scopolamine

(178; (±) R). Production of radio-active scopolamine and hyoscyamine on addition of [α-¹⁴C]tropine (179) and tropic acid to the culture medium has also been observed (94) with suspension cultures of D. stramonium, D. tatula and N. tabacum. It has furthermore been shown independently (61) that Datura cultures are capable of esterifying tropine with added tropic acid.

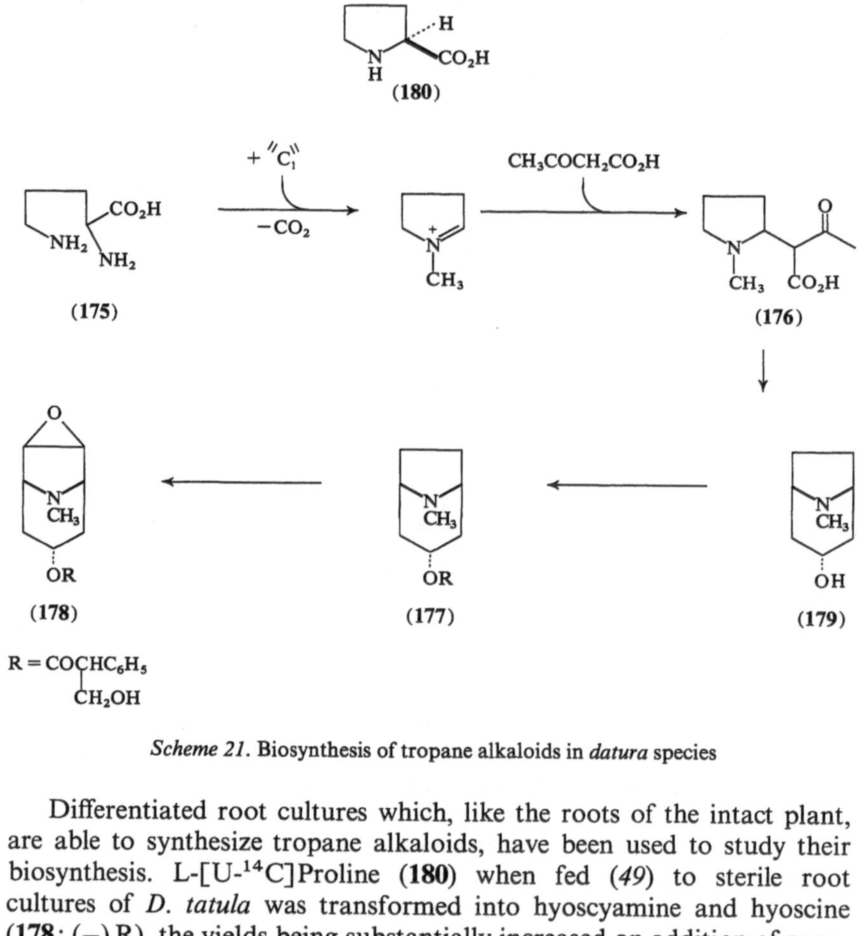

Scheme 21. Biosynthesis of tropane alkaloids in *datura* species

Differentiated root cultures which, like the roots of the intact plant, are able to synthesize tropane alkaloids, have been used to study their biosynthesis. L-[U-¹⁴C]Proline (180) when fed (49) to sterile root cultures of D. tatula was transformed into hyoscyamine and hyoscine (178; (−) R), the yields being substantially increased on addition of puromycin, an inhibitor of protein synthesis, thereby reserving proline for alkaloid synthesis. Competitive feeding experiments with [³H]sodium acetate and [3-¹⁴C]sodium acetoacetate in isolated root cultures of D. metel gave (66) preferential incorporation of radio-active carbon into the C_3 bridge of tropane alkaloids, showing acetoacetate to be the more direct precursor of the tropane portion of the skeleton. Surprisingly,

hygrine was less efficiently incorporated into hyoscyamine and scopolamine than was ornithine. Hygrine-α-carboxylate is postulated as the
key intermediate in the formation of the tropane alkaloids, but clearly
this needs to be confirmed by its isolation and feeding in labelled form
(see Scheme 21).

It has been found that, during root differentiation from cell cultures
of *D. innoxia,* scopolamine synthesis is initiated (*53*). A progressive
increase in alkaloid content was observed and the pattern of alkaloid
composition characteristic of the whole plant was restored in the majority
of regenerated plants.

References

1. ALLISON, A. J., D. N. BUTCHER, J. D. CONNOLLY, and K. H. OVERTON: Paniculides A,
B and C, Bisbolenoid Lactones from Tissue Cultures of *Andrographis paniculata.*
Chem. Comm. **1968,** 1493.

2. VON ARDENNE, M., G. OSSKE, K. SCHREIBER, K. STEINFELDER, and R. TEUMMLER:
Sterols and Triterpenoids IX. Sterols and Triterpenoids in *Solanum demissum* and
Solanum polyadenium. Kulturpflanze **13,** 115 (1965).

3. AUSTIN, D. J., and S. A. BROWN: Furanocoumarin Biosynthesis in *Ruta graveolens*
Cell Cultures. Phytochemistry **12,** 1657 (1973).

4. BARTON, D. H. R., J. G. T. CORRIE, P. J. MARSHALL, and D. A. WIDDOWSON: Biosynthesis of Terpenes and Steroids VII Unified Scheme for the Biosynthesis of
Ergosterol in *Saccharomyces cerevisiae.* Bioorganic Chemistry **2,** 363 (1973).

5. BARZ, W., F. MOHR, and E. TEUFEL: Catabolism of 4′,6′-dihydroxyaurones in Plant
Cell Suspension Cultures. Phytochemistry **13,** 1785 (1974).

6. BENVENISTE, P.: The Biosynthesis of Sterols in Tissues of Tobacco Grown *in vitro.*
Identification of Cycloeucalenol and Obtusifoliol. Phytochemistry **7,** 951 (1968).

7. BENVENISTE, P., M. A. HARTMANN, and F. DURST: Biosynthesis of Sterols in Jerusalem Artichoke Tuber Tissue. Phytochemistry **11,** 3003 (1972).

8. BENVENISTE, P., and R. HEINTZ: Biosynthesis of 24-methylene Cycloartanol from
Squalene-2(3)epoxide by Microsomes from Tissue Cultures of Bramble *(Rubus fruticosus).* Compt. Rend. D **274,** 947 (1972).

9. — — Plant Sterol Metabolism. Enzymatic Cleavage of the 9β,19β-Cyclopropane Ring
of Cyclopropyl Sterols in Bramble Tissue Cultures. J. Biol. Chem. **249,** 4267 (1974).

10. BENVENISTE, P., R. HEINTZ, W. H. ROBINSON, and R. M. COATES: Demonstration and
Identification of a Biosynthetic Intermediate between Farnesyl Pyrophosphate and
Squalene in a Higher Plant. Biochem. Biophys. Res. Comm. **49,** 1547 (1972).

11. BENVENISTE, P., M. J. E. HEWLINS, and B. FRITIG: Biosynthesis of Sterols in Cultured
Tobacco Tissues. Kinetics of Formation of Sterols and their Precursors. Eur. J.
Biochem. **9,** 526 (1969).

12. BENVENISTE, P., L. HIRTH, and G. OURISSON: Constituents of Plant Tissues Cultivated
in vitro I. Sterol Biosynthesis in Tobacco Tissues. Bull. Soc. Fr. Physiol. Veg. **11,**
252 (1965).

13. — — — Biosynthesis of Sterols in the Tissues of Tobacco Grown *in vitro* I. Isolation
of Sterols and Triterpenes. Phytochemistry **5,** 31 (1966).

14. — — — Biosynthesis of Sterols in Tobacco Tissue Culture II. Details of Biosynthesis of Phytosterols. Phytochemistry **5,** 45 (1966).

15. Benveniste, P., and R. A. Massy-Westropp: Demonstration of the 2,3-epoxide of Squalene in Tobacco Tissues *in vitro*. Tetrahedron Letters **1967**, 3553.
16. Berlin, J., and W. Barz: Metabolism of Isoflavones and Chalcones in Cell and Callus Suspension Cultures of *Phaseolus aureus*. Planta **98**, 300 (1971).
17. Berlin, J., W. Barz, H. Harms, and K. Haider: Degradation of Phenolic Compounds in Plant Cell Cultures. FEBS Letters **16**, 141 (1971).
18. Berlin, J., H. Harms, K. Haider, P. Kiss, and W. Barz: O-Demethylation and Decarboxylation of Benzoic Acids in Plant Cell Suspension Cultures. Planta **105**, 342 (1972).
19. Berlin, J., P. Kiss, D. Muller-Enoch, D. Gierse, W. Barz, and B. Janistyn: Degradation of Chalcones and Isoflavones in Plant Cell Suspension Cultures. Z. Naturforsch. C **29**, 374 (1974).
20. Carew, D. P., and E. J. Staba: Plant Tissue Culture. Its Fundamentals, Application and Relationships to Medicinal Plant Studies. Lloydia **28**, 1 (1965).
21. Chan, W., and E. J. Staba: Alkaloid Production by *Datura* Callus and Suspension Tissue Cultures. Lloydia **28**, 55 (1965).
22. Constabel, F., O. L. Gamborg, W. G. W. Kurz, and W. Steck: Production of Secondary Metabolites in Plant Cell Cultures. Planta Med. **25**, 158 (1974).
23. Constabel, F., J. P. Shyluk, and O. L. Gamborg: The Effects of Hormones on Anthocyanin Accumulation in Cell Cultures of *Haplopappus gracilis*. Planta **96**, 306 (1971).
24. Cotterell, G. P., T. G. Halsall, and M. J. Wriglesworth: The Chemistry of Triterpenes and Related Compounds Part XLVII Clarification of the Nature of the Tetracyclic Triterpene Acids of Elemi Resin. J. Chem. Soc. (C) **1970**, 739.
25. Durand, R., and M. H. Zenk: Enzymes of the Homogentisate Ring-Cleavage Pathway in Cell Suspension Cultures of Higher Plants. FEBS Letters **39**, 218 (1974).
26. — — The Homogentisate Ring-Cleavage Pathway in the Biosynthesis of Acetate-Derived Naphthoquinones of the *Droseraceae*. Phytochemistry **13**, 1483 (1974).
27. Ellis, B. E.: A Survey of Catechol Ring-Cleavage by Sterile Plant Tissue Cultures. FEBS Letters **18**, 228 (1971).
28. — Catabolic Ring-Cleavage of Tyrosine in Plant Cell Cultures. Planta **111**, 113 (1973).
29. — Degradation of Aromatic Compounds in Plants. Lloydia **37**, 168 (1974).
30. Ellis, B. E., and S. A. Brown: Isolation of Dimethylallyl Pyrophosphate: Umbelliferone Dimethylallyl Transferase from *Ruta graveolens*. Can. J. Biochem. **52**, 734 (1974).
31. Ellis, B. E., and G. H. N. Towers: Degradation of Aromatic Compounds by Sterile Plant Tissues. Phytochemistry **9**, 1457 (1970).
32. Elze, H., and E. Teuscher: Biochem. Physiol. Alkaloide, 4th Int. Symp., 1969, p. 239. Ed. K. Mothes. Berlin: Akademie Verlag. 1972.
33. Eppenberger, U., L. Hirth, and G. Ourisson: Anaerobic Cyclisation of Squalene Oxide to Cycloartenol in Tissue Cultures of *N. tabacum*. Eur. J. Biochem. **8**, 180 (1969).
34. Furuya, T., S. Ayabe, and K. Noda: Chrysophanol and Emodin from Callus Tissue of Rhubarb *(Rheum palmatum)*. Phytochemistry **14**, 1457 (1975).
35. Furuya, T., and A. Ikuta: Plant Tissue Cultures XV. Alkaloids from Callus Tissue of *Papaver somniferum*. Phytochemistry **11**, 3041 (1972).
36. Furuya, T., H. Kojima, and T. Katsuta: Plant Tissue Cultures XIV. 3-Methylpurpurin and Other Anthraquinones from Callus Tissue of *Digitalis lanata*. Phytochemistry **11**, 1073 (1972).
37. Furuya, T., K. Syono, and A. Ikuta: Plant Tissue Cultures XII. Isolation of Berberine from Callus Tissue of *Coptis japonica*. Phytochemistry **11**, 175 (1972).
38. Gamborg, O. L., F. Constabel, T. A. G. La Rue, R. A. Miller, and W. Steck: The

Influence of Hormones on Secondary Metabolite Formation in Plant Cell Cultures. Proc. Int. Symp. Plant Tissue Culture, Strasbourg, p. 335 (1970).

39. GARNIER, J., N. KUNESCH, E. SIOU, J. POISSON, G. KUNESCH, and M. KOCH: Study of Cultures of Tissues of *Vinca minor*. Isolation of a Lignan Lirioresorcinol B. Phytochemistry **14**, 1385 (1975).

40. GAUTHERET, R. J.: Sur la Possibilité de Realiser la Culture Indefinie des Tissues de Tubercules de Carotte. Compt. Rend. **208**, 118 (1939).

41. — La Culture des Tissus Végéteaux. Paris: Masson et Cie. 1959.

42. GOAD, L. J., and T. W. GOODWIN: The Biosynthesis of Sterols in Higher Plants. Biochem. J. **99**, 735 (1966).

43. GRISEBACH, H., and J. EBEL: Reduction of Cinnamic Acids to Cinamyl Alcohols with an Enzyme Preparation from Cell Suspension Cultures of Soyabean *(Glycine max)*. FEBS Letters **30**, 141 (1973).

44. GRISEBACH, H., J. EBEL, and K. HAHLBROCK: Purification and Properties of an o-dihydric Phenol meta-0-methyltransfcrase from Cell Suspension Cultures of Parsley and its Relation to Flavonoid Biosynthesis. Biochim. Biophys. Acta **268**, 313 (1972).

45. GRISEBACH, H., and H. FRITSCH: Biosynthesis of Cyanidin in Cell Cultures of *Haplopappus gracilis*. Phytochemistry **14**, 2437 (1975).

46. GRISEBACH, H., and K. HAHLBROCK: Formation of Coenzyme A esters of Cinnamic Acids with an Enzyme Preparation from Cell Suspension Cultures of Parsley. FEBS Letters **11**, 62 (1970).

47. — — Biosynthesis of Cyanidin in Cell Suspension Cultures of *Haplopappus gracilis*. Z. Naturforsch. **B26**, 581 (1971).

48. GRISEBACH, H., and L. SCHILL: Properties of a Phenolase Preparation from Cell Suspension Cultures of Parsley. Z. Physiol. Chem. **354**, 1555 (1973).

49. GUPTA, M. P., and M. R. GIBSON: *Datura stramonium* Sterile Root Cultures. J. Pharm. Sci. **61**, 1257 (1972).

50. HAHLBROCK, K., J. EBEL, R. ORTMANN, A. SUTTER, E. WELLMANN, and H. GRISEBACH: Regulation of Enzyme Activities Related to Biosynthesis of Flavone Glycosides in Cell Suspension Cultures of Parsley *(Petroselenium hortense)*. Biochim. Biophys. Acta **244**, 7 (1971).

51. HALL, J., A. R. H. SMITH, L. J. GOAD, and T. W. GOODWIN: The Conversion of Lanosterol, Cycloartenol and 24-Methylenecycloartanol into Poriferasterol by *Ochromonas malhamensis*. Biochem. J. **112**, 129 (1969).

52. HEWLINS, M. J. E., J. D. EHRHARDT, L. HIRTH, and G. OURISSON: The Conversion of [^{14}C]cycloartenol and [^{14}C]lanosterol into Phytosterols by Cultures of *Nicotiana tabacum*. Eur. J. Biochem. **8**, 184 (1969).

53. HIRAOKA, N., and M. TABATA: Alkaloid Production by Plants Regenerated from Cultured Cells of *Datura innoxia*. Phytochemistry **13**, 1671 (1974).

54. HIRAOKA, N., M. TABATA, and M. KONOSHIMA: Formation of Acetyltropine in *Datura* Callus Cultures. Phytochemistry **12**, 795 (1973).

55. HÖSEL, W., and W. BARZ: Enzymatic Transformation of Flavonols with a Cell Free Preparation from *Cicer arietinum*. Biochim. Biophys. Acta **261**, 294 (1972).

56. HÖSEL, W., P. D. SHAW, and W. BARZ: Degradation of Flavonols in Plant Cell Suspension Cultures. Z. Naturforsch. **B27**, 946 (1972).

57. IKUTA, A., K. SYONO, and T. FURUYA: Alkaloids of Callus Tissues and Redifferentiated Plantlets in the *Papaveraceae*. Phytochemistry **13**, 2175 (1974).

58. — — — Plant Tissue Cultures XXIV. Alkaloids in Plants Regenerated from *Coptis* Callus Cultures. Phytochemistry **14**, 1209 (1975).

59. IMPELLIZERI, G., and M. PIATELLI: Biosynthesis of Indicaxanthin in *Opuntia ficus-indica* Fruits. Phytochemistry **11**, 2499 (1972).

60. JANISTYN, B., W. BARZ, and R. POHL: Degradation of 2',4,4',6'-tetrahydroxychalcone-

2'-β-D-glucoside by Callus Suspension Cultures of *Pisum sativum*. Z. Naturforsch. **B 26**, 973 (1971).

61. Jindra, A., and E. J. Staba: *Datura* Tissue Cultures: Arginase, Transaminase and Esterase Activities. Phytochemistry **7**, 79 (1968).

62. Jurd, L.: The Acid Catalyzed Conversion of 3-Hydroxyflavanones to Anthocyanidins. Phytochemistry **8**, 2421 (1969).

63. Konoshima, M., M. Tabata, H. Yamamoto, and N. Hiraoka: Growth and Alkaloid Production of *Datura* Tissue Cultures. J. Pharm. Soc. Japan **90**, 370 (1970).

64. Leistner, E.: Mode of Incorporation of Precursors into Alizarin (1,2-Dihydroxy-9,10-Anthraquinone). Phytochemistry **12**, 337 (1973).

65. — Biosynthesis of Morindone and Alizarin in Intact Plants and Cell Suspension Cultures of *Morinda citrifolia*. Phytochemistry **12**, 1669 (1973).

66. Liebisch, H. W., K. Peisker, A. S. Radwan, and H. R. Schutte: Alkaloid Biosynthesis in *Datura metel* Isolated Root Cultures. Z. Pflanzenphysiol. **67**, 1 (1972).

67. Loewenberg, J. R.: Scopoletin and Scopolin Metabolism. Phytochemistry **9**, 361 (1970).

68. Miller, H. E., H. Rosler, A. Wohlpart, H. Wyler, M. E. Wilcox, H. Frohofer, T. J. Mabry, and A. S. Dreiding: Biogenesis of the Betalains. Helv. Chim. Acta **51**, 1470 (1968).

69. Moss, G. P., and S. A. Nicolaidis: Terpenoid Biosynthesis: the Stereochemistry of Squalene Cyclisation. Chem. Comm. **1969**, 1072.

70. Murashige, T., and F. Skoog: A Revised Medium for Rapid Growth and Bioassays with Tobacco Tissue Cultures. Physiol. Plantarum **15**, 473 (1962).

70a. Nag, K. K., and H. E. Street: Freeze Preservation of Cultured Plant Cells. Physiol. Plantarum **34**, 254 (1975).

71. Nettleship, L., and M. Slaytor: Ruine: a Glucosidic β-Carboline from *Peganum harmala*. Phytochemistry **10**, 231 (1971).

72. — — Limitations of Feeding Experiments in Studying Alkaloid Biosynthesis in *Peganum harmala* Callus Cultures. Phytochemistry **13**, 735 (1974).

73. Ourisson, G., B. Fritig, and L. Hirth: Biosynthesis of Scopoletin and Scopolin in Anergic Tobacco Tissue Cultured *in vitro*. Bull. Soc. Fr. Physiol. Veg. **13**, 51 (1967).

74. Overton, K. H., and D. J. Picken: Biosynthesis of Bisabolene by Callus Cultures of *Andrographis paniculata*. Chem. Comm. **1976**, 105.

75. Overton, K. H., and F. M. Roberts: Biosynthesis of *trans, trans*- and *cis, trans*-farnesols by Soluble Enzymes from Tissue Cultures of *Andrographis paniculata*. Biochem. J. **144**, 585 (1974).

76. — — Interconversion of *trans, trans*- and *cis, trans*-farnesol by Enzymes from *Andrographis*. Phytochemistry **13**, 2741 (1974).

77. Patterson, B. D., and D. P. Carew: Growth and Alkaloid Formation in *Catharanthus roseus* Tissue Cultures. Lloydia **32**, 131 (1969).

78. Petiard, V., and Y. Demarly: Presence of Glucosides and Alkaloids in Plant Tissue Cultures. Ann. Amelior. Plantes **22**, 361 (1972).

78a. Petiard, V., Y. Demarly, and R. R. Paris: Heterosides and Alkaloids in Tissue Cultures of Medicinal Plants. Plant. Med. Phytother. **6**, 41 (1972).

79. Puhan, Z., and S. M. Martin: The Industrial Potential of Plant Cell Culture. Progr. in Indust. Microbiol. **9**, 14 (1970).

80. Rai, P. P., T. D. Turner, and S. L. Greensmith: Anthracene Derivatives in Tissue Culture of *Cassia senna* L. J. Pharm. Pharmacol. **26**, 722 (1974).

81. Rink, E., and H. Böhm: Conversion of Reticuline into Scoulerine by a Cell Free Preparation from *Macleaya microcarpa* Cell Suspension Cultures. FEBS Letters **49**, 396 (1975).

82. Romeike, A., and H. Koblitz: Tissue Cultures of Alkaloid Bearing Plants II *Datura* Species. Kulturpflanze **18**, 165 (1970).

83. — — Tissue Cultures of Alkaloid Bearing Plants III Investigations Concerning the Esterification of Tropine. Kulturpflanze **20,** 165 (1972).

84. SAITOH, T., and S. SHIBATA: New Type Chalcones from Licorice Root. Tetrahedron Letters **1975,** 4461.

85. SAITOH, T., S. SHIBATA, U. SANKAWA, T. FURUYA, and S. AYABE: Biosynthesis of Echinatin. A New Biosynthetical Scheme of Retrochalcone. Tetrahedron Letters **1975,** 4463.

86. SCOTT, A. I., and S.-L. LEE: Biosynthesis of the Indole Alkaloids. A Cell-Free System from *Catharanthus roseus.* J. Amer. Chem. Soc. **97,** 6906 (1975).

86a. SMITH, A. R. H., L. J. GOAD, T. W. GOODWIN, and E. LEDERER: Phytosterol Biosynthesis: Evidence for a 24-ethylidine Intermediate During Sterol Formation in *Ochromonas malhamensis.* Biochem. J. **104,** 56c (1967).

87. STECK, W., B. K. BAILEY, J. P. SHYLUK, and O. L. GAMBORG: Coumarins and Alkaloids of Cell Cultures of *Ruta graveolens.* Phytochemistry **10,** 191 (1971).

88. STECK, W., D. BOULANGER, and B. K. BAILEY: Formation of Edulinine and Furoquinoline Alkaloids from Quinoline Derivatives by Cell Suspension Cultures of *Ruta graveolens.* Phytochemistry **12,** 2399 (1973).

89. STECK, W., and F. CONSTABEL: Biotransformations in Plant Cell Cultures. Lloydia **37,** 185 (1974).

90. STECK, W., O. L. GAMBORG, and B. K. BAILEY: Increased Yields of Alkaloids Through Precursor Biotransformations in Cell Suspension Cultures of *Ruta graveolens.* Lloydia **36,** 93 (1973).

91. STIENSTRA, T. M.: Formation of Mydriatic Alkaloids in Excised Root Cultures of *Datura stramonium* Grown on a Completely Synthetic Nutrient. Proc. K. Ned. Acad. Wet. **57,** 584 (1954).

92. STEWARD, F. C.: ed. Plant Physiology, Volume VB, Analysis of Growth: The Responses of Cells and Tissues in Culture. New York and London: Academic Press. 1969.

93. STEWARD, F. C., H. W. ISRAEL, and R. L. MOTT: Methods in Enzymology, eds. S. FLEISCHER and L. PACKER, Volume **32B,** p. 723. New York, San Francisco and London: Academic Press. 1974.

94. STOHS, S. J.: Production of Scopolamine and Hyoscyamine by *Datura stramonium* L. Suspension Cultures. J. Pharm. Sci. **58,** 703 (1969).

95. STOHS, S. J., B. KAUL, and E. J. STABA: *Dioscorea* Tissue Cultures II. Metabolism of ^{14}C Labelled Cholesterol by *Dioscorea deltoidea* Suspension Cultures. Phytochemistry **8,** 1679 (1969).

96. STOHS, S. J., and H. ROSENBERG: Steroids and Steroid Metabolism in Plant Tissue Cultures. Lloydia **38,** 181 (1975).

97. STOHS, S. J., J. J. SABATKA, and H. ROSENBERG: Incorporation of [4-^{14}C-22,23-^{3}H$_2$-] sitosterol into Diosgenin by *Dioscorea deltoidea* Tissue Suspension Cultures. Phytochemistry **13,** 2145 (1974).

98. TABATA, M., N. HIRAOKA, M. IKENOUE, Y. SANO, and M. KONOSHIMA: Production of Anthraquinones in Callus Cultures of *Cassia tora.* Lloydia **38,** 131 (1975).

99. TABATA, M., H. YAMAMOTO, N. HIRAOKA, and M. KONOSHIMA: Organisation and Alkaloid Production in Tissue Cultures of *Scopolia parviflora.* Phytochemistry **11,** 949 (1972).

100. TELLE, J., and R. J. GAUTHERET: On the Indefinite Culture of the Root of Jusquiame *(Hyoscyamus Niger* L.). Compt. Rend. **224,** 1653 (1947).

101. TEUSCHER, E.: Problems in the Production of Secondary Plant Metabolites with Cell Cultures. Pharmazie **28,** 6 (1973).

102. TOMITA, Y., and S. SEO: Biosynthesis of the Terpenes Maslinic Acid and 3-Epimaslinic Acid in Tissue Cultures of *Isodon japonicus* Hara. Chem. Comm. **1973,** 707.

103. TOMITA, Y. S. SEO, and K. TORI: Biosynthesis of Ursene Triterpenes from [1,2-^{13}C]-

acetate in Tissue Cultures of *Isodon japonicus*. Reassignment of ^{13}C NMR Signals in Urs-12-enes. Chem. Commun. **1975**, 954.

104. Tomita, Y., K. Tori, and S. Seo: Biosynthesis of Oleanene and Ursene-type Triterpenes from [4-^{13}C]mevalonic Acid in Tissue Cultures of *Isodon japonicus* Hara. Chem. Comm. **1975**, 270.

105. — — — Carbon-13 NMR Spectra of Urs-12-enes and Application to Structural Assignments of Components of *Isodon japonicus* Hara Tissue Cultures. Tetrahedron Letters **1975**, 7.

106. Tomita, Y., and A. Uomori: Mechanism of the Biosynthesis of the Ethyl Side Chain at C 24 of Stigmasterol in Tissue Cultures of *N. tabacum* and *D. tokoro*. Chem. Comm. **1970**, 1416.

107. — — Biosynthesis of Sapogenins in Tissue Cultures of *Dioscorea tokoro* Makino. Chem. Comm. **1971**, 284.

108. — — Biosynthesis of Isoprenoids Part III Mechanism of Alkylation During Biosynthesis of Stigmasterol in Tissue Cultures of Higher Plants. J. Chem. Soc. Perkin I **1973**, 2656.

109. — — Structure and Biosynthesis of Prototokoronin in Tissue Cultures of *Dioscorea tokoro*. Phytochemistry **13**, 729 (1974).

110. Tulecke, W.: Plant Tissue and Organ Culture, American Biology Teacher **25**, 90 (1963).

111. Veliky, I. A., and K. M. Barber: Biotransformation of Tryptophan by *Phaseolus vulgaris* Suspension Culture. Lloydia **38**, 125 (1975).

111a. Veliky, I. A.: Synthesis of Carboline Alkaloids by Plant Cell Cultures. Phytochemistry **11**, 1405 (1972).

112. White, P. R.: Potentially Unlimited Growth of Excised Plant Callus in an Artificial Nutrient. Amer. J. Botany **26**, 59 (1939).

113. — The Cultivation of Animal and Plant Cells, 2nd Edition. New York: The Ronald Press. 1963.

114. Yafin, Y., and I. Schechter: Comparison between Biosynthesis of *Ent*-kaurene in Germinating Tomato Seeds and Cell Suspension Cultures of Tobacco and Tomato. Plant Physiol. **56**, 671 (1975).

115. Yano, I., L. J. Morris, B. W. Nicholas, and A. T. James: Biosynthesis of Cyclopropane and Cyclopropene Fatty Acids in Higher Plants. Lipids **7**, 35 (1972).

(Received June 23, 1976)

Carbazole Alkaloids

By D. P. CHAKRABORTY, Bose Institute, Calcutta 700009, India

With 11 Figures

Contents

I. Introduction

Carbazole (1) was discovered in coal tar by GRAEBE and GLASER in 1872. The major development in the chemistry of carbazoles up to 1920 took place in Europe due to its importance in the European dye stuff industry. The use of vinyl carbazoles and its polymerisation products in industry gave further impetus to the studies of carbazole chemistry in the late thirties. Detailed references on the chemistry of carbazoles are available in previous monographs on the subject (56, 103).

References, pp. 366—371

Carbazoles were unknown as plant products except as degradation products of strychnine and aspidospermine until the discovery of ulein (**2**) and olivicine (**3**) (*77, 99, 100*) two novel pyridocarbazoles of the genus *Aspidosperma* (Fam. Apocynaceae) in 1957—58. Pyridocarbazoles constitute a novel group of indole bases lacking the conventional tryptamine fragment. Murrayanine (**4**) (*24*), the first member of the carbazole alkaloids without any basic nitrogen and distinctly different from the pyridocarbazoles (*58*), was reported in 1965*. Since then there has been a considerable progress in this group of alkaloids. Carbazole alkaloids are neutral or phenolic but this group of nitrogenous plant constituents originates from anthranilic acid like many alkaloids of Rutaceae. This justifies its inclusion in the alkaloidal group. Among the few members of these alkaloids there has been a considerable variation in structural complexity. Thus, murrayanine (**4**) is tricyclic and murrayazoline (mahanimbidine) (**5**) (*9, 72*) is hexacyclic. A review of carbazole alkaloids was first presented by KAPIL (*68*).

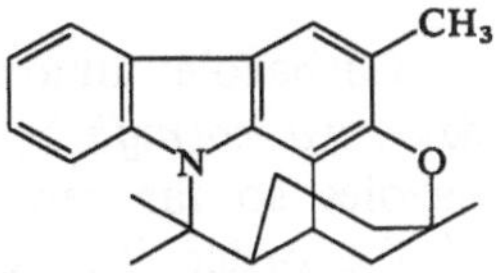

(**5**)

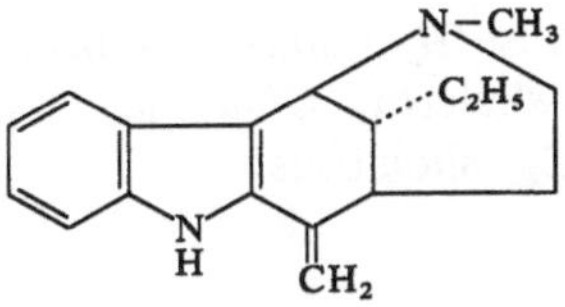

(**2**)

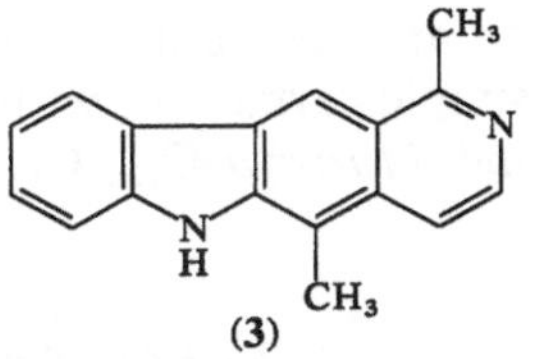

(**3**)

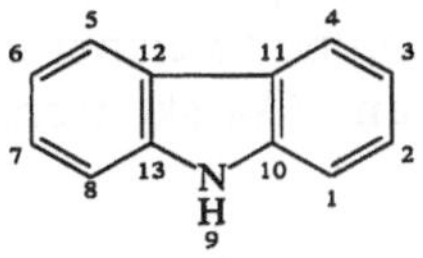

(**1**)

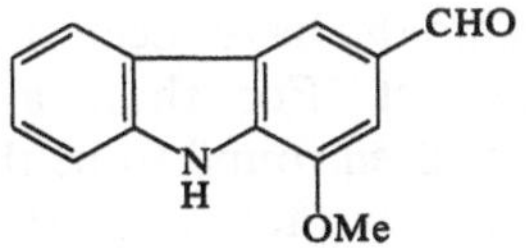

(**4**)

A. Nomenclature and Electronic Characteristics

The system of nomenclature used in this review is in accord with those followed in chemical abstracts (*85*) for carbazole (**1**). The electronic characteristics (*89*) of carbazoles are detailed below.

---

* In the report on the isolation of girinimbine (*23*), we reported 1-methoxy-3-formyl carbazole (murrayanine) as the first carbazole alkaloid.

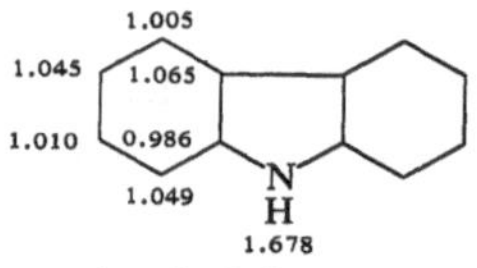

electrical charges

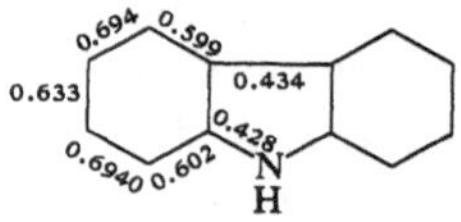

bond orders

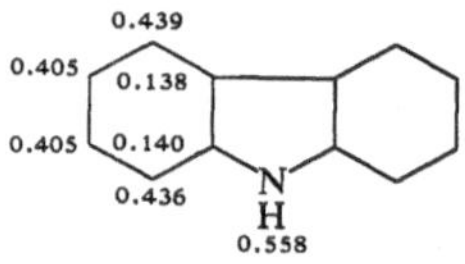

free valencies

B. Occurrence

Carbazole alkaloids were first reported from the stem bark of *Murraya koenigii* Spreng. (*69*) which is the richest source of phytocarbazoles so far reported. The plant belongs to the family Rutaceae of the order Rutales and is in the subtribe Clauseneae of the subfamily Aurantoidae (*51*). Subsequently they were isolated from the genera *Glycosmis* and *Clausena* belonging to the same subtribe of the family Rutaceae. So far (see Table 3), the genus *Glycosmis* has been found to elaborate simple carbazoles with a C_{13}-skeleton, *Clausena*, C_{13} and C_{18} skeletons and *Murraya* C_{13}, C_{18} and C_{23} skeletons.

C. Isolation

Several carbazole alkaloids have been isolated from the neutral fraction of the plant extract. For these alkaloids petroleum ether (boiling range 40—60°) has been found to be the best extracting solvent. From the petroleum ether extract, a mixture of several carbazole alkaloids crystallises readily. From this mixture pure compounds are obtained by chromatographic resolution over alumina. For phenolic or acidic compounds extraction using more polar solvents have been taken recourse to. After fractionation of the crude extract into phenolic or acidic fractions, purification by chromatography over silica gel is generally carried out. In some cases acid-catalyzed isomerisation of the natural compounds results in artifacts of isolation. Drastic acid treatment of the plant extract during fractionation procedure is not advisable, as it may lead to isolation of transformed natural products.

D. Detection of Carbazoles by Chromatographic Methods

Paper and thin layer chromatographic methods (33, 52) have been worked out to detect carbazoles and test their homogeneity. Separation of isomeric methyl carbazoles from mixtures was feasible by thin layer chromatography using appropriate solvent system.

In the developed chromatogram, carbazoles could be detected from fluorescence or quenching under uv light. Several spray reagents have been used to detect carbazoles on a developed chromatogram. Picric acid, antimony trichloride, antimony pentachloride (101), ferric chloride, concentrated hydrochloric acid and dichlorodicyanobenzoquinone (DDQ) (95) have been successfully used as spray reagents. The developed chromatogram gives brick red to red spots when sprayed with picric acid while with other spray reagents, greenish or bluish green or blue spots are formed depending both on reagents and compounds.

II. Methods of Structure Elucidation

A. Physical Methods

1. Ultraviolet Absorption Spectra

The ultraviolet absorption spectrum of carbazole (λ_{max} 233, 257, 293, 322, 336 nm; log ε 4.51, 4.18, 4.10, 3.46, 3.39) differs from its carbocyclic analogue fluorene due to the contribution of the lone pair of electrons of the heterocyclic nitrogen to the chromophoric system. This gives rise to the characteristic carbazole spectrum (46). Substitutions at different positions cause significant changes and in some cases a diagnostic spectral pattern is obtained. Substitutions of the methyl group causes a minor shift (Fig. 1). An approximate theoretical calculation of the extent and direction of the spectral shift with methyl substitutions has been made by BASU (6). The calculated shift has been found to agree with the experimental results (29). The significant and diagnostic spectra of substituted carbazoles have been recorded in the case of formyl (11) (Fig. 2) and methoxycarbazoles (29) (Figs. 3, 4, 5). Formyl-, methoxy- and pyranocarbazoles (Fig. 6) give rise to characteristic spectra which have been extensively utilized in the structure determination of carbazole and related alkaloids. The behaviour of compounds containing both formyl and other contributing substituents is interesting. Murrayanine and indizoline (6) which have a formyl group at the 3- and a methoxy at the 1-position show spectra characteristic of the 3-formyl carbazole chromophore. On reduction of the formyl group, uv spectra similar to that of 1-methoxycarbazole are obtained. The spectrum of the 3-formylcarbazole chromophore is readily discernible in murrayacine (7) and murrayacinine (8). They have a 2,2-substituted-Δ^3-pyran fragment besides the 3-formyl group. On reduction of the formyl group, they show spectra characteristic of a 2-substituted-Δ^3-pyran system. Heptazoline (9) and heptaphylline (10) having a 3-formyl group and hydroxyl groups at other positions show uv spectra characteristic of 3-formylcarbazole.

(6)

(7)

(8)

(9)

(10)

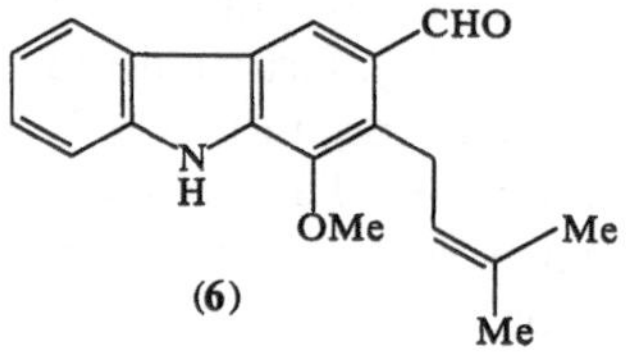

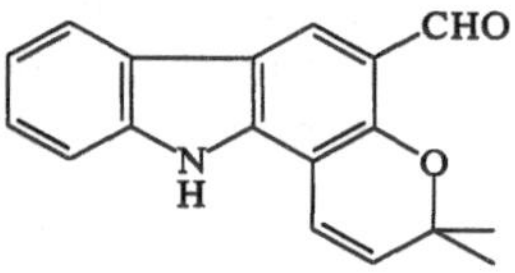

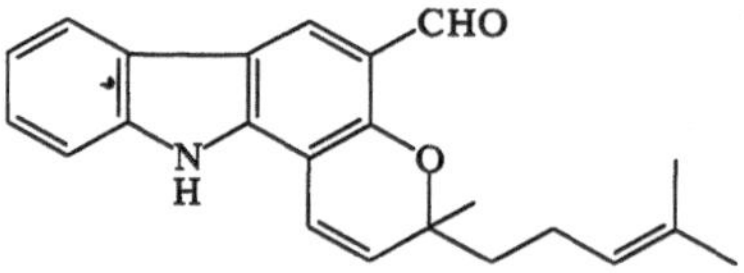

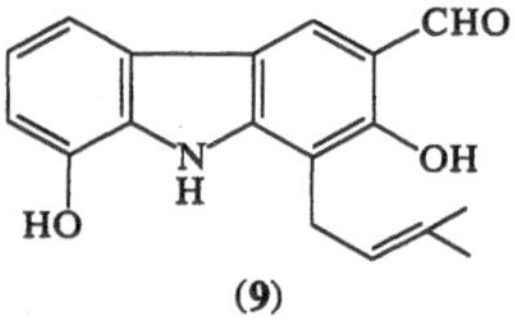

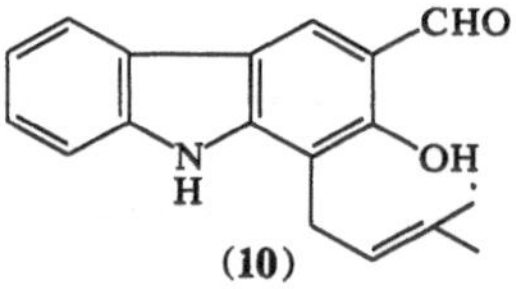

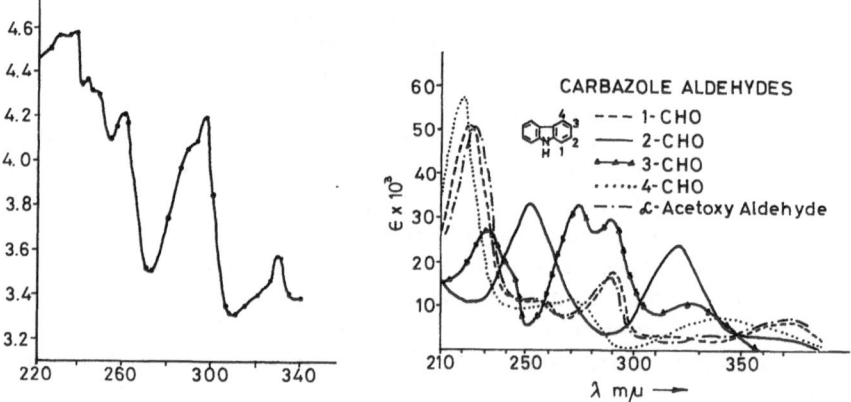

Fig. 1—8: Ultraviolet absorption spectra of carbazoles

Fig. 1. 3-Methylcarbazole

Fig. 2. Formylcarbazoles

References, pp. 366—371

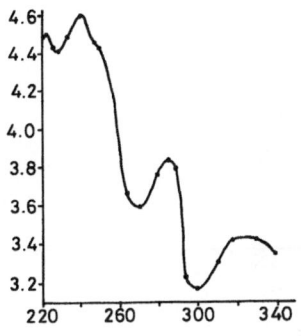

Fig. 3. 1-Methoxycarbazole

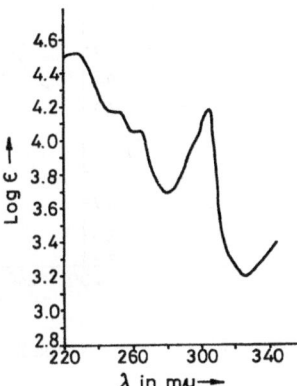

Fig. 4. 3-Methoxy-6-methylcarbazole

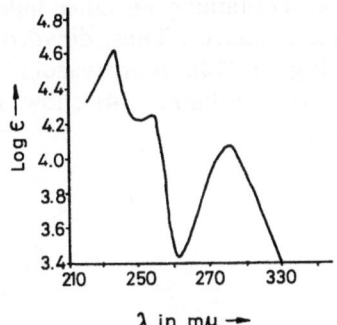

Fig. 5. 2-Methoxycarbazole

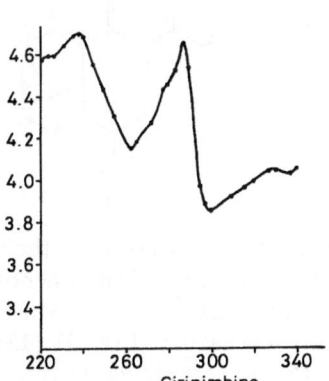

Girinimbine

Fig. 6. Girinimbine (Pyranocarbazole)

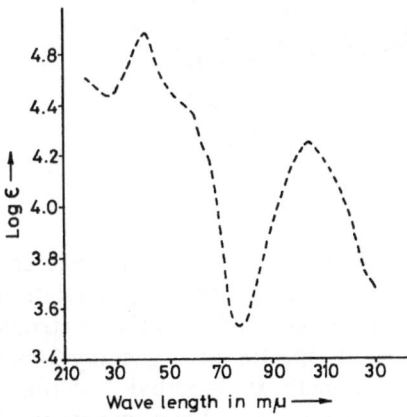

Fig. 7. Tetrahydromahanimbine

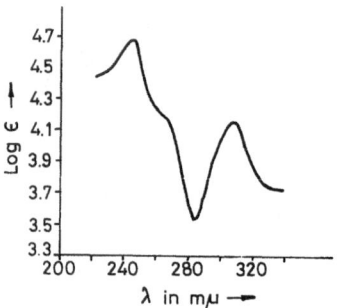

Fig. 8. Murrayazoline

When the double bond in 2,2-disubstituted-Δ^3-pyran is reduced the spectrum of the resulting alkaloid becomes that of the methoxy-carbazole chromophore containing an ether linkage in the appropriate position of the carbazole nucleus. Thus, dihydrogirinimbine (11), tetra-hydromahanimbine (Fig. 7) (12), murrayazoline (5) (Fig. 8), murray-azolidine (13) and murrayazolinine (14) show spectra similar to that of 2-methoxycarbazole.

(11)

(12)

(13)

(14)

The uv spectra of phenolic carbazoles are similar to those of their methoxy congeners. On acetylation, the oxygen lone pair of the phenol does not contribute to the electronic state of the carbazole system. Thus, the uv spectra of 2-acetoxy-3-methyl-1-formyl-carbazole (Fig. 2) (15) and 2-methoxy-3-methyl-6-acetoxycarbazole (16) are very similar to those of 1-formyl- and 2-methoxycarbazole, respectively. Similarly, heptazoline monoacetate assumes the spectrum of the heptaphylline chromophore.

The spectra of the acetylated products have been helpful in the structure determination of glycozolidine (17) and heptazoline (9). The uv spectra of carbazole alkaloids and related compounds are summarized in Table 1.

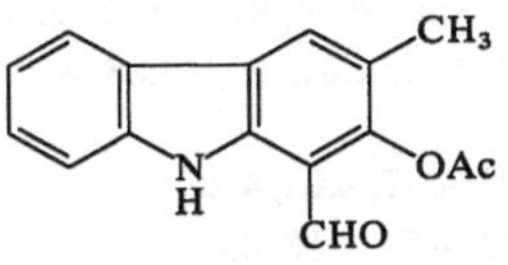

(15)

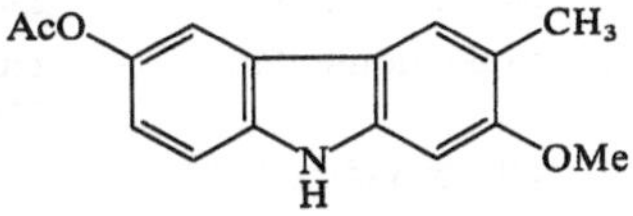

(16)

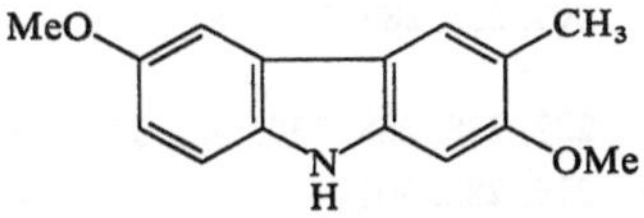

(17)

Table 1. *Ultraviolet Absorption Spectra of Carbazole Alkaloids and Some Related Compounds*

Compounds	λ_{max} in nm	log ϵ
1. Carbazole	233, 257, 293, 322–24, 336	4.5127, 4.1829, 4.1049, 3.4657, 3.3988
2. N-Methylcarbazole	235, 261, 293, 330, 336, 344	4.6743, 4.3654, 4.2509, 3.5655, 3.4930, 3.5977
3. 1-Methylcarbazole	214, 235, 259, 298, 322, 336	4.4153, 4.6684, 4.2898, 4.2238, 3.6098, 3.4338
4. 2-Methylcarbazole	235, 259, 296, 322, 334	4.5677, 4.1785, 4.1246, 3.5465, 3.4004
5. 3-Methylcarbazole	236, 260, 296, 328, 342	4.3960, 4.0267, 4.0267, 3.3088, 3.3088
6. 1-Methoxycarbazole	240, 252, 286, 320	4.5900, 4.3900, 3.8400, 3.4600
7. 2-Methoxycarbazole	235, 257, 303	4.5817, 4.2365, 4.0901,
8. 3-Methoxycarbazole	228–30, 251, 262–63, 300, 314, 329, 342	4.4917, 4.1838, 4.0816, 4.2365, 4.2365, 3.3914, 3.5053
9. 4-Methoxycarbazole	224, 243, 286, 327, 332	4.4568, 4.6658, 4.1558, 3.6344, 3.7555
10. 3-Methyltetrahydro-carbazole	232–33, 286	4.3326, 3.7258
11. 3-Methoxy-2-methyl-carbazole	232–34, 263, 303, 322, 336	4.4781, 4.0546, 4.2065, 3.4642, 3.5004
12. Glycozoline (19)	227, 252, 264, 304	4.16, 4.16, 4.06, 4.17
13. Murrayanine (4)	238, 247, 274, 289, 335	4.47, 4.30, 4.50, 4.56, 4.16

Table 1 (continued)

Compound	λ_{max} in nm	log ϵ
14. Mukoeic acid (38)	235, 270, 320	4.51, 4.58, 3.92
15. Mukonine (78)	236, 245, 266, 274, 306, 320	4.6, 4.57, 4.64, 4.73, 3.88, 3.78
16. Mukonidine (79)	230, 248, 266, 272, 308, 316 323	4.8, 4.57, 4.64, 4.73, 3.88, 3.78, 3.18
17. Glycozolidine (17)	233, 237, 260, 309	4.47, 4.47, 4.08, 4.20
18. Heptaphylline (10)	234, 278, 298, 346	4.42, 4.53, 4.58, 4.09
19. Indizóline (6)	237, 247, 275, 294, 330	4.61, 4.64, 4.78, 4.49, 4.16
20. Heptazoline (9)	240, 275, 298	4.66, 4.62, 4.40
21. 6-Methoxyhepta-phylline (129)	207, 235, 283, 311, 341	4.5, 4.35, 4.47, 4.57, 3.95
22. Girinimbine (18)	223, 238, 288, 330, 342, 358	4.49, 4.70, 4.65, 4.04, 3.89, 3.84
23. Murrayacine (7)	226, 282, 301	4.60, 4.57, 4.58
24. Koenimbine (76)	230, 240, 300, 340, 360	4.68, 4.69, 4.62, 4.24, 4.2
25. Heptazolidine (41)	230, 238, 278, 282, 335	4.68, 4.86, 4.29, 4.49, 3.70
26. Koenigicine (77)	239, 300, 361	4.56, 4.52, 3.97
27. Mahanimbine (27)	223, 239, 288, 330, 343	4.55, 4.06, 4.61, 3.90, 3.42
28. Mahanimbicine (82)	238, 281, 290, 335, 354	4.69, 4.39, 4.72, 3.89, 3.87
29. Mahanimbinine (193)	238, 288, 329, 344, 359	4.64, 4.61, 3.83, 3.87, 3.82
30. Murrayacinine (8)	234, 280, 301, 312	4.60, 4.5, 4.58, 4.50
31. Murrayazoline (5)	245, 257, 307	4.69, 4.37, 4.16
32. Murrayazolidine (13)	241, 260, 305	4.57, 4.23, 4.07
33. Bicyclomahanimbine (196)	242, 255, 260, 305, 331	4.61, 4.43, 4.40, 4.21, 3.62
34. Bicyclomahanimbicine (198)	243, 257, 263, 305	—
35. Murrayazolinine (14)	240, 253, 258, 304	—

2. IR Spectra

The ir spectra of carbazoles show bands characteristics for imino compounds. The peak of the NH bands shifts on change of phase. Thus, the –NH– bands of murrayanine, girinimbine and mahanimbine are different in the KBr phase as compared with those in the CHCl₃-phase due to intermolecular association in the solid phase. 1-, 2-, 3- and 4-methyl carbazoles have diagnostic ir bands in the region 850—800 cm⁻¹ (93). Richards' observations have been successfully utilized by Robinson and Pauscker (86) in the identification of methyl carbazoles obtained by degradation of strychnine. Chakraborty et al. utilized these data for the identification of carbazole, [ν_{max} 856 and

915 cm^{-1}] 3-methylcarbazole (v $_{max}$ 808 cm^{-1}) or a mixture of both in the degradation products of carbazole alkaloids. The unsaturation associated with 2 : 2-dimethyl-Δ^3-pyran shows an absorption band at 1645 cm^{-1}. Usually the aromatic aldehyde peaks of the carbazole alkaloids appear in the region 1680 cm^{-1}; with chelation as in hepta-phylline (10) and heptazoline (9), the carbonyl absorption moves to higher wave-length. The aromatic aldehyde band at 2870 cm^{-1} is also discernible in some cases.

3. NMR Spectra

The NMR spectra of the carbazole alkaloids (Table 2) have been very helpful in structure elucidation. The carbazole –NH– being acidic it absorbs near δ 10.0 but with substitution the signal for the –NH– proton has been found to be shifted. The signal for the –NH– proton has been registered at δ 7.6 to 12.01 in different carbazole alkaloids. The signal for the protons on C-4 and C-5 appear at lower field (around δ 7.17 to 7.5) as these two protons are phenanthrenic (78) and are mutually deshielded (13). The other aromatic protons of carbazole occur at higher frequency as complex multiplets. Resonances are shielded or deshielded according to the environment of the proton in question.

The downfield shift of the ethylenic protons of a chromene ring adjacent to the aromatic ring has been attributed to the deshielding effect of the heterocyclic nitrogen. This has been used to explain the angular fusion of the chromene ring in girinimbine (18) and related compounds. The ethylenic proton of the chromene adjacent to the aromatic ring shows a significant shift (δ 0.3 to 0.5) on changing the solvent from CDCl$_3$ to DMSO (84). This has been utilised in determining the angular fusion of the 2,2 substituted-Δ^3-pyran system in C$_{18}$ and C$_{23}$ carbazole alkaloids. The methyl protons of N–CH$_3$ carbazoles expe-rience a deshielding effect when an angular 2,2-dimethyl-Δ^3-pyran with an oxygen ether linkage at the 2-position is present (65).

The presence of a carbazole skeleton in murrayanine (4) was first detected from its NMR spectrum (Fig. 9).

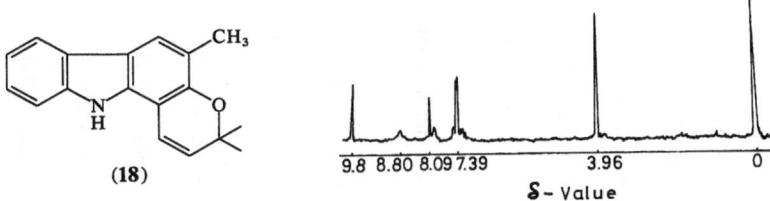

(18)

9.8 8.80 8.09 7.39   3.96   0

S - Value

Fig. 9. NMR spectrum of murrayanine

Table 2. *NMR Spectra of Carbazole Alkaloids*
(δ, ppm; TMS, O)

Compound	Aromatic protons				Protons of the substituent on the aromatic nucleus	Protons at other positions
	-NH-	C-4	C-5	Others		
1. Glycozoline (19)	7.8 (d)	7.5 (d)	7.5 (d)	6.8 (m); 7.18 (m)	3.9 (s, $-OCH_3$); 2.5 (s, $Ar-CH_3$)	—
2. Murrayanine (4)	8.80	7.39	7.39	8.09 (m)	3.96 (s, $-OCH_3$); 9.98 (s, $-CHO$)	
3. Mukoeic acid methylester (78)	8.53	8.15 (s)	7.65	7.3—7.5	4.07 (s, $-OCH_3$); 4.02 (s, $-COOCH_3$)	
4. Glycozolidine (17)	7.65 (m)	7.17 (m)	7.45 (m)	6.52 (m); 6.92—7.02	3.9 and 3.77 (s, $2-OCH_3$); 2.35 (s, $Ar-CH_3$)	
5. Heptaphylline (10)	10.3 (br)	8.25 (s)	8.0	7.1—8.3	9.9 (s, $-CHO$); 11.70 ($-OH$)	3.6 (d, $J = 6$ cps; benzylic; 5.35 (t, $J = 6$ cps; vinylic); 1.66 and 1.83 (d, $J = 1$ cps, *gem*-dimethyl group)
6. Indizoline (6)	8.8 (br)	8.4 (q)	8.02 (q)	7.4 (m)	3.29 (s, $-OCH_3$); 10.3 (s, $-CHO$)	3.9 (benzylic-); 5.25 (vinylic); 1.7 and 1.85 (d, $J=1$ cps *gem*-dimethyl groups)
7. 6-Methoxyheptaphylline (129)	8.15 (br)	7.92 (s)	7.45 (d) $J - 2$ Hz	7.0 (d, $J = 7.5$; 2 cps); 7.3 (d; $J = 7.5$ cps)	3.9 (s, $-OCH_3$); 9.9 (s, $-CHO$); 11.0 (s, $-OH$)	3.6 (d, $J = 8$ cps, benzylic); 5.3 t, $J = 8$ cps vinylic; 1.88, 1.75 (s, br, *gem*-dimethyl)
8. Heptazoline (9)	11.25	8.28	7.5 (dd) $J = 8, 2$ cps	6.7—7.16	9.91 ($-CHO$), 9.82 and 11.5 ($2-OH$)	3.63 (d, $J = 10$ cps, benzylic; 5.3 (m, vinylic); 1.65 & 1.81 (s, *gem*-dimethyl)
9. Girinimbine (18)	7.75	7.5		7.08	2.28 (s, $Ar-CH_3$)	1.42 (s, *gem*-dimethyl); 6.25 (d, $J = 10$ cps) (olefinic); 5.45 (d, $J = 10$ cps) (olefinic)
10. Murrayacine (7)	12.0	8.4	8.1—8.2	7.2—7.5	10.68 ($-CHO$)	1.40 (s, *gem*-dimethyl); 5.95 (d, $J = 10$ cps); 7.0 (d, $J = 10$ cps) (olefinic)
11. Koenine (171)		7.55	7.20 $J = 2$ cps	6.74 (dd, $J = 8$, 2 cps); 7.18 (d, $J = 8$ cps)		
12. Koenimbine (76)	7.75—6.85	7.61	7.4	6.91 (dd, $J = 8.5$, 2 cps); 7.18 (d, $J = 8.5$ cps)	3.91 (s, $-OCH_3$); 2.33 (s, $Ar-CH_3$)	1.46 (s, $O-C-CH_3$); 6.57 (d, $J = 10$ cps); 5.65 (d, $J = 10$ cps) (olefinic)

Table 2 (continued)

Compound	Aromatic protons				Protons of the substituent on the aromatic nucleus	Protons at other positions
	-NH-	C-4	C-5	Others		
13. Heptazolidine (**41**)	8.1 (br)	7.67 (s)	7.5 (d, J = 2 cps)	7.18 (d, J = 8.5 cps); 6.94 (d, J = 8.5 cps)	3.98 (s, -OCH₃); 2.37 (s, Ar-CH₃)	1.49 (s, *gem*-dimethyl); 6.63, (d, J = 10 cps), 5.65 (d, J = 10 cps) (olefinic)
14. Koenigicine (**77**)	7.81 (br)	6.83 (s)	7.4 (s)	7.5 (s)	3.86 and 3.96 (s, 2-OCH₃); 2.33 (s, Ar-CH₃)	1.5 (s, *gem*-dimethyl); 6.58 (d, J = 10 cps); 5.65 (d, J = 10 cps) (olefinic)
15. Koenigine (**170**)		7.52	7.42	6.81 (d, J = 2 cps)		5.65 (d, J = 10 cps); 6.55 (d, J = 10 cps) (olefinic)
16. Mahanimbine (**27**)	8.68	7.62 (s)	7.87 (complex)	7.05–7.35 (m)	2.29 (s, Ar-CH₃)	1.38 (s, -O-C-CH₃); 5.11 (d, J = 7 cps, vinylic); 1.54 and 1.64 (s, *gem*-dimethyl); 6.41 (d, J = 10 cps); 5.51 (d, J = 10 cps) (olefinic)
17. Mahanimbicine (**82**)	7.65 (m)	7.7 (d, J = 8.5 cps)	7.68	6.7 (d, J = 8.5 cps) 7.05 (d, J = 8 cps) 7.15 (d, J = 8 cps)	2.46 (s, Ar-CH₃)	1.44 (s, -O-C-CH₃); 1.7 to 2.4 (m, benzylic); 5.1 (t, vinylic); 1.55 & 1.65 (s, *gem*-dimethyl)
18. Mahanine (**189**)		7.58	7.54	6.54–6.74 (complex m)		
19. Mahanimbinine (**193**)		7.65	7.93	7.45–7.06	2.33 (s, Ar-CH₃)	1.41 (s, -O-C-CH₃); 1.86 to 1.35 (m, benzylic); 1.2 (s, *gem*-dimethyl); 6.60 (d, J = 10 cps); 5.58 (d, J = 10 cps)
20. Murrayazoline (**5**)		8.00	7.13		2.33 (s, Ar-CH₃)	1.43 (s, -O-C-CH₃); 3.33 (benzylic); 1.26 & 1.89 (s, *gem*-dimethyl)
21. Murrayazolidine (**13**)	7.8–7.9	7.65 (s)	7.2	7.35	2.35 (s, Ar-CH₃)	3.4 (benzylic), 2.00 (m); 1.45 (s, -O-C-CH₃); 4.8 (d, vinylic); 1.40 (s, *gem*-dimethyl)
22. Murrayazolinine (**14**)	7.9 (br)	7.63 (s)	7.45	7.00 (complex, m)	2.34 (s, Ar-CH₃); 1.62 (s, -OH)	1.22 (s, -O-C-CH₃); 1.4 (s, *gem*-dimethyl)
23. Bicyclomahanimbine (**196**)	8.08–7.05	7.74 (s)	8.08	7.05	2.36 (s, Ar-CH₃)	1.43 (s, O-C-CH₃); 3.26 (benzylic); 0.71 & 1.53 (*gem*-dimethyl)

Table 3. *Distribution of Carbazole Alkaloids*
(Fam. *Rutaceae;* subfam. *Aurantoidea;* subtribe *Clauseneae*)

Genus	C$_{13}$-Alkaloids	C$_{18}$-Alkaloids	C$_{23}$-Alkaloids
Glycosmis	Glycozoline Glycozolidine	—	—
Clausena	3-Methylcarbazole Murrayanine	Indizoline Heptazoline Heptaphylline 6-Methoxyheptaphylline Heptazolidine Girinimbine Murrayacine	—
Murraya	Murrayanine Mukoeic acid Mukonine Mukonidine	Girinimbine Koenimbine Koenigine Koenidine Koenine Murrayacine	Mahanimbine dl-Mahanimbine Mahanine Mahanimbinine Murrayazoline Mahanimbidine (Currayangin) Murrayazolinine Cyclomahanimbine Murrayazolidine Bicyclomahanimbine Mahanimbicine Iso mahanimbine Bicyclomahanimbicine Murrayacinine

4. Mass Spectra

Carbazole being a stable aromatic system, the mass spectral behaviour of the carbazole skeleton provides rather little information for structure elucidation. Substituted carbazoles behave very similar to substituted benzenes. Thus methoxycarbazoles lose a mass of 15 due to elimination of the methyl group. This is followed by ring contraction as shown below for glycozoline (**19**). Carbazole alkaloids with a carbomethoxy group lose a mass of 59 due to loss of the –COOCH$_3$ group (*2, 79*). A peak at (M-44) is obtained in the case of carbazole carboxylic acids. A formylcarbazole loses a mass of 28 due to loss of the –CO– group. Hydroxycarbazoles give a high intensity peak at (M-1) which is the reminiscent of the fragmentation behaviour of some phenolic compounds (*1*).

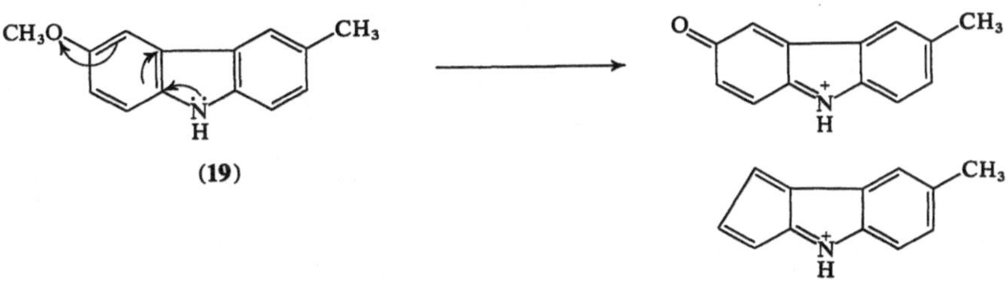

(19)

The mass spectral behaviour of carbazole alkaloids containing a 2,2-disubstituted-Δ^3-pyran or 2,2-disubstituted-pyran ring has been of much help in their structure elucidation. It is well known (106) that 2,2-dimethyl-Δ^3-pyran (M^+ 160) (20) gives rise to a high intensity peak at m/e 145 (M-15) due to the formation of the pyrylium ion (21).

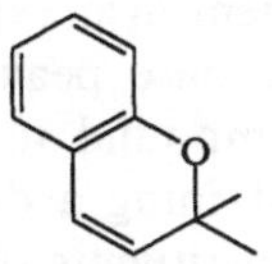

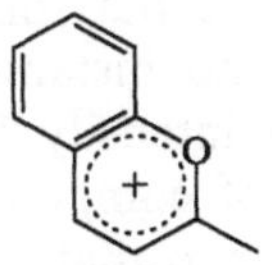

(20)   (21)

The dihydro compound (22) (M^+ 162) on the other hand, cleaves to give (23) (M-15) and (24) (M-55). The M-15 peak is not the base peak. The behaviour of girinimbine and dihydrogirinimbine is similar. Girinimbine (18) gives a high intensity base peak at m/e 248 (M-15) which could be represented by the carbazolopyrylium ion (25). In the dihydro series a peak at m/e 210 which corresponds to (26) is discernible. The high intensity peak (25) which has consistently been obtained from C_{18} and C_{23} alkaloids bearing this structural element is considered to be very diagnostic.

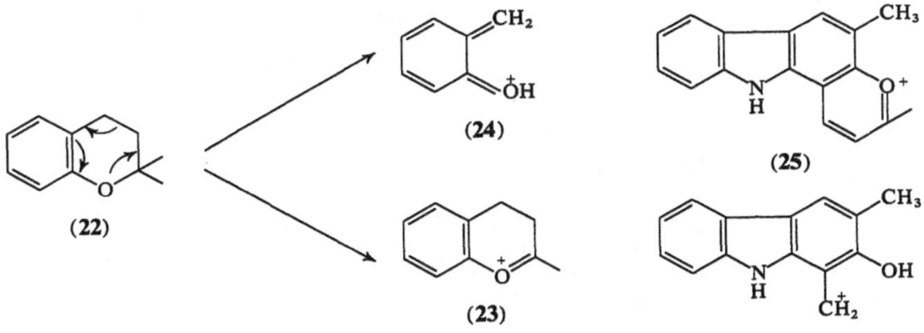

(22)   (24)   (25)

(23)   (26)

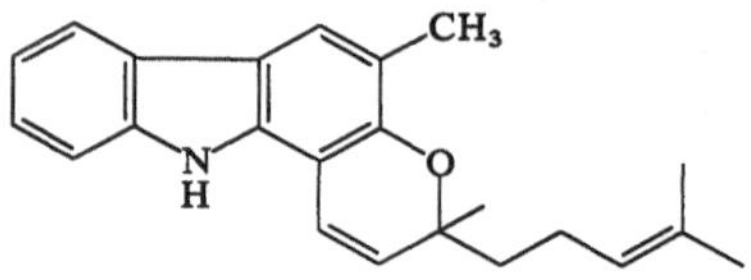

(27)

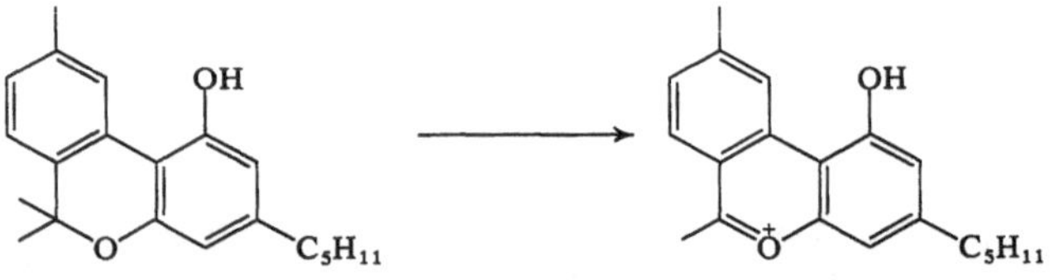

(28)   (29)

The behaviour of C_{23} alkaloids like mahanimbine (27) (Fig. 10) is similar. The latter gives a high intensity peak of type (25). C_{23} alkaloids with a saturated pyran ring behave essentially like dihydrogirinimbine. They show a peak at m/e 210 but some of them also exhibit the M-15 peak at m/e 248 which in no instance is the base peak. Further, the peak at m/e 210 differs in intensity from compound to compound. In this respect the behavior of compounds containing a dihydropyran or -Δ^3-pyran system is similar to that of the constituents of hashish (12) where the fragmentation depends on the nature of the other ring system attached to the 2,2-dimethyl-Δ^3-pyran system. The base peak in cannabinol (28) is at m/e 295 (29) as after expulsion of the methyl group, the compound assumes a compact aromatic system.

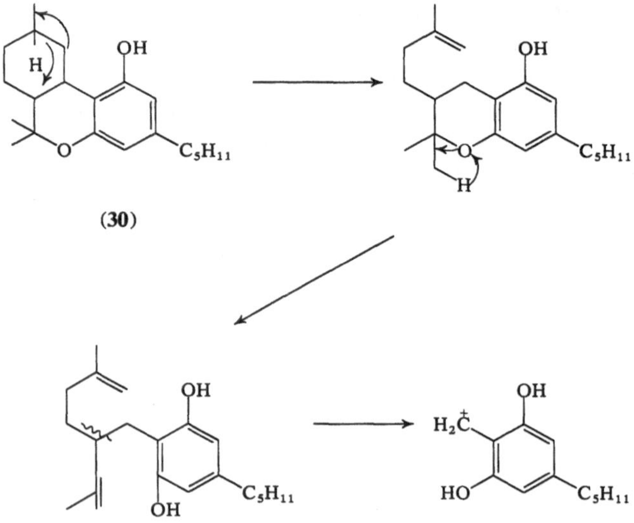

(30)

(31)

In the case of the hexahydrocannabinol (**30**) system, the fragmentation takes a different course and a high intensity peak at m/e 193 (**31**) is observed which corresponds to that of m/e 210 in the dihydro-girinimbine series.

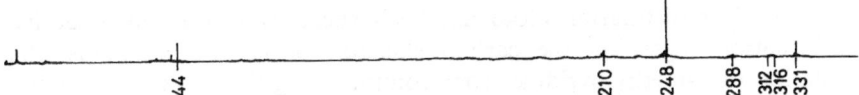

Fig. 10. Mass spectrum of mahanimbine

5. X-Ray Crystallographic Methods

X-ray crystallography was used by BORDNER *et al.* (*9*) to settle the structure of murrayazoline (**5**) unambiguously, the first carbazole alkaloid to have its structure determined in this manner (Fig. 11).

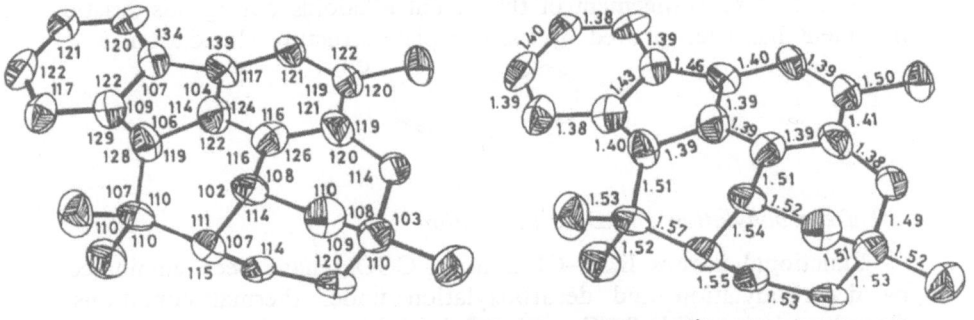

Fig. 11. X-ray crystal structure of murrayazoline. Bond distances (Å) and angles (degrees). The uncertainties in the C, N, O distances are about 0.01 Å. Uncertainties in the bond angles are about 0.5°. Angles not shown on the drawing are 9B-9-9A.111°; 8-9-9C.110°; 13-12-12A.114°; 1A-12-11.112°; 4-4A-7A.119°

B. Chemical Methods

Degradative Reactions

The degradation of the carbazole nucleus is difficult due to its high stability. In the structure elucidation of carbazole alkaloids, chemical degradations have been applied to get informations regarding the skeletal unit, transformation of the functional groups on the carbazole nucleus and informations regarding the nature of the vulnerable side chain or ring system fused to the carbazole nucleus. The following degradative methods have been successfully utilized in the structure elucidation of carbazole alkaloids: (a) Zinc dust distillation, (b) decarbonylation and decarboxylation, (c) demethylation, (d) ozonisation, (e) permanganate oxidation.

(a) Zinc Dust Distillation

Zinc dust distillation (*104*) has been successfully utilized to derive information regarding the carbon skeleton of various alkaloids, the isolation of 3-methylpyridine from coniine being the classical example. This drastic treatment sometime results into the isolation of rearranged products. The isolation of carbazole (**1**) from ajmaline may be cited as an example. In the case of carbazole alkaloids, zinc dust distillation has resulted in the isolation of carbazole (**1**), 3-methylcarbazole (**32**) or a mixture of both. Carbazole was obtained by zinc dust distillation of murrayanine (**4**), murrayacine (**7**) while 3-methylcarbazole was obtained from glycozoline (**19**), glycozolidine (**17**), girinimbine (**18**), mahanimbine (**27**) and the Wolff-Kishner reduction product of murrayanine. That the isolation of carbazole (**1**) or 3-methylcarbazole (**32**) is not due to rearrangement of the parent alkaloids during this drastic treatment has been proved by the complete structure elucidation.

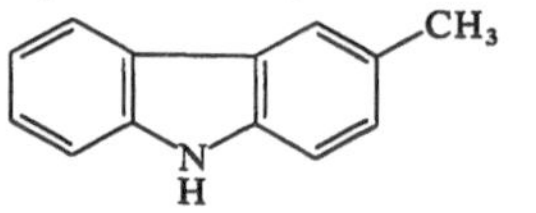

(32)

(b) Decarbonylation and Decarboxylation

Functional groups like –CHO and –COOH have been eliminated by decarbonylation and decarboxylation under thermal conditions. Decarbonylation with Pd/C at elevated temperatures has been used to convert murrayanine to 1-methoxycarbazole (**33**). The phenolic aldehyde (**34**) obtained by ozonisation of girinimbine (**18**) was decarbonylated to 2-hydroxy-3-methylcarbazole (**35**). Cycloheptaphylline (**36**) was decarbonylated to dihydropyranocarbazole (**37**).

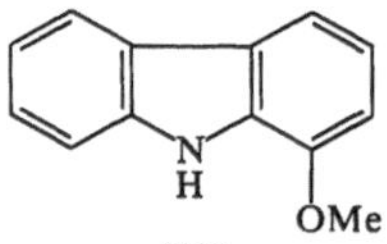

(33)

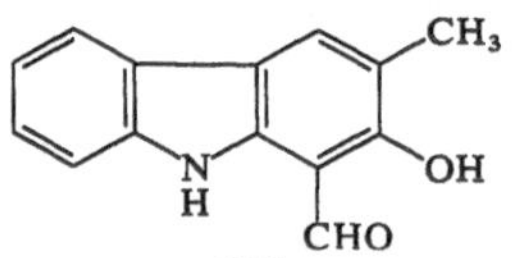

(34)

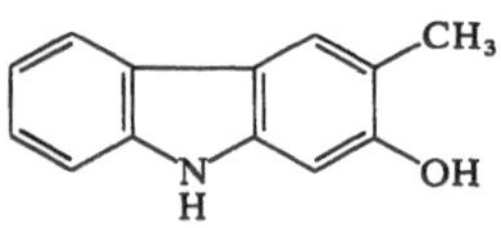

(35)

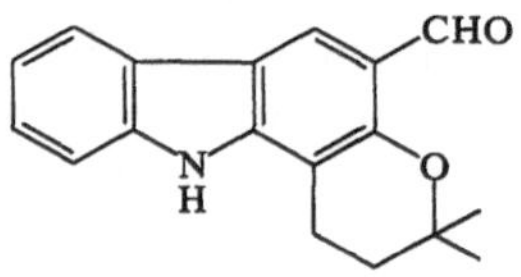

(36)

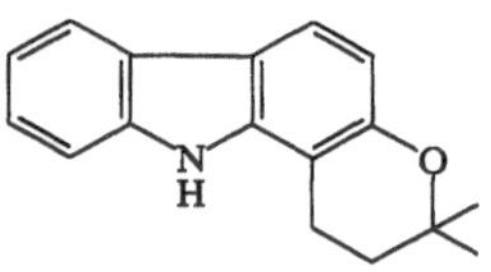

(37)

Decarboxylation of mukoeic acid (**38**) to 1-methoxycarbazole (**33**) was effected at 145° using antimony trichloride as a decarboxylating agent.

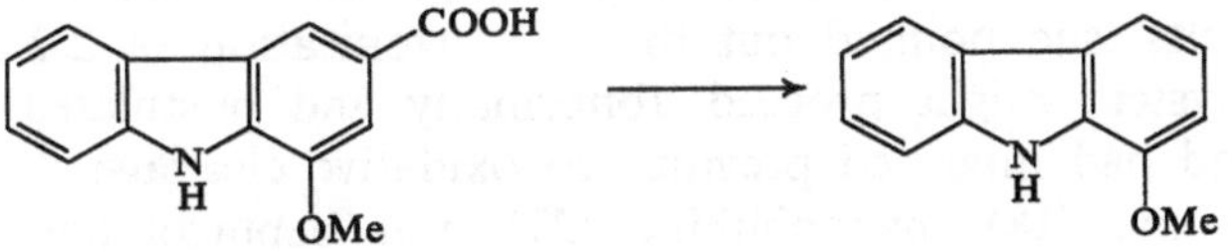
COOH
N
H
OMe
N
H
OMe

(**38**) (**33**)

(c) Demethylation

Demethylation with HBr (48%) has been used to demethylate the aromatic methoxyl group. The most interesting application has been in the structure elucidation of glycozolidine (**17**). Selective demethylation of glycozolidine (**17**) with HBr afforded a phenol (**39**) in which one of the methoxyl groups was retained. The uv spectrum of the O-acetate of the above phenol was similar to that of 2-methoxycarbazole showing that one methoxyl group was at the 2-position of the carbazole nucleus. Demethylation with pyridine hydrochloride however gave the dihydroxy compound.

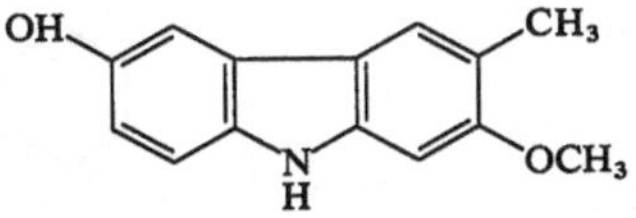
OH
CH₃
N
H
OCH₃

(**39**)

(d) Ozonisation

The nature of the fusion of the 2,2-dialkyl-Δ^3-pyran residue with the carbazole nucleus has been conclusively determined by the results of ozonisation. It is well known in the coumarin and alkaloidal series that ozonisation furnishes an aromatic α-hydroxyaldehyde and acetone. The characterisation of the α-hydroxy aldehyde has been utilized to

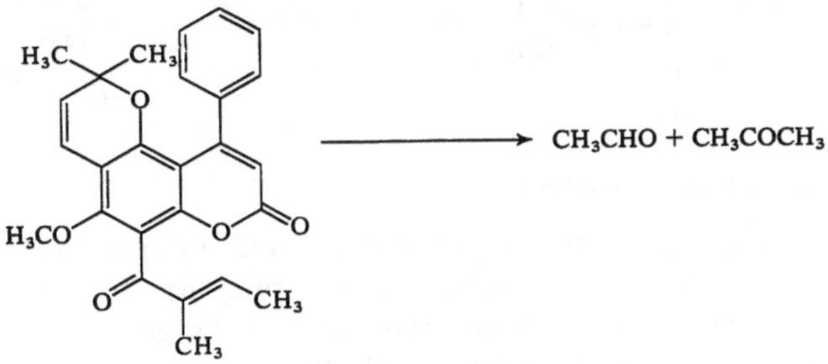
H₃C
CH₃
O
H₃CO
O
O
O
CH₃
CH₃
CH₃CHO + CH₃COCH₃

(**40**)

ascertain the point of fusion of the pyran ring and the final structure of the compounds. On the other hand, POLONSKY (*88*) obtained acetaldehyde and acetone by ozonisation of callophyllolide (**40**). On the basis of her results, she pointed out that the ozonisation of 2,2-dimethyl-Δ^3-pyran system might proceed abnormally and postulated that the double bond had migrated previous to oxidative cleavage.

Girinimbine (**18**) mahanimbine (**27**) and heptazolidine (**41**) are some of the alkaloids, whose ozonolysis resulted in assignment of the complete structure.

Girinimbine on ozonolysis furnished 2-hydroxy-1-formyl-3-methyl-carbazole (**34**) which settled the position of fusion of the pyran ring. Ozonisation of mahanimbine (**27**) yields (**34**) and a neutral product (**42**). Heptazolidine (**41**) on the other hand furnished the α-hydroxy-aldehyde (**43**) and a dialdehyde (**44**). Because of the isolation of the dialdehydes it was suggested (*38*) that ozonisation of 2,2-dimethyl-Δ^3-pyran proceeds in a normal fashion as follows. The first products of ozonisation are compounds of the dialdehyde type which undergo further cleavage and decarbonylation. This results in formation of an α-hydroxy aldehyde and acetone, the products isolated previously by ozonolysis of coumarins and alkaloids. The formation of acetone

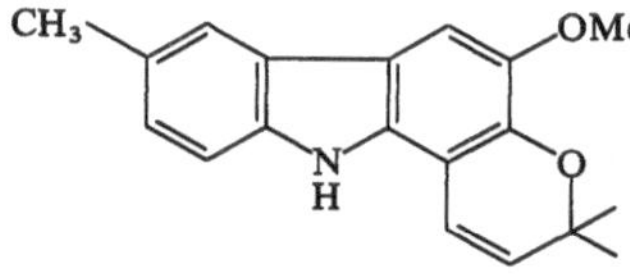

(**41**) 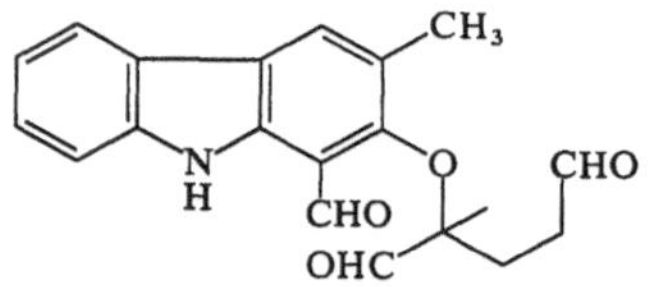   (**42**)

and α-hydroxyaldehyde (**43**) could be portrayed as follows. This concept also rationalises the formation of the trialdehyde (**42**) by ozonisation.

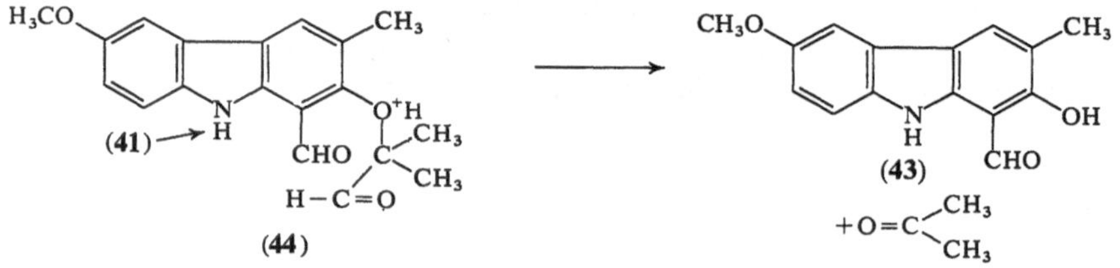

(**44**)   (**43**)

(e) Permanganate Oxidation

The isolation of α-hydroxyisobutyric acid by permanganate oxidation provides degradative proof for the presence of a 2,2-dimethyl-Δ^3-pyran system fused to an aromatic ring. Thus girinimbine (**18**) was found to give α-hydroxy-isobutyric acid (**45**) on permanganate oxidation.

References, pp. 366—371

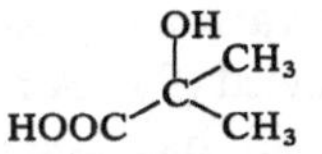

$$\underset{\text{(45)}}{\underset{\text{HOOC}}{\overset{\overset{\displaystyle OH}{\underset{\displaystyle |}{}}}{\underset{\displaystyle \diagdown}{\overset{\displaystyle C}{}}}}\overset{\displaystyle \diagup CH_3}{\underset{\displaystyle \diagdown CH_3}{}}}$$

C. Synthesis of Carbazoles

Various methods for the synthesis of carbazoles have been reviewed from time to time (*14, 56, 102, 103*). The methods successfully utilized in the synthesis of the alkaloids as also some newer methods of synthesis of carbazoles are briefly summarized below.

1. Graebe-Ullmann Synthesis

In the Graebe-Ullmann method of carbazole synthesis *o*-amino-diphenylamine (**46**) is treated with nitrous acid giving 1-phenyl 1,2,3-benzotriazole (**47**) which on heating gives carbazole.

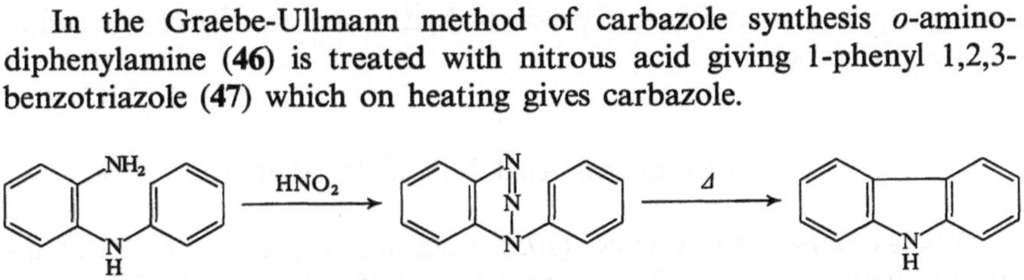

(46)   (47)   (1)

2. Borsche Synthesis

The Fischer-method (*94*) of indole synthesis was applied to the synthesis of tetrahydrocarbazoles by BORSCHE. The cyclohexanone-phenylhydrazone (**48**) obtained by reacting cyclohexanone (**49**) with phenylhydrazine (**50**) on indolisation furnished tetrahydrocarbazoles (**51**) which on dehydrogenation furnished carbazole. The success of the method is dependent upon the dehydrogenating agent and the reagent used for indolisation. Various condensing agents for indolisation and dehydrogenating agents for aromatisation have been used. The mechanism

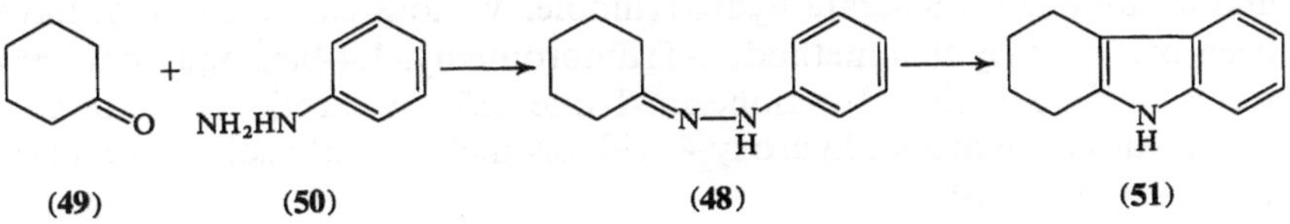

(49)   (50)   (48)   (51)

of the reaction and the use of various condensing agents and dehydrogenating methods have been reviewed (94). A mixture of glacial acetic acid and
concentrated hydrochloric acid has been successfully used for indolisation in our laboratory. The use of chloranil or Pd/C in p-cymene
has also found wide application for dehydrogenation in the synthesis of
carbazole alkaloids.

The preparation of the hydrazone required in the Borsche method
can be accomplished very conveniently using the Japp-Klingemann
reaction (87, 94). On reacting formylcyclohexanone (52) with a
diazotised aromatic amine (53) the hydrazone (54) is obtained. Cleaner

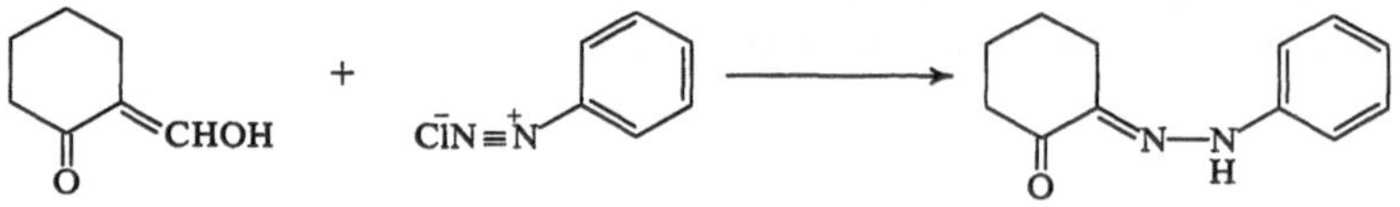

reaction products and better yields of hydrazone are the chief advantages of the process.

3. Watermann and Vivian's Method

WATERMANN and VIVIAN (105) obtained carbazole by reducing
nitrodiphenyl (55) with ferrous oxalate at 205—215° for half an hour.
This method has been improved using triethyl phosphite as reducing
agent (13a). POPLI et al. used the method in the synthesis of carbazole
alkaloids.

4. Nenitzescu Synthesis

The Nenitzescu method involves the condensation of an enamine
and a quinone to generate hydroxyindole. Various carbazoles (76) have
been prepared by this method. 2-Trifluoromethyl-1,4-benzoquinone (56)
reacts with 3-amino-2-cyclohexen-1-one (57) in acetic acid to give
4a,7,8,9a-tetrahydro-9a-hydroxy-4-trifluoromethyl carbazole 3—5 (4H,
6H) — dione (58).

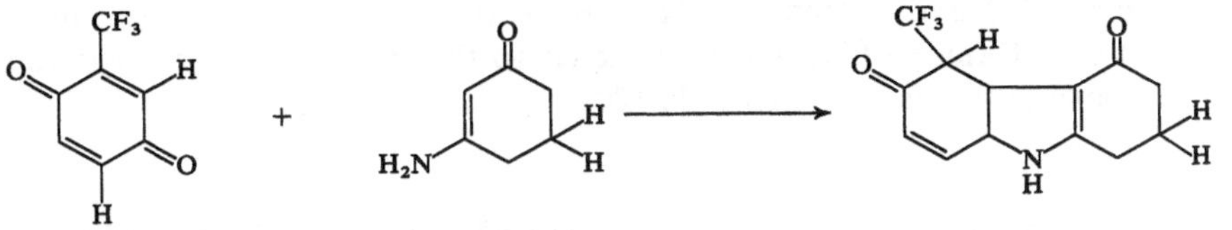

(56)   (57)   (58)

Two mechanisms have been proposed for the reaction which differ in the sequence of N–C and C–C condensation. The initial C–C condensation has been confirmed by recent studies and is compatible with certain aspects of enamine chemistry. It has further been established that the nitrogen-carbon condensation products are transformed into indoles only after conversion into compounds resulting from initial carbon-carbon condensation.

5. Thermal Synthesis of Carbazoles

Diphenylamine (59) was known to cyclise to carbazole (1) at red heat in very poor yield. CHAKRABORTY et al. (60) have recently shown that in the presence of elemental iodine and oxygen the reaction takes place at 350° in a sealed tube. A free radical mechanism proceeding via (60) by successive hydrogen abstraction for cyclisation was envisaged (90). Oxygen facilitates the breakdown of HI so as to form I_2.

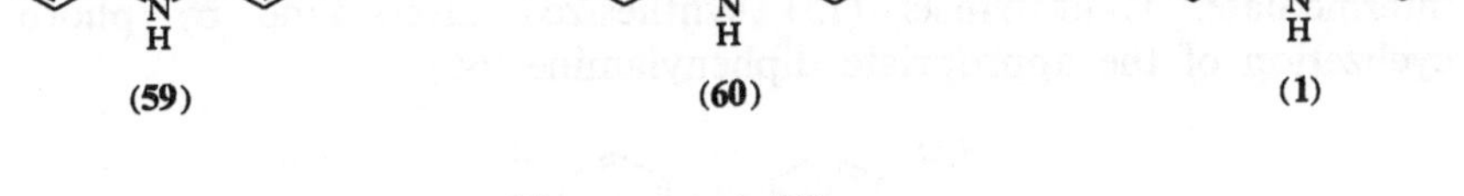

(59)   (60)   (1)

$$2I \longrightarrow 2I^*$$
$$2H + 2I \longrightarrow 2HI$$
$$2HI + \tfrac{1}{2}O_2 \longrightarrow H_2O + 2I$$

The reaction has been applied to the synthesis of carbazole, glycozoline, glycozolidine and 3-methylcarbazole. The yield in the case of carbazole and 3-methylcarbazole is fair but it is poor in case of glyco-

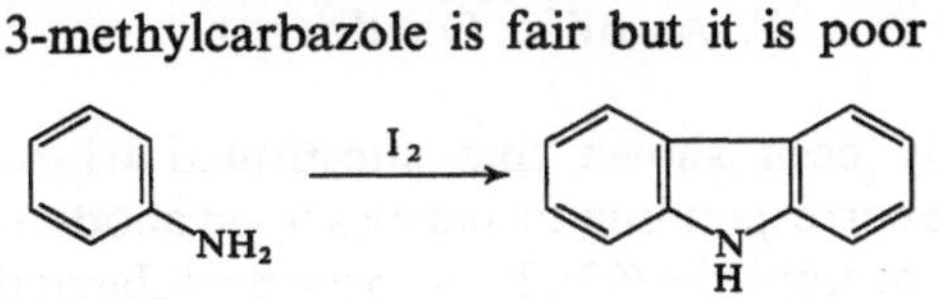

(61)   (1)

zolidine (17) and glycozoline (19). Recently CHAKRABORTY *et al.* have shown that aniline (61) when subjected to the iodine-catalysed reaction as above gives rise to carbazole (20).

6. Photocyclisation of Diphenylamine to Carbazole

Diphenylamine (59) and N-methyldiphenylamine (62) have recently been shown to yield carbazole and its N-methylderivative (63) photochemically (10, 59). The quantum yield of N-methylcarbazole is greater. A transient intermediate (65), formed *via* (64) with an absorption maximum at 610 nm has been proposed for the conversion of N-methyl-diphenylamine to N-methylcarbazole. It is obvious that carbazole is formed from diphenylamine by hydrogen abstraction through the transient

(59), R = H	(1), R = H
(62), R = CH₃	(63), R = CH₃

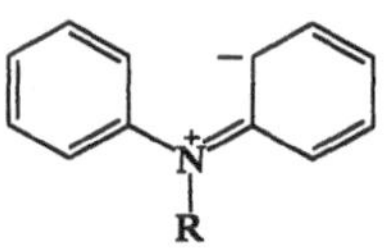

(64) 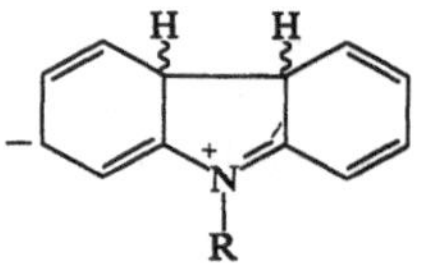   (65)

intermediate. CARRUTHERS (15) synthesized glycozoline by photocyclization of the appropriate diphenylamine (66).

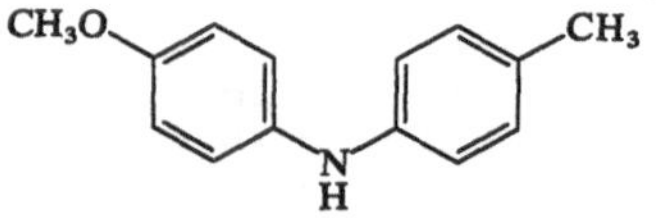

(66)

7. Anodic Oxidation

It has recently been shown that substituted triphenylamines (67) cyclize to form the corresponding carbazoles upon anodic oxidation at platinum electrode in acetonitrile (92). It has been postulated that the reaction proceeds through a stable dication radical (68) which cyclizes to (69).

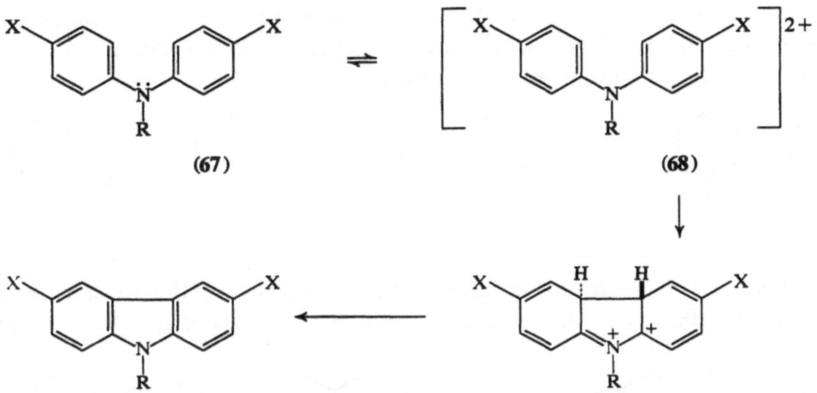

(67)   (68)

(69)

In the intermediate dihydrocarbazoles (69) the hydrogen would be *trans* leading to elimination of hydrogen and the formation of carbazoles.

III. Biogenesis of Carbazole Alkaloids

Because the carbazole alkaloids in Rutaceae have an anthranilic acid pattern, CHAKRABORTY (17) suggested an anthranilate origin for them. The co-occurrence of furanoquinoline and acridone bases together with glycozoline and glycozolidine in *Glycosmis pentaphylla* was considered circumstantial evidence in favour of this idea. Because of the ready photolytic cyclisation of diphenylamine to carbazole, CHAKRABORTY further considered the possibility that the carbazoles might be formed *via* diphenylamine as shown below. However, though the

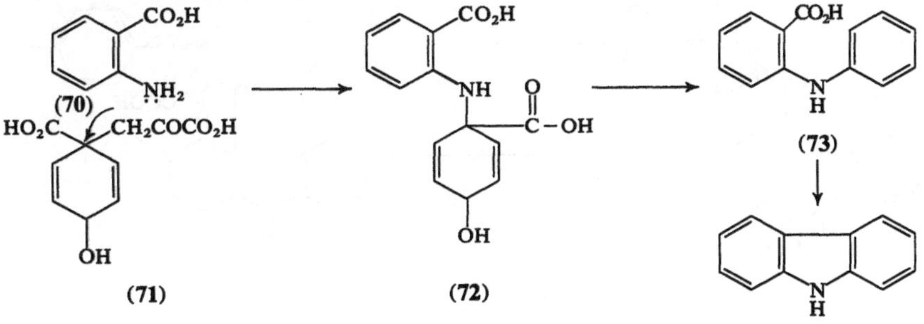

(70)

(71)   (72)

(73)

(1)

diphenylamine hypothesis was attractive it lacked circumstantial and biosynthetic evidence. Because mevalonic acid derived units participate in the biogenesis of indole bases and anthraquinones, CHAKRABORTY suggested that the five carbon atoms linked by bold lines in (4), (17) and (19) are derived from mevalonate.

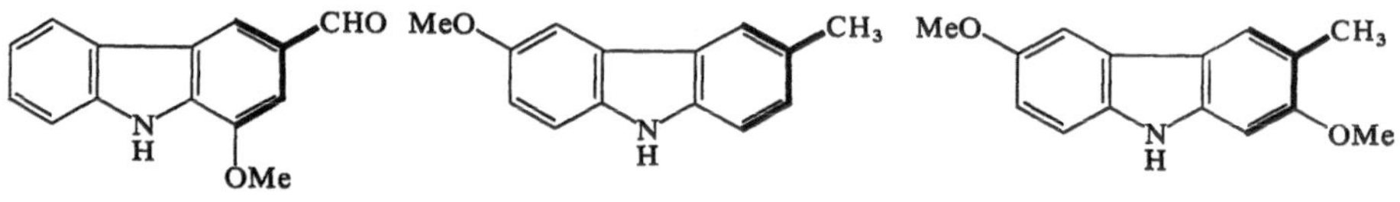

(4) (19) (17)

That mevalonate participates in the formation of the third ring of carbazole alkaloids has been advocated by KUREEL et al. (68) and by

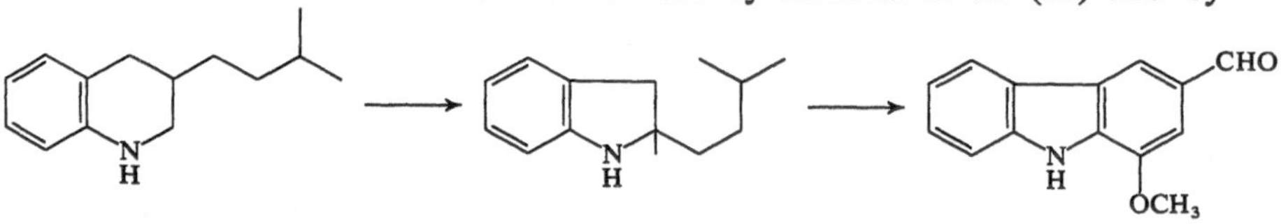

(74) (75) (4)

ERDTMAN (57). The latter pointed out that carbazoles could originate from a 3-prenylated quinoline (74) via a 2-prenylated indole (75).

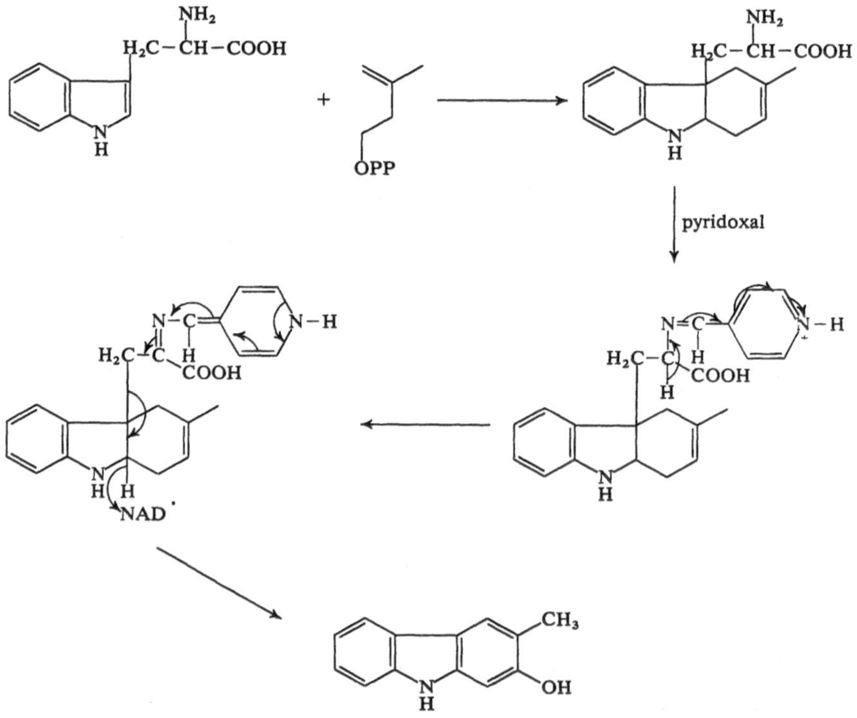

KUREEL et al. postulated that the indole ring could arise from anthranilic acid via dimethylallylquinolines and subsequent ring contraction. NARASHIMHAN (84), however, considers that tryptophan is the substrate to which the C_5-fragment is added after which a series of reactions eliminate the tryptophan side chain. The C_5-unit initially attacks the 3-position of the heterocyclic system; the subsequently cyclisation and loss of the serine residue in the presence of pyridoxal coenzyme gives a dihydrocarbazole which on dehydrogenation yields a 3-methyl carbazole.

Feeding experiments by POPLI and KAPIL (68) provided biosynthetic evidence in favour of the mevalonate origin of the ring carrying the extra methyl group. Feeding of $2\text{-}^{14}C$ and 2-^3H mevalonic acid lactone to M. koenigii resulted in isolation of highly radioactive koenimbine (76), koenigicine (77) and mahanimbine (27) though experiments establishing the location of the radioactivity are lacking.

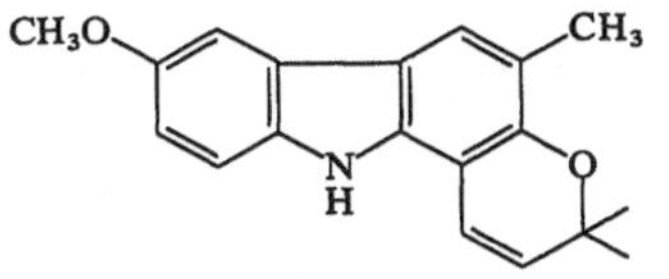

(76) 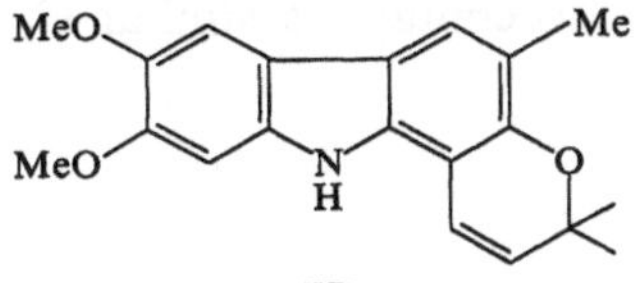   (77)

KAPIL (68) further pointed out that 3-methylcarbazole (32) is the key compound which gives rise to different structural patterns in carbazole alkaloids. The isolation of 3-methylcarbazole from the genus Clausena (62, 98) provides strong support for this idea. It has further been pointed out by CHOWDHURY and CHAKRABORTY (45) that the aromatic C-methyl groups in the carbazole alkaloids might become oxidized in vivo to formyl and carboxylic acid groups. The co-occurrence of 1-methoxy-3-methyl carbazole, murrayanine (4) and mukoeic acid (38) in Murraya koenigii Spreng. gives a strong support to this idea. The recent isolation of mukonine (78) and mukonidine (79) alkaloids with carbomethoxy groups shows that the carboxylic acid group can be further methylated to give the ester function.

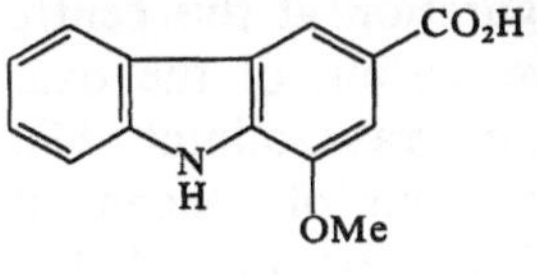

(38) 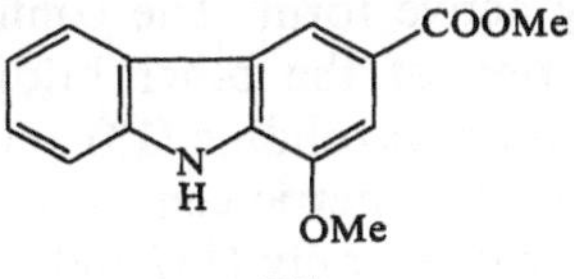   (78) 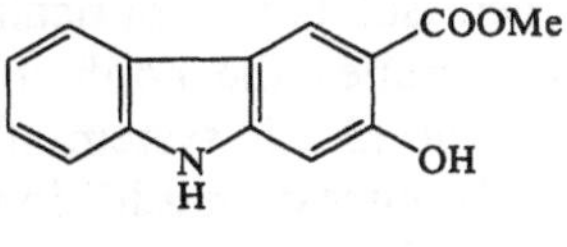   (79)

The formation of the pyran ring could be rationalized by assuming the incorporation of a mevalonate unit as has been proposed for many phenolic compounds. This gives rise to the group of carbazoles with

a C_{18} skeleton. Heptaphylline (10), girinimbine (18), murrayacine
(7) and their congeners are similar to typical plant phenolics with a
modified MVA unit. In fact the occurrence of heptaphylline and
girinimbine (18) in *Clausena heptaphylla* may be considered circumstantial
evidence in favour of the origin of the pyran ring from the prenylated
congener. Popli and Kapil (71) suggested that in the formation of
pyranocarbazoles, 2-hydroxy-3-methylcarbazole (35) plays a prominent
role. The recent isolation of 2-hydroxy-3-carbomethoxy-carbazole (mukonidine) (79) provides strong circumstantial evidence for the above idea.
Probably, the methyl group of 2-hydroxy-3-methylcarbazole has been
oxidized to the carbomethoxy group in mukonidine (79).

Mahanimbine (27) could be considered as the representative member
of carbazoles with a C_{23} skeleton. The incorporation of a monoterpene
unit as in mahanimbine (27) has precedences in the cannabichromene
(80) group of compounds derived from *Cannabis sp.* (80) as well as
in gambogic acid (91) obtained from *Garcinia xanthochymus*, which
also contains a similar C_{10} unit.

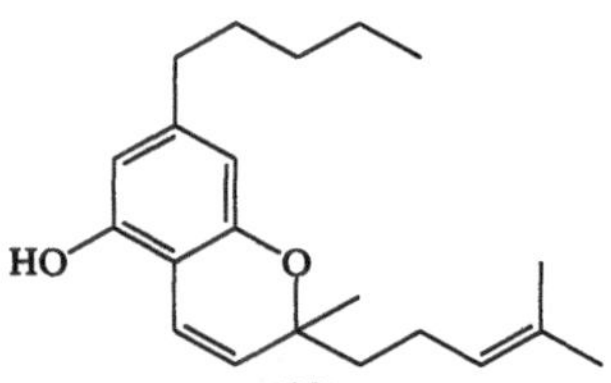

(80)

The monoterpene unit can undergo various transformations, thus
giving rise to isomeric compounds of the mahanimbine group such as
murrayazoline (5), murrayazolidine (13), murrayazolinine (14) and
bicyclomahanimbine (81).

Mahanimbine (27) and isomahanimbine (82) have the same C_{10}-unit
as cannabichromene (80) which contains an asymmetric centre. Unlike
cannabichromene, however, mahanimbine and isomahanimbine have
been isolated in optically active form. The configuration at this centre
determines the configuration at the other bridge carbon of the oxa-
bicyclo-(3,2,1)-system of murrayazolidine (13) and murrayazolinine (14);
both possess an additional asymmetric centre at the point of attachment
of the isopropyl group; however, only (13) and (5) have been isolated in
optically active form. While the relative configuration of the optically
active murrayazoline (5) is known from the results of the X-ray analysis,
the absolute configuration is unknown, as is the relative (when applicable)
and absolute configuration of the other carbazole alkaloids.

References, pp. 366—371

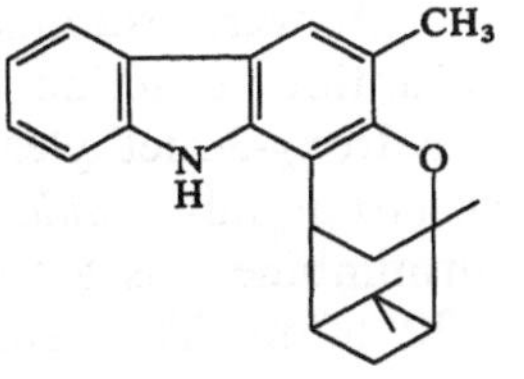

(81)

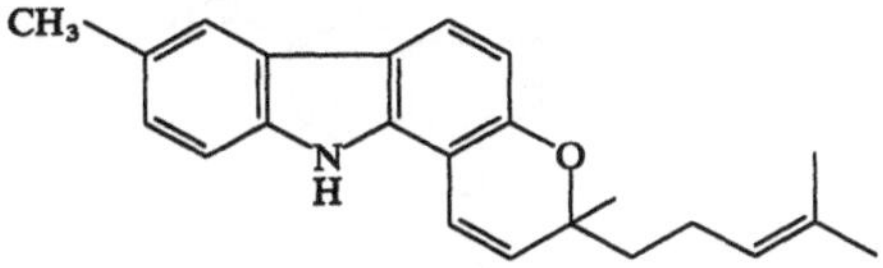

(82)

IV. Biochemical Properties of Carbazole Alkaloids

Glycozoline (19), mahanimbine (27), girinimbine (18) and glyco-zolidine (17) are feebly active when tested in the KB cell-culture. They were considered nontoxic.

The antibiotic properties (35, 50) of the carbazole alkaloids and some related products have been tested by the usual agar cup assay method using Sabouraud's medium against *Microsporum gypseum*, *Trichophyton rubrum*, *Epidermophyton floccusum*, *Candida albicans*, *Candida tropicalis*, *Staphyllococcus aureus* and *Escherichia coli*. Glycozoline, demethylated glycozoline, 1-methyl-6-hydroxycarbazole (83), 2-methyl-6-hydroxy-carbazole (84), glycozolidine (17), 1-hydroxycarbazole (85), 2,6-dihydroxy-3-methylcarbazole (86) murrayanine (4), girinimbine (18), mahanimbine

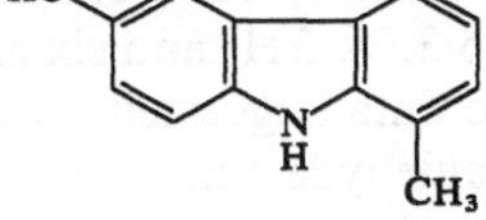

(83)

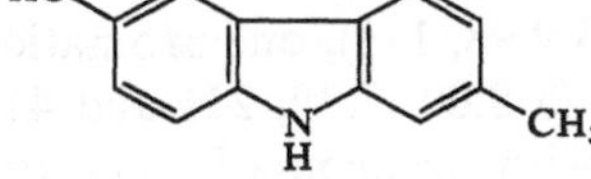

(84)

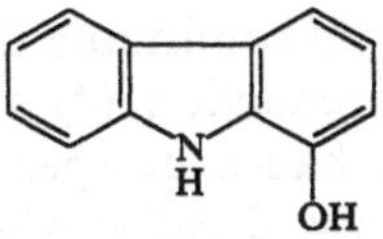

(85)

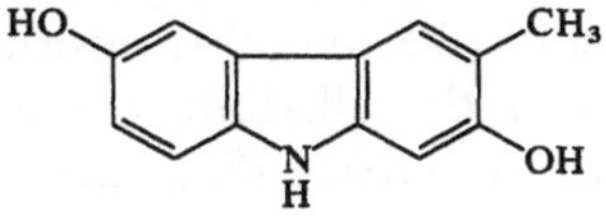

(86)

(27), and heptazoline (9) were examined against microorganisms. All of them have feeble antibiotic action at 100 μg/ml but the most significant is that of 6-hydroxy-3-methylcarbazole (87). It is active against *Trichophyton rubrum* (10 μg/ml). *Epidermophyton floccusum* and *Microsporum gypseum*. Girinimbine was active against *Nocardia asteroids* at a concentration of 30 μg/ml. The crude extract of *Murraya koenigii* was active against *Entamoeba histolytica* (7).

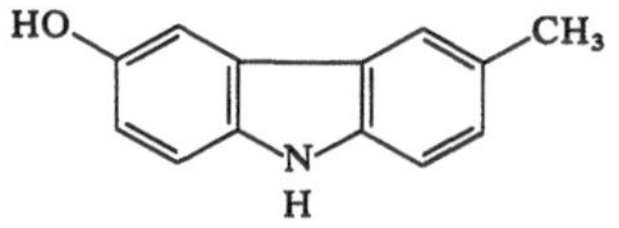

(87)

V. Chemistry of Carbazole Alkaloids

A. Members of the C_{13}-Skeleton Group

1. Murrayanine

Murrayanine, $C_{14}H_{11}NO_2$ (4), m.p. 168° (M^+ 225) was isolated from the petroleum ether (40—60°) extract of the stem bark of *Murraya koenigii* Spreng. by Chakraborty et al. (24). Later on, Bhattacharyya (8) also obtained (4) from *Clausena heptaphylla* Wt. and Arn.

The ir spectrum of (4) showed it to be an aromatic compound with –NH– (3450 cm^{-1}) aldehyde (2800, 1681 cm^{-1}) while the uv spectrum was strikingly similar to that of a 3-formylcarbazole (λ_{max} 238, 247, 274, 289, 335 nm; log ε 4.47, 4.30, 4.50, 4.56, 4.16) derivative suggesting the presence of such a chromophore. The nmr data showed the presence of an –NH– function (δ 8.80, 1H), one aldehyde group (δ 9.98, 1H), one aromatic methoxyl (δ 3.96, 3H) and six aromatic protons (δ 8.09, 7.39, 2H and 4H). All these data suggested a carbazole system with an aromatic methoxyl and an aldehyde function in (4).

The formation of an oxime, $C_{14}H_{12}N_2O_2$, m.p. 155—56°, and 2,4-dinitrophenylhydrazone (does not melt at 300°) were consistent with the above assignments; (4) also gave a malonic acid derivative. On zinc dust distillation (4) furnished carbazole (1) confirming the carbazole skeleton in the alkaloid. On borohydride reduction (4) furnished an alcohol, $C_{14}H_{13}NO_2$ (88), m.p. 127°, which had a uv spectrum similar to that of 1-methoxy-carbazole obtained by decarbonylation of (4) over Pd/C (10%). The oily Wolff-Kishner reduction product of

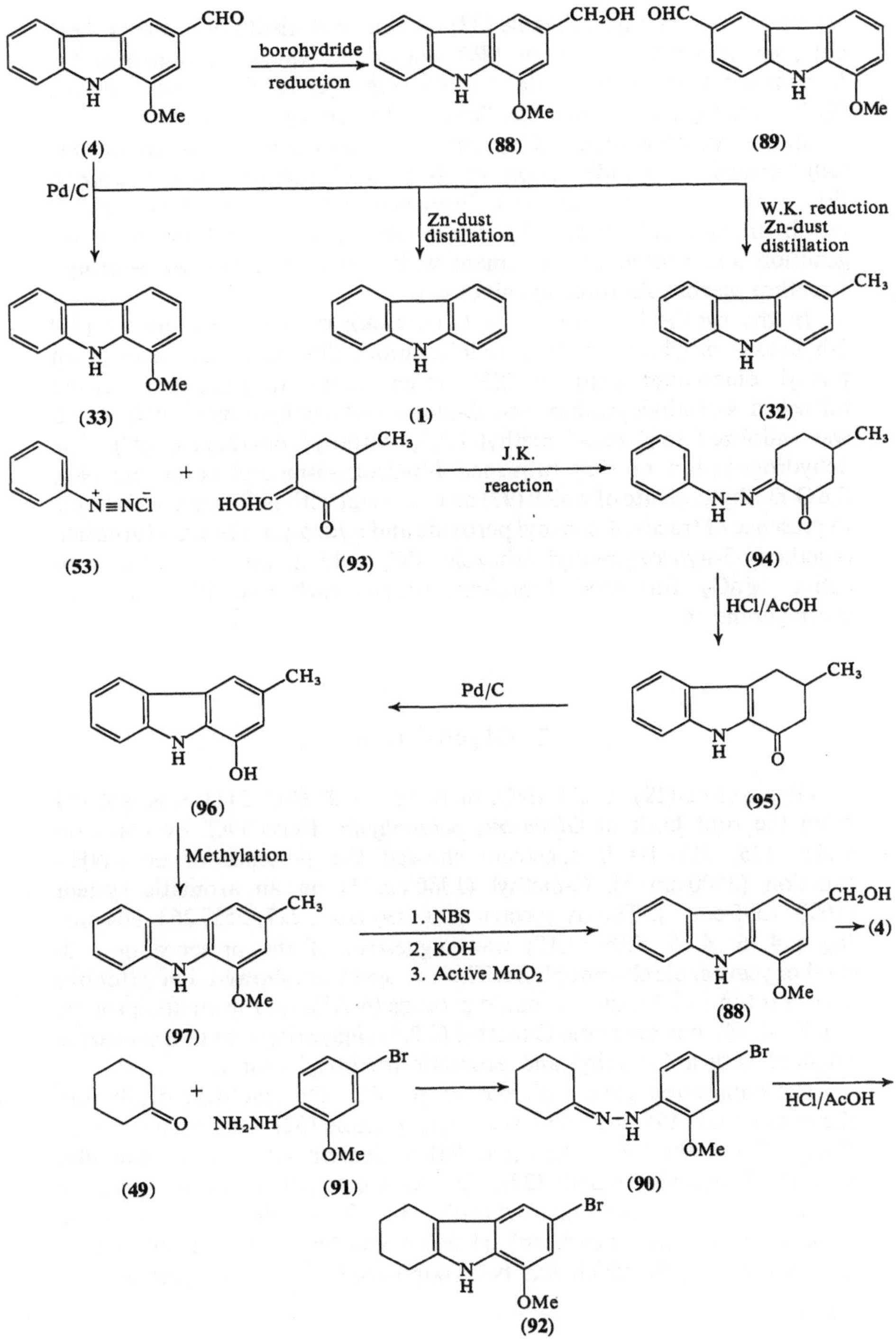

Chart 1. Reactions and synthesis of murrayanine

(4) furnished 3-methylcarbazole (32) on zinc dust distillation. From these data, either structures (4) or (89) were conceivable for murrayanine. In consideration of nmr data, Chakraborty et al. proposed structure (4) for murrayanine which has been confirmed by two syntheses.

In the synthesis reported by Crum and Sprague (49) the hydrazone (90) formed by condensation of 4-bromo-2-methoxyphenylhydrazine (91) with cyclohexanone (49) furnished 3-bromo-1-methoxy-5,6,7,8-tetrahydrocarbazole (92). The tetrahydrocarbazole (92) on dehydrogenation and subsequent treatment with formanilide yielded 3-formyl-1-methoxycarbazole (murrayanine) (4).

In the synthesis reported by Chakraborty and Choudhury (25) 2-hydroxymethylene-5-methyl cyclohexanone (93) on condensation with phenyl diazonium chloride (53) under Japp-Klingemann condition furnished 4-methylcyclohexane-1,2-dione-1-phenylhydrazone (94) which was indolized to 1-oxo-3-methyl-1,2,3,4-tetrahydrocarbazole (95). The dehydrogenation of (95) furnished 1-hydroxy-3-methyl carbazole (96), the O-methylderivate of which (97) on treatment with N-bromosuccinimide in presence of traces of benzoyl peroxide and hydrolysis *(in situ)* furnished 1-methoxy-3-hydroxymethylcarbazole (88) which on oxidation with active MnO_2 furnished 1-methoxy-3-formylcarbazole identical with murrayanine (4).

2. Glycozoline

Glycozoline (19), $C_{14}H_{13}NO$, m. p. 181—82° (M^+ 211), was isolated from the root bark of *Glycosmis pentaphylla* (Retz) DC. by Chakraborty (16, 18). Its ir spectrum showed the presence of an –NH– function (3500 cm^{-1}), C-methyl (1380 cm^{-1}) on an aromatic system (1600, 1595 cm^{-1}). The uv spectrum of (19) (λ_{max} 227, 252, 264, 304 nm; log ε 4.16, 4.16, 4.06, 4.17) was suggestive of the presence of a 3-methoxycarbazole chromophore. The nmr spectrum showed signals for one NH– proton (δ 7.8), two aromatic protons (δ 7.5), four aromatic protons (δ 6.8—7.18), one aromatic C-methyl (δ 2.5) suggesting it to be a carbazole alkaloid with a C-methyl and aromatic methoxyl group.

The compound gave a picrate, m. p. 182°. On zinc dust distillation the compound (19) furnished 3-methylcarbazole (32). On reduction with Raney nickel, the tosyl derivative (98) of demethylated glycozoline (87) afforded 3-methylcarbazole (32). The acetate of (87) had a uv spectrum similar to that of 3-methylcarbazole (32). This evidence showed the presence of a 3-methylcarbazole skeleton and led to the formulation of glycozoline as (19) which has been confirmed by several syntheses.

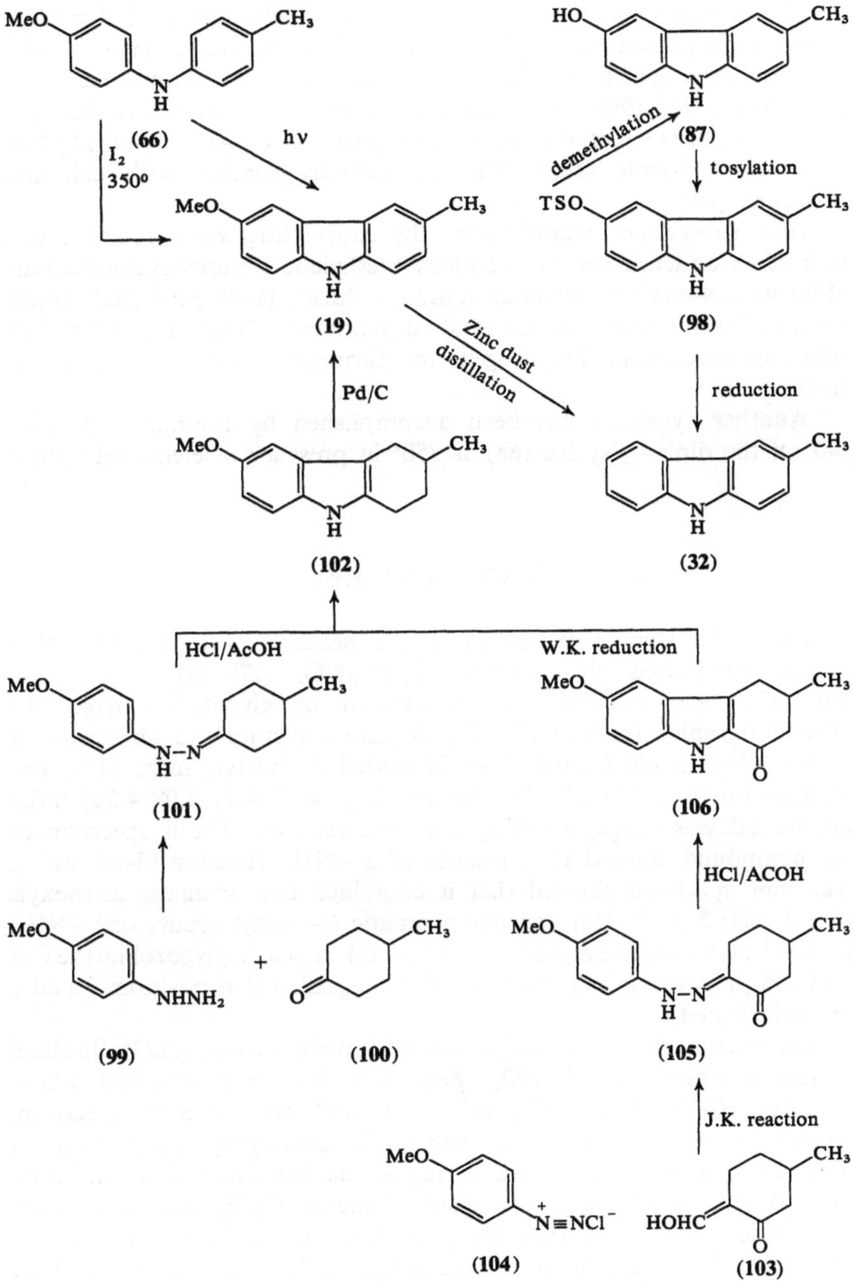

Chart 2. Reactions and synthesis of glycozoline

CARRUTHERS synthesized the alkaloid by photolysis of 4-methyl-4′-methoxydiphenylamine (66) in hexane (15). CHAKRABORTY et al. (31) provided three syntheses of this compound. Condensation of 4-methoxyphenylhydrazine (99) with 4-methylcyclohexanone (100) furnished the hydrazone (101) which on indolisation gave 3-methyl-1,2,3,4-tetrahydro-6-methoxycarbazole (102). This on dehydrogenation with chloranil furnished (19).

In a subsequent modification, the Japp-Klingemann reaction was utilized. Condensation of 2-hydroxymethylene-5-methylcyclohexanone (103) with 4-methoxyphenyldiazonium chloride (104) gave (105) which on indolisation gave the oxotetrahydrocarbazole (106). This on reduction and subsequent dehydrogenation furnished (102) and (19) respectively.

Another synthesis has been accomplished by thermal cyclisation (40) of the diphenylamine (66) at 350° in presence of elemental iodine.

3. Glycozolidine

From the stem bark of *Glycosmis pentaphylla* (Retz) DC., the second compound glycozolidine, $C_{15}H_{15}NO_2$ (17) (M$^+$ 241), m.p. 161—62°, was isolated after separation of glycozoline from the chromatographic fraction (26, 27). It gave a picrate and on treatment with methyl iodide furnished an N-methyl derivative, m.p. 140°. The uv spectrum [λ_{max} 233, 237, 260, 309 nm; log ε 4.47, 4.47, 4.08, 4.20] of the compound was suggestive of a carbazole skeleton. The ir spectrum of the compound showed the presence of a –NH– function (3450 cm^{-1}). The nmr spectrum showed that it contained two aromatic methoxyls (δ 3.9, 3 H; δ 3.77, 3 H) and one aromatic C-methyl group, one –NH– proton besides the five aromatic protons. Like those in glycozoline the C-4 and C-5 protons are not *ortho*-coupled suggesting that positions 3 and 6 are substituted.

On zinc dust distillation (17) furnished 3-methylcarbazole (32). On alkali fusion, a phenol $C_{13}H_{11}NO_2$ (86), m.p. 265° was obtained whose diacetate, $C_{17}H_{15}NO_4$ (107), m.p. 174—75°, showed a uv spectrum, similar to that of 3-methylcarbazole (32) confirming the position of a methyl group on the aromatic ring at the 3-position. Demethylation of (17) with HBr (48%) furnished a phenol $C_{14}H_{12}NO_2$ (39), m.p. 242°, whose acetate $C_{16}H_{15}NO_3$, m.p. 195° (M$^+$ 269), v_{max} 1755 cm^{-1}, showed uv data very similar to 2-methoxycarbazole, suggesting one of the methoxyl was at the 2- or 7-position. Reduction of the phenol *via* its tosyl derivative, m.p. 174—75°, afforded 2-methoxy-3-methylcarbazole

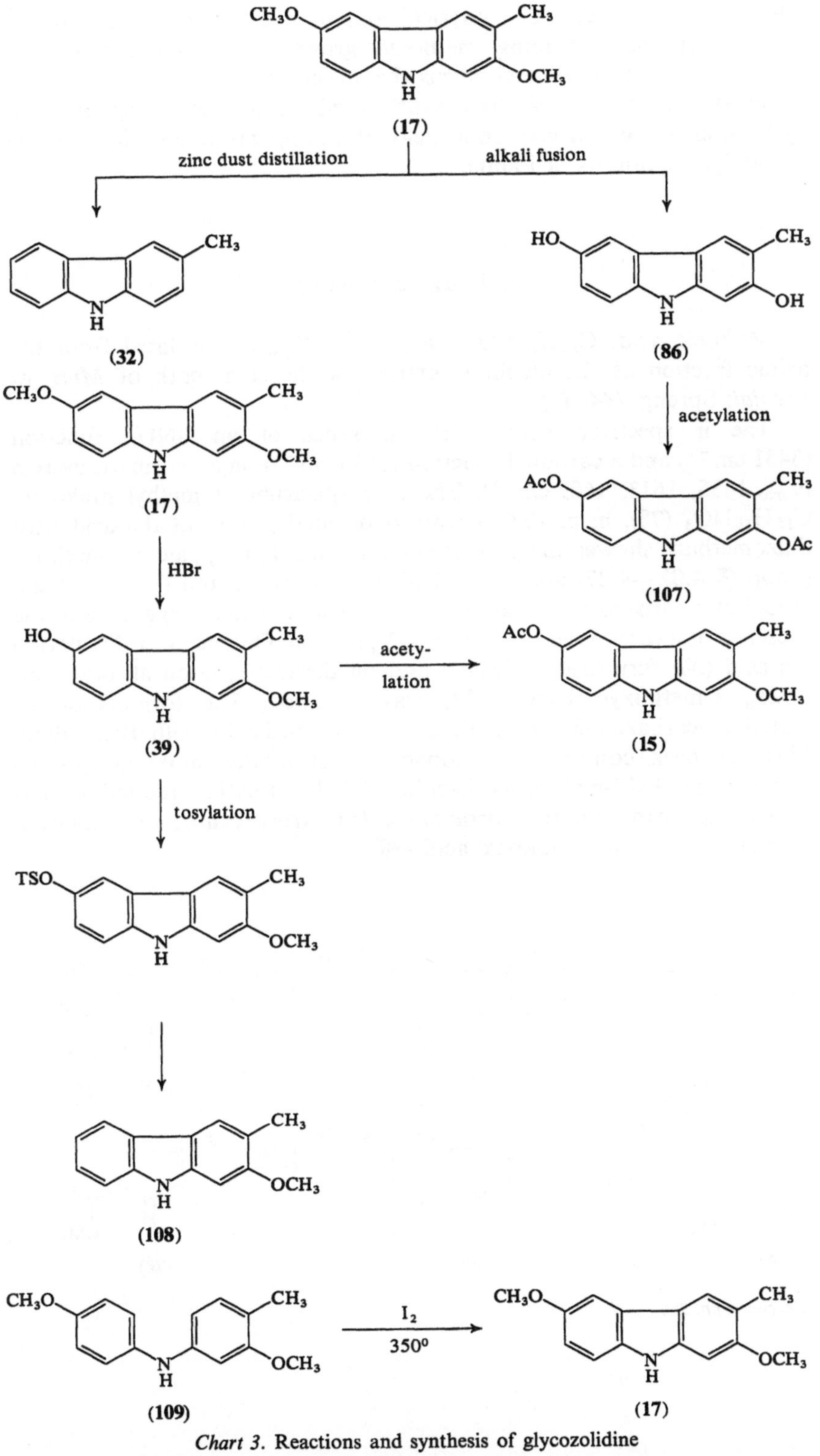

Chart 3. Reactions and synthesis of glycozolidine

(**108**), m.p. 224—25°. The physical and degradative data lead to the location of the remaining methoxyl group at the 6-position. The synthesis of glycozolidine confirms the structure.

CHAKRABORTY et al. (*60*) synthesized glycozolidine by thermal cyclisation of (**109**) in very poor yield, at a temperature 350° in presence of catalytic amounts of iodine.

4. Mukoeic Acid

Mukoeic acid, $C_{14}H_{11}NO_3$ (**38**), m.p. 242°, was isolated from the acidic fraction of the alcoholic extract of the stem bark of *Murraya koenigii* Spreng. (*44, 45*).

The ir spectrum showed the presence of an –NH– function (3431 cm^{-1}) and a carbonyl function (1690 cm^{-1}) on an aromatic system (v_{max} 1635, 1613, 1609 cm^{-1}). The nmr spectrum of methyl mukoate, $C_{15}H_{13}NO_3$ (**78**), m.p. 198°, obtained by methylation of the acid with diazomethane showed the presence of a methoxyl group, a carbomethoxy group (δ 4.02—4.07) and a deshielded aromatic proton (δ 8.08—8.20) at C-4 of a carbazole nucleus. These physical data were suggestive of the presence of a carboxylic group at the 3-position. On zinc dust distillation the acid (**38**) furnished carbazole and on decarboxylation at 145° with SbCl$_3$, 1-methoxycarbazole (**33**) was obtained. The formulation as 1-methoxycarbazole-3-carboxylic acid was consistent with these data. This has been confirmed by conversion of methyl mukoate (**78**) to 3-hydroxymethyl-1-methoxycarbazole (**88**) by LiAlH$_4$ reduction and subsequent oxidation to murrayanine (**4**). Murrayanine on oxidation with Ag$_2$O furnished mukoeic acid (**38**).

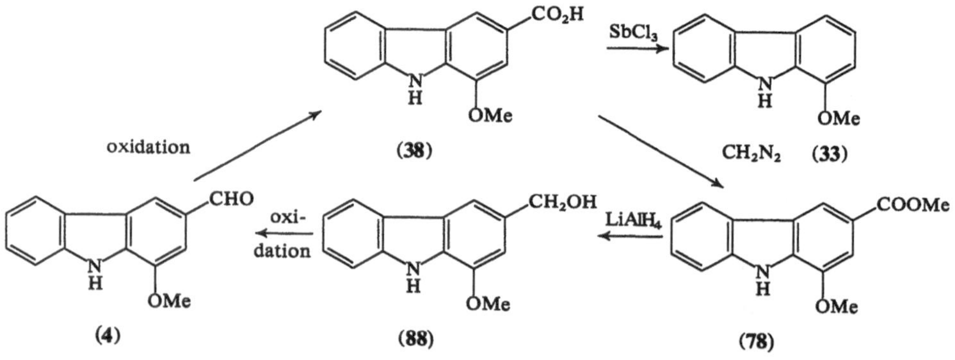

5. Mukonine

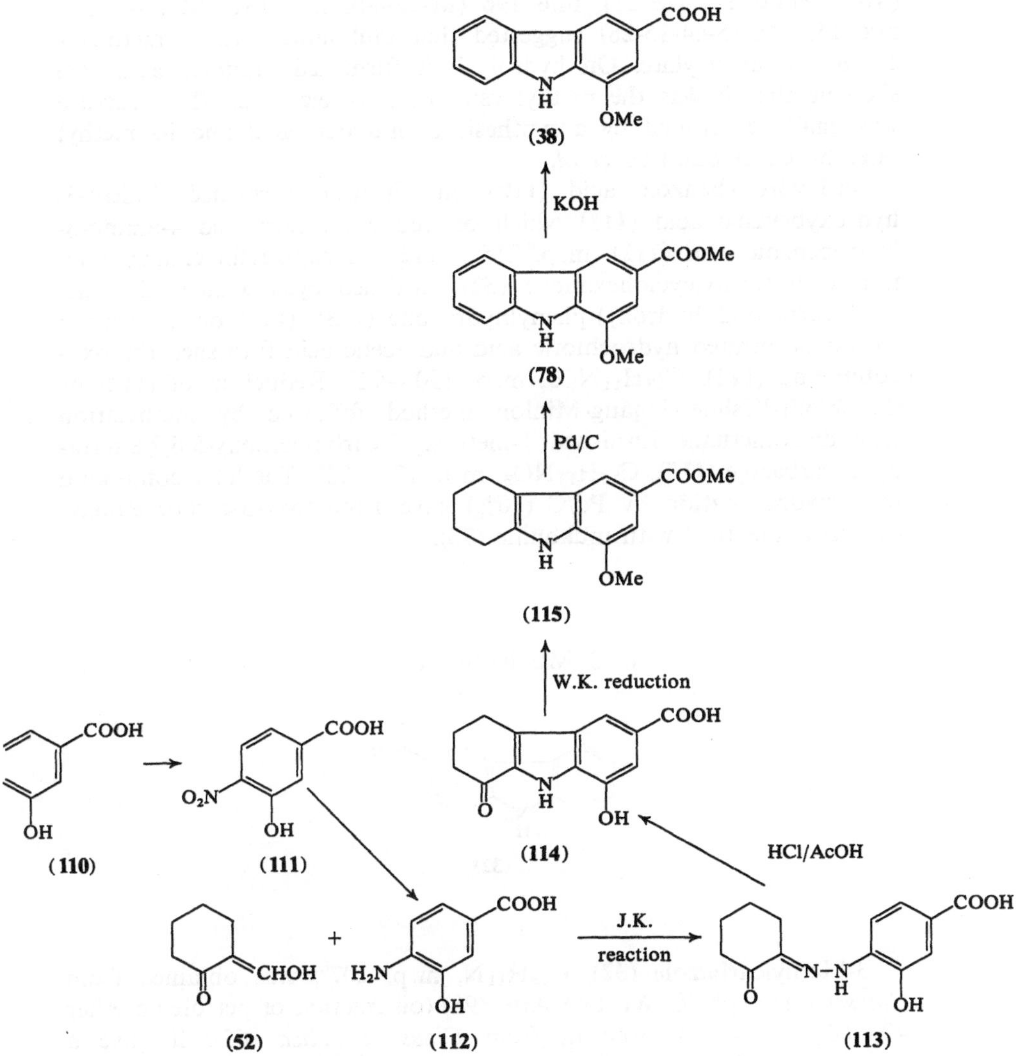

Chart 4. Reactions and synthesis of mukonine

Mukonine (**78**), $C_{15}H_{13}NO_3$, m.p. 195° (M$^+$ 255), was isolated from the stem bark of *Murraya koenigii* Spreng. by CHAKRABORTY *et al.* (*37*). The uv spectral data [λ_{max} 236, 245, 266, 274, 306, 320 nm; log ε 4.6, 4.57, 4.64, 4.73, 3.88, 3.78] of (**78**) were very similar to mukoeic acid (**38**). The ir spectrum showed an aromatic system (1600, 1510 cm^{-1}) with an –NH– function (3350 cm^{-1}) and an ester carbonyl function

(1700 cm^{-1}). The mass fragmentation pattern of (78), m/e 255, m/e 240 (M-15), m/e 224 (M-31), m/e 196 (M-15-44), m/e 181 (M-15-44-15), m/e 153 (M-15-44-15-28) suggested that mukonine was a methoxy-carbazole carboxylate. On hydrolysis it furnished mukoeic acid (38) showing that it was the methyl ester of mukoeic acid. The structure was finally confirmed by a synthesis of mukoeic acid and its methyl ester by Chakraborty et al.

m-Hydroxybenzoic acid (110) on nitration furnished 4-nitro-3-hydroxybenzoic acid (111) which on reduction furnished 4-amino-3-hydrobenzoic acid (112), m. p. 215°. (112) on Japp-Klingemann reaction with formylcyclohexanone (52) furnished cyclohexane-1,2-dione-1'(-4'-carboxy-2'-hydroxy)-phenylhydrazone (113). (113) on cyclization with concentrated hydrochloric acid and acetic acid furnished the oxo-compound (114), $C_{13}H_{11}NO_4$, m. p. 190—92°. Reduction of (114) by the Wolff-Kishner-Huang-Minlon method followed by methylation with diazomethane furnished 1-methoxy-3-carbomethoxy-5,6,7,8-tetra-hydrocarbazole (115), $C_{15}H_{17}NO_3$, m. p. 172—73°. The later compound on dehydrogenation by Pd/C (10%) gave 1-methoxy-3-carbomethoxy-carbazole identical with mukonine (78).

6. 3-Methylcarbazole

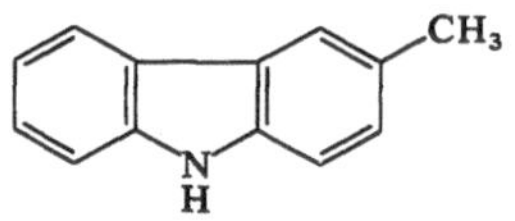

(32)

3-Methylcarbazole (32), $C_{13}H_{11}N$, m. p. 207°, was obtained from *Clausena heptaphylla* Wt. and Arn. (98) (oil fraction of petroleum ether 40—60° extract) as well as from *Clausena indica* (62). It gave a picrate m p. 176°. It has a uv spectrum (λ_{max} 236, 260, 296, 328, 342 nm; log ε 4.39, 4.02, 4.02, 3.30, 3.30) suggestive of a carbazole skeleton. The position of the methyl group at 3-position was readily discernible from the ir spectrum of the compound (ν_{max} 3450, 1600, 1493, 1390, 885 and 808 cm^{-1}), the band at 808 cm^{-1} being characteristic for 3-methylcarbazole (93). The identity of the compound was further established by direct comparison with an authentic specimen of 3-methyl-carbazole.

7. Mukonidine

Mukonidine, $C_{13}H_{11}NO_3$ (**79**) (M^+ 241, m.p. 245°), isolated from the stem bark of *Murraya koenigii* Spreng. by CHAKRABORTY *et al.* (*37*). The compound gave a positive $FeCl_3$ test. The ir spectrum showed it to be an aromatic substance with a $-NH-$ (3430 cm^{-1}), one chelated ester-carbonyl (1640 cm^{-1}) and one chelated hydroxyl (3200 cm^{-1}) groups while the uv spectrum (λ_{max} 230, 248, 266, 272, 308, 316, 323 nm; log ε 4.8, 4.57, 4.64, 4.73, 3.88, 3.78, 3.18) bears a strong similarity to that of 2-hydroxycarbazole with carboxyl substitution at the 3-position. The mass spectrum showed a high intensity base peak at M-1 (100%) consistent with the presence of a phenolic group. The other significant peaks obtained were at (M-1-15) due to the loss of a $-CH_3$, (M-1-15-44) for $-COO$ group and at (M-1-59) for $-COOCH_3$ group. In this respect the mass spectrum resembles that of the esters of aromatic acids. A peak at (M-1-59-29) is also informative as similar peak is found in other simple carbazoles like glycozoline (**19**) and mukonine (**78**). Mukonidine on methylation with diazomethane gave (**116**) the uv spectrum of which (λ_{max} 227, 250, 255.5, 285 nm; log λ 4.5, 4.28, 4.1, 4.13) was similar to that of 2-methoxycarbazole-3-methyl-carboxylate. On reduction with $LiAlH_4$ (**116**) furnished 2-methoxy-3-methylcarbazole (**108**), m.p. 225°, λ_{max} 238, 260, 302 nm; log ε 4.45, 4.34. On the basis of these physical and chemical data, mukonidine has been given the structure (**79**).

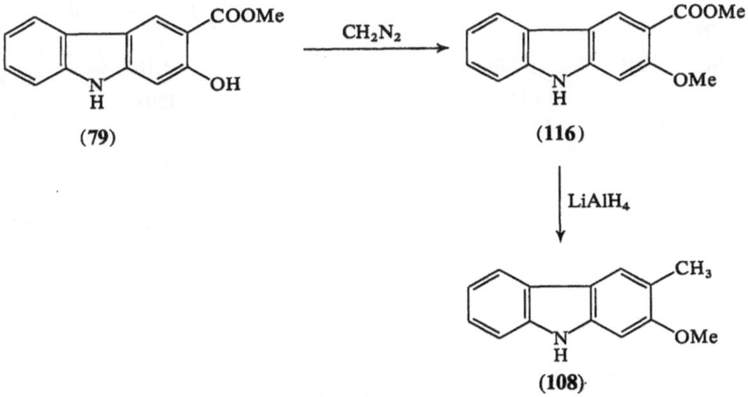

B. Members of the C_{18}-Skeleton Group

1. Indizoline

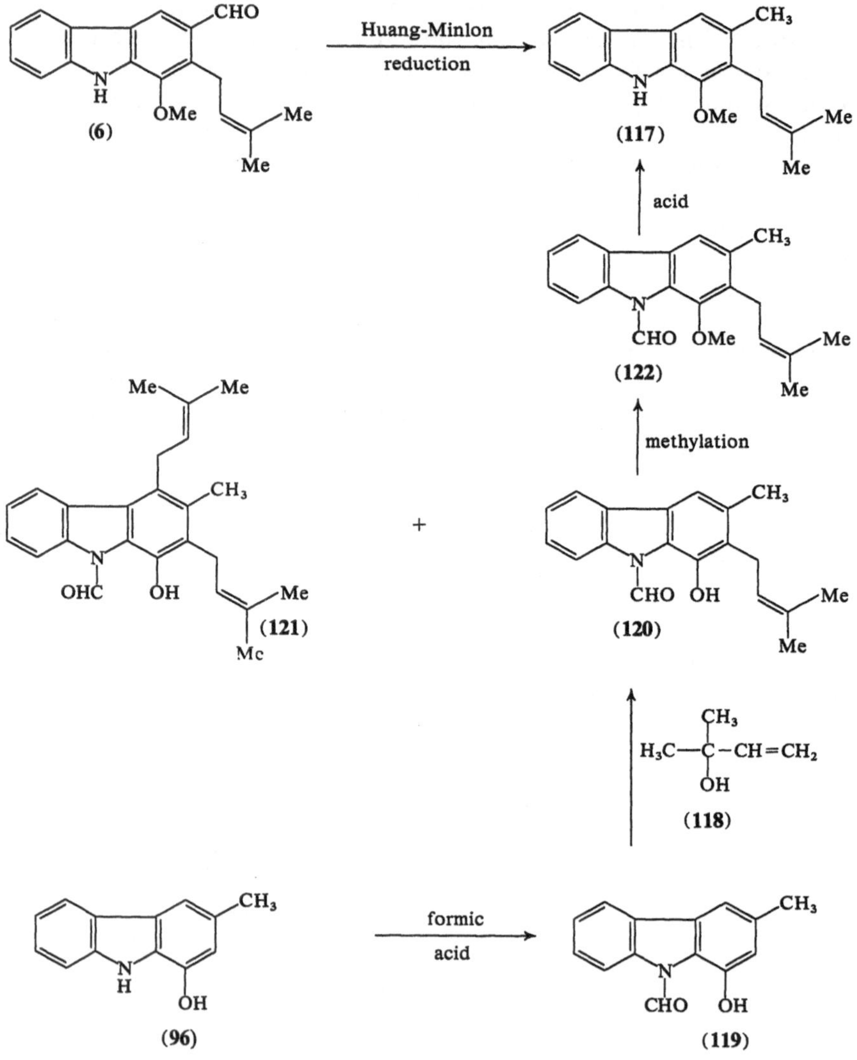

Indizoline, $C_{19}H_{19}NO_2$ (**6**) (M^+ 293), m.p. 171°, has been isolated from the roots of *Clausena indica* Oliv. (*62*). The ir spectrum of the compound showed v_{max} 1660 (CHO), 1620 and 1600 cm^{-1} (aromatic), the uv spectrum (λ_{max}^{EtOH} 237, 247, 275, 294 and 330 nm; log ε 4.61, 4.64, 4.78, 4.49 and 4.16). The nmr signals of indizoline (**6**) showed the presence of an aldehyde group (δ 10.3), $-NH-$ proton (δ 8.8), proton at C-4 (δ 8.4, singlet) and multiplets of 3 aromatic protons at δ 7.4 to 8.02, besides a vinylic proton (δ 5.25), benzylic methylene (δ 3.9 for 2 protons) and two methyl group (δ 1.7 and 1.85) which are parts of an isopentenyl chain. The proton at C-4 was deshielded due to the proximity of the aldehyde function at C-3.

On Huang-Minlon reduction indizoline (6) gave a compound having a uv spectrum very similar to that of 1-methoxycarbazole (33). In consideration of the singlet nature of C-4 proton and the substitution at the 1- and 3-positions, the isopentenyl side chain was located at the 2-position. On these basis indizoline was formulated as (6) and the reduction product as (117). Finally the structure was confirmed by the synthesis of the reduction product.

1-Hydroxy-3-methylcarbazole (96) was N-formylated with formic acid to give (119) which on reaction with 2-methyl-3-buten-2-ol (118) in the presence of BF_3/etherate gave (120) and (121). These were separated by column chromatography; (120) on methylation with $(CH_3)_2SO_4$ gave (122) which on acid treatment gave (117) identical with the Huang-Minlon reduction product of indizoline.

2. Heptaphylline

Heptaphylline, $C_{18}H_{17}NO_2$ (10), m.p. 170—71° (M^+ 279), was isolated from the roots of *Clausena heptaphylla* Wt. and Arn. by JOSHI *et al.* (*66, 63*). Its uv spectrum had λ_{max} 234, 278, 298 and 346 nm; log ε 4.42, 4.53, 4.58 and 4.09; ir bands appeared at 3300 (NH or OH), 2740, 1640 (weak-chelated aldehyde), 1618 (aromatic) cm^{-1} and its colour reactions showed it to be a 3-formylcarbazole derivative with a chelated aldehyde and phenolic group. The nmr spectrum showed the presence of a chelated hydroxyl (δ 11.70), and $-NH$ function (δ 10.3) and an aldehyde group (δ 9.9). The aromatic proton at C-4 was a singlet (δ 8.25) while the proton at C-5 was a doublet (δ 8.0, $J = 6.2$ respectively). The two methyl signals (δ 1.66 and 1.83), a benzylic methylene doublet at δ 3.6 ($J = 6$ cps) together with the signal for a vinylic proton could account for a $\gamma : \gamma$-dimethyl allyl residue in heptaphylline. On methylation heptaphylline gave a monomethoxy compound (123), m.p. 139—40°. On treatment with polyphosphoric acid or with boiling formic acid, it gave cycloheptaphylline (36). Its nmr spectrum showed it to be a 2,2-dimethyl chroman. On hydrogenation heptaphylline gave a dihydroderivative. On decarbonylation cycloheptaphylline gave a compound (37), m.p. 178°, the structure of which was established by synthesis. 2-Hydroxycarbazole (124) reacted with

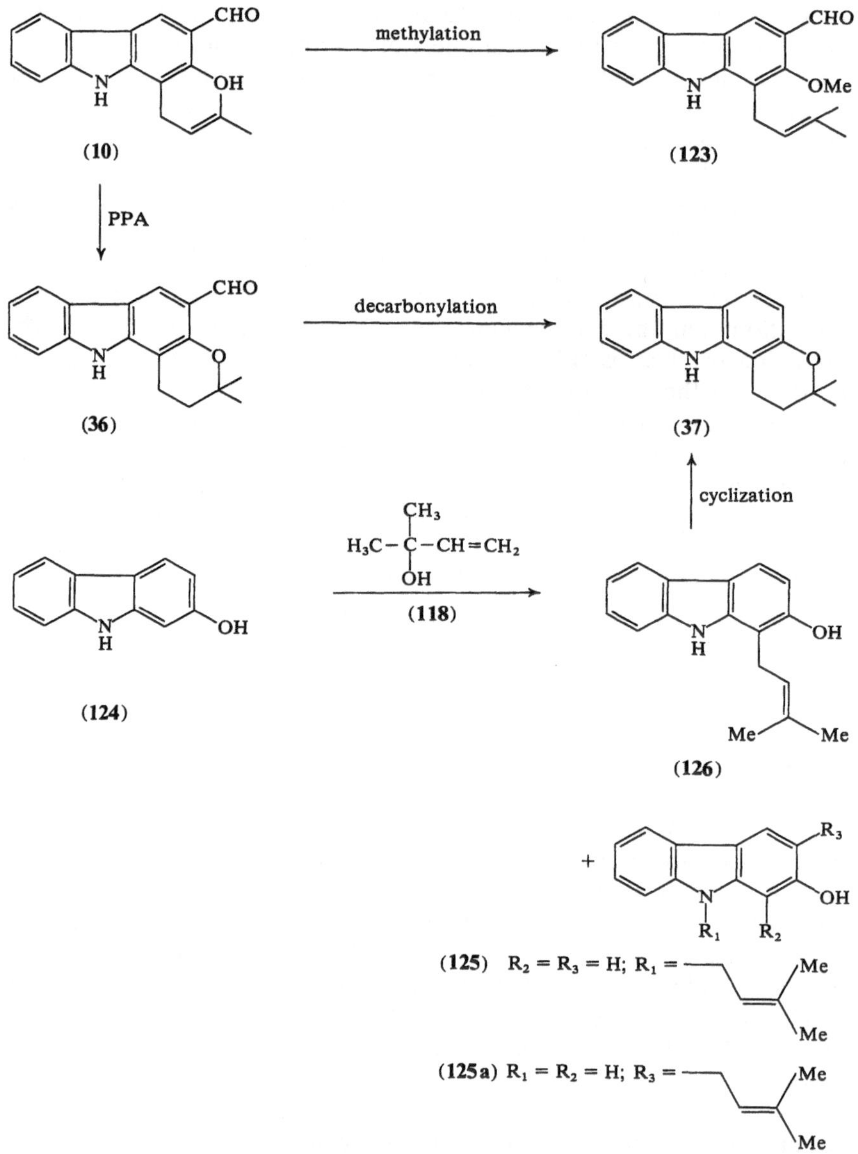

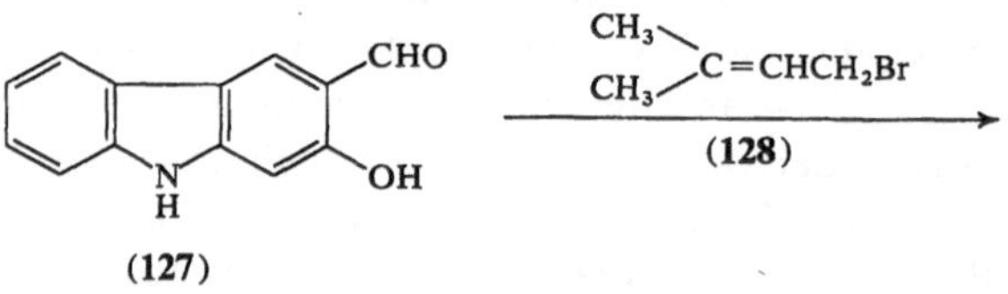

(10)

methylation →

(123)

PPA ↓

(36)

decarbonylation →

(37)

↑ cyclization

(124)

$H_3C-\overset{\overset{\displaystyle CH_3}{|}}{\underset{\underset{\displaystyle OH}{|}}{C}}-CH=CH_2$

(118)

(126)

+

(125) $R_2 = R_3 = H$; $R_1 =$ ⟍Me / Me

(125a) $R_1 = R_2 = H$; $R_3 =$ ⟍Me / Me

(127)

$\overset{\displaystyle CH_3}{\underset{\displaystyle CH_3}{\Big\rangle}} C=CHCH_2Br$

(128)

→ (10)

References, pp. 366—371

2-methyl-3-butene-2-ol (118) to give three isomeric compounds (125, 125a, 126). One of these (126) on acid catalyzed cyclisation furnished the compound (37). Heptaphylline was synthesized (61) in poor yield by reacting 2-hydroxy-3-formylcarbazole (127) with 3,3-dimethylallyl bromide (128) at 25° in presence of potassium hydroxide.

3. 6-Methoxyheptaphylline

6-Methoxyheptaphylline (129) was isolated from the hexane extract of the roots of *Clausena indica* by JOSHI *et al.* (64). The alkaloid, $C_{19}H_{19}NO_3$ (M^+ 309), m.p. 173—74° was phenolic and had uv spectrum (λ_{max} 207, 235, 283, 311, 341 nm; log ε 4.5, 4.35, 4.47, 4.57, 3.95) very similar to that of 3-formylcarbazole. Its nmr spectrum showed signals for chelated hydroxyl (δ–11), one aldehyde group (δ 9.9), an –NH– (δ 8.15), an aromatic proton singlet for C-4 (δ 7.92) and 3 aromatic protons at δ 6.9—7.5. The signals for two methyl groups on a double bond (δ 1.88, 1.75) together with signals at δ 5.3 for the vinylic proton and two benzylic protons at δ 3.6 accounted for the γ : γ-dimethyl allyl group on an aromatic system. The signal for the aromatic methoxyl group appeared at δ 3.9. On heating with formic acid the alkaloid gave an isomeric compound (130) (M^+ 309), m.p. 216—17°, which showed an ir band at 1640 cm^{-1}. From this, positions 2 for the hydroxyl and 1 for the isopentenyl chain were assigned. The formic acid cyclization product on Huang-Minlon reduction furnished dihydrokoenimbine (131), m.p. 250°. All these data were in conformity with structure of the alkaloid as 6-methoxyheptaphylline (129).

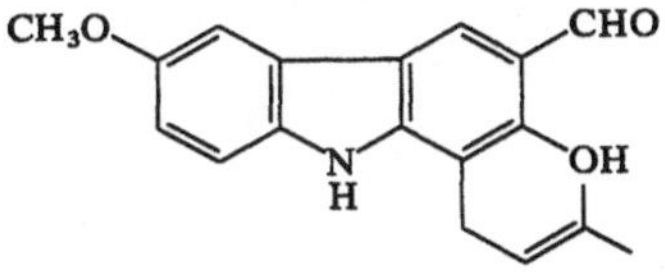

(129)

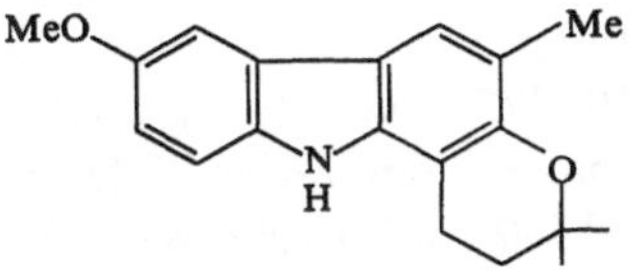

(131)

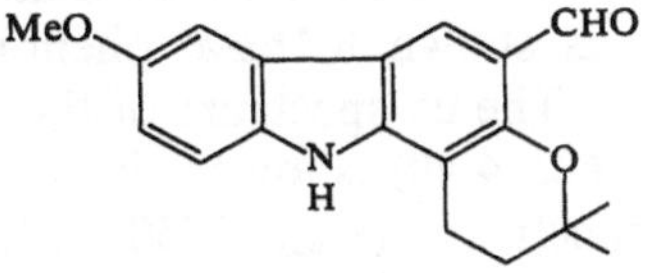

(130)

4. Heptazoline

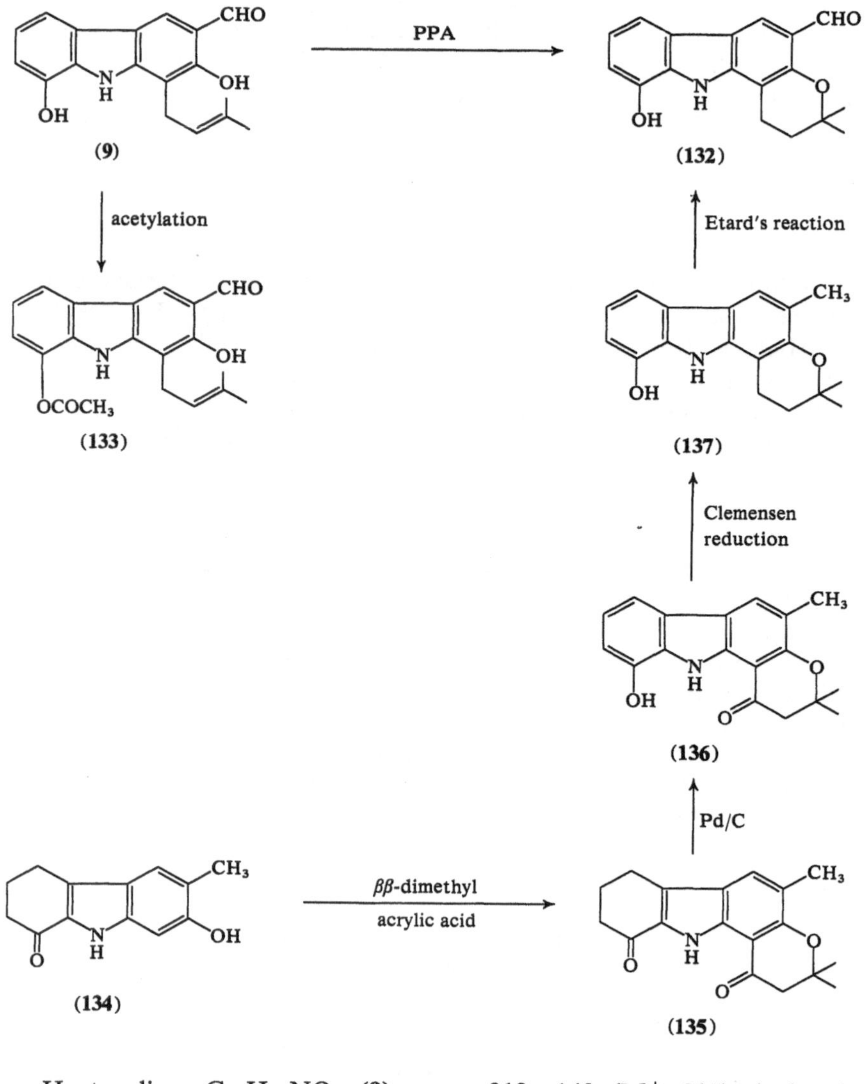
CHO
OH
N
H
OH
(9)
PPA
CHO
O
N
H
OH
(132)
acetylation
CHO
OH
N
H
OCOCH$_3$
(133)
Etard's reaction
CH$_3$
O
N
H
OH
(137)
Clemensen
reduction
CH$_3$
O
N
H
OH
O
(136)
Pd/C
CH$_3$
OH
N
H
O
(134)
ββ-dimethyl
acrylic acid
CH$_3$
O
N
H
O
O
(135)

Heptazoline, $C_{18}H_{17}NO_3$ (**9**), m.p. 212—14° (M^+ 295), isolated from the stem bark of *Clausena heptaphylla* by CHAKRABORTY *et al.* (*34*) was shown to have phenolic and aldehyde functions.

The uv spectrum of the compound (λ_{max} 240, 275, 298 nm; log ε 4.66, 4.62, 4.40) showed it to be a 3-formylcarbazole derivative. The ir spectrum (ν_{max} 3450, 3150, 1623, 1580, 1500 cm^{-1}) showed the presence of –NH–, chelated hydroxyl and aldehyde functions on an aromatic system, one of the phenolic hydroxyls being chelated to the aldehyde function at C-3.

Its nmr spectrum showed that the signal of the C-4 proton was deshielded (δ 8.28) due to a vicinal aldehyde function. The proton at C-5 was ortho- and meta-coupled (δ 7.5; $J = 8.2$ cps) showing that positions 6 and 7 were not substituted. The signals for the isopropylidene group (δ 1.65 and 1.81), vinylic proton (δ5.3) and benzylic methylene at (δ 3.63) showed it to have a $\gamma : \gamma$-dimethylallyl chain.

The isolation of carbazole by zinc dust distillation of (9) confirmed the carbazole skeleton of the alkaloid. On treatment with phosphoric acid heptazoline furnished cycloheptazoline (132) $C_{18}H_{17}NO_3$, m.p. 190°. Evidently the γ, γ-dimethylallyl chain occupies position 1. The acetylated heptazoline (133), m.p. 290° decomp. shows a uv spectrum very similar to that of heptaphylline suggesting the presence of the same chromophoric system in heptaphylline and heptazoline acetate. Since the protons at C-6 and C-7 are unsubstituted the second hydroxyl should occupy position 8. The nmr signal for the hydroxyl proton of cycloheptazoline was very similar to that of the 1-hydroxyl group in (85) which supports this conclusion. From these data the structure of cycloheptazoline was proposed as (132) which has been confirmed by a synthesis.

2-Hydroxy-3-methyl-8-oxo-5,6,7,8-tetrahydrocarbazole (134) was condensed with β,β-dimethylacrylic acid in the presence of fused zinc chloride and phosphorous oxychloride to yield 2,1-(2',2'-dimethylpyran-4'one)-3-methyl-8-oxo-5,6,7,8-tetrahydrocarbazole (135). This on dehydrogenation with Pd/C furnished 2:1-(2':2'-dimethylpyran-4'-one)-3-methyl-8-hydroxycarbazole (136). The chromanone on Clemensen reduction furnished the chroman (137) which on Etard's reaction furnished cycloheptazoline (132).

Thus the structure of heptazoline has been assigned as 1 (γ, γ-dimethylallyl)-2,8-dihydroxy-3-formylcarbazole (9).

5. Girinimbine

Girinimbine, $C_{18}H_{17}NO$ (18), m.p. 175° (M^+ 263), the first member of the group of carbazole alkaloids with a C_{18}-skeleton, was isolated from the stem bark of *Murraya koenigii* Spreng. by CHAKRABORTY *et al.* (23, 32). Later it was isolated from the root bark of *Clausena heptaphylla* Wt. and Arn. (63).

The uv spectrum of (18) (λ_{max} 223, 238, 288, 330, 342, 358 nm; log ε 4.49, 4.70, 4.65, 4.04, 3.89, 3.84), was suggestive of the presence of a carbazole system while the ir spectrum (v_{max} 3448, 1647, 1613 cm^{-1}) showed the presence of an –NH– function, unsaturation and an aromatic system. The nmr spectrum of (18) showed the presence of an NH proton (δ 7.75) two aromatic protons at δ 7.5, 3 aromatic protons (δ 7.08) and one aromatic C-methyl group (δ 2.28). The signal for the six protons singlet at δ 1.42 together with a symmetrical doublets at δ 5.45 and 6.25 ($J = 10$ cps each) suggested the presence of a 2,2-dimethyl-Δ^3-pyran ring fused to the carbazole system. The mass spectrum of girinimbine showed a molecular ion peak at m/e 263 and a high intensity peak at m/e 248 (M-15) which supports the presence of a 2,2-dimethyl-Δ^3-pyran system in girinimbine and could be represented by the carbazolopyrylium ion (25).

The 3-methylcarbazole skeleton of girinimbine was confirmed by the isolation of 3-methylcarbazole on zinc dust distillation of giri-nimbine and of the acid (138) obtained by ozonisation. On hydrogenation, girinimbine furnished dihydrogirinimbine, $C_{18}H_{19}NO$ (11), m. p. 176°. The uv spectrum of the compound (λ_{max} 240, 303 nm; log ε 4.65, 4.22), was very similar to that of 2-methoxycarbazole suggesting that the pyran oxygen is linked to the carbazole fragment at the 2-position. The nmr spectrum of dihydrogirinimbine (11) showed that the vinyl protons had disappeared. The mass spectrum of dihydrogirinimbine showed the peak at m/e 210.

Further proof for the presence of a 2,2-dimethyl-Δ^3-pyran ring was provided by chromic acid oxidation of girinimbine when acetone was obtained, and by alkaline permanganate oxidation when α-hydroxy-isobutyric acid (45) was obtained. Girinimbine (18) on ozonisation furnished an α-hydroxyaldehyde (34), m. p. 193°, and the corresponding acid (138). The O-acetate of the aldehyde had a uv spectrum (Fig. 2) very similar to that of 1-formylcarbazole suggesting that the aldehyde group was attached to C-1. On decarbonylation it afforded a phenolic carbazole (35), m. p. 243°, the methyl derivative (108) of which, m. p. 225°, had uv spectrum [λ_{max}^{EtOH} 237, 258, 300 nm; log ε 4.69, 4.36, 4.15] very similar to that of that of 2-methoxycarbazole. These data would be combined to show that the points of attachment of the 2,2-dimethyl-Δ^3-pyran ring were position 1 and 2 of the carbazole nucleus. Previously CHAKRABORTY considered the structure 6-methyl-2-hydroxyl-1-formyl-carbazole (139) for the aldehyde obtained on ozonolysis. The structure of the aldehyde was however revised to (34) on the basis of the following evidence.

On decarbonylation over Pd/C, the aldehyde furnished 2-hydroxy-3-methylcarbazole. From this the structure of the aldehyde was found to be 1-formyl-2-hydroxy-3-methylcarbazole (34) which JOSHI et al. (65) had shown to be the ozonisation product obtained from mahanimbine (27) and the structure of girinimbine turned out to be (18). This structure was also proposed by DUTTA et al. (53) on reinterpretation of the physical evidence especially the nmr data.

The structure of girinimbine has been confirmed by three syntheses. In two syntheses, the 2,2-dimethyl-Δ^3-pyran unit was built on using 2-hydroxy-3-methylcarbazole (35) as the phenolic substrate. CHAKRA-

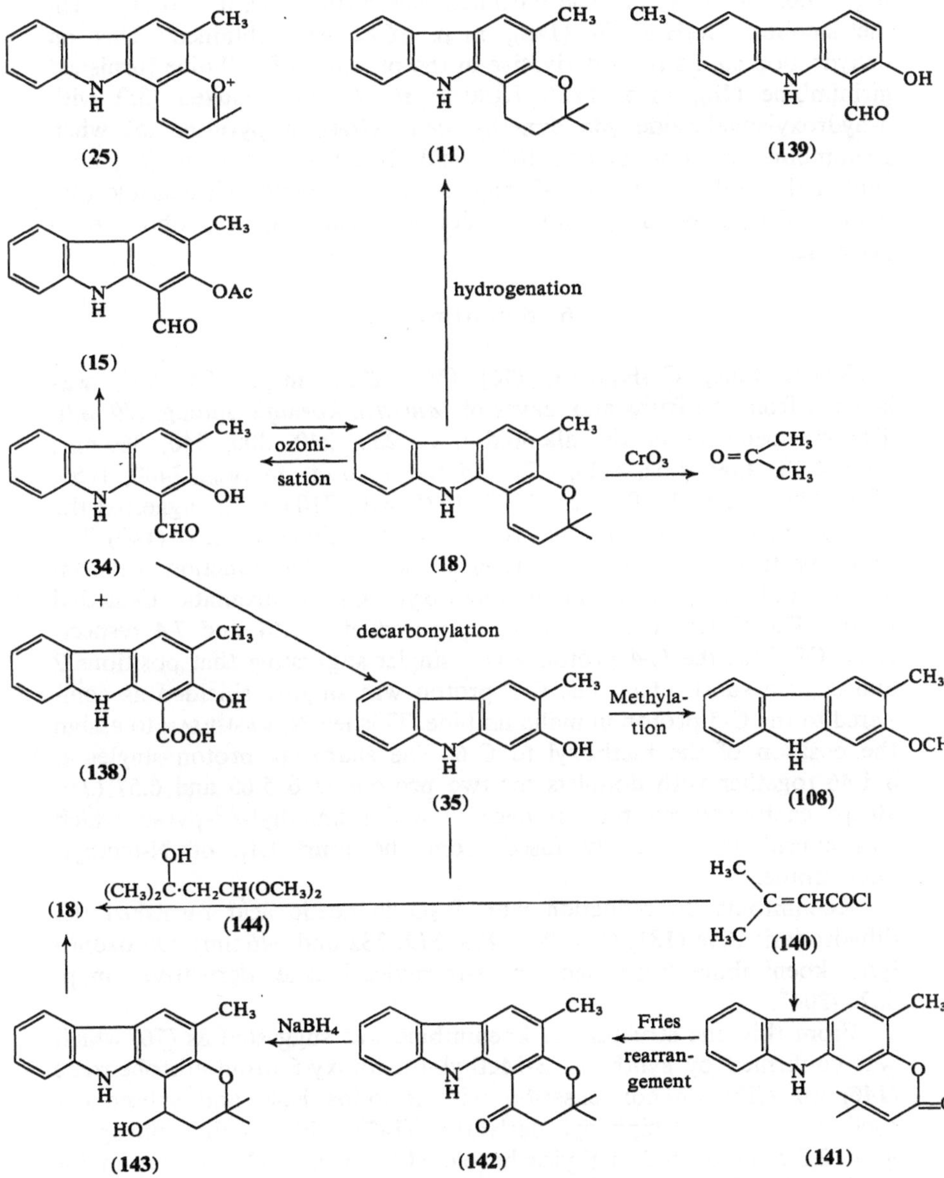

Chart 5. Reactions and synthesis of girinimbine

Borty and Islam (*19*) condensed β,β-dimethylacrylyl chloride (**140**) at 5° in presence of pyridine with (**35**) to form the acyl carbazole (**141**), m.p. 200°. This on Fries rearrangement and cyclisation gave the indolochromanone, $C_{18}H_{17}NO_2$ (**142**), m.p. 160—65° (λ_{max} 228, 282, 287 nm; log ε 4.65, 4.09, 4.22). The chroman was reduced with NaBH₄ when the alcohol, $C_{18}H_{19}NO_2$ (**143**), m.p. 160°, was obtained. This on dehydration *via* its tosyl derivative in the presence of collidine furnished girinimbine (**18**), m.p. 172°. Kureel et al. (*75*) refluxed (**35**) with 3-hydroxyisovaleraldehyde dimethylacetal (**144**) in pyridine (*6*) when girinimbine was obtained in 10% yield. Narashiman et al. (*81*) condensed the sodium salt of 1-formyl-2-hydroxy-3-methylcarbazole (**34**) with allyltriphenylphosphonium chloride when (**18**) was obtained in poor yield.

6. Koenimbine

Koenimbine, $C_{19}H_{19}NO_2$ (**76**) (M^+ 293), m.p. 194—95°, was isolated from the fruits and leaves of *Murraya koenigii* Spreng. (*70, 83*). The uv spectrum of the alkaloid (λ_{max} 230, 240, 300, 340, 360 nm; log ε 4.68, 4.69, 4.62, 4.24, 4.2) and the ir spectrum (ν_{max} 3400, 1650, 1580, 1470, 1210, 1140, 1125, 1110, 1025, 810, 710 cm^{-1}) suggested the presence of a carbazole nucleus. It gave an N-methyl derivative (**145**). The nmr spectrum showed the presence of an NH function between δ 7.75 and 6.85, an aromatic methoxyl and an aromatic C-methyl group. The C-4 and C-5 protons appeared at δ 7.61 and 7.4 respectively. Of these the C-4 proton was a singlet suggesting that positions 2 and 3 were substituted. The C-5 proton was slightly shielded as compared to the C-5 proton in mahanimbine. This led Narashiman to assign the position of the methoxyl to C-6. The sharp six proton singlet at δ 1.46 together with doublets for two protons at δ 5.65 and 6.57 (*J* = 10 cps each) showed the presence of a 2,2-dimethyl-Δ^3-pyran which was considered angularly fused from the nmr data of N-methyl-koenimbine.

Koenimbine on reduction with Pt₂O in acetic acid furnished the dihydroderivative (**131**) (λ_{max} 243, 258, 312, 332 and 346 nm). On ozonolysis koenimbine furnished an 1-formylcarbazole derivative, m.p. 302—304°.

From this, the structure of koenimbine was suggested as (**76**) which was confirmed by synthesis. 3-Methyl-4-methoxy-6-bromonitrobenzene (**146**) on Ullmann condensation with *m*-iodoanisole and subsequent cyclisation of the biphenyl derivative (**147**) with triethyl phosphite gave 2,6-dimethoxy-3-methylcarbazole (**17**), m.p. 169°, as the major product. The demethylated (**86**) product on condensation with 3-hydroxyisovaleraldehyde dimethylacetal (**144**) in pyridine furnished a phenolic compound (**148**) as one of the products. This on methylation with $(CH_3)_2SO_4$ in presence of K_2CO_3 yielded koenimbine (**76**).

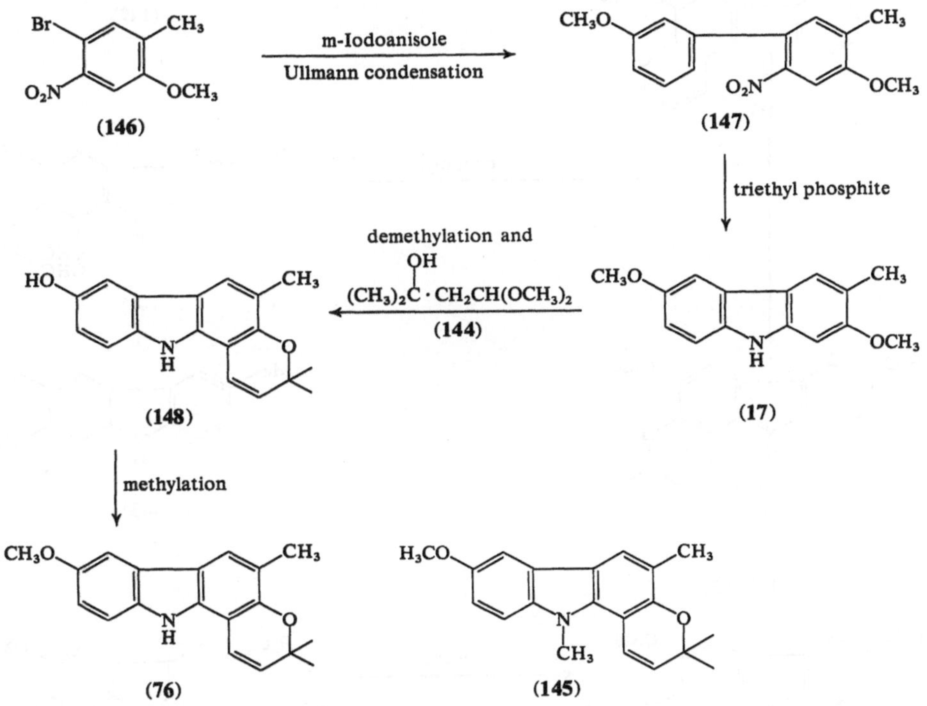
Br
CH3
O2N
OCH3
(146)
m-Iodoanisole
Ullmann condensation
CH3O
CH3
O2N
OCH3
(147)
triethyl phosphite
demethylation and
OH
(CH3)2C · CH2CH(OCH3)2
(144)
HO
CH3
N
H
O
(148)
CH3O
CH3
N
H
OCH3
(17)
methylation
CH3O
CH3
N
H
O
(76)
H3CO
CH3
N
CH3
O
(145)

7. Heptazolidine

Heptazolidine, $C_{18}H_{19}NO_2$ (**41**), m.p. 145° (M^+ 293), was isolated from root bark of *Clausena heptaphylla* by CHAKRABORTY *et al.* (*38*). The uv and ir spectra were suggestive of the presence of a pyranocarbazole system like that of girinimbine. The nmr spectrum of the alkaloid showed signals of an NH proton (δ 8.1), H-4 (δ 7.67), H-5 (δ 7.5 d, $J=2$ cps), one aromatic proton at δ 7.18 ($J=8.5$ cps) and 6.94 ($J=9.5$ cps). The 6 H singlet at δ 1.49 together with the doublets at δ 6.63 and 5.65 ($J=10$ cps each) for two vinylic protons showed the presence of a 2,2-dimethyl-Δ^3-pyran ring. In addition to these an aromatic C-methyl singlet at δ 2.37 and a methoxyl singlet at δ 3.98 showed the presence of methyl and methoxyl substitution on the carbazole nucleus. Since the protons at H-4 and H-5 were not ortho coupled, positions 3 and 6 were occupied.

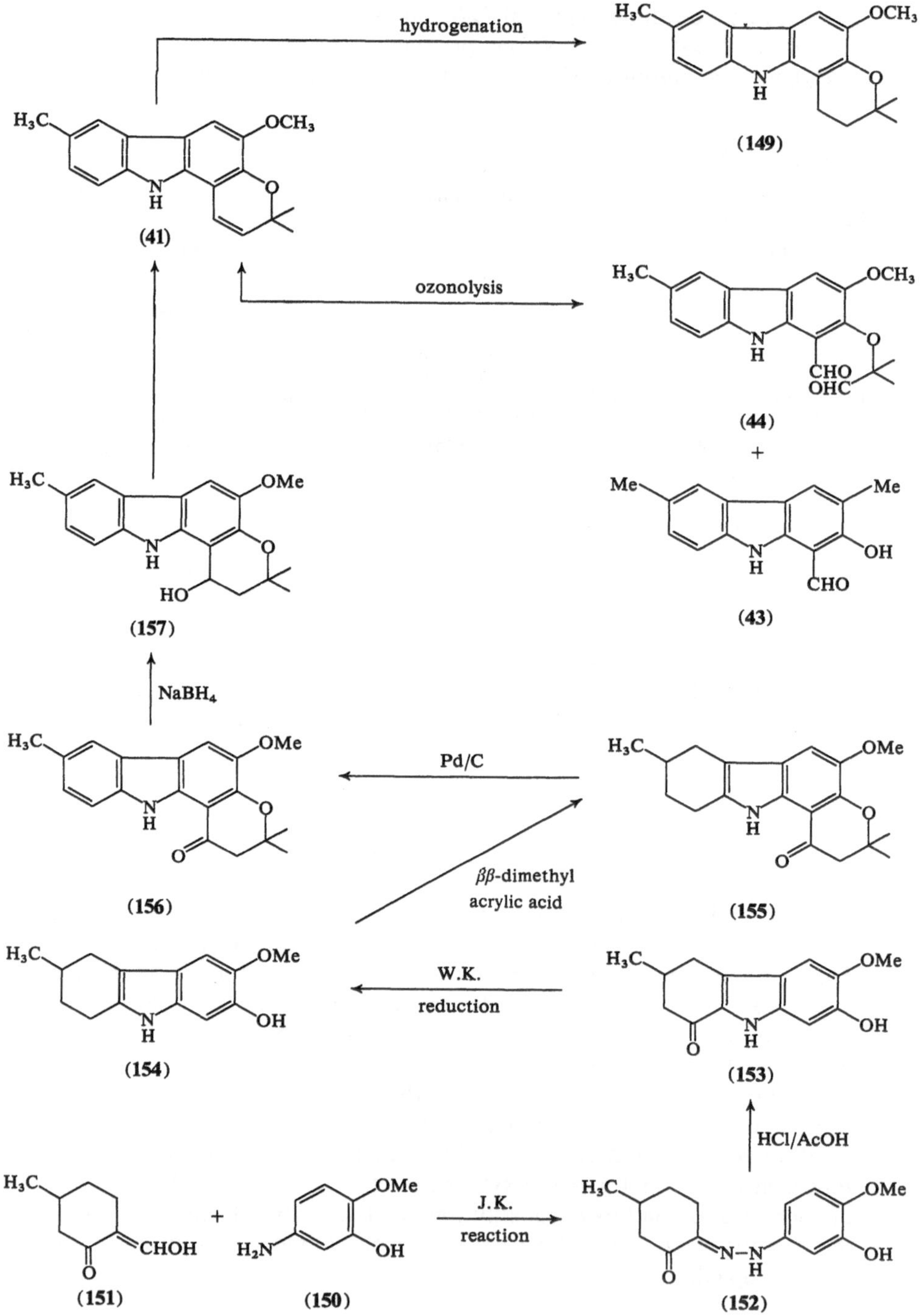

Chart 6. Reactions and synthesis of heptazolidine

Dihydroheptazolidine, $C_{19}H_{21}NO_2$ **(149)**, m.p. 130° (M^+ 295), obtained by hydrogenation over palladised charcoal had a uv spectrum very close to that of 6,7-dimethoxy-3-methylcarbazole which suggested that the methoxyl and the ether oxygen function were *ortho* with respect to each other. On ozonolysis heptazolidine **(41)** furnished a neutral product, $C_{19}H_{19}NO_4$ **(44)**, m.p. 150° (M^+ 325), considered to be a dialdehyde (v_{max} 1740, 1678 cm^{-1}) with the aromatic aldehydic function at C-1 as well as the α-hydroxyaldehyde **(43)** and acetone. From this evidence, the pyran ring was assumed to be attached to carbons at 1 and 2 with the pyran oxygen occupying the 2-position.

The structure of heptazolidine **(41)** has been confirmed by synthesis as follows. 3-Hydroxy-4-methoxyaniline **(150)** on Japp-Klingemann reaction with 2-formyl-3-methylcyclohexanone **(151)** furnished cyclohexane 1,2-dione-1'-(3'-hydroxy-4'-methoxy)-phenylhydrazone **(152)**, $C_{14}H_{18}N_2O_3$, m.p. 180—81°. The hydrazone **(152)** on cyclisation yielded 2 - hydroxy - 3 - methoxy - 6 - methyl - 8 - oxo - tetrahydrocarbazole **(153)**, $C_{14}H_{15}NO_3$, m.p. 152—53°. Wolff-Kishner reduction of the oxo compound furnished gummy 2-hydroxy-3-methoxy-6-methyl-5,6,7,8-tetrahydrocarbazole **(154)** which on treatment with β : β-dimethylacrylic acid in the presence of $POCl_3$ and $ZnCl_2$ yielded the chromanone **(155)**, $C_{19}H_{23}NO_3$, m.p. 160—61°. The chromanone **(155)** was dehydrogenated to indolochromanone **(156)**, $C_{19}H_{19}NO_3$, m.p. 150—51°. On borohydride reduction **(156)** furnished the alcohol **(157)**, $C_{19}H_{21}NO_3$, m.p. 165—67°, whose tosyl derivative was dehydrotosylated by heating with collidine, to heptazolidine **(41)**.

8. Murrayacine

Murrayacine, $C_{18}H_{15}NO_2$ **(7)**, m.p. 244—45°, M^+ 277, was isolated by CHAKRABORTY and DAS from the stem-bark of *Murraya koenigii* Spreng. (*30, 32*) and by ROY and CHAKRABORTY from *Clausena heptaphylla* (*96*).

The uv absorption spectrum of murrayacine (λ_{max} 226, 282, 301 nm; log ε 4.60, 4.57, 4.58) showed it to be a 3-formylcarbazole derivative. This was also supported by its ir spectrum (v_{max}^{KBr} 3250, 1675, 1040, 1600). The alcohol $C_{18}H_{17}NO_2$ **(158)**, m.p. 176°, obtained by sodium borohydride reduction of **(7)** had a uv spectrum strikingly similar to that of girinimbine suggesting the presence of a formyl-pyranocarbazole skeleton in **(7)**. The nmr spectrum of **(7)** showed signals for an aldehyde function (δ 10.68) and an NH-group (δ 12.0). The aromatic proton at C-4 appeared as singlet at δ 8.4 which was deshielded due to the proximity of aldehyde function at C-3, while the other four protons appeared as multiplets at δ 8.2—7.2. The sharp six-proton singlet at δ 1.4 together with two symmetrical one-proton doublets at δ 7.00 and 5.95 ($J = 10$ cps each) confirmed the presence of a 2,2-dimethyl-Δ^3-pyran system in murrayacine. The high intensity mass spectral peaks at m/e 262 and 234 could be represented by the species (159) and (160).

The isolation of carbazole by zinc dust distillation of the alkaloid confirmed the presence of a carbazole skeleton. Murrayacine on hydrogenation over Pt$_2$O in alcohol furnished a dihydro derivative, $C_{18}H_{17}NO_2$ (161), m.p. 176° (M^+ 279) which on reduction with LiAlH$_4$ furnished dihydrogirinimbine (26). The structure has been confirmed by two syntheses as follows.

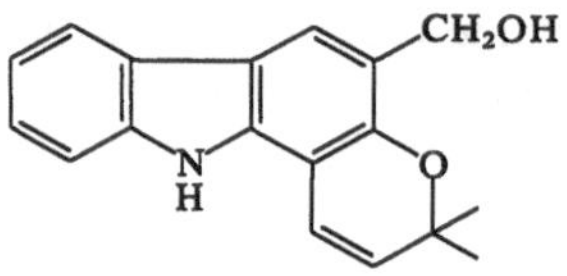

(158)

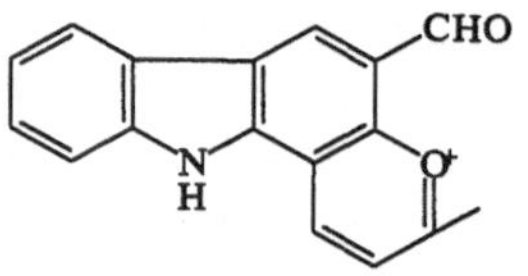

(159)

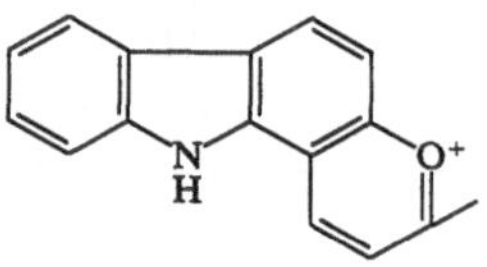

(160)

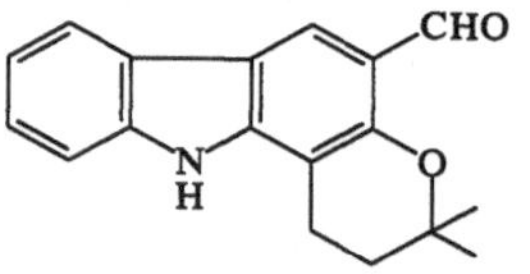

(161)

(i) In the synthesis reported by CHAKRABORTY et al. (21) 4-hydroxy-methyl-3-hydroxyaniline (162), m.p. 110°, on treatment with formyl cyclohexanone (52) under Japp-Klingemann condition furnished the hydrazone (163), m.p. 99—100°, which on cyclisation with a mixture of glacial acetic acid and hydrochloric acid furnished 2-hydroxy-3-hydroxymethyl-8-oxo-5,6,7,8-tetrahydrocarbazole (164), m.p. 113 — 115°. Wolff-Kishner reduction of (164) gave an alcohol (165) which on acetylation with 2,2-dimethylacrylyl chloride at 5° in presence of pyridine furnished the compound (166), m.p. 147—50°. The phenol ester (166) on Fries migration and subsequent treatment with hydrochloric acid furnished the chromanone (167), m.p. 125° which on dehydrogenation with Pd/C gave indolochromanone (168), m.p. 160—62°. Borohydride reduction of (168) gave alcohol (169), m.p. 114—15°, which on tosylation followed by dehydrotosylation in presence of collidine furnished chromenoindole (158), m.p. 199—208°. Oxidation of the chromenoindole (158) with active MnO$_2$ furnished murrayacine (7).

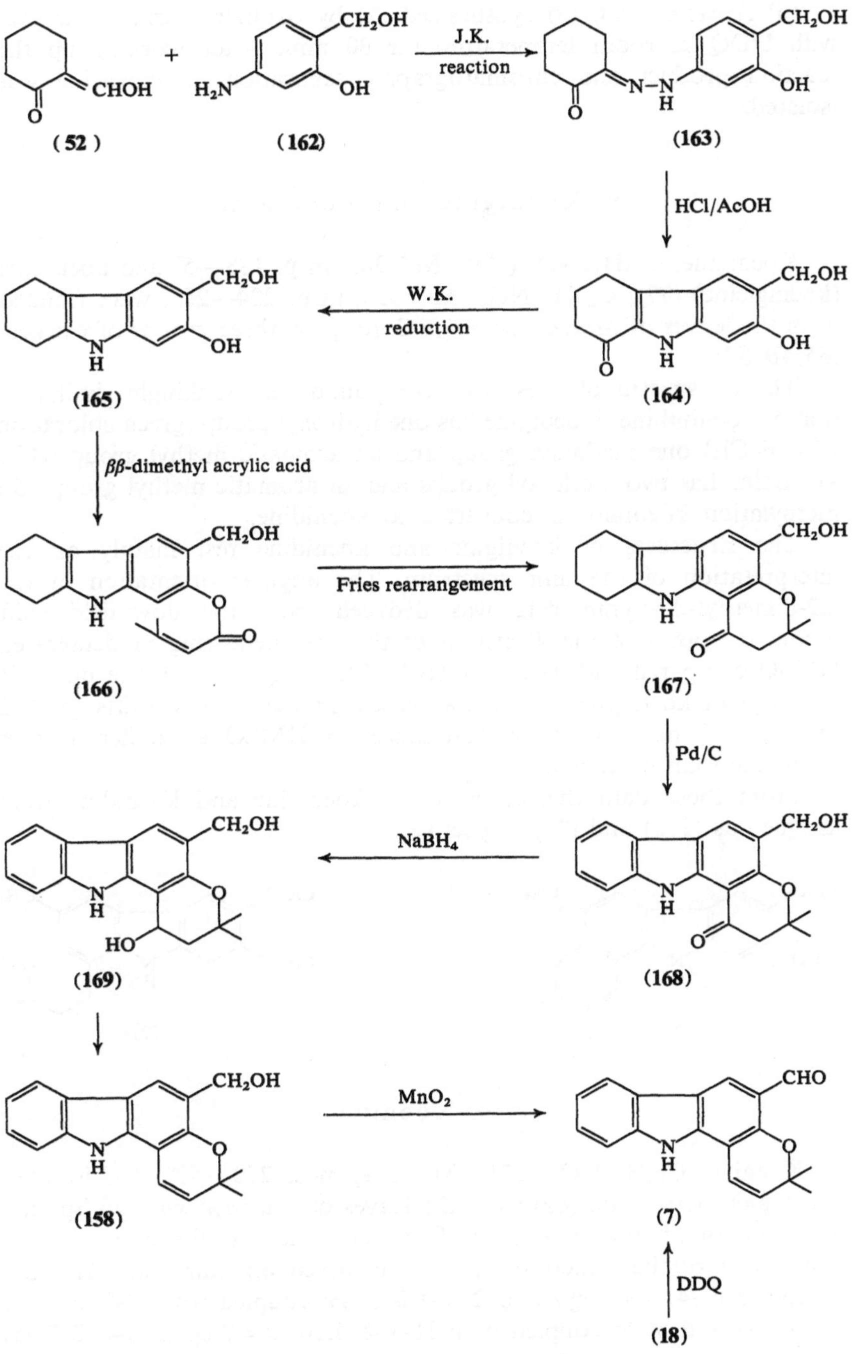

Chart 7. Synthesis of murrayacine

(ii) ANWER *et al.* (*3*) synthesized (**7**) by oxidizing girinimbine (**18**) with DDQ at room temperature for 30 min. After working up the reaction product and chromatographic separation, murrayacine was isolated.

9. Koenigine and Koenidine

Koenigine, $C_{19}H_{19}NO_3$ (**170**), M^+ 309, m.p. 183—5° and koenidine (koenigicine) (**77**), $C_{20}H_{21}NO_3$, M^+ 323, m.p. 224—25°, were isolated from the leaves of *Murraya koenigii* Spreng. by three groups of workers (*65, 70, 82*).

The uv spectra of these two compounds are strikingly similar to that of koenimbine. Koenigine has one hydroxyl group (green coloration with $FeCl_3$) one methoxyl group and an aromatic methyl group while koenidine has two methoxyl groups and an aromatic methyl group. On methylation koenigine is converted to koenidine.

The structures of koenigine and koenidine rest mainly on the interpretation of the nmr spectrum. The angular orientation of the 2,2-dimethyl-Δ^3-pyran ring was deduced from the downfield shift (about 0.3 ppm) of the 4'-proton of the chromene ring in deuterated DMSO compared with that in $CHCl_3$. The assignment of the phenolic hydroxyl in koenigine to C-7 was based on the relative shifts [δ 6.92 to 6.82] of the H-8 proton resonances in DMSO at higher field in koenigine than in koenidine.

From these data the structures of koenigine and koenidine were assigned as (**170**) and (**77**) respectively.

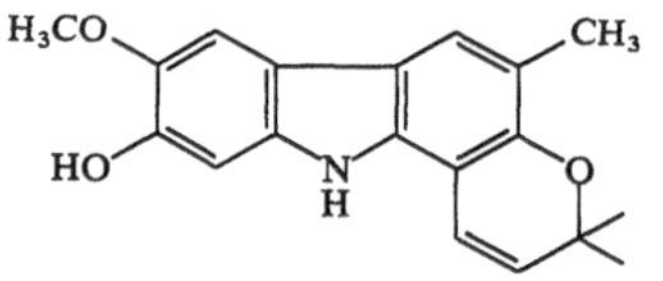

(170) 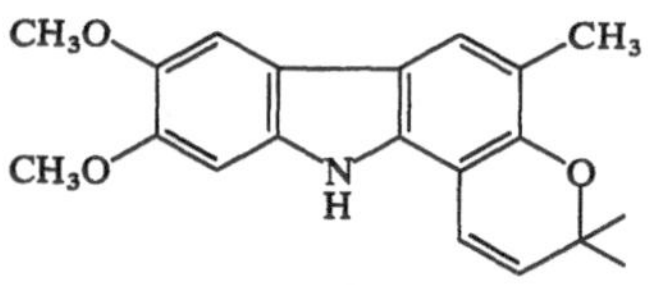   (77)

10. Koenine

Koenine, $C_{18}H_{17}NO_2$ (**171**), M^+ 279, m.p. 250—52°, was isolated by NARASHIMAN *et al.* (*82*) from the leaves of *Murraya koenigii* Spreng. The compound gave a positive ferric chloride test for phenols. The assignment of the structure was mainly based on nmr data. The C-7 proton (δ 6.74; $J = 8$ cps and 2 cps) is *ortho* coupled with H-8 (δ 7.18; $J = 8$ cps) and *meta* coupled with H-5 (δ 7.20; $J = 2$ cps). H-4 (δ 7.55) was a singlet. The presence of an aromatic C-methyl group is also deduced from nmr spectrum. The downfield shift of the 4'-proton in deuterated DMSO as compared with that in chloroform has been considered by NARASHIMHAN as characteristic for angular fusion of the 2,2-dimethyl-Δ^3-pyran ring to the carbazole nucleus.

From these data and the conversion of koenine to koenimbine by methylation, NARASHIMHAN proposed structure (171) for koenine.

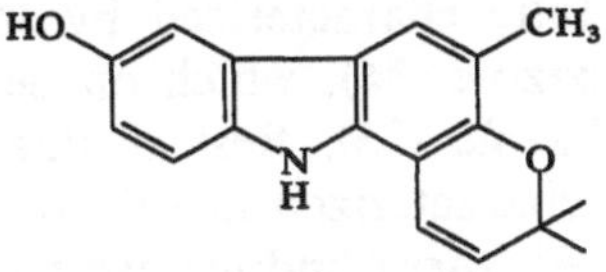

(171)

C. Members of the C$_{23}$-Skeleton Group

1. Mahanimbine

Mahanimbine, $C_{23}H_{25}NO$ (27), m. p. 94—95° (M^+ 331), $[\alpha]_D^{CHCl_3}$ + 45.4°, the first member of the C$_{23}$ carbazole alkaloid group, was isolated by CHAKRABORTY et al. (36), from the stem bark of Murraya koenigii Spreng. Later on it was isolated from the leaves and the fruits of the plant (83, 84). The compound gave a picrate, m.p. 142°. It also gave an N-methyl derivative (172).

The ir spectrum of mahanimbine showed bands for an –NH– function, an aromatic C-Me group, a double bond and an aromatic residue ($v_{max}^{CHCl_3}$ 3450, 1650, 1620, 1490, 1379, 1365; v_{max}^{KBr} 3350, 1647, 1613, 1575, 1493 cm^{-1}). The uv absorption spectrum (λ_{max} 223, 239, 288, 330, 343 nm; log ε 4.55, 4.06, 4.61, 3.90 and 3.42) was very similar to that of girinimbine (18) showing the presence of the same pyrano-carbazole chromophore in the two alkaloids. The mass spectrum of mahanimbine (27) showed a high intensity peak at m/e 248 which could be represented by the same carbazolopyrylium ion (25) as that formed from girinimbine. On hydrogenation mahanimbine gave dihydro derivative (173), tetrahydro derivative (12), $C_{23}H_{29}NO$, m. p. 108° $[\alpha]_D$—21°.

From this and the nmr data of the compound CHAKRABORTY et al. concluded that it had a 2,2-dimethyl-Δ^3-pyran skeleton like girinimbine and C_5H_9 residue containing an unsaturation. The 3-methyl carbazole skeleton in (27) was established by isolation of 3-methyl carbazole by zinc dust distillation of mahanimbine.

NARASIMHAN *et al.* made a more precise study of the nmr spectrum
of mahanimbine and assigned the signals at δ 7.62 and 7.87 to H-5 and
H-4. From the singlet nature of the signal for C-4 proton they concluded
that positions 2 and 3 were substituted. Because of the uv data of
tetrahydromahanimbine and the isolation of a hydroxyaldehyde, m.p.
193°, on ozonisation of mahanimbine, the pyran ring was considered
to be fused to the carbazole nucleus at the 2:1 position like that
in girinimbine (18). The α-hydroxy aldehyde obtained by ozonisation
of mahanimbine was fully characterized by JOSHI *et al.* as 1-formyl-
2-hydroxy-3-methylcarbazole (34), which on decarbonylation furnished
2-hydroxy-3-methylcarbazole (35). Besides this ozonolysis furnished a
trialdehyde which was characterized as (42) by JOSHI *et al.* (65) on the
basis of spectral data. All these findings were consistent with structure
(27) for mahanimbine proposed by NARASIMHAN *et al.* (83). The structure
was confirmed by four almost simultaneous syntheses of racemic material.

In all syntheses, 2-hydroxy-3-methylcarbazole (35) was condensed
with citral (174) (47, 67) leading to the formation of mahanimbine and
other products. 2-Hydroxy-3-methylcarbazole (35) was obtained by
different methods by different workers. KUREEL *et al.* (71) obtained the
phenolic substrate by Ullmann condensation of 2-nitrobromobenzene
(175) with 4-bromo-2-methylanisole (176). The biphenyl (177) so obtained was reduced with triethyl phosphite to isomeric methoxy methyl-
carbazoles. 2-Methoxy-3-methylcarbazole (108) on demethylation furnished 2-hydroxy-3-methylcarbazole (35).

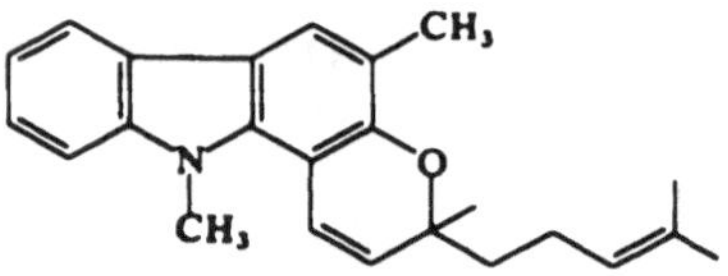

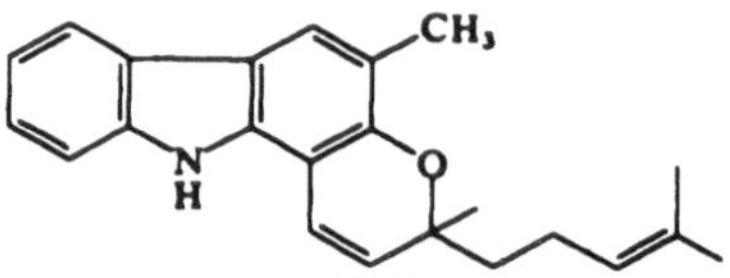

(27)

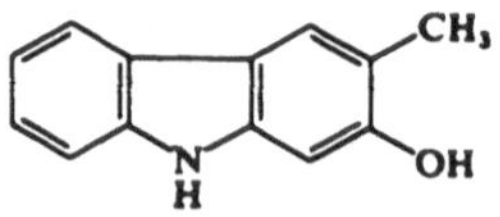

(172)

(35)

ozonolysis

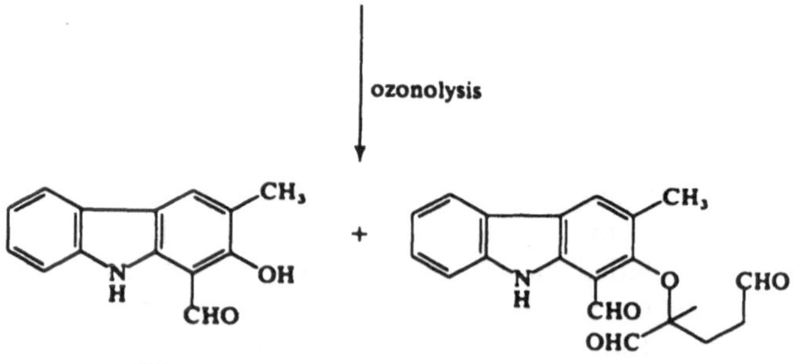

(34)

+

(42)

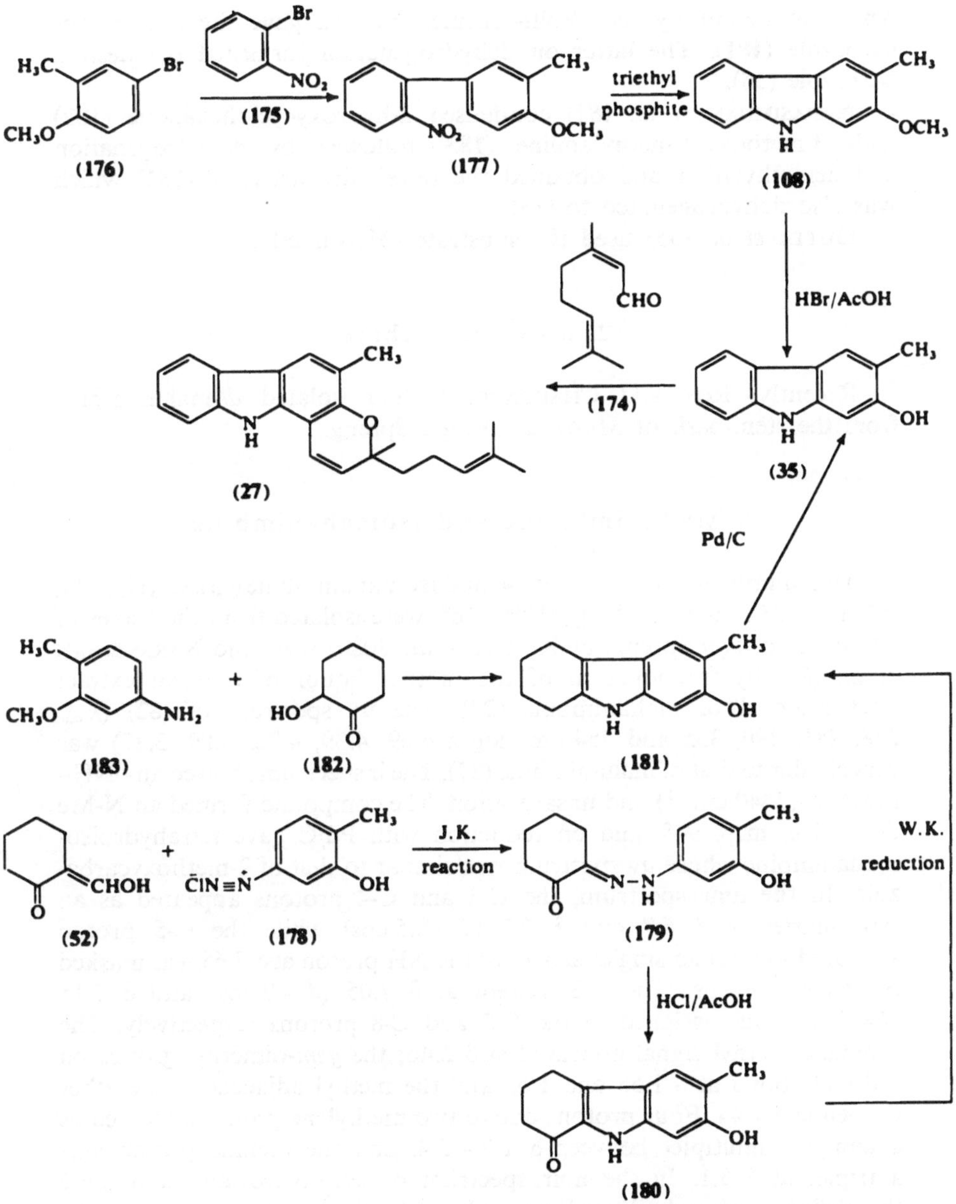

Chart 8. Reactions and synthesis of mahanimbine

CHAKRABORTY et al. (28) reacted diazotised 5-amino-2-cresol (178) with 2-hydroxymethylenecyclohexanone (52) to form the hydrazone (179). This (179) on indolisation furnished oxotetrahydrocarbazole (180),

23*

whose reduction by the Wolff-Kishner method gave the tetrahydro-carbazole (181). The latter on dehydrogenation furnished the desired carbazole (35).

Narasimhan et al. (81) condensed 2-hydroxycyclohexanone (182) with 3-methoxy-4-methylaniline (183) followed by dehydrogenation and demethylation and obtained the tetrahydrocarbazole (181) which was also dehydrogenated to (35).

Dutta et al. (55) used the substrate (35) directly.

2. dl-Mahanimbine

Recently, Roy and Chakraborty (97) isolated dl-mahanimbine from the stem-bark of Murraya koenigii Spreng.

3. Mahanimbicine and Isomahanimbine

Mahanimbicine $[\alpha]_D + 18.6°$ (74) and Isomahanimbine (65), $C_{23}H_{25}NO$, (82) (M$^+$ 331, m.p. 142°), $[\alpha]_D^{CHCl_3} -6°$, were isolated from the leaves of Murraya koenigii Spreng. by Kureel et al., Joshi et al. and Narasimhan et al. (84) by fractionation of the mother liquor of the leaf extract after removal of mahanimbine (27). The uv spectrum of (82) (λ_{max} 238, 281, 290, 335 and 354 nm; log ε 4.69, 4.39, 4.72, 3.89, 3.87) was very similar to that of mahanimbine (27). The ir spectrum showed an $-NH-$ function (3440 cm^{-1}) and unsaturation. The compound formed an N-Me derivative, m.p. 94°, and on reduction with Pd/C gave tetrahydroiso-mahanimbine whose uv spectrum was similar to that of 2-methoxycarba-zole. In the nmr spectrum, the C-3 and C-4 protons appeared as an AB quartet at δ 6.7 and δ 7.7 ($J = 8.5$ cps) while the C-5 proton appeared as a broad singlet at δ 7.68; the NH proton at δ 7.65 was masked by other protons. The AB system at δ 7.05 ($J = 8$ cps) and δ 7.15 ($J = 8$ cps) was assigned to the C-7 and C-8 protons respectively. The aromatic methyl signal appeared at δ 2.46; the gem-dimethyl groups on a double bond at δ 1.55 and 1.65 and the methyl adjacent to the ether oxygen at δ 1.44. Four protons due to two methylene groups appeared as a complex multiplet between δ 1.7—2.4, and the olefinic proton was a triplet at δ 5.1. In the nmr spectrum of tetrahydroisomahanimbine the olefinic proton doublets were replaced by the resonance of a benzylic methylene system. The AB spectrum due to the C-7 and C-8 protons showed a better resolved quartet centred at δ 7.15.

Ozonolysis of isomahanimbine furnished a phenolic aldehyde (184), m.p. 198—200°, with a uv spectrum very similar to that of 1-formyl-

carbazole. The neutral product, m.p. 194—95° obtained by ozonolysis
was assigned structure (185) on the basis of physical data. All the
evidence indicate structure (82) for isomahanimbine which was confirmed by synthesis of the racemic substance by KUREEL et al.

Ullmann condensation of 4-methoxy-2-nitrobromobenzene (186)
with m-iodotoluene and subsequent cyclization of the biphenyl derivative
(187) with triethyl phosphite and demethylation gave 2-hydroxy-6-methyl-
carbazole (188). The latter on condensation with citral for 30 min,
gave (dl)-mahanimbicine in low yield.

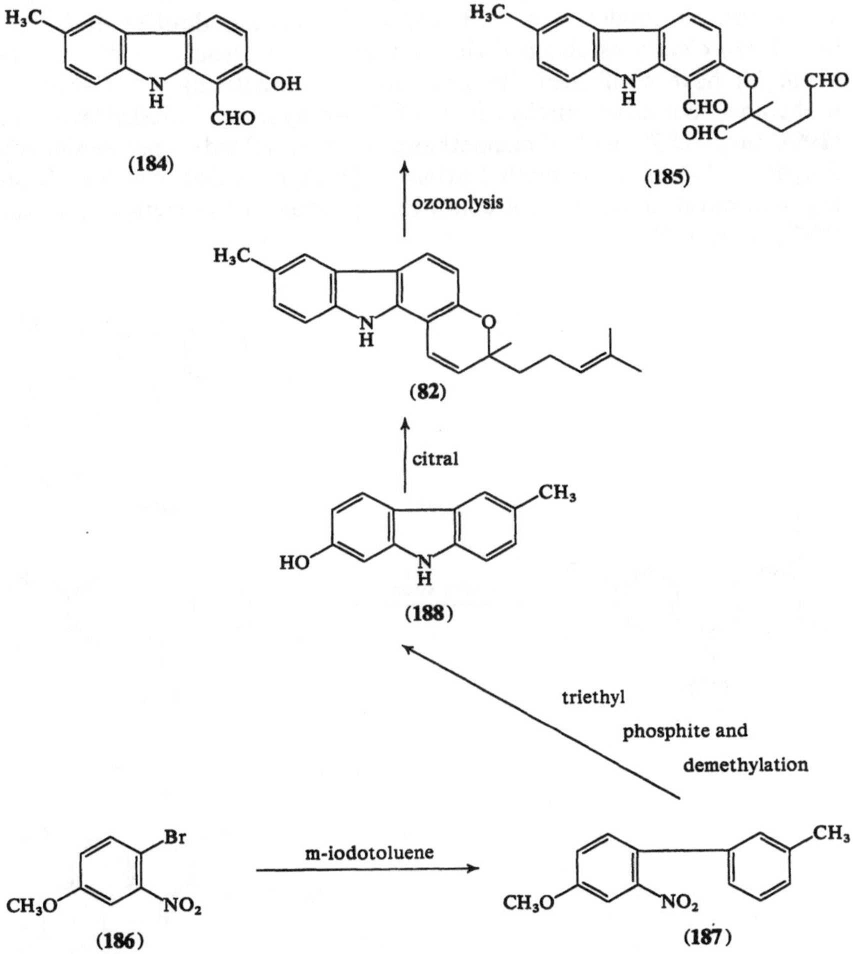

(184)

(185)

↑
ozonolysis

(82)

↑
citral

(188)

triethyl
phosphite and
demethylation

m-iodotoluene

(186)

(187)

NARASIMHAN et al. have recently isolated (+)-mahanimbicine from
the leaves of Murraya koenigii Spreng. (84), differing in optical rotation.

4. Mahanine

Mahanine, $C_{23}H_{23}NO_2$ (189), M^+ 247, m.p. 100°, $[\alpha]_D^{CHCl_3}$ — 24.4°, was isolated by NARASIMHAN *et al.* (*82, 84*) from the leaves of *Murraya koenigii* Spreng. The uv and ir spectra suggested the presence of the pyranocarbazole system.

The nmr spectrum showed that like mahanimbine it had three highfield methyl singlets indicating the presence of a monoterpene unit substituted on the pyran ring. The further appearance of a C-6 proton quartet at δ 6.54 ($J = 7.8$ cps), a C-8 proton doublet at 6.74 ($J = 2$ cps), a C-4 proton singlet at δ 7.58 and a C-5 proton doublet at δ 7.54 ($J = 8$ cps) clearly established that mahanine has structure (189). POPLI *et al.* (*4*) have confirmed the structure by synthesizing (±)-O-methylmahanine. Selective methylation of 2,7-dihydroxy-3-methylcarbazole (190), m.p. 285°, with diazomethane in ether afforded predominantly 2-hydroxy-7-methoxy-3-methylcarbazole (191), m.p. 240° which on heating with citral under normal conditions yielded (±)-O-methylmahanine (192), m.p. 180°.

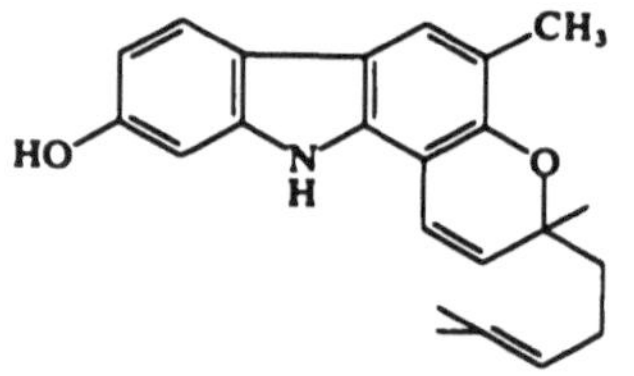

(189)

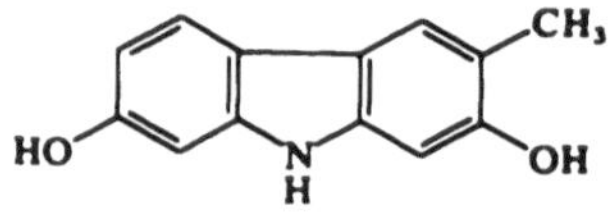

(190)   methylation   (191)

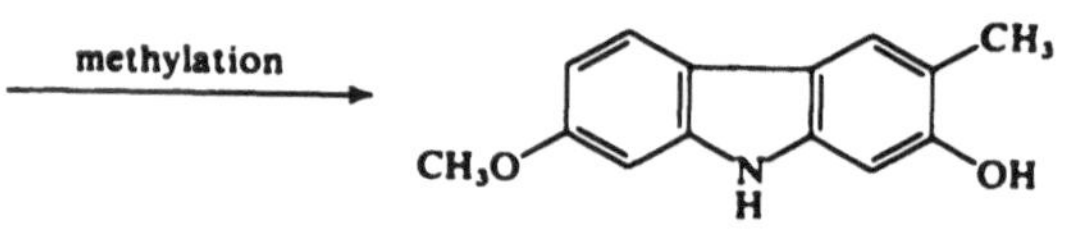

citral

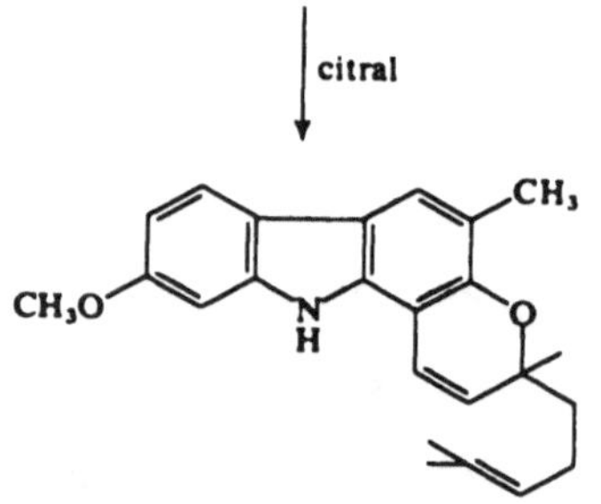

(192)

5. Mahanimbinine

Mahanimbinine, $C_{23}H_{27}NO_2$ (193) (M^+ 349), m.p. 179°, was isolated from the leaves of *Murraya koenigii* Spreng. (73). The ir spectrum of the compound showed the presence of an alcoholic hydroxyl (3580 cm^{-1}) and an imino (3450 cm^{-1}) function. The uv spectrum indicated the presence of a mahanimbine-like chromophore (λ_{max} 238, 288, 329, 344 and 359 nm). The nmr spectrum of mahanimbinine was similar to that of mahanimbine except for the absence of the signal of the vinylic proton and the replacement of the two vinyl methyl resonances two aliphatic methyl signals as a singlet at δ 1.2. Structure (193) was proposed for mahanimbine and has been confirmed by partial synthesis. Mahanimbine (27) on treatment with *m*-chloroperbenzoic acid furnished the epoxide (194) which on reduction with LiAlH$_4$ furnished mahanimbinine (193), m.p. 148°.

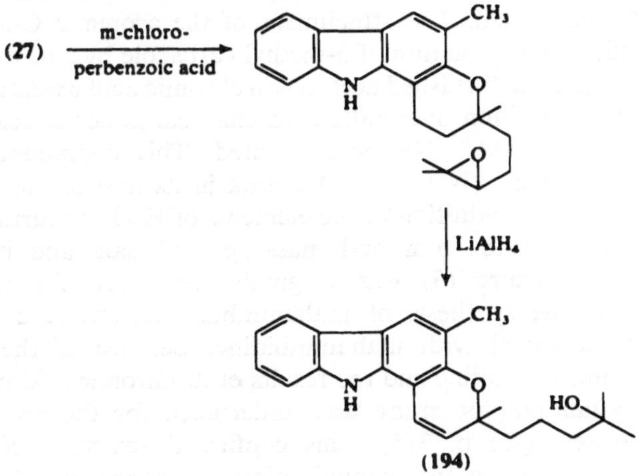

6. Murrayazoline

Murrayazoline (5), $C_{23}H_{25}NO$ (M^+ 331), $[\alpha]_D^{CHCl_3}$ — 10°, m. p. 260—62°, was isolated by CHAKRABORTY and CHOWDHURY in 1966 from the alcoholic extract of the stem bark of *Murraya koenigii* Spreng. (9). Later, its racemate named variously mahanimbidine (70) and curryangin (54) was isolated from the leaves and stem bark of the plant. Its uv. spectrum (λ_{max} 245, 257, 307 nm; log ε 4.69, 4.37, 4.16) were suggestive of a 2-methoxycarbazole chromophore. The ir spectrum of (5) lacked an NH peak. The mass spectrum of murrayazoline showed the base peak at m/e 331 and lacked the high intensity peak at m/e 248 represented by the carbazolopyrylium ion (25). Instead it had a prominent peak at m/e 210 (26) characteristic of the 2,2-dimethylpyranocarbazole system.

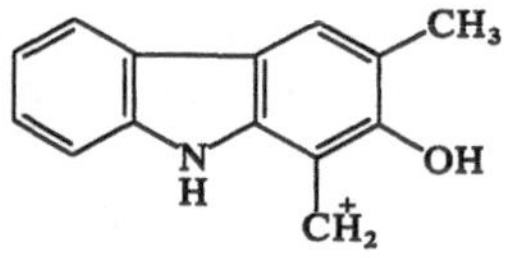

(26)

The nmr spectrum of murrayazoline (or mahanimbidine) shows the presence of five aromatic protons. The environment of the aromatic protons was very similar to that of mahanimbine and related compounds suggesting that positions 3, 2 and 1 of the carbazole nucleus are similarly substituted. The attachment of the aromatic C-methyl to C-3 was confirmed by isolation of 3-methyl carbazole by zinc dust distillation. Murrayazoline furnished acetone on chromic acid oxidation. In an attempted hydrogenation over palladised charcoal in acetic acid a compound $C_{23}H_{27}NO_2$ (M^+ 349) was isolated. This compound, unlike murrayazoline, had an –NH– and OH peak in its ir spectrum and was apparently formed by addition of the elements of H_2O to murrayazoline.

On the basis of the nmr and mass spectral data and biogenetic consideration structure (5) was originally advanced for mahanimbidine. During the synthesis of mahanimbine, DUTTA et al. isolated a compound identical with mahanimbidine. Because of the unusual hydration of murrayazoline and the results of its chromic acid oxidation, an x-ray crystallographic study was undertaken for the proof of the hexacyclic system (see p. 315). This confirmed structure (5) for the alkaloids, murrayazoline and mahanimbidine (currayangin). The hydration of murrayazoline was however clarified during the structural studies on murrayazolinine (14).

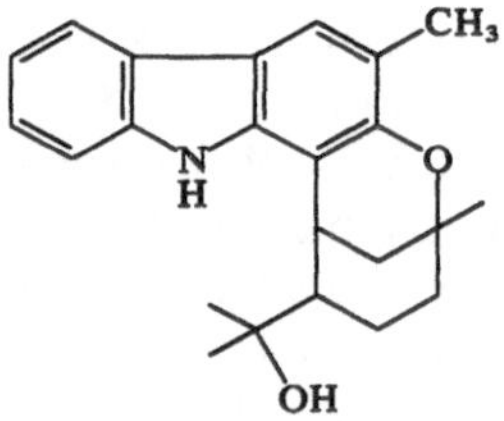

(14)

7. Murrayazolinine

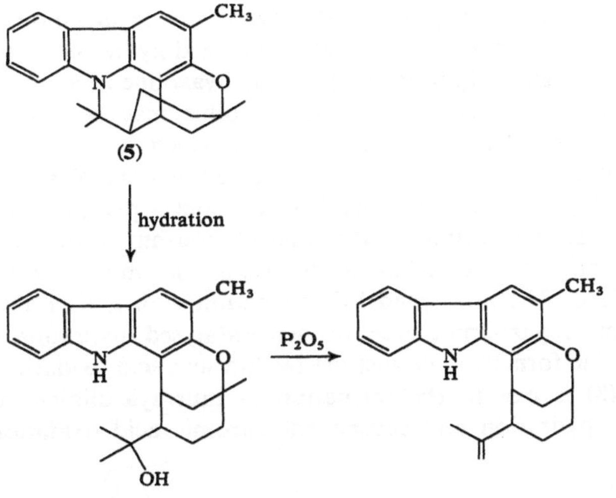

(5)

hydration

(14)   (13)

Murrayazolinine (14), $C_{23}H_{27}NO_2$, m.p. 184°, was isolated by CHAKRABORTY et al. (42). The infrared data showed the presence of an amino and hydroxy function and an aromatic system in (14). The ultraviolet absorption spectrum showed bands (λ_{max}240, 253, 258, 304 nm) characteristic for a 2-methoxycarbazole chromophore. The nmr spectrum showed signals at δ 7.9 for an NH proton, one aromatic singlet at δ 7.63, 7.45—7.0 (4H complex aromatic) protons, δ 3.68 (1H, benzylic methine), δ 2.34 (3H, s, aromatic C-methyl), δ 1.62 (1H, s, OH, exchangeable with D_2O), δ 1.4 (6H, s, gem-dimethyl shielded by oxygen ring.

The presence of the grouping $(CH_3)_2C-OH$ was confirmed by the isolation of acetone during chromic acid oxidation of the compound. The mass spectrum of murrayazolinine showed the base peak at m/e 349 and two other significant peaks at m/e 248 (74%) and 210 (17%) which could be represented by the species (25) and (26).

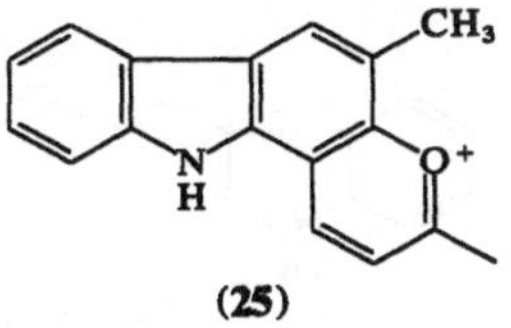

(25) 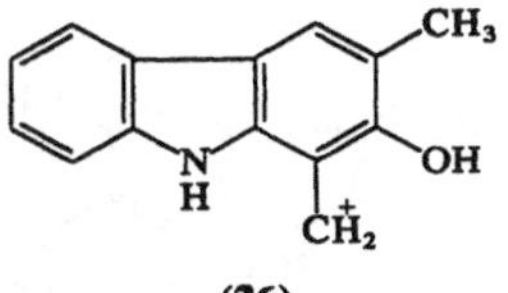   (26)

The evidence suggested that murrayazolinine is a pentacyclic alkaloid built on a 2-hydroxy-3-methylcarbazole skeleton to which a mono-

terpene unit is attached through an ether linkage. In turn the terpenoid
fragment constitutes a bicyclic system with an hydroxyisopropyl group
and a benzylic hydrogen. Treatment of murrayazolinine with $POCl_3$
furnished murrayazoline while acid-catalyzed hydration (aqueous acetic,
hydrochloric and sulphuric acid) of murrayazoline gave murrayazolinine.
The facile N–C cleavage of murrayazoline may be due to the ready
formation of a tertiary carbonium ion which has some rather bulky
groups attached to it. The situation is reminiscent of the hydrolysis of
tertiary alkyl halides where back strain effects play an important role.

With the clarification of the interrelationship of murrayazoline and
murrayazolinine the unusual reactions of murrayazoline can be
rationalized. The attempted hydrogenation with glacial acetic acid furnished murrayazolinine due to acid-catalysed hydration of murrayazoline. The formation of acetone by chromic acid oxidation of murrayazoline (5) is due to the formation of murrayazolinine (14) by acid-
catalysed hydration and subsequent chromic acid oxidation.

8. Murrayazolidine

Murrayazolidine, $[\alpha]_D^{30}$ $CHCl_3 + 20°$, $C_{23}H_{25}NO$ (13) (M^+ 331),
m. p. 143°, was isolated from the stem bark of *Murraya koenigii* Spreng.
by Chakraborty *et al.* (22). The ir spectrum showed it to have an
–NH– function, an aromatic system and unsaturation. The uv spec-
trum of (12) (λ_{max} 241, 260, 305 nm; log ε 4.57, 4.23, 4.07) was consistent
with a 2-methoxycarbazole chromophore. On hydrogenation it gave a
dihydro compound, $C_{23}H_{27}NO$, m. p. 176°, which did not show any
shift of the absorption maxima suggesting that the double bond in the
alkaloid is not conjugated with the carbazole system. The nmr spectrum
of murrayazolidine showed it to have an aromatic substitution pattern
similar to those of mahanimbine, murrayazoline and murrayazolinine.
It had signals for two vinylic protons (δ 4.8) and for a vinyl methyl
group on a double bond at (δ 2.35) besides a benzylic methine proton at
δ 3.4. Like in many carbazoles with a 2,2-substituted pyran system the

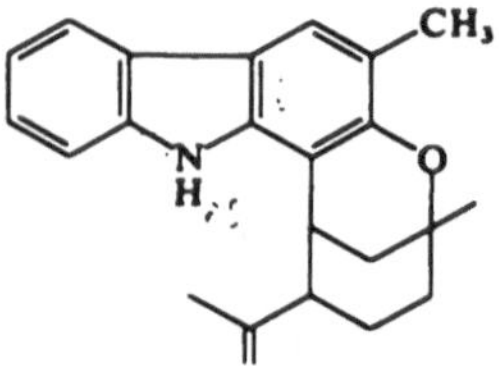

(13) 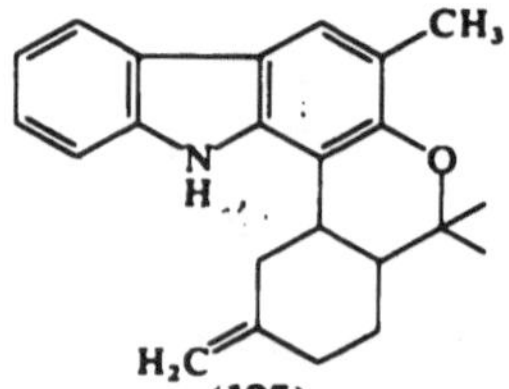   (195)

molecular ion peak was the base peak while the other peaks were at m/e 248 and at m/e 210. From the molecular formula and the functionalities of the compound two probable structures (13) or (195) of murrayazolidine were conceivable.

Structure (13) has been confirmed by partial synthesis (39). Murrayazoline (5) on acid-catalyzed hydration afforded murrayazolinine (14) which on dehydration with P_2O_5 afforded murrayazolidine. This synthesis conclusively proved the structure of murrayazolidine as (13) except for its absolute configuration.

9. Cyclomahanimbine

The racemate of murrayazolidine called cyclomahanimbine (72) or currayanine (55) was isolated by KUREEL et al. and DUTTA et al. respectively from the leaves of Murraya koenigii Spreng. Cyclomahanimbine, $C_{23}H_{25}NO$ (M$^+$ 331), m. p. 146° $[\alpha]_D \pm 0°$, was shown to have (λ_{max} 251, 257, 307 and 241 nm) very similar to that of 2-methoxycarbazole indicating the ether linkage at the 2-position of the carbazole nucleus. The nmr spectrum of the compound was very similar to that of murrayazolidine. From the physical data KUREEL et al. considered structure (13) for cyclomahanimbine. This has been supported by conversion of mahanimbine (27) to cyclomahanimbine by shaking with hydrochloric acid.

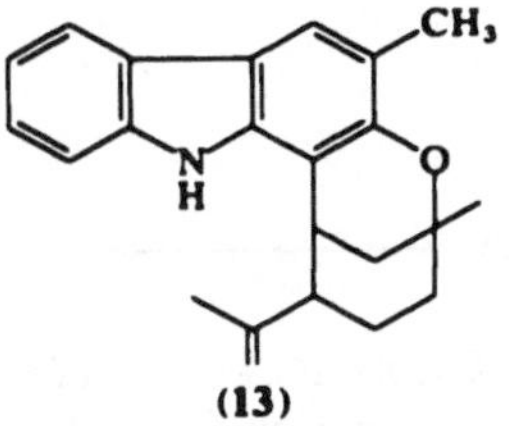

(13)

10. Bicyclomahanimbine

Bicyclomahanimbine, $C_{23}H_{25}NO$ (81) (M$^+$ 331), m. p. 145°, $[\alpha]_D$ CHCl$_3$ — 1.2°, has been isolated by KUREEL et al. (72) from the petroleum ether extract of M. koenigii leaves. The uv spectrum (λ_{max} 242, 255, 260, 305, 331 nm; log ε 4.61, 4.43, 4.40, 4.21, 3.62) indicative of the presence of a 2-methoxycarbazole chromophore as in tetrahydromahanimbine, murrayazoline and its congeners. It gave an N-methyl derivative.

The nmr spectrum of bicyclomahanimbine showed an aromatic substitution pattern similar to that of mahanimbine, murrayazoline or
murrayazolidine. The high field methyl signal at δ 0.71 was attributed
to attachment to a cyclobutane system. In consideration of the molecular
formula and the absence of double bonds the compound could be
considered a hexacyclic base. The data was consistent with either of
two structures (81) or (196) proposed by Kureel et al. Structure (81)
was preferred because of the conversion of mahanimbine (27) to
bicyclomahanimbine on silica gel or ion exchange resin (H⁺) treatment.
The conversion was visualised as follows.

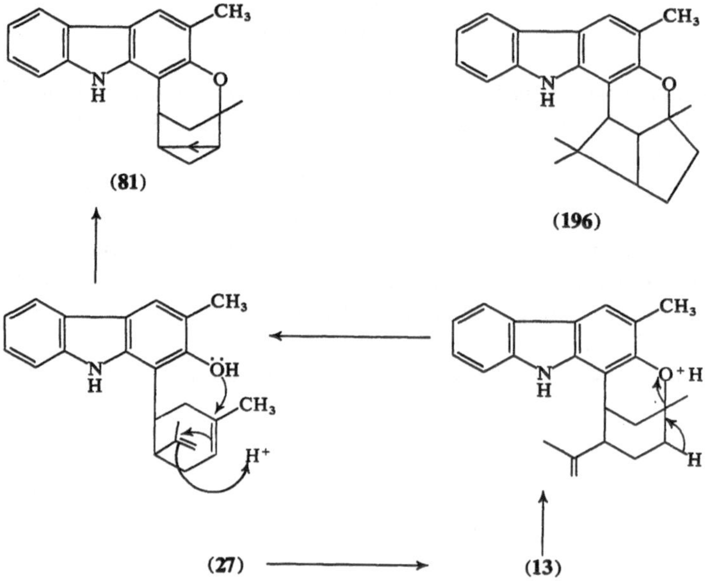

(81)

(196)

(27) ──────────→ (13)

On the other hand, Crombie et al. (48) in consideration of the
structure of cannabicyclol for which formula (197) had been determined
by x-ray analysis came to the conclusion that structure (196) was
correct for bicyclomahanimbine.

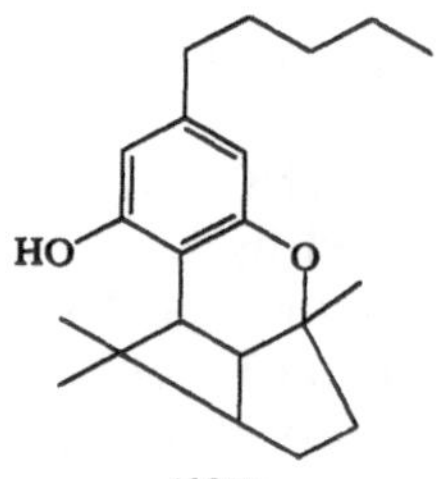

(197)

Bicyclomahanimbine was also obtained during the synthesis of mahanimbine by KUREEL et al. (71).

11. Murrayacinine

Murrayacinine, $C_{23}H_{23}NO_2$ (8), m.p. 105° (M^+ 345), was isolated from the stem bark of *Murraya koenigii* Spreng. by CHAKRABORTY et al. (41). The uv and ir spectral data of murrayacinine suggested a 3-formyl-carbazole derivative, while the uv spectrum of the alcohol $C_{23}H_{25}NO_2$ obtained by borohydride reduction of the alkaloid showed it to be a pyranocarbazole like girinimbine (18) or mahanimbine (27). The mass spectrum of the compound showed the base peak at m/e 262 (M-15-68) and another at m/e 234 (M-15-68-28) like those of murrayacine which could be represented by the species (159) and (160). It could therefore be concluded that murrayacinine was a 3-formyl pyranocarbazole deriva-tive like murrayacine with a C_5H_8 residue on the pyran ring. On reduction with LiAlH₄ murrayacinine furnished mahanimbine (27). This suggested structure (8) which has been confirmed by two syntheses.

2-Hydroxy-3-methylcarbazole (35) on chromyl chloride oxidation furnished 2-hydroxy-3-formylcarbazole (127), m.p. 240—41°, which

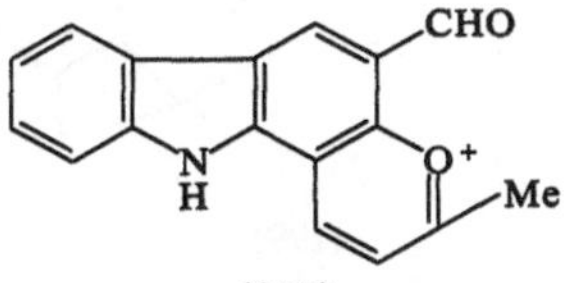

(159) 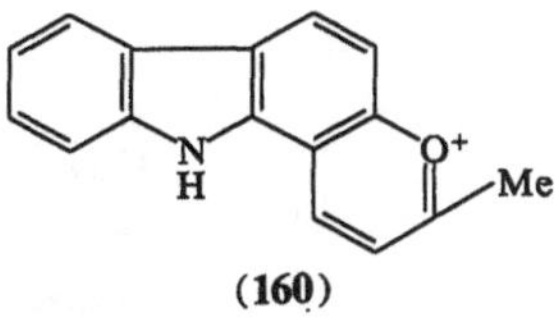   (160)

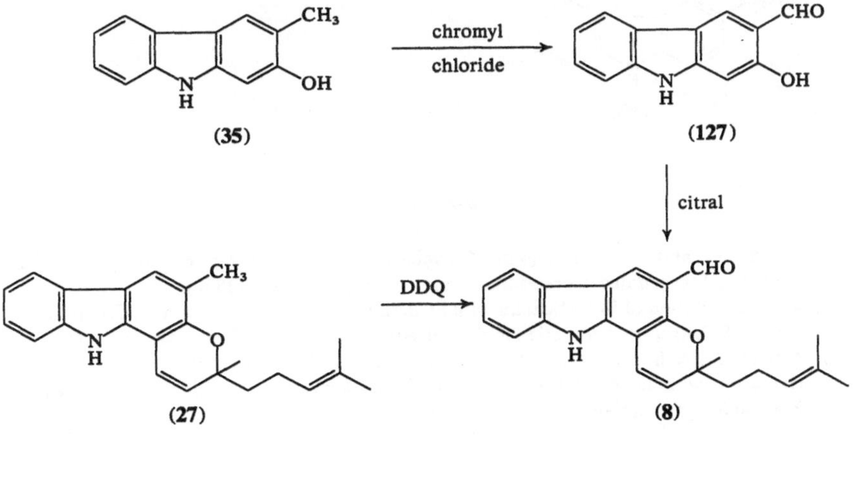

(35)   chromyl chloride →   (127)

citral

(27)   DDQ →   (8)

on condensation with citral in presence of pyridine at room temperature gave murrayacinine (8).

In another synthesis, mahanimbine (27) on oxidation with DDQ furnished murrayacinine (8) (43).

12. Bicyclomahanimbicine

Bicyclomahanimbicine, $C_{23}H_{25}NO$ (198), m.p. 218°, was isolated in poor yield from the amorphous fraction left after the separation of mahanimbicine (74). The uv spectrum was similar to that of tetra-hydromahanimbine. The nmr data indicate the close relationship of the base to mahanimbicine. Two structures (198) and (199) were therefore considered for bicyclomahanimbicine.

KUREEL et al. obtained (±)-bicyclomahanimbicine by condensing (188) with citral in pyridine and refluxing for 6 hr. A benezene solution of mahanimbicine on shaking with ion exchange resin, also furnished bicyclomahanimbicine. On the basis of this structure (198) for bicyclo-mahanimbicine was favored.

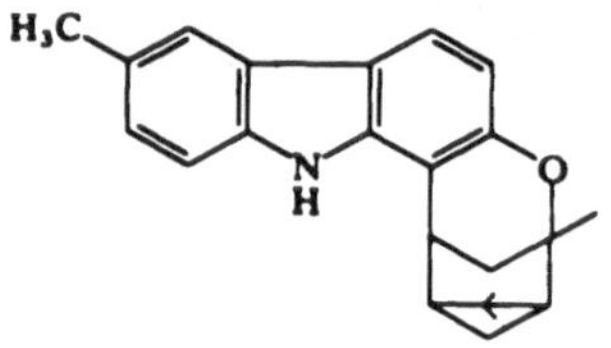

(198)

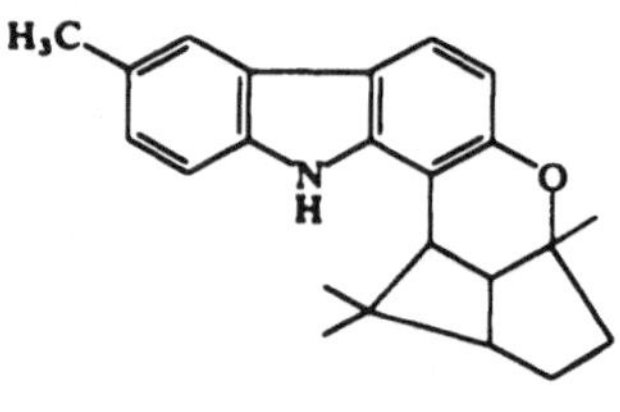

(199)

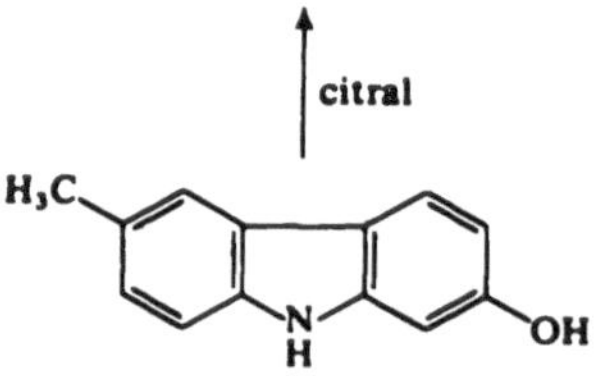

citral

(188)

References

1. ACZEL, T., and H. E. LUMPKIN: Correlation of Mass Spectrum with Structure in Aromatic Oxygenated Compounds. Anal. Chem. 32, 1819 (1960).
2. — — Correlation of Mass Spectra with Structure in Aromatic Oxygenated Compounds: Methyl Substituted Aromatic Acids and Aldehydes. Anal. Chem. 33, 387 (1961).
3. ANWER, F., A. S. MASALDAN, R. S. KAPIL, and S. P. POPLI: Synthesis of Murrayacine; Oxidation with DDQ of the Activated Methyl Group of the Alkaloids of Murraya koenigii Spreng. Indian J. Chem. 11, 1314 (1973).

4. ANWER, F., R. S. KAPIL, and S. P. POPLI: Terpenoid Alkaloids from *Murraya koenigii* Spreng. VII: Synthesis of DL-O-Methyl Mahanine and Related Carbazoles. Experientia **28**, 769 (1972).

5. BANDARNAYAKE, W. M., L. CROMBIE, and D. A. WHITING: 3-Hydroxy-3-methyl-1,1-dimethoxybutane, a New Reagent for Dimethylchromenylation. Synthesis of Lonchocarpin, Jacareubin, Evodionol Methyl Ether and Other Chromenes. J. Chem. Soc. (C) **8**, 11 (1971).

6. BASU, R.: Electronic Spectra of Carbazole Derivatives. J. Indian Chem. Soc. **44**, 580 (1967).

7. BHAKUNI, D. S., M. L. DHAR, M. M. DHAR, B. N. DHAWAN, and B. N. MEHROTRA: Screening of Indian plants for Biological activity: Part II. Indian J. Exptl. Biol. **7**, 250 (1969).

8. BHATTACHARYYA, P., and D. P. CHAKRABORTY: Murrayanine and Dentatin from *Clausena heptaphylla* Wt. and Arn. Phytochemistry **12**, 1831 (1973).

9. BORDNER, J., D. P. CHAKRABORTY, B. K. CHOUDHURY, S. N. GANGULY, K. C. DAS, and B. WEINSTEIN: The X-ray Crystal Structure of Murrayazoline (Mahanimbidine and Currayangin). Experientia **28**, 1406 (1972).

10. BOWEN, E. J., and J. H. D. ELAND: Photochemistry of Diphenylamine Solutions. Proc. Chem. Soc. 202 (1963).

11. BÜCHI, G., and E. W. WARNHOFF: The Structure of Uleine: J. Am. Chem. Soc. **81**, 4433 (1959). And private communications from G. BÜCHI.

12. BUDZIKIEWIEZ, H., R. T. ALPIN, D. A. LIGHTNER, C. DJERASSI, R. MECHOULAM, and Y. GAONI: Massenspektroskopie und ihre Anwendung auf strukturelle und stereochemische Probleme. Massenspektroskopische Untersuchung der Inhaltstoffe von Haschisch. Tetrahedron **21**, 1881 (1965).

13. BURNELL. R. H., and D. CASA: Alkaloids of *Aspidosperma vargasii*. Can. J. Chem. **45**, 89 (1967).

13a. CADOGAN, J. I. G., and M. CAMERONWOOD: Reaction of Nitrocompounds by Triethylphosphite: A New Cyclisation Reaction. Proc. Chem. Soc. 361 (1962).

14. CAMPBELL, N., and B. M. BARCLAY: Recent Advances in the Chemistry of Carbazole. Chem. Rev. **40**, 359 (1947).

15. CARRUTHERS, W.: Photocyclisation of Diphenylamines: Synthesis of Glycozoline. Chem. Commun. 272 (1966).

16. CHAKRABORTY, D. P.: Glycozoline. a Carbazole Derivative from *Glycosmis pentaphylla* (Retz.) DC. Tetrahedron Letters 661 (1966).

17. — On the Biogenesis of the Carbazoles of Rutaceae. J. Indian Chem. Soc. **46**, 177 (1969).

18. — Glycozoline, a Carbazole Derivative from *Glycosmis pentaphylla*. Phytochem. **8**, 769 (1969).

19. CHAKRABORTY, D. P., and A. ISLAM: Synthesis of Girinimbine. J. Indian Chem. Soc. **48**, 91 (1971).

20. CHAKRABORTY, D. P., A. ISLAM, and (Miss) M. SARKAR: Iodine-Promoted Thermal Synthesis of Carbazole from Aniline. Proc. of the 5th International Congress of Heterocyclic Chemistry, held at Ljubljana, p. 341 (1975).

21. CHAKRABORTY, D. P., A. ISLAM, and P. BHATTACHARYYA: Synthesis of Murrayacine. J. Organ. Chem. **38**, 2783 (1973).

22. CHAKRABORTY, D. P., A. ISLAM, S. P. BASAK, and R. DAS: Murrayazolidine, the First Pentacyclic Carbazole from *Murraya koenigii* Spreng. Chem. and Ind. (London) 593 (1970).

23. CHAKRABORTY, D. P., B. K. BARMAN, and P. K. BOSE: On the Structure of Girinimbine, a Pyranocarbazole Derivative Isolated from *Murraya koenigii* Spreng. Science and Cult. (India) **30**, 445 (1964).

24. CHAKRABORTY, D. P., B. K. BARMAN, and P. K. BOSE: On the Constitution of Murrayanine, a Carbazole Derivative Isolated from *Murraya koenigii* Spreng. Tetrahedron **21**, 681 (1965).
25. CHAKRABORTY, D. P., and B. K. CHOWDHURY: Synthesis of Murrayanine. J. Organ. Chem. **33**, 1265 (1968).
26. CHAKRABORTY, D. P., and B. P. DAS: Glycozolidine, a New Carbazole Derivative from *Glycosmis pentaphylla* (Retz) DC. Science and Cult. (India) **32**, 181 (1966).
27. CHAKRABORTY, D. P., B. P. DAS, and S. P. BASAK: Structure of Glycozolidine. The Plant Biochemical Journal (India) **1**, 73 (1974).
28. CHAKRABORTY, D. P., D. CHATTERJEE, and S. N. GANGULI: Synthesis of Mahanimbine. Chem. and Ind. (London) 1662 (1969).
29. CHAKRABORTY, D. P., J. DUTTA, and A. GHOSH: The Ultraviolet Absorption Spectra of Some Simple Carbazole Derivatives. Science and Cult. (India) **31**, 529 (1965).
30. CHAKRABORTY, D. P., and K. C. DAS: Structure of Murrayacine. Chem. Commun. 967 (1968).
31. CHAKRABORTY, D. P., K. C. DAS, and B. K. CHOWDHURY: Synthesis of Glycozoline. Phytochem. **8**, 773 (1969).
32. — — — Structure of Murrayacine. J. Organ. Chem. **36**, 725 (1971).
33. CHAKRABORTY, D. P., K. C. DAS and B. P. DAS: Paper Chromatographic Separation of Some Carbazole Derivatives. Indian J. Chem. **4**, 416 (1966).
34. CHAKRABORTY, D. P., K. C. DAS, and A. ISLAM: Heptazoline, a New Carbazole Alkaloid from *Clausena heptaphylla* Wt. and Arn. J. Indian Chem. Soc. **47**, 1197 (1970).
35. CHAKRABORTY, D. P., K. C. DAS, B. P. DAS, and B. K. CHOWDHURY: On the Antibiotic Properties of some Carbazole Alkaloids. Trans. Bose Res. Inst. **38**, 1 (1975).
36. CHAKRABORTY, D. P., K. C. DAS, and P. K. BOSE: Structure of Mahanimbine, a Pyranocarbazole Derivative from *Murraya koenigii* Spreng. Science and Cult. (India) **32**, 83 (1966).
37. CHAKRABORTY, D. P., P. BHATTACHARYYA, S. ROY, R. GUHA, and S. P. BHATTACHARYYA: Some Minor Carbazole Alkaloids of *Murraya koenigii* Spreng. Paper presented at the 4th Indo Soviet Symposium of the Chemistry of Natural Products, held in Feb. 1976 at Lucknow (India).
38. CHAKRABORTY, D. P., P. BHATTACHARYYA, A. ISLAM, and S. ROY: Heptazolidine, a Carbazole Alkaloid from *Clausena heptaphylla* Wt. and Arn. Chem. and Ind. 303 (1974).
39. CHAKRABORTY, D. P., P. BHATTACHARYYA, and A. R. MITRA: Murrayazolidine. Chem. and Ind. 260 (1974).
40. — — — Thermal Synthesis of Glycozoline. J. Indian Chem. Soc. **53**, 321 (1976).
41. CHAKRABORTY, D. P., P. BHATTACHARYYA, A. ISLAM, and S. ROY: Structure of Murrayacinine, a New Carbazole Alkaloid from *Murraya koenigii* Spreng. Chem. and Ind. 165 (1974).
42. CHAKRABORTY, D. P., S. N. GANGULY, P. N. MAJI, A. R. MITRA, K. C. DAS, and B. WEINSTEIN: Murrayazolinine, a Carbazole Alkaloid from *Murraya koenigii* Spreng. Chem. and Ind. 322 (1973).
43. CHAKRABORTY, D. P., and S. ROY: Unpublished data.
44. CHOWDHURY, B. K., and D. P. CHAKRABORTY: Mukoeic Acid, the First Carbazole Carboxylic Acid From Plant Source. Chem. and Ind. 549 (1969).
45. — — Mukoeic Acid, the First Carbazole Carboxylic Acid from Plant Source. Phytochem. **10**, 1967 (1971).
46. CLEMO, G. R., and D. G. FELTON: The Influence of Structure on the UV Absorption Spectra of Heterocyclic Systems. J. Chem. Soc. 1658 (1952).
47. CROMBIE, L., and R. PONSFORD: Synthesis of Hashish Cannabinoids by Terpenic Cyclisation. Chem. Commun. 894 (1968).

48. CROMBIE, L., D. A. WHITING, D. G. CLARKE, and M. J. BEGLEY: The X-ray Structure of Dibromocannabicyclol: Structure of Bicyclomahanimbine. Chem. Commun. 1547 (1970).
49. CRUM, J. D., and P. W. SPRAGUE: Synthesis of Murrayanine. Chem. Commun. 417 (1966).
50. DAS, K. C., D. P. CHAKRABORTY, and P. K. BOSE: Antifungal Activity of Some Constituents of Murraya koenigii Spreng. Experientia 21, 340 (1965).
51. DREYER, D. L., MICHAEL V. PICKERING, and P. COHAN: Distribution of Limonoids in the Rutaceae. Phytochem. 11, 705 (1972).
52. DUTTA, J., M. HOQUE, and D. P. CHAKRABORTY: Thin-layer Chromatography of Carbazole Derivatives. J. Chromatogr. 42, 555 (1969).
53. DUTTA, N. L., and C. QUASSIM: Constituents of Murraya koenigii. Structure of Girinimbine. Indian J. Chem. 7, 307 (1969).
54. DUTTA, N. L., C. QUASSIM, and M. S. WADIA: Constituents of Murraya koenigii, Structure of Curryangin. Indian J. Chem. 7, 1061 (1969).
55. — — — Synthesis of Mahanimbine and Curryangin. Indian J. Chem. 7, 1168 (1969).
56. ELDERFIELD, R. C.: Heterocyclic Compounds, Vol. 3, pp. 291, 2nd printing 1960. London-New York: J. Wiley and Sons, Inc.
57. ERDTMAN, H.: In "Perspectives in Phytochemistry" (J. B. HARBORNE and T. SWAIN, Eds.), p. 107. New York: Academic Press. Inc. 1969.
58. GASKELL, A. J., and J. A. JOULE: Subincanine, a C_{22}-Carbazole Alkaloid. Tetrahedron Letters 77 (1970).
59. GRELMAN, K. A., G. M. SHERMAN, and H. LINSCHITZ: Photoconversion of Diphenylamines to Carbazoles and Accompanying Transient Species. J. Amer. Chem. Soc. 85, 1881 (1963).
60. ISLAM, A., P. BHATTACHARYYA, and D. P. CHAKRABORTY: Thermal Cyclisation of Diphenylamine to Carbazole: Synthesis of Natural Product Glycozolidine. Chem. Commun. 537 (1972).
61. JOSHI, B. S., and D. F. RANE: Synthesis of Heptaphylline. Chem. and Ind. 685 (1968).
62. JOSHI, B. S., and D. H. GAWAD: Isolation and Structure of Indizoline, a New Carbazole Alkaloid from Clausena indica Oliv. Indian J. Chem. 12, 437 (1974).
63. JOSHI, B. S., D. GAWAD, V. N. KAMAT, and T. R. GOVINDACHARI: Structure and Synthesis of Heptaphylline. Phytochem. 11, 2065 (1972).
64. JOSHI, B. S., D. H. GAWAD, and V. N. KAMAT: 6-Methoxyheptaphylline, a New Carbazole Alkaloid from Clausena indica Oliv. Indian J. Chem. 10, 1123 (1972).
65. JOSHI, B. S., V. N. KAMAT, and D. H. GAWAD: On the Structure of Girinimbine, Mahanimbine, Isomahanimbine, Koenimbidine and Murrayacine. Tetrahedron 26, 1475 (1970).
66. JOSHI, B. S., V. N. KAMAT, A. K. SAKSENA, and T. R. GOVINDACHARI: Structure of Heptaphylline, a Carbazole Alkaloid from Clausena heptaphylla Wt. and Arn. Tetrahedron Letters 4019 (1967).
67. KANE, V. V., and R. K. RAZDAN: Hashish II: Reaction of Substituted Resorcinols with Citral in the Presence of Pyridine — A proposed Mechanism. Tetrahedron Letters 591 (1969).
68. KAPIL, R. S.: Carbazole Alkaloids, in Alkaloids, Vol. 13 (R. H. F. MANSKE, Ed.), pp. 273. New York-London: Academic Press, Inc. 1971.
69. KIRTIKAR, K. R., and B. D. BASU: Indian Medicinal Plants, 2nd ed., Vol. 1, p. 472, Basu, Allahabad, India (1933).
70. KUREEL, S. P., R. S. KAPIL, and S. P. POPLI: New Alkaloids from Murraya koenigii Spreng. Experientia 25, 790 (1969).

71. Kureel, S. P., R. S. Kapil, and S. P. Popli: The Synthesis of (±) Mahanimbine and Bicyclomahanimbine. Chem. Commun. 1120 (1969).

72. — — — Terpenoid Alkaloids from *Murraya koenigii* Spreng-11. The Constitution of Cyclomahanimbine, Bicyclomahanimbine and Mahanimbidine. Tetrahedron Letters 3857 (1969).

73. — — — Terpenoid Alkaloids and Synthesis of Mahanimbine. Experientia **26**, 1055 (1970).

74. — — — Two Novel Alkaloids from *Murraya koenigii*. Mahanimbicine and Bicyclomahanimbicine. Chem. and Ind. 958 (1970).

75. — — — Synthesis of Koenine, Koenimbine and Girinimbine. Chem. and Ind. 1262 (1970).

76. Littell, R., G. O. Morton, and G. R. Allen, Jr.: Observation on the Mechanism of the Nenitzescu Indole Synthesis and its Utilisation for the Preparation of Carbazoles. J. Amer. Chem. Soc. **17**, 3740 (1970).

77. Marini-Bettolo, G. B., and J. Schmutz: Zur Konstitution der Alkaloide Olivacin Guatambuin und U-Alkaloid. Helv. Chim. Acta **42**, 2146 (1959).

78. Martin, R. H., N. Defay, and F. Geeris-Irvard: Applications de la spectrographie de résonance magnétique nucléaire (RMN) dans le domaine des dérives polycycliques — a caractère aromatique XI. Tetrahedron **21**, 2421 (1965).

79. McLafferty, F. W., and R. S. Gohelle: Mass Spectrometric Analysis of Aromatic Acids and Esters. Anal. Chem. **31**, 2076 (1959).

80. Mechoulam, R., and Y. Gaoni: Recent Advances in the Chemistry of Hashish. Fortschr. Chem. Organ. Naturstoffe **25**, 175 (1967).

81. Narasimhan, N. S., M. V. Paradkar, and A. M. Gokhale: Synthesis of Girinimbine and (±) Mahanimbine. Tetrahedron Letters 1605 (1970).

82. Narasimhan, N. S., M. V. Paradhkar, and S. L. Kelkar: Alkaloids of *Murraya koenigii*. Structure of Mahanine, Koenine, Koenigine, and Koenidine. Indian J. Chem. **8**, 473 (1970).

83. Narasimhan, N. S., M. V. Paradhkar, and V. P. Chitguppi; Structures of Mahanimbine and Koenimbine. Tetrahedron Letters 5502 (1968).

84. Narashimhan, N. S., M. V. Paradkar, V. P. Chitguppi, and S. L. Kelkar: Alkaloids of *Murraya koenigii*: Structures of Mahanimbine, Koenimbine, Mahanine, Koeniginine, Koenine and (±) Isomahanimbine. Indian J. Chem. **13**, 993 (1975).

85. Patterson, A. M., and L. T. Capell: The Ring Index, p. 229, No. 1675. New York: Reinhold. 1940.

86. Pausacker, K. H., and R. Robinson: Strychnine and Brucine. Degradation of the Strychnineacetic Acid Prepared from Pseudostrychnine. J. Chem. Soc. 1557 (1947).

87. Phillips, R. R.: Organic Reactions, Vol. 10, p. 143 (edited by R. Adams). New York: J. Wiley and Sons. 1959.

88. Polonsky, J.: Structures du Calophyllolide de l'Inophyllolide et de l'Acide Calophyllique. VI — Synthèse du Dihydroinophyllolide et de l'Acide Dihydrocalophyllique. Bull. Soc. Chim. Fr. 929 (1957).

89. Pullman, B., and A. Pullman: In Quantum Biochemistry, p. 325. New York-London: Interscience Publishers. 1963.

90. Raley, J. H., R. D. Mullineaux, and C. W. Bittner: High Temperature Reactions of Iodine with Hydrocarbons. J. Amer. Chem. Soc. **85**, 3174 (1963).

91. Ramsay, M. V. J., W. D. Ollis, and I. O. Sutherland: The Constitution of Gambogic Acid. Tetrahedron **21**, 1453 (1965).

92. Reynolds, R., L. L. Larry, and R. F. Nelson: Electrochemical Generation of Carbazoles from Aromatic Amines. J. Amer. Chem. Soc. **96**, 1087 (1974).

93. Richards, R. E.: Infrared Spectrum and Structural Diagnosis: Substituted Carbazoles. J. Chem. Soc. 778 (1947).

94. ROBINSON, B.: Fischer Indole Synthesis. Chem. Rev. **63**, 373 (1963).
95. ROY, S., and D. P. CHAKRABORTY: DDQ as a Spray Reagent for Carbazoles, J. Chromatogr. **96**, 266 (1974).
96. — — Murrayacine from *Clausena heptaphylla*. Phytochem. **15**, 356 (1976).
97. — — Mahanimbine from *Murraya koenigii* Spreng. Phytochem. **13**, 2893 (1974).
98. ROY, S., P. BHATTACHARYYA, and D. P. CHAKRABORTY: 3-Methyl Carbazole from *Clausena heptaphylla*. Phytochem. **13**, 1017 (1974).
99. SCHMUTZ, J., and F. HUNZICKER: Alkaloids of *Aspidosperma olivaceum*. Pharm. Acta Helv. **33**, 341 (1958).
100. SCHMUTZ, J., F. HUNZICKER, and R. HIRT: Ulein, das Hauptalkaloid von *Aspidosperma* ulei. Aspidosperma-Alkaloid, I. Mitteilung. Helv. Chim. Acta **40**, 1189 (1957).
101. STAHL, E.: Thin-layer Chromatography, A Laboratory Handbook, p. 500. London: G. Allen and Unwin Ltd.; Berlin-Heidelberg-New York: Springer. 1969.
102. STEVENS, T. S.: In Chemistry of Carbon Compounds (E. H. RODD, ed.), Vol. IVA, pp. 120. Elsevier Publishing Co. 1957.
103. SUMPTER, W. C., and F. M. MILLER: In Heterocyclic Compounds with Indole and Carbazole System, pp. 70. New York: Interscience Publishers, Inc. 1954.
104. VALENTA, Z.: Zinc Dust Distillation in Technique of Organic Chemistry (A. WEISS-BERGER, ed.), Vol. XI, p. 643. New York: Interscience Publishers, Inc. 1963.
105. WATERMAN, H. C., and D. L. VIVIAN: Direct Ring-Closure Through a Nitro Group in Certain Aromatic Compounds with the Formation of Nitrogen Heterocycles: A New Reaction. J. Organ. Chem. **14**, 289 (1949).
106. WILLHALM, B., A. F. THOMAS, and F. GAUTSCHI: Mass Spectra and Organic Analysis – II. Mass Spectra of Aromatic Ethers in Which the Oxygen Forms Part of a Ring. Tetrahedron **20**, 1185 (1964).

(Received March 23, 1976)

Bürzeldrüsenlipide

Von J. JACOB, Biochemisches Institut für Umweltcarcinogene,
Ahrensburg/Holstein, und Universität Hamburg, Fachbereich Biologie,
Bundesrepublik Deutschland

Mit 12 Abbildungen

Inhalt

Verwendete Abkürzungen

TLC = Dünnschichtchromatographie
SC = Säulenchromatographie
TMS = Trimethylsilyl
GLC = Gaschromatographie

I. Einleitung

Soweit heute bekannt, ist das Integument bei allen Tierarten mit einer mehr oder minder starken Lipidschicht überzogen (*1*), die je nach Tierklasse vornehmlich zu bestimmten Lipidklassen gehört, z. B. aliphatische Kohlenwasserstoffe bei Insekten (*2*), Wachse bei Vögeln, freie Fettsäuren bei Säugetieren (*3*). Ihre Funktion ist immer noch Gegenstand von Spekulationen, jedoch scheint die Schutzwirkung gegen Durchnässung und Austrocknung unbestritten. Daneben werden heute Bakteriostase und Fungistase diskutiert.

Die Hautlipide werden in holokrinen Drüsen produziert (*4*), d. h. die Drüsenzellen platzen nach vollständiger Ausreifung und entleeren ihren Inhalt über Drüsengänge auf die Körperoberfläche, die Haut.

Die Vogelhaut nimmt innerhalb der höher entwickelten Tiere insofern eine Sonderstellung ein, als ihre Lipidversorgung praktisch ausschließlich über eine einzige Drüse, die Bürzeldrüse, die sich oberhalb der Schwanzfederkiele befindet, erfolgt. Über ihre Funktion ist schon vor 400 Jahren spekuliert worden (*5*). Nach neueren Untersuchungen kommen daneben allerdings auch andere Lipidgenese-Orte in der Vogelhaut in Betracht, die jedoch nur für einzelne Arten von Bedeutung zu sein scheinen (näheres Kap. VI). Auch besitzen Vögel keine Schweißdrüsen und zeichnen sich durch eine weitere Anpassung der Haut, die Federbildung, aus.

Literaturverzeichnis: SS. 431—438

Die Sekrete der Bürzeldrüse *(glans uropygialis)* bestehen in erster Linie aus kompliziert zusammengesetzten Esterwachsen; hierbei überwiegen Monoesterwachse, an deren Aufbau mehr oder minder stark alkylverzweigte Fettsäuren und Alkohole beteiligt sind. Daneben kommen bei manchen Arten auch Diesterwachse oder Triglyceride vor. Ferner wurden Squalen, Sterole, Triesterwachse, freie Alkohole, Diole und freie Fettsäuren nachgewiesen.

Die in den letzten Jahrzehnten verbesserten analytischen Verfahren haben eine intensive Beschäftigung mit der Struktur von Hautlipiden möglich gemacht und bewirkt. Vor allem die Kombination der Gaschromatographie mit der Massenspektrometrie ermöglichte die Strukturaufklärung von isomeren Fettsäure- und Alkoholgemischen und stellt heute das tragende Verfahren auf diesem Gebiet dar.

II. Anatomie und Histologie der Bürzeldrüse

Von wenigen Ausnahmen abgesehen, besitzen alle Vögel eine Bürzeldrüse *(glans uropygialis,* engl.: uropygial gland, preen gland), die als paariges Organ direkt oberhalb der Federkiele den Schwanzfedern aufliegt. Beide Drüsenloben *(lobus glandis uropygialis)* besitzen je einen Drüsengang *(ductus glandis uropygialis),* der in eine gemeinsame Öffnung *(porus ductus uropygialis)* einmündet, und sind durch einen Muskelgewebestreifen *(isthmus glandarius)* miteinander verbunden. Um den Drüsenausgang findet man meistens einen Kranz *(circulus uropygialis)* von feinen Federn (Pinselfedern, Bürzeldocht). Anatomische Details finden sich bei PARIS *(6),* GRASSE *(7)* sowie sehr ausführlich bei LUCAS und STETTENHEIM *(8)* und sind andeutungsweise in Abb. 1 wiedergegeben.

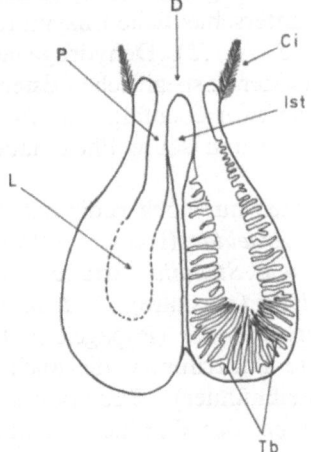

Abb. 1. Schematischer Querschnitt durch eine Bürzeldrüse (*L,* Lobus; *Tb,* Tubuli (im linken Lobus fortgelassen); *Ist,* Isthmus; *P,* Porus; *D,* Ductus; *Ci,* Circulus)

Das Bürzeldrüsensekret wird in den zahllosen Tubuli produziert und wird nach Ausreifung und Auflösung der Zellen in die Loben entleert. Die Tubuli lassen sich elektronenoptisch in 3 voneinander unterscheidbare Zonen unterteilen, die auch im Lichtmikroskop durch ihre unterschiedliche Anfärbbarkeit auszumachen sind (Abb. 2) (*9, 10, 11*).

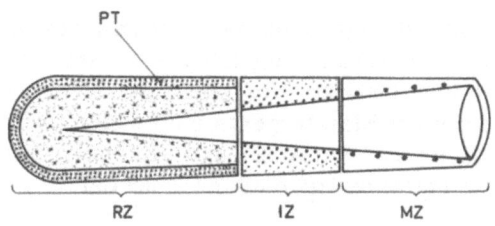

Abb. 2. Schematischer Längsschnitt durch einen Tubulus nach Lennert und Weitzel (*11*). (*RZ*, Rinden- oder Außenzone; *PT*, Peripherer Teil der Rindenzone; *IZ* Intermediäre Zone; *MZ*, Mark- oder Innenzone; Schwärzung zeigt Osmiophilie an)

Der periphere Teil der Rindenzone sowie die Intermediärzone zeigen starke Osmiophilie, während die Markzone nur schwach OsO_4-positiv ist. Untersuchungen von Lennert und Weitzel (*11*) haben gezeigt, daß die Rindenzone vornehmlich Wachse enthält, die intermediäre Zone dagegen reich an Triglyceriden ist. Die Markzone besteht hauptsächlich aus Tubuliausgängen und spielt für die Lipidproduktion offenbar keine wesentliche Rolle.

Topographische Untersuchungen der Enzymaktivitäten innerhalb der Drüse haben gezeigt, daß unterschiedliche Enzymmuster in den einzelnen Zellen vorliegen (*11, 12, 13, 14, 15*). Dehydrogenasen und saure Phosphatase dominieren neben einer unspezifischen Esterase in der Markzone, während in der Rindenzone zwar ebenfalls eine Reihe von Dehydrogenasen, dagegen keine Esterase und saure Phosphatase nachweisbar sind (*14, 15*).

Einige Vogelarten besitzen nur noch rudimentäre Bürzeldrüsen, z. B. *Columbiformes* (Tauben), *Ardeidae* (Reiher), *Rhynchochitidae* (Kagus), andere überhaupt keine, wie *Struthio* (Strauß), *Rheiformes* (Nandus), *Dromaeus* (Emu), *Casuariidae* (Kasuare), *Otididae* (Trappen) und einige Spezies der Ordnung *Psittaciformes* (Papageien). Diese Arten besitzen meistens stark ausgeprägte Puderdunen, die auch bei Arten der Ordnungen *Tinamiformes* (Steißhühner), *Ciconiiformes* (Stelzvögel), *Gruiformes* (Rallenartige) und bei der Familie *Artamidae* (Schwalbenstare) aus der Ordnung *Passeriformes* gefunden wurden, und die möglicherweise die Funktion des Bürzeldrüsensekretes übernehmen.

Literaturverzeichnis: SS. 431—438

III. Bürzeldrüsensekrete

1. Analytische Methoden

a) Isolierung und Reinigung

Das Bürzeldrüsensekret kann durch vorsichtiges Abdrücken am lebenden Tier gewonnen werden, was vor allem bei in Gefangenschaft lebenden Arten mit hinreichend großen Drüsen angewendet wird (16, 17). Der Vorteil dieser Methode liegt in der Tatsache, daß so gewonnenes Material praktisch frei von Lipidverunreinigungen aus dem umgebenden Gewebe ist. Sie ist aus technischen Gründen jedoch bei Kleinvögeln nicht anwendbar und birgt überdies ein hohes Risiko für das Tier in sich. Eine zweite, meistens angewendete Methode ist die sorgfältige Ektomie der gesamten Drüse eines frischtoten Tieres und die nachfolgende Extraktion der Lipide mit $CHCl_3/CH_3OH$ (2:1; v/v) (18). Hierbei werden alle Bürzellipide extrahiert. Nach Zugabe von Wasser (1 v) enthält die schwere Phase die Rohlipide. Kann eine Extraktion nicht sofort erfolgen, läßt sich die enzymatische Aktivität und damit die Bildung von Artefakten (z. B. Lipolyse) im Drüsengewebe durch Aufbewahren der Drüse in abs. Aceton reduzieren bzw. blockieren.

Für die Auftrennung der Rohlipide werden dünnschicht (TLC)- und säulenchromatographische (SC) Verfahren benutzt. Bei Verwendung von SiO_2-Fertigplatten (E. Merck) trennt das System i-Octan Paraffine, Olefine und aromatische Kohlenwasserstoffe. $CCl_4/CHCl_3$ (1:1; v/v) trennt Sterolester, Mono-, Di- und Triesterwachse von Triglyceriden. Für polare Lipide sind die Systeme (a) $CHCl_3$, (b) Benzol/Essigester (9:1; v/v) bzw. (c) die doppelte Entwicklung der Platte mit 1. Äther/Hexan/Eisessig (90:10:1; v/v/v) und 2. Äther/Hexan/25% NH_3 (80:20:1; v/v/v) erfolgreich (19). Threo- und erythro-Diole, die als Bestandteile von Diesterwachsen (z. B. bei Galliformes) eine Rolle spielen, können an mit 5% Borsäure-imprägnierten Platten getrennt werden (20, 21, 22, 23, 24).

Bei Sekretmengen von weniger als 1 mg empfiehlt sich die präparative TLC in obigen Systemen, bei größeren Mengen die SiO_2-SC mit Mischsystemen (25) oder mit Lösungsmittelgradienten (26, 27, 28).

Die Strukturaufklärung von Bürzelwachsen ohne vorherige Spaltung in Fettsäuren und Alkohole gelang wegen der zahlreichen möglichen Isomerien nur in wenigen Fällen (29, 30). Methanolyse in 5% methanolischer HCl (31) oder BF_3/Methanol (32) führt zu Fettsäuremethylestern und Alkoholen, die sich säulenchromatographisch leicht trennen lassen. Infolge sterischer Hinderung kann die Spaltung erschwert sein, wenn größere Substituenten (äthyl-, propyl-, butyl- usw.) sich am C-2 der Säure oder des Alkohols befinden (33). Da Alkohole schwer deutbare Massenspektren liefern, werden diese durch CrO_3-Oxidation in die korres-

pondierenden Fettsäuren überführt, die als Methylester massenspektrometrisch untersucht werden (29). Alkandiole werden in die Acetonide (34)
oder TMS-Derivate (35) überführt oder nach oxidativer Spaltung als
Aldehyde (36) bzw. als Fettsäuremethylester (37, 38) bestimmt.

Die Umesterung von Diesterwachsen, die 3-Hydroxyfettsäuren (wie
z. B. das Bürzelwachs von *Columba palumbus* (Ringeltaube) (39, 40)] oder
Alkylhydroxymalonsäuren (41) enthalten, liefert leicht Artefakte infolge
Wasserabspaltung und Methoxylierung (42, 43, 44), weshalb die alkalische Verseifung in 1 n methanolischer NaOH vorzuziehen ist.

b) Gaschromatographie (GLC)

Bei den Fettsäure- und Alkoholfraktionen aus Bürzelwachsen handelt
es sich im allgemeinen um sehr komplizierte Gemische, für deren gas-
chromatographische Auftrennung Säulen mit hoher Trennleistung er-

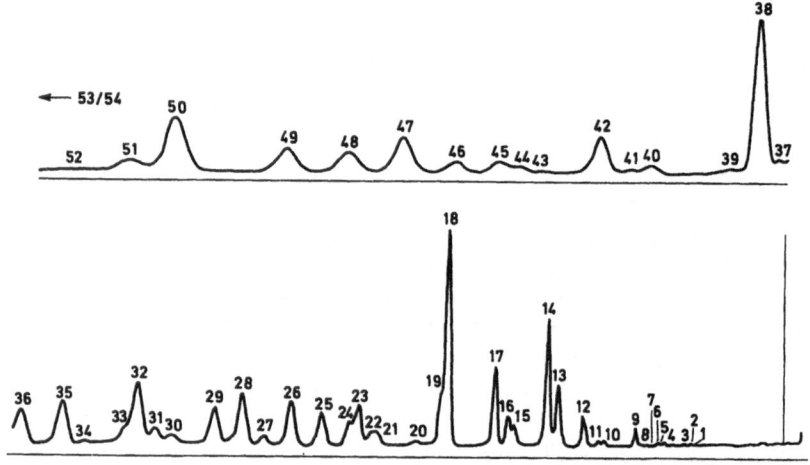

Abb. 3. GLC der Fettsäuremethylester aus dem Bürzelwachs von *Fulica atra* (Bläßhuhn).
50 m Kapillarsäule mit OV 101-Imprägnierung, Teilung 1:20, Säulentemperatur 180°,
1,4 ml N_2/Min. (46)

Nr. 3, 9, 13, 19, 27 = $2\text{-}C_{11-15}$
Nr. 4, 10, 15, 21, 30, 40, 50 = $2,6\text{-}C_{11}\text{-}C_{17}$
Nr. 5, 11, 16, 22, 30, 40 = $2,8\text{-}C_{11-16}$
Nr. 8, 12, 17, 23, 31, 41, 50 = $2,10\text{-}C_{11-17}$
Nr. 25, 34, 44 = $2,12\text{-}C_{14-16}$
Nr. 14, 18, 26, 35, 45, 53 = $2,6,10\text{-}C_{12-17}$
Nr. 28, 37, 46 = 2,6,12- und $2,8,12\text{-}C_{14-16}$
Nr. 24, 32, 42, 51 = 4,8- und $4,10\text{-}C_{14-17}$
Nr. 33, 42, 51 = $4,12\text{-}C_{15-17}$
Nr. 29, 38, 47, 54 = 4, 8, $12\text{-}C_{14-17}$
Nr. 49 = $4,8,14\text{-}C_{16}$
Nr. 1, 2, 6, 7, 20, 36, 39, 43, 48, 52 = nicht identifiziert

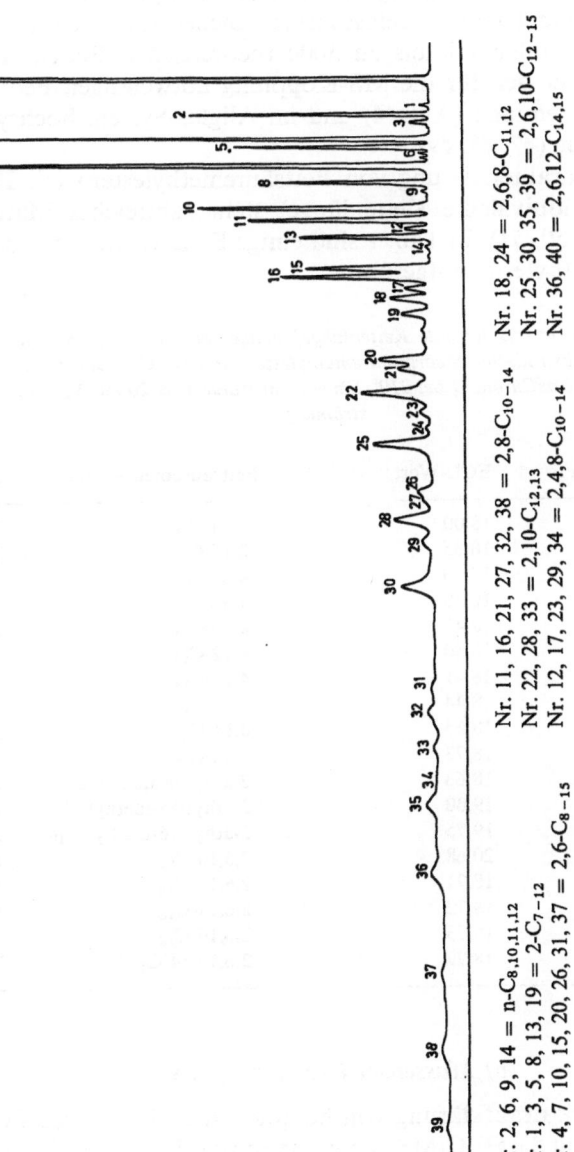

Abb. 4. GLC der Fettsäuremethylester aus dem Bürzelwachs von *Larus argentatus* (Silber-möwe), 10 m Glassäule, 5% OV 101, Säulentemperatur 180°, 20 ml N₂/Min.

Nr. 2, 6, 9, 14 = n-C$_{8,10,11,12}$
Nr. 1, 3, 5, 8, 13, 19 = 2-C$_{7-12}$
Nr. 4, 7, 10, 15, 20, 26, 31, 37 = 2,6-C$_{8-15}$

Nr. 11, 16, 21, 27, 32, 38 = 2,8-C$_{10-14}$
Nr. 22, 28, 33 = 2,10-C$_{12,13}$
Nr. 12, 17, 23, 29, 34 = 2,4,8-C$_{10-14}$

Nr. 18, 24 = 2,6,8-C$_{11,12}$
Nr. 25, 30, 35, 39 = 2,6,10-C$_{12-15}$
Nr. 36, 40 = 2,6,12-C$_{14,15}$

forderlich sind. Diese Forderung ist bei Kapillarsäulen mit Polypropylen-glykol- (45) bzw. OV 101-Imprägnierungen (46) weitgehend erfüllt (Abb. 3). Aufgrund der mangelhaften Belastbarkeit solcher Säulen ist man aber zu gepackten Glassäulen mit bis zu 3000 theoretischen Böden/m übergegangen, die unter den für die MS-Kopplung notwendigen Bedingungen noch gute Auftrennung (Abb. 4) und im Allglas-System hochsymmetrische Peaks geben (47, 48, 49).

Für die Charakterisierung von Fettsäuremethylestern mit Hilfe von ECL-Werten (äquivalente Kettenlänge) steht zahlreiches Material zur Verfügung (25, 50, 51). In Tab. 1 sind einige ECL-Werte für substituierte Stearinsäuremethylester aufgeführt.

Tabelle 1. *ECL-Werte (äquivalente Kettenlänge) einiger als Bürzelwachsbestandteile nach-gewiesener alkyl-substituierter Stearinsäuremethylester für 10 m Glassäulen mit 3% OV 101-Imprägnierung auf GasChrom Q bei 210° Säulentemperatur und 20 ml N_2/Min. Trägergas-strömung*

Fettsäuremethylester	ECL-Wert	Fettsäuremethylester	ECL-Wert
n-C_{18}	18,00	2,14-C_{18}	18,84
2-C_{18}	18,35	2,16-C_{18}	19,07
3-C_{18}	18,40	4,6-C_{18}	18,70
4-C_{18}	18,50	4,8-C_{18}	18,76
6-C_{18}	18,41	4,10-C_{18}	18,80
8-C_{18}	18,40	4,12-C_{18}	18,88
10-C_{18}	18,41	4,14-C_{18}	18,96
12-C_{18}	18,44	4,16-C_{18}	19,20
14-C_{18}	18,55	10,14-C_{18}	18,89
16-C_{18} (anteiso)	18,73	10,16-C_{18}	19,12
17-C_{18} (iso)	18,63	2-äthyl-6-methyl-C_{18}	19,45
2-äthyl-C_{18}	19,00	2-äthyl-14-methyl-C_{18}	19,60
2-propyl-C_{18}	19,75	2-äthyl-16-methyl-C_{18}	19,80
2-butyl-C_{18}	20,68	2,6,10-C_{18}	19,08
2,6-C_{18}	18,71	2,6,12-C_{18}	19,10
2,8-C_{18}	18,72	2,6,14-C_{18}	19,22
2,10-C_{18}	18,73	2,6,16-C_{18}	19,46
2,12-C_{18}	18,77	2,6,10,14-C_{18}	19,77

c) Massenspektrometrie (MS)

Für die Strukturaufklärung von homologen und isomeren Fettsäure-methylestern stellt die GLC/MS-Kombination das Verfahren der Wahl dar. Massenspektren von Lipiden (52, 53, 54) sowie speziell von Fettsäure-methylestern (25, 55, 56, 57, 58, 59) sind in der Literatur beschrieben und die charakteristischen massenspektrometrischen Fragmente tabellarisch aufgeführt (33, 60, 61, 62, 63).

Literaturverzeichnis: SS. 431—438

Allgemein lassen sich Fettsäuremethylester am Ion m/e 74 erkennen, das durch McLafferty-Umlagerung gemäß:

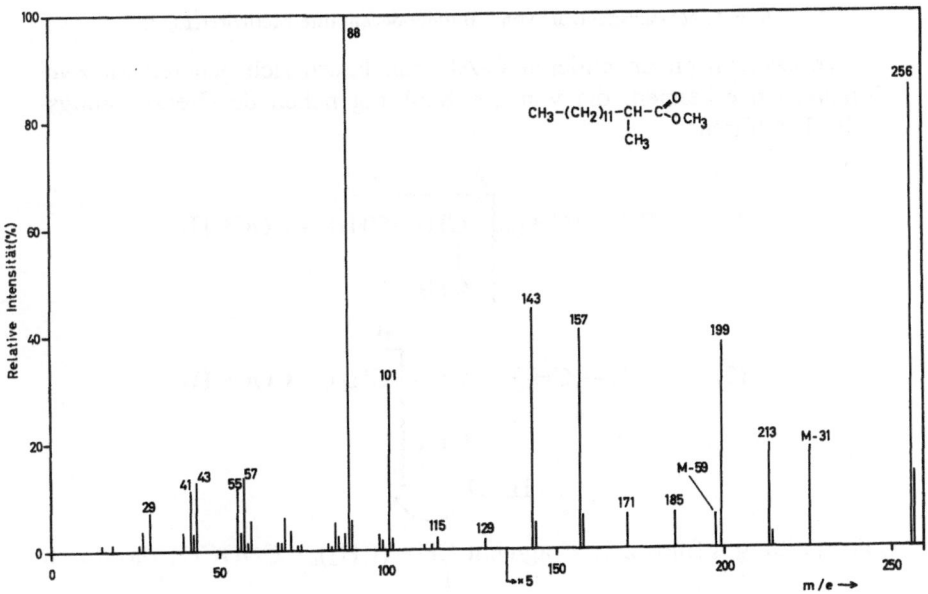

entsteht. Die in Bürzelwachsen häufig vorkommenden C-2-substituierten Fettsäuren führen zu entsprechend verschobenen McLafferty-Ionen, z. B. m/e 88 bei 2-Methyl-, m/e 102 bei 2-Äthyl-, m/e 116 bei 2-Propyl- und m/e 130 bei 2-Butyl-Fettsäuremethylestern (Abb. 5 und 6). Gleichzeitig wird eine Olefin-Elimination beobachtet, wenn der Substituent größer

Tabelle 2. *Schlüsselionen für 2-alkyl-substituierte Fettsäuremethylester*

Substituent am C-2	McLafferty-Ion	Olefin-Elimination
H	74	—
CH_3	88	—
C_2H_5	102	M-28 (= $M-C_2H_4$)
C_3H_7	116	M-42 (= $M-C_3H_6$)
C_4H_9	130	M-56 (= $M-C_4H_8$)

Abb. 5. Massenspektrum von 2-Methyltetradecansäure-Methylester (*79*)

als CH_3 ist (vgl. Tabelle 2). Die Substitution am C-2 führt weiterhin zur
bevorzugten Abspaltung von $-COOCH_3$ (= M-59), die auch beobachtet
wird, wenn der Substituent =O, $-OH$ oder OCH_3 ist (*64*).

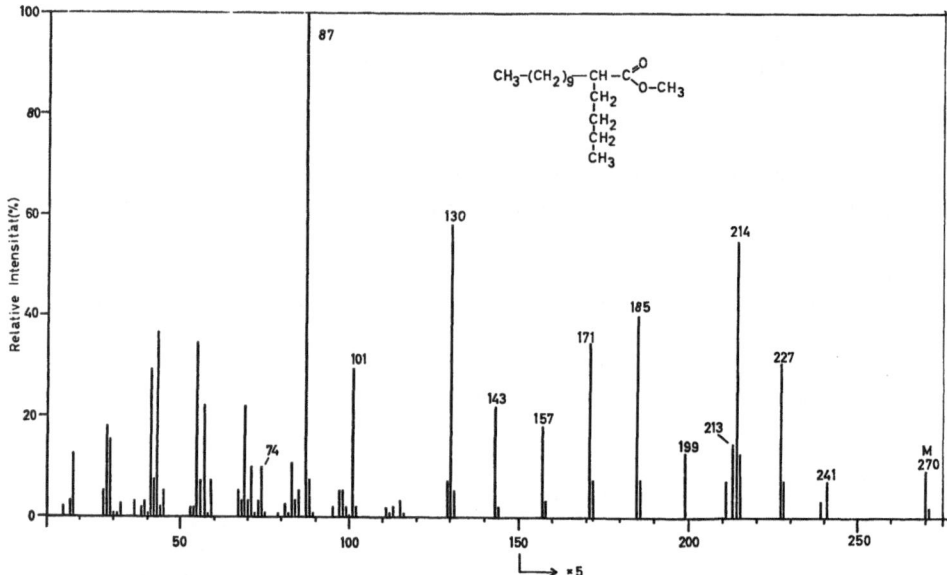

Abb. 6. Massenspektrum von 2-Butyldodecansäure-Methylester (*33*)

Verzweigungen an anderen C-Atomen lassen sich generell an zwei
Ionenserien erkennen, die von der Spaltung neben der Verzweigungsstelle herrühren:

$$
\begin{array}{cc}
& A \\
(1) & CH_3-(CH_2)_m - CH - (CH_2)_n - COOCH_3 \\
& \mid \\
& CH_3
\end{array}
$$

$$
\begin{array}{cc}
& B \\
(2) & CH_3-(CH_2)_m - CH - (CH_2)_n - COOCH_3 \\
& \mid \\
& CH_3 \\
& H, 2\,H
\end{array}
$$

Die Spaltung A führt zum Fragment $CH-(CH_2)_n - COOCH_3$, das durch
$\mid$
CH_3

Literaturverzeichnis: SS. 431—438

Methanol- sowie nachfolgender Wasserabspaltung die Ionenserie
A → (A-32) → (A-32-18) liefert, z. B. im Fall von 10-Methylfettsäure-
Methylester (199 → 167 → 149). Die Spaltung B liefert ein Triplett B,
B + 1, B + 2, z. B. für den obigen Fall 171, 172, 173.

Die in Bürzelwachsen von zahlreichen Arten vorkommenden 3-
Methylfettsäuren (u. a. *Sphenisciformes (65), Procellariiformes (66, 67),
Piciformes (60), Cuculiformes (68), Strigiformes (33), Passeriformes (29,
62, 69)*] zeigen praktisch kein m/e 87, aber intensive Ionen M-15 und
M-74, die für die Erkennung dieses Estertyps nützlich sind (Abb. 7).

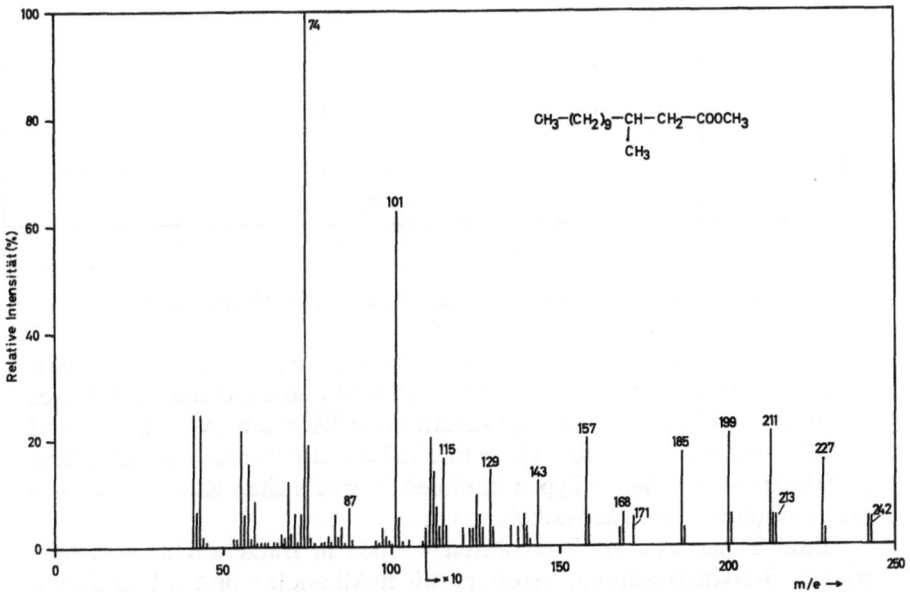

Abb. 7. Massenspektrum von 3-Methyltridecansäure-Methylester (*29*)

4-Methyl-substituierte Ester können am Intensitätsverhältnis m/e
87 > 74 und den intensiven Ionen M-73 [= M − (CH$_2$ − COOCH$_3$)] und
M-49 [= M − (OCH$_3$ + H$_2$O)] erkannt werden. Sie sind als Bürzeldrüsen-
Wachsbestandteile weit verbreitet und wurden bei Arten der folgenden
Ordnungen gefunden: *Sphenisciformes (65), Procellariiformes (66, 67),
Gruiformes (46, 70), Lariformes (63), Charadriiformes (71), Psittaciformes
(72), Anseriformes (73, 76, 77), Phoenicopteriformes (78)* und *Passeri-
formes (74, 75)*.

6-Methyl-substituierte Methylester zeigen neben den Spaltungen an
der Verzweigungsstelle ein intensives (M-76)-Ion, an dem sie leicht
erkannt werden können; sie treten in Bürzelwachsen auf, bei denen auch
4-Methylfettsäuren nachgewiesen wurden.

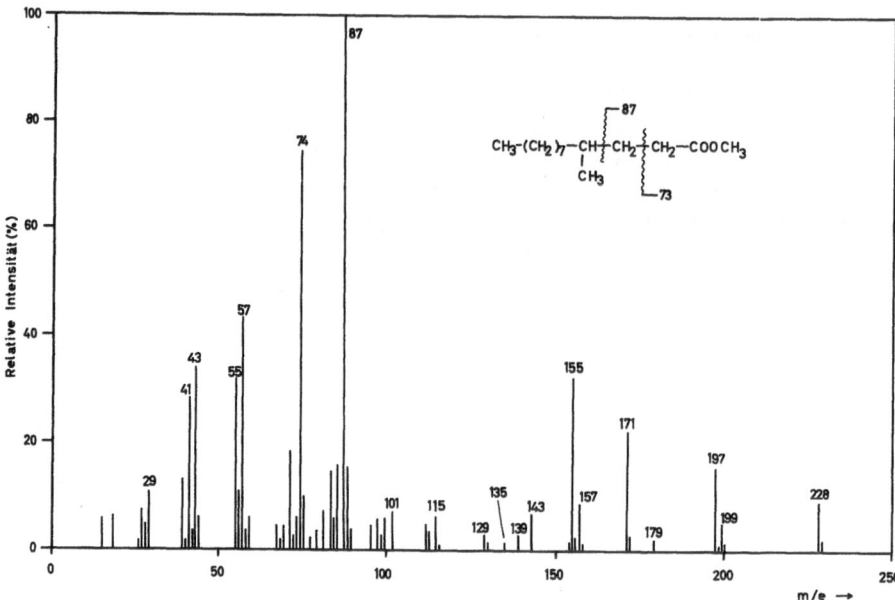

Abb. 8. Massenspektrum von 4-Methyldodecansäure-Methylester (*79*)

8-Methyl-substituierte Fettsäuremethylester zeigen m/e 143 als Basision sowie ein Intensitätsverhältnis m/e 87 > 74 und ein charakteristisches (M-49)-Ion, während 10-methyl-substituierte Ester am (M-132)-Ion und am Intensitätsverhältnis m/e 129 ≈ m/e 130 erkannt werden können. Diese mittelkettig-verzweigten Typen kommen in erheblichen Konzentrationen als Bestandteile der Wachsalkohole vor.

Zwei Typen von Hydroxysäuren wurden in Bürzelwachsen nachgewiesen. β-Hydroxysäuren, verestert mit n-Alkanolen und n-Fettsäuren, sind Hauptbestandteil des Bürzeldrüsensekretes von *Columba palumbus* (*39, 40*). Das Massenspektrum von β-Hydroxysäure-Methylestern zeigt kein Molekülion, dagegen ein schwaches (M-1)-Ion und ein intensives Ion

m/e 103, das dem Fragment $\overset{\oplus}{\underset{\underset{OH}{|}}{CH}} - CH_2 - COOCH_3$ zugeordnet werden

kann. Daneben werden die Fragmente (M-18) (= Wasserabspaltung), (M-31) (= OCH₃-Abspaltung) sowie das Dublett 152/153 (= M−CH₃ OH−H₂O und M−OCH₃−H₂O) beobachtet (Abb. 9).

Alkyl-hydroxymalonsäuren treten als Bestandteile von Bürzeldrüsenlipiden bei Arten aus unterschiedlichen Ordnungen auf (*41*). In den Massenspektren ihrer Methylester wird kein Molekülion, dagegen aber

Literaturverzeichnis: SS. 431—438

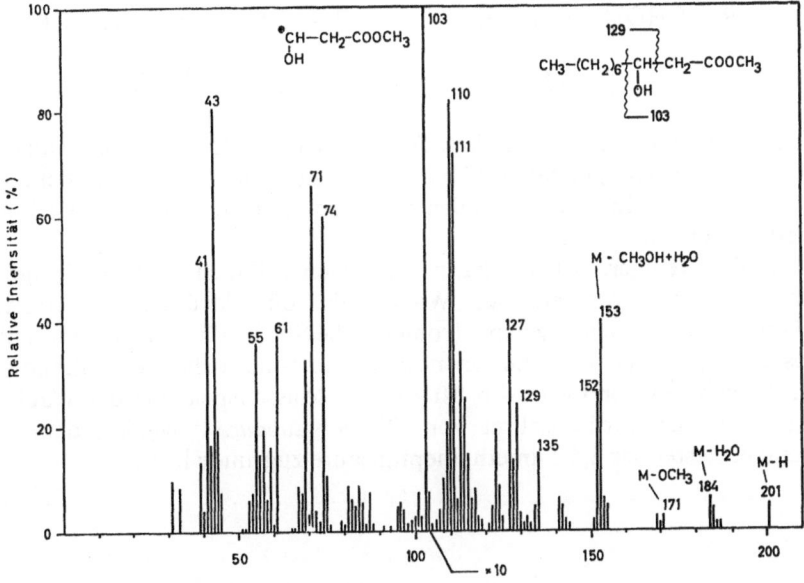

Abb. 9. Massenspektrum von 3-Hydroxydecansäure-Methylester (*39*)

ein schwaches (M-32)- (Methanolabspaltung) und das für die Substitution am C-2 charakteristische (M-59)-Ion beobachtet. Als Basision tritt M-119 auf, das aus dem (M-59)-Ion durch Eliminierung von HCOOCH$_3$ entsteht. Daneben zeigen solche Ester ein McLafferty-Ion m/e 148 entsprechend:

$$HO-C \overset{\displaystyle C\diagdown OH}{\diagup \diagdown} OCH_3$$
$$\underset{\underset{O}{\displaystyle \|}}{C-OCH_3}$$

Bei den Wachsalkoholen dominieren einwertige, primäre Alkohole mit und ohne Verzweigungen, deren Strukturaufklärung, nach CrO$_3$-Oxidation und nachfolgender Veresterung, als Methylester massenspektrometrisch möglich ist. Die Massenspektren der freien Alkohole liefern dagegen uncharakteristische Fragmente. Als Bestandteile von Diesterwachsen wurden Alkan-2,3-diole bei verschiedenen *Galliformes*-Arten nachgewiesen (*22, 24, 34, 35, 38, 80, 81, 82*). In den Bürzellipiden einiger Arten wurden Alkan-1,2-diole identifiziert (*83*). Die OH-Positionen lassen sich aus den Massenspektren der freien Diole aufgrund der Spaltung zwischen den OH-tragenden C-Atomen (*34, 38*)

$$R - CHOH -\{CH_2OH \qquad\qquad R - CHOH -\{CHOH - CH_3$$

$$\{M\text{-}31 \qquad\qquad\qquad\qquad\qquad \{M\text{-}45$$

oder vorteilhafter aus denen ihrer Acetonide (*81*) oder TMS-Derivate (*35*) erkennen. Verzweigungen der Kette lassen sich allerdings am besten nach Oxidation mit MnO_4^-/JO_4^- aus den Spektren der resultierenden Fettsäure-methylester nachweisen.

In den wenigen bisher bekannt gewordenen Fällen, in denen lediglich ein einzelnes isomerenfreies Wachs als Bürzeldrüsensekret auftritt (*Ploceidae*), ist die massenspektrometrische Strukturaufklärung mit Hilfe des direkten Einlasses in die Ionenquelle und ohne vorherige Methanolyse erfolgreich gewesen (*30*). Abb. 10 gibt das Massenspektrum des Wachses aus dem Bürzeldrüsensekret von *Ploceus subaureus* wieder, dem der Octadecylester der 2,4-Dimethylheptansäure zugrunde liegt.

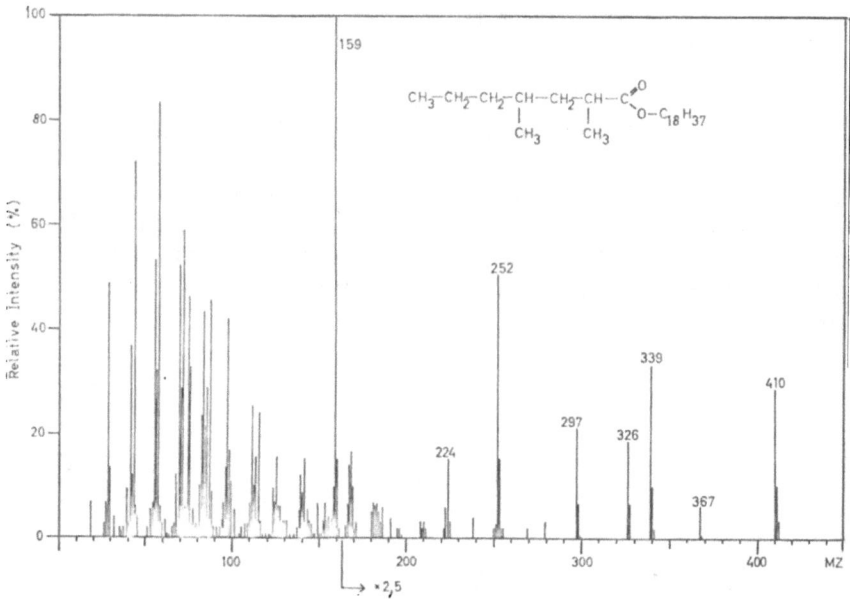

Abb. 10. Massenspektrum des Bürzelwachses von *Ploceus subaureus* (Direkteinlaß) bei 80 eV (*30*)

Das Spektrum zeigt ein intensives Molekülion (m/e 410). Die Abspaltung von C_3H_7 (m/e 367), die Abwesenheit von m/e 353 und das intensive Ion bei m/e 339 zeigen eine Verzweigung in 4-Position der Wachssäure an. Die 2-Position trägt ebenfalls eine Methylverzweigung wie aus dem McLafferty-Ion m/e 326 entsprechend

Literaturverzeichnis: SS. 431—438

$$CH_3-CH=C\overset{\displaystyle OH}{\diagup}O-C_{18}H_{37}$$

erkannt werden kann. Die Kettenlänge des Alkohols ergibt sich aus der Olefinabspaltung $(M-C_{18}H_{36})$ bei m/e 252, das von einer weiteren C_2H_4-Abspaltung m/e 224 begleitet wird. Das Basision m/e 159, das durch Abspaltung von Wasser in m/e 141 übergeht, zeigt schließlich die Kettenlänge der Säure $(C_8H_{17}COOH_2^{\oplus})$ an.

2. Zusammensetzung der Bürzellipide bei Arten unterschiedlicher Ordnungen

a) Systematik

Die Klasse *Aves* (Vögel) stellt im Vergleich zu anderen Tierklassen eine anatomisch-morphologisch verhältnismäßig homogene Tiergruppe dar. Diese vergleichsweise große Ähnlichkeit der einzelnen Arten hat Anlaß zu zahlreichen Meinungsverschiedenheiten der Zoologen bezüglich ihrer Eingliederung in Ordnungen, Familien, Unterfamilien usw. gegeben und zu taxonomischen Systemen mit sogar unterschiedlichen Anzahlen von Ordnungen geführt (*84, 85, 86, 87*). Das auf WETMORE (*88*) zurückgehende von CUISIN (*86*) vorgeschlagene System steht, soweit die experimentellen Ergebnisse bis heute erkennen lassen, in Übereinstimmung mit den vorliegenden chemotaxonomischen Befunden, und ihm entspricht daher die Gliederung dieses Kapitels (Tab. 3). Von den 33 Ordnungen dieser Systematik sind 5 fossil und von den verbleibenden 28 rezenten Ordnungen liegen aus 16 Ordnungen Untersuchungen an Bürzeldrüsensekreten vor. Sie lassen bereits heute deutlich erkennen, daß im allgemeinen die qualitative Zusammensetzung dieser Sekrete signifikant von Ordnung zu Ordnung variiert, während innerhalb einer Ordnung lediglich quantitative Variation beobachtet wird (vgl. Kap. VII).

Die folgenden Abschnitte geben einen Überblick über die Zusammensetzung der Bürzeldrüsensekrete der untersuchten 16 Ordnungen.

b) Einzelne Ordnungen

1. Sphenisciformes (Pinguine)

Die Zusammensetzung der Bürzeldrüsensekrete von 3 der insgesamt 15 Pinguinarten wurden untersucht (*65*). In dieser Ordnung treten 2 unterschiedliche Wachsmuster auf: bei *Pygoscelis* und *Eudyptes* dominieren 3- bzw. 3,x-dimethyl-substituierte Fettsäuren, bei *Spheniscus*

Tabelle 3. *Systematik der Klasse Aves [nach* Cuisin *(86), basierend auf* Wetmore]

1. Unterklasse: Archaeornithes (fossil) (Urvögel)
   1. Ordnung: Archaeopterygiformes (fossil) (Echsenvögel)
2. Unterklasse: Neornithes (eigentliche Vögel)
   1. Gruppe: Odontognathae (Zahnvögel)
     1. Ordnung: Hesperornithiformes (fossil)
     2. Ordnung: Ichtyornithiformes (fossil)
     3. Ordnung: Sphenisciformes (Pinguine)
   2. Gruppe: Neognathae (neuzeitliche Vögel)
     1. Ordnung: Struthioniformes (Strauße)
     2. Ordnung: Rheiformes (Nandus)
     3. Ordnung: Casuariiformes (Kasuare)
     4. Ordnung: Aepyornithiformes (fossil) (Elefantenvögel)
     5. Ordnung: Dinornithiformes (fossil) (Moas)
     6. Ordnung: Apterygiformes (Kiwis)
     7. Ordnung: Tinamiformes (Steißhühner)
     8. Ordnung: Gaviiformes (Seetaucher)
     9. Ordnung: Procellariiformes (Röhrennasen)
     10. Ordnung: Pelecaniformes (Pelikanartige)
     11. Ordnung: Ciconiiformes (Stelzvögel)
     12. Ordnung: Phoenicopteriformes (Flamingos)
     13. Ordnung: Anseriformes (Gänseartige)
     14. Ordnung: Falconiformes (Greifvögel)
     15. Ordnung: Galliformes (Hühnervögel)
     16. Ordnung: Gruiformes (Rallen)
     17. Ordnung: Charadriiformes (Regenpfeifer und Schnepfenartige)
     18. Ordnung: Lariformes (Möwen- und Alkenartige)
     19. Ordnung: Columbiformes (Tauben)
     20. Ordnung: Psittaciformes (Papageien)
     21. Ordnung: Cuculiformes (Kuckucke)
     22. Ordnung: Strigiformes (Eulen)
     23. Ordnung: Caprimulgiformes (Nachtschwalben)
     24. Ordnung: Apodiformes (Segler)
     25. Ordnung: Coliiformes (Mausvögel)
     26. Ordnung: Trogoniformes (Nageschnäbler)
     27. Ordnung: Coraciiformes (Rackenartige)
     28. Ordnung: Piciformes (Spechte)
     29. Ordnung: Passeriformes (Sperlingsvögel)

dagegen 2-, 4- und 6-methyl- sowie 2,x- und 4,x-dimethyl-substituierte Fettsäuren. Allen Arten gemeinsam ist die Ähnlichkeit der Wachs-Alkohol-Zusammensetzung, bei der mittelkettige Verzweigungen (z. B. 10-Methyl-alkanole) oder Verzweigungen im aliphatischen Ende des Alkohols (z. B. 14-C_{16}, 16-C_{20}, 20-C_{22} usw.) vorherrschen (Tab. 4).

Der Vergleich mit den an Procellariiformes-Arten gewonnenen Ergebnissen zeigt eine nahe Verwandtschaft dieser beiden Ordnungen.

Literaturverzeichnis: SS. 431—438

Tabelle 4. *Quantitative Zusammensetzung der Bürzelwachsbestandteile einiger Pinguinarten (65)*

	Pygoscelis papua	Eudyptes crestatus	Spheniscus magellanicus
Fettsäuren			
3-methyl-subsituiert	67,3	67,2	—
3,x-dimethyl-substituiert (3,7-; 3,9-; 3,11-)	9,0	8,1	—
2-methyl-substituiert	1,7	1,8	19,1
2,x-dimethyl-substituiert			
(2,6-; 2,8-; 2,10-; 2,12-; 2,14-)	9,9	5,6	25,4
2,x,y-trimethyl-substituiert			
(2,6,10-; 2,6,12-; 2,6,14-)	—	1,2	6,9
4- und 6-methyl-substituiert	6,7	10,7	37,6
4,x-dimethyl-substituiert	3,5	—	9,0
andere (n-; 2-äthyl-; x-methyl-)	—	2,1	2,0
nicht identifiziert	1,9	3,3	—
Alkohole			
n-Alkanole	—	1,0	1,4
2-methyl-substituiert	6,2	1,7	2,7
3-methyl-substituiert	0,2	0,5	—
4-; 6-; 8-; 10-; 12-methyl-substituiert	18,7	26,7	28,7
14-; 16-; 18-; 20-methyl-substituiert	10,5	14,6	28,2
2,x-dimethyl-substituiert			
(2,6-; 2,10-; 2,12-; 2,14-; 2,16-)	13,8	4,5	2,5
2,x,y-trimethyl-substituiert (2,6,12-; 2,10,12-)	1,4	0,8	—
4,x-dimethyl-substituiert	—	—	0,9
x,14- und x,16-dimethyl-substituiert			
(6,14-; 8,14-; 8,16-; 10,14-; 10,16-; 12,16-)	24,3	16,3	9,4
x,18-dimethyl-substituiert			
(6,18-; 8,18-; 10,18-; 12,18-)	15,6	17,9	12,2
x,20-dimethyl-substituiert (10,20-; 16,20-)	7,5	15,8	12,8
unidentifiziert	1,8	0,2	1,2

2. Procellariiformes (Röhrennasenvögel)

Diese phylogenetisch alte Vogelgruppe, die 4 Familien umfaßt (*Diomedeidae* = Albatrosse, *Procellariidae* = Sturmvögel, *Hydrobatidae* = Sturmschwalben und *Pelecanoidae* = Sturmtaucher) und von denen mindestens je eine Art untersucht worden ist (*66, 67*), zeigt chemotaxonomisch sehr große Ähnlichkeit mit den Pinguinen, mit denen sie offenbar verwandt sind. Ihre Bürzelwachse sind durch das Vorkommen von vor allem 2-, 3-, 4-, 2,x-, 4,x-substituierten Fettsäuren charakterisiert. Lediglich *Diomedea* weist einen hohen Anteil an 2,4,x-trimethylsubstituierten Säuren auf und leitet vielleicht zu einer anderen Ordnung über. Auch in dieser Ordnung sind, wie bei den *Sphenisciformes* 2 Typen unterscheidbar: a) Arten mit hauptsächlich 3-methyl-substituierten

Tabelle 5. *Quantitative Zusammensetzung der Bürzelwachsbestandteile einiger Procellariiformes-Arten (66, 67)*

	Diomedea melanophris	*Procellaria aequinoctialis*	*Puffinus griseus*	*Fulmarus glacialis*	*Pelecanoides urinatrix*	*Pachyptila belcheri*	*Garrodia nereis*
Fettsäuren							
n-	—	—	—	—	—	—	16,7
2-methyl-substituiert	—	Spur	—	0,4	—	0,3	25,4
4-methyl-substituiert	—	4,1	—	22,6	6,0	43,7	36,7
6-methyl-substituiert	—	2,0	—	4,0	—	2,0	6,7
3-methyl-substituiert	—	22,2	65,9	53,3	50,0	31,7	2,6
2,x-dimethyl-substituiert (2,4-; 2,6-; 2,8-; 2,12-)	19,3	5,3	1,3	6,5	—	7,1	4,3
3,x-dimethyl-substituiert (3,5-; 3,7-; 3,9-; 3,11-)	2,9	30,9	23,1	8,8	29,4	2,7	—
4,x-dimethyl-substituiert (4,6-; 4,8-; 4,10-)	—	—	2,2	0,4	8,4	12,5	7,1
2,x,y-trimethyl-substituiert (2,4,6-; 2,4,8-; 2,6,8-; 2,6,10-)	67,4	26,0	—	—	—	—	—
3,x,y-trimethyl-substituiert (3,5,7-)	—	7,7	—	—	—	—	—
unidentifiziert	10,4	1,8	7,5	4,0	6,2	—	0,5
Alkohole							
n-	5,2	—	2,9	40,5	43,2	43,2	69,2
2-methyl-substituiert	4,5	3,5	7,8	5,0	9,7	7,8	2,9
4-methyl-substituiert (neben 6-; 8-; 10-; 12-; 14-methyl-substituiert)	32,9	24,3	33,0	36,7	21,6	40,9	26,8
3-methyl-substituiert	—	2,0	3,2	—	12,6	3,8	1,1
2,x-dimethyl-substituiert (2,6-; 2,8-; 2,10-; 2,12-; 2,14-)	1,8	5,8	10,3	—	2,8	1,2	—
3,x-dimethyl-substituiert (3,5-; 3,7-; 3,9-; 3,11-)	—	2,7	—	—	1,5	0,3	—
4,x-dimethyl-substituiert (4,6-; 4,8-; 4,10-; 4,12-; 4,14-; 4,16-)	27,5	37,4	24,5	1,4	5,5	1,9	—
6,x- und 10,x-dimethyl-substituiert	9,5	—	—	—	—	—	—
2,x,y-trimethyl-substituiert	3,1	8,0	4,1	—	—	—	—
4,x,y-trimethyl-substituiert	7,8	10,8	—	—	—	—	—
3,x,y-trimethyl-substituiert	—	1,4	—	—	—	—	—
unidentifiziert	7,7	4,1	14,2	16,4	3,1	0,9	—

Literaturverzeichnis: SS. 431—438

(Procellaria, Puffinus, Fulmarus, Pelecanoides) und b) mit vornehmlich 4- und 6-methyl-substituierten Fettsäuren *(Pachyptila, Garrodia)*.

Die Wachsalkohole dieser Ordnung unterscheiden sich von denen der Pinguine. Bei den Röhrennasen dominieren 4- und 6-methyl-substituierte und bei einigen Arten unverzweigte Alkanole (Tab. 5).

Authentische Vergleichssubstanzen (3-Methylhexansäure, 4-Methyl-heptansäure sowie höhere Homologe) wurden durch Kondensation von Na-Malonester mit z. B. 2-Jodpentan bzw. 1-Jod-2-methyl-pentan hergestellt *(26, 29, 66, 89)*. Die entsprechenden Alkohole lassen sich leicht durch Reduktion mit LiAlH₄ gewinnen. 3,7-Dimethyloctansäure wurde nach Hydrierung von Geraniol mit Pd/H₂ in Cyclohexan durch Oxidation mit CrO₃ in Eisessig/tert. Butanol aus dem 3,7-Dimethyloctanol hergestellt *(66)*.

3. Ciconiiformes (Stelzvögel)

Diese Ordnung mit 111 Arten vereinigt die 6 ziemlich heterogenen Familien *Ardeidae* (Reiher und Dommeln), *Ciconiidae* (Störche und Marabus), *Threskiornithidae* (Löffler und Ibisse), sowie die jeweils nur mit einer Art vertretenen *Cochleariidae* (Kahnschnabel), *Balaenicipitidae* (Schuhschnabel) und *Scopidae* (Schattenvogel = Hammerkopf).

Das Bürzeldrüsensekret des Graureihers *(Ardea cinerea)* enthält Esterwachse und Triglyceride (Tab. 6), an deren Aufbau gesättigte, unverzweigte Fettsäuren beteiligt sind. Die Spuren von ungesättigten Fett-

Tabelle 6. *Zusammensetzung der Triglyceride und Esterwachse des Bürzeldrüsensekretes vom Graureiher (Ardea cinerea)*

Kettenlänge: Anzahl der Doppelbindungen	Triglyceridfettsäure	Wachsfettsäure	Wachsalkohol
6:0	0,4	—	—
8:0	1,2	—	1,7
9:0	0,1	—	2,1
10:0	14,8	0,6	78,5
11:0	2,0	0,1	3,6
12:0	29,4	3,1	5,4
13:0	1,5	0,3	2,5
14:0	29,6	18,1	2,4
15:0	0,4	2,7	1,4
16:0	16,6	67,8	2,4
16:1	3,0	—	—
17:0	0,1	0,5	—
18:0	—	6,8	—
20:1	0,1	—	—
20:polyen	0,1	—	—
22:polyen	0,7	—	—

säuren stammen vermutlich aus Verunreinigungen durch Depotfett (90), das 26% Polyenfettsäuren enthält.

Untersuchungen an anderen Reihern (Nycticorax) (91), Ibissen (92) und am Hammerkopf (93) deuten auch chemotaxonomisch auf die Heterogenität dieser Ordnung.

4. Phoenicopteriformes (Flamingos)

Aus dieser Ordnung wurde erst eine Art, *Phoenicopterus ruber* (Flamingo) untersucht (78). Die Wachsfettsäuren des Bürzeldrüsensekretes sind di- bzw. trimethyl-substituiert und gehören 5 homologen Reihen an, während die Alkohole hauptsächlich unverzweigt sind (Tab. 7). Die Verzweigungspositionen der Alkohole sind unbekannt. Das Bürzeldrüsensekret dieser Art zeigt deutlichere Beziehungen zu den Wachsen der *Anseriformes*-Arten als zu denen der *Ciconiiformes*-Arten.

Tabelle 7. *Zusammensetzung des Bürzeldrüsensekretes von Phoenicopterus ruber (Flamingo)* (78)

Fettsäuren	%
2,6-dimethyl-substituiert	35,7
4,6-dimethyl-substituiert	3,8
2,4,6-trimethyl-substituiert	30,6
2,4,8-trimethyl-substituiert	22,8
2,6,8-trimethyl-substituiert	7,1
Alkohole	
n-Alkanole	87,3
verzweigte Alkanole	12,7

5. Anseriformes (Gänseartige)

Die Ordnung *Anseriformes,* die Gänse, Schwäne, Säger und Enten umfaßt, ist mit 36 untersuchten von unsgesamt 148 Arten die am gründlichsten bearbeitete. Über sie wurde wiederholt zusammenfassend berichtet (17, 59, 100, 101). Die Systematik innerhalb dieser Ordnung ist umstritten. Hier wird der Einteilung von BERNDT und MEISE (84), die 12 Unterfamilien aufführt, gefolgt, ohne die chemotaxonomischen Befunde zu berücksichtigen, die zu einer etwas abweichenden Einteilung führen (Tab. 8).

Die Bürzelwachse der *Anseriformes*-Arten sind mehr oder minder stark methylverzweigt. Während bei den Wachsalkoholen n-Alkanole überwiegen, kommen bei den Wachssäuren homologe Reihen mit einer bis

Tabelle 8. *Systematik der Ordnung Anseriformes [nach* BERNDT *und* MEISE *(84)] mit den bislang untersuchten Arten*

Systematik	Untersuchte Arten (Literatur)

Ordnung: Anseriformes (ca. 148 Arten, 36 untersucht)

1. Familie: Anhimidae

2. Familie: Anatidae

1. Unterfamilie: Anseranatinae	Anseranas semipalmata (*94*)
2. Unterfamilie: Dendrocygninae	Dendrocygna viduata (*76*)
3. Unterfamilie: Coscorobinae	Coscoroba coscoroba (*17*)
4. Unterfamilie: Cygninae	Cygnus olor (*45*)
	Cygnus atratus (*17, 94*)
	Cygnus cygnus (*17*)
	Cygnus columbianus (*17*)
	Cygnus melanocoryphus (*17*)
5. Unterfamilie: Anserinae	Anser anser (*73*)
	Anser a. f. domesticus (*16, 95, 96*)
	Anser indicus (*73*)
	Anser fabalis (*73*)
	Anser caerulescens (*73*)
	Branta leucopsis (*76*)
6. Unterfamilie: Anatinae	Anas domesticus (*77, 89, 96, 97*)
	Anas platyrhynchos (*26, 77*)
	Anas strepera (*73*)
	Anas clypeata (*73*)
	Tadorna tadorna (*98*)
	Tadorna ferruginea (*98*)
	Tadorna tadornoides (*94*)
	Stictonetta naevosa (*94*)
	Tachyeres pteneres (*102*)
	Tachyeres patachonicus (*102*)
7. Unterfamilie: Cairininae	Cairina moschata (*76*)
	Chenonetta jubata (*94*)
	Nettapus pulchellus (*94*)
8. Unterfamilie: Cereopsinae	Cereopsis novaehollandiae (*94*)
9. Unterfamilie: Aythyinae	Aythya fuligula (*99*)
	Aythya ferina (*73*)
10. Unterfamilie: Somateriinae	Somateria mollissima (*76*)
	Melanitta nigra (*47*)
11. Unterfamilie: Oxyurinae	Biziura lobata (*94*)
12. Unterfamilie: Merginae	Mergus serrator (*76*)
	Mergus albellus (*73*)
	Bucephala clangula (*96*)

vier Methylverzweigungen vor, die sich alle an geradzahligen C-Atomen befinden. Hierbei ist der alternierende Rhythmus (2,4,6- oder 2,4,6,8-) gegenüber dem statistischen (z. B. 2,6,8- neben 2,4,8- usw.) bevorzugt.

Tabelle 9. Zusammensetzung der Bürzelwachs-Komponenten einiger Anseriformes-Spezies. (˙ bei Tachyeres pteneres kommen nur trimethyl-substituierte Fettsäuren vom Typ 4,x,y- vor.)

Spezies (Lit.)	Substitution der Fettsäuren									Alkohole		
	n-	2-	4-	andere mono-s.	2,x-	4,x-	2,x,y-	2,4,6,8-	unidenti-fiziert	n-	verzweigt	unidenti-fiziert
Anseranas semipalmata (94)	93,5	6,5	—	—	—	—	—	—	—	97,8	2,2	—
Dendrocygna viduata (76)	—	—	—	—	—	—	56,8	43,2	—	100,0	—	—
Coscoroba coscoroba (17)	—	—	—	—	—	—	100,0	—	—	100,0	—	—
Cygnus cygnus (17)	—	—	—	—	—	—	100,0	—	—	79,6	20,4	—
Cygnus columbianus (17)	—	—	—	—	—	—	100,0	—	—	74,8	25,2	—
Cygnus melanocoryphus (17)	—	—	—	—	—	—	44,5	54,2	1,3	80,3	19,7	—
Cygnus olor (45)	—	—	—	—	—	—	86,5	13,5	—	73,0	27,0	—
Cygnus atratus (17, 94)	—	—	—	—	78,3	—	21,7	—	—	78,1	21,9	—
Anser anser (73)	—	—	—	—	—	—	—	100,0	—	99,1	0,9	—
Anser a. f. domesticus (16)	—	—	—	—	—	—	—	100,0	—	100,0	—	—
Anser fabalis (73)	—	—	—	—	—	—	89,3	10,7	—	88,4	11,6	—
Anser caerulescens (73)	—	—	—	—	—	—	100,0	—	—	87,1	12,9	—
Anser indicus (73)	—	—	—	—	2,0	—	87,3	10,7	—	95,0	5,0	—
Branta leucopsis (76)	—	—	—	—	—	—	100,0	—	—	98,4	1,6	—
Tadorna ferruginea (98)	—	—	—	—	—	—	95,4	4,6	—	98,3	1,7	—
Tadorna tadorna (98)	—	—	—	—	—	—	96,3	3,7	—	98,6	1,4	—
Anas platyrhynchos (26, 77)	12,7	43,3	43,9	—	—	—	—	—	0,1	62,0	37,3	0,7
A. platyrh. f. domesticus (77, 89, 96, 97)	2,5	44,2	50,2	—	—	—	—	—	3,1	40,9	58,1	1,0
Anas strepera (73)	—	72,5	27,5	—	—	—	—	—	—	65,6	34,4	—

Literaturverzeichnis: SS. 431—438

Tabelle 9 (*Fortsetzung*)

Substitution der Fettsäuren

Spezies (Lit.)	n-	2-	4	andere mono-s.	2,x-	4,x-	2,x,y-	2,4,6,8-	unidentifiziert	*Alkohole* n-	verzweigt	unidentifiziert
Anas clypeata (73)	—	94,5	—	—	2,9	2,6	—	—	—	60,2	39,8	—
Tachyeres pteneres (*102*)	34,4	3,9	12,5	10,6	2,4	15,5	3,5*	—	17,2	64,5	35,2	0,3
Tachyeres patachonicus (*102*)	13,6	9,0	7,3	—	12,0	14,1	27,6	—	16,4	52,3	47,3	0,4
Cairina moschata (76)	—	—	—	—	—	—	100,0	—	—	100,0	—	—
Aythya ferina (73)	—	—	—	—	1,3	—	84,8	13,9	—	88,5	11,5	—
Aythya fuligula (99)	—	6,3	—	—	23,4	—	62,9	3,6	3,8	72,7	23,9	3,4
Somateria mollissima (76)	4,3	—	—	—	33,1	4,9	48,2	3,6	5,9	94,9	5,1	—
Mergus serrator (76)	—	—	—	—	—	—	100,0	—	—	100,0	—	—
Mergus albellus (73)	—	—	—	—	2,6	—	97,4	—	—	99,0	1,0	—
Melanitta nigra (47)	—	—	—	—	5,0	—	95,0	—	—	77,6	22,4	—

Der Grad der Verzweigung nimmt von den Gänsen und Schwänen
(tetramethyl-verzweigt) über die Säger (trimethyl-verzweigt) zu den
Enten (mehrheitlich mono- und dimethyl-verzweigt) hin ab. Obgleich
nur für wenige Fettsäuren bewiesen, scheint die D-Konfiguration bei
den natürlich vorkommenden Säuren vorzuherrschen, wie durch Vergleich der Retentionszeiten an Kapillarsäulen und durch Synthese nachgewiesen wurde (16, 103).

Bei 2 Arten wurden neben Esterwachsen beträchtliche Mengen (bis
zu 87%) Squalen im Bürzeldrüsensekret *(Anseranas semipalmata (94),
Cairina moschata (76))* nachgewiesen. Ferner wurden bei zahlreichen
Anseriformes-Arten Triesterwachse aufgefunden, die aus mit n-Alkanolen
und n-Fettsäuren veresterten Alkyl-hydroxymalonsäuren bestehen (41).
Dieser Wachstyp scheint jedoch in Bürzeldrüsensekreten weit verbreitet
zu sein und ist nicht auf die Ordnung *Anseriformes* beschränkt.

Tab. 9 gibt einen Überblick über die Zusammensetzung der Bürzellipide von *Anseriformes*-Arten, wobei nur diejenigen Untersuchungen
berücksichtigt wurden, bei denen die Wachsbestandteile hinreichend
sicher charakterisiert wurden (GLC/MS).

6. Falconiformes (Greifvögel)

Aus dieser 274 Arten umfassenden Ordnung, die Neuweltgeier,
Sekretär, Adler, Bussarde und Falken vereint, wurden bisher 5 Arten
untersucht (104): *Falco tinnunculus* (Turmfalke), *Falco columbarius*
(Merlin), *Accipiter nisus* (Sperber), *Accipiter gentilis* (Habicht) und
Buteo buteo (Mäusebussard). Die Bürzeldrüsensekrete aller Arten bestehen aus sehr kompliziert zusammengesetzten Monoesterwachsen. Die
Wachsfettsäuren mit Kettenlängen zwischen C_{12} und C_{27} sind vornehmlich dimethyl-substituiert mit der ersten Methylverzweigung an C-2
und einer weiteren nahe dem Methylende des Moleküls. Daneben kommen
Mono- und Trimethylfettsäuren vor. Die Wachsalkohole, die hauptsächlich mono- und dimethylverzweigt sind, zeigen eine gewisse strukturelle Verwandtschaft zu den Fettsäuren; ihre Verzweigungen befinden
sich ebenfalls nahe dem Methylende, die Methylverzweigung am C-2
fehlt dagegen (Tab. 10).

Innerhalb der Familien der Ordnung *Falconiformes* wird eine
stufenweise Zunahme des Verzweigungsgrades von *Falco* über *Accipiter*
zu *Buteo* (phylogenetische Reihe) beobachtet, wie sie auch in anderen
Ordnungen nachgewiesen wurde *(Anseriformes, Charadriiformes, Lariformes)*.

Literaturverzeichnis: SS. 431—438

Tabelle 10. *Zusammensetzung der Bürzelwachsbestandteile einiger Falconiformes-Arten*

Wachskomponente	Vorkommende Typen	Falco tinnunculus	Falco columbarius	Accipiter nisus	Accipiter gentilis	Buteo buteo
Fettsäuren						
unverzweigt (n-)	6-; 12-; 16-	0,9	5,3	—	1,6	0,5
2-methyl-substituiert		37,1	29,1	20,7	12,7	0,8
4-methyl-substituiert		—	1,2	—	0,7	—
andere monomethyl-substituierte		—	0,5	—	0,9	0,1
2,x-dimethyl-substituiert	2,6-; 2,8-; 2,10-; 2,12-; 2,14-; 2,16-; 2,20-; 2,24-	45,0	47,4	70,7	55,6	40,7
4,x-dimethyl-substituiert und andere	4,6-; 4,8-; 4,10-; 6,14-; 10,14-; 10,16-	—	0,3	—	1,1	—
2,x,y-trimethyl-substituiert	2,6,8-; 2,6,10-; 2,6,12-; 2,6,14-; 2,6,16-; 2,8,16-; 2,10,14-; 2,10,16-; 2,10,18-; 2,10,20-; 2,14,16-; 2,14,18-; 2,14,20-; 2,14,22-; 2,16,18-; 2,16,20-; 2,16,22-; 2,18,20-; 2,18,22-; 2,20,24-	15,7	16,2	8,6	23,8	49,9
2,x,y,z-tetramethyl-substituiert	2,10,12,14-; 2,10,14,16-; 2,14,18,22-				—	3,9
nicht identifiziert		1,3	—	—	3,6	4,1
Alkohole						
unverzweigt (n-)	2-; 6-; 8-; 10-; 12-; 14-; 15-; 16-; 18-; 20-; 22-; 24-; 2-äthyl-	32,1	10,3	13,8	3,4	1,0
monomethyl-substituiert		56,6	68,5	65,1	41,3	41,6
dimethyl-substituiert	2,12-; 2,16-; 2,18-; 4,16-; 6,14-; 10,14-; 10,16-; 10,20-; 10,22-; 12,16-; 12,18-; 12,20-; 12,22-; 14,16-; 14,18-; 14,20-; 14,22-; 14,24-; 16,20-; 16,22-; 16,24-; 18,22-	10,3	21,2	19,1	52,0	48,0
trimethyl-substituiert	2,18,22-	—	—	—	—	4,0
nicht identifiziert		1,0	—	2,0	3,3	5,4

7. Galliformes (Hühnervögel)

Die Bürzeldrüsenlipide der bislang untersuchten *Galliformes*-Arten zeichnen sich durch das Vorkommen von Diesterwachsen aus, die aus mit n-Fettsäuren veresterten Alkan-2,3-diolen bestehen und sich durch die allgemeine Formel

$$
\begin{array}{l}
CH_3 \qquad O \\
| \qquad\quad \| \\
CH-O-C-(CH_2)_x-CH_3 \\
| \\
CH-O-C-(CH_2)_y-CH_3 \\
| \qquad\quad \| \\
| \qquad\quad O \\
(CH_2)_z \\
| \\
CH_3
\end{array}
$$

wiedergeben lassen. Hierbei können die Diole, die gelegentlich auch als Uropygiole, heute aber mit dem systematischen Namen bezeichnet werden, threo- oder erythro-Konfiguration haben. Die Konfiguration läßt sich nach der alkalischen Hydrolyse der Diesterwachse aus der TLC an Borsäure-imprägnierten SiO_2-Platten sowie aus der GLC der Acetonide erkennen. Nur zwei Arten, *Gallus domesticus* (Haushuhn) (*22, 34*) und *Coturnix pectoralis* (Grauwachtel) (*82*) weisen einen wesentlichen Anteil an threo-Isomeren auf. Die 4 anderen untersuchten Arten, *Phasianus colchicus* (Fasan) (*38, 81*), *Perdix perdix* (Rebhuhn) (*24*), *Meleagris pavo* (Truthuhn) (*22*), *Leipoa ocellata* (Thermometerhuhn) (*82*) enthalten ausschließlich erythro-Diole (Tab. 11). Die Kettenlängenverteilung der Diole bei *Coturnix coturnix japonica* (japanische Wachtel) (*35*), deren Stereochemie jedoch nicht untersucht wurde, weicht deutlich von der von *Coturnix pectoralis* ab. Bei *Leipoa ocellata* (*82*) und *Phasianus colchicus* (*105*) wurden außerdem geringe Mengen unverzweigter Monoesterwachse nachgewiesen.

8. Gruiformes (Rallenartige)

Die Ordnung *Gruiformes* ist eine heterogene Vogelgruppe, die recht verschiedene Familien, die teilweise nur durch eine Art repräsentiert sind, zusammenfaßt. Einige Arten besitzen keine Bürzeldrüse *(Otididae)*, zahlreiche Arten besitzen ausgeprägte Puderdunen (z. B. *Mesitornithidae*, der Kagu, die Sonnenralle). Die Bürzeldrüsensekrete von 4 Spezies (alle zur Familie *Rallidae* gehörend) wurden näher untersucht: *Rallus aquaticus* (Wasserralle) (*70*), *Porzana porzana* (Tüpfelsumpfhuhn) (*70*),

Tabelle 11. Zusammensetzung der Diole aus den Bürzelwachsen verschiedener Galliformes-Arten

Diol-Kettenlänge	Gallus domesticus		Coturnix pectoralis		Perdix perdix	Leipoa ocellata	Phasianus colchicus	Meleagris pavo
	erythro	threo	erythro	threo	erythro	erythro	erythro	erythro
C_{12}	—	—	—	—	3,8	—	—	—
C_{13}	—	—	—	—	1,3	—	—	—
C_{14}	—	—	6,3	—	18,3	—	—	—
C_{15}	—	—	—	—	0,8	3,7	—	—
C_{16}	—	—	5,8	1,4	12,6	—	1,4	—
C_{17}	—	—	—	—	1,5	—	93,6	—
C_{18}	—	—	17,7	7,4	21,5	92,8	1,5	—
C_{19}	—	—	0,5	0,2	5,8	—	3,5	1,2
C_{20}	—	—	2,6	1,0	16,8	3,5	—	15,5
C_{21}	—	—	2,1	0,5	4,1	—	—	27,0
C_{22}	13,8	7,8	23,7	2,4	11,6	—	—	41,2
C_{23}	18,5	15,7	17,6	1,3	1,9	—	—	15,1
C_{24}	20,3	23,9	9,3	0,2	—	—	—	—

Gallinula chloropus (Teichhuhn) (*70*) und *Fulica atra* (Bläßhuhn) (*46*). Ihre Strukturen sind durch das Vorkommen von Fettsäuren charakterisiert, bei denen vorzugsweise jedes vierte C-Atom eine Methylgruppe trägt, wobei sich die erste Verzweigung an C-2 oder C-4 befinden kann (z. B. 2,6- oder 4,8-dimethyl- bzw. 2,6,10- oder 4,8,12-trimethyl- bzw. 2,6,10,14-tetramethyl-substituierte Säuren). Von *Rallus* über *Gallinula* zu *Fulica* nimmt der Anteil an 4-substituierten Säuren zu und der an 2-substituierten ab (Tab. 12). Bei den Wachsalkoholen dominieren n-Alkanole (>50%); daneben kommen monomethyl-substituierte und Spuren dimethyl-substituierter Alkanole vor, deren Substituenten sich an geradzahligen C-Atomen befinden.

Tabelle 12. *Zusammensetzung der Bürzelwachsbestandteile einiger Gruiformes-Arten* (*46, 70*)

Wachsbestandteil	Rallus aquaticus	Gallinula chloropus	Porzana porzana	Fulica atra
Fettsäuren				
2-methyl-substituiert	4,9	12,9	3,4	4,9
4-methyl-substituiert	—	1,1	0,6	—
2,6-dimethyl-substituiert	18,1	4,9	7,4	4,6
2,8-dimethyl-substituiert	—	—	—	2,9
2,10-dimethyl-substituiert	2,4	0,4	1,7	13,2
2,12-dimethyl-substituiert	14,0	7,4	1,1	2,0
2,14-dimethyl-substituiert	3,3	3,1	1,3	
2,16-dimethyl-substituiert	—	0,6	0,9	—
4,8-dimethyl-substituiert	0,7	35,1	48,0	9,5
4,10-dimethyl-substituiert	—	—	—	0,9
4,12-dimethyl-substituiert	—	—	—	2,9
4,14-dimethyl-substituiert	—	0,4	—	—
2,6,10-trimethyl-substituiert	41,9	7,8	5,3	20,3
2,6,12-trimethyl-substituiert	7,4	3,8	1,9	4,8
2,6,14-trimethyl-substituiert	6,1	1,5	5,3	—
4,8,12-trimethyl-substituiert	—	17,0	16,6	25,0
4,8,14-trimethyl-substituiert	—	4,0	6,5	3,2
2,6,10,14-tetramethyl-substituiert	1,2	—	—	—
nicht identifiziert	—	—	—	5,8
Alkohole				
unverzweigt	54,0	76,3	80,8	60,6
2-methyl-substituiert	1,2	0,2	—	1,5
4-methyl-substituiert	—	0,4	0,6	—
6-,8-,10-methyl-substituiert	22,3	8,2	4,7	14,5
12-methyl-substituiert	0,2	1,7	3,2	—
14-methyl-substituiert	15,7	8,0	4,8	9,6
16-methyl-substituiert	1,6	2,5	4,4	—
10,14- und 10,16-dimethyl-substituiert	4,3	2,7	1,5	—
nicht identifiziert	0,7	—	—	13,8

Literaturverzeichnis: SS. 431—438

Bei einigen Arten dieser Ordnung wurden im Verseifungsprodukt des Bürzeldrüsensekretes 1,2-Diole nachgewiesen (83). Ihre Struktur wurde durch Massenspektrometrie der Isopropylidenderivate (Acetonide) aufgeklärt, die Kettenlängen von $C_{16} - C_{22}$ ergab.

9. Charadriiformes (Regenpfeifer und Schnepfenartige)

Die Ordnung *Charadriiformes*, die insgesamt 12 Familien mit etwa 184 Arten vereint, wird häufig mit der folgenden Ordnung *Lariformes* zu einer Überordnung zusammengefaßt (85). In der Tat sind auch chemotaxonomisch die Übergänge sowohl zu den Alken als auch zu den näher untersuchten Möwen fließend. Während in den bislang analysierten Bürzelwachsen von *Charadriiformes*-Arten mono- (vornehmlich 2-methyl-substituierte) und dimethyl-verzweigte (2,x- und 4,x-) Fettsäuren domi-

Tabelle 13. *Bürzelwachszusammensetzung einiger Charadriiformes-Arten (71, 79)*

Wachsbestandteil	Haematopus ostralegus	Tringa totanus	Calidris canutus	Calidris alpina
Fettsäuren				
unverzweigt	1,3	0,9	Spur	1,3
2-methyl-substituiert	60,9	13,3	30,5	11,3
4-methyl-substituiert	1,3	0,3	11,8	0,6
6-methyl-substituiert	1,0	—	0,2	0,4
2,6-dimethyl-substituiert	10,4	29,7	16,5	14,9
2,8-dimethyl-substituiert	15,1	13,0	13,3	16,9
2,10-dimethyl-substituiert	5,9	17,4	11,3	19,5
2,12-dimethyl-substituiert	1,3	1,2	4,2	9,1
2,14-dimethyl-substituiert	0,5	0,1	0,4	1,0
4,8-dimethyl-substituiert	1,6	2,6	5,9	0,3
4,10-dimethyl-substituiert	0,2	—	—	—
2,6,10-trimethyl-substituiert	—	13,4	2,3	13,4
nicht identifiziert	0,5	8,1	3,6	11,3
Alkohole				
unverzweigt	49,6	46,6	57,8	47,1
2-methyl-substituiert	—	18,7	16,2	17,9
4-methyl-substituiert	—	9,9	13,5	10,8
6-methyl-substituiert	—	1,7	2,2	5,1
10-methyl-substituiert	13,4	—	—	—
12-methyl-substituiert	3,1	—	—	—
14-methyl-substituiert	15,6	2,1	3,8	4,3
16-methyl-substituiert	16,0	—	—	—
2,6-dimethyl-substituiert	—	3,2	1,4	2,3
2,8-dimethyl-substituiert	—	2,4	0,6	2,3
2,10-dimethyl-substituiert	—	5,7	1,3	2,9
2,12-dimethyl-substituiert	—	5,3	0,4	1,3
nicht identifiziert	2,3	4,4	2,8	6,0

nieren, herrschen bei den *Lariformes*-Arten n-Fettsäuren und trimethyl-substituierte homologe Reihen vor. Die Alkohole zeigen dagegen in beiden Ordnungen weitgehende Ähnlichkeiten.

Es wurden 3 Spezies aus der Familie *Scolopacidae,* nämlich *Tringa totanus* (Rotschenkel) (*71*), *Calidris canutus* (Knutt) (*71*) und *Calidris alpina* (Alpenstrandläufer) (*71*) und eine Art aus der Familie *Haematopodidae, Haematopus ostralegus* (Austernfischer) (*71, 79*), untersucht. Die Ergebnisse sind in Tab. 13 wiedergegeben.

10. Lariformes (Möwenartige)

Von den fünf Familien [*Stercorariidae* (Raubmöwen), *Laridae* (Möwen), *Sternidae* (Seeschwalben), *Rhynchopidae* (Scherenschnäbler), *Alcidae* (Alken)] mit insgesamt 132 Arten sind die Familien *Laridae* (*106, 107, 108*) und *Alcidae* (*23, 63*) näher untersucht. Aus Tab. 14 läßt sich auch hier eine chemotaxonomische Verwandtschaftsreihe erkennen, die von unsubstituierten Bürzelwachskomponenten zu zunehmend verzweigteren Bestandteilen führt. Während *Fratercula arctica* (Papageientaucher) (*23*) ausschließlich n-Fettsäuren und n-Alkohole aufweist, ist bei *Uria aalge* (*63*), von der 2 Rassen untersucht wurden (Trottel- und Ringellumme) und bei *Cepphus grylle* (Grylteiste) (*63*), der Anteil an n-Fettsäuren auf $40-50\%$ reduziert. Das Bürzeldrüsensekret von *Alca torda* (Tordalk) (*63*) enthält nur noch 15% n-Fettsäuren und leitet damit zu den *Laridae* über, an deren Bürzellipiden n-Fettsäuren mit $8,5-38\%$ beteiligt sind. Die übrigen Wachssäuren gehören zu mehreren homologen Reihen von methyl-substituierten Säuren (2-, 4-mono-, 2,6-, 2,8-, 2,10-, 4,6-, 4,8-, 4,10-, 4,12-di- und 2,4,8-, 2,6,8-, 2,6,10-, 2,6,12-, 4,6,10- und 4,8,12-trimethyl-substituierten).

Die Alkohole zeigen den gleichen Trend, sind aber allgemein weniger stark verzweigt. Lediglich *Fratercula arctica* enthält ausschließlich n-Alkanole, während die Möwen den höchsten Verzweigungsgrad aufweisen (Tab. 15).

11. Columbiformes (Tauben)

Das Bürzeldrüsensekret einer Art aus der Familie *Columbidae* (Tauben) der insgesamt 3 Familien [*Pteroclidae* (Flughühner), *Raphidae* (Dronten, ausgestorben), *Columbidae* (Tauben)], wurde untersucht (*Columba palumbus,* Ringeltaube) (*39*). Dabei handelt es sich um ein Diesterwachs, bei dem 3-Hydroxyfettsäuren mit n-Alkanolen und n-Fettsäuren verestert sind:

Literaturverzeichnis: SS. 431—438

Tabelle 14. *Bürzelwachs-Fettsäuren einiger Lariformes-Arten (23, 63, 106, 107, 108)*

Fettsäure	Fratercula arctica (23)	Uria aalge (63)	Cepphus grylle (63)	Alca torda (63)	Rissa tridactyla (106)	Larus ridibundus (107, 108)	Larus fuscus (107)	Larus argentatus (107)
unverzweigt (n-)	100,0	41,8	49,2	15,0	8,5	37,6	25,4	15,7
2-methyl-substituiert	—	0,4	2,5	9,7	46,6	5,2	10,2	14,4
4-methyl-substituiert	—	2,0	1,5	8,1	—	—	—	—
6-methyl-substituiert	—	0,7	0,2	12,9	—	—	—	—
8-, 10- und 12-methyl-substituiert	—	0,7	0,2	2,3	—	—	—	—
2,6-dimethyl-substituiert	—	—	5,6	—	13,8	14,1	21,5	28,6
2,8-dimethyl-substituiert	—	—	4,9	—	9,9	8,3	12,8	15,2
2,10-dimethyl-substituiert	—	—	7,3	—	8,7	2,5	7,2	8,6
4,6-dimethyl-substituiert	—	7,7	—	8,3	—	—	—	—
4,8-dimethyl-substituiert	—	4,3	4,6	7,1	—	—	—	—
4,10-dimethyl-substituiert	—	7,7	4,0	7,6	—	—	—	—
4,12-dimethyl-substituiert	—	1,5	—	—	—	—	—	—
2,4,8-trimethyl-substituiert	—	0,4	1,1	—	—	5,6	9,6	3,0
2,6,8-trimethyl-substituiert	—	—	13,2	—	—	13,5	2,4	1,8
2,6,10-trimethyl-substituiert	—	—	3,5	—	—	11,9	10,0	10,1
2,6,12-trimethyl-substituiert	—	—	—	17,0	—	1,3	0,9	2,6
4,6,10-trimethyl-substituiert	—	11,3	1,7	—	—	—	—	—
4,8,12-trimethyl-substituiert	—	5,4	0,5	—	—	—	—	—
nicht identifiziert	—	16,1	—	12,0	12,5	—	—	—

Tabelle 15. *Bürzelwachs-Alkohole einiger Lariformes-Arten (23, 63, 106, 107, 108)*

Alkohol	Fratercula arctica (23)	Uria aalge (63)	Cepphus grylle (63)	Alca torda (63)	Rissa tridactyla (106)	Larus ridibundus (107, 108)	Larus fuscus (107)	Larus argentatus (107)
unverzweigt	100,0	59,7	49,0	43,4	46,3	64,7	54,1	49,4
2-methyl-substituiert	—	10,1	7,6	7,0	34,7	10,7	14,2	10,9
4-methyl-substituiert	—	9,9	8,2	9,8	—	7,3	3,9	5,5
6-methyl-substituiert	—	9,5	—	8,4	—	—	—	—
8-methyl-substituiert	—	—	—	1,0	—	—	—	—
12-methyl-substituiert	—	0,1	—	—	—	—	—	—
2,6-dimethyl-substituiert	—	—	3,1	—	3,2	3,7	5,7	7,4
2,8-dimethyl-substituiert	—	—	6,9	—	2,8	4,3	5,1	7,3
2,10-dimethyl-substituiert	—	—	5,6	—	4,9	3,9	4,9	6,2
2,12-dimethyl-substituiert	—	—	—	—	4,0	1,4	3,9	3,9
4,6-dimethyl-substituiert	—	0,7	—	1,9	—	—	—	—
4,8-dimethyl-substituiert	—	1,4	1,9	2,1	—	—	—	—
4,10-dimethyl-substituiert	—	2,1	5,0	4,1	—	—	—	—
4,12-dimethyl-substituiert	—	1,0	—	5,1	—	—	—	—
2,6,10-trimethyl-substituiert	—	—	—	—	—	4,0	5,9	7,2
2,6,12-trimethyl-substituiert	—	—	—	—	—	—	2,3	2,2
nicht identifiziert	—	5,5	12,7	17,2	4,1	—	—	—

Literaturverzeichnis: SS. 431—438

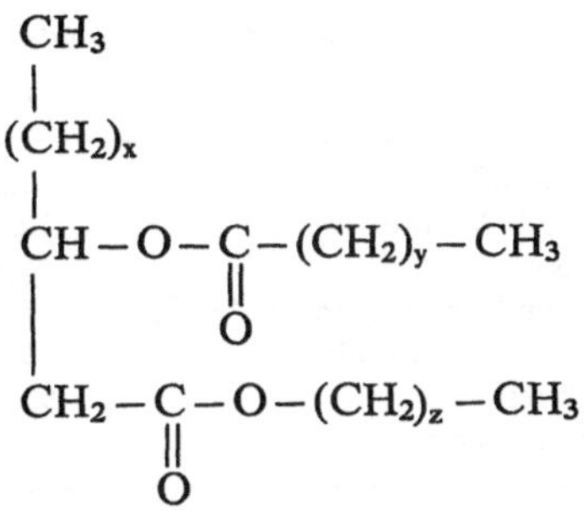

$$
\begin{array}{l}
CH_3 \\
| \\
(CH_2)_x \\
| \\
CH - O - \underset{\substack{\| \\ O}}{C} - (CH_2)_y - CH_3 \\
| \\
CH_2 - \underset{\substack{\| \\ O}}{C} - O - (CH_2)_z - CH_3
\end{array}
$$

Die quantitative Zusammensetzung gibt Tab. 16 wieder.

Tabelle 16. *Quantitative Zusammensetzung der Diesterwachs-Komponenten aus dem Bürzeldrüsensekret der Ringeltaube (Columba palumbus) (39)*

Fettsäuren		Hydroxysäuren		Alkohole	
$8:0$	0,6	$3\text{-}OH - C_8$	Spur	$8:0$	0,6
$9:0$	Spur	$3\text{-}OH - C_9$	86,0	$9:0$	Spur
$10:0$	15,4	$3\text{-}OH - C_{10}$	14,0	$10:0$	12,2
$11:0$	Spur			$11:0$	Spur
$12:0$	10,8			$12:0$	23,4
$14:0$	2,5			$14:0$	11,5
$16:0$	70,7			$16:0$	52,3

Tauben besitzen gut ausgeprägte Puderdunen und nur rudimentäre Bürzeldrüsen, deren Sekretmenge schwerlich zur Gefiederimprägnierung ausreichen dürfte. Tatsächlich hat die Analyse der Gefiederlipide bei *Columba palumbus* (40), deren Zusammensetzung sich deutlich von der des Bürzeldrüsensekretes unterscheidet (Tab. 17), gezeigt, daß andere Lipidquellen in der Taubenhaut eine Rolle spielen, was auch durch frühere histologische Untersuchungen (109, 8) bereits zu vermuten war. Hauptbestandteil der Gefiederlipide sind freie Fettsäuren (55,6%) mit Kettenlängen von $C_{12} - C_{30}$, wobei geradzahlige Individuen vorherrschen (Hauptkomponenten C_{16}, C_{18}, C_{20}, C_{22}, C_{24}, C_{26}). Daneben wurden u. a. freies Cholestanol, freie Alkohole (geradzahlige bevorzugt), Cholestanolester, Monoesterwachse sowie Kohlenwasserstoffe aufgefunden. Die Hauptkomponenten letzterer zeigen eine um 1 C kürzere Kette als die Fettsäuren, so daß ein biogenetischer Zusammenhang (Decarboxylierung) zu vermuten ist. Das Bürzeldrüsensekret ist mit 6,7% nur ein geringer Bestandteil der Gefiederlipide. Da die an ihrem Aufbau beteiligten 3-Hydroxysäuren fungizide Eigenschaften besitzen (110), ist eine ähnliche physiologische Funktion hier zu erwägen.

Tabelle 17. *Zusammensetzung der Gefiederlipide der Ringeltaube (Columba palumbus) (40)*

Lipid	Gewichts-%
Kohlenwasserstoffe	3,3
Sterolester und Wachse	10,7
Diesterwachse	6,7
Triglyceride und ähnlich polare Stoffe	5,0
Freie Alkohole und Sterole	5,0
Freie Fettsäuren	55,6
Polare Verbindungen	13,7

12. Psittaciformes (Papageien)

In dieser großen, über 300 Arten umfassenden Ordnung sind Bürzeldrüsen häufig nur schwach entwickelt bzw. rudimentär vorhanden, einigen Arten fehlt ein solches Organ völlig (Aras, Amazonen). Das Bürzeldrüsensekret nur einer Art, *Melopsittacus undulatus* (Wellensittich) wurde untersucht (72). Es handelt sich um ein aus vorwiegend (ω-1)-, (ω-2)- und (ω-3)-methyl-verzweigten Fettsäuren und n-Alkoholen zusammengesetztes Esterwachs (Tab. 18).

Tabelle 18. *Zusammensetzung der Bürzelwachsbestandteile von Melopsittacus undulatus (72)*

Fettsäuren		Alkohole	
unverzweigt	20,9	unverzweigt	93,0
2-methyl-substituiert	15,2	6-methyl-substituiert	1,1
(ω-1)-, (ω-2)- und (ω-3)-substituiert	32,7	10-methyl-substituiert	0,8
andere monomethyl-substituierte	14,1	12-methyl-substituiert	1,5
2,x-dimethyl-substituiert	5,0	14-methyl-substituiert	2,2
4,x- und 6,x-dimethyl-substituiert	12,1	16-methyl-substituiert	1,4

13. Cuculiformes (Kuckucke)

Von der 2 Familien [*Musophagidae* (Turakos), *Cuculidae* (Kuckucke)] umfassenden und 146 Arten enthaltenden Ordnung wurde bisher das Bürzeldrüsensekret einer Art, *Cuculus canorus* (europäischer Kuckuck) (68) untersucht. Das Monoesterwachs (Tab. 19) ist zu über 80% aus 3-methyl-substituierten Fettsäuren und Alkoholen aufgebaut und läßt sich durch die Formel

$$R-\underset{\underset{CH_3}{|}}{CH}-(CH_2)_x-\underset{\underset{O}{\|}}{C}-O-(CH_2)_y-\underset{\underset{CH_3}{|}}{CH}-R' \qquad \begin{matrix} R = C_8-C_{16} \\ R' = C_6-C_{16} \\ x = 1,2 \\ y = 2\,(1,3) \end{matrix}$$

beschreiben.

Literaturverzeichnis: SS. 431—438

3-Methyl-verzweigte Fettsäuren und Alkohole als Hauptbestandteile von Bürzelwachsen wurden auch in den Ordnungen *Piciformes* (Spechte), *Passeriformes* (Sperlingsvögel) sowie bei einer Eulenart *(Tyto alba)* als wesentliche Bestandteile bei *Procellariiformes-* und *Spheniseiformes-* Arten nachgewiesen, sind also ein weit verbreitetes Merkmal.

Tabelle 19. *Zusammensetzung der Bürzelwachs-Komponenten des europäischen Kuckucks (Cuculus canorus) (68)*

Verzweigungsstelle	Fettsäuren	Alkohole
unverzweigt	5,6	2,8
3-methyl-substituiert	76,8	89,0
4-methyl-substituiert	8,9	Spur
2-methyl-substituiert	—	Spur
3,x-dimethyl-substituiert	4,1	4,4
nicht identifiziert	4,6	3,8

Tabelle 20. *Bürzelwachszusammensetzung von Asio otus und Bubo bubo (33)*

Wachskomponente	Asio otus	Bubo bubo
Fettsäuren		
unverzweigt	—	2,6
2-methyl-substituiert	—	1,7
2-äthyl-substituiert	24,4	42,3
2-propyl-substituiert	10,0	—
2-butyl-substituiert	55,6	—
2,6-dimethyl-substituiert	—	13,0
2,8-dimethyl-substituiert	—	12,8
2,10-dimethyl-substituiert	—	0,7
2-äthyl-6-methyl-substituiert	—	9,9
2,6,12-trimethyl-substituiert	—	1,2
2,6,14-trimethyl-substituiert	—	1,3
nicht identifiziert	10,0	14,5
Alkohole		
unverzweigt	28,8	23,0
2-methyl-substituiert	3,2	13,9
4-methyl-substituiert	3,9	0,3
6-methyl-substituiert	20,9	1,0
8-methyl-substituiert	4,1	—
10-methyl-substituiert	22,3	32,0
12-methyl-substituiert	5,7	0,3
2-butyl-substituiert	5,4	—
2,6-dimethyl-substituiert	—	2,0
2,8-dimethyl-substituiert	—	3,3
2,10-dimethyl-substituiert	—	21,5
2,16-dimethyl-substituiert	—	0,8
2,18-dimethyl-substituiert	—	1,9
nicht identifiziert	5,7	—

14. Strigiformes (Eulen)

Die zwei Familien der Ordnung *Strigiformes (Strigidae* und *Tytonidae)* unterscheiden sich in ihren Bürzeldrüsensekreten signifikant. Bei den *Strigidae,* von denen drei Arten untersucht wurden [*Asio otus* (Waldohreule) *(33)*, *Strix aluco* (Waldkauz) *(111)*, *Bubo bubo* (Uhu) *(33)*] kommen 2-alkyl-substituierte Fettsäuren als Hauptbestandteile vor (2-methyl-, 2-äthyl-, 2-propyl-, 2-butyl-, 2,6-, 2,8-, 2,10-dimethyl- und 2-äthyl-6-methyl- sowie 2,6,12- und 2,6,14-trimethyl-substituierte), die mit vornehmlich unverzweigten bzw. monomethyl-substituierten Alkanolen verestert sind. Demgegenüber ist das Bürzeldrüsensekret von *Tyto alba* (Schleiereule) *(33)* hauptsächlich aus 3-methyl-substituierten Fettsäuren und Alkoholen aufgebaut. Die Zusammensetzungen von 3 Arten sind in den Tabellen 20 und 21 wiedergegeben. Eine dünnschichtchromatographische Untersuchung *(96)* deutet auch bei *Aegolius funereus* (Rauhfußkauz) auf ein Monoesterwachs.

Tyto setzt sich damit deutlich von den *Strigidae* ab.

Tabelle 21. *Zusammensetzung der Wachskomponenten des Bürzeldrüsensekretes von Tyto alba (Schleiereule) (33)*

Wachskomponente	Tyto alba
Fettsäuren	
3-methyl-substituiert	59,3
3,5-dimethyl-substituiert	1,5
3,7-dimethyl-substituiert	10,2
3,9-dimethyl-substituiert	6,2
3,11-dimethyl-substituiert	11,9
3,13-dimethyl-substituiert	8,5
3,15-dimethyl-substituiert	1,4
3,9,11-trimethyl-substituiert	1,0
Alkohole	
unverzweigt	5,0
2-methyl-substituiert	11,7
3-methyl-substituiert	35,5
4-methyl-substituiert	12,2
3,7-dimethyl-substituiert	1,3
3,9-dimethyl-substituiert	1,2
3,11-dimethyl-substituiert	3,0
3,13-dimethyl-substituiert	4,8
3,15-dimethyl-substituiert	2,9
4,6-dimethyl-substituiert	4,0
4,8-dimethyl-substituiert	5,8
4,14-dimethyl-substituiert	3,1
4,16-dimethyl-substituiert	4,0
4,6,14-trimethyl-substituiert	0,6
nicht identifiziert	4,9

Literaturverzeichnis: SS. 431—438

15. Piciformes (Spechte)

Von den auf 6 Familien verteilten 377 Arten der Ordnung *Piciformes* wurden die Bürzeldrüsensekrete zweier Arten, *Dryocopus martius* (Schwarzspecht) (*60*) und *Picus viridis* (Grünspecht) (*60*) untersucht. Die Esterwachse sind durch die hohen Anteile an 3-monomethyl- und 3,x-dimethyl-substituierten Komponenten charakterisiert (Tab. 22). Derartige Wachse kommen auch bei *Tyto alba, Cuculus canorus* sowie bei zahlreichen *Passeriformes*-Arten vor. Ferner wurden 1,2-Diole im Bürzeldrüsensekret von *Piciformes*-Arten beobachtet (*83*).

Tabelle 22. *Zusammensetzung des Bürzeldrüsensekretes von Dryocopus martius (Schwarzspecht) und Picus viridis (Grünspecht) (60)*

Wachskomponente	Dryocopus martius	Picus viridis
Fettsäuren		
3-methyl-substituiert	54,2	64,0
3,7-dimethyl-substituiert	14,4	9,4
3,9-dimethyl-substituiert	3,4	12,1
3,11-dimethyl-substituiert	4,7	9,3
3,13-dimethyl-substituiert	10,5	3,7
3,15-dimethyl-substituiert	6,0	—
3,17-dimethyl-substituiert	0,5	—
3,7,11-trimethyl-substituiert	2,7	0,8
nicht identifiziert	3,6	0,7
Alkohole		
unverzweigt	70,9	4,6
3-methyl-substituiert	11,6	53,0
3,7-dimethyl-substituiert	—	8,2
3,9-dimethyl-substituiert	—	5,4
3,11-dimethyl-substituiert	—	2,1
3,13-dimethyl-substituiert	1,3	11,7
3,15-dimethyl-substituiert	1,2	6,6
anteiso-substituiert	12,0	5,4
andere monomethyl-substituierte	3,0	3,0

16. Passeriformes (Sperlingsvögel)

Die Ordnung *Passeriformes* ist die bei weitem umfangreichste der Klasse Aves und vereinigt in vier Unterordnungen (Tab. 23) etwa 5100 Spezies und damit über die Hälfte aller rezenten Arten.

Tabelle 23. *Systematik der Ordnung Passeriformes*

1. Unterordnung: Eurylaimi (14 Spezies)
2. Unterordnung: Tyranni (1110 Spezies)
3. Unterordnung: Menurae (4 Spezies)
4. Unterordnung: Oscines (= Passeres; 3990 Spezies)

Die Unterordnung Oscines, aus der allein bislang Untersuchungen an Bürzeldrüsensekreten vorliegen, umfaßt 47 Familien; von diesen wurden 10 mehr oder minder intensiv bearbeitet. Die Untersuchungen zeigen, daß qualitativ signifikante Unterschiede der Wachskompositionen bereits auf Familienebene erkennbar sind, so daß eine Entscheidungshilfe bei der in vielen Fällen strittigen Zuordnung von Arten zu einzelnen Familien möglich ist.

16.1 Corvidae (Rabenvögel)

Aus der 102 Arten enthaltenden Familie *Corvidae* liegen Bürzelwachsanalysen von 8 Arten vor: *Corvus frugilegus* (Saatkrähe) (*112*), *Coloeus monedula* (Dohle) (*75*), *Corvus corone corone* (Rabenkrähe) (*75*), *Corvus corone cornix* (Nebelkrähe) (*75*), *Garrulus glandarius* (Eichelhäher) (*113*), *Pica pica* (Elster), (*113*), *Pyrrhocorax graculus* (Alpendohle) (*113*) und *Corvus corax* (Kolkrabe) (*113*).

Hauptsächliche Bestandteile der Wachssäuren sind 2-monomethyl-, 2,x-dimethyl- und 2,x,y-trimethyl-substituierte Fettsäuren (Tab. 24). Bei *Coloeus* kommen daneben n-Fettsäuren vor. Die beiden Rassen *Corvus corone* enthalten dagegen vor allem 4-methyl-substituierte Säuren, jedoch scheint die Biosynthese von 4-substituierten Fettsäuren genetisch eng mit der von 2-substituierten zusammenzuhängen; beide Komponenten kommen in verschiedenen Familien und sogar bei zahlreichen Arten nebeneinander vor (z. B. *Anseriformes, Sphenisciformes*). Bei den Wachsalkoholen dominieren n- und Monomethyl-alkanole. Auch hier ist bei den beiden Krähen die 4-Substitution bevorzugt (Tab. 25).

16.2. Sylviidae (Grasmücken)

8 Arten der insgesamt 321 Sylviidae-Arten sind bisher untersucht worden (*74*). Die Bürzelwachs-Säuren gehören homologen Reihen methylsubstituierter Fettsäuren an, deren erste Verzweigung sich in 2-Position befindet. Weitere Methylgruppen befinden sich vorzugsweise an jedem weiteren 4. C-Atom, so daß 2,6-dimethyl-, 2,6,10-trimethyl- und 2,6,10,14-tetramethyl-substituierte Säuren dominieren (Tab. 26). Bei den Alkoholen sind n-, 2- und 4-Methylalkanole bevorzugt, doch kommen auch höher substituierte Alkohole vor (Tab. 27). Ähnliche Wachsmuster, speziell mit Methylverzweigungen an jedem 4. C-Atom, wurden bei den Rallen und einigen *Charadriiformes*-Arten aufgefunden.

Tabelle 24. Zusammensetzung der Bürzelwachs-Fettsäuren einiger Corvidae-Arten (75, 112, 113)

Fettsäure	Coloeus monedula	Corvus frugilegus	Corvus corax	Corvus corone corone	Corvus corone conix	Pyrrhocorax graculus	Pica pica	Garrulus glandarius
unverzweigt	35,7	—	—	—	—	—	1,7	0,5
2-methyl-substituiert	64,3	100,0	85,3	—	—	37,4	34,2	9,8
4-methyl-substituiert	—	—	—	34,8	34,2	—	—	—
6-methyl-substituiert	—	—	—	3,2	2,7	—	—	—
2,6-, 2,8- und 2,10-dimethyl-substituiert	—	—	13,0	—	—	61,3	45,1	59,1
2,12-dimethyl-substituiert	—	—	1,4	—	—	1,0	0,2	—
2,14- und 2,16-dimethyl-substituiert	—	—	—	—	—	—	1,7	—
4,8-dimethyl-substituiert	—	—	—	27,2	27,0	—	—	—
4,10-dimethyl-substituiert	—	—	—	15,1	16,0	—	—	—
2,6,10-trimethyl-substituiert	—	—	—	—	—	—	5,9	11,2
2,6,12-, 2,8,12-trimethyl-substituiert	—	—	—	—	—	—	3,7	5,9
2,6,14-trimethyl-substituiert	—	—	—	—	—	—	5,6	7,8
2,6,16-trimethyl-substituiert	—	—	—	—	—	—	1,3	5,1
2,6,18-trimethyl-substituiert	—	—	—	—	—	—	0,6	0,2
4,6,10-trimethyl-substituiert	—	—	—	6,4	6,5	—	—	—
4,6,12-trimethyl-substituiert	—	—	—	4,8	4,6	—	—	—
nicht identifiziert	—	—	0,3	8,5	9,0	0,3	—	0,4

Tabelle 25. Zusammensetzung der Bürzelwachs-Alkohole einiger Corvidae-Arten (75, 112, 113)

Alkohole	Coloeus monedula	Corvus frugilegus	Corvus corax	Corvus corone corone	Corvus corone cornix	Pyrrhocorax graculus	Pica pica	Garrulus glandarius
unverzweigt	87,5	100,0	6,9	59,7	58,2	87,0	67,1	11,1
2-methyl-substituiert	12,5	—	88,6	—	—	9,1	12,4	64,7
4-, 6- und 8-methyl-substituiert	—	—	—	28,7	29,0	1,6	13,3	—
10-methyl-substituiert	—	—	—	10,6	11,7	0,2	—	—
12-methyl-substituiert	—	—	—	—	—	Spur	—	—
14-methyl-substituiert	—	—	—	Spur	Spur	1,4	2,0	—
16-methyl-substituiert	—	—	—	—	—	0,4	1,7	—
18-methyl-substituiert	—	—	—	—	—	—	0,4	—
2,6-, 2,8- 2,10-dimethyl-substituiert	—	—	3,6	—	—	—	2,7	12,9
2,12-dimethyl-substituiert	—	—	—	—	—	—	0,2	0,6
2,14-dimethyl-substituiert	—	—	—	—	—	—	—	4,1
2,16-dimethyl-substituiert	—	—	—	—	—	—	—	5,5
2,18-dimethyl-substituiert	—	—	—	—	—	—	—	0,5
nicht identifiziert	—	—	0,9	1,0	1,1	0,3	0,2	0,6

Tabelle 26. Zusammensetzung der Bürzelwachs-Fettsäuren einiger Sylviidae-Arten (74). (Verzweigungstypen in Klammern kommen nur in Spuren vor.)

Fettsäuren	Acrocephalus melanopogon	A. schoe-nebaenus	A. palu-stris	A. scir-paceus	A. arundi-naceus	Sylvia atricapilla	S. curruca	S. communis
unverzweigt	0,4	0,7	1,1	0,9	0,8	1,4	2,7	1,2
2-methyl-substituiert	5,9	8,2	1,6	3,7	—	10,6	45,2	1,5
2,6- (2,8-, 2,10-, 2,12-) dimethyl-substituiert	64,6	48,5	13,6	18,6	21,3	39,6	50,6	42,0
2,14-dimethyl-substituiert	—	—	7,4	1,0	0,4	0,8	—	—
2,16-dimethyl-substituiert	—	0,7	3,0	—	2,1	6,4	—	—
2,4,6-, 2,4,8- und 2,4,10-trimethyl-substituiert	—	—	—	—	—	—	—	1,0
2,6,10- (2,6,12-, 2,6,14-) trimethyl-substituiert	27,5	34,3	24,7	29,3	34,8	24,5	1,5	37,2
2,6,16-trimethyl-substituiert	1,1	3,9	22,1	11,0	8,8	7,1	—	10,2
2,6,18-trimethyl-substituiert	0,3	0,6	4,0	2,9	0,9	2,1	—	0,8
2,6,10,14-tetramethyl-substituiert	—	—	12,0	26,3	26,6	2,0	—	3,5
2,6,10,16-tetramethyl-substituiert	—	—	6,2	4,0	3,4	5,2	—	2,2
2,6,10,18-tetramethyl-substituiert	—	—	1,8	1,1	—	—	—	—
nicht identifiziert	0,2	3,1	2,5	1,2	0,9	0,3	—	0,4

Tabelle 27. Zusammensetzung der Bürzelwachs-Alkohole einiger Sylviidae-Arten (74). (Verzweigungstypen in Klammern kommen nur in Spuren vor.)

Alkohole	Acrocephalus melanopogon	A. schoe-nebaenus	A. palu-stris	A. scir-paceus	A. arundi-naceus	Sylvia atricapilla	S. curruca	S. communis
unverzweigt	70,8	12,8	16,5	25,6	22,8	17,4	8,9	12,7
2-methyl-substituiert	15,2	9,6	1,6	–	8,1	69,2	75,7	58,0
andere monomethyl-substituierte	6,5	5,1	16,9	19,0	17,6	2,3	0,5	1,0
2,6- (2,8-, 2,10-, 2,14-, 2,16-, 2,18-)dimethyl-substituiert	7,0	30,6	0,6	–	10,5	9,9	14,8	25,9
4,8-, 4,10- und 4,12-dimethyl-substituiert	–	0,7	39,0	32,8	6,3	–	–	–
andere dimethyl-substituierte (6,10-, 6,14-, 10,14-)	–	–	–	0,8	3,9	–	–	–
2,6,10-(2,6,8-, 2,6,12-, 2,6,14-, 2,6,16-)trimethyl-substituiert	–	40,8	1,3	18,8	24,6	–	–	–
4,8,12- und 4,8,14-trimethyl substituiert	–	–	21,9	2,5	–	–	–	–
2,6,10,14- und 2,6,10,16-tetramethyl-substituiert	–	–	–	–	4,9	–	–	–
nicht identifiziert	0,5	0,4	2,2	0,5	1,3	1,2	0,1	2,4

Tabelle 28. *Bürzelwachsbestandteile einiger Paridae-Arten (61)*

Wachsbestandteil	Parus ater	P. melano-lophus	P. ater/P. melanol.	P. maior	P. caeruleus	P. atricapillus montana	P. a. rhe-nanus	P. xantho-genys
Fettsäuren								
unverzweigt	5,9	6,5	1,3	2,1	1,7	0,8	0,2	1,0
2-äthyl-substituiert	43,9	75,2	60,3	68,7	49,2	56,3	70,0	74,2
2-äthyl-6-methyl-, -8-methyl-, -10-methyl-, -12-methyl-, -14-methyl-substituiert	48,8	14,1	34,4	28,6	46,3	40,3	28,5	24,3
2-äthyl-16-methyl-substituiert	0,3	0,2	0,4	—	0,5	0,6	0,4	Spur
2-äthyl-18-methyl-substituiert	—	0,2	0,6	—	—	—	—	0,1
2,6- (2,8-, 2,10-)dimethyl-substituiert	1,0	2,3	1,5	—	0,5	0,9	0,7	0,2
2-äthyl-6,14-, -8,14-, -8,16-, -10,14-dimethyl-substituiert	—	—	0,6	0,2	1,6	1,1	0,2	—
nicht identifiziert	0,1	1,5	0,9	0,4	0,2	—	—	0,2
Alkohole								
unverzweigt	83,3	95,7	90,5	96,6	91,7	84,6	89,5	80,4
4-, 6-, 8-monomethyl-substituiert	14,8	1,6	8,4	2,7	7,7	13,8	7,6	15,9
anteiso-monomethyl-substituiert	1,8	1,9	1,0	0,7	0,6	1,5	2,1	2,9
nicht identifiziert	0,1	0,8	0,1	—	—	0,1	0,8	0,8

16.3. Paridae (Meisen)

Chemotaxonomisch ist diese Familie innerhalb der *Passeriformes* klar abgegrenzt durch das Vorkommen von 2-äthyl- und 2-äthyl-x-methyl-substituierten Fettsäuren, die vorzugsweise mit n-Alkanolen verestert sind (*61*). Die quantitative Zusammensetzung ihrer Wachsbestandteile ist in der Tab. 28 wiedergegeben. Ein Vergleich mit den Bürzelsekreten von *Aegithalos caudatus* (Schwanzmeise) (*114*), *Remiz pendulinus* (Beutelmeise) (*114*) und *Panurus biarmicus* (Bartmeise) (*114*), die häufig zu den Meisen gezählt werden, zeigt jedoch klar, daß diese Arten eher bei den *Sylviidae* (vgl. 16.2) anzusiedeln sind (Tab. 29).

Tabelle 29. *Quantitative Zusammensetzung der Bürzelwachs-Bestandteile von Remiz pendulinus (Beutelmeise), Aegithalos caudatus (Schwanzmeise) und Panurus biarmicus (Bartmeise) (114)*

Wachsbestandteil	Remiz pendulinus	Aegithalos caudatus	Panurus biarmicus
Fettsäuren			
unverzweigt	0,9	14,2	0,3
2-methyl-substituiert	13,2	31,6	2,6
2,6-dimethyl-substituiert	49,0	45,1	42,3
2,12-dimethyl-substituiert	1,8	2,6	1,1
2,14-dimethyl-substituiert	3,5	2,3	6,4
2,6,10-trimethyl-substituiert	11,0	1,3	14,3
2,6,12-trimethyl-substituiert	7,8	1,3	13,4
2,6,14-trimethyl-substituiert	10,5	1,6	14,0
2,6,16-trimethyl-substituiert	2,3	—	4,6
2,6,18-trimethyl-substituiert	—	—	1,0
Alkohole			
unverzweigt	78,7	33,4	67,2
2-methyl-substituiert	—	60,3	14,1
4-,6-,8-,10-,12-methyl-substituiert	14,0	—	8,7
14-methyl-substituiert	6,5	—	6,8
16-methyl-substituiert	—	—	0,6
2,6-dimethyl-substituiert	0,8	3,8	2,6
2,14-dimethyl-substituiert	—	2,5	—

16.4. Fringillidae (Finken und Ammern)

Das Bürzeldrüsensekret dieser großen Vogelgruppe ist durch das Vorkommen von Esterwachsen charakterisiert, die aus 3-Methylfettsäuren und n- bzw. 3-Methylalkanolen aufgebaut sind. 11 Arten [*Emberiza schoeniclus* (Rohrammer), *E. bruniceps* (Braunkopfammer), *Fringilla coelebs* (Buchfink), *F. montifringilla* (Bergfink), *Serinus serinus* (Girlitz), *Carduelis spinus* (Zeisig), *C. chloris* (Grünling), *C. flammea*

Tabelle 30. Zusammensetzung der Bürzelwachsbestandteile einiger Fringillidae-Arten (29, 62, 69)

Wachskomponente	Emberiza schoeniclus	E. bruniceps	Fringilla coelebs	F. montifringilla	Serinus serinus	Carduelis spinus	C. chloris	C. flammea	C. cannabina	Loxia curvirostris	Pyrrhula p.	Passer domesticus	Passer montanus
Fettsäuren													
unverzweigt	—	—	8,4	Spur	—	2,4	—	1,6	—	0,8	—	—	—
3-methyl-substituiert	89,4	87,3	88,6	94,2	96,6	88,5	100,0	96,3	100,0	97,3	100,0	100,0	76,4
3,x-dimethyl-subst.	10,3	12,7	3,0	5,4	1,4	9,1	—	2,1	—	1,9	—	—	17,7
nicht identifiziert	0,3	—	—	0,4	2,0	—	—	—	—	—	—	—	5,9
Alkohole													
unverzweigt	51,1	47,7	67,8	70,6	70,2	50,8	61,7	81,9	81,0	78,4	34,0	100,0	78,4
3-methyl-substituiert	46,9	52,3	28,0	26,8	29,8	48,6	38,3	16,8	19,0	20,7	66,0	—	19,5
nicht identifiziert	2,0	—	4,2	2,6	—	0,6	—	1,3	—	0,9	—	—	2,1

(Birkenzeisig), *C. cannabina* (Hänfling), *Loxia curvirostris* (Fichten-kreuzschnabel), *Pyrrhula pyrrhula* (Gimpel)] wurden untersucht (*62, 69*), jedoch müssen zwei, gewöhnlich zu den *Ploceidae* gezählte Arten, nämlich *Passer montanus* (Feldsperling) (*69*) und *Passer domesticus* (Haussperling) (*29*) aufgrund ihrer Wachszusammensetzung zu den *Fringillidae* gestellt werden (Tab. 30).

Bei einigen Arten dieser Familie wurde das Vorkommen von Alkan-1,2-diolen [Finken und Ammern (*83*), *Zonotrichia leucophrys* (*115*)] mit Kettenlängen zwischen C_{14} und C_{18} nachgewiesen.

16.5. Ploceidae (Webervögel)

Chemotaxonomisch deutlich separiert von den *Fringillidae* durch die Struktur ihrer Bürzelwachse sind die *Ploceidae*. Die Monoester-wachse sind, soweit bisher bekannt, einheitliche Wachse, d. h. praktisch ausschließlich aus einer Fettsäure und einem Alkohol aufgebaut; dies erlaubt eine direkte massenspektrometrische Strukturaufklärung ohne vorherige Spaltung der Wachse in ihre Bestandteile. Bei den Fettsäuren wurden 2,4-dimethyl- und 2,4,6- bzw. 2,4,8-trimethyl-substituierte Indi-viduen nachgewiesen. Unter den Alkoholen dominieren n-Alkanole (Tab. 31). Die von zahlreichen Zoologen zu den *Ploceidae* gestellten *Passer*-Arten zeigen von diesen abweichende Bürzelwachszusammen-setzungen und wurden daher unter den *Fringillidae* erwähnt (vgl. 16.4.).

Tabelle 31. *Zusammensetzung der Bürzelwachse einiger Ploceus-Arten (30)*

Wachskomponente	*Ploceus cucullatus*	*Ploceus subaureus*	*Ploceus galbula*	*Ploceus spec.*	*Quelea quelea*
Fettsäuren					
2,4-Dimethylheptansäure	100,0	100,0	100,0	95,9	0,4
2,4-Dimethyloctansäure	—	—	—	1,0	Spur
2,4-Dimethylnonansäure	—	—	—	3,1	—
2,4,6-Trimethyloctansäure	—	—	—	—	0,6
2,4,6-Trimethylnonansäure	—	—	—	—	90,9
2,4,8-Trimethylundecansäure	—	—	—	—	8,1
Alkohole					
unverzweigt	100,0	100,0	100,0	100,0	98,9
4-methyl-verzweigt	—	—	—	—	0,3
12-methyl-verzweigt	—	—	—	—	0,2
14-methyl-verzweigt	—	—	—	—	0,6

16.6. Certhiidae (Baumläufer) und andere

Neben den erwähnten *Passeriformes*-Arten wurden einige andere Familien untersucht, so z. B. *Regulus regulus Santae Mariae* (Santa-Maria-

Tabelle 32. Zusammensetzung der Bürzelwachse einiger Certhiidae-, Paradoxornithidae- und Timaliidae-Arten (114) sowie einer Regulus-Art

Wachskomponente	Certhia familiaris	C. brachy-dactyla	Leiothrix lutea	Mesia ar-gentauris	Paradoxor-nis gularis	P. webbi-anus	Regulus regulus	Siva cyano-uroptera
Fettsäuren								
unverzweigt	5,1	3,8	0,5	47,1	2,2	—	1,3	Spur
2-methyl-substituiert	33,2	75,2	73,5	51,5	12,4	1,1	26,8	16,0
2,6-dimethyl-substituiert	6,1	1,1	21,8	1,4	41,7	38,3	68,3	48,2
2,10-dimethyl-substituiert	20,1	14,0	—	—	1,3	2,0	—	4,5
2,12-dimethyl-substituiert	17,0	4,4	—	—	—	—	—	1,5
2,14-dimethyl-substituiert	10,4	0,6	—	—	1,0	—	0,2	—
2,18-dimethyl-substituiert	0,4	—	—	—	—	—	—	—
2,6,10-trimethyl-substituiert	0,4	0,3	2,2	—	13,9	8,9	0,9	16,7
2,6,12-trimethyl-substituiert	0,5	—	1,7	—	9,9	12,0	0,8	10,7
2,6,14-trimethyl-substituiert	4,3	0,6	0,3	—	4,3	12,1	0,9	1,0
2,6,16-trimethyl-substituiert	—	—	—	—	8,2	15,0	0,8	1,4
2,14,16-trimethyl-substituiert	2,5	—	—	—	—	—	—	—
2,6,10,14-tetramethyl-substituiert	—	—	—	—	5,1	10,6	—	—
Alkohole								
unverzweigt	67,7	58,9	3,2	20,7	2,3	57,2	64,9	13,2
2-methyl-substituiert	5,1	37,5	89,9	79,3	74,4	8,0	20,4	71,1
andere monomethyl-substituierte	21,5	—	—	—	—	28,9	14,7	3,2
dimethyl-substituiert	5,7	3,6	6,9	—	21,2	5,9	—	12,5
trimethyl-substituiert	—	—	—	—	2,1	—	—	—

Tabelle 33. *Zusammensetzung der Triglyceridfettsäuren des Depotfettes einiger Arten mit unterschiedlichen Nahrungsgewohnheiten*

Triglycerid-fettsäure	*Fratercula arctica* (Papageien-taucher) (*23*)	*Ardea cinerea* (Graureiher) (*90*)	*Anser anser* (Graugans) (*73*)	*Phasianus colchicus* (Fasan) (*38*)	*Columba palumbus* (Ringeltaube) (*39*)	*Pyrrhula pyrrhula* (Gimpel) (*62*)	*Passer domesticus* (Haussperling) (*29*)
12:0	—	0,1	—	—	—	0,2	0,2
14:0	5,0	2,1	0,4	0,8	1,2	0,2	1,5
16:0	25,6	24,3	18,2	28,5	18,3	12,7	33,1
16:1	6,9	8,3	3,9	9,0	7,4	1,4	6,8
17:0	0,1	0,3	0,3	Spur	Spur	Spur	Spur
18:0	8,6	7,2	5,8	6,7	2,8	5,1	6,1
18:1	22,5	28,1	57,5	48,1	44,5	21,3	33,1
18:2	1,0	5,3	13,0	6,3	25,3	58,5	17,8
18:3	2,6		0,9	0,6	0,5	Spur	1,0
20:polyen	15,6	3,2	—	—	—	—	—
22:polyen	11,7	10,2	—	—	—	—	—
andere	0,4	3,6	—	—	—	0,6	0,4

Literaturverzeichnis: SS. 431—438

Azoren-Goldhähnchen) aus der Familie *Regulidae* (Goldhähnchen)
(114), *Certhia familiaris* (Waldbaumläufer) *(114)* und *Certhia brachy-
dactyla* (Gartenbaumläufer) *(114)* aus der Familie *Certhiidae* (Baumläufer),
Paradoxornis gularis und *Paradoxornis webbianus* aus der Familie
Paradoxornithidae (114), sowie die Arten *Mesia argentauris* (Silberohr-
sonnenvogel), *Siva cyanouroptera* und *Leiothrix lutea* (Chinasonnenvogel)
aus der Familie *Timaliidae*. Alle Arten zeigen die für die *Sylviidae*
charakteristischen Bürzelwachsmuster, 2-, 2,x-, 2,x,y- und 2,x,y,z-
methyl-substituierte Fettsäuren, die mit n- bzw. Monomethylalkanolen
verestert sind (Tab. 32).

IV. Andere Vogellipide

Neben den erwähnten Bürzeldrüsensekreten, an deren Aufbau sich,
ähnlich wie bei Talgdrüsensekreten anderer Tierarten, ungewöhnliche
Fettsäuren beteiligen, wurden auch andere Vogellipide wie die Depotfette zahlreicher Arten und die Magenöle einiger *Procellariiformes*-
Arten untersucht.

1. Depotfette

Alle bislang analysierten Depotfette aus der Klasse Aves sind Triglyceride mit gesättigten und ungesättigten Fettsäuren der Kettenlänge $C_{12}-C_{22}$, deren Zusammensetzung von der eingenommenen
Nahrung abhängig ist. So ist der Anteil an Linol- und Linolensäure
(18:2 und 18:3) bei körnerfressenden Arten, der Anteil an Polyen-
säuren der Kettenlänge C_{20} und C_{22} bei fischfressenden Arten erhöht,
während insekten- und aasfressende Arten einen höheren Anteil an
Palmitin- und Ölsäure (16:0 und 18:1) aufweisen. Die Tab. 33 gibt
einen Überblick über Arten mit unterschiedlichen Nahrungsgewohnheiten (Fischnahrung: Papageientaucher, Graureiher; reine Pflanzen-
kost: Graugans; Pflanzen- und Körnernahrung: Fasan; Körnerkost:
Ringeltaube, Gimpel; gemischte Kost: Haussperling).

Triglyceride wurden auch als regelmäßige Bestandteile in Lipidex-
trakten von Bürzeldrüsengewebe gefunden; ihre Zusammensetzungen gleichen oder ähneln denjenigen der Depotfette. Neben diesen, als Reserve-
lipid zu bezeichnenden Triglyceriden kommen bei einigen Arten deutlich anders zusammengesetzte Triglyceride als Bestandteile der Bürzeldrüsensekrete vor *(Fratercula, Ardea)*.

Außerdem wurden die Depotfette folgender Arten untersucht (Tab. 34).

Tabelle 34. *Bislang untersuchte Depotfette verschiedener Vogelarten*

Art	Literatur
Fulmarus glacialis (Eissturmvogel)	(116)
Sula bassana (Baßtölpel)	(116)
Megalestris catarrhactes (Raubmöwe)	(116)
Dromaius novaehollandiae (Emu)	(117)
Phoenicopterus chilensis (Flamingo)	(118)
Larus argentatus (Silbermöwe)	(116)
Fratercula arctica (Papageientaucher)	(23)
Struthio camelus (Strauß)	(118)
Ardea cinerea (Graureiher)	(90)
Anser anser (Graugans)	(73)
Anser fabalis (Saatgans)	(73)
Anser caerulescens (Schneegans)	(73)
Anser indicus (Streifengans)	(73)
Aythya ferina (Tafelente)	(73)
Mergus albellus (Zwergsänger)	(73)
Anas clypeata (Löffelente)	(73)
Anas strepera (Schnatterente)	(73)
Apterix australis (Streifenkiwi)	(119)
Phasianus colchicus (Fasan)	(38)
Perdix perdix (Rebhuhn)	(24)
Gallus domesticus (Haushuhn)	(120)
Columba palumbus (Ringeltaube)	(39)
Pyrrhula pyrrhula (Gimpel)	(62)
Passer domesticus (Haussperling)	(29)

2. Magenöle

Arten der Ordnung *Procellariiformes* (Röhrennasenvögel) besitzen stark riechende Magenöle, die sie bei Gefahr Feinden entgegenspeien, die sie aber vermutlich auch an ihre Jungen verfüttern. Hierbei handelt es sich nicht um Magensekrete, wie vielfach angenommen wurde, sondern um nicht oder nur partiell abgebaute Nahrungsrückstände (*121, 122, 123, 124, 125*).

So wurden im Magenöl von *Puffinus griseus* (Dunkler Sturmtaucher) Esterwachse (*126, 127, 128, 129*), bei *Oceanodroma leucorhoa* (Wellenläufer) Diglyceridäther (*124, 125*) aufgefunden, die zugleich auch Bestandteile der Nahrung dieser Art sind (Anchovies, Krabben, Zooplankton). Die Magenöle von *Puffinus pacificus* (Keilschwanzsturmtaucher) und *Pterodroma macroptera* (Langflügelsturmvogel) zeigen einen hohen Gehalt an Cholesterinestern, Triglyceriden und freiem Cholesterin (*123*), während das Magenöl von *Puffinus tenuirostris*

(Kurzschwanzsturmtaucher) zu 80% aus Esterwachsen, 10% Triglyceriden und 10% Mono- und Diglyceriden besteht (*130*). Wie sehr die Zusammensetzung der Magenöle von der aufgenommenen Nahrung abhängt, zeigen die unterschiedlichen Analysenergebnisse der Magenöl-Lipide von *Fulmarus glacialis* (Eissturmvogel), einer weltweit verbreiteten *Procellariiformes*-Art. ROSENHEIM und WEBSTER (*131*) fanden Esterwachse, während CHEAH und HANSEN (*130*) Triglyceride als Hauptlipide des Magenöls bei dieser Art nachweisen konnten.

V. Biosynthese

Das umfangreiche Analysenmaterial über die Zusammensetzungen von Bürzeldrüsensekreten läßt erkennen, daß ein breites Spektrum von verzweigten Fettsäuren und Alkoholen sowie Hydroxysäuren und Diole, die teilweise erstmalig an dieser Stelle in der Natur aufgefunden wurden, in der Bürzeldrüse des Vogels synthetisiert wird. Dennoch sind hinsichtlich der Biosynthese der Wachskomponenten viele Fragen offen. Bisher wurden keine Enzyme in reiner Form isoliert.

Es wird allgemein angenommen, obgleich nicht bewiesen, daß die auch in Bürzeldrüsensekreten weit verbreiteten n-Fettsäuren durch einen Multienzymkomplex biosynthetisiert werden, wie er etwa aus der Hefe (*132, 133*) oder der Taubenleber (*134, 135*) isoliert wurde und der die Kondensation von Malonyl-CoA zu mittelkettigen Fettsäuren mit Acetyl-CoA als Starter gemäß:

$$CH_3CO - SCoA + n\,HOOC - CH_2 - CO - SCoA + 2n\,NADPH + 2n\,H^+$$
$$\rightarrow CH_3(CH_2 - CH_2)_n - CO - SCoA + n\,CO_2 + n\,HSCoA + 2n\,NADP^+ +$$
$$+ n\,H_2O \; (n = 7,8)$$

katalysiert.

Die Kettenlängenverteilungen der n-Fettsäuren aus Bürzeldrüsensekreten, die von denen anderer Organlipide abweichen, lassen sich durch das variable Substratverhältnis Acetyl-CoA/Malonyl-CoA erklären (*136*).

Daß die Biosynthese von 2,4,6,8-tetramethyl-substituierten Fettsäuren gemäß obiger Gleichung mit Propionyl-CoA als Substrat verläuft, konnte durch Injektion von Propionat-[3-^{14}C] in die Bürzeldrüse von *Anser anser* gezeigt werden, wobei eine hohe Einbaurate in die 2,4,6,8-Tetramethyl-decansäure und -undecansäure beobachtet wurde (*137*). Es ist zu vermuten, daß die Biosynthese dieser Säuren via Methylmalonyl-CoA abläuft. Im Gegensatz dazu konnte eine Methylübertragung aus L-Methionin-[^{14}C] ausgeschlossen werden.

Ein kritischer Blick auf die Strukturen der Bürzelwachsfettsäuren läßt jedoch erkennen, daß zumindest bei manchen Arten Synthetase-Systeme vorhanden sein müssen, die sowohl Malonyl- als auch Methyl-malonyl-CoA als Substrat akzeptieren. Hierfür spricht, daß z. B. bei *Anseriformes*-Arten neben 2-methyl- auch 4- bzw. 6-methyl-substituierte Fettsäuren und andererseits auch polymethyl-verzweigte Fettsäuren vorkommen, deren Methylsubstituenten (an geraden C-Atomen) statistisch über das Molekül verteilt sind (z. B. 2,4,6-, 2,4,8-, 2,6,8-, 2,6,10- und 2,6,12-trimethyl-verzweigte Fettsäuren nebeneinander). Es wurden allerdings bislang keine Fettsäuren mit mehr als vier Methylverzweigungen sicher nachgewiesen.

Abweichend von der statistischen Verteilung der Methylgruppen findet man bei einigen Ordnungen Wachsfettsäuren, bei denen jedes 4. C-Atom eine Methylverzweigung trägt (z. B. 2,6-, 2,6,10- und 2,6,10,14-). Solche Säuren dominieren bei *Ralliformes, Sylviidae, Lariformes* und *Charadriiformes*. Für diese Arten liegen keine Untersuchungen bezüglich der Fettsäure-Biosynthese vor.

Für Arten, bei denen ausschließlich 2-methyl-substituierte Wachssäuren nachgewiesen wurden [z. B. *Corvus frugilegus (112)*], wurde vermutet, daß die Kondensation mit Methylmalonyl-CoA der finale Schritt der Biosynthese ist, der die Ablösung von der Enzymoberfläche zugleich bewirkt. Experimentelle Beweise für diese Hypothese liegen jedoch nicht vor.

2- und 4-methyl-substituierte Fettsäuren kommen in Bürzeldrüsensekreten häufig nebeneinander vor [*Alcidae (63)*, *Anseriformes (73)*], und sicherlich nahe verwandte Arten unterscheiden sich chemotaxonomisch gelegentlich lediglich dadurch, daß die Wachskomponenten der einen Art 2-, die der anderen 4-methyl-substituiert sind [z. B. die Fettsäuren bei den *Ralliformes (70)* oder die Alkohole bei den *Sylviidae (74)*]. Dies legt die Vermutung für eine sehr ähnliche Biosynthese nahe; so wurde vermutet, daß 4-Methylfettsäuren durch C_2-Verlängerung mit Malonyl-CoA aus 2-Methylfettsäuren entstehen (*63, 74, 75, 113*).

Über die Biosynthese der in Bürzeldrüsensekreten weit verbreiteten 3-Methylfettsäuren *(Passeriformes, Cuculiformes, Piciformes, Sphenisciformes, Procellariiformes, Tyto alba)* ist nichts bekannt. Ebenfalls ungeklärt ist die Biosynthese von 2-alkyl-substituierten Fettsäuren mit Substituenten, die mehr als ein C-Atom enthalten (z. B. 2-äthyl-, 2-propyl-, 2-butyl-verzweigt), wie sie in den Bürzeldrüsensekreten von *Eulen (33)*, *Meisen (61)* und in Spuren auch in denen von *Pinguinen (65)* nachgewiesen wurden. Obgleich auch hier experimentelle Beweise ausstehen, wurde vermutet, daß Butyryl-, Valeroyl- und Caproyl-CoA möglicherweise über die entsprechenden Alkyl-Malonyl-CoA-Verbindungen (Äthyl-, Propyl-, Butylmalonyl-CoA) als Substrate vorkommen

Literaturverzeichnis: SS. 431—438

(33). Nicht bekannt ist ferner die Biosynthese von 3-Hydroxyfettsäuren, die in den Sekreten der Ringeltaube *(Columba palumbus)* *(39, 40)* vorkommen, sowie derjenigen der weit verbreiteten Alkylhydroxymalonsäuren *(41)*.

Nicht untersucht ist die Biosynthese der Wachsalkohole. Bei einigen Arten zeigt sich eine deutliche Strukturverwandtschaft zwischen den Wachsfettsäuren und den Wachsalkoholen, z. B. bei *Alcidae (23, 63)*, *Ardeidae (90)*, *Charadriiformes (71)*, einigen *Corvidae*-Arten *(75, 113)*, *Cuculiformes (68)*, *Emberizidae (62, 69)*, *Falconiformes (104)*, *Fringillidae (62, 69)*, *Lariformes (106, 107, 108)*, *Sylviidae (74, 114)*, *Strigiformes (33)*. Bei diesen Arten kann man eine Reduktion der Fettsäuren zu den korrespondierenden Alkoholen vermuten. Andere Arten zeigen dagegen von den Fettsäuren klar unterschiedene Alkoholmuster, wie z. B. *Anseriformes (47, 73, 99)*, *Paridae (61)*, bestimmte *Corvidae*-Arten *(112)*, *Ploceidae (30)*, *Procellariiformes (66, 67)*, *Psittaciformes (72)* und *Ralliformes (46, 70)*.

Besser untersucht ist die Biosynthese der Diole. Die Biogenese von Alkan-2,3-diolen, die als alkoholische Komponenten der Diesterwachse in den Bürzeldrüsensekreten von *Galliformes*-Arten vorkommen, wurde von mehreren Autoren untersucht *(35, 80, 138)*. TANG und HANSEN *(80)* haben den Einbau von Acetat-[^{14}C], Pyruvat-[^{14}C] und Glucose-[^{14}C] in die Fettsäuren und Diole des Bürzeldrüsensekretes von Hühnern nach Injektion in die Bürzeldrüse beschrieben. Propionat wurde in die ungeradzahligen Komponenten eingebaut. Die *in vitro* Fettsäuresynthese aus Acetat und Propionat durch Zentrifugationsüberstände aus Bürzeldrüsenhomogenaten lieferte die gleichen Fettsäuremuster wie *in vivo*.

SAWAYA und KOLATTUKUDY *(35, 139)* untersuchten den Einbau von Acetat-[1-^{14}C], Palmitinsäure-[1-^{14}C] und Stearinsäure-[1-^{14}C] in das Bürzeldrüsensekret des Fasans *(Phasianus colchicus)* nach Injektion unter die Bürzeldrüse. Acetat wurde in annähernd gleicher Menge in die Fettsäuren und Diole eingebaut. Das C-1 der Palmitinsäure, die ebenfalls in beide Wachskomponenten inkorporiert wurde, konnte als C-3 in der Hauptkomponente der Diole (Octadecan-2,3-diol) wiedergefunden werden. Analog hierzu liefert Stearinsäure-[1-^{14}C] hauptsächlich ein in 3-Position markiertes Eicosan-2,3-diol. Diese Versuche beweisen, daß die Diole durch C_2-Verlängerung aus Fettsäuren entstehen können. Der Mechanismus dieser Elongation ist gegenwärtig noch ungeklärt. Versuche, das C-1 oder C-2 der Diole durch Angebot von Pyruvat-[2-^{14}C], Alanin-[U-^{14}C] oder Lactat-[2-^{14}C] zu markieren, schlugen fehl.

Die Biosynthese von Alkan-1,2-diolen verläuft wahrscheinlich über die Hydrierung von 2-Hydroxyfettsäuren. Ein 16.000-g-Überstand aus dem Bürzeldrüsen-Homogenat von *Zonotrichia leucophrys* (Dachsammerfink) katalysiert die Reduktion von 2-Hydroxyfettsäuren zu 1,2-Diolen in Anwesenheit von ATP, CoASH, NADH und NADPH (*115*).

VI. Physiologische Funktion

Es wird allgemein angenommen, daß eine Hauptfunktion des Bürzeldrüsensekretes, das der Vogel mit dem Schnabel gleichmäßig über sein Gefieder verteilt, in der Reinigung und Imprägnierung der Federn besteht. Gleichzeitig wurde über eine Verbesserung der Schwimmlage bei Wasservögeln berichtet (*6, 140*). Dies setzt einen niedrigen Schmelzpunkt, Wasserunlöslichkeit, hohes Spreitungsvermögen gegenüber Wasser und Beständigkeit gegenüber Oxidation durch Luftsauerstoff voraus. Alle diese Voraussetzungen sind in der Tat bei Bürzeldrüsensekreten erfüllt. Dennoch ist schwer zu verstehen, weshalb die Sekrete einerseits so ungeheuer kompliziert zusammengesetzt sind – einfachere Kompositionen würden den gleichen Zweck einfacher erfüllen, und tatsächlich werden ja auch solche einfach zusammengesetzte Sekrete z. B. bei *Fratercula arctica* und *Ardea cinerea* sowie bei zahlreichen *Anseriformes*-Arten gebildet – und weshalb andererseits die Zusammensetzungen einen so hohen Grad von Ordnungs-, Familien- oder sogar Artspezifität aufweisen. Es liegt daher nahe, nach anderen Funktionen des Bürzeldrüsensekretes zu suchen, wobei man am ehesten noch an Wehrfunktionen denken könnte. Tatsächlich haben alkylsubstituierte Fettsäuren bakterizide und fungizide Eigenschaften (*141– 147*). Über die Abhängigkeit der antibakteriellen und hämolytischen Wirkung von der Struktur verschiedener Fettsäuren wurde berichtet (*148, 149*). Die im Bürzeldrüsensekret der Ringeltaube *(Columba palumbus)* (*39, 40*) nachgewiesenen 3-Hydroxyfettsäuren wurden auch in freier Form in Ernte- und Blattschneiderameisen beobachtet, denen sie wegen ihrer fungiziden Eigenschaften als Wachstumsregulatoren bzw. Herbizide dienen (hauptsächlich D-β-Hydroxydecansäure = Myrmicacin) (*110*). Auch einige alkyl-substituierte Fettsäuren haben neben bakteriziden antimykotische Wirkung (*146*).

Von PUGH wurde auf die Wechselwirkung zwischen Federlipiden und Dermatophyten, die offenbar sehr artspezifisch sind, hingewiesen (*150, 151, 152*). Hierbei zeigte sich allerdings bei manchen Arten eine wachstumsfördernde Wirkung der Gefiederlipide gegenüber den Dermatophyten.

Gegenwärtig liegen keine experimentellen Untersuchungen über die Wirkung von Bürzeldrüsensekreten auf Mallophagen (Federlinge) vor, die im hohen Maße wirtspezifisch sind (*153, 154*).

Literaturverzeichnis: SS. 431—438

Von Hou (*155, 156, 157*) wurde auf die Synthese von Provitamin D in der Bürzeldrüse hingewiesen, das unter Lichteinwirkung auf das Gefieder in Vitamin D umgewandelt werden soll. Rachitische Hühner normalisierten sich unter UV-Bestrahlung, wenn eine Bürzeldrüse vorhanden war, drüsenektomierte Tiere zeigten dagegen keine positiven Befunde. Andererseits führt die Exstirpation der Bürzeldrüse bei verschiedenen Arten, wie z. B. Tauben (*6, 158, 159*), Hühnern (*6, 160*), Enten (*6*) und *Passeriformes*-Arten (*6, 161*) zu keinen nachteiligen Konsequenzen für die Tiere. Tauben scheinen jedoch für diese Experimente wenig geeignete Arten zu sein, da die Lipidversorgung des Gefieders nur unwesentlich durch die Bürzeldrüse als vielmehr durch die Haut direkt erfolgt (*40*).

VII. Chemotaxonomie und Systematik

Viele Einzelergebnisse haben gezeigt, daß die Kompositionen von Bürzeldrüsensekreten signifikant von Ordnung zu Ordnung variieren; d. h. Arten, die zu unterschiedlichen Ordnungen gehören, unterscheiden sich in der qualitativen Zusammensetzung ihrer Bürzeldrüsensekrete, Arten der gleichen Ordnung unterscheiden sich lediglich in quantitativer Hinsicht. Damit kann das Bürzelwachs in gewissen Grenzen als ein chemotaxonomischer Parameter für die Bestimmung der Verwandtschaft von Arten benutzt werden, zumal es in vielen Fällen bereits eine Differenzierung zwischen einzelnen Vogelfamilien erlaubt. Einzelne Arten und Rassen lassen sich dagegen nicht immer sicher unterscheiden. Obgleich das Bürzelwachsmuster nur eines von vielen genetisch fixierten Merkmalen und dadurch mit anderen z. B. anatomischen, morphologischen oder ethologischen Merkmalen verglichen werden kann, ist die hohe Konstanz dieses Parameters und seine exakte Bestimmbarkeit durch die GLC/MS-Technik der wesentliche Vorteil und Nutzen für systematische Untersuchungen. Eine Einschränkung findet das Verfahren durch die Tatsache, daß die quantitative Zusammensetzung innerhalb ein und derselben Art in Grenzen schwanken kann (was aber auch für andere z. B. anatomische Messungen zutrifft). Daneben ließ sich für Hühner auch eine Abweichung der quantitativen Zusammensetzung der Wachskomponenten zwischen Kücken und ausgewachsenen Tieren feststellen, d. h. für chemotaxonomische Untersuchungen sollen adulte Tiere herangezogen werden (*162*). Trotz dieser Einschränkung erbrachten chemotaxonomische Untersuchungen eine Reihe von in systematischer Hinsicht interessanten Ergebnissen:

a) Es besteht eine nahe Verwandtschaft zwischen *Sphenisciformes* (Pinguine) und *Procellariiformes* (Röhrennasenvögel) (*65, 66, 67*).

Innerhalb der Procellariiformes nimmt *Diomedea* eine Sonderstellung ein und verbindet diese Ordnung möglicherweise mit den *Larolimicolae* oder einer anderen Vogelgruppe.

b) die Ordnung *Ciconiiformes* (Stelzvögel) ist als heterogene Sammelordnung aufzufassen. *Ardea* (Reiher) und *Nycticorax* (Nachtreiher) zeigen keine klare chemotaxonomische Verwandtschaft und setzen sich beide deutlich von *Scopus* (Hammerkopf) und *Threskiornis* (Ibis) ab (*90, 91, 92*).

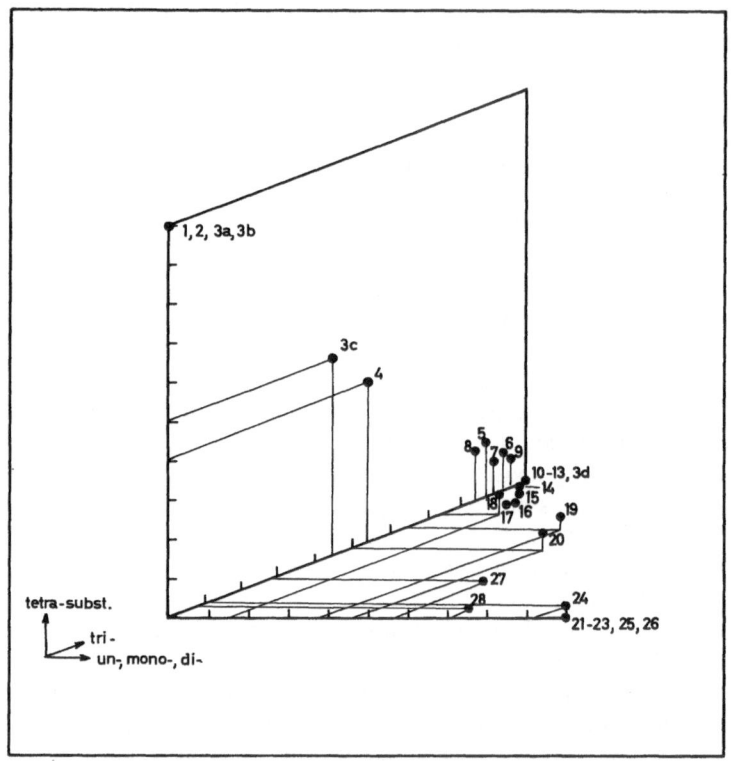

Abb. 11. Chemotaxonomische Verwandtschaft einiger *Anseriformes*-Arten, basierend auf dem Verzweigungsgrad der Fettsäuren des Bürzeldrüsensekretes.

1 Anser anser, 2 Anser anser f. domesticus, 3 Anser caerulescens, 3a Cygnus cygnus, 3b Cygnus columbianus, 3c Cygnus melanocoryphus, 3d Coscoroba coscoroba, 4 Dendrocygna viduata, 5 Cygnus olor, 6 Anser fabalis, 7 Anser indicus, 8 Aythya ferina, 9 Tadorna (Casarca) tadornoides, 10 Cereopsis novaehollandiae, 11 Branta leucopsis, 12 Cairina moschata, 13 Mergus serrator, 14 Mergus albellus, 15 Tadorna tadorna, 16 Tadorna ferruginea, 17 Melanitta nigra, 18 Stictonetta naevosa, 19 Aythya fuligula, 20 Somateria mollissima, 21 Chenonetta jubata, 22 Anas strepera, 23 Anas clypeata, 24 Cygnus atratus, 25 Anas platyrhynchos f. domesticus, 26 Anas platyrhynchos, 27 Tachyeres patachonicus, 28 Tachyeres pteneres

Literaturverzeichnis: SS. 431—438

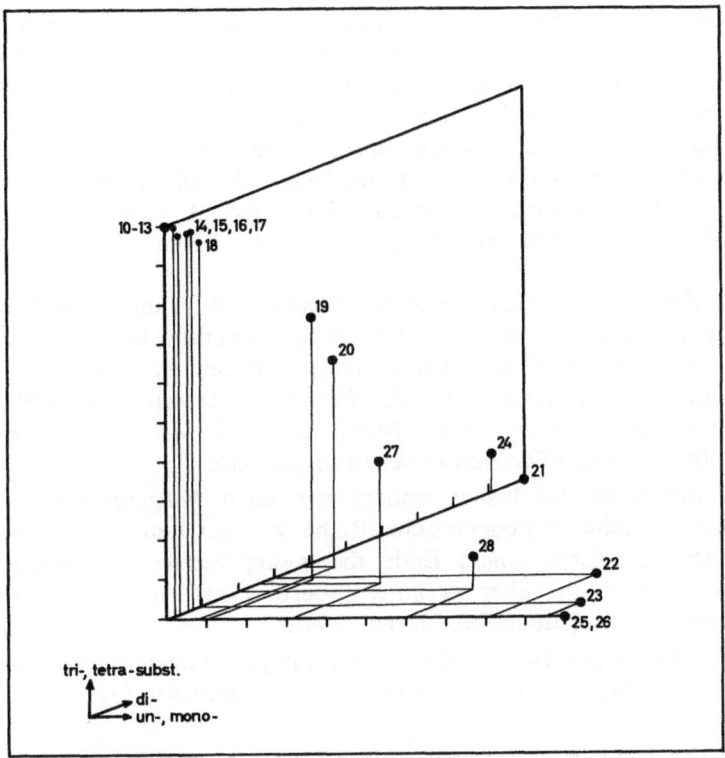

Abb. 12. Chemotaxonomische Verwandtschaft einiger *Anseriformes*-Arten, basierend auf dem Verzweigungsgrad der Fettsäuren des Bürzeldrüsensekretes. (Nummern bedeuten die gleichen Arten wie in Abb. 11. Arten 1–9 weggelassen.)

c) *Phoenicopterus* (Flamingo) (*78*) scheint eher mit den *Anseriformes* (Gänseartige) als mit den *Ciconiiformes* (Stelzvögel) verwandt zu sein.

d) Die Ordnung *Anseriformes* (Gänseartige) zeigt eine klare chemotaxonomische Gliederung, die sich graphisch im n-dimensionalen Raum darstellen läßt. Hierbei werden die verschiedenen Parameter (Fettsäuren und Alkohole mit unterschiedlich hohem Verzweigungsgrad) quantitativ auf Raumachsen aufgetragen. Die Bürzelwachs-Kompositionen erscheinen hierbei als Punkt. Punktwolken fassen ähnliche Arten zusammen. Der besseren Anschaulichkeit und Darstellbarkeit halber beschränkt man sich auf einen dreidimensionalen Raum, wobei dann mehrere Parameter zusammengelegt und eine zunehmende Ungenauigkeit der Aussage in Kauf genommen werden müssen. Die Abb. 11 und 12 geben die Ergebnisse aus der Ordnung *Anseriformes* wieder (*73*).

Beide Abbildungen berücksichtigen nur die Wachsfettsäuren. Während in Abb. 11 für tetra- und tri-substituierte Fettsäuren je eine Achse zur Verfügung steht, sind un-, mono- und di-substituierte Säuren gemeinsam auf einer Achse untergebracht. Eine bessere Einsicht in die Verwandtschaft der Entenvögel (*Anas, Aythya, Somateria, Cairina* usw.) und Säger *(Mergus)* ermöglicht die Abb. 12, bei der tetra- und trisubstituierte Fettsäuren auf einer Achse vereint, dafür aber die di- und die Summe von mono- und un-substiuierten Säuren gestreckt sind.

Hierbei erweisen sich Gänse und Schwäne aufgrund des hohen Verzweigungsgrades als untereinander nahe verwandt. Einige Entenarten (z. B. *Melanitta, Cairina, Branta* usw.) scheinen eng mit den Sägern *(Mergus)* verwandt zu sein. Die *Anas*-Arten setzen sich deutlich als gering verzweigte Arten von den vorgenannten ab. Mit ihnen scheinen die Dampfschiffenten *(Tachyeres)* verwandt zu sein.

e) Innerhalb der bisher isoliert stehenden *Falconiformes* (Greife) ist eine ähnliche phylogenetische Reihe wie bei den *Anseriformes* zu erkennen, an deren einem Ende die wenig verzweigten *Falco*-Arten (Falken) stehen, die über *Accipiter* (Sperber) zu den stärker verzweigten *Buteo*-Arten (Bussarde) führen (*104*).

f) *Galliformes* (Hühner) sind eine homogene Vogelgruppe, die bisher zu keiner anderen Ordnung Verbindungen aufweist (*22, 34, 35, 38, 81, 82*).

g) Die wenigen untersuchten *Ralliformes*-Arten (Rallen) zeigen gewisse Ähnlichkeiten mit den *Sylviidae,* jedoch sollte es sich hier um konvergente Entwicklungen bei der sonst erheblichen Verschiedenheit dieser Ordnungen handeln (*46, 70*).

h) Die Ordnung *Charadriiformes* (Regenpfeifer und Schnepfen) zeigt nahe Verwandtschaft mit der Ordnung *Lariformes* (Möwenartige). Faßt man beide Ordnungen zusammen, läßt sich wiederum eine Reihe abgestufter Verwandtschaften aufstellen, an deren einem Ende die wenig verzweigten *Alcidae* (Alken) stehen, die über *Cepphus grylle* (Gryllteiste) mit den *Laridae* (Möwen) verbunden sind; bei den *Charadriiformes* ist schließlich der Anteil an unverzweigten Wachsen vollständig durch solche mit mono-, di- und trisubstituierten Komponenten ersetzt (*23, 63, 71, 79, 106, 107, 108*).

i) *Columbiformes* stehen isoliert. Das Bürzeldrüsensekret macht nur einen geringen Prozentsatz der Gefiederlipide aus (*39, 40*).

j) *Psittaciformes* (Papageien) scheint eine klar abgegrenzte Ordnung zu sein (*72*).

k) Innerhalb der *Strigiformes* (Eulen), die sich wiederum deutlich von anderen Ordnungen separieren, bildet *Tyto alba* (Schleiereule) eine

Ausnahme. Sie zeigt Ähnlichkeit mit zahlreichen anderen Arten aus verschiedenen Ordnungen *(Cuculiformes, Piciformes, Passeriformes)* *(33, 111)*.

l) *Cuculiformes* (Kuckucke) und *Piciformes* (Spechte) sind chemotaxonomisch untereinander sowie von manchen *Passeriformes*-Arten nicht unterscheidbar.

m) Innerhalb der *Passeriformes* (Sperlingsvögel) existieren eine Reihe klar abgrenzbarer Familien. Die *Corvidae* (Rabenvögel) *(75, 112, 113)*, *Paridae* (Meisen) *(61)*, *Fringillidae* (Finken und Ammern) *(29, 62, 69)*, *Ploceidae* (Webervögel) *(30)* und *Sylviidae* (Grasmücken) *(74)* sind klar unterscheidbar.

Remiz pendulinus (Beutelmeise), *Aegithalos caudatus* (Schwanzmeise) und *Panurus biarmicus* (Bartmeise) sind keine Meisen, sondern gehören wie *Regulus regulus* (Goldhähnchen), die *Certhiidae* (Baumläufer) und die *Timaliidae* (Timalien) in die Nähe der *Sylviidae* *(114)*.

Passer montanus und *Passer domesticus* sind vermutlich keine *Ploceidae*, sondern gehören zu den *Fringillidae* *(29, 69)*.

Neben dem hier erwähnten Verfahren sind andere chemotaxonomische Methoden mit unterschiedlichem Erfolg auf die Klasse Aves angewendet worden. Besonders die Arbeiten von SIBLEY *(163—177)* über die unterschiedlichen Eiweißmuster verschiedener Arten bei Betrachtung der Vogelei-Eiweiße, aber auch der Augenlinsenproteine und Hämoglobine nach qualitativer Elektrophorese haben Entscheidungshilfen für die Vogelsystematik geliefert.

Literatur

1. KOLATTUKUDY, P. E. (Hrsg.): Chemistry and Biochemistry of Natural Waxes. Amsterdam: Elsevier Scientif. Publ. 1976.
2. JACKSON, L. L., and G. L. BAKER: Cuticular Lipids of Insects. Lipids **9**, 239 (1974).
3. HAAHTI, E. O. A.: Major Lipid Constituents of Human Skin Surface. Helsinki: Mercatorin Kirjapaino. 1961.
4. WELSCH, U., und V. STORCH: Einführung in Cytologie und Histologie der Tiere. Stuttgart: G. Fischer. 1973.
5. FRIEDRICH II.: De arte venandi cum avibus, Augusta vindelicorum, Augsburg 1596.
6. PARIS, P.: Récherches sur la Glande Uropygienne des Oiseaux. Arch. zool. exp. gén. **53**, 139 (1913).
7. GRASSÉ, P. P.: Traité de Zoologie, Band XV, Oiseaux, Paris 1950.
8. LUCAS, A. M., and P. R. STETTENHEIM: Avian Anatomy Integument, Part II, S. 613. Agriculture Handbook 362. Washington: U.S. Dept. of Agriculture. 1972.
9. STERN, M.: Histologische Beiträge zur Sekretion der Bürzeldrüse. Arch. mikrosk. Anat. **66**, 299 (1905).

10. Lennert, K., und G. Weitzel: Untersuchung über die Bürzeldrüse der Vögel, II. Mitt. Morphologie und Histochemie der Bürzeldrüsen von Enten. Hoppe-Seyler's Z. physiol. Chem. **288**, 266 (1951).

11. — — Morphologie und Histochemie der Bürzeldrüse von Enten. Z. mikr. anat. Forsch. **58**, 208 (1952).

12. Cater, D. B., and N. R. Lawrie: Some histochemical and biochemical observations on the preen gland. J. Physiol. **111**, 231 (1950).

13. — — A histochemical study of the developing preen glands of chicks from 14th day of incubation until 14 days after hatching. J. Physiol. **112**, 405 (1951).

14. Ishida, K., S. Kusuhara, and M. Yamaguchi: Histochemical studies of preen glands. Nipp. Chik. Gakkai-Ho **42**, 544 (1971).

15. Ishida, K., S. Kusuhari, T. Suzuki, and M. Yamaguchi: Histochemical demonstration of enzymes in the uropygial gland of fowl. Brit. Poult. Sci. **14**, 179 (1973).

16. Odham, G.: Studies on feather waxes of birds. I. On the chemical composition of the wax in the free flowing secretion from the preen gland of domestic geese. Ark. kemi **21**, 379 (1963).

17. Bertelsen, O.: The Chemical composition of the Wax in the Free-flowing Secretion of Waterfowls of the Genus *Cygnus* (Swans). Chem. Scr. **4**, 163 (1973).

18. Sperry, W. M.: Lipid Analysis, in: Methods of Biochemical Analysis, Vol. 2, S. 83. New York/London: Interscience Publ. 1955.

19. Nikkari, T., and E. O. A. Haahti: Isolation and Analysis of Two Types of Diester Waxes from the Skin Surface Lipids of the Rat. Biochim. Biophys. Acta **164**, 294 (1968).

20. Morris, L. J.: Separation of isomeric long-chain polyhydroxy acids by thin-layer chromatography. J. Chromatogr. **12**, 321 (1963).

21. Thomas III, A. E., J. E. Sharoun, and H. Ralston: Quantitative estimation of isomeric monoglycerides by thin-layer chromatography. J. Amer. Oil Chem. Soc. **42**, 789 (1965).

22. Hansen, I. A., B. K. Tang, and E. Edkins: Erythro-Diols of wax from the uropygial gland of the turkey. J. Lipid Res. **10**, 267 (1969).

23. Jacob, J., und G. Grimmer: Das Bürzeldrüsensekret des Papageientauchers *(Fratercula arctica)*. Z. Naturforsch. **25b**, 54 (1970).

24. — — Die Bürzellipide des Rebhuhns *(Perdix perdix)*. Z. Naturforsch. **25b**, 689 (1970).

25. Jacob, J.: TLC, GLC and MS of Complex Lipid Mixtures from Uropygial Secretions. J. Chromatogr. Sci. **13**, 415–422 (1975).

26. Odham, G.: Studies on feather waxes of birds. II. On the chemical composition of the wax in the free fllowing secretion from the preen gland of Peiping ducks *(Anas platyrhynchos L.)*. Ark. kemi **22**, 417 (1964).

27. Grimmer, G., J. Jacob, and J. Kimmig: Difference between the composition of positional isomeric fatty acids from psoriatic scales and normal human skin. Z. klin. Chem. klin. Biochem. **9**, 111 (1971).

28. Jacob, J., und G. Grimmer: Lipidzusammensetzung gesunder und pathologischer Hautbezirke bei Psoriasis vulgaris. Z. klin. Chem. klin. Biochem. **11**, 297 (1973).

29. Jacob, J., und A. Zeman: Die Bürzeldrüsenlipide des Haussperlings *(Passer domesticus)*. Z. Naturforsch. **25b**, 984 (1970).

30. Poltz, J., und J. Jacob: Bürzeldrüsensekrete von Webervögeln *(Ploceidae)*. Z. Naturforsch. **28c**, 449 (1973).

31. Jacob, J., and G. Grimmer: Occurrence of positional isomers of octadecenoic and hexadecenoic acids in human depot fat. J. Lipid Res. **8**, 308 (1967).

32. Morrison, W. L., and L. M. Smith: Preparation of fatty acid methyl esters and dimethylacetals from lipids with boronfluoride-methanol. J. Lipid Res. **5**, 600 (1964).

33. Jacob, J., and J. Poltz: Chemical composition of uropygial gland secretions of owls. J. Lipid Res. **15**, 243 (1974).

34. HAAHTI, E. O. A., and H. M. FALES: The uropygiols: identification of the unsaponifiable constituents of a diester wax from chicken preen glands. J. Lipid Res. **8**, 131 (1967).

35. SAWAYA, W., and P. E. KOLATTUKUDY: Structure and Biosynthesis of Diesters of Alkane-2,3-diols of the Uropygial Glands of Ring-Necked Pheasants. Biochem. **11**, 4398 (1972).

36. HUBER, W. F.: A study of n-Octadecenoic Acids. I. Synthesis of *cis*- and *trans*-7-through 12- and of 17-Octadecenoic Acids. J. Amer. Chem. Soc. **73**, 2730 (1951).

37. GRIMMER, G., und J. JACOB: Optimierung der oxidativen Spaltung mit MnO_4'/JO_4'. I. Mitt.: Monoenfettsäuren. Z. Naturforsch. **24b**, 565 (1969).

38. JACOB, J., und G. GRIMMER: Das Bürzeldrüsensekret des Fasans *(Phasianus colchicus)*. Z. Naturforsch. **25b**, 577 (1970).

39. JACOB, J., und A. ZEMAN: Das Bürzeldrüsensekret der Ringeltaube *(Columba palumbus)*. Hoppe-Seyler's Z. physiol. Chem. **353**, 492 (1972).

40. JACOB, J., und G. GRIMMER: Gefiederlipide der Ringeltaube *(Columba palumbus)*. Z. Naturforsch. **30c**, 363 (1975).

41. — — Vorkommen von Alkyl-hydroxymalonsäuren in Bürzeldrüsensekreten. Hoppe-Seyler's Z. physiol. Chem. **354**, 1648 (1973).

42. LOUGH, A. K.: The Production of Methoxy-Substituted Fatty Acids as Artefacts during Eesterification of Unsaturated Fatty Acids with Methanol containing Boron Trifluoride. Biochem. J. **90**, 4C (1964).

43. — Use of methanol containing boron trifluoride for esterification of unsaturated fatty acids. Nature **202**, 795 (1964).

44. HANSEN, R. P., and J. F. SMITH: The occurrence of methyl methoxystearate isomers in the methyl esters prepared from sheep perinephric fat. Lipids **1**, 316 (1966).

45. ODHAM, G.: Studies on feather waxes of birds. III. The chemical composition of the wax in the free flowing secretion from the preen gland of the mute swan *(Cygnus olor)*. Ark. kemi **23**, 431 (1965).

46. JACOB, J., und A. ZEMAN: Das Bürzeldrüsensekret des Bläßhuhns *(Fulica atra)*. Z. Naturforsch. **26b**, 1344 (1971).

47. — — Das Bürzeldrüsensekret der Trauerente *(Melanitta nigra)*. Z. Naturforsch. **27b**, 695 (1972).

48. GRIMMER, G., A. HILDEBRANDT und H. BÖHNKE: Probenahme und Analytik polycyclischer aromatischer Kohlenwasserstoffe in Kraftfahrzeugabgasen. Erdöl und Kohle **25**, 531 (1972).

49. — — — Sampling and Analytics of Polycyclic Aromatic Hydrocarbons in Automobile Exhaust Gas. 2. Enrichment of the PAH and Separation of the Mixture of all PAH. Zbl. Bakt. Hyg., I. Abt. Orig. Ber. **158**, 22 (1973).

50. NICOLAIDES, N.: Structure of branched fatty acids in the wax esters of vernix caseosa. Lipids **6**, 901 (1971).

51. ACKMAN, R. G.: Influence of Methyl Substituent Position on Retention Times in the GLC of Higher Monomethylbranched Fatty Acid Esters and Hydrocarbons. J. Chromatogr. Sci. **10**, 243 (1972).

52. McCLOSKEY, J. A.: Mass Spectrometry of Lipids and Steroids. Methods of Enzymology, **XIV**, Lipids, J. M. LOWENSTEIN (Hrsg.), S. 382, New York: Academic Press. 1969.

53. ZEMAN, A., und H. SCHARMANN: Massenspektrometrie von Lipiden. I. Fette, Seifen, Anstrichm. **74**, 509 (1972).

54. — — Massenspektrometrie von Lipiden II. Fette, Seifen, Anstrichm. **75**, 32 (1973).

55. STENHAGEN, E., and H. A. BOEKENOOGEN: Analyses and Characterisation of Oils, Fats and Fat Products. London: Interscience Publ. 1968.

56. RYHAGE, R., and E. STENHAGEN: in F. W. McLAFFERTY, Mass Spectrometry of Organic Ions. New York: Academic Press. 1963.

57. Ryhage, R., and E. Stenhagen: Mass spectrometric studies. IV. Esters of mono-methyl-substituted long chain carboxylic acids. Ark. kemi **15**, 291 (1960).

58. — — Mass spectrometric studies. V. Methyl esters of monoalkyl-substituted acids with ethyl or longer side chain and methyl esters of di- and polyalkyl-substituted acids. Ark. kemi **15**, 333 (1960).

59. Zeman, A., und J. Jacob: Massenspektrometrische Identifizierung von verzweigten Fettsäuren und Alkoholen aus Bürzeldrüsenlipiden. Fette, Seifen, Anstrichm. **75**, 667 (1973).

60. Jacob, J., und J. Poltz: Bürzeldrüsensekrete von Spechten *(Piciformes)*. Z. Natur-forsch. **29c**, 236 (1974).

61. Poltz, J., and J. Jacob: Waxes of the uropygial gland secretion of birds of the genus *Parus*. Biochim. Biophys. Acta **360**, 348 (1974).

62. Jacob, J., und A. Zeman: Über Bürzeldrüsensekrete von Finkenvögeln. Vergleichende Untersuchung der Bürzellipide vom Grünfink *(Carduelis chloris)*, Gimpel *(Pyrrhula pyrrhula)* und Hänfling *(Carduelis cannabina)*. Z. Naturforsch. **26b**, 1352 (1971).

63. — — Zur Chemotaxonomie der Alkenvögel. Die Zusammensetzung der Bürzel-lipide des Tordalk *(Alca torda)*, der Trottel- und Ringellumme *(Uria aalge)* und der Gryllteiste *(Cepphus grylle)*. Z. Naturforsch. **28c**, 78 (1973).

64. Ryhage, R., and E. Stenhagen: Mass spectrometric studies. VI. Methyl esters of normal chain oxo-, hydroxy-, methoxy- and epoxy-acids. Ark. kemi **15**, 545 (1960).

65. Jacob, J.: Uropygial Gland Lipids of Penguins *(Sphenisciformes)*. Biochem. System. **4**, 209 (1976).

66. Jacob, J., und A. Zeman: Das Bürzeldrüsensekret des Eissturmvogels *(Fulmarus glacialis)*. Z. Naturforsch. **26b**, 33 (1971).

67. Jacob, J.: Chemotaxonomical relationship between Penguins and Tubenoses. Biochem. System. **4**, 215 (1976).

68. Jacob, J., und J. Poltz: Das Bürzeldrüsensekret des Kuckucks *(Cuculus canorus)*. Hoppe-Seyler's Z. physiol. Chem. **353**, 1657 (1972).

69. Poltz, J., und J. Jacob: Bürzeldrüsensekrete bei Ammern *(Emberizidae)*, Finken *(Fringillidae)* und Webern *(Ploceidae)*. J. Ornithol. **115**, 119 (1974).

70. Jacob, J., and J. Poltz: The chemical composition of uropygial gland secretions from *Ralliformes*. Biochem. System. **3**, 263 (1975).

71. — — Chemotaxonomische Untersuchungen an Limikolen. Die Zusammensetzung des Bürzeldrüsensekretes von Austernfischer, Rotschenkel, Knutt und Alpen-strandläufer. Biochem. System. **1**, 169 (1973).

72. — — Zur systematischen Stellung der Papageien. Chemotaxonomische Untersuchungen am Wellensittich *(Melopsittacus undulatus* Shaw). J. Ornithol. **115**, 454 (1974).

73. Jacob, J., and A. Glaser: Chemotaxonomy of *Anseriformes*. Biochem. System. **2**, 215 (1975).

74. Poltz, J., and J. Jacob: The uropygial gland secretions of birds of the family *Sylviidae*. Biochem. System. **3**, 57 (1975).

75. Jacob, J., und G. Grimmer: Zur Chemotaxonomie der Rabenvögel. Zusammen-setzung der Bürzeldrüsensekrete der Dohle *(Coloeus monedula)*, Raben- *(Corvus corone corone)* und Nebelkrähe *(Corvus corone cornix)*. Z. Naturforsch. **28c**, 75 (1973).

76. Odham, G.: Studies on feather waxes of birds. VI. Further investigation of the free flowing preen gland secretion from species within the family of *Anatidae*. Ark. kemi **27**, 263 (1967).

77. — Studies on feather waxes of birds. VII. A comparison between the preen gland wax of the Peiping duck and its wild ancestor, the mallard. Ark. kemi **27**, 289 (1967).

78. Bertelsen, O.: The chemical composition of the wax in the free flowing preen gland secretion of the flamingo *(Phoenicopterus ruber* L.*)*. Ark. kemi **32**, 17 (1970).

79. Karlsson, H., and G. Odham: Studies on feather waxes of birds. VIII. The chemical

composition of the wax in the free flowing secretion from the preen gland of the oystercatcher *(Haematopus ostralegus* L.*)*. Ark. kemi **31**, 143 (1969).

80. TANG, B. K., and I. A. HANSEN: Lipid Synthesis in Chicken Preen Glands. Proc. Australian Biochem. Soc. **3**, 84 (1970).

81. SAITO, K., and M. GAMO: The Occurrence of Diester of 2,3-Dihydroxyoctadecane in the Preen Gland of Green Pheasant *(Phasianus colchicus)*. J. Biochem. **67**, 841 (1970).

82. EDKINS, E., and I. A. HANSEN: Diol Esters from the Uropygial Glands of Mallee Fowl and Stubble Quail. Comp. Biochem. Physiol. **39B**, 1 (1971).

83. SAITO, K., and M. GAMO: The Occurrence of 1,2-Diols in Preen Glands of some birds. Biochim. Biophys. Acta **260**, 164 (1972).

84. BERNDT, R., und W. MEISE: Naturgeschichte der Vögel, S. 7. Stuttgart: Franckh'sche Verlagshandlg. 1962.

85. MAYR, E., and D. AMADON: A classification of recent birds. Amer. Mus. Novitates **1496**, 1 (1951).

86. CUISIN, M.: in J. DORST: Das Leben der Vögel I und II. Enzyklopädie der Natur, Bd. 12/13. Lausanne: Edition Rencontre. 1972.

87. FREYE, H. A.: Vögel, Das Tierreich VII/5, S. 104. Berlin: W. de Gruyter. 1960.

88. WETMORE, A.: A revised classification for the birds of the world. Smithsonian Misc. Coll. **117**, 22 (1951).

89. WEITZEL, G., A.-M. FRETZDORFF und J. WOJAHN: Untersuchungen über die Bürzeldrüse der Vögel. III. Strukturaufklärung der optisch aktiven Heptansäure aus Bürzeldrüsen von Enten. Hoppe-Seyler's Z. physiol. Chem. **291**, 29 (1952).

90. POLTZ, J., und J. JACOB: Das Bürzeldrüsensekret vom Graureiher *(Ardea cinera)*, J. Ornithol. **115**, 103 (1974).

91. JACOB, J.: Waxes Containing Secondary Alcohols from the Uropygial Gland Secretion of *Nycticorax nycticorax* (Night heron). Hoppe-Seyler's Z. physiol. Chem. **356**, 1823 (1975).

92. — unveröffentlicht.

93. — unveröffentlicht.

94. EDKINS, E., and I. A. HANSEN: Wax Esters Secreted by the Uropygial Glands of some Australian Waterfowl, Including the Magpie Goose. Comp. Biochem. Physiol. **41B**, 105 (1972).

95. WEITZEL, G., A.-M. FRETZDORFF und J. WOJAHN: Untersuchungen über die Bürzeldrüse der Vögel. IV. Die Fettstoffe der Bürzeldrüsen von Gänsen. Hoppe-Seyler's Z. physiol. Chem. **291**, 46 (1952).

96. HAAHTI, E. O. A., K. LAGERSPETZ, T. NIKKARI, and H. M. FALES: Lipids of the Uropygial Gland of Birds. Comp. Biochem. Physiol. **12**, 435 (1964).

97. WEITZEL, G., und K. LENNERT: Untersuchungen über die Bürzeldrüse der Vögel. I. Die Fettstoffe der Bürzeldrüsen von Enten. Hoppe-Seyler's Z. physiol. Chem. **288**, 251 (1951).

98. ODHAM, G.: Studies on feather waxes of birds. IV. The chemical composition of the wax in the flowing secretion from the preen gland of the ruddy shelduck *(Tadorna ferruginea* Pall.*)* and the common shelduck *(Tadorna tadorna* L.*)*. Ark. kemi **25**, 543 (1966).

99. JACOB, J., und A. ZEMAN: Das Bürzeldrüsensekret der Reiherente *(Aythya fuligula)*. Z. Naturforsch. **25b**, 1438 (1970).

100. ODHAM, G.: Studies on feather waxes of waterfowl. Ark. kemi **27**, 295 (1967).

101. — Studies on the Fatty Acids in the Feather Waxes of some Water-Birds. Fette, Seifen, Anstrichm. **69**, 164 (1967).

102. JACOB, J.: unveröffentlicht.

103. ODHAM, G.: Studies on feather waxes of birds. V. Note on the steric configuration

of the 2,4,6-trimethylnonanoic acid present in the preen gland wax of several species of waterfowl. Ark. kemi **27**, 251 (1967).

104. Jacob, J., and J. Poltz: Composition of uropygial gland secretions of birds of prey. Lipids **10**, 1 (1975).

105. Kolattukudy, P. E.: persönliche Mitteilung.

106. Jacob, J., und A. Zeman: Das Bürzeldrüsensekret der Dreizehmöwe *(Rissa tridactyla)*. Z. Naturforsch. **27b**, 691 (1972).

107. Zeman, A., und J. Jacob: Vergleichende Untersuchung der Bürzeldrüsenlipide der Lachmöwe *(Larus ridibundus)*, Heringsmöwe *(Larus fuscus)* und Silbermöwe *(Larus argentatus)*. Z. anal. Chem. **261**, 306 (1972).

108. Zeman, A., and J. Jacob: The preen lipids of sea birds. Petrolio e ambiente **1973**, 123.

109. Lucas, A. M.: Lipoid secretion in the avian epidermis. Anat. Rec. **160**, 386 (1968).

110. Schildknecht, H., und K. Koob: Myrmicacin, das erste Insekten-Herbizid. Ang. Chem. **83**, 110 (1971).

111. Jacob, J.: unveröffentlicht.

112. Jacob, J., und A. Glaser: Das Bürzeldrüsensekret der Saatkrähe *(Corvus frugilegus)*. Z. Naturforsch. **25b**, 1435 (1970).

113. Poltz, J., und J. Jacob: Zur Chemotaxonomie der Corviden II. Bürzeldrüsensekrete von Eichelhäher *(Garrulus glandarius)*, Elster *(Pica pica)*, Alpendohle *(Pyrrhocorax graculus)* und Kolkrabe *(Corvus corax)*. Z. Naturforsch. **29c**, 239 (1974).

114. Jacob, J., and G. Grimmer: On the classification of some passerine birds by chemotaxonomical methods including *Aegithalos, Certhia, Leiothrix, Mesia, Panurus, Paradoxornis, Regulus, Remiz* and *Siva*. Biochem. System. **3**, 267 (1975).

115. Kolattukudy, P. E.: Structure and cell-free synthesis of alkane-1,2-diols of the uropygial gland of white crowned sparrow *(Zonotrichia leucophrys)*. Biochem. biophys. Res. Commun. **49**, 1376 (1972).

116. Lovern, J. A.: Body Fats of some Seabirds. Biochem. J. **32**, 2142 (1938).

117. Hilditch, T. P., I. C. Sime, and L. Maddison: The Component Acids of some Wild Animal and Bird Fats. Biochem. J. **36**, 98 (1942).

118. Gunstone, F. D., and W. C. Russell: Animal Fats. 3. The Component Acids of Ostrich Fat. Biochem. J. **57**, 459 (1954).

119. Shorland, F. B., and J. P. Gass: Fatty Acid Composition of the Depot Fats of the Kiwi. J. Sci. Food Agric. **12**, 174 (1961).

120. Hilditch, T. P., E. C. Jones, and A. J. Rhead: The Body Fats of the Hen. Biochem. J. **28**, 786 (1934).

121. Hagerup, O.: Communities of birds in the North Atlantic Ocean. Vidensk. Meddr. dansk naturh. Foren, Kobenhavn **82**, 127 (1926).

122. Kritzler, H.: Observations on behaviour in captive fulmars. Condor **50**, 5 (1948).

123. Cheah, C. C., and I. A. Hansen: Stomach Oil and tissue lipids of the petrels *Puffinus pacificus* and *Pterodroma macroptera*. Int. J. Biochem. **1**, 203 (1970).

124. Lewis, R. W.: Studies of the glyceryl ethers of stomach oils. Diss. Abstr. **26**, 5012 (1966).

125. — Studies of the glyceryl ethers of the stomach oil of Leach's petrel *Oceanodroma leucorhoa*. Comp. Biochem. Physiol. **19**, 363 (1966).

126. Smith, L. H.: Some constants of mutton bird oil. J. Soc. chem. Ind. London **30**, 405 (1911).

127. Carter, C. L.: A chemical investigation of mutton-bird oil. J. Soc. chem. Ind. London **40**, 220T (1921).

128. — A chemical investigation of mutton-bird oil. Part II. Comparison of stomach oil and body fat. J. Soc. chem. Ind. London **47**, 26T (1928).

129. Carter, C. L., and J. Malcolm: Observations on the biochemistry of mutton-bird oil. Biochem. J. **21**, 484 (1927).

130. CHEAH, C. C., and I. A. HANSEN: Wax Esters in the Stomach Oil of Petrels. Int. J. Biochem. 1, 198 (1970).
131. ROSENHEIM, O., and T. A. WEBSTER: The stomach oil of the fulmar petrel (Fulmarus glacialis). Biochem. J. 21, 111 (1927).
132. LYNEN, F., I. HOPPER-KESSEL und H. EGGERER: Zur Biosynthese der Fettsäuren. III. Die Fettsäuresynthetase der Hefe und die Bildung enzymgebundener Acetessigsäure. Biochem. Z. 340, 95 (1964).
133. LYNEN, F.: Fatty Acid Synthesis from Malonyl-CoA. Methods in Enzymology 5, 443 (1964).
134. HSU, R. Y., G. WASSON, and J. W. PORTER: The Purification and Properties of the Fatty Acid Synthetase of Pigeon Liver. J. Biol. Chem. 240, 3736 (1965).
135. HSU, R. Y., P. H. W. BUTTERWORTH, and J. W. PORTER: Pigeon Liver Fatty Acid Synthase. Methods in Enzymology 14, 33 (1969).
136. SUMPER, M., D. OESTERHELT, C. RIEPERTINGER und F. LYNEN: Die Synthese verschiedener Carbonsäuren durch den Multienzymkomplex der Fettsäuresynthese aus Hefe und die Erklärung ihrer Bildung. Europ. J. Biochem. 10, 377 (1969).
137. NOBLE, R. E., R. L. STJERNHOLM, D. MERCIER, and E. LEDERER: Incorporation of Propionic Acid into a Branched-chain Fatty Acid of the Preen Gland of the Goose. Nature 199, 600 (1963).
138. TANG, B. K., and I. A. HANSEN: Lipogenesis in chicken uropygial glands. Europ. J. Biochem. 31, 372 (1972).
139. SAWAYA, W. N.: Structure and Biosynthesis of Diesters of Alkane-2,3-diols of the Uropygial Gland of Ring-necked Pheasants (Phasianus colchicus). Diss. Abstr. Int. B. 33, 5662 (1973).
140. WEITZEL, G.: Beziehungen zwischen Struktur und Funktion beim Bürzeldrüsenfett. Fette, Seifen, Anstrichm. 53, 667 (1951).
141. STANLEY, W. M., M. S. JAY, and R. ADAMS: The preparation of certain octadecanoic acids and their bactericidal action toward B. Leprae. J. Amer. Chem. Soc. 51, 1261 (1929).
142. STANLEY, W. M., and R. ADAMS: The surface tension of various aliphatic acids previously studied for bactericidal action to Mycobacterium Leprae. J. Amer. Chem. Soc. 54, 1548 (1932).
143. WEITZEL, G.: Verzweigte Fettsäuren und Tuberkulose. Ang. Chem. 60, 263 (1948).
144. WEITZEL, G., und E. SCHRAUFSTÄTTER: Bakteriostatische und fungistatische Wirkung aliphatischer Carbonsäuren und Alkohole. Hoppe-Seyler's Z. physiol. Chem. 285, 172 (1950).
145. BUU-HOI, N. P., und P. CAGNIANT: Zur Kenntnis der biologischen Bedeutung der höheren verzweigten Fettsäuren. Hoppe-Seyler's Z. physiol. Chem. 279, 76 (1943).
146. JACOB, J.: unveröffentlichte Ergebnisse.
147. — Diesterwachse und ungewöhnliche Fettsäuren aus Bürzeldrüsenlipiden. Fette, Seifen, Anstrichm. 76, 241 (1974).
148. BREUSCH, F. L.: 1. Welt-Fett-Kongreß 1964, Hamburg. Abstract S. 45.
149. — Homologe und isomere Reihen. Fortschr. chem. Forsch. 12, 119 (1969).
150. PUGH, G. J. F., and M. D. EVANS: Keratinophilic Fungi Associated with Birds. I. Fungi isolated from feathers, nests and soils. Trans. Br. mycol. Soc. 54, 233 (1970).
151. — — Keratinophilic Fungi Associated with Birds. II. Physiological Studies. Trans. Br. myol. Soc. 54, 241 (1970).
152. PUGH, G. J. F.: The Contamination of Bird's Feathers by Fungi. Ibis 114, 172 (1972).
153. TIMMERMANN, G.: Studien zu einer vergleichenden Parasitologie der Charadriiformes oder Regenpfeifervögel. Teil I: Mallophaga. Jena: VEB G. Fischer. 1957.
154. — Die Federlingsfauna der Sturmvögel und die Phylogenese des procellariiformen Vogelstammes. Hamburg: Kommissionsverlag Cram, de Gruyter & Co. 1965.

155. Hou, H. C.: Über die Funktion der Bürzeldrüse. Chin. J. Physiol. **2**, 345 (1928).
156. — Beobachtungen über die Beziehung der Bürzeldrüse der Vögel zur Rachitis. Chin. J. Physiol. **3**, 171 (1929).
157. — Weitere Beobachtungen über die Beziehung der Bürzeldrüse der Vögel zur Rachitis. Chin. J. Physiol. **4**, 79 (1930).
158. Kossmann, R.: Über die Talgdrüsen der Vögel. Z. wiss. Zool. **21**, 568 (1871).
159. Esther, K. H.: Über Bau, Entwicklung und Funktion der Bürzeldrüse *(Glandula uropygii)* der Tauben. Morph. Jahrb. **82**, 321 (1938).
160. Lunghetti, B.: Konformation, Struktur und Entwicklung der Bürzeldrüse bei verschiedenen Vogelarten. Arch. mikros. Anat. **69**, 264 (1907).
161. Jacob, J.: unveröffentlichte Ergebnisse.
162. Kolattukudy, P. E., and W. N. Sawaya: Age dependent structural changes in the diol esters of uropygial glands of chicken. Lipids **9**, 290 (1974).
163. Sibley, C. G.: The evolutionary and taxonomic significance of sexual and hybridization in birds. Condor **59**, 166 (1957).
164. — The electrophoretic patterns of egg-white proteins as taxonomic characters. Ibis **102**, 215 (1960).
165. — The comparative morphology of protein molecules as data for classification. Syst. Zool. **11**, 108 (1962).
166. — The characteristics of specific peptides from single proteins as data for classification, in Leone (ed.): Taxonomic biochemistry and serology. New York: Ronald Press. 1964.
167. — Molecular systematics: new techniques applied to old problems. Oiseau Rev. Franc. Orn. **35**, 112 (1965).
168. Sibley, C. G., and A. H. Brush: An electrophoretic study of avian eye-lense proteins. Auk **84**, 203 (1967).
169. Sibley, C. G., and H. T. Hendrickson: A comparative electrophoretic study of avian plasma proteins. Condor **72**, 43 (1970).
170. Sibley, C. G.: A comparative study of the egg-white proteins of passerine birds. Peabody Museum of Natural History, Yale Univ. New Haven, Bulletin 32 (1970).
171. Sibley, C. G., and J. E. Ahlquist: A comparative study of the egg-white proteins of nonpasserine birds. Peabody Museum of Natural History, Yale Univ. New Haven, Bulletin 39 (1972).
172. Sibley, C. G., K. W. Corbin, and J. E. Ahlquist: The relationship of the seedsnipe *(Thinocoridae)* as indicated by their egg-white proteins and hemoglobins. Bonn. Zool. Beitr. **19**, 235 (1968).
173. Sibley, C. G., K. W. Corbin, and J. H. Haavie: The relationship of the flamingos as indicated by the egg-white proteins and hemoglobins. Condor **71**, 155 (1969).
174. Sibley, C. G.: The Relationships of *Picathartes.* Bull. Brit. Orn. Cl. **93**, 23 (1973).
175. Sibley, C. G., and J. E. Ahlquist: The relationship of the hoatzin. Auk **90**, 1 (1973).
176. Sibley, C. G.: The relationships of the silky flycatchers. Auk **90**, 394 (1973).
177. Sibley, C. G., and J. E. Ahlquist: The relationships of the African sugarbirds *(Promerops).* Ostrich **45**, 22 (1974).

(Eingegangen am 14. September 1975)

Hypothalamus-Regulationshormone

Von W. VOELTER, Institut für Organische Chemie,
Eberhard-Karls-Universität, Tübingen, Bundesrepublik Deutschland

Mit 33 Abbildungen

Inhaltsübersicht

Häufig verwendete Abkürzungen

Ac-	Acetyl
Boc-	tert-Butyloxycarbonyl-
Bzl-	Benzyl-

DCC	Dicyclohexylcarbodiimid
Dnp-	2,4-Dinitrophenyl-
Et$_3$N	Triäthylamin
HOBt	1-Hydroxybenzotriazol
Iboc-	Isobornyloxycarbonyl-
Mbh-	Dimethoxybenzhydryl-
Me-	Methyl-
MeOH	Methanol
-OBu^t	tert-Butylester
-OBzl	Benzylester
-OMe	Methylester
-ONP	p-Nitrophenylester
-ONSu	N-Hydroxysuccinimidester
-OPcp	Pentachlorphenylester
-OTcp	Trichlorphenylester
Polym.	Polymeres Harz
Pyr	Pyroglutaminsäure
TFA	Trifluoressigsäure
Z-	Benzyloxycarbonyl-

I. Einführung

Durch anatomische, klinische, physiologische, biochemische und chemische Studien (*1—35*) gelang es, den Bildungsort der Hypothalamus-Hormone zu bestimmen. Die Biosynthese von Oxytocin und Vasopressin erfolgt in den Zellen des nucleus supraopticus und nucleus paraventricularis. Die eminentia mediana ist die Produktionsstätte der Hypothalamus-Regulationshormone.

In Tab. 1 sind die Strukturen und biologischen Wirkungen von Hypothalamus-Hormonen mit Hormon-freisetzender Wirkung zusammengestellt. Tab. 2 gibt eine Übersicht über Hypothalamus-Hormone mit Hormonausschüttung-hemmender Wirkung.

Die Existenz eines Corticotropin-freisetzenden Hormons kann auf Grund zahlreicher physiologischer Studien als sicher angenommen werden; Reindarstellung und Strukturaufklärung dieses Hormons sind jedoch noch nicht gelungen (*36—39*).

Unmittelbar nach Ermittlung der Sequenz vom Follikel-stimulierenden Hormon-freisetzenden Hormon bzw. luteinisierenden Hormon-freisetzenden Hormon und dessen Totalsynthese wurden zahlreiche Untersuchungen an Mensch (*40—78*) und Tier (*79—125*) mit diesem Peptidhormon begonnen. Wie die biologischen Tests zeigen, reguliert das Dekapeptidamid (vgl. Tab. 1) die Sekretion vom luteinisierenden und Follikel-stimulierenden Hormon (*126, 127*). Vermutlich wird LH/FSH – RH in den Neuronen des Nucleus arcuatus des Hypothalamus gebildet (*128*). Durch Untersuchun-

Tabelle 1. *Strukturen und biologische Wirkungen von Hypothalamus-Hormonen mit Hormon-freisetzender Wirkung*

Name des Hormons	Struktur(vorschlag)	biologische Wirkung

Corticotropin-freisetzendes Hormon (CRH)	Ac–Ser–Tyr–Cys–Phe–His–(Asn, Gln)–
	    |
	    –Cys–(Pro, Val)–Lys–Gly–NH$_2$
		verursacht Adreno-corticotropin-ausschüttung

Follikel-stimulie-rendes Hormon-freisetzendes Hormon (FSH–RH)	Pyr–His–Trp–Ser–Tyr–Gly–Leu–Arg–Pro–Gly–NH$_2$	verursacht Aus-schüttung des Follikel-stimulie-renden Hormons
Luteinisierendes Hormon-freisetzen-des Hormon (LH–RH)	Pyr–His–Trp–Ser–Tyr–Gly–Leu–Arg–Pro–Gly–NH$_2$	verursacht Aus-schüttung des luteinisierenden Hormons
Melanotropin-frei-setzendes Hormon (MRH)	Cys–Tyr–Ile–Gln–Asn–Cys	verursacht Aus-schüttung von Melanotropin
Prolactin-frei-setzendes Hormon (PRH)		verursacht Aus-schüttung von Prolactin
Thyreotropin-freisetzendes Hormon (TRH)	Pyr–His–Pro–NH$_2$	verursacht Aus-schüttung von Thyreotropin
Wachstumshormon-freisetzendes Hormon (GRH)	Val–His–Leu–Ser–Ala–Glu–Glu–Lys–Glu–Ala (nicht bestätigt) Pyr–Ser–Gly–NH$_2$ (nicht bestätigt)	verursacht Aus-schüttung von Wachstumshormon

gen an Ratten, welchen die Schilddrüse entfernt wurde, konnte gezeigt werden, daß die LH/FSH – RH-Produktion im Hypothalamus über einen negativen „Feedback"-Mechanismus von der Hypophyse kontrolliert wird (*129*). LH/FSH – RH kann in biologischem Material direkt durch Radioimmunoassays bestimmt werden (*130—132*). Nach Verabreichung von LH – RH ist eine Erhöhung (*133*) des Wachstumshormonspiegels im Blutplasma beobachtet worden; eine Beeinflussung des Plasmaprolactinspiegels dagegen wird nicht festgestellt (*134*).

Obgleich für das Melanotropin-freisetzende Hormon schon Strukturvorschläge gemacht worden sind, müssen diese Untersuchungen noch erhärtet werden (*39, 135, 136*).

Tabelle 2. *Strukturen und biologische Wirkungen von Hypothalamus-Hormonen mit Hormon-ausschüttung-hemmender Wirkung*

Hormon	Struktur(vorschlag)	biologische Wirkung
Melanotropinfreisetzung-hemmendes Hormon (MRIH)	Pro–Leu–Gly–NH$_2$ Pro–His–Phe–Arg–Gly–NH$_2$ S————————S | | Cys–Tyr–Ile–Gln–Asn–Cys	hemmt die Ausschüttung von Melanotropin
Prolactinfreisetzunghemmendes Hormon (PIH)		hemmt die Ausschüttung von Prolactin
Wachstumshormonfreisetzung-hemmendes Hormon (GIH), Somatostatin	S—————————————————————S | | Ala–Gly–Cys–Lys–Asn–Phe–Phe–Trp–Lys–Thr–Phe–Thr–Ser–Cys	hemmt die Ausschüttung von Wachstumshormon

Von der Existenz des Prolactin-freisetzenden Hormons ist man überzeugt, es ist jedoch noch keine Struktur des Naturstoffs bekannt. Vermutlich ist das PRH ebenfalls ein Polypeptidhormon (*39, 137*).

Nachdem synthetisches Thyreotropin-freisetzendes Hormon zur Verfügung stand, wurde durch zahlreiche klinische Studien (*138—232*) und Tierexperimente (*233—281*) die Wirkung dieses Naturstoffs untersucht. Schon relativ früh ist dem TRH von Prange antidepressive Wirkung (*282—283*) zugeschrieben worden; andere Arbeitskreise jedoch bezweifeln diese Wirkung (*284—285*). Auch das Verhalten Schizophrener nach Gabe von TRH ist beschrieben worden (*286*). Tierexperimente deuten darauf hin, daß TRH bei der Thermoregulation eine Rolle spielt (*287*). TRH verursacht nicht nur die Ausschüttung von Thyreotropin (TSH), sondern auch von Prolactin (*288, 289*). Im allgemeinen wird die biologische Aktivität von TRH und seiner Derivate über deren TSH-ausschüttende Wirkung bestimmt (vgl. Zitate, oben); es ist jedoch auch gelungen, Radioimmunoassays zu entwickeln, mit welchen die TRH-Konzentration im Blut oder Urin direkt bestimmt werden kann (*290, 291*). Intravenös verabreichtes TRH wird bei Warmblütlern rasch wieder ausgeschieden: Bei Ratten findet man eine Stunde nach der Injektion von [^{3}H]-TRH 38% der Gesamtradioaktivität in der Blase (*292*). In Gegenwart von

Methylenblau und Sauerstoff nimmt die biologische Aktivität von TRH bei Lichteinstrahlung rasch ab (*293*).

Man hat seither viele Anstrengungen unternommen, das Wachstums-hormon-freisetzende Hormon (GRH) zu isolieren (*39, 294—297*). Von SCHALLY *et al.* (*298*) wurde für GRH die Sequenz H − Val − His − Leu − Ser − Ala − Glu − Glu − Lys − Glu − Ala − OH vorgeschlagen; das synthe-tisierte Peptid (*299—301*) zeigte allerdings keine Wachstumshormon-freisetzende Wirkung. Vor kurzem wurde von YUDAEV und UTESHEVA (*302*) für die Struktur von GRH ein Tripeptid der Sequenz Pyr − Ser − Gly − NH_2 angenommen. Synthesen des Tripeptiamids und biologische Aktivitäts-messungen zeigten jedoch, daß die Sequenz Pyr − Ser − Gly − NH_2 ebenfalls kein Wachstumshormon freizusetzen vermag (*303—306*).

Auch für das Melanotropinfreisetzung-hemmende Hormon (MRIH) sind verschiedene Strukturen vorgeschlagen worden (*39*): H − Pro − Leu − Gly − NH_2, H − Cys − Tyr − Ile − Gln − Asn − Cys − OH und H − Pro −

$$\text{S}\underline{\qquad\qquad\qquad\qquad\qquad\qquad}\text{S}$$

His − Phe − Arg − Gly − NH_2. Die beiden ersten Peptide sind Bruchstücke des Oxytocins. Obgleich von dem Tripeptidamid H − Pro − Leu − Gly − NH_2 eine Reihe physiologischer Wirkungsstudien (*307—319*) vorliegen, ist noch umstritten, ob die oben angegebenen Peptidsequenzen als eigentliche Strukturen für MRIH angesehen werden können (*320—322*). Von der Se-quenz H − Pro − Leu − Gly − NH_2 liegen auch schon Untersuchungen zur Tertiärstruktur vor (*323—324*).

Obgleich man an der Isolierung des Prolactinfreisetzung-hemmenden Hormons (PIH) arbeitet, ist seine Reindarstellung bisher noch nicht gelun-gen (*39, 325—328*).

Anfang 1973 wurde von der Guillemin-Arbeitsgruppe die Struktur des Somatostatins publiziert und wie aus den Zitaten *329* bis *370* hervor-geht, sind seither schon eine Fülle von Studien mit dem synthetischen Hormon durchgeführt worden.

II. Thyreotropin-freisetzendes Hormon

A. Strukturaufklärung

Jahrzehntelang wurde an der Strukturaufklärung von TRH gearbeitet; da das Hormon im Hypothalamus nur in Nanogrammengen vorkommt und lange Zeit keine verläßlichen biologischen Tests verfügbar waren, war man vor eine besonders schwierige Aufgabe gestellt. Schließlich gelang jedoch der Arbeitsgruppe um SCHALLY (*371*) die Reindarstellung von

Schweine-TRH; BURGUS, GUILLEMIN *et al.* (*372*) berichteten als erste die Reinisolierung von TRH aus Schafen.

Die Isolierungsschritte der Schallyschen Aufarbeitung (*373*) sind in Tab. 3 zusammengefaßt: aus 165 000 Hypothalamusfragmenten können

Tabelle 3. *Reinigungsschritte zur Isolierung von Schweine-TRH nach* SCHALLY *et al.* (*373*)
(*Ausgangsmaterial 165 000 Hypothalamusfragmente*)

Reinigungsschritte	Menge in g
1. Lyophylisiertes Material	2 500
2. Nach Entfetten des Gewebes	2 200
3. Lyophylisierter Extrakt mit 2 N CH$_3$COOH	1 075
4. Lyophylisiertes Eisessig-Konzentrat	665
5. Sephadex-G-25-Trennung	380
6. Phenolextraktion	83
7. Cellulose-Chromatographie	16,6
8. Gegenstromverteilung [0,1-proz. CH$_3$COOH/1-Butanol/Pyridin (11:5:3)]	4,98
9. Elektrophorese	0,538
10. Verteilungschromatographie	0,0469
11. Verteilungschromatographie	0,0449
12. Adsorptionschromatographie	0,028
13. Adsorptionschromatographie	0,0151
14. Chromatographie auf Sephadex G-25	0,0071
15. Papierchromatographie	0,00505

Tabelle 4. *Schema zur Isolierung von Schaf-TRH nach* BURGUS *et al.* (*372*)

Isolierungsschritte	Zahl der Hypothalamus- fragmente	Menge	TRH-Ein- heiten/mg (*374*)
1. Lyophylisiertes Rohmaterial	294 000	25 kg	
2. Alkohol-Chloroform-Extrakt	294 000	294 g	1
3. Ultrafiltration „UM-3"	294 000	71 g	3
4. Gelfiltration (Sephadex G-25, 0,5 M Essigsäure)			
5. Gelfiltration (Sephadex G-25, 0,5 M Essigsäure)	286 000	16 g	16
6. Verteilungschromatographie [0,01% CH$_3$COOH/n-Butanol/Pyridin (11:5:3)]			
7. Verteilungschromatographie [0,01% CH$_3$COOH/n-Butanol/Pyridin (11:5:3)]	280 000	246 mg	800
8. Adsorptionschromatographie (Norit/H$_2$O-Äthanol-Phenol)			
9. Adsorptionschromatographie (Norit/H$_2$O-Äthanol-Phenol)	275 000	4,2 mg	30 500
10. Verteilungschromatographie [n-Butanol/CH$_3$COOH/H$_2$O (4:1:5)]	273 000	2,0 mg	58 500
11. Verteilungschromatographie [n-Butanol/CH$_3$COOH/H$_2$O (4:1:5)]	270 000	1,0 mg	57 000

ungefähr 5 mg reines Hormon gewonnen werden. In Tab. 4 sind die Isolierungsschritte zur Darstellung von TRH nach BURGUS *et al.* *(372)* zusammengestellt: Nach 11 Stufen werden aus 270 000 Schaf-Hypothalami 1 mg reines Hormon gewonnen.

Nach der Isolierung von TRH konnte die Aminosäurezusammensetzung des Hormons bestimmt werden: Nach der Hydrolyse mit 6 N Salzsäure werden Histidin, Prolin und Glutaminsäure im Verhältnis 1:1:1 gefunden. Da weder Carboxypeptidase mit dem natürlichen Hormon reagiert, noch ein Dansylderivat gebildet wird, mußte gefolgert werden, daß der Naturstoff weder eine freie COOH- noch NH$_2$-Gruppe besitzt.

Obgleich es sich beim TRH um einen Naturstoff mit relativ kleinem Molekulargewicht handelt, war eine Strukturidentifizierung durch Massenspektrometrie anfangs nicht möglich. Daher versuchte man das Strukturproblem zunächst durch Synthese zu lösen. Auf Grund der Aminosäureanalyse sollte TRH aus den Aminosäuren His, Pro und Glu aufgebaut sein.

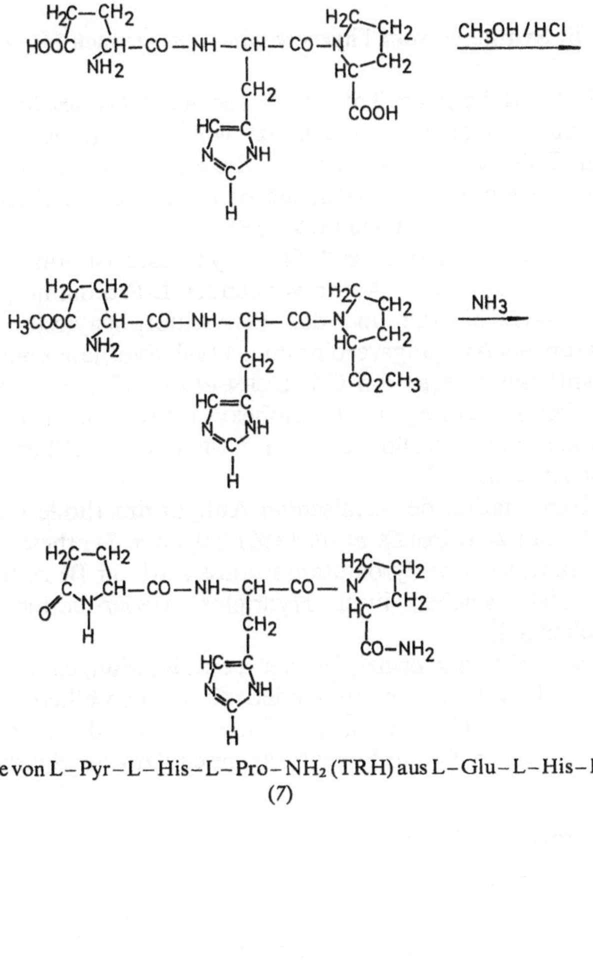

Abb. 1. Synthese von L–Pyr–L–His–L–Pro–NH$_2$ (TRH) aus L–Glu–L–His–L–Pro–OH

(7)

Die Synthese sämtlicher möglicher Tripeptidkombinationen (H − L −
His − L − Glu − L − Pro − OH, H − L − Pro − L − Glu − L − His − OH, H −
L − Glu − L − Pro − L − His − OH, H − L − Glu − L − His − L − Pro − OH,
H − L − His − L − Pro − L − Glu − OH, H − L − Pro − L − His − L − Glu −
OH) erlaubte jedoch zu zeigen, daß keine davon biologisch aktiv ist
(*375, 376*).

Im Zusammenhang mit diesen Studien wurde das Peptid H − L − Glu −
L − His − L − Pro − OH mit Methanol/HCl zum entsprechenden Di-
methylester umgesetzt. Anschließende Reaktion mit Ammoniak führt
den Methylester des Prolins in das entsprechende Säureamid über; außerdem aber erfolgt am N-terminalen Ende des Peptids ein Ringschluß zur
Pyroglutaminsäure. Obgleich die Reaktion in nicht besonders hohen
Ausbeuten verläuft, war damit, fast zufällig, die erste TRH-Synthese
gelungen (Abb. 1) (*7, 377, 378*); das erhaltene Produkt stimmt chromato-
graphisch mit dem natürlichen Hormon überein und zeigt die erwartete
biologische Aktivität (*7*).

B. Synthesen vom Thyreotropin-freisetzenden Hormon

Um 1970 sind die ersten Totalsynthesen von TRH beschrieben worden.
Da TRH auch medizinisch eingesetzt wird, eignen sich Synthesen an
polymeren Trägern nur wenig; bei diesen Syntheseprinzipien bilden
sich Fehl- und Rumpfsequenzen, die sich vom gewünschten Peptid nur
unvollständig separieren lassen (*379—383*).

Eine der ersten „klassischen" TRH-Synthesen ist von Flouret (*384*)
beschrieben worden. Der Autor verwendet L-Prolinamid, tert − Butyl-
oxycarbonyl − L − histidin und den Pentachlorphenylester der L-Pyroglutaminsäure als Ausgangsverbindungen (vgl. Syntheseschema 1).

Die Synthesestrategie von Gillessen et al. (*385*) benutzt zur Knüp-
fung der Peptidbindungen Dicyclohexylcarbodiimid (DCC) und die
Azidmethode, die ursprünglich von Curtius eingeführt wurde (vgl.
Syntheseschema 2).

Eine Kombination der gemischten Anhydridmethode und der DCC-
Knüpfung benutzten Bøler et al. (*386*) bei ihrer Synthese. Zum Schutz
der Aminofunktion der Pyroglutaminsäure wird der Benzyloxycarbonylrest verwendet, welcher durch Hydrieren wieder entfernt wird (vgl.
Syntheseschema 3).

Analoge Knüpfungsprinzipien der Peptidbindungen werden in zwei
weiteren Synthesestrategien verwendet, die in demselben Laboratorium
entwickelt wurden (*387, 388*), allerdings werden die Seitenketten der
Aminosäuren durch Z- und Benzylreste geschützt (vgl. Syntheseschemata
4 und 5).

Die Peptidsequenz von TRH wird nach der Methode von INOUYE (*389*) über den p-Nitrophenylester von Glutamin und das Azid von Histidin hergestellt. Der Glutaminrest der Zwischenstufe L–Gln–L–His–L–Pro–NH$_2$ wird zum Schluß durch Erhitzen in Essigsäure zum Pyroglutaminsäurerest cyclisiert (vgl. Syntheseschema 6).

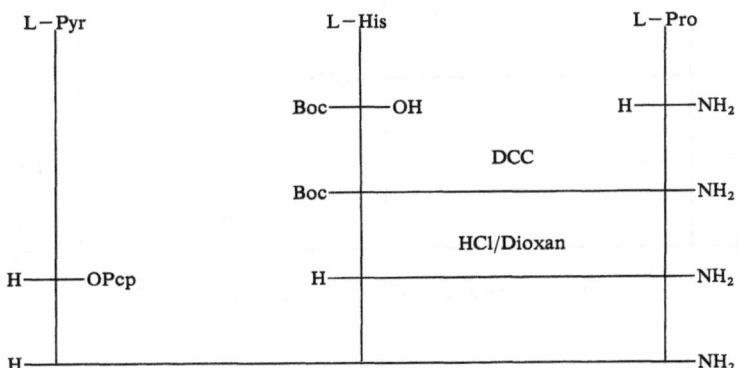

Syntheseschema 1. TRH-Synthese nach FLOURET (*384*)

Sämtliche in diesem Artikel genauer beschriebenen Peptidsynthesen sind analog zu Syntheseschema 1 dargestellt: Aus der obersten Zeile ist die Sequenz des Peptids ersichtlich. Den folgenden Zeilen können die einzelnen Syntheseschritte zur Darstellung des Peptids entnommen werden. Die einzelnen Aminosäurereste sind ab der 2. Zeile nur noch durch kreuzende Linien angegeben. Substituenten der α-NH$_2$-Gruppen sind links, solche der α-COOH-Gruppen sind rechts mit dem entsprechenden Aminosäurerest durch waagerechte Striche verbunden. Die Verknüpfung der Seitenkettenschutzgruppen wird durch einen Schrägstrich (nach oben) angegeben.

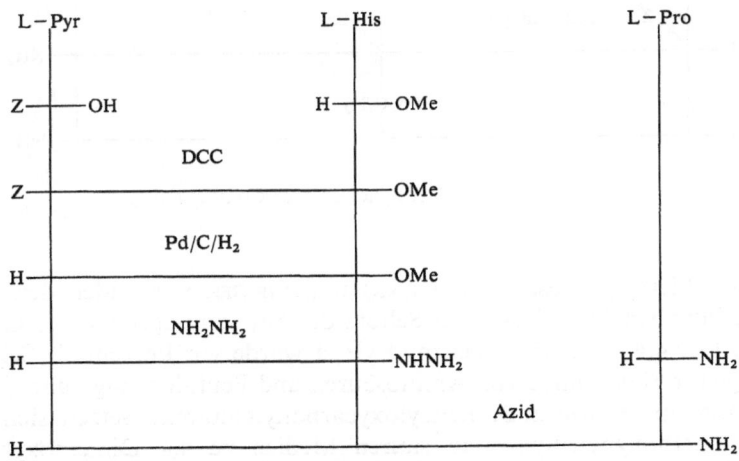

Syntheseschema 2. TRH-Synthese nach GILLESSEN et al. (*385*)

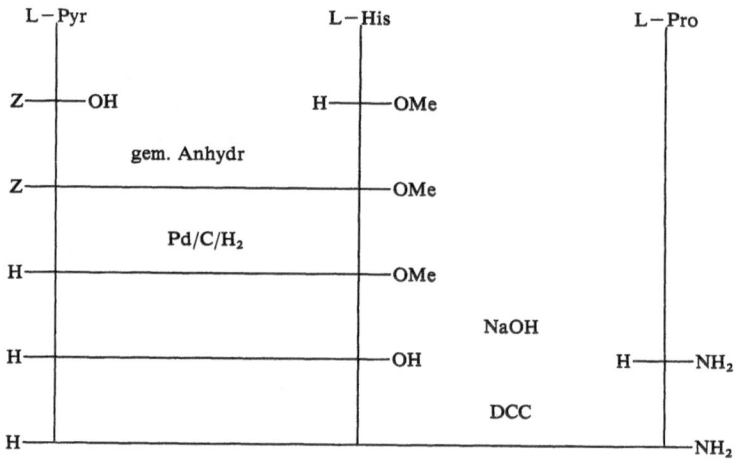

Syntheseschema 3. TRH-Synthese nach BøLER *et al.* (*386*)

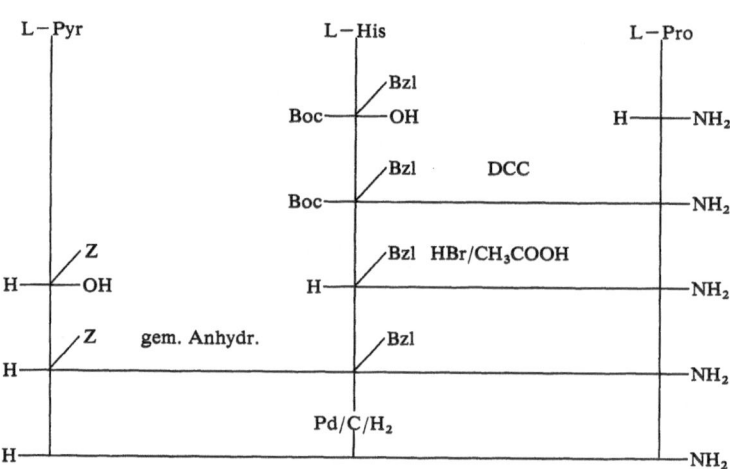

Syntheseschema 4. TRH-Synthese nach CHANG *et al.* (*387*)

Die TRH-Synthesen von KÖNIG und GEIGER verwenden den 4,4'-Dimethoxybenzhydrylrest zum Schutz der Säureamidgruppe des Glutamins. Der 4,4'-Dimethoxybenzhydrylrest wurde vor kurzem als Schutzgruppe für Säureamide von Aminosäuren und Peptiden eingeführt (*390*).

Säureamide, wie z. B. Benzyloxycarbonylglutamin, setzen sich mit 4,4'-Dimethoxybenzhydrol in sauren Medien zu 4,4'-Dimethoxybenzhydrylamiden um (*390*) (vgl. Abb. 2).

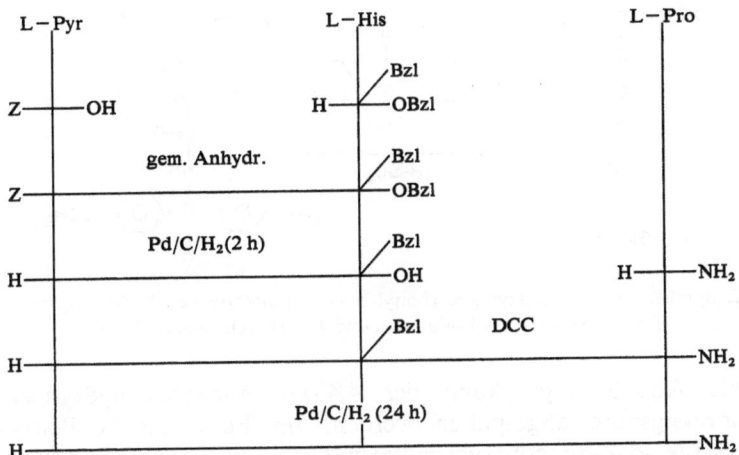

Syntheseschema 5. TRH-Synthese nach CHANG *et al.* (*388*)

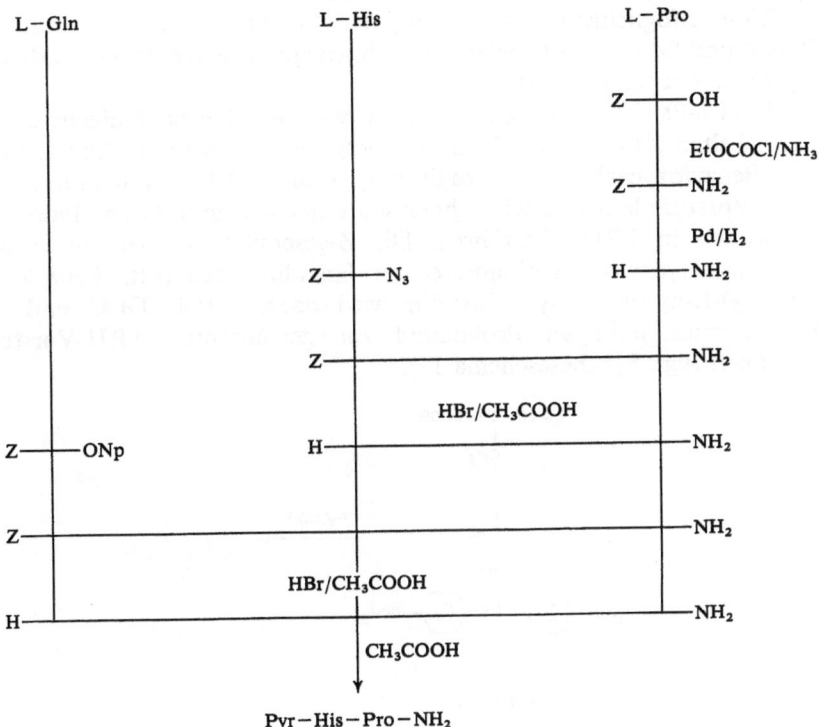

Pyr−His−Pro−NH₂

Syntheseschema 6. TRH-Synthese von INOUYE *et al.* (*389*)

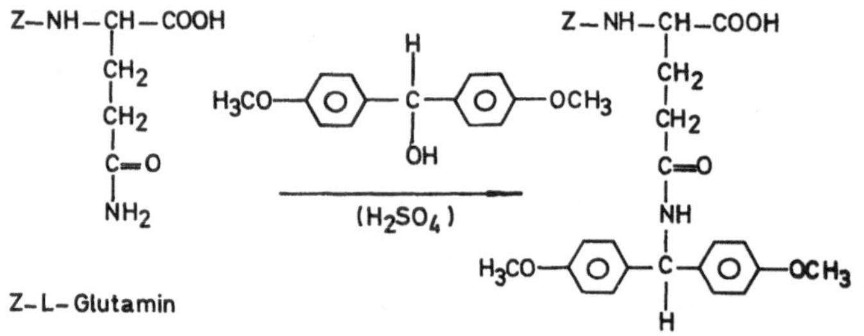

Abb. 2. Synthese von N$^\alpha$-Benzyloxycarbonyl-N$^\gamma$-4,4'-dimethoxybenzhydryl-L-glutamin aus
Benzyloxycarbonyl-L-glutamin und 4,4'-Dimethoxybenzhydrol

Wie Abb. 3 zeigt, kann der 4,4'-Dimethoxybenzhydrylrest mit
Trifluoressigsäure abgespalten werden. Im Falle des N$^\gamma$-Benzyloxy-
carbonyl-N$^\alpha$-4,4'-dimethoxybenzhydryl-L-glutaminesters werden in sie-
dender Trifluoressigsäure der Benzyloxycarbonylrest und der 4,4'-Dime-
thoxybenzhydrylrest abgespalten; außerdem wird der Glutaminsäurerest
zu dem der Pyroglutaminsäure cyclisiert.

Diese Möglichkeit der Bildung der Pyroglutaminsäure benutzten
KÖNIG und GEIGER (391) bei den von ihnen entwickelten TRH-Synthesen
(Syntheseschemata 7—10).

TRH läßt sich durch Umkristallisieren nur schlecht reinigen; daher
entwickelten KURATH und THOMAS (392) eine Synthese zur Darstellung
von Benzyloxycarbonyl-L-pyroglutamyl-L-histidyl-L-prolinamid. Diese
TRH-Vorstufe läßt sich leicht hochrein darstellen und durch Hydrieren
quantitativ in TRH überführen. Die Z-geschützte Pyroglutaminsäure
wird als Hydroxysuccinimidester an Histidin gekuppelt. Benzyloxycarbonyl-L-pyroglutamyl-L-histidin wird dann mittels DCC und N-
Hydroxysuccinimid an Prolinamid zur gewünschten TRH-Vorstufe
gekuppelt (vgl. Syntheseschema 11).

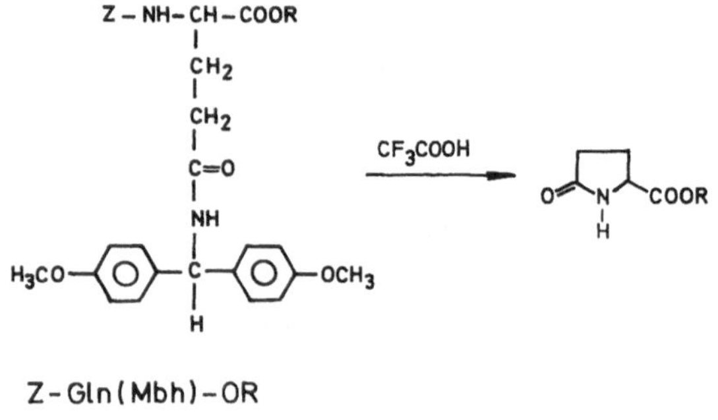

Z – Gln (Mbh) – OR

Abb. 3. Synthese von Pyroglutaminsäureestern aus N$^\alpha$-Benzyloxycarbonyl-N$^\gamma$-4,4'-di-
methoxybenzhydrylglutaminestern

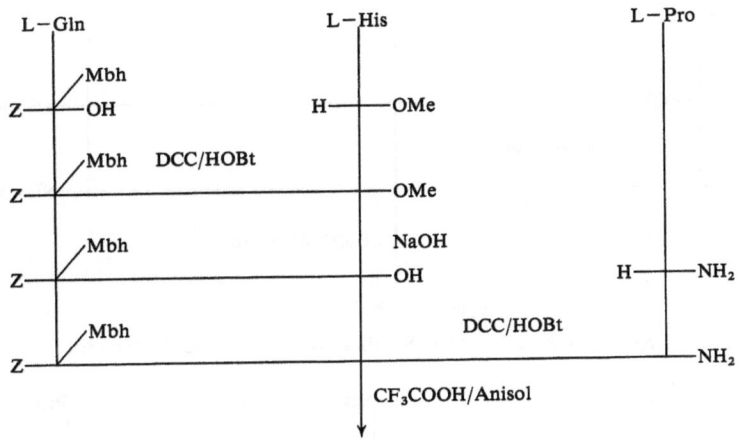

Syntheseschema 7. TRH-Synthese von KÖNIG und GEIGER (391)

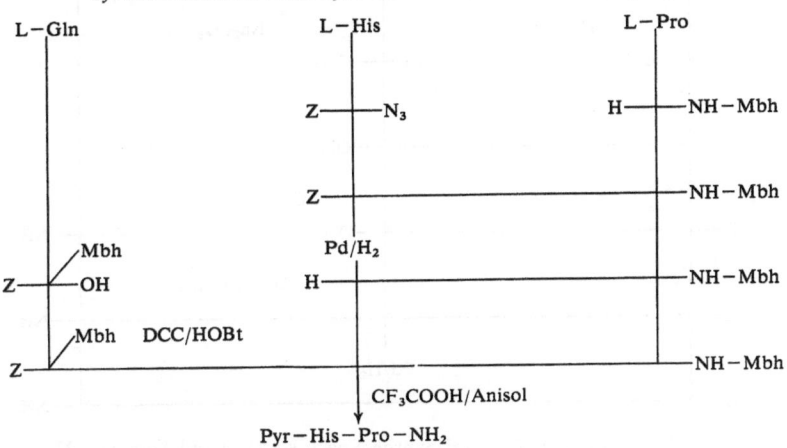

Syntheseschema 8. TRH-Synthese von KÖNIG und GEIGER (391)

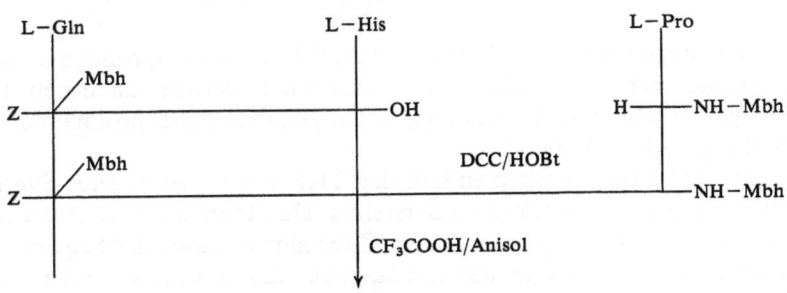

Syntheseschema 9. TRH-Synthese von KÖNIG und GEIGER (391)

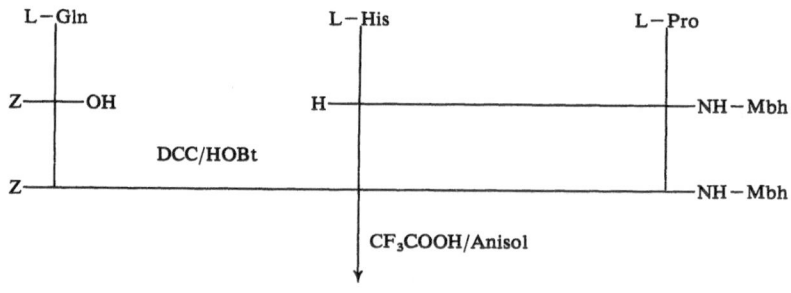

Syntheseschema 10. TRH-Synthese von KÖNIG und GEIGER (391)

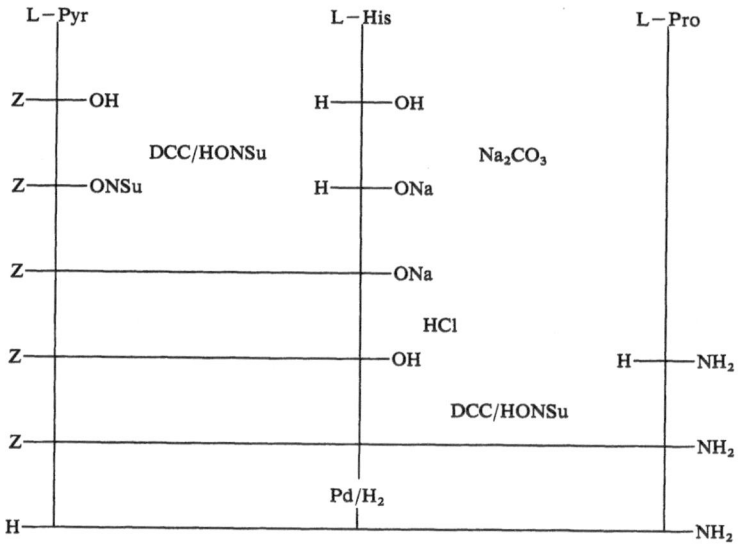

Syntheseschema 11. TRH-Synthese nach KUPATH und THOMAS (392)

N-Benzyloxycarbonyl-L-pyroglutamyl-L-histidyl-L-prolinamid sollte allerdings nicht in Methanol umkristallisiert werden, da neben TRH Benzyloxycarbonyl-L-(O^5-methyl)glutamyl-L-histidyl-L-prolinamid entsteht (vgl. Abb. 4) (392).

In recht hohen Ausbeuten läßt sich TRH nach einer neueren Synthese von BAJUSZ und FAUSZT (393) darstellen. Das Tripeptid wird stufenweise aufgebaut; begonnen wird mit dem C-terminalen Ende der Sequenz. Die α-Aminofunktionen und der Imidazolrest des Histidins werden durch den Benzyloxycarbonylrest geschützt. Zum Schutz der C-terminalen Amidgruppe wird der p,p'-Dimethoxybenzhydrylrest verwendet. Zur

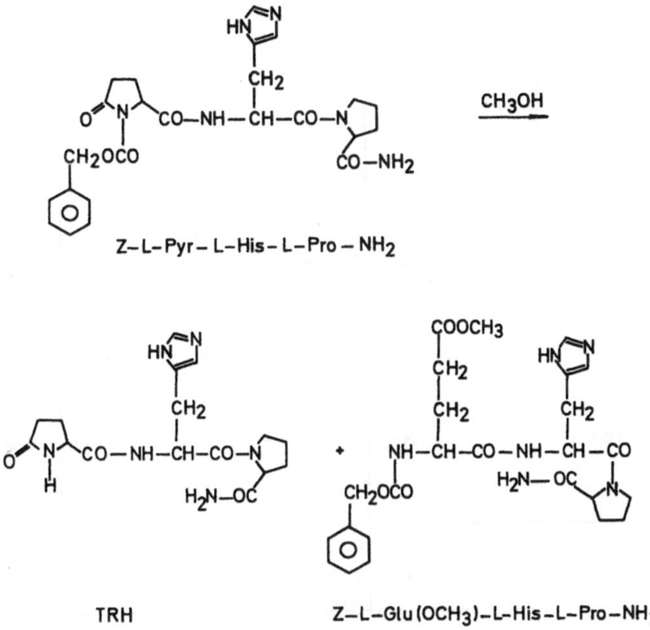

Abb. 4. Methanolyse von Benzyloxycarbonyl-L-pyroglutamyl-L-histidyl-L-prolinamid (*392*)

Herstellung der Peptidbindungen werden Trichlorphenylester eingesetzt (vgl. Reaktionsschema 12).

In einer weiteren TRH-Synthese, welche mit sehr hohen Ausbeuten verläuft, wird zunächst Benzyloxycarbonyl-L-pyroglutamyl-L-histidin aus Benzyloxycarbonyl-L-pyroglutaminsäure-N-hydroxy-5-norbornen-2,3-dicarboximidester (Z–L–Pyr–ONB) und L-Histidin durch eine Aktivesterkupplung hergestellt. Die weiteren Reaktionsschritte zum TRH sind aus Syntheseschema 13 zu entnehmen. Als Tartrat kann TRH in schönen Kristallen erhalten werden (*394*).

Wie zu Beginn dieses Abschnitts bereits erwähnt wurde, sind TRH-Festphasensyntheseprodukte für medizinische Zwecke ungeeignet. Trotzdem wurden bald nach Bekanntwerden der Struktur des Hormons einige TRH-Synthesen am polymeren Träger beschrieben.

Bei der Beyermanschen Synthese wird Boc-Prolin an chlormethyliertes Polystyrol-Divinylbenzol geknüpft und nach Abspaltung der Boc-Gruppe wird Boc-L-His(Bzl)-OH mit DCC an das Prolin-Harz gekuppelt. Die zweite Kupplung wird mit dem 2,4,5-Trichlorphenylester der Pyroglutaminsäure durchgeführt. Das geschützte Peptid wird mit Triäthylamin/CH₃OH vom Harz gespalten, hydriert und durch Ammonolyse in TRH überführt (vgl. Syntheseschema 14) (*395*).

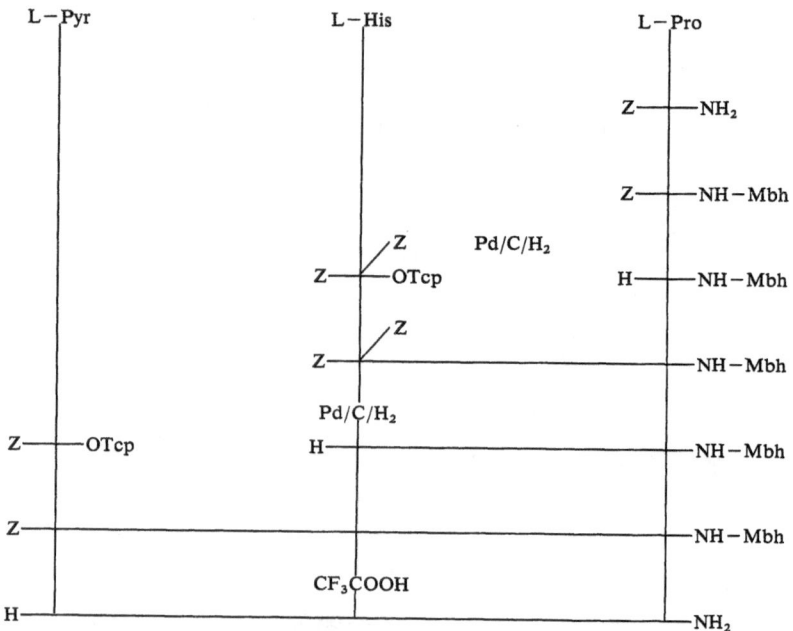

Syntheseschema 12. TRH-Synthese nach BAJUSZ und FAUSZT (*393*)

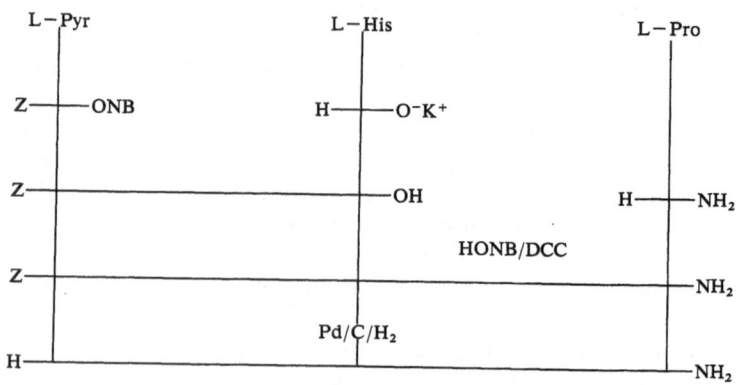

Syntheseschema 13. TRH-Synthese nach HATANAKA *et al.* (*394*)

Im Zusammenhang mit ihrer TRH-Synthese untersuchten BEYERMAN *et al.* (*396, 397*) DCC-Kupplungsreaktionen mit Histidin. Die Autoren konnten zeigen, daß bei Einsatz von N^{im}-Benzylhistidinderivaten eine beträchtliche Razemisierung stattfindet. Durch Zusatz von 1-Hydroxy-

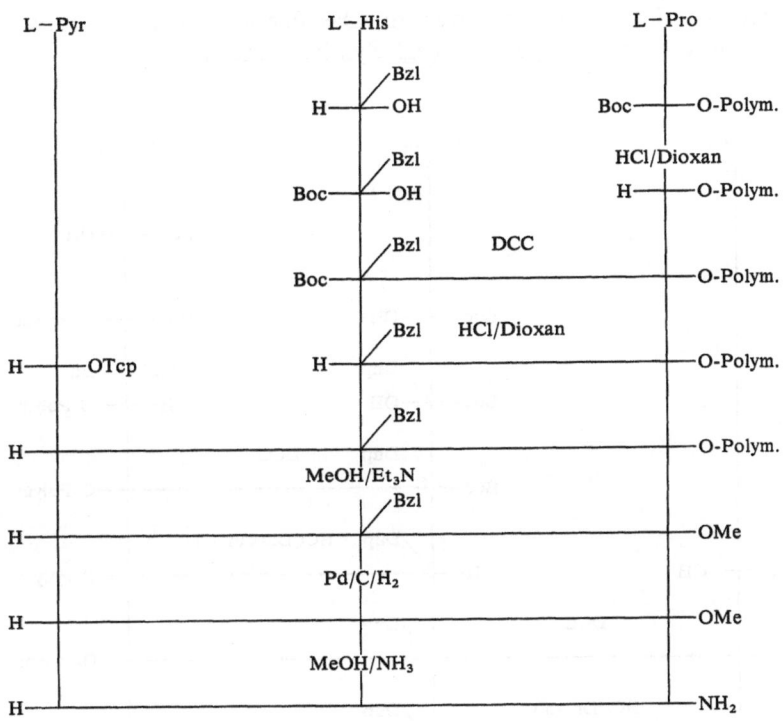

Syntheseschema 14. TRH-Synthese nach BEYERMAN *et al.* (*395*)

benzotriazol kann die Razemisierung jedoch weitgehend verhindert werden. Nach weiteren Studien der Beyerman-Gruppe sollte der Imidazolrest des Histidins bei gemischten Anhydridkupplungen durch geeignete Reste geschützt werden, um hohe Ausbeuten und racematfreie Produkte zu erhalten. tert-Butyloxycarbonyl-, Isobutyloxycarbonyl-, Benzyloxycarbonyl- und p-Toluolsulfonylreste sind z. B. als Schutzgruppen geeignet.

Zwei weitere Festphasensynthesen wurden von RIVAILLE und MILHAUD beschrieben (*398*): tert-Butyloxycarbonyl-L-prolin wird mit chlormethyliertem Styrol-Divinylbenzol-Harz umgesetzt. Nach Abspalten des Boc-Rests mit HCl/Dioxan wird N(im)-o,p-Dinitrophenyl-N-tert-butyloxycarbonyl-L-histidin an das Prolinpolymere mit DCC gekuppelt. Die Pyroglutaminsäure wird entweder mit DCC oder über eine Aktivesterkupplung mit dem Dipeptidpolymeren verknüpft (vgl. Syntheseschemata 15 und 16).

PIETTA *et al.* (*399*) benutzen zu ihrer TRH-Synthese ein Benzhydrylaminharz als polymeren Träger; Boc-L-prolin kann an die Aminogruppe dieses

Harzes mit DCC gekuppelt werden. Das Peptid wird mit Fluorwasserstoff/Anisol vom Harz getrennt (vgl. Syntheseschema 17).

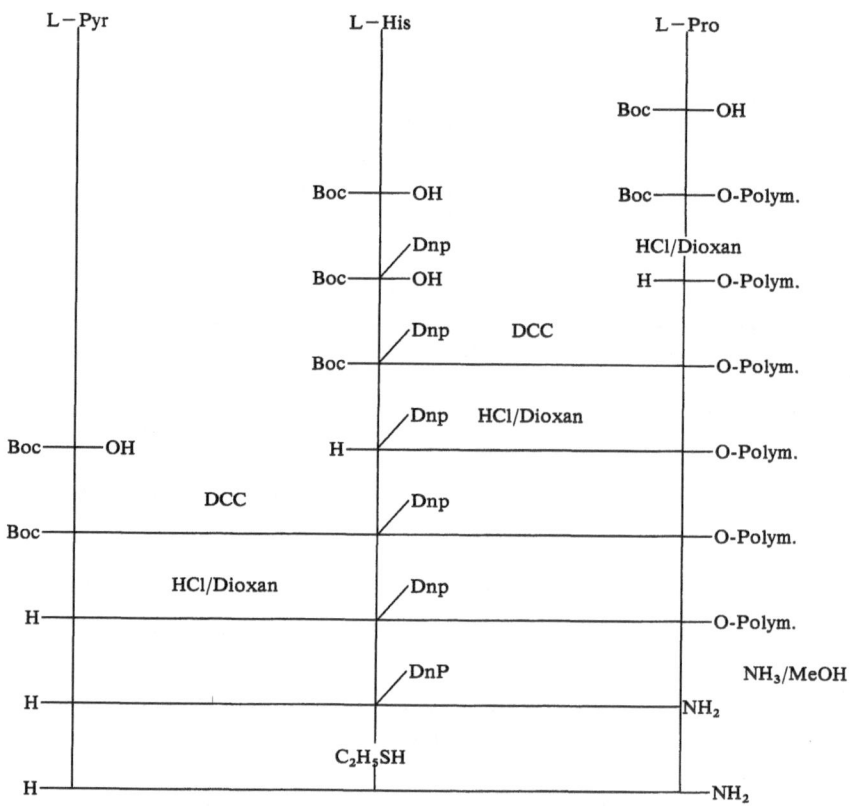

Syntheseschema 15. TRH-Synthese nach RIVAILLE und MILHAUD (*398*)

C. Synthesen von TRH-Analoga. Struktur-Aktivitätsbeziehungen beim Thyreotropin-freisetzenden Hormon

Im Zusammenhang mit der Strukturaufklärung von TRH synthetisierten GILLESEN et al. (*385*) folgende Modellpeptide: L-Histidyl-L-prolyl-L-glutaminsäure, L-Histidyl-L-glutamyl-L-prolin, L-Prolyl-L-histidyl-L-glutaminsäure, L-Prolyl-L-glutamyl-L-histidin, L-Glutamyl-L-prolyl-L-histidin, L-Glutamyl-L-histidyl-L-prolin, L-Histidyl-L-γ-glutamyl-L-prolin, N^α-Acetyl-L-glutamyl-L-histidyl-L-prolin, L-Pyroglutamyl-L-histidyl-L-prolin und L-Pyroglutamyl-L-histidyl-L-prolinamid.

Zur Synthese des ersten dieser Peptide werden N-tert-Butyloxycarbonyl-L-prolin und L-Glutaminsäuredibenzylester durch die gemischte An-

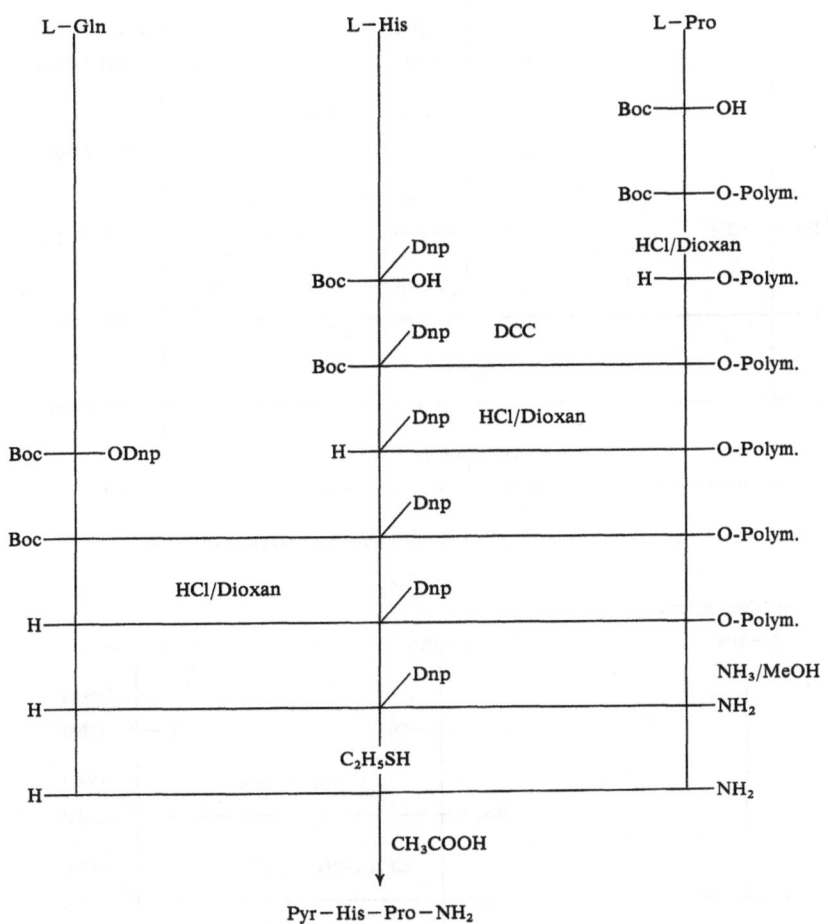

Syntheseschema 16. TRH-Synthese nach RIVAILLE und MILHAUD *(398)*

hydridmethode miteinander verknüpft. Vom Dipeptidderivat wird die Boc-Gruppe mit Trifluoressigsäure abgespalten. Benzyloxycarbonyl-L-histidinazid wird mit L-Prolyl-L-glutaminsäuredibenzylester zum

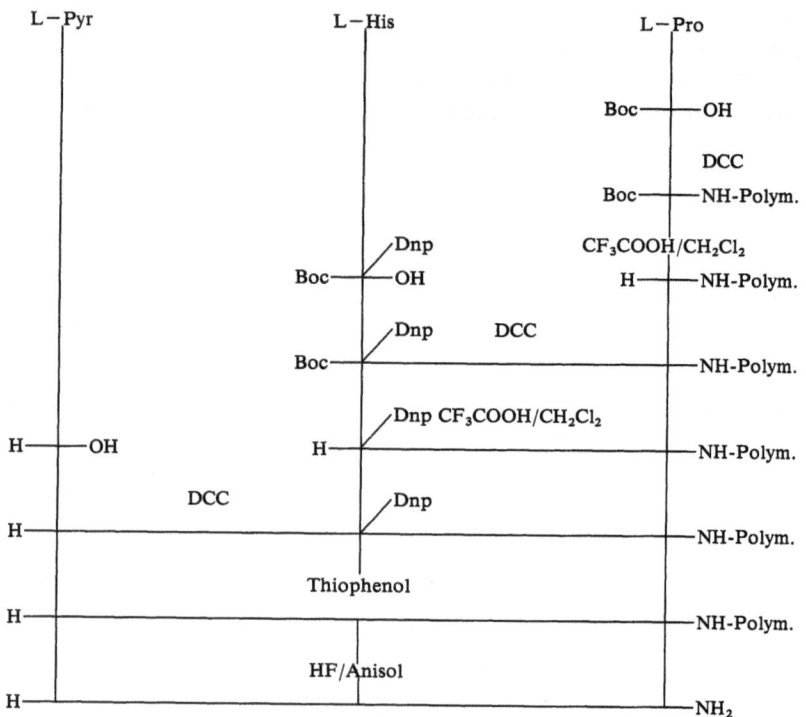

Syntheseschema 17. TRH-Synthese nach PIETTA *et al.* (*399*)

Syntheseschema 18.

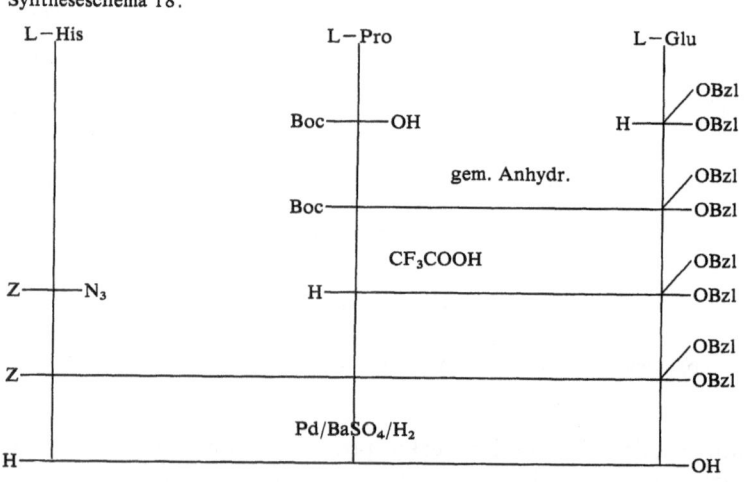

Syntheseschema 18. Darstellung von L−His−L−Pro−L−Glu−OH nach GILLESSEN *et al.* (*385*)

Literaturverzeichnis: SS. 534—564

Benzyloxycarbonyl-L-histidyl-L-prolyl-L-glutaminsäuredibenzylester umgesetzt. Hydrierung in Eisessig mit Pd/BaSO₄ liefert das gewünschte Peptid L-Histidyl-L-prolyl-L-glutaminsäure (vgl. Syntheseschema 18).

Nach einer analogen Strategie wird die Darstellung von L-Histidyl-L-glutamyl-L-prolin beschrieben (vgl. Syntheseschema 19) (*385*).

Zur Darstellung der Sequenz L-Prolyl-L-histidyl-L-glutaminsäure werden als Knüpfungsprinzipien ebenfalls die gemischte Anhydridmethode in Kombination mit der Azidkupplung benutzt (vgl. Syntheseschema 20) (*385*).

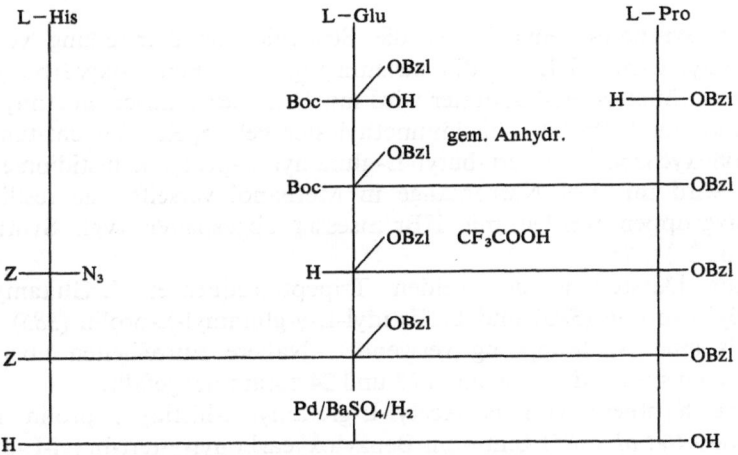

Syntheseschema 19. Darstellung von L−His−L−Glu−L−Pro−OH nach GILLESSEN *et al.* (*385*)

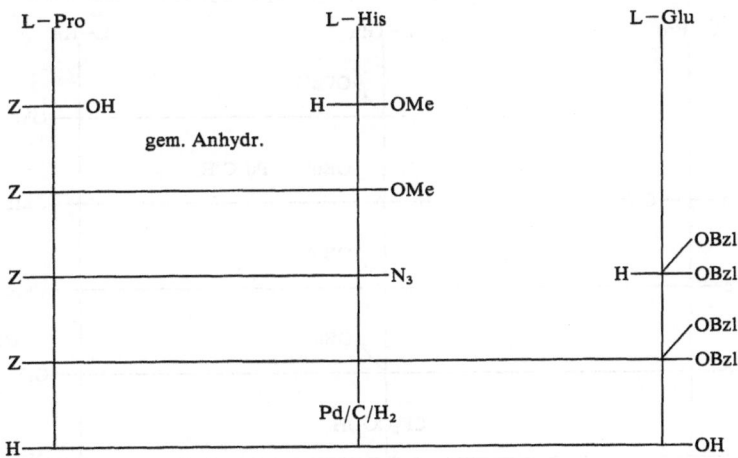

Syntheseschema 20. Darstellung von L−Pro−L−His−L−Glu−OH nach GILLESSEN *et al.* (*385*)

Das Tripeptid L-Prolyl-L-glutamyl-L-histidin wird über den geschützten Dipeptidmethylester Benzyloxycarbonyl-γ-tert-butyl-L-glutamil-L-histidinmethylester hergestellt. Von Z−L−Glu(OBuᵗ)−L−His−OMe wird die Benzyloxycarbonylschutzgruppe durch Hydrieren spezifisch entfernt. Der so erhaltene γ-tert-Butyl-L-glutamyl-L-histidinmethylester wird mit Boc-L-prolin-N-hydroxysuccinimidester zum Boc-L-prolyl-γ-tert-butyl-L-glutamyl-L-histidinmethylester umgesetzt. Nach Verseifung des Methylesters mit 1 N NaOH werden die restlichen Schutzgruppen mit Trifluoressigsäure abgespalten (vgl. Syntheseschema 21) (385).

Im Syntheseschema 22 ist die Strategie zur Darstellung von L-Glutamyl-L-prolyl-L-histidin zusammengefaßt: Benzyloxycarbonyl-L-glutaminsäure-γ-tert-butylester wird mit Hilfe der gemischten Anhydridmethode an L-Prolyl-L-histidinmethylester gekuppelt. Der entstandene Benzyloxycarbonyl-γ-tert-butyl-L-glutamyl-L-prolyl-L-histidinmethylester wird mit 1 N Natronlauge in Methanol verseift; die restlichen Schutzgruppen werden mit HBr/Eisessig abgespalten (vgl. Syntheseschema 22) (385).

Zur Darstellung der beiden Tripeptidsequenzen L-Glutamyl-L-histidyl-L-prolin (385) und L-Histidyl-L-γ-glutamyl-L-prolin (385) wird jeweils eine Azidkupplung verwendet. Nähere Einzelheiten über die Synthesen sind in den Schemata 23 und 24 zusammengefaßt.

Die Synthese von Nᵅ-Acetyl-L-glutamyl-L-histidyl-L-prolin nach Gillessen et al. (385) geht vom Benzyloxycarbonyl-γ-tert-butyl-L-glutamyl-L-histidyl-L-prolinmethylester aus. Die Z-Schutzgruppe wird in Methanol mit Pd/C abhydriert und anschließend durch die Acetylgruppe ersetzt, welche durch den Essigsäure-p-nitrophenylester eingeführt wird.

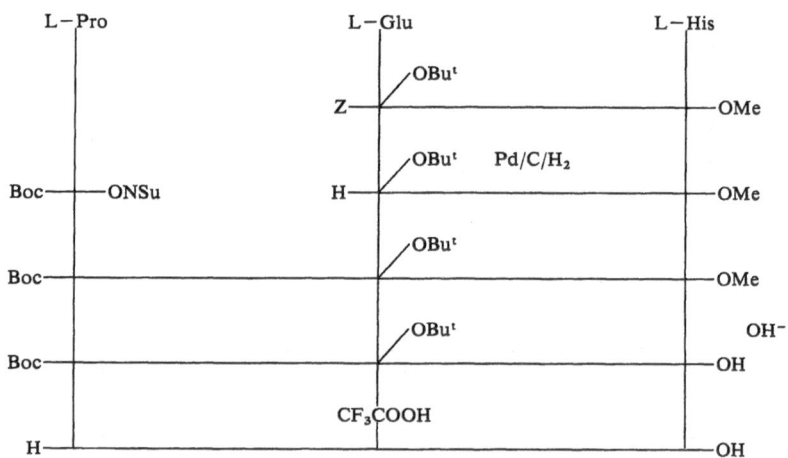

Syntheseschema 21. Darstellung von L−Pro−L−Glu−L−His−OH nach Gillessen *et al.* (385)

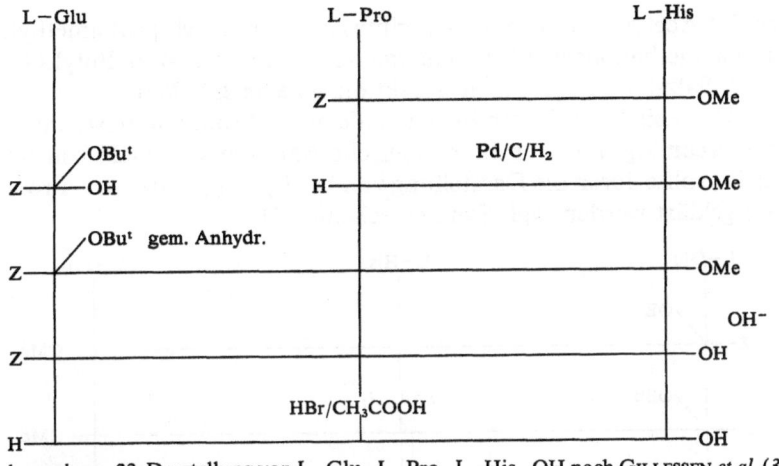

Syntheseschema 22. Darstellung von L–Glu–L–Pro–L–His–OH nach GILLESSEN *et al.* (*385*)

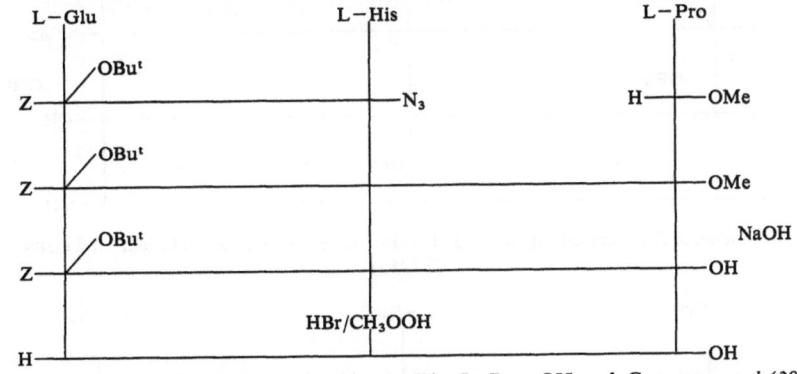

Syntheseschema 23. Darstellung von L–Glu–L–His–L–Pro–OH nach GILLESSEN *et al.* (*385*)

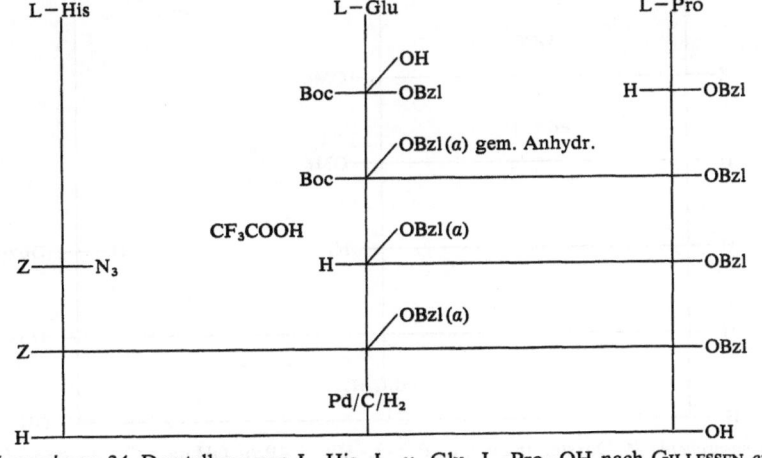

Syntheseschema 24. Darstellung von L–His–L–γ–Glu–L–Pro–OH nach GILLESSEN *et al.* (*385*)

Der N^α-Acetyl-γ-tert-butyl-L-glutamyl-L-histidyl-prolinmethylester wird mit methanolischer Natronlauge verseift und die γ-tert-Butyl-Gruppe mit 2N Salzsäure abgespalten (Reaktionsschema 25) (*385*).

TRH besitzt am C-terminalen Ende einen Prolinamidrest; inwieweit die Säureamidgruppe für die biologische Aktivität von TRH von Bedeutung ist, sollte durch die Darstellung von L−Pyr−L−His−L−Pro−OH (*385*) geklärt werden (vgl. Syntheseschema 26).

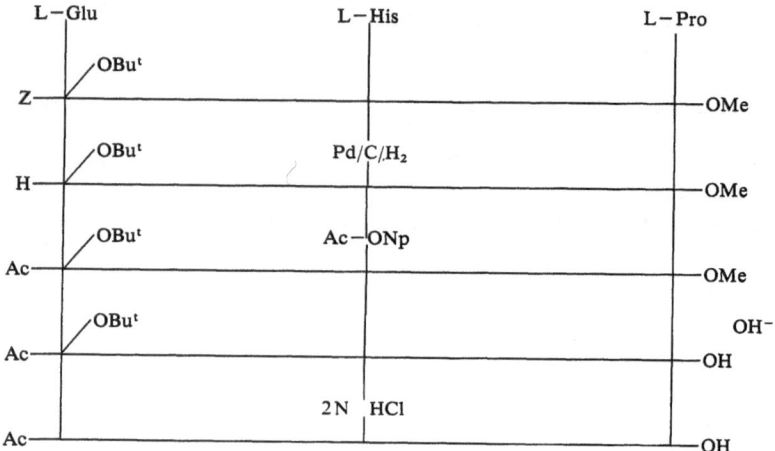

Syntheseschema 25. Darstellung von Ac−L−Glu−L−His−L−Pro−OH nach Gillessen *et al.*
(*385*)

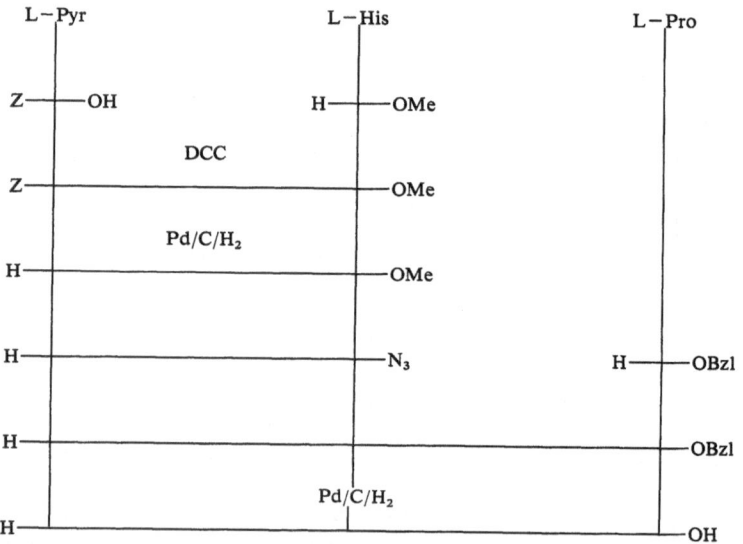

Syntheseschema 26. Darstellung von L−Pyr−L−His−L−Pro−OH nach Gillessen *et al.* (*385*)

Literaturverzeichnis: SS. 534—564

HOFMANN und BOWERS (*400*) untersuchten den Einfluß des Austausches des Imidazolrestes von TRH durch den Pyrazolrest; L-Pyroglutamyl-L-β-(pyrazolyl-3)-alanyl-L-prolinamid besitzt im Vergleich zum natürlichen Hormon nur 5% biologische Aktivität.

Mehr als 20 TRH-Analoga wurden von CHANG et al. (*388*) hergestellt. Für ihre Synthesen benutzen die Autoren Pyroglutamyl-N^{im}-benzyl-histidin als Zwischenstufe, da sich daraus leicht zu reinigende N^{im}-Benzyl-tripeptide herstellen lassen. Der eingeschlagene Syntheseweg für L – Pyr – L – His(Bzl) – OH folgt aus Syntheseschema 27.

Um die Bedeutung des Histidins für die biologische Aktivität von TRH zu untersuchen, synthetisierten GILLESSEN et al. (*401*) TRH-Analoga, bei welchen L-Histidin durch die körpereigenen Aminosäuren L-Arginin und L-Lysin bzw. die körperfremden Aminosäuren L-α,γ-Diamino-buttersäure und β-(3-Pyrazolyl)-L-alanin ersetzt ist.

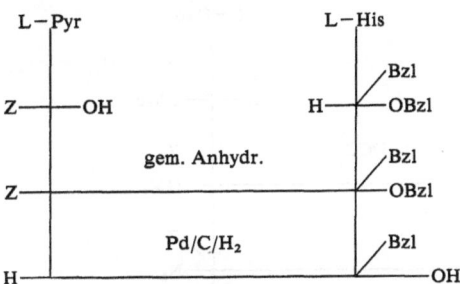

Syntheseschema 27. Darstellung von L–Pyr–L–His(Bzl)–OH nach CHANG et al. (*388*)

Die Darstellungen für L-Pyroglutamyl-L-lysyl-L-prolinamid, L-Pyro-glutamyl-L-α,γ-diaminobutyryl-L-prolinamid und L-Pyroglutamyl-β-(3-pyrazolyl)-L-alanyl-L-prolinamid erfolgen nach einer analogen Strategie: Pyroglutaminsäure wird durch eine DCC-Kupplung mit den mittelständigen Aminosäuren verknüpft; die zweite Peptidbindung wird durch eine Azidkupplung hergestellt (vgl. Syntheseschemata 28—30) (*401*).

Zur Darstellung von L-Pyroglutamyl-L-arginyl-L-prolinamid wird L-Pyroglutaminsäure mit Dicyclohexylcarbodiimid an den N^G-Nitro-L-argininmethylester gekuppelt. Der entstandene L-Pyroglutamyl-N^G-nitro-L-argininmethylester wird mit Natronlauge verseift und die freie Säure mit DCC an Prolinamid geknüpft. Die Nitrogruppe wird mit Raney-Nickel abhydriert (vgl. Reaktionsschema 31) (*401, 402*).

Um die Bedeutung der Carbonamidgruppe des Prolinamids zur TRH-Wirkung zu ermitteln, synthetisierten KÖNIG und GEIGER (*391*) Des-carbamoyl-TRH (vgl. Reaktionsschema 32), welches noch beträchtliche biologische Aktivität besitzt.

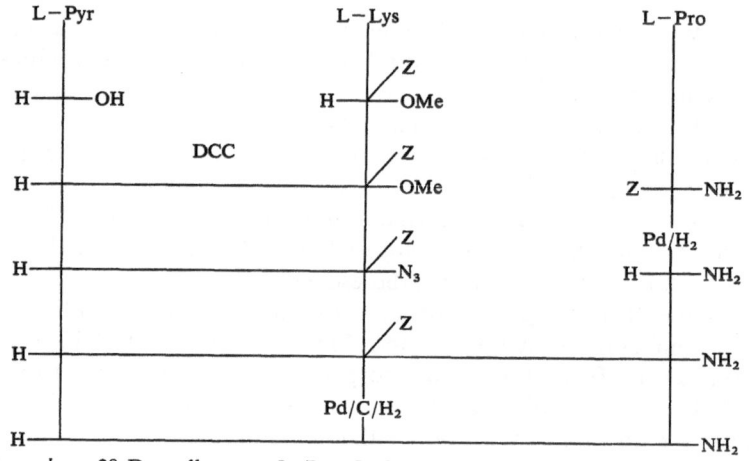

Syntheseschema 28. Darstellung von L–Pyr–L–Lys–L–Pro–NH$_2$ nach Gillessen *et al. (401)*

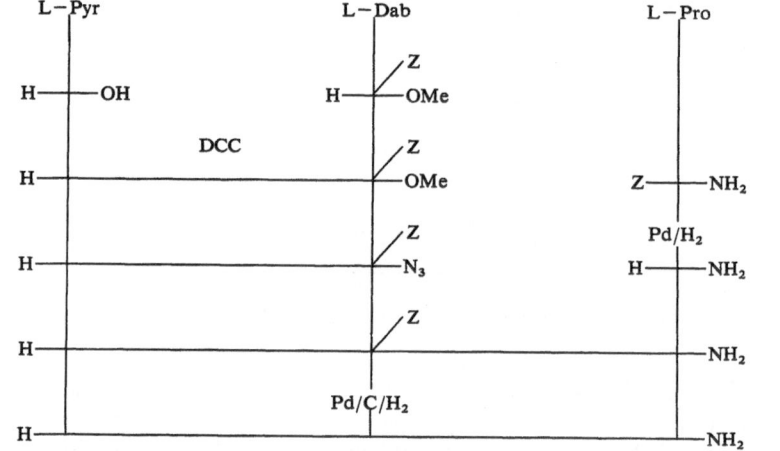

Syntheseschema 29. Darstellung von L–Pyr–L–Dab–L–Pro–NH$_2$ nach Gillessen *et al.*

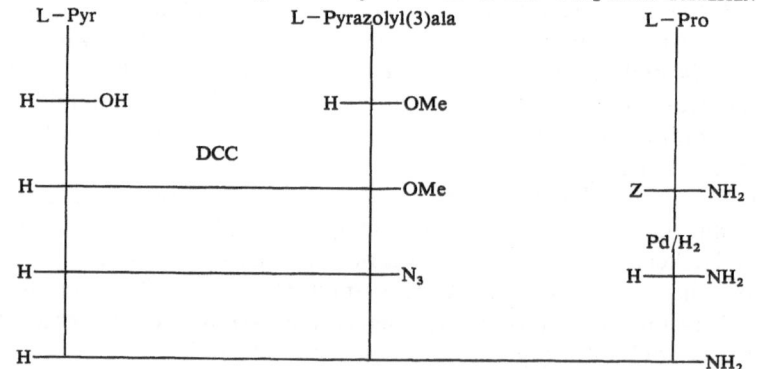

Syntheseschema 30. Darstellung von L–Pyr–L–Pyrazolyl(3)ala–L–Pro–NH$_2$ nach Gillessen *et al. (401)*

Literaturverzeichnis: SS. 534—564

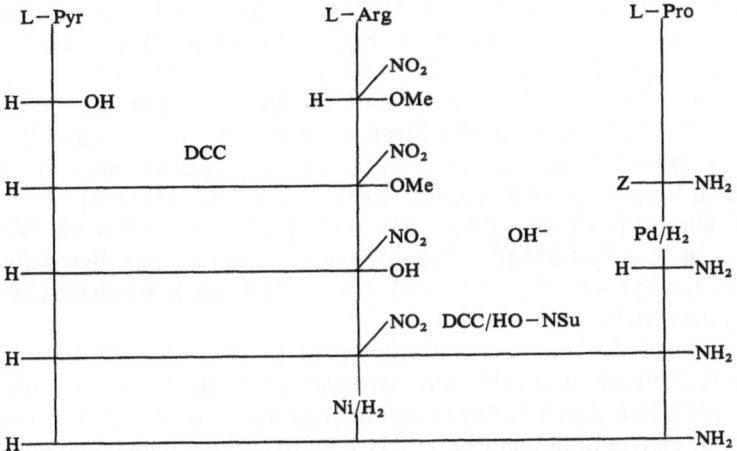

Syntheseschema 31. Darstellung von L–Pyr–L–Arg–L–Pro–NH$_2$ nach GILLESSEN *et al.* (*401*)

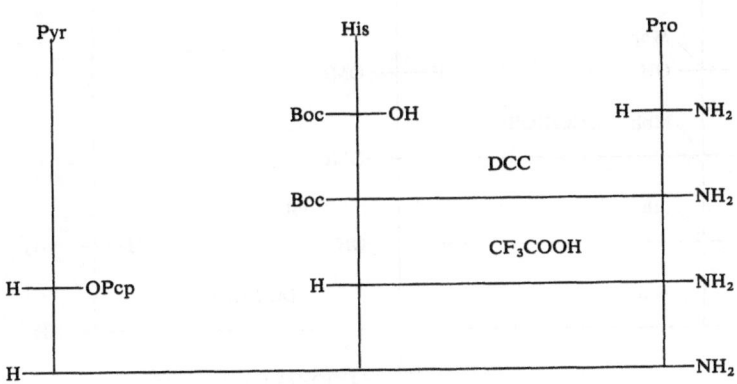

Syntheseschema 32. Darstellung von Des-carbamoyl-TRH nach KÖNIG und GEIGER (*391*)

Reaktionsschema 33. Darstellungen von D–Pyr–L–His–L–Pro–NH$_2$,
L–Pyr–D–His–L–Pro–NH$_2$, L–Pyr–L–His–D–Pro–NH$_2$ und
D–Pyr–D–His–D–Pro–NH$_2$ nach FLOURET *et al.* (*403*)

Aktivitätsänderungen beim Austausch von natürlichen L- durch D-Aminosäuren untersuchten G. R. Flouret et al. (403). Zur Darstellung von D – Pyr – L – His – L – Pro – NH$_2$, L – Pyr – D – His – L – Pro – NH$_2$, L – Pyr – L – His – D – Pro – NH$_2$ und D – Pyr – D – His – D – Pro – NH$_2$ benutzten sie die Strategie von Reaktionsschema 33 (403).

L – Pyr – D – His – L – Pro – NH$_2$ ist vor kurzem auch auf einem anderen Weg dargestellt worden (Reaktionsschema 34) (404).

Während L – Tyr^2 – TRH noch geringe TSH-ausschüttende Wirkung zeigt, ist L – Tyr(OMe)^2-TRH biologisch inaktiv; zur Beweisführung wurde L – Pyr – L – Tyr(OMe) – L – Pro – NH$_2$ nach Syntheseschema 35 (405) hergestellt.

Inwieweit der heteroaromatische Imidazolrest des Histidins für die biologische Aktivität von TRH verantwortlich ist, sollte durch den Austausch von L-Histidin durch L-Phenylalanin ermittelt werden. Einen gangbaren Weg zur Herstellung von Phe^2-TRH zeigt Syntheseschema 36 (406).

In einer umfangreichen Studie haben Sievertsson et al. (407) hauptsächlich TRH-Analoga auf ihre TSH-ausschüttende Wirkung untersucht, in welchen Histidin durch eine andere Aminosäure ersetzt ist. Von den Autoren der Referenz 407 wurden die Peptide Pyr – Gly – Pro – OCH$_3$ und Pyr – Gly – Pro – NH$_2$ nach ein und demselben Schema aufgebaut (Syntheseschema 37).

Leu^2 – TRH und dessen Methylester kann nach Sievertsson et al. (407) durch zweifache Aktivesterkupplung dargestellt werden. Benzyloxy-carbonyl-L-leucin-p-nitrophenylester wird zunächst mit Prolinmethyl-ester zum Benzyloxycarbonyl-L-leucyl-L-prolinmethylester umgesetzt.

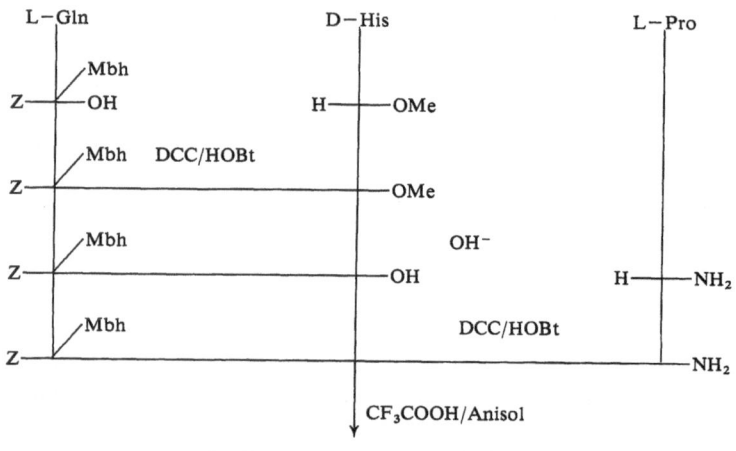

Reaktionsschema 34. Darstellung von L – Pyr – D – His – L – Pro – NH$_2$ nach Voelter et al. (404)

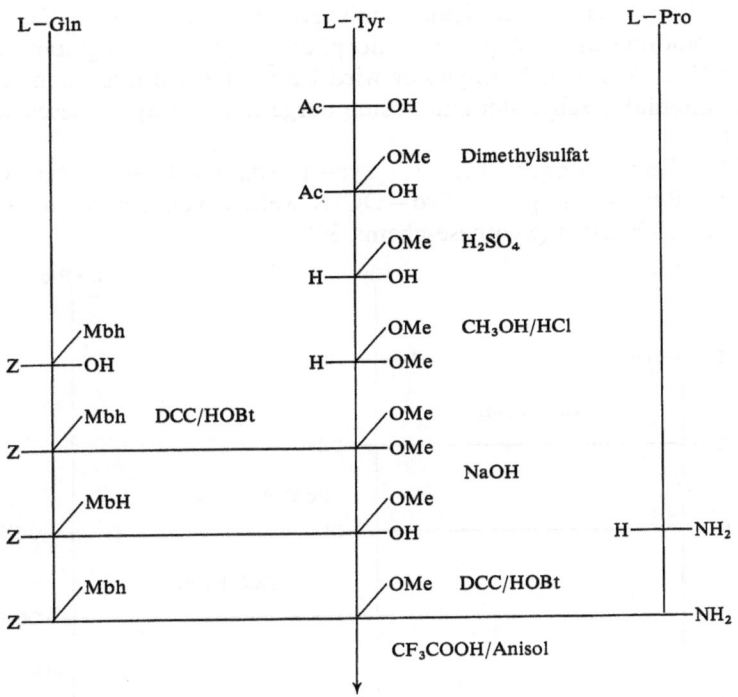

Reaktionsschema 35. Darstellung von L–Pyr–L–Tyr(OMe)–L–Pro–NH₂ nach VOELTER und HORN *et al.* (*405*)

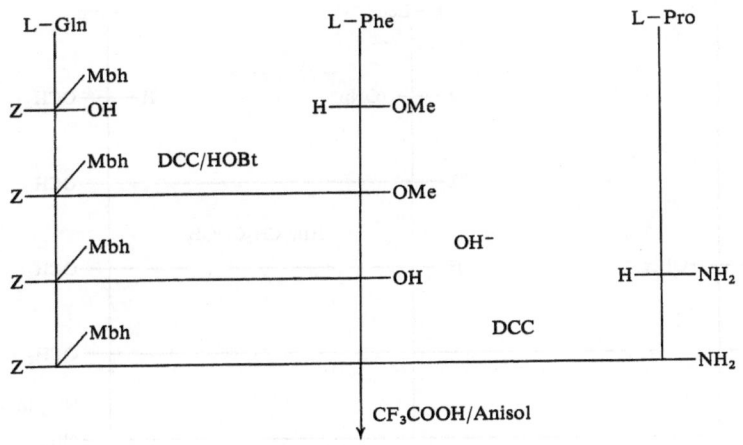

Syntheseschema 36. Darstellung von L–Pyr–L–Phe–L–Pro–NH₂ nach ZECH *et al.* (*406*)

30*

Nach Abspalten der Z-Gruppe mit HBr/Essigsäure wird die zweite Peptidbindung durch den Pentachlorphenylester der Pyroglutaminsäure hergestellt. Aus dem Methylester wird $Leu^2 - TRH$ durch Ammonolyse mit ammoniakalischer Methanollösung dargestellt (vgl. Syntheseschema 38) (*407*).

Die Darstellungen von $L - Pyr - L - Arg(NO_2) - L - Pro - OCH_3$ und $L - Pyr - L - Arg - L - Pro - OCH_3$ werden von derselben Arbeitsgruppe beschrieben (Syntheseschema 39).

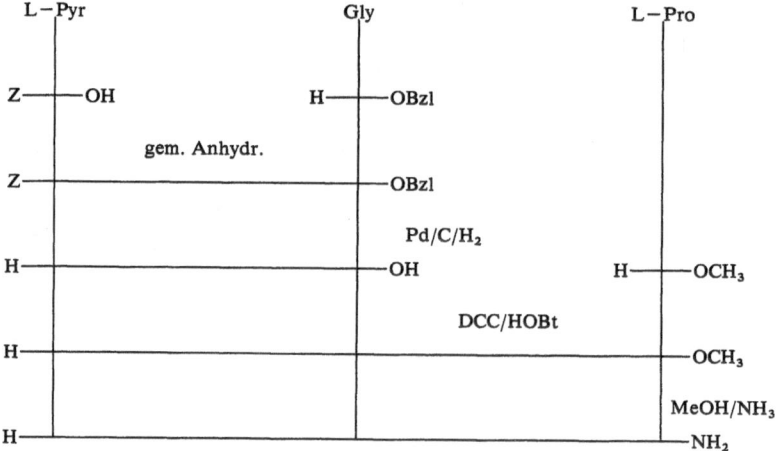

Syntheseschema 37. Darstellungen von $L-Pyr-Gly-L-Pro-OCH_3$ und $L-Pyr-Gly-L-Pro-NH_2$ nach Sievertsson *et al.* (*407*)

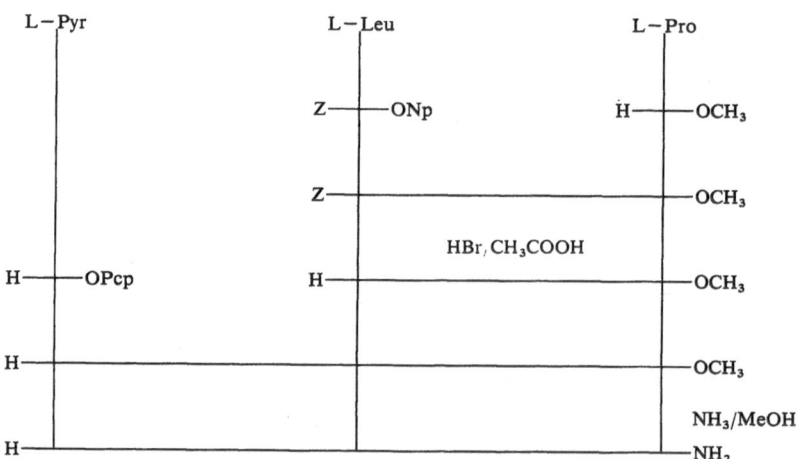

Syntheseschema 38. Darstellungen von $L-Pyr-L-Leu-L-Pro-OCH_3$ und $L-Pyr-L-Leu-Pro-NH_2$ nach Sievertsson *et al.* (*407*)

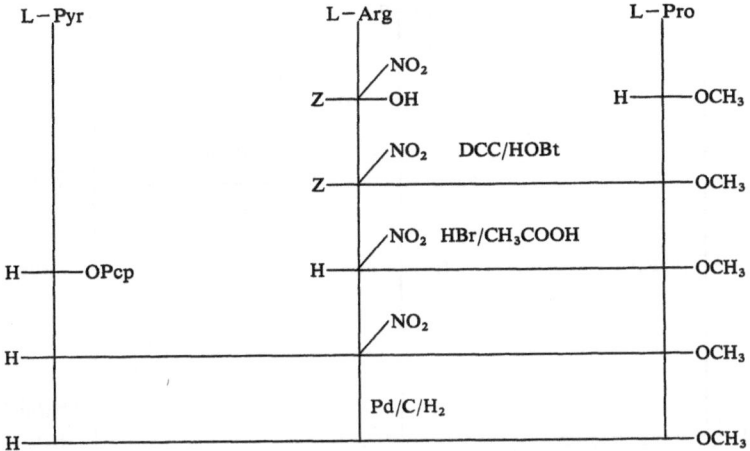

Syntheseschema 39. Darstellungen von L−Pyr−L−Arg(NO₂)−L−Pro−OCH₃ und L−Pyr−L−Arg−L−Pro−OCH₃ nach SIEVERTSSON *et al.* (*407*)

Zu den Synthesen von Pyroglutamyl-β-(2-thienyl)-D-alanyl-L-prolin-methylester und Pyroglutamyl-β-(2-thienyl)-L-alanyl-L-prolinmethylester und den entsprechenden Prolinamiden wird zunächst Formyl-β-(2-thienyl)-D,L-alanin in die optischen Antipoden aufgespalten. SIEVERTSSON *et al.* stellten die Verbindungen nach Syntheseschema 40 dar (*407*).

Die biologischen Tests der von SIEVERTSSON *et al.* (*407*) hergestellten Verbindungen zeigen, daß das Leu² − TRH noch beträchtliche TSH-ausschüttende Wirkung besitzt. Dieser Sachverhalt beweist, daß weder die basischen noch die heteroaromatischen Eigenschaften des Histidins unbedingte Voraussetzungen für die Hormonwirkung sind. Ebenso liefert der Austausch des Imidazolrestes von TRH durch den Thienylrest Hormonderivate mit beträchtlicher biologischer Aktivität. Da Gly² − TRH biologisch völlig inaktiv ist, ist die Seitenkette der zweiten Aminosäure für die TSH-ausschüttende Wirkung eines Peptids notwendig (*407*).

Inwieweit Kettenverlängerung die biologische Aktivität von TRH beeinflußt, untersuchten SIEVERTSSON *et al.* durch Synthese und Tests einiger Tetrapeptide (*408*).

C-terminale Verlängerung des TRH-Restes um Glycinamid liefert ein Peptid (L−Pyr−L−His−L−Pro−Gly−NH₂), welches im Vergleich zu TRH noch 30-proz. biologische Aktivität besitzt. Die Tetrapeptide L-Pyroglutamyl-L-histidyl-glycyl-L-prolinamid und L-Pyroglutamyl-glycyl-L-histidyl-L-prolinamid sind TRH-Derivate, bei welchen die ursprüngliche Sequenz um einen mittelständigen Glycinrest verlängert ist. Beide Verbindungen zeigen keine TSH-ausschüttende Wirkung.

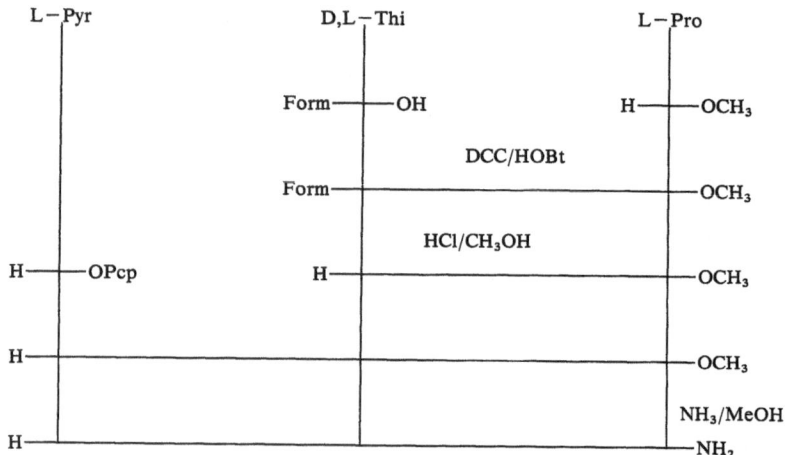

Syntheseschema 40. Darstellungen von L–Pyr–β–(2-thienyl)–L–Ala–L–Pro–OCH₃,
L–Pyr–β–(2-thienyl)–L–Ala–L–Pro–NH₂, L–Pyr–β–(2-thienyl)–D–Ala–L–Pro–OCH₃
und L–Pyr–β–(2-thienyl)–D–Ala–L–Pro–NH₂ nach SIEVERTSSON *et al.* (*407*)

Die Darstellungsmethoden der Verbindungen L−Pyr−L−His−
L−Pro−Gly−NH₂, L−Pyr−L−His(Bzl)−L−Pro−Gly−NH₂, L−
Pyr−L−His−Gly−L−Pro−NH₂ und L−Pyr−Gly−L−His−L−
Pro−NH₂ sind in den Syntheseschemata 41—43 zusammengefaßt (*408*).
Zur Knüpfung der Peptidbindungen wird durchwegs die Dicyclohexyl-
carbodiimidmethode verwendet (*408*).

Von den vier TRH-Derivaten L-Pyroglutamyl-L-histidyl-pyrrolidin,
Cyclopentylcarbonyl-L-histidyl-L-prolinamid, Cyclopentylcarbonyl-L-
histidyl-pyrrolidin und L-Prolyl-L-histidyl-L-prolinamid zeigt keines
TSH-ausschüttende Wirkung (*409*). Durch Cyclopentylcarbonyl-L-histi-
dyl-pyrrolidin allerdings läßt sich die TRH-induzierte Freisetzung von
TSH ebenso hemmen, wie durch L-Trijodthyronin, L-Thyroxin, D-Tri-
jodthyronin und D-Thyroxin (*409, 410*).

Für zahlreiche physiologische Untersuchungen ist radioaktiv markiertes TRH besonders geeignet. Jodierung von TRH und anschließender
Austausch der Jodatome durch Tritium liefert ein sehr reines, hoch-
markiertes Hormon (*411*).

GH₃-Zellen sind hypophysäre Ratten-Tumor-Zellen, welche zwar
Prolactin und Wachstumshormone, aber kein Thyreotropin in die Nährlösung abgeben. TRH wird spezifisch an Membran-Rezeptoren von
GH₃-Zellen gebunden und verursacht dadurch eine Zunahme der
Synthese und Freisetzung von Prolactin; die Wachstumshormonpro-
duktion dagegen wird gehemmt. In einer Untersuchung von HINKLE *et al.*
(*412*) wird von TRH-Analoga die Prolactin-freisetzende Wirkung, die

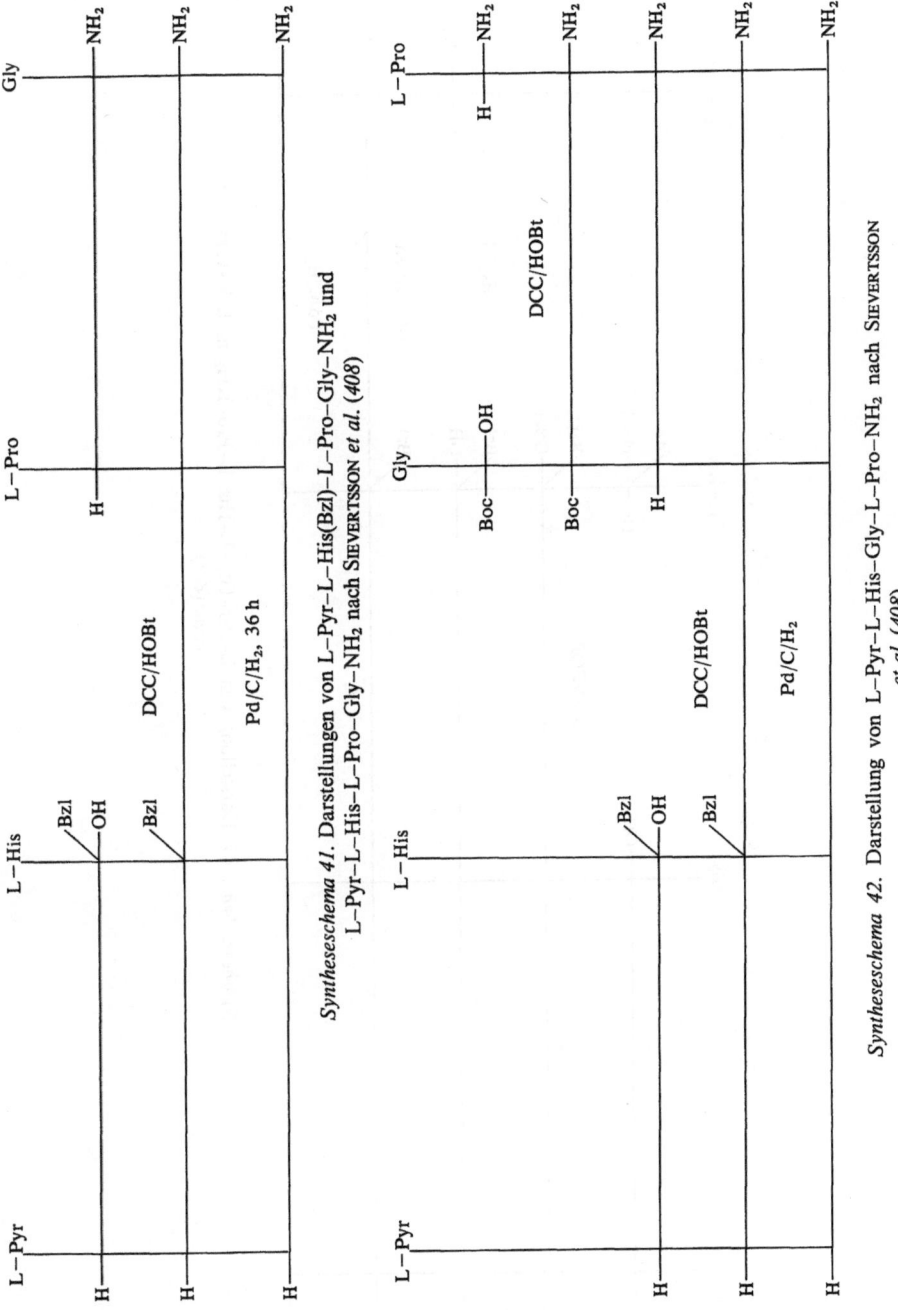

Syntheseschema 41. Darstellungen von L–Pyr–L–His(Bzl)–L–Pro–Gly–NH₂ und L–Pyr–L–His–L–Pro–Gly–NH₂ nach SIEVERTSSON *et al.* (408)

Syntheseschema 42. Darstellung von L–Pyr–L–His–Gly–L–Pro–NH₂ nach SIEVERTSSON *et al.* (408)

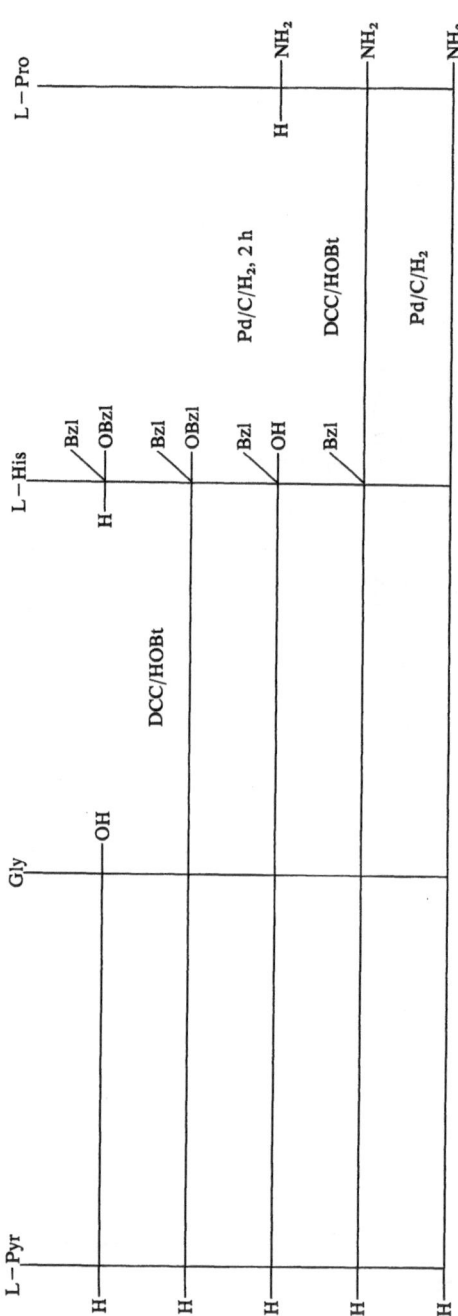

Syntheseschema 43. Darstellung von L–Pyr–Gly–L–His–L–Pro–NH₂ nach Sievertsson *et al.* (408)

Wachstumshormonfreisetzung-hemmende Wirkung und die TRH-Rezeptor-Affinität bei GH₃-Zellen bestimmt.

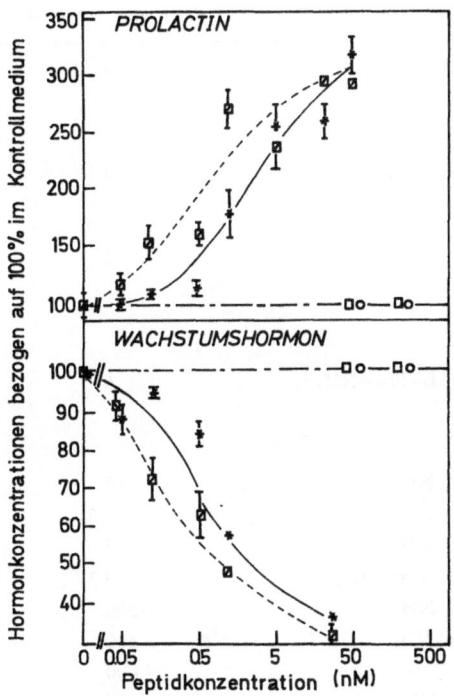

Abb. 5. Prolactin- und Wachstumshormonkonzentrationen in Nährmedien von GH₃-Zellen nach Gabe verschiedener Konzentrationen von TRH bzw. TRH-Analoga. Die Bestimmungen von Prolactin und Wachstumshormon erfolgen 3 Tage nach Zugabe der Peptide. ✴ ✴ ✴ TRH, ▨ ▨ ▨ [3–Me–L–His]²–TRH, o o o Lys²–TRH, ☐ ☐ ☐ Arg²–TRH (412)

Abb. 5 zeigt die Prolactin-freisetzende und Wachstumshormon-freisetzung-hemmende Wirkung von TRH, [3-Me – L – His]² – TRH, Lys² – TRH und Arg² – TRH. Die Ergebnisse von 26 weiteren TRH-Analoga sind in Tab. 5 zusammengefaßt. Es werden jeweils die Prolactin- und Wachstumshormonkonzentrationen im Nährmedium von GH₃-Zellen nach dreitägiger Einwirkung der Peptide bestimmt. Als Maß für die biologische Aktivität eines Peptids, wird die Hälfte derjenigen Konzentration angegeben, welche maximale Prolactinsynthese verursacht.

Da [³H]TRH spezifisch von GH₃-Zellen gebunden wird, kann durch Zugabe von TRH oder dessen Analoga die Verdrängung des markierten Hormons vom Rezeptor untersucht werden. In Abb. 6 ist die Verdrängung

Tabelle 5. *Peptidkonzentration (nM) zur Stimulation der halben maximalen Prolactinausschüttung und Dissoziationskonstanten der Peptid-Rezeptor-Komplexe bei GH$_3$-Zellen (K$_{Diss}$) für TRH und verschiedene Analoga (412)*

Peptid	Konzentration (nM)	K$_{Diss}$(nM)
L–Pyr–L–His–L–Pro–NH$_2$	2	25
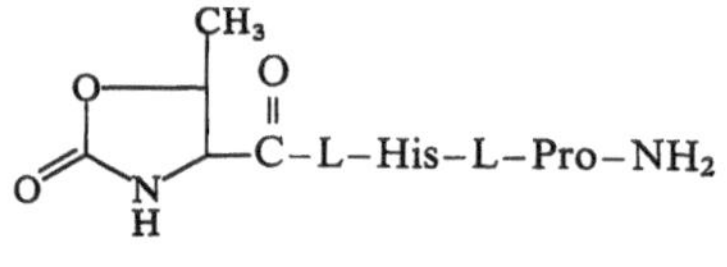 (5-methyl-oxazolidin-2-on-4-carbonyl)–L–His–L–Pro–NH$_2$	4	120
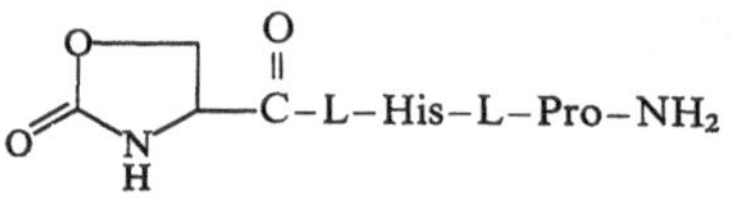 (oxazolidin-2-on-4-carbonyl)–L–His–L–Pro–NH$_2$	6	200
L–Pro–L–His–L–Pro–NH$_2$	250	1 000
L–Glu(OCH$_3$)–L–His–L–Pro–NH$_2$	2,5	35
L–Gln–L–His–L–Pro–NH$_2$	7	5 400
L–Pyr–3–Me–L His–L–Pro–NH$_2$	0,3	4,7
L–Pyr–L–Lys–L–Pro–NH$_2$	> 500	> 2 500
L–Pyr–L–Arg–L–Pro–NH$_2$	> 500	> 2 500
L–Pyr–L–His–L–Pro–OH	4 000	13 000
L–Pyr–L–His–L–Pro–NH–CH$_3$	50	1 000
L–Pyr–L–His–L–Pro–NH–CH$_2$–CH$_3$	40	500
L–Pyr–L–His–L–Pro–N(CH$_2$)$_2$	> 1 000	> 30 000
L–Pyr–L–His–NH$_2$	> 500	> 2 500
L–Pyr–L–His–Gly–NH$_2$	> 500	> 2 500
L–Pyr–L–His–L–Leu–NH$_2$	> 500	> 2 500
L–Pyr–L–His–L–Met–NH$_2$	> 500	> 2 500
L–Pyr–L–His–L–Ser–NH$_2$	> 500	> 2 500
L–Pyr–L–His–L–Val–NH$_2$	> 500	> 2 500
L–Pyr–L–His–L–Phe–NH$_2$	> 500	> 2 500
L–Pyr–L–His–L–Trp–NH$_2$	> 500	> 2 500
L–Pyr–L–His–L–Tyr–NH$_2$	> 500	> 2 500
L–Pyr–L–His–L–Pro–L–Ala–NH$_2$	> 500	> 2 500
L–Pyr–D–His–L Pro–NH$_2$	100	600
D–Pyr–L–His–L–Pro–NH$_2$	> 4 000	> 20 000
L–Pyr–L–His–D–Pro–NH$_2$	> 4 000	> 20 000
D–Pyr–D–His–D–Pro–NH$_2$	> 4 000	> 20 000

Literaturverzeichnis: SS. 534—564

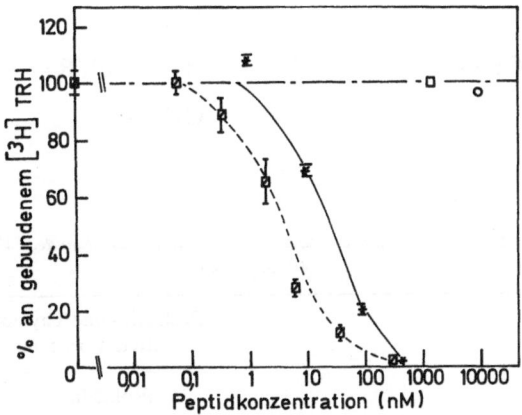

Abb. 6. Beeinflussung der [^{3}H]TRH-Bindung an GH$_3$-Zellen durch die Gegenwart von TRH und einigen Analoga bei verschiedenen Konzentrationen. Die GH$_3$-Zellen werden vor den Experimenten 1 Stunde lang mit [^{3}H]TRH-haltigen (10 nM) Kulturmedium inkubiert. ✱ ✱ ✱ TRH, ▨▨▨ [3–Me–L–His]^2–TRH, ○○○ Lys^2–TRH, □□□ Arg^2–TRH

von an GH$_3$-Zellen gebundenes [^{3}H]TRH bei steigender Konzentration von TRH- bzw. TRH-Analoga dargestellt. Die GH$_3$-Zellen wurden 1 Stunde lang mit Nährmedium, welches 10 nM [^{3}H]TRH enthält, inkubiert. Die Dissoziationskonstanten der TRH-Analoga-Rezeptor-Komplexe sind in Tab. 5 zusammengestellt.

Die Untersuchungen von HINKLE et al. (412) zeigen, daß für die biologische Wirkung von TRH die Fünfringstruktur der Pyroglutaminsäure ganz entscheidend ist, da das TRH-Derivat L – Gln – L – His – L – Pro – NH$_2$ mit N-terminalem Glutaminrest extrem geringe Rezeptoraffinität besitzt. Auch das C-terminale Prolinamidende ist für die Rezeptorbindung notwendig: Von 14 Peptiden mit substituiertem Prolinamid zeigen nur die Prolinmethylamid- und Prolinäthylamidderivate noch 1% der Bindungsaktivität des natürlichen Hormons.

Wie aus Tab. 6 hervorgeht, haben Verbindungen mit großer TSH-ausschüttender auch starke Prolactin-freisetzende Wirkung.

Die Ergebnisse der Tab. 6 lassen den Schluß zu, daß die Rezeptoren für die Prolactin- und Thyreotropinfreisetzung in ihren Strukturen ähnlich oder gar identisch sind.

Inwieweit sich Substitutionen der Amidwasserstoffatome des Prolins auf die biologische Aktivität von TRH auswirken, sollte nach der Darstellung verschiedener [Pro-alkylamid]^3 – TRH-Verbindungen überprüft werden (413—415). Die TRH-Derivate der allgemeinen Form L – Pyr – L – His – L – Pro – NHR (R = – CH$_3$ (413), R = – CH$_2$ – CH$_3$ (414)

$$R = -CH_2-CH_2-CH_3 \ (416), \quad R = -CH\begin{smallmatrix} CH_3 \\ \diagup \\ \diagdown \\ CH_3 \end{smallmatrix} \ (416), \quad R = -CH_2-CH_2-$$

Tabelle 6. *Vergleich der Prolactin- und Thyreotropin-freisetzenden Wirkung von TRH-Analoga (412)*

Peptid	Prolactin- und Thyreotropin-freisetzende Aktivität bezogen auf TRH = 100%	
	Prolactin	Thyreotropin
L–Pyr–L–His–L–Pro–NH₂	100	100
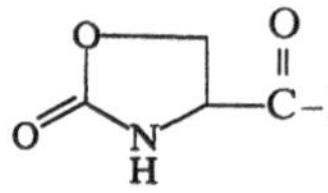 –C–L–His–L–Pro–NH₂	33	38
L–Pro–L–His–L–Pro–NH₂	0,8	0,01
L–Gln–L–His–L–Pro–NH₂		< 5
L–Pyr–3–Me–L–His–L–Pro–NH₂	670	800
L–Pyr–L–Lys–L–Pro–NH₂	< 0,4	0,02
L–Pyr–L–Arg–L–Pro–NH₂	< 0,4	0,05
L–Pyr–L–His–L–Pro–OH	0,05	0,02—0,1
L–Pyr–L–His–L–Pro–NH–CH₃	4	6—9
L–Pyr–L–His–L–Pro–NH–CH₂–CH₃	5	14
L–Pyr–L–His–L–Pro–N(CH₃)₂	< 0,2	0,5
L–Pyr–L–His–L–Leu–NH₂	< 0,4	0,04—0,1
L–Pyr–L–His–L–Val–NH₂	< 0,4	0,1—0,2
L–Pyr–L–His–L–Pro–L–Ala–NH₂	< 0,4	0,5
L–Pyr–L–His–NH₂	< 0,4	< 0,02
L–Pyr–L–His–Gly–NH₂	< 0,4	< 0,02
L–Pyr–L–His–L–Met–NH₂	< 0,4	< 0,02
L–Pyr–L–His–L–Phe–NH₂	< 0,4	< 0,02
L–Pyr–L–His–L–Tyr–NH₂	< 0,4	< 0,02
L–Pyr–L–His–L–Trp–NH₂	< 0,4	< 0,02
L–Pyr–D–His–L–Pro–NH₂	2	3
D–Pyr–L–His–L–Pro–NH₂	< 0,05	0,01
L–Pyr–L–His–D–Pro–NH₂	< 0,05	0,01
D–Pyr–D–His–D–Pro–NH₂	< 0,05	< 0,01

$$CH_2 - CH_3 \; (417), \; R = - \overset{\overset{\displaystyle CH_3}{\displaystyle |}}{\underset{\underset{\displaystyle CH_3}{\displaystyle |}}{C}} - CH_3 \; (417), \; R = - CH_2 - CH_2 - CH_2 - CH_2 - $$

CH_3 *(417)* und $R = - CH_2 - CH_2 - CH_2 - CH_2 - CH_2 - CH_2 - CH_3$ *(418)*
sind nach Syntheseschema 44 hergestellt worden.

Die Hormonderivate (vgl. Syntheseschema 44) wurden mit einem modifizierten McKenzie-Test an Mäusen auf ihre biologische Aktivität überprüft. Dabei zeigt sich ein konstantes Ansteigen der Thyreotropin-freisetzenden Wirkung vom Methyl- zum n-Pentyl-Derivat. Stark verminderte bzw. keine Aktivität zeigen N^{amid}-Heptyl-TRH, N^{amid}-tert-Butyl-TRH und N^{amid}-Isopropyl-TRH.

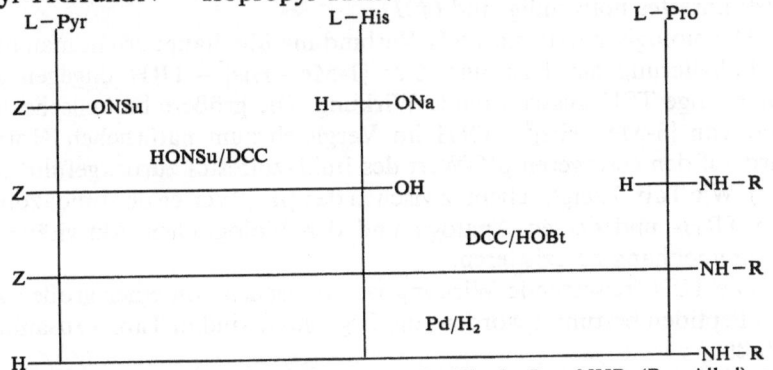

Syntheseschema 44. Darstellungen von L—Pyr—L—His—L—Pro—NHR (R = Alkyl) nach
VOELTER *et al. (413—418)*

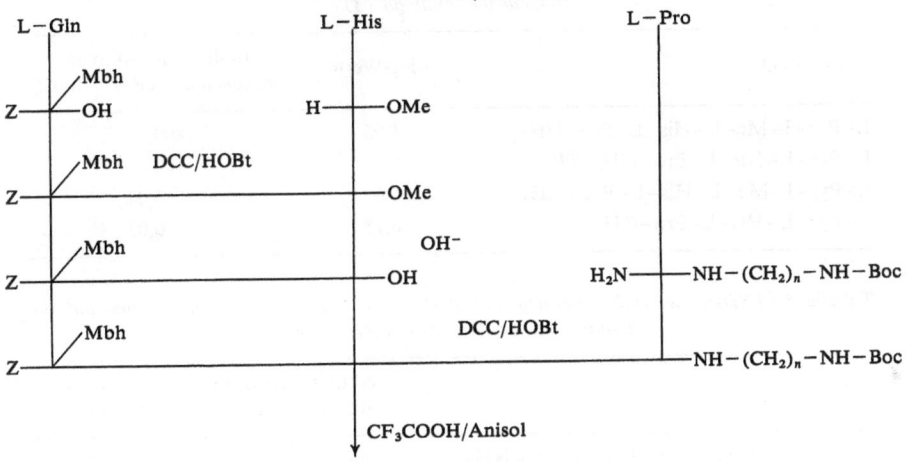

Syntheseschema 45. Darstellungen von L—Pyr—L—His—L—Pro—NH–(CH$_2$)$_n$–NH$_2$ für
n = 4, 7, 10 nach FUCHS *et al. (419, 420)*

Vor kurzem ist auch eine Synthesestrategie für TRH-Derivate der allgemeinen Form $L - Pyr - L - His - L - Pro - NH - (CH_2)_n - NH_2$ beschrieben worden. Nach Syntheseschema 45 sind drei Hormonderivate (n = 4,7 und 10) hergestellt worden (419, 420).

Einige weitere wichtige Untersuchungen beschäftigen sich mit der Bedeutung des Histidinrestes für die biologische Aktivität von TRH (421—423).

Wie schon erwähnt, besitzt Phe^2-TRH noch beträchtliche TSH-ausschüttende Wirkung; das Hormonderivat wird, ebenso wie TRH selbst, durch Blutserum desaktiviert und seine biologische Wirkung durch Trijodthyronin inhibiert. Daraus wird der Schluß gezogen, daß für die volle Hormonwirkung sowohl die Π-Elektronen als auch die Basizität des Histidinrestes notwendig sind (421).

Die biologisch aktivste TRH-Verbindung überhaupt erhält man durch 3-Methylierung des Histidinrestes; $[1-Me - His]^2 - TRH$ dagegen zeigt nur geringe TSH-ausschüttende Wirkung. Die größere biologische Aktivität von $[3-Me - His]^2 - TRH$ im Vergleich zum natürlichen Hormon wird auf den geringeren pK-Wert des Imidazolrestes zurückgeführt (422, 423). Wie Tab. 7 zeigt, scheint zwischen den pK_a-Werten der Imidazolreste von TRH- und dessen Analoga und den biologischen Aktivitäten ein Zusammenhang zu existieren.

Die TSH-freisetzende Wirkung ist inzwischen von einer großen Zahl von Peptiden bestimmt worden; die Ergebnisse sind in Tab. 8 zusammengefaßt.

Tabelle 7. *pK$_a$-Werte der Imidazolreste und biologische Aktivitäten von TRH und verschiedenen Analoga (31)*

Verbindung	pK$_a$-Werte	Biologische Aktivität bezogen auf TRH = 100%
L–Pyr–3–Me–L–His–L–Pro–NH$_2$	5.95	800
L–Pyr–L–His–L–Pro–NH$_2$ (TRH)	6.25	100
L–Pyr–1–Me–L–His–L–Pro–NH$_2$	6.6	0,04
L–Pyr–L–His–L–Pro–OH	6.75	0,02

Tabelle 8. *TSH-freisetzende Wirkung von Peptiden und Peptidderivaten bezogen auf die biologische Aktivität von TRH = 100%*

Verbindung	Aktivität bezogen auf TRH = 100%	Literaturzitate
Cyclopentylcarbonyl–L–His–L–Pro–NH$_2$	inaktiv	(409)
Cyclopentylcarbonyl–L–His–L–pyrrolidin	inaktiv	(409)
L–Gln–L–His–L–Pro–NH$_2$	<5	(412)
L–Glu–L–His–L–Pro–OH	inaktiv	(373)

Literaturverzeichnis: SS. 534—564

Tabelle 8 *(Fortsetzung)*

Verbindung	Aktivität bezogen auf TRH = 100%	Literatur-zitate
L–Glu–L–His–L–Pro–OH	inaktiv	*(385)*
L–Glu–L–His–L–Pro–OH	inaktiv	*(375)*
Ac–L–Glu–L–His–L–Pro–OH	inaktiv	*(375)*
Ac–L–Glu–L–His–L–Pro–OH	inaktiv	*(385)*
γ–L–Glu–L–His–L–Pro–OH	inaktiv	*(375)*
Ac–γ–L–Glu–L–His–L–Pro–OH	inaktiv	*(375)*
L–Glu–L–Pro–L–His–OH	inaktiv	*(385)*
L–Glu–L–Pro–L–His–OH	inaktiv	*(375)*
L–Glu–L–Pro–L–His–OH	inaktiv	*(373)*
L–His–L–Glu–L–Pro–OH	inaktiv	*(373)*
L–His–L–Glu–L–Pro–OH	inaktiv	*(385)*
L–His–L–Glu–L–Pro–OH	inaktiv	*(375)*
Ac–L–His–L–Glu–L–Pro–OH	inaktiv	*(375)*
L–His–L–γ–Glu–L–Pro–OH	inaktiv	*(385)*
L–His–L–γ–Glu–L–Pro–OH	inaktiv	*(375)*
Ac–L–His–L–γ–Glu–L–Pro–OH	inaktiv	*(375)*
L–His–L–Pro–L–Gln–OH	inaktiv	*(373)*
L–His–L–Pro–L–Glu–OH	inaktiv	*(373)*
L–His–L–Pro–L–Glu–OH	inaktiv	*(385)*
L–His–L–Pro–L–Glu–OH	inaktiv	*(375)*
Ac–L–His–L–Pro–L–Glu–OH	inaktiv	*(375)*
L–Pro–L–Glu–L–His–OH	inaktiv	*(373)*
L–Pro–L–Glu–L–His–OH	inaktiv	*(385)*
L–Pro–L–Glu–L–His–OH	inaktiv	*(375)*
Ac–L–Pro–L–Glu–L–His–OH	inaktiv	*(375)*
L–Pro–L–His–L–Gln–OH	inaktiv	*(373)*
L–Pro–L–His–L–Glu–OH	inaktiv	*(373)*
L–Pro–L–His–L–Glu–OH	inaktiv	*(385)*
L–Pro–L–His–L–Glu–OH	inaktiv	*(375)*
Ac–L–Pro–L–His–L–Glu–OH	inaktiv	*(375)*
L–Pro–L–His–L–Pro–NH_2	0,01	*(23)*
L–Pro–L–His–L–Pro–NH_2	inaktiv	*(409)*
L–Pyr–L–Ala–L–His–NH_2	inaktiv	*(424)*
L–Pyr–L–Arg–L–Pro–NH_2	inaktiv	*(401, 402)*
L–Pyr–L–Arg–L–Pro–NH_2	0,05	*(31)*
L–Pyr–L–Arg(NO_2)–L–Pro–OCH_3	inaktiv	*(407)*
L–Pyr–L–α,γ–diaminobutyryl–L–Pro–NH_2	inaktiv	*(401, 402)*
L–Pyr–Gly–L–Pro–OCH_3	inaktiv	*(407)*
L–Pyr–Gly–L–Pro–NH_2	inaktiv	*(407)*
L–Pyr–Gly–L–His–L–Pro–NH_2	inaktiv	*(408)*
L–Pyr–Gly–L–His(Bzl)–L–Pro–NH_2	inaktiv	*(408)*
L–Pyr–Gly–L–Tyr–NH_2	inaktiv	*(424)*
L–Pyr–1–Me–L–His–OMe	<0,0025	*(423)*
L–Pyr–1–Me–L–His–OMe	<0,0025	*(422)*
L–Pyr–3–Me–L–His–OMe	0,02	*(423)*
L–Pyr–3–Me–L–His–OMe	0,02	*(422)*
L–Pyr–1–Me–L–His–L–Pro–NH_2	0,04	*(423)*

Tabelle 8 *(Fortsetzung)*

Verbindung	Aktivität bezogen auf TRH = 100%	Literatur-zitate
L–Pyr–1–Me–L–His–L–Pro–NH$_2$	0,04	*(422)*
L–Pyr–1–Me–L–His–L–Pro–NH$_2$	0,04	*(31)*
L–Pyr–3–Me–L–His–L–Pro–NH$_2$	800	*(423)*
L–Pyr–3–Me–L–His–L–Pro–NH$_2$	800	*(422)*
L–Pyr–3–Me–L–His–L–Pro–NH$_2$	800	*(31)*
L–Pyr–L–His–NH$_2$	inaktiv	*(424)*
L–Pyr–L–His–NH$_2$	<0,02	*(412)*
L–Pyr–L–His–OMe	<0,0005	*(423)*
L–Pyr–L–His–OMe	<0,0005	*(422)*
L–Pyr–L–His–OMe	inaktiv	*(375)*
L–Pyr–L–His–Abu–NH$_2$	0,09	*(388)*
L–Pyr–L–His–L–Ala–NH$_2$	<0,09	*(388)*
L–Pyr–L–His(Bzl)–L–Ala–NH$_2$	inaktiv	*(388)*
L–Pyr–L–His–Gly–NH$_2$	<0,02	*(23)*
L–Pyr–L–His–Gly–NH$_2$	inaktiv	*(388)*
L–Pyr–L–His–Gly–L–Pro–NH$_2$	inaktiv	*(408)*
L–Pyr–L–His(Bzl)–Gly–NH$_2$	inaktiv	*(388)*
L–Pyr–L–His(Bzl)–Gly–L–Pro–NH$_2$	inaktiv	*(408)*
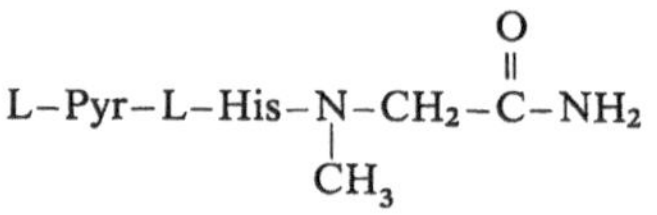 L–Pyr–L–His–N–CH$_2$–C–NH$_2$ (with O double bond, CH$_3$)	0,32	*(23)*
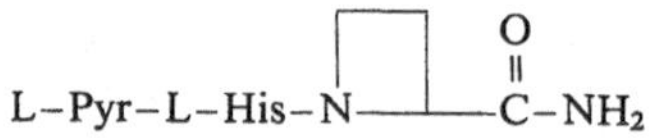 L–Pyr–L–His–N⎡⎤–C–NH$_2$ (with O double bond)	1,6	*(23)*
L–Pyr–L–His–L–Ile–OCH$_3$	inaktiv	*(388)*
L–Pyr–L–His–L–Ile–NH$_2$	inaktiv	*(388)*
L–Pyr–L–His–L–Leu–NH$_2$	0,09	*(388)*
L–Pyr–L–His–L–Leu–NH$_2$	0,04	*(23)*
L–Pyr–L–His–L–Leu–NH$_2$	0,04—0,1	*(412)*
L–Pyr–L–His(Bzl)–L–Leu–NH$_2$	inaktiv	*(388)*
L–Pyr–L–His–L–Met–NH$_2$	inaktiv	*(388)*
L–Pyr–L–His–L–Met–NH$_2$	<0,02	*(412)*
L–Pyr–L–His–L–Phe–NH$_2$	inaktiv	*(388)*
L–Pyr–L–His–L–Phe–NH$_2$	<0,02	*(412)*
L–Pyr–L–His(Bzl)–L–Phe–NH$_2$	inaktiv	*(388)*
L–Pyr–L–His(Bzl)–L–Pro–NH$_2$	0,09	*(388)*
L–Pyr–D–His–L–Pro–NH$_2$	3	*(412)*
L–Pyr–D–His–L–Pro–NH$_2$	2—3	*(403)*
L–Pyr–D–His–L–Pro–NH$_2$	30	*(404)*
D–Pyr–L–His–L–Pro–NH$_2$	0,01	*(412)*
D–Pyr–L–His–L–Pro–NH$_2$	0,1	*(403)*
D–Pyr–D–His–D–Pro–NH$_2$	<0,01	*(412)*
D–Pyr–D–His–D–Pro–NH$_2$	inaktiv	*(403)*
L–Pyr–L–His–D–Pro–NH$_2$	0,01	*(412)*
L–Pyr–L–His–D–Pro–NH$_2$	0,1	*(403)*

Literaturverzeichnis: SS. 534—564

Tabelle 8 *(Fortsetzung)*

Verbindung	Aktivität bezogen auf TRH = 100%	Literaturzitate
L–Pyr–L–His–L–Pro–OH	0,02	*(31)*
L–Pyr–L–His–L–Pro–OH	gering	*(385)*
L–Pyr–L–His–L–Pro–OH	gering	*(375)*
L–Pyr–L–His–L–Pro–OH	0,02	*(23)*
L–Pyr–L–His–L–Pro–OH	0,1	*(412)*
L–Pyr–L–His–L–Pro–OMe	10	*(31)*
L–Pyr–L–His–L–Pro–OMe	9	*(388)*
L–Pyr–L–His–L–Pro–OMe	stark aktiv	*(375)*
L–Pyr–L–His–L–Pro–OMe	10	*(23)*
L–Pyr–L–His(Bzl)–L–Pro–OMe	inaktiv	*(388)*
L–Pyr(Me)–L–His–L–Pro–NH$_2$	1,7	*(23)*
L–Pyr–L–His–L–Pro–NH–NH$_2$	14	*(23)*
L–Pyr–L–His–L–Pro–NH–CH$_3$	6	*(412)*
L–Pyr–L–His–L–Pro–NH–CH$_3$	9	*(388)*
L–Pyr–L–His–L–Pro–NH–CH$_3$	beträchtliche Aktivität	*(413)*
L–Pyr–L–His(Bzl)–L–Pro–NH–CH$_3$	inaktiv	*(388)*
L–Pyr–L–His–L–Pro–NH–CH$_2$–CH$_3$	14	*(23)*
L–Pyr–L–His–L–Pro–NH–CH$_2$–CH$_3$	beträchtliche Aktivität	*(414)*
L–Pyr–L–His–L–Pro–NH–(CH$_2$)$_2$–CH$_3$	beträchtliche Aktivität	*(416)*

$$\text{L–Pyr–L–His–L–Pro–NH–CH} \begin{array}{c} \text{CH}_3 \\ | \\ \\ | \\ \text{CH}_3 \end{array}$$

L–Pyr–L–His–L–Pro–NH–CH(CH$_3$)–CH$_3$	kaum Aktivität	*(416)*
L–Pyr–L–His–L–Pro–NH–(CH$_2$)$_3$–CH$_3$	beträchtliche Aktivität	*(417)*

$$\text{L–Pyr–L–His–L–Pro–NH–C} \begin{array}{c} \text{CH}_3 \\ | \\ \text{–CH}_3 \\ | \\ \text{CH}_3 \end{array}$$

L–Pyr–L–His–L–Pro–NH–C(CH$_3$)$_2$–CH$_3$	kaum Aktivität	*(417)*
L–Pyr–L–His–L–Pro–NH–(CH$_2$)$_4$–CH$_3$	beträchtliche Aktivität	*(417)*
L–Pyr–L–His–L–Pro–NH–(CH$_2$)$_6$–CH$_3$	kaum Aktivität	*(418)*
L–Pyr–L–His–L–Pro–NH–(CH$_2$)$_4$–NH$_2$	7,5	*(419)*
L–Pyr–L–His–L–Pro–NH–(CH$_2$)$_7$–NH$_2$	4	*(419)*
L–Pyr–L–His–L–Pro–NH–(CH$_2$)$_{10}$–NH$_2$	0,2	*(419)*
L–Pyr–L–His–L–Pro–NH–CH$_2$–CH$_2$OH	16	*(23)*
L–Pyr–L–His–L–Pro–N(CH$_3$)$_2$	0,5	*(23)*
L–Pyr–L–His–L–Pro–N(CH$_2$–CH$_3$)$_2$	0,05	*(23)*
L–Pyr–L–His–L–Pro–N(piperidinyl) 	0,2	*(31)*
L–Pyr–L–His–L–Pro–N(piperidinyl) 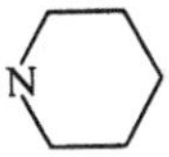	0,6	*(23)*
L–Pyr–L–His–L–Pro–NH–(phenyl) 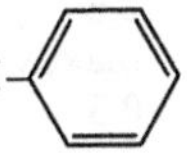	16	*(23)*

Tabelle 8 *(Fortsetzung)*

Verbindung	Aktivität bezogen auf TRH = 100%	Literatur-zitate
L–Pyr–L–His–N◁ (Azetidinring) 	beträchtliche Aktivität	(391)
L–Pyr–L–His–N⬠ (Pyrrolidinring) 	0,3	(23)
L–Pyr–L–His–pyrrolidin	inaktiv	(409)
L–Pyr–L–His–N⬡ (Piperidinring) 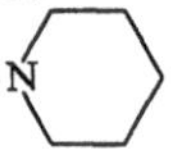	0,04	(23)
L–Pyr–L–His–N⬠—CH₂OH 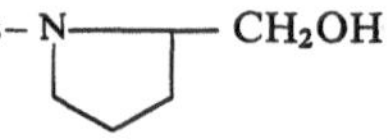	1,2	(23)
L–Pyr–L–His–N⬠—C(=O)–NH₂, OH 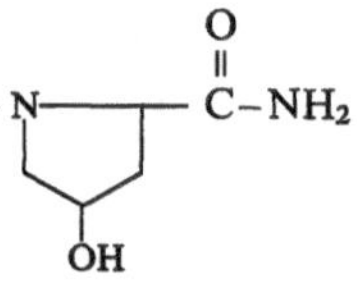	0,14	(23)
L–Pyr–L–His–L–Pro–L–Ala–NH₂	0,5	(23)
L–Pyr–L–His–L–Pro–Gly–NH₂	30	(408)
L–Pyr–L–His–L–Pro–Gly–NH₂	35	(23)
L–Pyr–L–His(Bzl)–L–Pro–Gly–NH₂	geringe Aktivität	(408)
L–Pyr–L–His–L–Thr–NH₂	inaktiv	(388)
L–Pyr–L–His–L–Val–NH₂	0,18	(388)
L–Pyr–L–His–L–Val–NH₂	0,1—0,2	(412)
L–Pyr–L–His(Bzl)–L–Val–NH₂	inaktiv	(388)
L–Pyr–L–His–L–Trp–NH₂	<0,02	(23)
L–Pyr–L–His–L–Trp–NH₂	inaktiv	(388)
L–Pyr–L–His–L–Tyr–NH₂	<0,02	(412)
L–Pyr–L–Leu–L–Pro–NH₂	0,5	(407)
L–Pyr–L–Lys–L–Pro–NH₂	0,02	(423)
L–Pyr–L–Lys–L–Pro–NH₂	inaktiv	(401, 402)
L–Pyr–L–Lys–L–Pro–NH₂	0,02	(31)
L–Pyr–L–Met–L–Pro–NH₂	1	(31)
L–Pyr–L–Met–L–Pro–NH₂	1	(423)
L–Pyr–L–Orn–L–Pro–NH₂	0,02	(31)
L–Pyr–L–Orn–L–Pro–NH₂	0,025	(423)
L–Pyr–L–Phe–L–His–NH₂	inaktiv	(424)
L–Pyr–L–Phe–L–3 Hyp–NH₂	inaktiv	(421)
L–Pyr–L–Phe–L–Pro–NH₂	10	(421)
L–Pyr–L–Phe–L–Pro–NH₂	<10	(425)
L–Pyr–L–Phe–L–Pro–NH₂	<10	(406)
L–Pyr–β–(3-pyrazolyl)–L–Ala–L–Pro–NH₂	5	(400)
L–Pyr–β–(3-pyrazolyl)–L–Ala–L–Pro–NH₂	2,6	(401, 402)
L–Pyr–β–(2-thienyl)–L–Ala–L–Pro–OCH₃	inaktiv	(407)
L–Pyr–β–(2-thienyl)–L–Ala–L–Pro–NH₂	0,5	(407)

Literaturverzeichnis: SS. 534—564

Tabelle 8 *(Fortsetzung)*

Verbindung	Aktivität bezogen auf TRH = 100%	Literatur-zitate
L–Pyr–β–(2-thienyl)–D–Ala–L–Pro–OCH₃	inaktiv	*(407)*
L–Pyr–β–(2-thienyl)–D–Ala–L–Pro–NH₂	0,1	*(407)*
L–Pyr–L–Trp–L–His–NH₂	inaktiv	*(424)*
L–Pyr–L–Trp–L–Pro–NH₂	inaktiv	*(421)*
L–Pyr–L–Trp–L–Tyr–NH₂	inaktiv	*(424)*
L–Pyr–L–Tyr–NH₂	inaktiv	*(424)*
L–Pyr–L–Tyr–L–His–NH₂	inaktiv	*(424)*
L–Pyr–L–Tyr–L–Phe–NH₂	inaktiv	*(424)*
L–Pyr–L–Tyr–L–Pro–NH₂	inaktiv	*(421)*
L–Pyr–L–Tyr–L–Pro–NH₂	<0,1	*(423)*
L–Pyr–L–Tyr–L–Pro–NH₂	0.08	*(31)*
L–Pyr–L–Tyr(OMe)–L–Pro–NH₂	inaktiv	*(405)*
L–Pyr–L–Tyr(OMe)–L–Pro–NH₂	inaktiv	*(393)*
L–Pyr–L–Tyr–L–Trp–NH₂	inaktiv	*(424)*
L–Pyr–L–Tyr–L–Tyr–NH₂	inaktiv	*(424)*

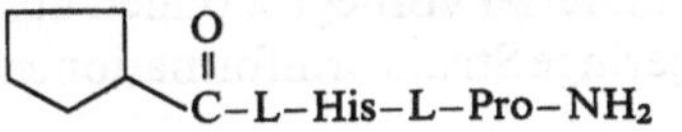 [structure] C–L–His–L–Pro–NH₂	<0,01	*(23)*
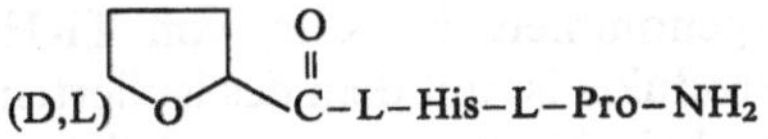 (D,L) [structure] C–L–His–L–Pro–NH₂	0,01	*(23)*
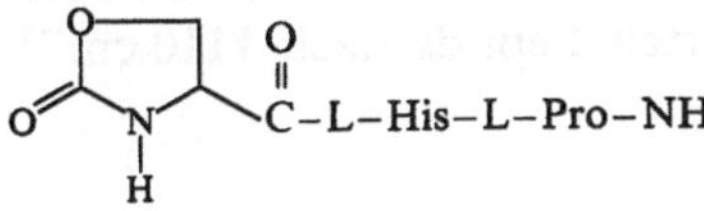 [structure] C–L–His–L–Pro–NH₂	38	*(412)*
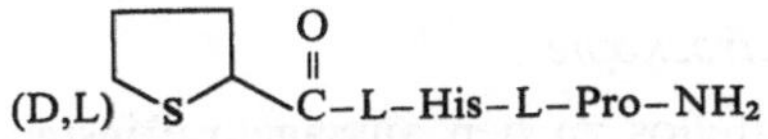 (D,L) [structure] C–L–His–L–Pro–NH₂	0,2	*(23)*

D. Spektroskopische Untersuchungen von TRH und seinen Derivaten

Die moderne Naturstoffchemie setzt die verschiedenen spektroskopischen Methoden einmal zur Strukturaufklärung und zum anderen zur Korrelation spektroskopischer Daten mit der biologischen Aktivität von bioaktiven Molekülen ein.

D 1. Infrarotspektroskopie

Die Infrarotspektroskopie leistet wertvolle Dienste zum Nachweis identischer Verbindungen. Da IR-Banden charakteristisch sind für bestimmte Atomschwingungen im Molekül, erlaubt das IR-Spektrum meist

31*

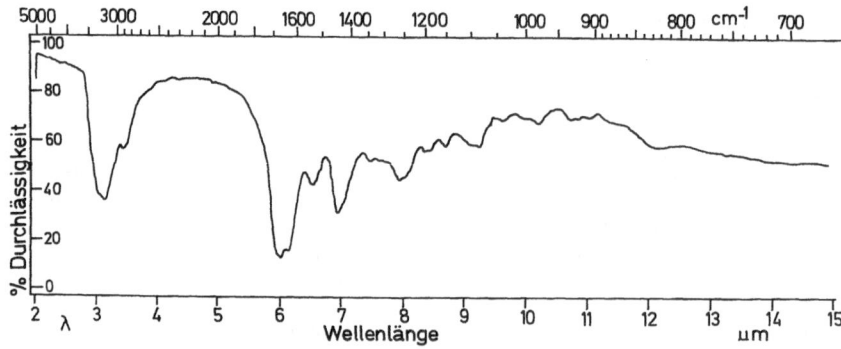

Abb. 7. IR-Spektrum von TRH, aufgenommen in KBr (426)

nur Atomgruppen, nicht aber die Molekülklasse einer unbekannten Verbindung zu identifizieren.

Obgleich im Zusammenhang mit TRH-Synthesen bzw. TRH-Analoga-Synthesen (419, 426) zahlreiche IR-Spektren aufgenommen worden sind und daher viel spektroskopisches Vergleichsmaterial vorliegt, können aus dem IR-Spektrum von TRH (7, 372, 426) nur geringe Strukturinformationen entnommen werden.

Abb. 7 zeigt das IR-Spektrum, aufgenommen in KBr, von TRH (426). Das Spektrum des synthetischen Produkts ist mit dem des isolierten Naturstoffs nahezu identisch. Die Bande bei 1121 cm^{-1} des natürlichen Hormons ist im Spektrum des synthetisierten Peptids nach 1110 cm^{-1} verschoben (372).

D 2. Massenspektroskopie

Die Massenspektroskopie zählt zweifellos zu den aussagekräftigsten Untersuchungsmethoden der Naturstoffchemie (427—431). Auch die Strukturaufklärung von TRH wurde durch die Massenspektroskopie wesentlich unterstützt (432, 433).

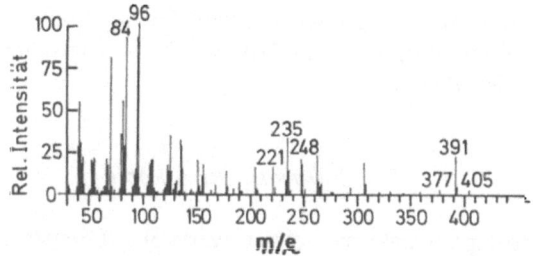

Abb. 8. Massenspektrum von mit Diazomethan methyliertem L–Pyr–L–His–L–Pro–OH (LKB 9 000, 70 eV) (433)

Literaturverzeichnis: SS. 534—564

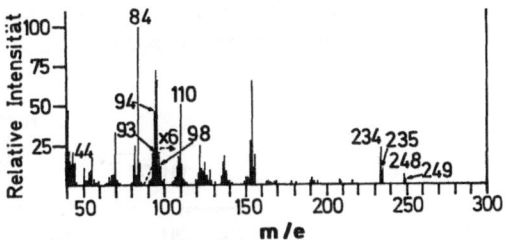

Abb. 9. Massenspektrum von aus natürlichem Material isoliertem TRH nach Derivatisierung mit Diazomethan (LKB 9 000, 70 eV) (*433*)

Durch Massenspektrometrie konnte z. B. gezeigt werden, daß L − Glu − L − His − L − Pro − OH bei der Acetylierung L − Pyr − L − His − L − Pro − OH bildet, welches schwache biologische Aktivität besitzt (*433*).

Durch Umsetzung von L − Pyr − L − His − L − Pro − OH mit Diazomethan bildet sich der Methylester, außerdem erfolgt teilweise Methylierung an den Seitenketten des Histidins und der Pyroglutaminsäure. Im Massenspektrum des methylierten Produkts von L − Pyr − L − His − L − Pro − OH wird der Methylester durch einen Peak bei m/e 377, das methylierte Histidinderivat durch einen solchen bei m/e 391 und das doppelt methylierte Produkt (Histidin- und Pyroglutaminsäurerest) durch das massenhöchste Ion bei m/e 405 angezeigt (vgl. Abb. 8). Obgleich Abb. 8 das Massenspektrum eines Gemisches zeigt, kann auf die Peptidsequenz geschlossen werden. Der Peak bei m/e 221 wird durch ein Dipeptidfragment und die Fragmentpeaks bei m/e 84, 98, 81 und 95 werden durch die Ionen von Pyr, Pyr + CH$_2$, His und His + CH$_2$ verursacht. Abb. 8 zeigt das Massenspektrum von mit Diazomethan methyliertem L − Pyr − L − His − L − Pro − OH.

Selbst nach Derivatisierung mit Diazomethan konnten von natürlichem und synthetischem TRH zunächst nur Massenspektren ohne sichtbaren Molekülpeak aufgenommen werden (Abb. 9) (*433*). Immerhin konnte aus dem Massenspektrum auf die Sequenz geschlossen werden: Die Peaks bei m/e 84, 98, 81, 95 und 235 können den Fragmentionen Pyr, Pyr + CH$_2$, His, His + CH$_2$ und Pyr − His zugeordnet werden.

Die Massenspektren von TRH sind von der Temperatur der Ionenquelle abhängig: Bei einer Ionenquellentemperatur von 160° C treten folgende Fragmentpeaks auf: bei m/e 84 (Pyrrolidonfragment), m/e 81 (Imidazolylmethylenion) und m/e 70 (Prolinfragment). Wird die Ionenquellentemperatur auf 200° C erhöht, dann bilden sich, bedingt durch thermische Reaktionen, neue Fragmentationen. Der Peak bei 234, 1129 wird einem Diketopiperazin, welches sich aus der Histidylprolinsequenz bildet, zugeordnet (*387, 434*).

Die wichtigsten Massenpeaks des Spektrums von TRH sind in Tab. 9 zusammengefaßt (*387*), die wichtigsten Fragmentionen zeigt Abb. 10 (*434*).

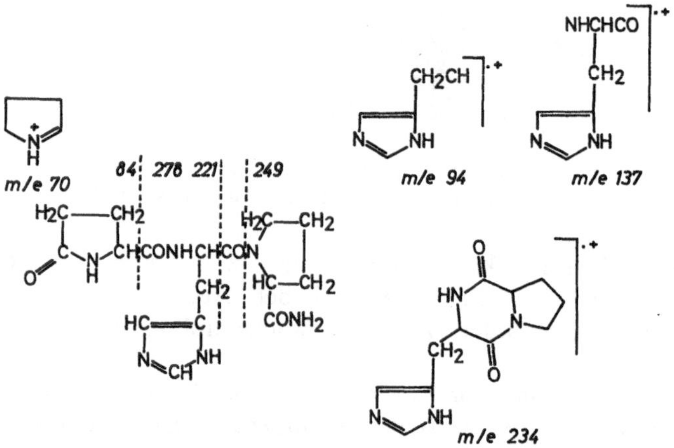

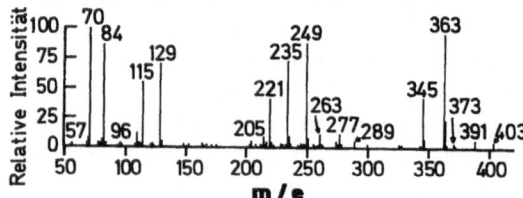

Abb. 10. Die wichtigsten Fragmentationen bei der massenspektroskopischen Fragmentierung von TRH (*434*)

Bessere Ergebnisse liefert die chemische Ionisationsmassenspektrometrie. Abb. 11 zeigt das chemische Ionisationsmassenspektrum von synthetischem TRH. Die Ionen bei m/e 363 (M + 1), 391 (M + 29) und 403 (M + 41) werden durch Protonen-, Äthyl- bzw. Allylanlagerung gebildet (*433*). Die Dipeptidfragmente Pyr − His (m/e 221) und Pyr − His − CO (m/e 249) verursachen charakteristische Peaks im Spektrum und sind für die Sequenzanalyse von Bedeutung.

Abb. 11. Chemisches Ionisationsmassenspektrum von synthetischem TRH (MS 9, 70 eV) (*433*)

In neuester Zeit sind auch Felddesorptionsmassenspektren von TRH und seinen Analoga aufgenommen worden (*435*). Die ersten Untersuchungen zeigen, daß diese Methode eindeutig zu interpretierende Spektren mit intensivem Molekülpeak liefert.

Die Massenspektren von TRH, (N-Butylprolinamid)[3] − TRH, (N-tert-Butylprolinamid)[3]-TRH und (N-Pentylprolinamid)[3]-TRH wurden kürzlich ohne Derivatisierung mit einem MS 9-Gerät vermessen. Die Proben wurden als Hydrochloride über das Direkteinlaßsystem eingeführt (Ionenquellentemperatur ca. 200° C, 70 eV). Dabei ergeben sich außer beim unsubstituierten TRH keine Hinweise auf thermische Umwandlungen vor der Ionisierung. Unter diesen Bedingungen werden ausgeprägte Molekülionen bei m/e 362 (TRH), 418 ((N-Butylprolinamid)^3-TRH, N-tert-Butylprolinamid)^3-TRH) bzw. 432 ((N-Pentylprolinamid)^3-TRH) erhalten. Die wichtigsten Fragmentionen sind in den Tab. 10—13 erfaßt. Abb. 12 zeigt die Strukturen der wichtigsten Fragmentionen (*417*).

Tabelle 9. *Die wichtigsten Massenpeaks (m/e-Werte) mit entsprechenden relativen Intensitäten im Massenspektrum von TRH (387)*

m/e	relative Intensität	m/e	relative Intensität
362	0,2	278	1,4
249	2,9	248	2,4
235	5,7	234	7,6
221	3,8	154	15,7
153	8,6	137	10,0
136	7,2	122	8,6
110	11,4	109	10,9
94	10,0	84	61,0
82	71,5	81	32,0
70	100,0		

Gut interpretierbare Spektren erhält man auch von den Trimethylsilyl(TMS)derivaten von TRH und dessen Teilsequenzen. Die TMS-Verbindungen werden durch Umsatz der Peptide mit N-Methyl-N-trimethylsilyltrifluoracetamid bei Raumtemperatur gewonnen. Abb. 13 und 14 zeigen die Massenspektren der Trimethylsilylderivate von L – Pyr – L – His – OMe und TRH. Die Interpretationen der Massenpeaks sind in den Tab. 14 und 15 zusammengestellt (*436*).

Auch die Trimethylsilylderivate der TRH-Analoga eignen sich hervorragend zur massenspektroskopischen Untersuchung. Als Beispiele sind die Massenspektren der TMS-Derivate von Phe^2-TRH (*437*) und (N-Äthylprolinamid)^3-TRH (*414*) in Abb. 15 und 16 angegeben. Interpretationen der wichtigsten Fragmentpeaks des Massenspektrums vom TMS-Derivat von (N-Äthylprolinamid)^3-TRH sind in Tab. 16 zusammengestellt (*414*).

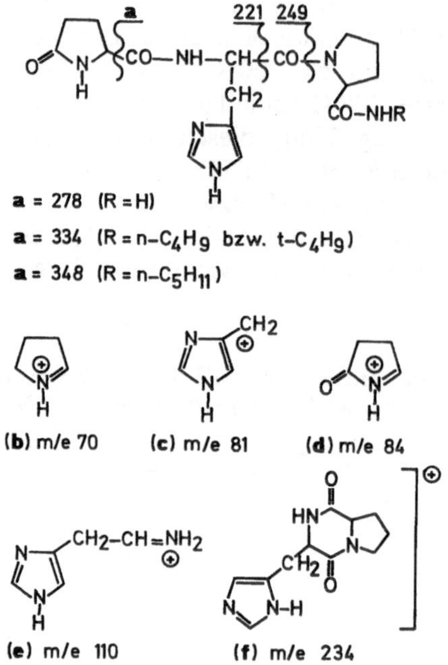

$a = 278 \quad (R = H)$

$a = 334 \quad (R = n\text{-}C_4H_9 \text{ bzw. } t\text{-}C_4H_9)$

$a = 348 \quad (R = n\text{-}C_5H_{11})$

(b) m/e 70   (c) m/e 81   (d) m/e 84

(e) m/e 110   (f) m/e 234

Abb. 12. Die wichtigsten Fragmentationen bei den massenspektroskopischen Fragmentierungen von TRH, (N-Butylprolinamid)³-TRH, (N-tert-Butylprolinamid)³-TRH und (N-Pentylprolinamid)³-TRH (*417*)

Tabelle 10. *Wichtigste Fragmentationen von TRH (417)*

m/e	relative Intensität	Erklärung*
362	4	M⁺
278	12	+
249	17	+
234	18	<u>f</u>
221	32	+
154	28	—
136	19	—
110	30	<u>e</u>
84	75	<u>d</u>
82	90	—
81	33	<u>c</u>
70	100	<u>b</u>

* Die als Buchstaben angegebenen Ionen sind in Abb. 12 dargestellt.

⁺ Vgl. Abb. 12.

Literaturverzeichnis: SS. 534—564

Tabelle 11. *Wichtigste Fragmentationen von (N-Butylprolinamid)^3-TRH (417)*

m/e	relative Intensität	Erklärung*
418	7	M$^+$
334	12	+
319	5	334−CH$_3$
249	17	+
221	18	+
110	20	e
84	14	d
82	17	−
81	11	e
70	100	b

* Die als Buchstaben angegebenen Ionen sind in Abb. 12 dargestellt.
$^+$ Vgl. Abb. 12.

Tabelle 12. *Wichtigste Fragmentationen von (N-tert-Butylprolinamid)^3-TRH (417)*

m/e	relative Intensität	Erklärung*
418	3	M$^+$
334	4	+
319	5	334−CH$_3$
249	41	+
221	32	+
110	37	e
84	22	d
82	20	−
81	13	e
70	100	b

* Die als Buchstaben angegebenen Ionen sind in Abb. 12 dargestellt.
$^+$ Vgl. Abb. 12.

Tabelle 13. *Wichtigste Fragmentationen von (N-Pentylprolinamid)³-TRH (417)*

m/e	relative Intensität	Erklärung*
432	4	M⁺
348	7	⁺
319	4	$348 - C_2H_5$
304	3	$319 - CH_3$
249	37	⁺
221	35	⁺
195	13	−
190	8	−
184	8	−
136	9	−
121	12	−
110	35	<u>e</u>
84	36	<u>d</u>
82	25	−
81	19	<u>c</u>
70	100	<u>b</u>

* Die als Buchstaben angegebenen Ionen sind in Abb. 12 dargestellt.

⁺ Vgl. Abb. 12.

Tabelle 14. *Interpretation der wichtigsten Ionen im Massenspektrum des Trimethylsilyl-derivates von L−Pyr−L−His−OMe (vgl. Abb. 13) (436)*

m/e	Interpretation
424	M⁺
409	$M - CH_3$
393	$M - OCH_3$
365	$M - COOCH_3$
268	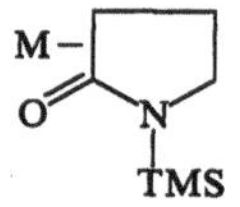 M− (structure: pyrrolidinone ring with O, N–TMS)
240	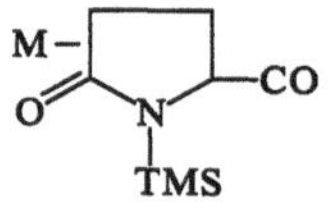 M− (structure: pyrrolidinone ring with O, N–TMS, –CO)
224	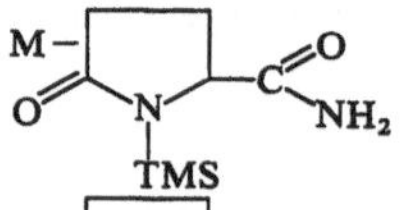 M− (structure: pyrrolidinone ring with O, N–TMS, C=O, NH₂)
156	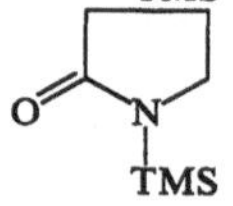 (structure: pyrrolidinone ring with O, N–TMS)

Literaturverzeichnis: SS. 534—564

Tabelle 15. *Interpretation der wichtigsten Ionen im Massenspektrum des Trimethylsilylderivates von TRH (vgl. Abb. 14) (436)*

m/e	Interpretation
578	M^+ (3 TMS)
563	$M - CH_3$
560	$M - H_2O$
545	$M - (CH_3 + H_2O)$
506	M^+ (2 TMS)
488	$506 - H_2O$
463	$M - TMS - N = C = O$
422	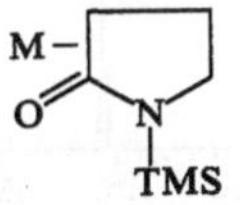 $M-$ [pyrrolidinone ring, $O=$, N–TMS]
394	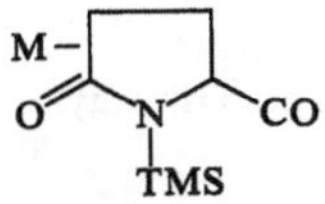 $M-$ [pyrrolidinone ring, $O=$, N–TMS, –CO]
365	$393 - CO$
156	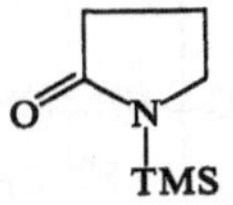 [pyrrolidinone ring, $O=$, N–TMS]

Tabelle 16. *Interpretation der wichtigsten Ionen im Massenspektrum des Trimethylsilylderivates von (N-Äthylprolinamid)³-TRH (vgl. Massenspektrum Abb. 16) (414)*

m/e	Interpretation
534	M^+ [2(CH_3)$_3$Si]
519	$M - CH_3$
516	$M - H_2O$
501	$M - (CH_3 + H_2O)$
463	$M - CH_3CH_2 - N = C = O$
393	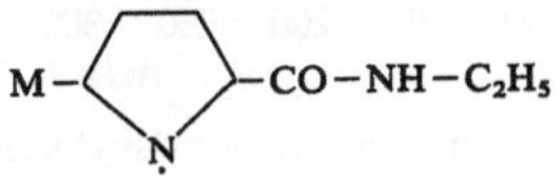 $M-$ [pyrrolidine ring, N] $-CO - NH - C_2H_5$
378	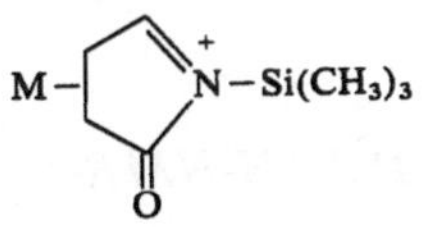 $M-$ [ring, $\overset{+}{N} - Si(CH_3)_3$, O]
365	$393 - CO$
156	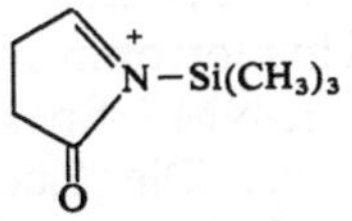 [ring, $\overset{+}{N} - Si(CH_3)_3$, O]
142	$Pro - NH - C_2H_5$

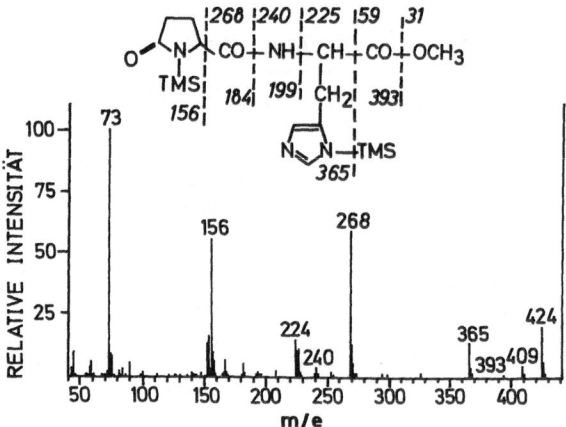

Abb. 13. Massenspektrum des Trimethylsilylderivates von L–Pyr–L–His–OMe (LKB 9 000, 70 eV) (*436*)

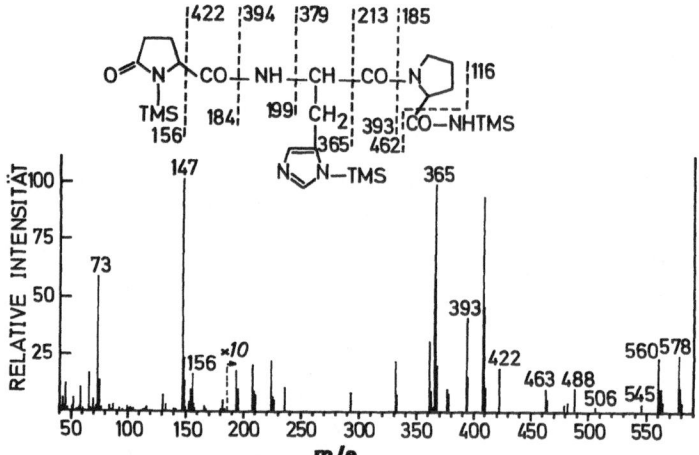

Abb. 14. Massenspektrum des Trimethylsilylderivates von TRH (LKB 9 000, 70 eV) (*436*)

D 3. ¹H-NMR-Spektroskopie

Auf Grund der Daten eines 250-MHz-¹H-NMR-Spektrums wurde ein Raummodell vom Thyreotropin-freisetzenden Hormon entwickelt. Abb. 17 zeigt das Protonen-NMR-Spektrum von TRH und Abb. 18 das Raummodell des Hormons. Die Zuordnungen der Protonen-NMR-Signale von TRH sind in Tab. 17 zusammengestellt (*438*).

Literaturverzeichnis: SS. 534—564

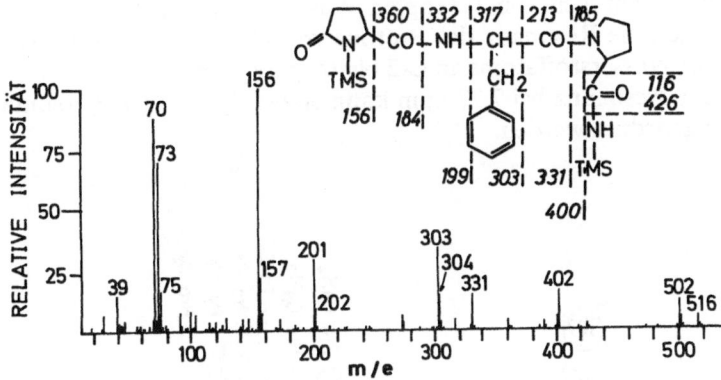

Abb. 15. Massenspektrum des Trimethylsilylderivates von Phe²-TRH (LKB 9 000, 70 eV) (437)

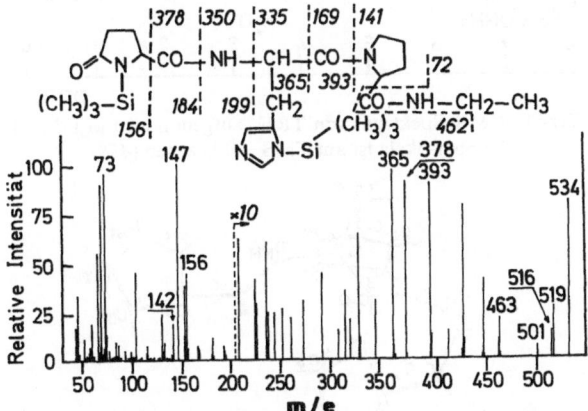

Abb. 16. Massenspektrum des Trimethylsilylderivates von (N-Äthylprolinamid)³-TRH (LKB 9 000, 70 eV) (414)

Auf die Konformation des Histidinrestes kann aus der Abhängigkeit zwischen der vicinalen Kopplungskonstanten J(NH−C$_\alpha$H) und dem Diederwinkel(C$_\alpha$H − NH) geschlossen werden. ∅ läßt sich aus θ= (∅ −60) errechnen (439, 440).

Abb. 17 läßt folgende Zusammenhänge erkennen (438):

1. Die trans- und cis-Protonen der Säureamidgruppe des Prolins verursachen Signale bei 7,94 bzw. 6,89 ppm.

2. Bei 7,51 und 6,89 ppm liegen die Resonanzen der Imidazolprotonen C$_2$H und C$_4$H.

3. Das Signal des Protons an C-4 des Imidazolrestes überlappt mit der Resonanz des cisständigen Wasserstoffatoms der CONH$_2$-Gruppe.

4. Beim Zufügen von 1 Tropfen Wasser zu der DMSO-Lösung verbreitert sich die Resonanz des Protons von C-4 des Imidazolrestes, diejenige des Wasserstoffatoms an C-2 bleibt jedoch unverändert.

5. Die Resonanz bei 7,77 ppm kann einem Proton der Pyroglutaminsäure zugeordnet werden.

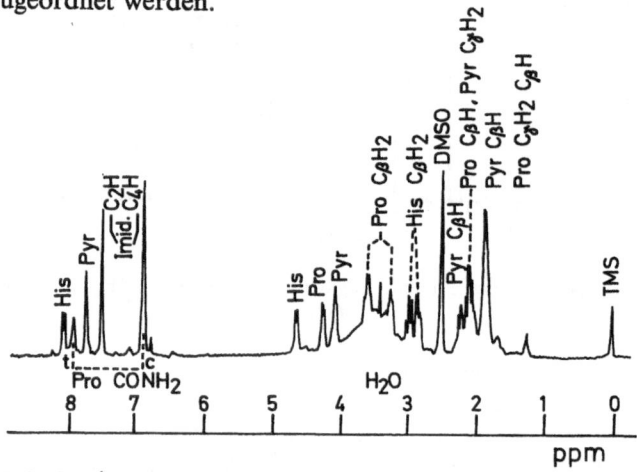

Abb. 17. 250-MHz-¹H-NMR-Spektrum von TRH, aufgenommen in DMSO-D₆ bei 20° C. Die ppm-Skala ist auf TMS = 0 bezogen (*438*)

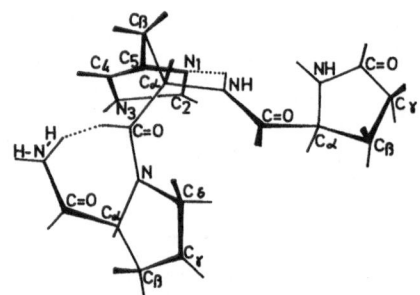

Abb. 18. Strukturmodell für TRH (*438*)

6. Das Amidproton des Histidinrestes verursacht ein Dublett bei 8.07 ppm mit einer Kopplungskonstanten von 7,5 bis 8 Hz.

7. Durch spektralen Vergleich und Entkopplungsexperimente können folgende Signale zugeordnet werden: $Pro - C_\alpha H$: 4,22 ppm, $Pro - C_\beta H_2$: 2,10 und 1,82 ppm, $Pro - C_\gamma H_2$: 1,82 ppm, $Pyr - C_\alpha H$: 4,04 ppm und $Pyr - C_\beta H$: 1,82 und 2,02 ppm.

Auf Grund des Wertes der Kopplungskonstanten $J(C_\alpha H - NH)$ von 7,5 bis 8 Hz des Histidinrestes müssen drei Möglichkeiten für $\emptyset$ diskutiert werden: $\emptyset = -150°$, $\emptyset = -90°$ und $\emptyset = +60°$. Anhand von Dreidingmodellen kann gezeigt werden, daß das TRH-Molekül für $\emptyset = -150°$ maximal stabil ist (*438*).

Tabelle 17. *Zuordnung der Protonenresonanzsignale von TRH (DMSO-D$_6$, 20° C, ppm-Werte bezogen auf TMS = 0) (438)*

	C$_\alpha$H	N–H	C$_\beta$H	C$_\gamma$H	C$_\delta$H
Pyr	4.04	7.77	1.82	2.10	
			2.20		
	4.01	7.82	1.86	2.10	
			2.24		
His	4.64	8.07	2.89		
	4.68	8.46	3.10		
	3.85		3.13		
	4.07		3.19		
Pro	4.22		2.10	1.82	3.57
			1.82		3.23
	4.33		1.60	1.80	3.59
			1.80		≈3.30
	4.59 cis		≈2.10	≈2.10	3.47
	4.39 trans		≈2.10	≈2.10	3.47
	4.25		1.90	1.90	3.35
			2.35		

Die extrem starke Tieffeldverschiebung der Resonanz des C-terminalen transständigen Amidprotons läßt den Schluß zu, daß dieses Wasserstoffatom zu der C=O-Gruppe des Histidins eine Wasserstoffbrücke ausbildet (*438*).

Weitere gründliche Untersuchungen zur Konformation von TRH kommen allerdings zu dem Schluß, daß sich in DMSO kein Siebenring mit einer Wasserstoffbrücke ausbildet und daß das Hormon als *cis*- und *trans*-Isomer vorliegt (*441—447*).

D 4. ^{13}C-NMR-Spektroskopie

Durch die Entwicklung der Impuls-Fourier-Transform-NMR-Spektroskopie gelang der ^{13}C-Kernresonanz der entscheidende Durchbruch zu einer der wichtigsten Strukturuntersuchungsmethoden der Naturstoffchemie (*448—451*).

Verschiedene ^{13}C-NMR-Untersuchungen von TRH sind schon in der Literatur erschienen (*452—456*).

Abb. 19 zeigt das Impuls-Fourier-Transform-^{13}C-NMR-Spektrum von TRH.

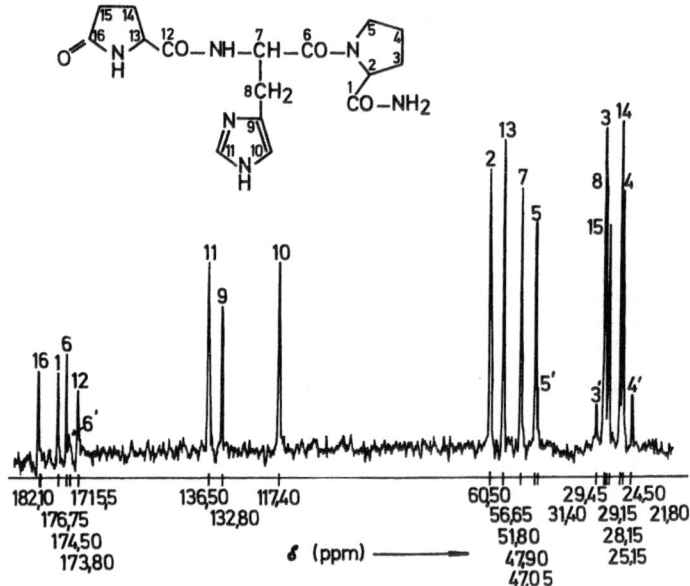

Abb. 19. Impuls-Fourier-Transform-^{13}C-NMR-Spektrum von TRH (protonenbreitband-entkoppelt; 22,63 MHz; 200 mg in 1,5 ml D$_2$O; Akkumulation von 16 384 Pulsinterfero-grammen; ppm-Werte sind auf TMS = 0 bezogen; 3′, 4′, 5′ und 6′ vgl. Text) (455)

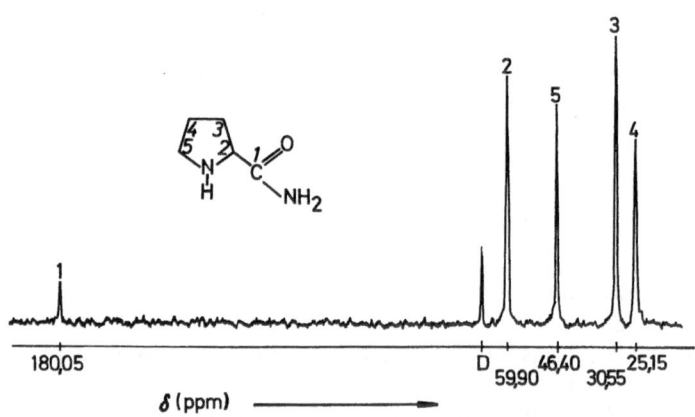

Abb. 20. Impuls-Fourier-Transform-^{13}C-NMR-Spektrum von L-Prolinamid (protonen-breitbandentkoppelt; 22,63 MHz; 150 mg in 1,5 ml D$_2$O; Akkumulation von 2 048 Puls-interferogrammen; ppm-Werte sind auf TMS = 0 bezogen; D = Dioxansignal) (467)

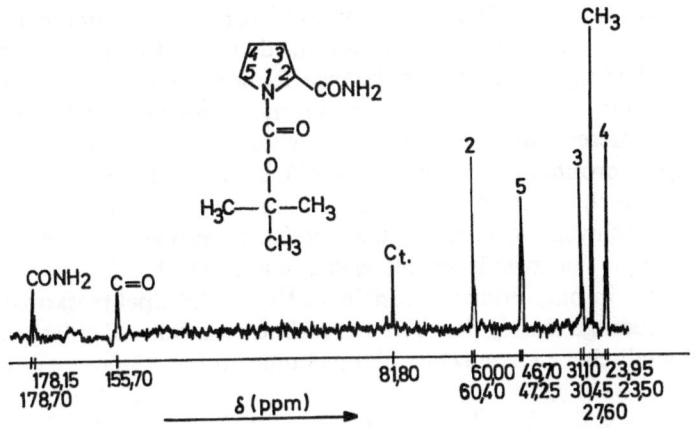

Abb. 21. Impuls-Fourier-Transform-^{13}C-NMR-Spektrum von tert-Butyloxycarbonyl-L-prolinamid (protonenbreitbandentkoppelt; 22,63 MHz; 80 mg in 1,5 ml D_2O; Akkumulation von 1 024 Pulsinterferogrammen; ppm-Werte sind auf TMS = 0 bezogen) (455)

Zur Signalidentifizierung des Spektrums von Abb. 19 werden folgende Zuordnungshilfen herangezogen:

1. Spektroskopischer Vergleich mit den Molekülbruchstücken von TRH.

2. Durch „Off-Resonance"-Teilentkopplung kann zwischen primären (Quartett), sekundären (Triplett), tertiären (Dublett) und quartären (Singulett) Kohlenstoffatomen unterschieden werden.

3. Die Anwendung allgemeiner chemischer Verschiebungsregeln.

Wie sich herausstellte, eignet sich die ^{13}C-NMR-Spektroskopie hervorragend zum Nachweis der cis-trans-Isomerie von Prolinpeptiden (426, 457—466).

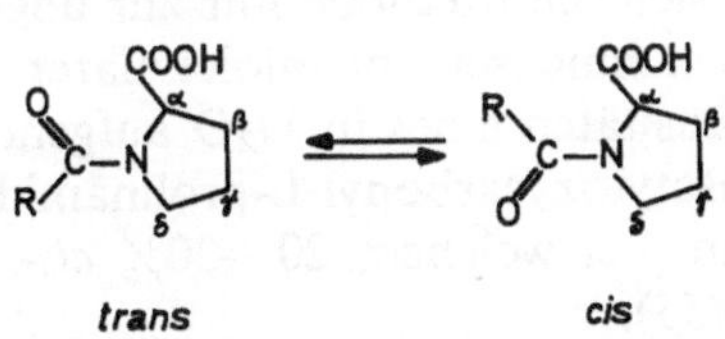

Abb. 22. cis-trans-Isomerie von Prolinderivaten

Unter anderem waren die Aufnahme des ^{13}C-NMR-Spektrums von L-Prolinamid und seine Interpretation zur Signalzuordnung des TRH-Spektrums (vgl. Abb. 19) notwendig. Abb. 20 zeigt das Impuls-Fourier-Transform-^{13}C-NMR-Spektrum von L-Prolinamid. Die Resonanzidentifizierung gelingt auf Grund allgemeiner chemischer Verschiebungsregeln (467).

Ersetzt man das NH-Ringproton des Prolinamids durch eine Schutzgruppe vom Urethantyp, z. B. durch den tert-Butyloxycarbonylrest,
dann ist, bedingt durch die behinderte Rotation um die C – N-Bindung
eine *cis*- oder *trans*ständige Verknüpfung möglich. Das Vorliegen der
beiden Isomeren (*cis* und *trans*) in Lösung läßt sich für tert-Butyloxycarbonyl-L-prolinamid durch ^{13}C-NMR-Spektroskopie experimentell
eindeutig, nachweisen (Abb. 21): Den Ringkohlenstoffatomen des Prolins
müssen durchweg Doppelsignale zugeordnet werden, die sich nur durch
das Vorliegen von zwei Isomeren erklären lassen (Abb. 22).

In der Impuls-Fourier-Transform-^{13}C-NMR-Spektroskopie verursachen, bedingt durch den Kern-Overhauser-Effekt, Kohlenstoffatome
unterschiedlicher chemischer Umgebung Intensitäten verschiedener

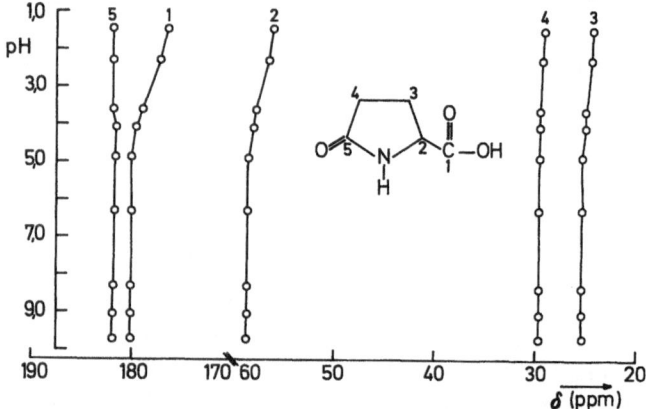

Abb. 23. pH-Abhängigkeit der ^{13}C-NMR-Signale von L-Pyroglutaminsäure. Die ppm-
Werte sind auf TMS = 0 bezogen (*467*)

Größe; daher eignet sich die Methode nur zur ungefähren Abschätzung
der prozentualen Zusammensetzung molekularer Gleichgewichte. Auf
Grund der Signalintensitäten eines in D$_2$O aufgenommenen ^{13}C-NMR-
Spektrums von tert-Butyloxycarbonyl-L-prolinamid stellt sich in Wasser
ein Gleichgewicht ein, bei welchem 20—30% *cis*- und 80—70% *trans*-
Isomeres vorliegen (*455*).

Für die ^{13}C-NMR-Spektreninterpretation von TRH war ferner die
eindeutige Signalzuordnung im Spektrum der Pyroglutaminsäure notwendig. Auf Grund allgemeiner chemischer Verschiebungsregeln erwartet
man im Spektrum der Pyroglutaminsäure für C-1 und C-5 (vgl. Abb. 23)
zwei Resonanzen bei tiefem Feld. Diese Resonanzen liegen jedoch so nahe
beieinander, daß die Zuordnung der Signale von C-1 und C-5 zunächst
nicht getroffen werden kann (Tab. 18). Wird jedoch das Spektrum von
Pyroglutaminsäure bei verschiedenen pH-Werten aufgenommen, dann

sollte die Resonanz der Carboxylgruppe (C-1) eine viel stärkere pH-Abhängigkeit zeigen, als das C=O-Signal des Pyrrolidonrings (C-5), da sich durch die Abdissoziation des Protons die Elektronendichte an C-1 viel stärker ändert als an C-5. Wie Abb. 23 zeigt, wird diese Vermutung durch das Experiment voll bestätigt (467).

Tabelle 18. ^{13}C-Chemische Verschiebungen von L-Pyroglutaminsäure, aufgenommen in D_2O (467)

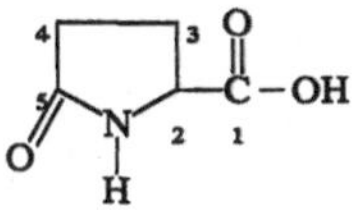

Kohlenstoffatom	ppm-Werte, bezogen auf TMS = 0
C-1	180,20
C-2	58,35
C-3	25,35
C-4	29,65
C-5	181,20

Durch die Ermittlung der pH-Abhängigkeit der Kohlenstoffresonanzen von Pyroglutaminsäure gelingt die eindeutige Signalzuordnung für C-1 und C-5: Im Vergleich zu C-5 ist die Resonanz von C-1 tieffeldverschoben (467).

Eine weitere Hilfe für die Signalzuordnung im ^{13}C-NMR-Spektrum von TRH liefert die Interpretation des Kohlenstoffresonanzspektrums vom L-Histidinmethylester. Die in Tab. 19 zusammengestellten Signalzuordnungen wurden auf Grund allgemeiner chemischer Verschiebungsregeln und durch Protonen-„Off-Resonance"-Teilentkopplung getroffen (467).

Im ^{13}C-NMR-Spektrum von TRH (Abb. 19) liegen in unmittelbarer Nachbarschaft (1—2 ppm) der Resonanzen der Prolinkohlenstoffatome Signale, die nur 10—20% der Intensität der übrigen Resonanzen aufweisen. Durch die ^{13}C-NMR-Studien von tert-Butyloxycarbonyl-L-prolinamid (vgl. Abb. 21) kann die Ursache dieser Nebenresonanzen gedeutet werden: TRH liegt in wäßriger Lösung zu 10—20% als cis- und 80—90% als trans-Isomeres vor (vgl. Abb. 24) (455).

Substituiert man ein Wasserstoffatom der C-terminalen Säureamidgruppe von TRH durch Alkylreste, dann erhält man Derivate des Hormons mit interessanten biologischen Eigenschaften (vgl. Tab. 8). Von Interesse war, zu ermitteln, inwieweit Substitution der Prolinamidwasserstoffatome durch Alkylreste das cis-trans-Isomeriegleichgewicht von TRH beeinflußt. Zur Klärung der Frage wurden von (Propylprolinamid)^3-TRH-Verbindungen ^{13}C-NMR-Spektren aufgenommen (468).

Tabelle 19. ^{13}C-Chemische Verschiebungen von L-Histidinmethylester, aufgenommen in DMSO (467)

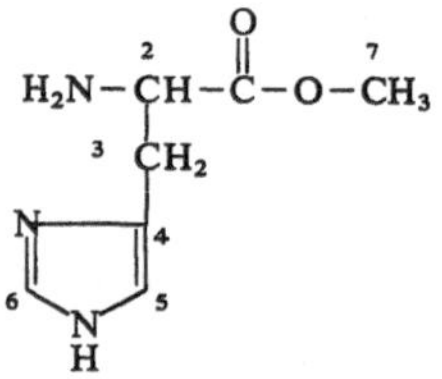

Kohlenstoffatom	ppm-Werte, bezogen auf TMS = 0
C-1	168,75
C-2	53,40
C-3	25,35
C-4	126,75
C-5	118,35
C-6	134,35
C-7	51,35

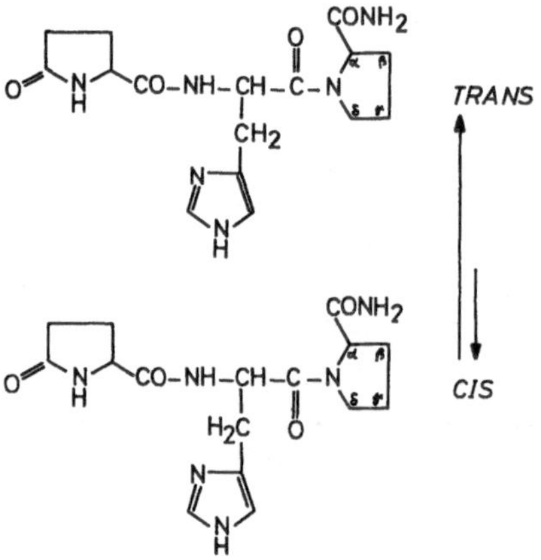

Abb. 24. cis-trans-Isomerie des Thyreotropin-freisetzenden Hormons

Abb. 25 zeigt das Impuls-Fourier-Transform-^{13}C-NMR-Spektrum von [n-Propylprolinamid]^3-TRH in D$_2$O. Anhand von vergleichenden Strichspektren (Abb. 26) mit TRH können die Resonanzzuordnungen für die n-Propylamid und Isopropylamidreste der TRH-Derivate getroffen werden.

Obgleich die ^{13}C-Signallagen von TRH und den Derivaten erwartungsgemäß weitgehend identisch sind, verschieben sich die Prolin-C=O- (induktiver Effekt) und Histidin-C=O-Resonanzen der Derivate gegenüber der reinen Verbindung nach höherem Feld. Die Resonanzverschiebungen könnten durch Wechselwirkung zwischen der Amidgruppe und der Histidin-C=O-Gruppe verursacht werden.

Literaturverzeichnis: SS. 534—564

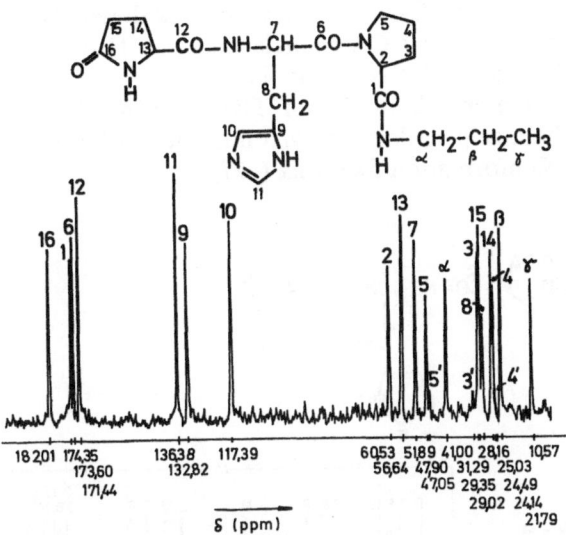

Abb. 25. Impuls-Fourier-Transform-^{13}C-NMR-Spektrum von (Pro−NH−CH$_2$−CH$_2$−CH$_3$)^3−TRH (protonenbreitbandentkoppelt; 22,63 MHz; 230 mg in 1,2 ml D$_2$O; Akkumulation von 16 384 Pulsinterferogrammen: ppm-Werte sind auf TMS = 0 bezogen) (468)

Sowohl das reine TRH, als auch die amidsubstituierten Derivate (Abb. 25) zeigen in wäßriger Lösung neben den intensiven Signalen der Prolin-C-Atome 3,4 und 5 im Abstand von 1—2 ppm noch weitere Resonanzen, die nur 10—15% der Intensität der übrigen Signale aufweisen. Diese Signalverdoppelung beruht auf dem Vorliegen eines cis-trans-Isomerengleichgewichts, das hier zu 85—90% auf der Seite der trans-Verbindung liegt. Die Lage des Gleichgewichts ist lösungsmittelabhängig, wird jedoch, wie die ^{13}C-NMR-Messungen zeigen, nur geringfügig vom Substituenten der Amidgruppe beeinflußt (468).

Ebenso wie zum Nachweis von cis-trans-Isomeren eignet sich die ^{13}C-NMR-Spektroskopie zur Unterscheidung von Diastereomeren: Racemisierung von TRH an C-α von Histidin ist im ^{13}C-NMR-Spektrum durch Doppelresonanzen der α-Kohlenstoffatome des Histidin- und Prolinrestes und Doppelsignale von C-β, γ und δ des Prolinringes nachweisbar (454).

Zur Bestimmung der Beweglichkeit von Peptidfragmenten werden in neuester Zeit Messungen von Spin-Gitter-Relaxationszeiten (T$_1$) herangezogen (454, 469, 470). Aus den T$_1$-Messungen kann z. B. abgeleitet werden, daß der Pyrrolidonring der Pyroglutaminsäure in verschiedenen Konformationen vorliegt, da die T$_1$-Werte für C-β und C-γ verschieden sind. Die T$_1$-Werte für die α-Kohlenstoffatome von TRH nehmen in der

Reihe Pyr — C-α, His — C-α, Pro — C-α ab. Aus diesem Sachverhalt kann gefolgert werden, daß das C-terminale Ende des Hormons geringere Beweglichkeit besitzt als das N-terminale.

^{13}C-NMR-Untersuchungen von TRH, welches 85% angereichertes L-Prolin enthält, zeigen, daß der Pyrrolidinring im Hormon in einer ganz bestimmten Konformation fixiert ist (471).

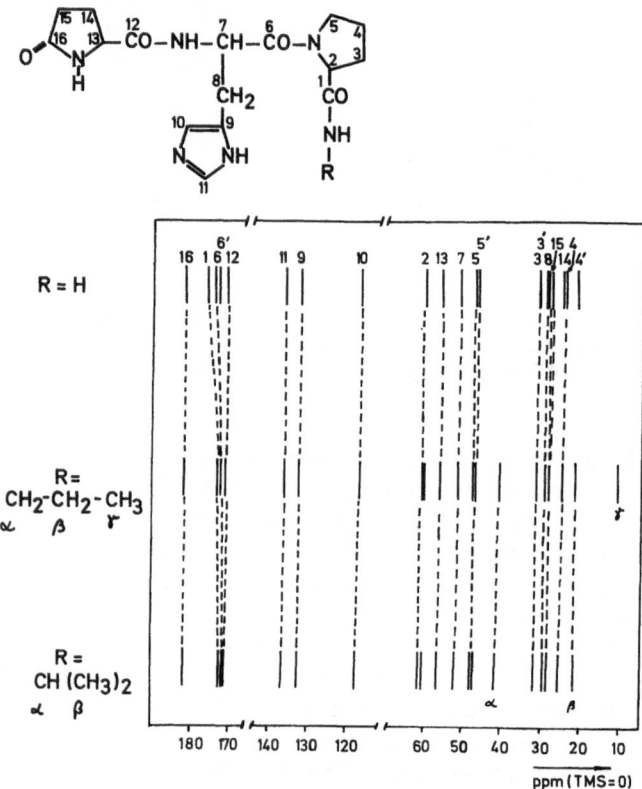

Abb. 26. ^{13}C-Signalzuordnungen von (Propylprolinamid)[3]-TRH-Verbindungen durch spektroskopischen Vergleich mit TRH (468)

D 5. Circulardichroismus

Zur Zeit liegen noch wenig Circulardichroismusuntersuchungen über TRH vor (472, 473). Wie Abb. 27 zeigt, ist das CD-Spektrum von TRH stark vom pH-Wert der Lösung abhängig (472).

Literaturverzeichnis: SS. 534—564

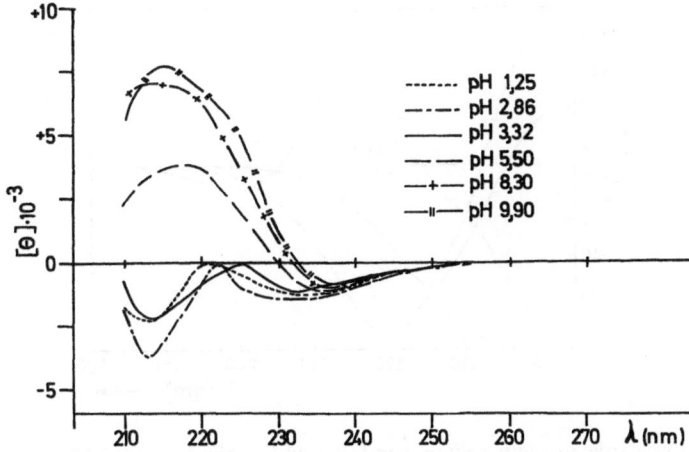

Abb. 27. Circulardichroismusspektren von TRH bei verschiedenen pH-Werten (*472*)

Zur Deutung der starken pH-Abhängigkeit des Circulardichroismus (CD) von TRH sind auch die CD-Spektren von L-Histidin und L-Alanyl-L-prolin bei verschiedenen pH-Werten aufgenommen worden (Abb. 28 und 29) (*472*).

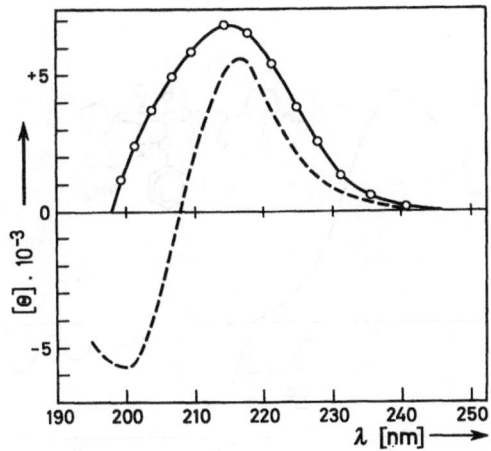

Abb. 28. Circulardichroismusspektren von L-Histidin (–o–o–o pH 1,30; – – – pH 11,10) (*472*)

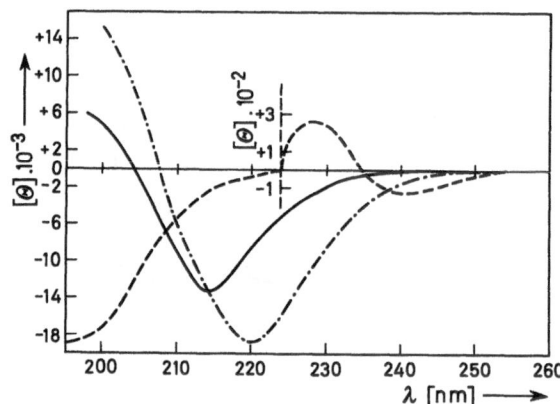

Abb. 29. Circulardichroismusspektren von L-Alanyl-L-prolin (– – –pH 0,78; ──── pH 8,94; –·–·–·– Methanol) (*472*)

   Die CD-Untersuchungen von TRH und einigen Analoga in verschiedenen Lösungsmitteln lassen den Schluß zu, daß sich besonders in Dioxanlösung zwischen einem Säureamidproton des Prolins und der Carbonylgruppe des Histidins eine Wasserstoffbrücke ausbildet, welche einen starken negativen Cottoneffekt bei 226 nm verursacht (*473*).
   Die starke pH-Abhängigkeit des Circulardichroismus von TRH wird auf Protonierung und/oder Konformationsänderungen des Hormons zurückgeführt (*472*). Im Gegensatz dazu wird das CD-Spektrum von Phe² – TRH vom pH-Wert der Lösung kaum beeinflußt. Am Phenylrest

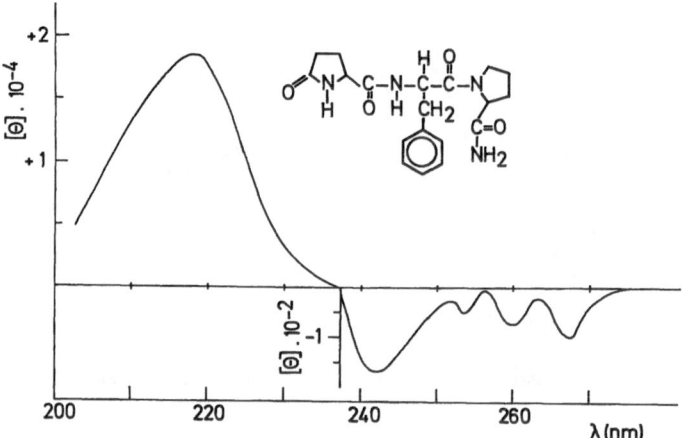

Abb. 30. Circulardichroismusspektrum von Phe²-TRH in H_2O (*474*)

Literaturverzeichnis: SS. 534—564

von Phe² – TRH können keine Protonen angelagert werden; die Konformation des TRH-Derivates scheint durch die Wasserstoffionenkonzentration nicht verändert zu werden. Das in Wasser aufgenommene CD-Spektrum von Phe² – TRH zeigt Abb. 30 (*474*).

III. Luteinisierendes/Follikel-stimulierendes Hormon-freisetzendes Hormon

A. Strukturaufklärung

Die Isolierung des zweiten Hypothalamus-Hormons mit Hormonfreisetzender Wirkung gelang den Arbeitsgruppen von SCHALLY (*475*) und GUILLEMIN (*476*).

Zur Bestimmung der Aminosäurezusammensetzung wurde das aus Schweinehypothalami isolierte Material mit 6 N Salzsäure hydrolysiert; die Tryptophanbestimmung wurde nach alkalischer Hydrolyse durchgeführt. Nach diesen Bestimmungen enthält das Hormon folgende Aminosäuren: Arg(1), Glu(1), Gly(2), His(1), Leu(1), Pro(1), Ser(1), Trp(1) und Tyr(1) (*477*).

Da das Peptidhormon nicht mit Dansylchlorid reagiert und sein Massenspektrum Peaks bei m/e 111, 1181 und 112, 1248 zeigt, kann geschlossen werden, daß Pyroglutaminsäure das N-terminale Ende bildet (*477*).

Chymotrypsin- und Thermolysinspaltungen sowie Edman-Dansyl-Abbau führten zur Strukturaufklärung des Hormons (vgl. Abb. 31 und 32) (*477*).

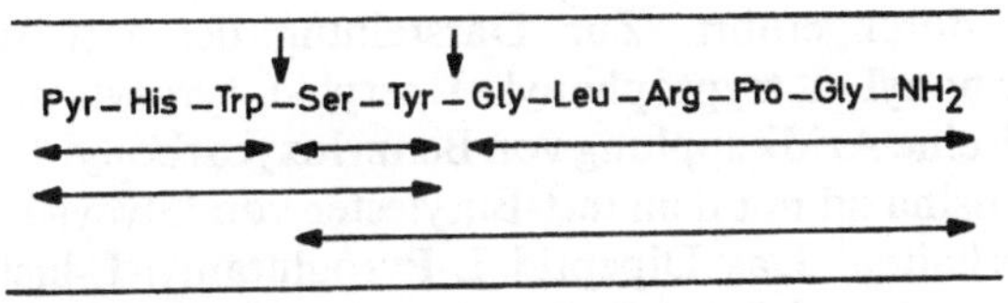

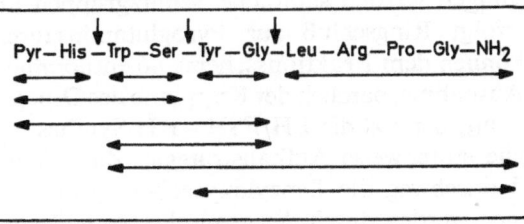

Abb. 31. Schematische Darstellung der Chymotrypsinspaltung von LH/FSH–RH (*477*)

Abb. 32. Schematische Darstellung der Thermolysinspaltung von LH/FSH–RH (*477*)

Durch chemische und enzymatische Inaktivierungen des Hormons kann ebenfalls geschlossen werden, daß Pyroglutaminsäure, Arginin-, Tyrosin- und Tryptophanreste Bestandteile des Hormons sind (478, 479).

Nach der Isolierung des Hormons ist in einigen Arbeiten gezeigt worden, daß die Biosynthese von LH/FSH − RH in den Mitochondrien stattfindet (480—483).

Es liegt bisher nur eine ^{1}H-NMR und ^{13}C-NMR-Untersuchung von LH/FSH − RH vor. Nach dieser Studie soll das Hormon in Wasser Knäuelstruktur haben; der Prolinrest soll in trans-Konfiguration vorliegen (484).

Auch Circulardichroismusuntersuchungen bestätigen, daß das LH/ FSH − RH zumindest in sauren pH-Bereichen Knäuelstruktur einnimmt (485).

B. Synthesen vom Luteinisierenden/Follikel-stimulierenden Hormon-freisetzenden Hormon

Unmittelbar nach Bekanntwerden der Struktur von LH/FSH − RH sind eine Reihe von Totalsynthesen beschrieben worden (486—501). Obgleich die Nachteile der Festphasensynthese zur Darstellung medizinisch anwendbarer Peptidhormone bekannt sind (vgl. II. B.), wurde das Dekapeptidamid auch nach dem Merrifield-Syntheseverfahren hergestellt.

In der ersten klassischen Synthese von LH/FSH − RH (486) werden sämtliche Peptidbindungen mit Ausnahme der Knüpfung von Tyrosin und Glycin mit Dicyclohexylcarbodiimid in Gegenwart von 1-Hydroxybenzotriazol durchgeführt. Zur Darstellung des tert-Butylesters von Benzyloxycarbonyl - L tryptophanyl - L seryl - L tyrosyl - glycyl - L - leucin werden durch eine Azidkupplung von Benzyloxycarbonyl-L-tryptophanyl-L-seryl-L-tyrosinazid mit dem tert-Butylester von Glycyl-L-leucin bessere Ausbeuten erhalten. Das Dipeptid L-Pyroglutamyl-L-histidin wird aus N^α-Benzyloxycarbonyl-N^γ-4,4′-dimethoxybenzhydryl-L-glutaminyl-L-histidin durch Erhitzen mit Trifluoressigsäure/Anisol gewonnen: unter diesen Bedingungen werden sämtliche Schutzgruppen abgespalten und gleichzeitig erfolgt Ringschluß zur Pyroglutaminsäure. Die näheren Einzelheiten können dem Reaktionsschema 46 entnommen werden (486).

Mit einer Ausnahme, nämlich der Kupplung des Dipeptidazids Z − L − Ser − L − Tyr − N$_3$, benutzt die LH/FSH − RH-Synthese von Yanaihara et al. (499) eine stufenweise Aufbaustrategie. Zum Schutz der Aminofunktion wird durchweg die Benzyloxycarbonylgruppe verwendet. Die Peptidbindungen werden durch die gemischte Anhydrid-, die Azid- und die aktivierte Estermethode geknüpft (vgl. Syntheseschema 47) (499).

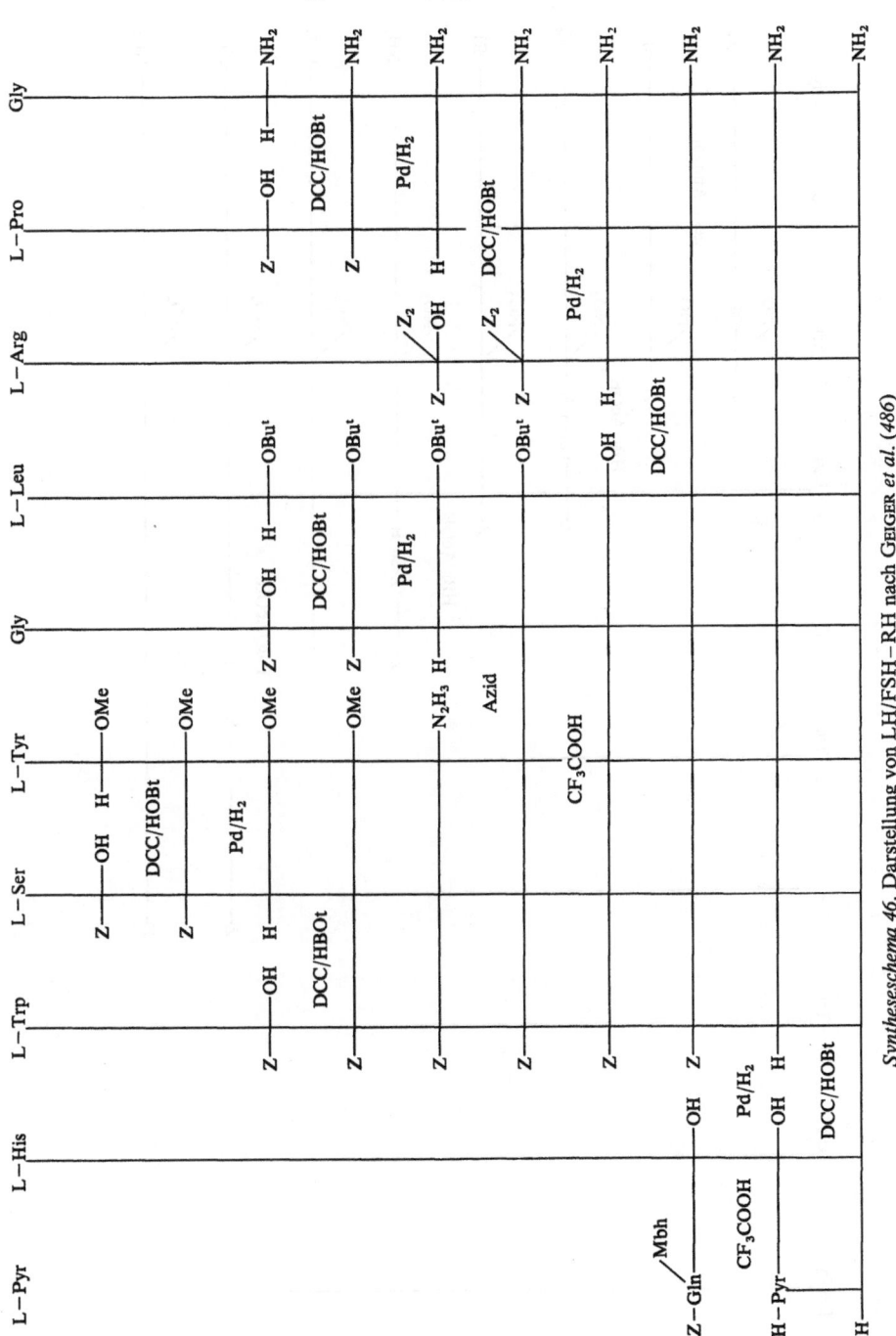

Syntheseschema 46. Darstellung von LH/FSH–RH nach GEIGER *et al.* (486)

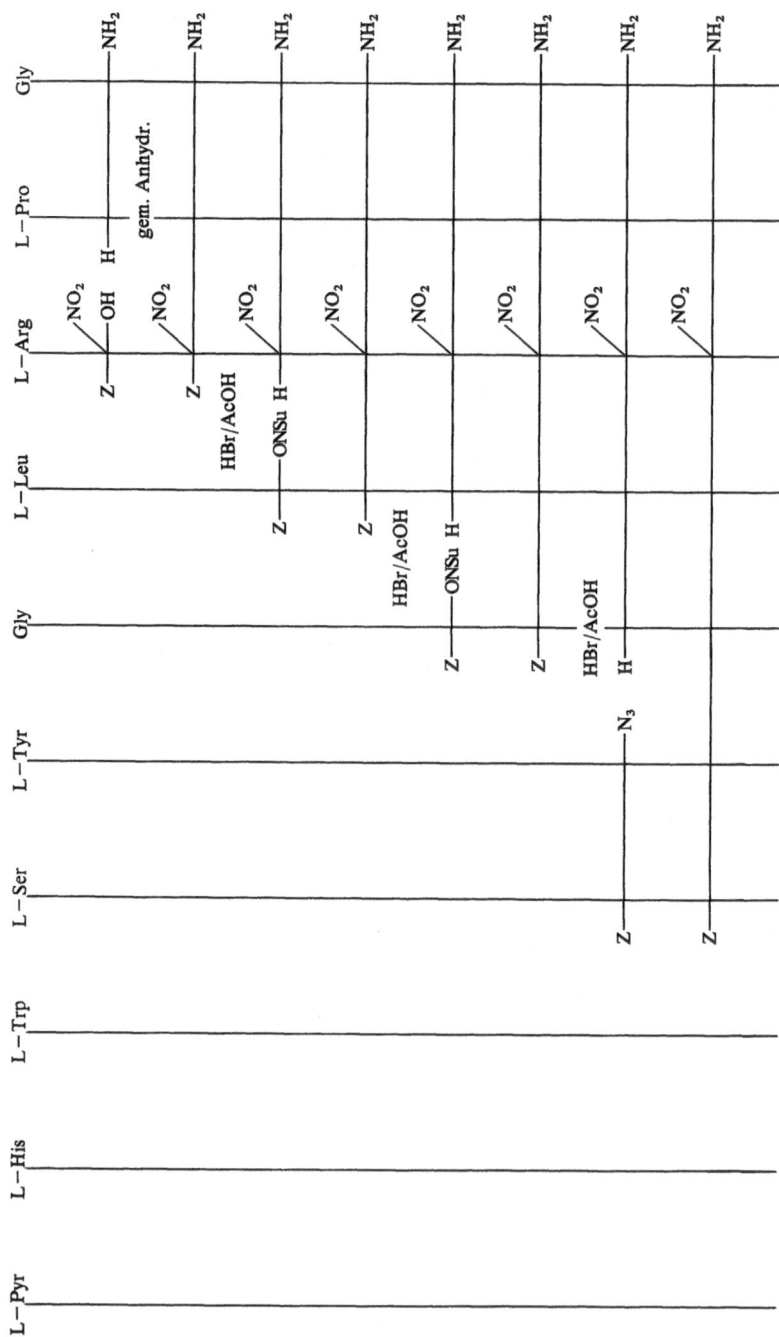

Literaturverzeichnis: SS. 534—564

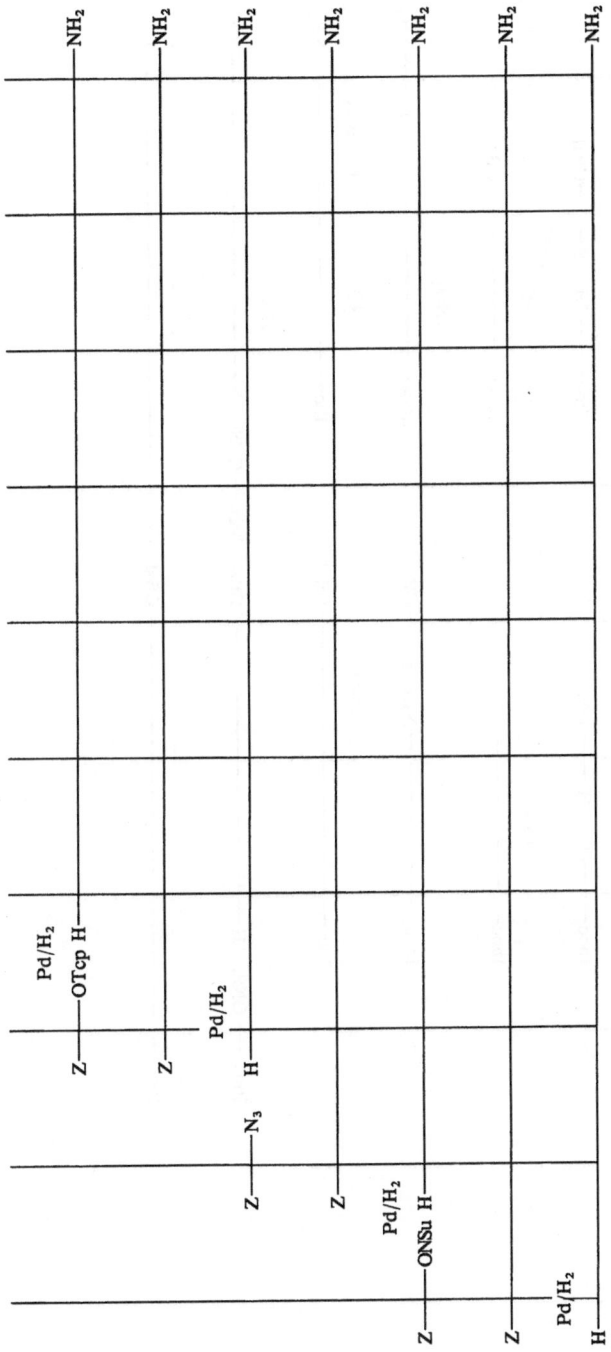

Syntheseschema 47. Darstellung von LH/FSH–RH nach YANAIHARA et al. (499)

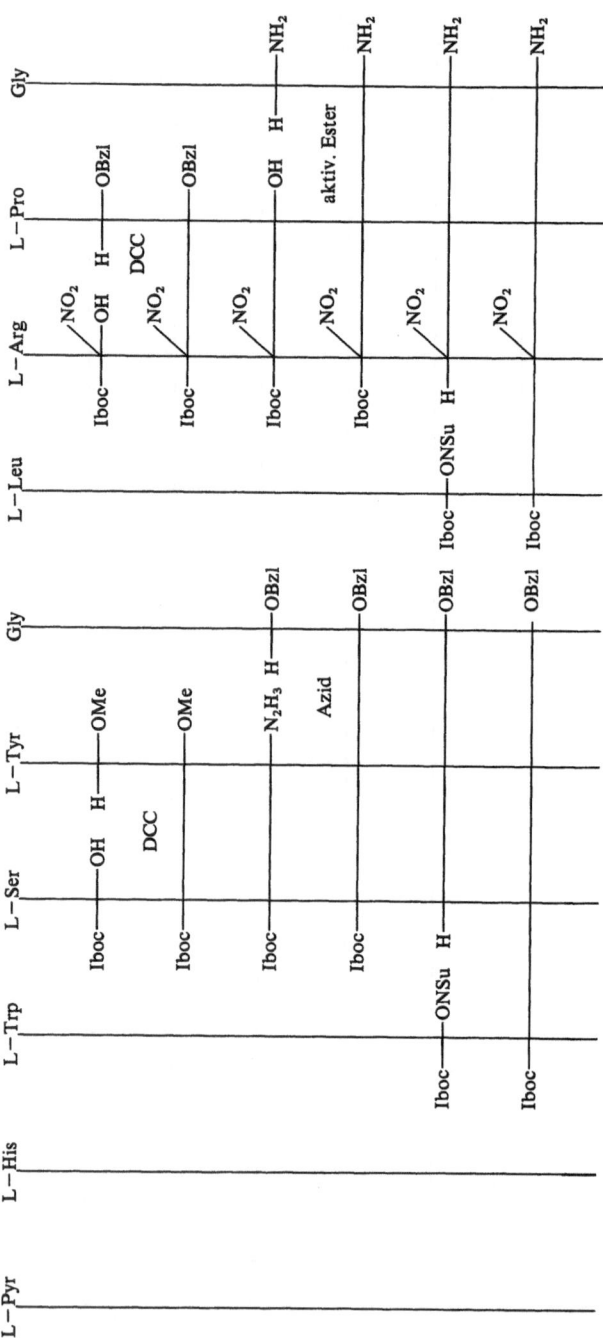

Literaturverzeichnis: SS. 534—564

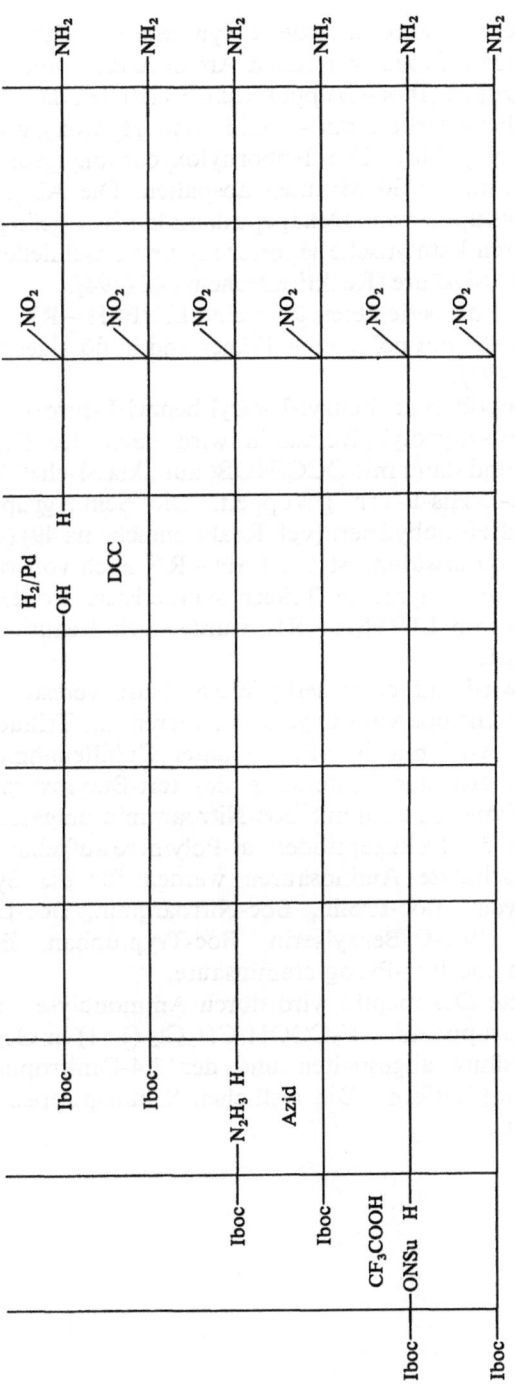

Syntheseschema 48. Darstellung von LH/FSH–RH nach FUJINO et al. (494)

Die in demselben Jahre publizierte Synthese von Fujino et al. (494) benutzt mit Erfolg zur Blockierung der α-Aminofunktion die Isobornyloxy-carbonylschutzgruppe (Iboc-Gruppe). Zum Herstellen der Peptidbindung werden Dicyclohexylcarbodiimid-, Azid- und Hydroxysuccinimidester-kupplungen durchgeführt. Die Isobornyloxycarbonylgruppe läßt sich bei Raumtemperatur in 30 Minuten abspalten. Die Abspaltungen der Iboc- und Nitrogruppe vom Dekapeptidamidderivat gelingen entweder mit HF oder durch katalytische Hydrierung und anschließende Behand-lung mit Trifluoressigsäure (Reaktionsschema 48 (494).

Sievertsson et al. berichteten über eine LH/FSH−RH-Synthese, bei welcher Kupplungen am polymeren Träger und in flüssiger Phase durch-geführt werden (491).

Das Heptapeptidamid Benzyl-L-seryl-benzyl-L-tyrosyl-glycyl-L-leu-cyl-nitro-L-arginyl-L-prolyl-glycinamid wird nach der Festphasensyn-these hergestellt und dann mit DCC/HOBt auf „klassische" Weise an das Tripeptid L-Pyr-L-His-L-Trp gekuppelt. Die Schutzgruppen werden schließlich mit Pd/H$_2$ abhydriert (vgl. Reaktionsschema 49) (491).

Wie oben schon erwähnt, ist LH/FSH−RH auch von verschiedenen Arbeitsgruppen am polymeren Träger synthetisiert worden. Eine der ersten Synthesen von LH/FSH−RH wurde nach Reaktionsschema 50 (492) durchgeführt.

Boc-Glycin wird mit chlormethyliertem Harz verestert. Nach Ab-spaltung der Boc-Gruppe vom Glycin-Polymeren mit Trifluoressigsäure/ Methylenchlorid wird mit Boc-Prolin unter Zuhilfenahme von DCC gekuppelt. Nach erneuter Entfernung des tert-Butyloxycarbonylrestes vom Dipeptidpolymeren wird mit Boc-Nitroarginin umgesetzt und dann in analoger Weise das Dekapeptidderivat-Polymere aufgebaut.

Folgende geschützte Aminosäuren werden für die Synthese ein-gesetzt: Boc-Glycin, Boc-Prolin, Boc-Nitroarginin, Boc-Leucin, Boc-O-Benzyltyrosin, Boc-O-Benzylserin, Boc-Tryptophan, Boc-2,4-Dini-trophenylhistidin und Boc-Pyroglutaminsäure.

Das geschützte Dekapeptid wird durch Ammonolyse vom Harz ge-trennt, die Boc-Gruppe mit CF$_3$COOH/CH$_2$Cl$_2$ (1 : 1) in Gegenwart von 10% Thioglykolsäure abgespalten und der 2,4-Dinitrophenylrest mit 2-Mercaptoäthanol entfernt. Die restlichen Schutzgruppen werden mit H$_2$/Pd abhydriert.

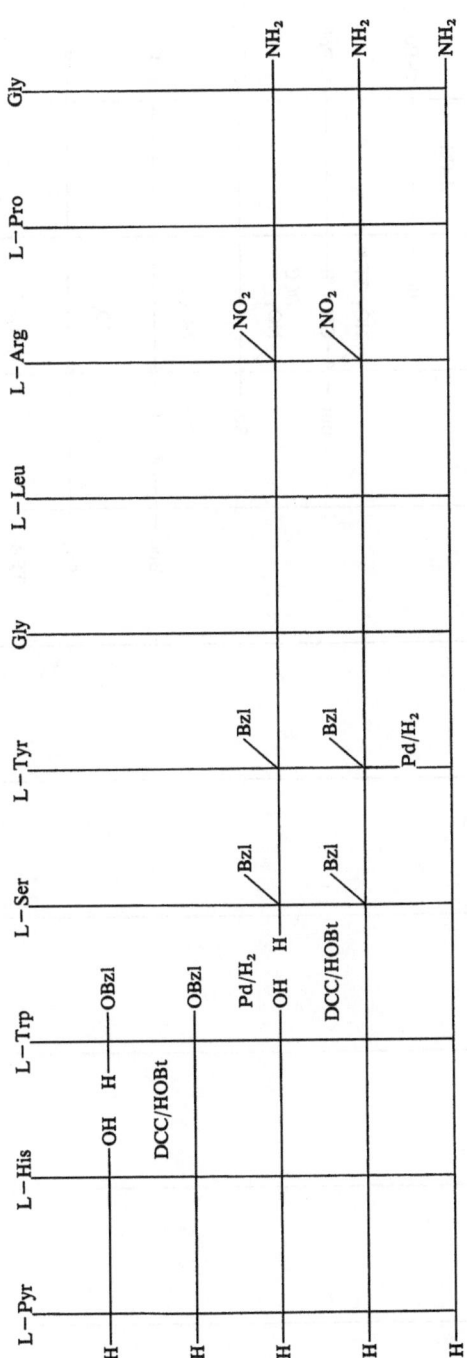

Syntheseschema 49. Darstellung von LH/FSH–RH nach SIEVERTSSON *et al.* (491)   I

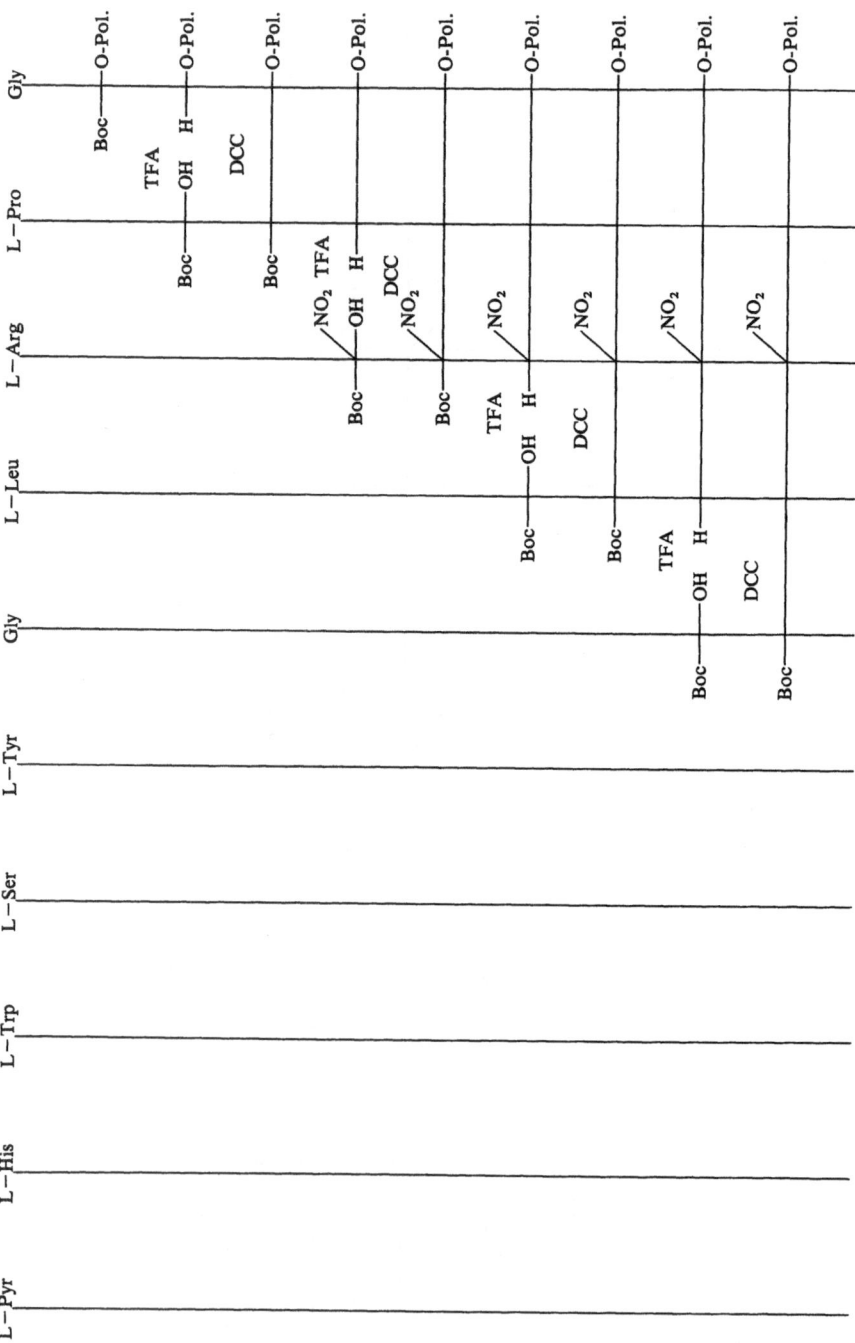

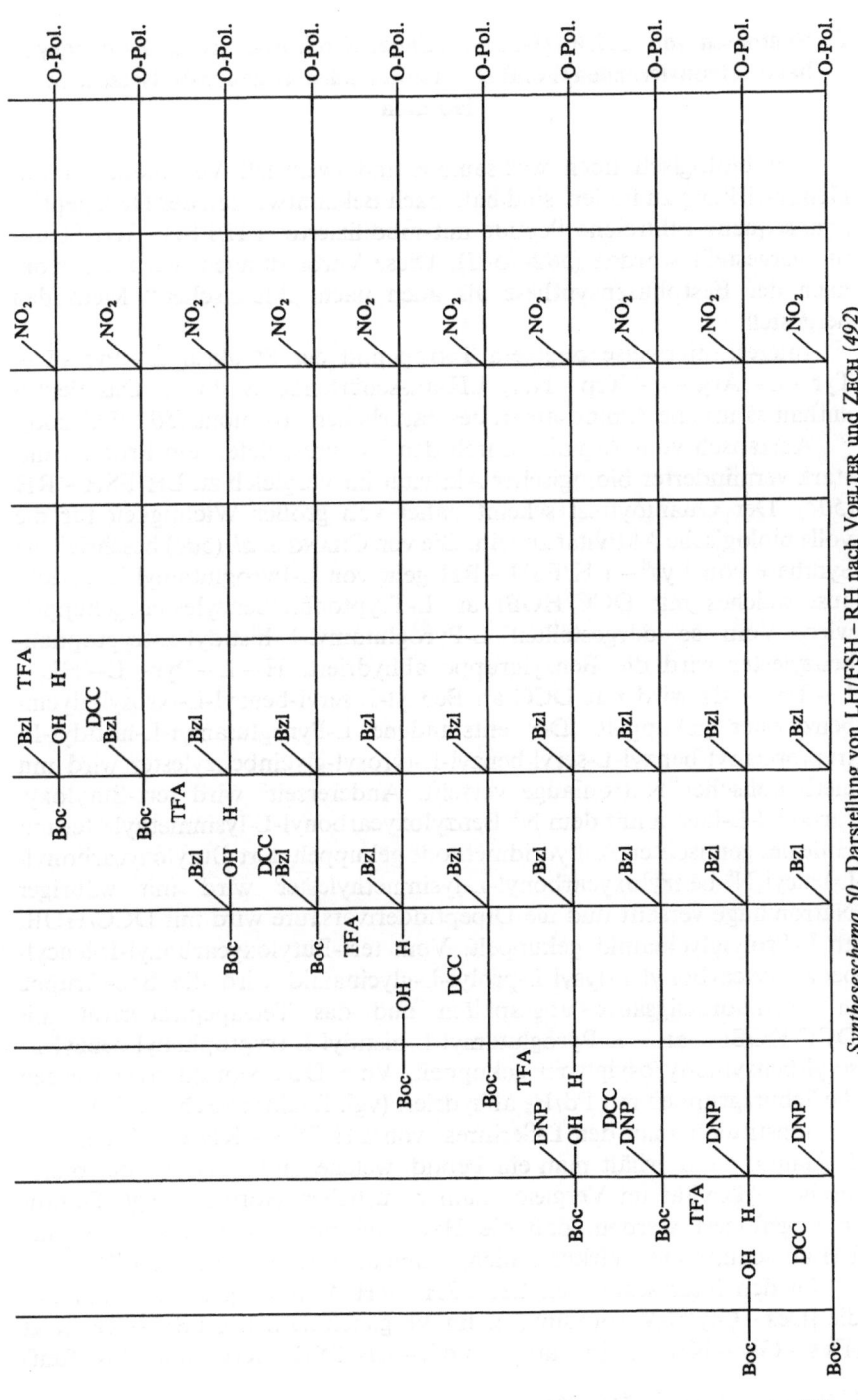

Syntheseschema 50. Darstellung von LH/FSH – RH nach VOELTER und ZECH (*492*)

C. Synthesen von LH/FSH-RH-Analoga. Struktur-Aktivitätsbeziehungen beim luteinisierenden/Follikel-stimulierenden Hormon-freisetzenden Hormon

Um biologisch noch wirksamere und eventuell Verbindungen mit Hemmwirkung zu finden, sind bald nach Bekanntwerden der Dekapeptid-amidsequenz zahlreiche Peptide mit modifizierter LH/FSH−RH-Struktur hergestellt worden (502—541). Diese Verbindungen wurden sowohl nach der Festphasensynthese als auch nach „klassischen" Methoden hergestellt.

Interessanterweise zeigt ein Tetrapeptid der Struktur L−Pyr−L−Tyr−L−Arg−L−Trp−NH$_2$ LH-ausschüttende Wirkung. Das Peptid enthält sämtliche Aminosäuren des natürlichen Hormons (502, 503, 508).

Austausch vom Arginin- durch den Lysinrest liefert ein Produkt mit stark verminderter biologischer Aktivität im Vergleich zu LH/FSH−RH (504). Der Guanidylrest scheint daher von großer Wichtigkeit für die volle biologische Aktivität zu sein. Die von CHANG et al. (504) beschriebene Synthese von Lys^8−LH/FSH−RH geht von L-Pyroglutamyl-L-histidin aus, welches mit DCC/HOBt an L-Tryptophanbenzylester gekuppelt wird. Vom so dargestellten L-Pyroglutamyl-L-histidyl-L-tryptophan-benzylester wird die Benzylgruppe abhydriert. H−L−Pyr−L−His−L−Trp−OH wird mit DCC an Benzyl-L-seryl-benzyl-L-tyrosyl-glycin-benzylester gekuppelt. Der entstandene L-Pyroglutamyl-L-histidyl-L-tryptophanyl-benzyl-L-seryl-benzyl-L-tyrosyl-glycinbenzylester wird mit methanolischer Natronlauge verseift. Andererseits wird tert-Butyloxy-carbonyl-L-leucin mit dem N$^\varepsilon$-Benzyloxycarbonyl-L-lysinmethylester mit Hilfe der gemischten Anhydridmethode gekuppelt. tert-Butyloxycarbonyl-L-leucyl-N$^\varepsilon$-benzyloxycarbonyl-L-lysinmethylester wird mit wäßriger Natronlauge verseift und die Dipeptidderivatsäure wird mit DCC/HOBt an L-Prolylglycinamid gekuppelt. Vom tert-Butyloxycarbonyl-L-leucyl-benzyloxycarbonyl-L-lysyl-L-prolyl-L-glycinamid wird die Boc-Gruppe mit Trifluoressigsäure abgespalten und das Tetrapeptidderivat mit DCC/HOBt an L-Pyroglutamyl-L-histidyl-L-tryptophanyl-benzyl-L-seryl-benzyl-L-tyrosylglycin gekuppelt. Vom Decapeptidderivat werden die Schutzgruppen mit Pd/H$_2$ abhydriert (vgl. Reaktionsschema 51) (504).

Substituiert man den L-Serinrest von LH/FSH−RH durch den von L-Alanin, dann erhält man ein Peptid, welches nur noch 5% der biologischen Aktivität im Vergleich zum natürlichen Hormon zeigt. Daraus muß gefolgert werden, daß die Hydroxylgruppe von L-Serin für die LH-ausschüttende Wirkung nicht unbedingt notwendig ist (507).

Zu den interessantesten LH/FSH−RH-Analoga gehören zweifellos die [Des−Gly^{10}]-Verbindungen. Im Vergleich zum LH/FSH−RH zeigt (Des−Gly−NH$_2$^{10}, Pro-äthylamid^9)−LH/FSH−RH drei- bis fünf-

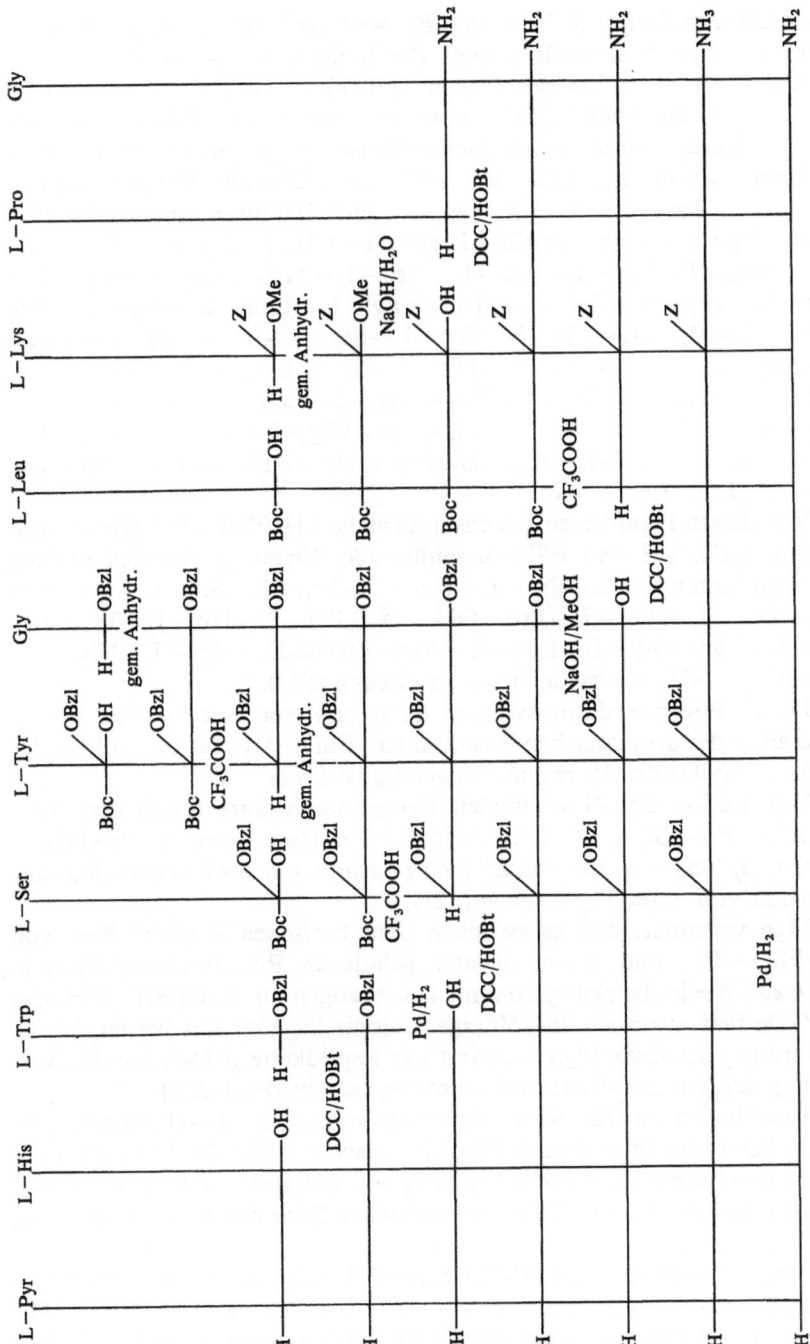

Syntheseschema 51. Darstellung von Lys[8]–LH/FSH–RH nach CHANG *et al.* (504)

fache LH-ausschüttende Wirkung. Bei weiterer Verlängerung des Alkyl-restes am Prolinamidende nimmt die biologische Wirkung allerdings wieder ab. Substitution des Glycinamidendes durch Alkylreste scheint daher die Bindung des Peptids an den Rezeptor zu vergrößern (511, 512).

Um Rezeptorstudien an menschlichem Hypophysenzellen durch-zuführen, wurde die LH- und FSH-ausschüttende Wirkung einiger Peptide an humanen fetalen Hypophysenzellkulturen untersucht. Fol-gende Peptide wurden geprüft: LH/FSH − RH, L − Pyr − L − Gln − L − Ala − NH$_2$, L − Pyr − L − Tyr − L − Arg − L − Trp − NH$_2$, L − Pyr − L − His − L − Trp − L − Ser − L − Tyr − Gly − L − Leu − L − Arg − L − Pro − OH, L − Pyr − L − His − L − Trp − L − Ser − L − Tyr − Gly − L − Leu − L − Arg − OH, L − Pyr − L − His − L − Trp − L − Ser − L − Tyr − Gly − OH, L − Trp − L − Ser − L − Tyr − Gly − L − Leu − L − Arg − L − Pro − Gly − NH$_2$, L − Trp − L − Ser − L − Tyr − Gly − L − Leu − L − Arg − L − Pro − OH, L − Leu − L − Arg − L − Pro − Gly − NH$_2$ und L − Pyr − L − His − L − Pro − Gly − NH$_2$.

Von diesen Peptiden zeigen das natürliche LH/FSH − RH erwartungs-gemäß starke LH und FSH-ausschüttende Wirkung; von den übrigen Peptiden haben nur L − Pyr − L − His − L − Trp − L − Ser − L − Tyr − Gly − L − Leu − L − Arg − L − Pro − OH, L − Pyr − L − His − L − Trp − L − Ser − L − Tyr − Gly − L − Leu − L − Arg − OH und L − Pyr − L − His − L − Pro − Gly − NH$_2$ schwache Hormonwirkung (513).

Durch Photooxydationsstudien kann bewiesen werden, daß wenig-stens einer der aromatischen Reste im LH/FSH − RH für die volle biolo-gische Aktivität des Hormons notwendig ist (514).

Substitution der N-terminalen Pyroglutaminsäure durch den Ac − L − Ala-, Ac − Gly-, D − Pyr- oder L − Pro-Rest führt zu Produkten, welche, im Vergleich zum natürlichen Hormon, nur noch eine biologische Aktivität von unter 1% besitzen (521).

Man vermutet, daß es zwischen dem basischen Arginin[8]-Rest von LH/FSH − RH und einem negativ geladenen Rezeptormolekülbruch-stück zur Wechselwirkung kommt. Da, bezogen auf LH/FSH − RH, das Har[8]-Derivat nur noch 25—50%ige biologische Aktivität besitzt, ist die Entfernung des Guanidylrestes von der Peptidkette jedoch für die Aus-bildung des Hormon-Rezeptorkomplexes entscheidend (522).

Substitution des His[2]-Restes durch den von (2 − L − β − (Pyrazolyl − 3)-alanin liefert ein Produkt mit 19% biologischer Aktivität. Man vermutet daher, daß für die hormonelle Wirkung von LH/FSH − RH an Position 2 ein Aminosäurerest mit heteroaromatischer Seitenkette von Bedeutung ist (523).

Pentamethylphenylalanin[3]-LH/FSH-RH zeigt noch die erstaunlich hohe biologische Aktivität von 34—70% (bezogen auf LH/FSH − RH). Sowohl Tryptophan als auch Pentamethylphenylalanin besitzt die Fähig-

Literaturverzeichnis: SS. 534—564

keit, mit aromatischen MolekülenΠ-Komplexe zu bilden. Die Ausbildung solcher Komplexe zwischen der Seitenkette der 3. Aminosäure von LH/FSH − RH und dem Rezeptor scheinen notwendig für die physiologische Wirkung zu sein (*524*).

Wird die Position 6 von LH/FSH−RH durch D-Aminosäurereste, wie z. B. D-Alanin ausgetauscht, dann erhält man Derivate mit bis zu 50facher biologischer Aktivität im Vergleich zum natürlichen Hormon (*525, 526, 534*).

Erstaunlicherweise führt Substitution eines Glycinamidwasserstoffatoms von LH/FSH − RH durch den $NH_2 − (CH_2)_4$-Rest zu einem Derivat, welches keine biologische Aktivität zeigt (*530*).

Für die LH/FSH-ausschüttende Wirkung des Dekapeptidamids scheint an Position 5 eine Aminosäure mit einem isocyclischen Seitenkettenrest von Bedeutung zu sein: $(Tyr(Me)^5 − LH/FSH − RH$ und $Phe^5 − LH/FSH − RH$ mit aromatischen Seitenkettenresten an Position 5 zeigen 8,6 bzw. 42,5% biologische Aktivität, während das Cyclohexyl-L-alanin-haltige Derivat $(Cha^5) − LH/FSH − RH$ noch zu 22,4% biologisch aktiv ist (*531*).

Verkürzt man vom N-terminalen Ende ausgehend die LH/FSH − RH-Dekapeptidamidsequenz um jeweils eine Aminosäure, dann erhält man 8 Peptide mit geringer biologischer Aktivität. Alle 9 möglichen Nonapeptidamide, welchen an irgend einer Position eine Aminosäure in der Sequenz des LH/FSH − RH fehlt, zeigen mit Ausnahme von $(Des − Gly − NH_2^{10}$, $Pro − NH_2^9) − LH/FSH − RH$ äußerst geringe LH-ausschüttende Wirkung (*533*).

Substituiert man $[D − Phe^2] − LH/FSH − RH$ in Position 6 durch D − Ala, D − Leu, D − Arg, D − (Ph)Gly-, D − Phe oder 2Me − Ala, dann erhält man Verbindungen mit unterschiedlichen antiovulatorischen Eigenschaften (*536*).

Zur Darstellung der Derivate von $(Des − Gly − NH_2^{10}$, $Pro − äthyl-amid^9) − LH/FSH − RH$ ist vor kurzem eine klassische Synthese beschrieben worden, welche mit relativ guter Ausbeute abläuft (vgl. Reaktionsschema 52) (*537*).

Obgleich $(L − Ala^4) − LH/FSH − RH$ noch beträchtliche biologische Aktivität zeigt, führt Blockierung der OH-Gruppe von Serin im LH/FSH−RH durch den Bu^t-Rest zum Verlust der LH-freisetzenden Wirkung. Einen Weg zur Darstellung von $(L − Ser(Bu^t)^4) − LH/FSH − RH$ zeigt Syntheseschema 53 (*539*).

Durch Festphasensynthese wurden vor kurzem 9 Peptide mit der natürlichen Sequenz von LH/FSH − RH hergestellt, bei welchen jeweils ein Aminosäurerest durch einen Prolinrest substituiert ist. Keines der Peptide zeigt jedoch große biologische Aktivität (*541*).

Die LH- bzw. FSH-ausschüttende Wirkung einer großen Zahl von Peptiden ist in Tab. 20 zusammengestellt.

Tabelle 20. *Biologische Aktivität von Peptiden und Peptidderivaten bezogen auf LH/FSH–RH = 100%*

Verbindung	Aktivität bezogen auf LH/FSH–RH = 100%	Literatur- zitate
(Ac–L–Ala^1)–LH/FSH–RH	~0,25(LH); 0,1—0,25 FSH	(521)
(L–Ala^3)–LH/FSH–RH	<0,001	(31)
(L–Ala^4)–LH/FSH–	5	(507)
(L–Ala^4)–LH/FSH–RH	schwach aktiv	(530)
(L–Ala^4)–LH/FSH–RH	3—6 (LH); 16 (FSH)	(506)
(L–Ala^5)–LH/FSH–RH	schwach aktiv	(530)
(L–Ala^6)–LH/FSH–RH	1	(31)
(D–Ala^6)–LH/FSH–RH	≈200^7	(519)
(D–Ala^6)–LH/FSH–RH	570 (LH); 475 (FSH)	(525)
(D–Ala^6)–LH/FSH–RH	570 (LH); 475 (FSH)	(526)
(D–Ala^6, Des–Gly–NH$_2^{10}$)–LH/FSH–RH–äthylamid	200^7	(519)
(L–Ala^7)–LH/FSH–RH	5—6 (LH); 3—5 (FSH)	(506)
(L–Ala^9)–LH/FSH–RH	0,85	(31)
(L–Ala^{10})–LH/FSH–RH	10 (LH)	(506)
(L–Ala^{10})–LH/FSH–RH	6,0 (LH); 3,5 (FSH)	(511)
(L–Ala^4, L–Phe^5)–LH/FSH–RH	schwach aktiv	(530)
(L–Ala^4, L–Phe^5, L–Lys^8)–LH/FSH–RH	schwach aktiv	(530)
(L–Pyr(1)–Ala^2)–LH/FSH–RH	1	(538)
(L–Arg^2)–LH/FSH–RH	<0,1 (LH); 0,1 (FSH)	(506)
(L–Arg^2)–LH/FSH–RH	≈0,03 (LH); 0,01 (FSH)	(521)
(L–Arg^2)–LH/FSH–RH	0,005	(532)
(D–Arg^2)–LH/FSH–RH	>0,05<0,5	(532)
Ac–L–Arg–L–Pro–Gly–NH$_2$	<0,001 (in vitro)	(533)
(L–Cit^8)–LH/FSH–RH	6—33	(522)
(L–Cit^8)–LH/FSH–RH	7,5 (in vivo)	(535)
(L–Cha^5)–LH/FSH–RH^8	22,4 (in vivo)	(531)
(Dab^8)–LH/FSH–RH	2 (in vivo)	(535)
Des–(L–Arg^8–Gly^{10})–LH/FSH–RH^6	<0,01	(517)

Verbindung	Aktivität	Lit.
$(Des-L-Arg^8-L-Pro^9-Gly^{10})-LH/FSH-RH$	inaktiv	(31)
$(Des-Gly^6)-LH/FSH-RH$	~5 (LH); 5—7 (FSH)	(521)
$(Des-Gly^{10})-LH/FSH-RH$	11 (LH)	(511)
$(Des-Gly^{10})-LH/FSH-RH$	11	(517)
$Des-(Gly^6-Gly^{10})-LH/FSH-RH^6$	<0,01	(517)
$(Des-Gly-NH_2^{10}, L-Pro-OEt^9)-LH/FSH-RH$	11,5 (LH); 6,4 (FSH)	(511)
$(Des-Gly-NH_2^{10}, L-Pro-NH_2^9)-LH/FSH-RH$	<5 (LH); 3 (FSH)	(521)
$(Des-Gly-NH_2^{10}, L-Pro-n-butylamid^9)-LH/FSH-RH$	7,2 (LH); 4,3 (FSH)	(511)
$(Des-Gly-NH_2^{10}, L-Pro-cyclohexylamid^9)-LH/FSH-RH$	0,7 (LH); 0,6 (FSH)	(511)
$(Des-Gly-NH_2^{10}, L-Pro-dimethylamid^9)-LH/FSH-RH$	15 (LH); 9 (FSH)	(511)
$(Des-Gly-NH_2^{10}, L-Pro-äthanolamid^9)-LH/FSH-RH$	100—150	(505)
$(Des-Gly-NH_2^{10}, L-Pro-äthanolamid^9)-LH/FSH-RH$	210 (LH); 220 (FSH)	(511)
$(Des-Gly-NH_2^{10}, L-Pro-äthylamid^9)-LH/FSH-RH$	500	(512)
$(Des-Gly-NH_2^{10}, L-Pro-äthylamid^9)-LH/FSH-RH$	120 (LH); 158 (FSH)	(521)
$(Des-Gly-NH_2^{10}, L-Pro-äthylamid^9)-LH/FSH-RH$	500	(505)
$(Des-Gly-NH_2^{10}, L-Pro-äthylamid^9)-LH/FSH-RH$	300 (LH); 280 (FSH)	(511)
$(Des-Gly-NH_2^{10}, L-Pro-äthylamid^9)-LH/FSH-RH$	550 (LH); 700 (FSH)	(520)
$(Des-Gly-NH_2^{10}, L-Pro-äthylamid^9)-LH/FSH-RH$	300 (LH); 280 (FSH)	(525)
$(Des-Gly-NH_2^{10}, L-Pro-äthylamid^9)-LH/FSH-RH$	300 (LH); 280 (FSH)	(526)
$(Des-Gly-NH_2^{10}, L-Pro-äthylamid^9)-LH/FSH-RH$	stark aktiv	(540)
$(Des-Gly-NH_2^{10}, L-Pro-äthylamid^9, L-Ala^6)-LH/FSH-RH$	22 (LH)	(526)
$(Des-Gly-NH_2^{10}, L-Pro-äthylamid^9, D-Ala^6)-LH/FSH-RH$	180 (LH); 300 (FSH)	(526)
$(Des-Gly-NH_2^{10}, L-Pro-äthylamid^9, D-Ala^6)-LH/FSH-RH$	180 (LH); 300 (FSH)	(525)
$(Des-Gly-NH_2^{10}, L-Pro-äthylamid^9, \alpha-Aminoisobuttersäure^6)-LH/FSH-RH$	470 (LH); 300 (FSH)	(526)
$(Des-Gly-NH_2^{10}, L-Pro-äthylamid^9, D-Ala^6, L-Phe^5)-LH/FSH-RH$	270 (LH); 300 (FSH)	(525)
$(Des-Gly-NH_2^{10}, L-Pro-äthylamid^9, D-Ala^6, L-Phe^5)-LH/FSH-RH$	270 (LH); 300 (FSH)	(526)
$(Des-Gly-NH_2^{10}, L-Pro-äthylamid^9, D-Ala^6, L-Ile^5)-LH/FSH-RH$	160 (LH); 350 (FSH)	(525)
$(Des-Gly-NH_2^{10}, L-Pro-äthylamid^9, D-Ala^6, L-Ile^5)-LH/FSH-RH$	160 (LH); 350 (FSH)	(526)
$(Des-Gly-NH_2^{10}, L-Pro-äthylamid^9, D-Abu^6)-LH/FSH-RH$	180 (LH); 170 (FSH)	(525)
$(Des-Gly-NH_2^{10}, L-Pro-äthylamid^9, D-Nva^6)-LH/FSH-RH$	90—400 (LH); 160—290 (FSH)	(525)
$(Des-Gly-NH_2^{10}, L-Pro-äthylamid^9, D-Leu^6)-LH/FSH-RH$	275 (LH); 280 (FSH)	(525)
$(Des-Gly-NH_2^{10}, L-Pro-äthylamid^9, D-Leu^6)-LH/FSH-RH$	5 360 (LH); 1 450 (FSH)	(534)
$(Des-Gly-NH_2^{10}, L-Pro-äthylamid^9, D-Leu^6, L-Phe^5)-LH/FSH-RH$	100 (LH); 90 (FSH)	(525)

Tabelle 20 *(Fortsetzung)*

Verbindung	Aktivität bezogen auf LH/FSH-RH = 100%	Literatur-zitate
(Des-Gly-NH$_2^{10}$, L-Pro-äthylamid^9, D-Leu^6, L-Ile^5)-LH/FSH-RH	85 (LH); 67 (FSH)	(525)
(Des-Gly-NH$_2^{10}$, L-Pro-äthylamid^9, D-Phe^6, L-Phe^3)-LH/FSH-RH	140 (LH); 120—520 (FSH)	(525)
(Des-Gly-NH$_2^{10}$, L-Pro-äthylamid^9, D-Ser^6)-LH/FSH-RH	100 (LH); 100 (FSH)	(525)
(Des-Gly-NH$_2^{10}$, L-Pro-isobutylamid^9)-LH/FSH-RH	3 (LH); 2,8 (FSH)	(511)
(Des-Gly-NH$_2^{10}$, L-Pro-isopropylamid^9)-LH/FSH-RH	150 (LH); 100 (FSH)	(511)
(Des-Gly-NH$_2^{10}$, L-Pro-methylamid^9)-LH/FSH-RH	80—100	(505)
(Des-Gly-NH$_2^{10}$, L-Pro-methylamid^9)-LH/FSH-RH	47 (LH); 56 (FSH)	(511)
(Des-Gly-NH$_2^{10}$, L-Pro-OMe^9)-LH/FSH-RH	6,2 (LH); 2,9 (FSH)	(511)
(Des-Gly-NH$_2^{10}$, L-Pro-morpholinamid^9)-LH/FSH-RH	20—30	(505)
(Des-Gly-NH$_2^{10}$, L-Pro-morpholinamid^9)-LH/FSH-RH	17 (LH); 16 (FSH)	(511)
(Des-Gly-NH$_2^{10}$, L-Pro-piperidinamid^9)-LH/FSH-RH	1,5 (LH); 2,0 (FSH)	(511)
(Des-Gly-NH$_2^{10}$, L-Pro-propylamid^9)-LH/FSH-RH	200—300	(505)
(Des-Gly-NH$_2^{10}$, L-Pro-propylamid^9)-LH/FSH-RH	550 (LH); 940 (FSH)	(520)
(Des-Gly-NH$_2^{10}$, L-Pro-propylamid^9)-LH/FSH-RH	190 (LH); 210 (FSH)	(511)
(Des-Gly-NH$_2^{10}$, L-Pro-pyrrolidinamid^9)-LH/FSH-RH	70—80	(505)
(Des-Gly-NH$_2^{10}$, L-Pro-pyrrolidinamid^9)-LH/FSH-RH	110 (LH); 92 (FSH)	(511)
(Des-His^2)-LH/FSH-RH	<0,001	(31)
(Des-His^2)-LH/FSH-RH	0,001	(532)
(Des-L-His^2, Des-L-Pro^9)-LH/FSH-RH	≪0,1 (LH); ≪0,1 (FSH)	(521)
Des-(L-His^2-Gly^{10})-LH/FSH-RH$^{\underline{6}}$	<0,01	(517)
Des-L-Leu^7-L-Arg^8)-LH/FSH-RH	inaktiv	(31)
Des-(L-Leu^7-Gly^{10})-LH/FSH-RH$^{\underline{6}}$	<0,01	(517)
Des-L-Leu^7-L-Arg^8-L-Pro^9-Gly^{10})-LH/FSH-RH	schwach aktiv	(31)
Des-(L-Pro^9-Gly^{10})-LH/FSH-RH$^{\underline{6}}$	<0,01	(517)
(Des-L-Pro^9)-LH/FSH-RH	~1 (LH); 0,8 (FSH)	(521)
(L-Pro^1)-LH/FSH-RH	0,02 (LH); <0,5 (FSH)	(521)
(D-Pyr^1)-LH/FSH-RH	~1 (LH); 0,6 (FSH)	(521)
(Des-L-Pyr^1)-LH/FSH-RH	<0,002	(509)

Literaturverzeichnis: SS. 534—564

$(\text{Des-L-Pyr}^1\text{-Des-L-His}^2)\text{-LH/FSH-RH}$	inaktiv	(509)
$\text{Des-}(\text{L-Ser}^4\text{-Gly}^{10})\text{-LH/FSH-RH}^{\underline{6}}$	<0,01	(517)
$\text{Des-}(\text{L-Trp}^3\text{-Gly}^{10})\text{-LH/FSH-RH}^{\underline{6}}$	<0,01	(517)
$(5\text{-F-L-Trp}^3)\text{-LH/FSH-RH}$	6	(524)
$\text{Des-}(\text{L-Tyr}^5\text{-Gly}^{10})\text{-LH/FSH-RH}^{\underline{6}}$	<0,01	(517)
$(\text{L-Glu}^8)\text{-LH/FSH-RH}$	schwach aktiv	(530)
$(\text{L-Gln}^4)\text{-LH/FSH-RH}$	8 (LH); 6 (FSH)	(506)
$\text{Ac-Gly-L-Leu-L-Arg-L-Pro-Gly-NH}_2$	<0,001 (in vitro)	(533)
$(\text{Ac-Gly}^1)\text{-LH/FSH-RH}$	>0,25 (LH); 0,1—0,5 (FSH)	(521)
$(\text{Gly}^3)\text{-LH/FSH-RH}$	<0,001	(31)
$(\text{Gly}^4)\text{-LH/FSH-RH}$	1,5	(31)
$(\text{Gly}^5)\text{-LH/FSH-RH}$	0,1	(31)
$(\text{Gly}^5)\text{-LH/FSH-RH}$	0,1 (in vivo)	(531)
$(\text{Gly}^7)\text{-LH/FSH-RH}$	3 (LH); 5 (FSH)	(506)
$(\text{Gly}^7)\text{-LH/FSH-RH}$	0,2	(31)
$(\text{Gly}^8)\text{-LH/FSH-RH}$	0,1	(31)
$(\text{Gly}^9)\text{-LH/FSH-RH}$	0,2	(31)
$(\text{Gly-NHCH}_3^{10})\text{-LH/FSH-RH}$	1,7 (LH); 1,9 (FSH)	(521)
$((\text{Gly-NH(CH}_2)_4\text{-NH}_2^{10})\text{-LH/FSH-RH}$	inaktiv	(530)
$(\text{Gly-NMe}_2^{10})\text{-LH/FSH-RH}$	14	(31)
$(\text{Gly-OH}^{10})\text{-LH/FSH-RH}$	0,5 (LH); 0,5 (FSH)	(521)
$(\text{Gly-OCH}_3^{10})\text{-LH/FSH-RH}$	1,4 (LH); 1,0 (FSH)	(521)
$(\text{Gly}^{10})\text{-LH/FSH-RH}$	0,003 (LH); 0,003 (FSH)	(529)
$(\text{Gly}^{10}, \text{Gly}^{2a})\text{-LH/FSH-RH}$	0,0003 (LH); 0,0003 (FSH)	(529)
$(\text{Gly}^{10}, \text{Gly}^{1a})\text{-LH/FSH-RH}$	<0,0003 (LH); <0,0003 (FSH)	(529)
$(\text{Gly}^{10}, \text{L-Tyr}^3, \text{L-Trp}^5)\text{-LH/FSH-RH}$	0,0003 (LH); 0,00015 (FSH)	(529)
$(\text{CH}_3\text{CH}_2\text{-}\overset{\text{O}}{\text{C}}\text{-Gly}^1)\text{-LH/FSH-RH}$	0,2	(31)
$(\text{L-Har}^8)\text{-LH/FSH-RH}$	25—50	(522)
$(\text{L-Har}^8)\text{-LH/FSH-RH}$	21,7	(527)
$(\text{L-Har}^8)\text{-LH/FSH-RH}$	12,6 (in vivo)	(535)
$\text{Ac-L-His-L-Trp-L-Ser-L-Tyr-Gly-L-Leu-L-Arg-L-Pro-Gly-NH}_2$	0,2	(518)
$\text{Ac-L-His-L-Trp-L-Ser-L-Tyr-Gly-L-Leu-L-Arg-L-Pro-Gly-NH}_2$	0,2 (in vitro)	(533)

Tabelle 20 *(Fortsetzung)*

Verbindung	Aktivität bezogen auf LH/FSH−RH = 100%	Literaturzitate
(L−His^3)−LH/FSH−RH	<0,05	(524)
(D−His^2)−LH/FSH−RH	10,0	(532)
(3−Me−L−His^2)−LH/FSH−RH	1 (LH); 1—2 (FSH)	(506)
(N$^\tau$−Me−L−His^2)−LH/FSH−RH	6	(31)
(N$^\tau$−Me−L−His^2)−LH/FSH−RH	2	(31)
(L−Ile^2)−LH/FSH−RH	0,03	(528)
(L−Ile^6)−LH/FSH−RH	0,034	(31)
(L−Ile^7)−LH/FSH−RH	45 (LH); 33 (FSH)	(506)
L−Leu−L−Arg−L−Pro−Gly−NH$_2$	inaktiv	(31)
Ac−L−Leu−L−Arg−L−Pro−Gly−NH$_2$	<0,001 (in vitro)	(533)
(L−Leu^1)−LH/FSH−RH	0,003	(31)
(L−Leu^2)−LH/FSH−RH	0,5	(532)
(D−Leu^2)−LH/FSH−RH	>0,01, <0,05	(532)
(L−Leu^{2,3})−LH/FSH−RH	inaktiv	(528)
(L−Leu^4)−LH/FSH−RH	inaktiv	(539)
(D−Leu^6)−LH/FSH−RH	110 (LH); 100 (FSH)	(525)
(D−Leu^6)−LH/FSH−RH	900 (LH); 500 (FSH)	(534)
(L−Leu^8)−LH/FSH−RH	<1	(527)
(L−Lys^2)−LH/FSH−RH	<0,1 (LH); <0,1 (FSH)	(506)
(L−Lys^8)−LH/FSH−RH	beträchtlich	(504)
(L−Lys^8)−LH/FSH−RH	10—30	(522)
(L−Lys^8)−LH/FSH−RH	7,6	(527)
(L−Lys^8)−LH/FSH−RH	11—28 (LH); 25 (FSH)	(506)
(L−Lys^8)−LH/FSH−RH		(515)
(L−Lys^8)−LH/FSH−RH	schwach aktiv	(530)
(L−Lys^8)−LH/FSH−RH	6,0 (in vivo)	(535)
(Nar^8)−LH/FSH−RH	15,1 (in vivo)	(535)
(L−Nle^7)−LH/FSH−RH^5	30 (LH); 22 (FSH)	(506)

Literaturverzeichnis: SS. 534—564

Verbindung	Aktivität	Lit.
(L–Orn⁸)–LH/FSH–RH	6—12 (LH); 5 (FSH)	(506)
(L–Orn⁸)–LH/FSH–RH	aktiv	(515)
(L–Orn⁸)–LH/FSH–RH	5,5	(527)
(L–Orn⁸)–LH/FSH–RH	2,0 (in vivo)	(535)
(Nᵟ-Ac–L–Orn⁸)–LH/FSH–RH	5,1 (in vivo)	(535)
(Nᵟ-Bz–L–Orn⁸)–LH/FSH–RH	1,1 (in vivo)	(535)
(L–Phe²)–LH/FSH–RH	4	(31)
(L–Phe²)–LH/FSH–RH	4—7 (LH); 2 (FSH)	(506)
(D–Phe²)–LH/FSH–RH	<0,001	(532)
(L–Phe³)–LH/FSH–RH	2	(31)
(L–Phe⁵)–LH/FSH–RH	≈50	(530)
(L–Phe⁵)–LH/FSH–RH	42,5 (in vivo)	(531)
(L–Phe⁵)–LH/FSH–RH	64	(31)
(D–Phe²)–LH/FSH–RH	0,0001	(536)
(D–Phe²–D–Ala⁶)–LH/FSH–RH	0,1	(536)
(D–Phe²–L–Leu⁶)–LH/FSH–RH	0,01	(536)
(D–Phe²–D–Arg⁶)–LH/FSH–RH	0,001	(536)
(D–Phe²–2–Me–L–Ala⁶)–LH/FSH–RH	0,0001	(536)
(p–Aminophenylalanin³)–LH/FSH–RH	0,59	(524)
(p–Aminophenylalanin⁵)–LH/FSH–RH	37	(510)
(D,L–Chlorphenylalanin⁵)–LH/FSH–RH	inaktiv	(515)
(p–Methoxyphenylalanin⁵)–LH/FSH–RH	24	(510)
(Pentamethylphenylalanin³)–LH/FSH–RH	69	(524)
(p–Nitrophenylalanin³)–LH/FSH–RH	0,01	(524)
(p–Nitrophenylalanin⁵)–LH/FSH–RH	5	(510)
Ac–L–Pro–Gly–NH₂	<0,001 (in vitro)	(533)
(L–Pro¹)–LH/FSH–RH	<0,1 (LH); <0,1 (FSH)	(506)
(L–Pro¹)–LH/FSH–RH	geringe Aktivität	(541)
(L–Pro³)–LH/FSH–RH	geringe Aktivität	(541)
(L–Pro³)–LH/FSH–RH	geringe Aktivität	(541)
(L–Pro⁴)–LH/FSH–RH	geringe Aktivität	(541)
(L–Pro⁵)–LH/FSH–RH	geringe Aktivität	(541)
(L–Pro⁵)–LH/FSH–RH	geringe Aktivität	(541)
(L–Pro⁶)–LH/FSH–RH	geringe Aktivität	(541)

Tabelle 20 *(Fortsetzung)*

Verbindung	Aktivität bezogen auf LH/FSH-RH = 100%	Literaturzitate
(L-Pro^7)-LH/FSH-RH	geringe Aktivität	(541)
(L-Pro^8)-LH/FSH-RH	geringe Aktivität	(541)
(L-Pro^9)-LH/FSH-RH	45	(541)
(L-Pro^{10})-LH/FSH-RH	geringe Aktivität	(541)
(L-Pro^{10})-LH/FSH-RH	0,18 (LH); 0,14 (FSH)	(511)
(L-Pyr(3)Ala^2)-LH/FSH-RH	19	(523)
L-Pyr-L-Gln-L-Ala-NH$_2$	inaktiv	(509)
L-Pyr-L-His-NH$_2$	0,01	(31)
L-Pyr-L-His-L-Phe-L-Ala-L-Tyr-Gly-L-Leu-L-Arg-L-Pro-NHEt	7,5 (in vivo); 6,04 (in vitro)	(537)
L-Pyr-L-His-L-Phe-Gly-L-Tyr-Gly-L-Leu-L-Arg-L-Pro-NHEt	5,1 (in vivo); 2,81 (in vitro)	(537)
L-Pyr-L-His-L-Phe-L-Ala-L-Tyr-Gly-L-Leu-L-Lys-L-Pro-NHEt	0,6 (in vivo); 0,12 (in vitro)	(537)
L-Pyr-L-His-L-Phe-Gly-L-Tyr-Gly-L-Leu-L-Lys-L-Pro-NHEt	0,091 (in vitro)	(537)
L-Pyr-L-His-L-Phe-Gly-L-Tyr-Gly-L-Leu-L-Har-L-Pro-NHEt	0,0013 (in vitro)	(537)
L-Pyr-L-His-L-Phe-L-Ala-L-Tyr-Gly-L-Leu-L-Har-L-Pro-NHEt	0,0019 (in vitro)	(537)
L-Pyr-L-His-L-Phe-L-Ala-L-Tyr-Gly-L-Phe-L-Arg-L-Pro-NHEt	8,1 (in vivo); 1,59 (in vitro)	(537)
L-Pyr-L-His-L-Phe-L-Ala-L-Tyr-D-Ala-L-Leu-L-Arg-L-Pro-NHEt	28,3 (in vivo); 2,13 (in vitro)	(537)
L-Pyr-L-His-L-Phe-D-Ala-L-Tyr-Gly-L-Leu-L-Arg-L-Pro-NHEt	0,3 (in vivo); 0,24 (in vitro)	(537)
L-Pyr-L-His-L-Phe-Gly-L-Tyr-Gly-L-Leu-L-Arg-L-Pro-AMT-Me^2	1,4 (in vivo); 1,18 (in vitro)	(537)
L-Pyr-L-His-L-Pro-NH$_2$(TRH)	inaktiv	(509)
L-Pyr-L-His-L-Ser-L-Tyr-Gly-L-Leu-L-Arg-L-Pro-Gly-NH$_2$	<0,001 (in vitro)	(533)
L-Pyr-L-His-L-Trp-NH$_2$	0,1	(31)
L-Pyr-L-His-L-Trp-NH$_2$	inaktiv	(509)
L-Pyr-L-His-L-Trp-OH	inaktiv	(509)
L-Pyr-L-His-L-Trp-OH	<0,0003 (LH); <0,0003 (FSH)	(529)
L-Pyr-L-His-L-Trp-OH	1	(539)
L-Pyr-L-His-L-Trp-L-Ser-NH$_2$	<0,01	(31)
L-Pyr-L-His-L-Trp-L-Ser-NH$_2$	inaktiv	(509)
L-Pyr-L-His-L-Trp-L-Ser-Gly-L-Leu-L-Arg-L-Pro-Gly-NH$_2$	<0,001 (in vitro)	(533)

Peptid	Aktivität	Lit.
L-Pyr-L-His-L-Trp-L-Ser-L-Tyr-NH$_2$	0,02	*(31)*
L-Pyr-L-His-L-Trp-L-Ser-L-Tyr-Gly-NH$_2$	<0,01	*(31)*
L-Pyr-L-His-L-Trp-L-Ser-L-Tyr-Gly-OH	<0,0003 (LH); <0,0003 (FSH)	*(529)*
L-Pyr-L-His-L-Trp-L-Ser-L-Tyr-Gly-OH	inaktiv	*(31)*
L-Pyr-L-His-L-Trp-L-Ser-L-Tyr-Gly-NH$_2$	1,8 (LH); 1,3 (FSH)	*(521)*
L-Pyr-L-His-L-Trp-L-Ser-L-Tyr-Gly-L-Arg-L-Pro-Gly-NH$_2$	0,005 (in vitro)	*(533)*
L-Pyr-L-His-L-Trp-L-Ser-L-Tyr-Gly-L-Leu-NH$_2$	<0,01	*(31)*
L-Pyr-L-His-L-Trp-L-Ser-L-Tyr-Gly-L-Leu-L-Arg-NH$_2$	0,01	*(31)*
L-Pyr-L-His-L-Trp-L-Ser-L-Tyr-Gly-L-Leu-L-Arg-Gly-NH$_2$	0,11 (in vitro)	*(533)*
L-Pyr-L-His-L-Trp-L-Ser-L-Tyr-Gly-L-Leu-L-Arg-L-Pro-NH$_2$	11 (in vitro)	*(533)*
L-Pyr-L-His-L-Trp-L-Ser-L-Tyr-Gly-L-Leu-L-Arg-L-Pro-Gly-NH$_2$	10	*(31)*
L-Pyr-L-His-L-Trp-L-Ser-L-Tyr-Gly-L-Leu-L-Arg-L-Pro-Gly-NH$_2$	<0,001 (in vitro)	*(533)*
L-Pyr-L-His-L-Trp-L-Ser-L-Tyr-L-Leu-L-Arg-L-Pro-Gly-NH$_2$	0,1 (in vitro)	*(533)*
L-Pyr-L-His-L-Trp-L-Tyr-Gly-L-Leu-L-Arg-L-Pro-Gly-NH$_2$	<0,001 (in vitro)	*(533)*
L-Pyr-L-Ser-L-Val-NH$_2$	<0,004	*(509)*
L-Pyr-L-His-L-Tyr-L-Ala-L-Phe-Gly-L-Leu-L-Arg-L-Pro-NHEt	0,24 (in vitro)	*(537)*
L-Pyr-L-Trp-L-Ser-L-Tyr-Gly-L-Leu-L-Arg-L-Pro-Gly-NH$_2$	<0,001 (in vitro)	*(533)*
L-Pyr-L-Tyr-L-Arg-L-Trp-NH$_2$	≈0,001	*(508)*
L-Pyr-L-Tyr-L-Arg-L-Trp-NH$_2$	0,013	*(509)*
L-Pyr-L-Tyr-L-Arg-L-Trp-Gly-L-His-L-Leu-NH$_2$	aktiv	*(31)*
L-Pyr-L-Val-L-Ser-NH$_2$	inaktiv	*(509)*
Ac-L-Ser-L-Tyr-Gly-L-Leu-L-Arg-L-Pro-Gly-NH$_2$	<0,001 (in vitro)	*(533)*
(O=^{c}Ser^1)LH/FSH-RH^1	5—25 (LH); 6 (FSH)	*(506)*
(L-Ser(Bu^t)^4)-LH/FSH-RH	inaktiv	*(539)*
(L-Ser^5)-LH/FSH-RH	~0,5 (LH); ~0,5 (FSH)	*(521)*
(L-Ser^7)-LH/FSH-RH	schwach aktiv	*(530)*
(Sar^6)-LH/FSH-RH	~2 (LH); 2 (FSH)	*(521)*
(Sar^6),Des-L-His^2)-LH/FSH-RH	<0,1 (LH); <0,1 (FSH)	*(521)*
(Sar^6,Des-L-Tyr^5)-LH/FSH-RH	<0,1 (LH); <0,1 (FSH)	*(521)*
(L-Thr^2)-LH/FSH-RH	0,1	*(532)*
(L-Thr^4)-LH/FSH-RH	4 (LH);17 (FSH)	*(506)*
(L-Thr^4)-LH/RH	19	*(31)*
(O=^{c}Thr^1)LH/FSH-RH^2	<5 (FSH)	*(506)*

Tabelle 20 (Fortsetzung)

Verbindung	Aktivität bezogen auf LH/FSH-RH = 100%	Literaturzitate
Ac-L-Trp-L-Ser-L-Tyr-Gly-L-Leu-L-Arg-L-Pro-Gly-NH$_2$	<0,001 (in vitro)	(533)
(L-Trp^2)-LH/FSH-RH	>5,0 <50	(532)
(D-Trp^2)-LH/FSH-RH	<0,001	(532)
L-Trp-L-Ser(Bu^t)-L-Tyr-Gly-L-Leu-L-Arg-L-Pro-Gly-NH$_2$	inaktiv	(539)
Ac-L-Tyr-Gly-L-Leu-L-Arg-L-Pro-Gly-NH$_2$	<0,001 (in vitro)	(533)
(L-Tyr^2)-LH/FSH-RH	5,0	(532)
(L-Tyr^3)-LH/FSH-RH	0,013	(31)
(L-Tyr^{11})-LH/FSH-RH	0,08 (LH); 0,1 (FSH)	(521)
(L-Tyr(Me)^3)-LH/FSH-RH	~2 (LH); ~2 (FSH)	(521)
(L-Tyr(Me)^5)-LH/FSH-RH	8,6 in vivo)	(531)
(L-Tyr(Me)^{3,5})-LH/FSH-RH	0,5 (LH); 0,5 (FSH)	(521)
(Cl-L-Tyr^5)-LH/FSH-RH	8 (LH); 5 (FSH)	(506)
(Di-Cl-L-Tyr)^5-LH/FSH-RH^4	<1 (LH); <1 (FSH)	(506)
(L-Tyr^3,L-Trp^5)-LH/RH	0,003	(528)
(D-Val^2)-LH/FSH-RH	2,0	(532)
(L-Val^7)-LH/FSH-RH	16 (LH); 20—35 (FSH)	(506)

1: O=cSer: 2-Oxo-oxazolidin-4-carbonsäure
2: O=cThr: 2-Oxo-5-methyl-oxazolidin-4-carbonsäure
3: Cl-Tyr: ortho-Chlortyrosin
4: Di-Cl-Tyr: ortho, ortho'-Dichlortyrosin
5: Nle: Norleucin

6: Bindestrich in der Klammer bedeutet bis
7: nach 30 Minuten bestimmt
8: Cha: Cyclohexyl-L-alanin
9: AMT-Me: 1-Methyl-5-aminomethyltetrazol

Literaturverzeichnis: SS. 534—564

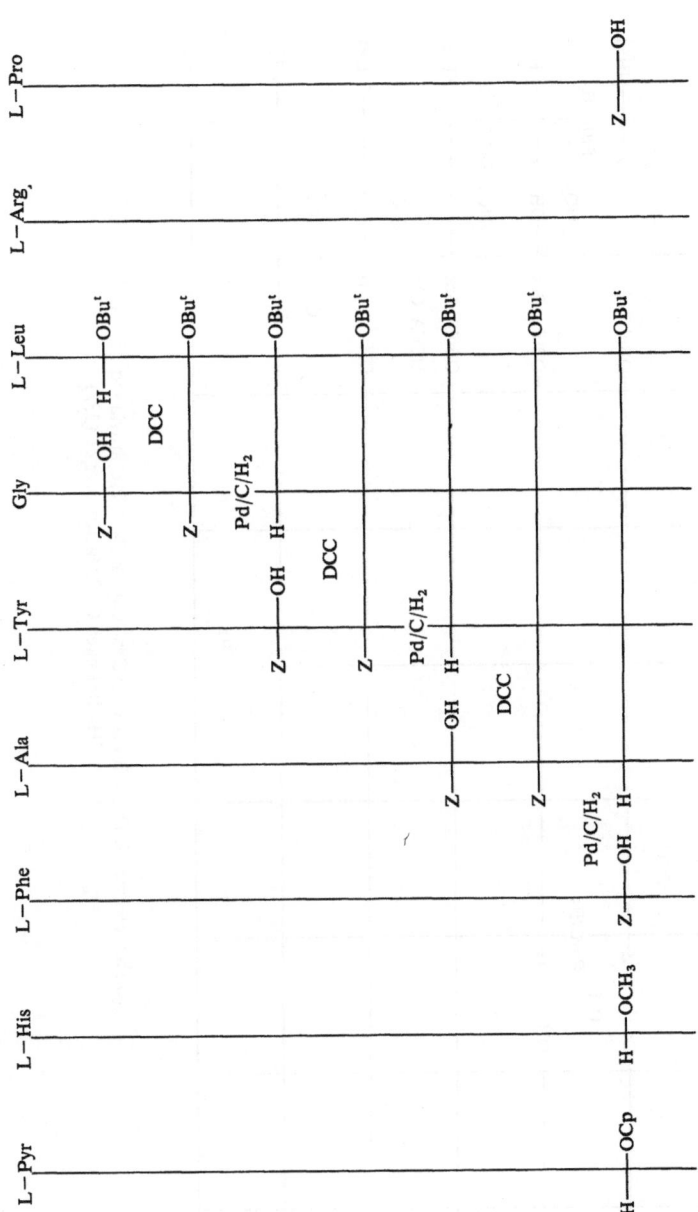

(Fortsetzung S. 530)

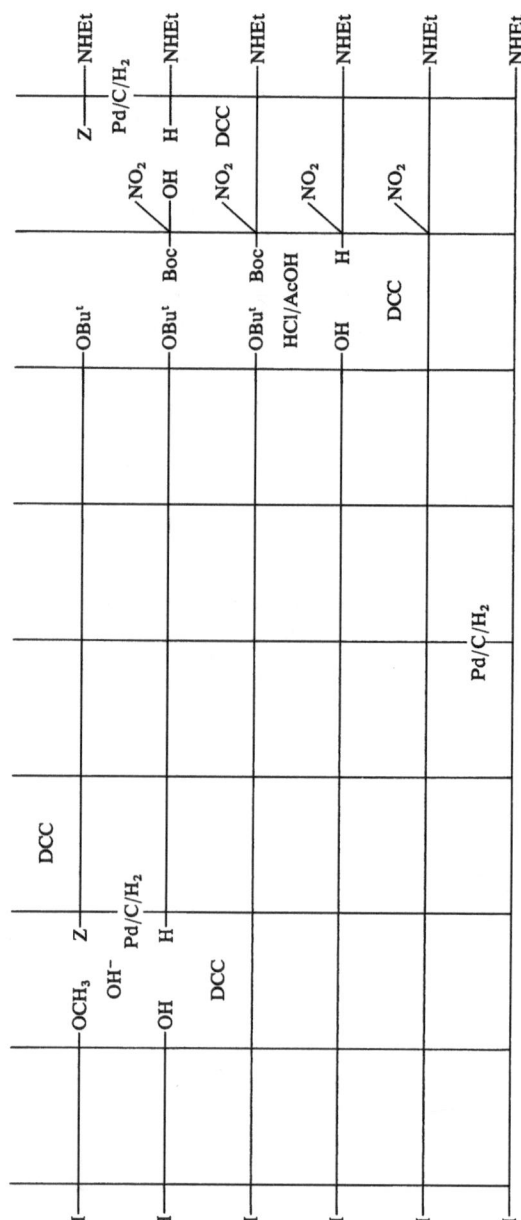

Syntheseschema 52. Darstellung von (Des–Gly–NH$_2$[10], Pro-äthylamid[9], L–Ala[4], L–Phe[3])–
LH/FSH–RH nach C. R. Beddell et al. (537)

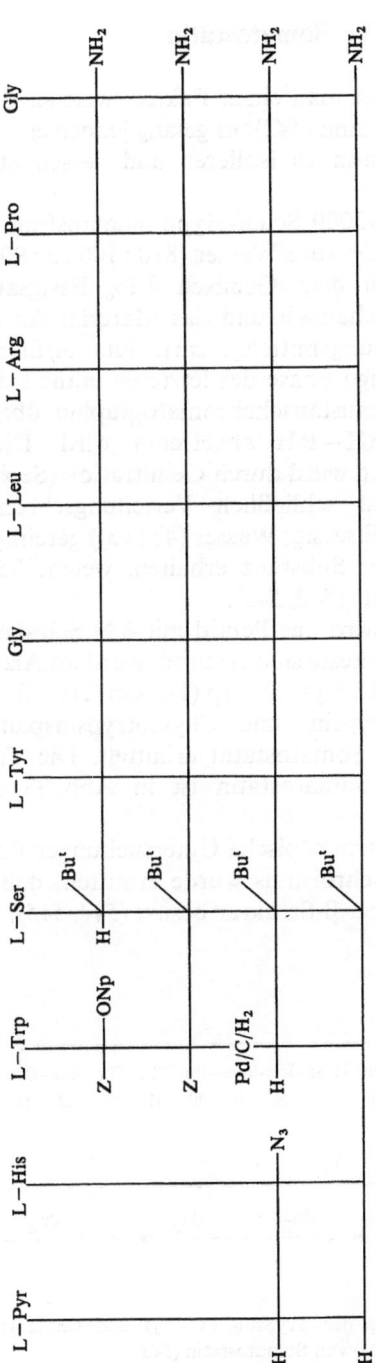

Syntheseschema 53. Darstellung von (L—Ser(Bu^t)^4)—LH/FSH—RH nach Schafer *et al.* (539)

IV. Somatostatin

Seit längerer Zeit vermutet man einen Faktor, welcher die Sekretion von Wachstumshormonen hemmt (542); es gelang jedoch erst vor kurzem, dieses sogenannte Somatostatin zu isolieren und dessen Struktur aufzuklären (542, 543).

Zur Isolierung werden 500 000 Schaf-Hypothalamusfragmente (2 kg) mit Äthanol/Chloroform/Essigsäure/Wasser (810 : 100 : 5 : 90) extrahiert. Der Extrakt wird dann mit dem Gemisch 0,1% Essigsäure/n-Butylalkohol/Pyridin (11 : 5 : 3) behandelt und das Material der organischen Phase anschließend in das Lösungsmittelsystem n-Butanol/Eisessig/Wasser (4 : 1 : 5) gebracht. Die wäßrige Phase des letzteren enthält das Somatostatin, welches durch Ionenaustauschchromatographie über Carboxymethylcellulose von LH/FSH−RH abgetrennt wird. Die so angereicherte Somatostatinfraktion wurd durch Gelfiltration (Sephadex G-25, 0,5 molare Essigsäure) und schließlich Verteilungschromatographie (Sephadex G-25, n-Butanol, Eisessig, Wasser (4 : 1 : 5)) gereinigt. Als Endprodukt werden 8,5 mg einer Substanz erhalten, welche 75 Gewichtsprozente Aminosäuren enthält (542, 543).

Zur Aminosäureanalyse wird das Peptid mit 6 N Salzsäure hydrolysiert und folgende Aminosäurereste sind ermittelt worden: Ala (1), Gly (1), Thr (2), Lys (2), Phe (3), Ser (1), Cys (2), Trp (1), Asp (1) (542, 543).

Durch Edmanabbau, Trypsin- und Chymotrypsinspaltung wurde schließlich die Sequenz von Somatostatin ermittelt. Die Trypsin- und Chymotrypsinspaltung von Somatostatin ist in Abb. 33 schematisch dargestellt (543).

Es liegen noch wenig spektroskopische Untersuchungen über Somatostatin vor. Durch Circulardichroismus wurde ermittelt, daß das Tetradecapeptid keine α-helicale aber β-Struktur besitzt (544, 545).

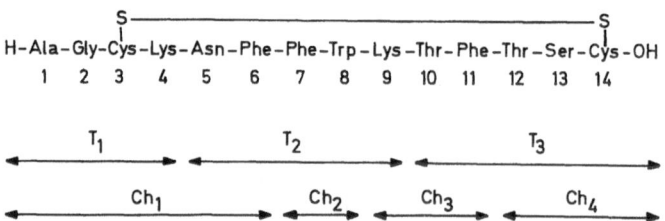

Abb. 33. Schematische Darstellung der Trypsin- (T_1–T_3) und Chymotrypsinspaltung (Ch$_1$–Ch$_4$) von Somatostatin (543)

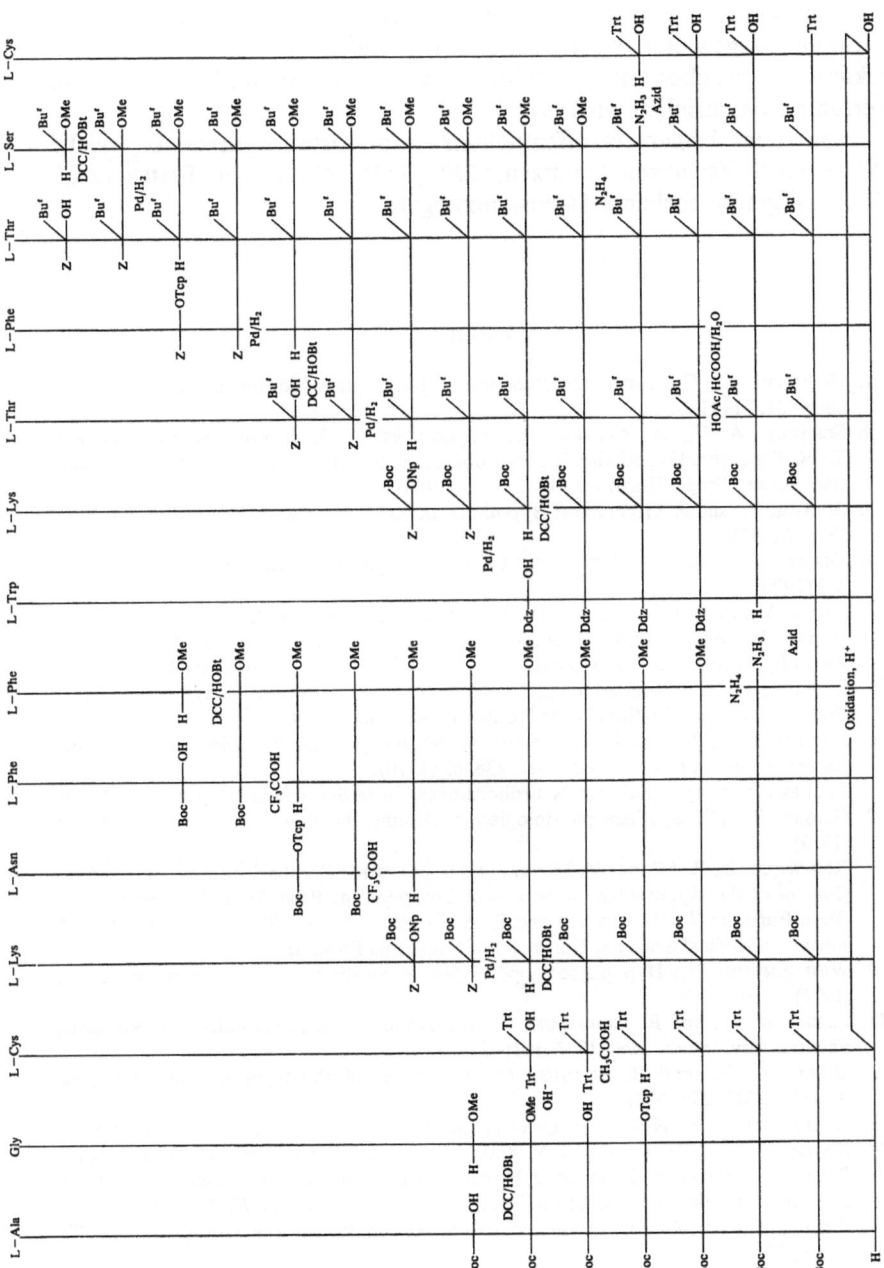

Syntheseschema 54. Darstellung von Somatostatin nach IMMER et al. (550)

Obgleich in diesem Artikel schon mehrmals auf die Nachteile der Festphasensynthese zur Darstellung medizinisch anwendbarer Peptidhormone hingewiesen wurde (vgl. II B), sind schon mehrere Arbeiten bekannt, in welchen die Darstellung des Tetradecapeptids nach diesem Verfahren beschrieben wird (*546—548*).

Von den bisher veröffentlichten Somatostatinsynthesen, welche „klassische" Verfahren benutzen (*549—551*), sei die von Immer et al. (*550*) im Syntheseschema 54 zusammengefaßt.

Literatur

1. Hofmann, K.: Chemistry and Function of Polypeptide Hormones. Ann. Rev. Biochem. **31**, 213 (1962).

2. Schally, A. V., A. Arimura, C. Y. Bowers, A. J. Kastin, S. Sawano, and T. W. Redding: Hypothalamic Neurohormones Regulating Anterior Pituitary Function. Recent Progr. Hormone Res. **24**, 497 (1968).

3. Burgus, R., and R. Guillemin: Hypothalamic Releasing Factors. Ann. Rev. Biochem. **39**, 499 (1970).

4. Diassi, P. A., and Z. P. Horovitz: Endocrine Hormones. Ann. Rev. Pharmacol. **10**, 219 (1970).

5. Jutisz, M., M. P. de la Llosa, A. Berault et B. Kerdelhué: Le Rôle des Hormones Hypothalamiques dans la Régulation de l'Excrétion et de la Synthèse des Hormones Adénohypophysaires. In: Colloques Natiaux du Centre National de la Recherche Scientifique, Neuroendocrinologie. (J. Benoit et C. Kordon, eds.), p. 287. Paris: Editions du Centre National de la Recherche Scientifique. 1970.

6. Pittman, J. A., Jr., and J. M. Hershman: Physiology of the Thyroid Feedback Loop. Excerpta Medica Intern. Congr. Ser. **238**, 69 (1970).

7. Folkers, K.: Hypothalamic Neurohormones. Intra-Sci. Chem. Rep. **5**, 263 (1971).

8. Geiger, R.: Die Synthese physiologisch wirksamer Peptide. Angew. Chem. **83**, 155 (1971).

9. Kniewald, Z., R. Massa, M. Motta, and L. Martini: Feedback Mechanisms and the Control of the Hypothalamohypophysial Complex. In: Proc. Steroid Hormones and Brain Function. (C. H. Sawyer and R. A. Gorski, eds.), p. 289. Los Angeles: UCLA Forum in Medical Sciences, University of California Press. **15**, 1971.

10. Winkelmann, W.: Hypophyseotrope Hypothalamushormone. Der Internist **12**, 193 (1971).

11. Fleischer, N., and R. Guillemin: Clinical Applications of Hypothalamic-Releasing Factors. Adv. Intern. Med. **18**, 303 (1972).

12. Guillemin, R., and R. Burgus: The Hormones of the Hypothalamus. Scientific American **227**, 24 (1972).

13. Hollander, C. S.: Newer Aspects of Hyperthyroidism. Hospital Practice **7**, 87 (1972).

14. Kastin, A. J., C. Gual, and A. V. Schally: Clinical Experience with Hypothalamic Releasing Hormones. Luteinizing Hormone-Releasing Hormone and Other Hypophysiotropic Releasing Hormones. Recent Progr. Hormone Res. **28**, 201 (1972).

15. Retiene, K.: Die Rolle des Zwischenhirns in der endokrinen Regulation. Umschau **72**, 114 (1972).

16. Schally, A. V., A. J. Kastin, and A. Arimura: The Hypothalamus and Reproduction. Am. J. Obstet. Gynecol. **114**, 423 (1972).

17. SIEVERTSSON, H.: Chemistry of Hypothalamic Neurohormones. Acta Pharm. Suecica 9, 19 (1972).

18. STUDER, R. O., und H. STEINER: Hypothalamische Releasing-Hormone. Schweiz. med. Wschr. 102, 1270 (1972).

19. BLACKWELL, R. E., and R. GUILLEMIN: Hypothalamic Control of Adenohypophysial Secretions. Ann. Rev. Physiol. 35, 357 (1973).

20. FOLKERS, K., N.-G. JOHANSSON, F. HOOPER, B. L. CURRIE, H. SIEVERTSSON, J.-K. CHANG und C. Y. BOWERS: Chemie und Biosynthese der freisetzenden und hemmenden Hypothalamus-Neurohormone. Angew. Chem. 85, 271 (1973).

21. GEIGER, R.: Chemische Synthese von Peptidhormonen. Chemie in unserer Zeit 7, 131 (1973).

22. PRANGE, A. J., Jr., I. C. WILSON, G. R. BREESE, N. P. PLOTNIKOFF, P. P. LARA, and M. A. LIPTON: Hypothalamic Releasing Hormones and Catecholamines: A New Interface. Frontiers in Catecholamine Res. 1149 (1973).

23. VALE, W., G. GRANT, and R. GUILLEMIN: Chemistry of the Hypothalamic Releasing Factors — Studies on Structure-Function Relationships. In: Frontiers in Neuroendocrinology (W. F. GANONG and L. MARTINI, eds.), p. 375. New York: Oxford University Press. 1973.

24. CZYGAN, P.-J.: Regulationsprinzipien der weiblichen Keimdrüsenfunktion. Fortschr. Geburtsh. Gynäkol. 52, 121 (1974).

25. GUILLEMIN, R.: The Hormones of the Hypothalamus. Amer. J. Med. 57, 591 (1974).

26. MARTINI, L.: Gonadotropin Releasing Factors: Recent Physiological Findings. In: Recent Progress in Reproductive Endocrinology. (F. G. CROSIGNANI and V. H. T. JAMES, eds.), p. 295. London: Academic Press. 1974.

27. SCHALLY, A. V., A. ARIMURA, A. J. KASTIN, D. H. COY, J. VILCHEZ-MARTINEZ, T. W. REDDING, R. GEIGER, W. KÖNIG, and H. WISSMAN: Physiological, Veterinary and Clinical Effects of Hypothalamic Releasing Hormones (RH), Especially LH−RH. Pure Appl. Chem. 37, 315 (1974).

28. SCHWARTZ, I. L., and R. WALTER: Neurohypophyseal Hormones as Precursors of Hypophysiotropic Hormones. Israel J. Med. Sci. 10, 1288 (1974).

29. STUDER, R. O.: Chemische und biochemische Aspekte der Releasing Hormone. Wiener klin. Wschr. 86, 57 (1974).

30. VOELTER, W.: Hypothalamus-Peptidhormone. Chemiker-Ztg. 98, 554 (1974).

31. GUPTA, D., and W. VOELTER: Hypothalamic Hormones — Structure, Synthesis and Biological Activity. Weinheim: Verlag Chemie. 1975.

32. LABRIE, F., P. BORGEAT, A. LEMAY, S. LEMAIRE, N. BARDEN, J. DROUIN, I. LEMAIRE, P. JOLICOEUR, and A. BÉLANGER: Role of Cyclic AMP in the Action of Hypothalamic Regulatory Hormones. Adv. Cyclic Nucleotide Res. 5, 787 (1975).

33. MOSS, R. L., S. M. MCCANN, and C. A. DUDLEY: Releasing Hormones and Sexual Behavior. Progr. Brain Res. 42, 37 (1975).

34. SCHÖNEBERG, D. K.: Dynamics of Hypothalamic-Pituitary Function during Puberty. Clin. Endocr. Metab. 4, 57 (1975).

35. VOELTER, W.: Struktur, biologische Wirkung und Strategie zur Synthese von Peptidhormonen. Chemiker-Ztg. 100, 130 (1976).

36. KAPPELER, H., and R. SCHWYZER: Synthetic Peptides Related to the Corticotropins (ACTH) and the Melanophore Stimulating Hormones (MSH) Possessing Corticotropin Releasing Activity (CRF-Activity). Experientia 16, 415 (1960).

37. FINK, G., J. R. SMITH, and J. TIBBALLS: Corticotrophin Releasing Factor in Hypophysical Portal Blood of Rats. Nature 230, 467 (1971).

38. FEHM, H. L., K. H. VOIGT und E. F. PFEIFFER: Nebennierenrindeninsuffizienz als Folge eines isolierten Mangels an Corticotropin-Releasing-Hormon (CRH). Dtsch. med. Wschr. 98, 2066 (1973).

39. SCHALLY, A. V., A. ARIMURA, and A. J. KASTIN: Hypothalamic Regulatory Hormones. Science **179**, 341 (1973).

40. MILHAUD, G., P. RIVAILLE, P. E. GARNIER, J.-L. CHAUSSAIN, E. BINET et J.-C. JOB: Synthèse de la LH − RH et effet sélectif sur la libération de l'hormone lutéotrophique chez l'Homme adulte. Compt. Rend. Acad. Sci. 1858 (1975).

41. AKANDE, E. O., P. J. CARR, A. DUTTON, J. BONNAR, C. S. CORKER, P. C. B. MACKINNON, and D. ROBINSON: Effect of Synthetic Gonadotrophin-Releasing Hormone in Secondary Amenorrhoea. Lancet **112** (1972).

42. JOB, J. C., P. E. GARNIER, J. L. CHAUSSAIN, and G. MILHAUD: Elevation of Serum Gonadotropins (LH and FSH) after Releasing Hormone (LH − RH) Injection in Normal Children and in Patients with Disorders of Puberty. J. Clin. Endocr. Metab. **35**, 473 (1972).

43. JOB, J. C., P. E. GARNIER, J. L. CHAUSSAIN, E. BINET, P. RIVAILLE, and G. MILHAUD: Effects of Synthetic Luteinizing Hormone-Releasing Hormone (LH − RH) on Serum Gonadotropins (LH and FSH) in Normal Children and Adults. Eur. J. Clin. Biol. Res. **17**, 411 (1972).

44. WAGNER, H., K. BÖCKEL, M. HRUBESCH, G. GROTE, and W. H. HAUSS: On the Effect of Synthetic LH-Releasing Hormone on the Blood Levels of LH, FSH, HGH, TSH, ACTH, Insulin and Blood Sugar in Man. Horm. Metab. Res. **4**, 403 (1972).

45. WAGNER, H., K. BÖCKEL, M. HRUBESCH, and G. GROTE: Examination of the Pituitary-Gonadal Relationship in Man with Synthetic LH/FSH-Releasing Hormone. Excerpta Medica Intern. Congr. Ser. **263**, 257 (1972).

46. HAUG, E., and P. TORJESEN: Effect of Synthetic Luteinizing Hormone- and Follicle Stimulating Hormone-Releasing Hormone (LH/FSH − RH) on Serum Levels of LH, FSH, Thyrotrophin (TSH) and Growth Hormone (HGH) in Normal Male Subjects. Acta Endocr. **73**, 465 (1973).

47. JEPPSSON, S., S. KULLANDER, G. RANNEVIK, and J. THORELL: Intranasal Administration of Synthetic Gonadotrophin-Releasing Hormone. Brit. Med. J. **4**, 231 (1973).

48. JOB, J. C., P. E. GARNIER, J. L. CHAUSSAIN, and P. CANLORBE: Effect of Synthetic Luteinizing Hormone-Releasing Hormone (LH − RH) on the Release of Gonadotropins in Hypophyso-Gonadal Disorders of Children and Adolescents. Biomedicine **19**, 77 (1973).

49. LUNDBERG, P. O.: Clinical Evaluation of the Luteinizing Hormone-Releasing Hormone (LH − RH) Test in Cases with Anatomically Verified Disorders of the Hypothalamo-Pituitary Region. Acta Neurol. Scand. **49**, 461 (1973).

50. LUNDBERG, P. O., and L. WIDE: The Effect of Synthetic Luteinizing Hormone-Releasing Hormone on Blood Levels of Luteinizing Hormone and Follicle Stimulating Hormone in Klinefelter Patients. Int. J. Fertil. **18**, 97 (1973).

51. SCHNEIDER, H. P. G., and H. G. DAHLÉN: Studies with Synthetic LH-Releasing Hormone in the Human. Neuroendocr. **11**, 328 (1973).

52. SOLBACH, H. G., W. WIEGELMANN, H. K. KLEY, H. ZIMMERMANN und H. L. KRÜS-KEMPER: Die diagnostische Bedeutung des synthetischen LH − RH (Luteinisierungs-hormon-Releasinghormon) für die Überprüfung der gonadotropen Funktion des Hypophysenvorderlappens. Dtsch. med. Wschr. **98**, 2114 (1973).

53. SOLBACH, H. G., and W. WIEGELMANN: Intranasal Application of Luteinising-Hormone Releasing Hormone. Lancet 1259 (1973).

54. ARIMURA, A., A. J. KASTIN, D. GONZALEZ-BARCENA, J. SILLER, R. E. WEAVER, and A. V. SCHALLY: Disappearance of LH-Releasing Hormone in Man as Determined by Radioimmunoassay. Clin. Endocr. **3**, 421 (1974).

55. ARIMURA, A., A. J. KASTIN, and A. V. SCHALLY: Immunoreactive LH-Releasing Hormone in Plasma: Midcycle Elevation in Women. J. Clin. Endocr. Metab. **38**, 510 (1974).

56. BRECKWOLDT, M., P.-J. CZYGAN, F. LEHMANN, and G. BETTENDORF: Synthetic LH − RH as a Therapeutic Agent. Acta Endocr. **75**, 209 (1974).

57. CHAUSSAIN, J. L., P. E. GARNIER, E. BINET, J. VASSAL, R. SCHOLLER, and J. C. JOB: Effect of Synthetic Luteinizing Hormone-Releasing Hormone (LH – RH) on the Release of Gonadotropins in Hypophyso-Gonadal Disorders of Children and Adolescents. III. Hypopituitarism. J. Clin. Endocr. Metab. **38**, 58 (1974).

58. CZYGAN, P.-J., M. BRECKWOLDT, F. LEHMANN, R. LANGEFELD, and G. BETTENDORF: LH – RH-Test in 100 Patients with Ovarian Insufficiency. Acta Endocr. **75**, 428 (1974).

59. FINK, G., G. GENNSER, P. LIEDHOLM, J. THORELL, and J. MULDER: Comparison of Plasma Levels of Luteinizing Hormone Releasing Hormone in Men After Intravenous or Intranasal Administration. J. Endocr. **63**, 351 (1974).

60. GENNSER, G., P. LIEDHOLM, and J. THORELL: Pituitary Responses to Continuous Administration of LRH in Human Males with Oligozoospermia. Horm. Metab. Res. **6**, 79 (1974).

61. GONZALEZ-BARCENA, D., A. J. KASTIN, D. S. SCHALCH, M. C. MILLER, L. A. LEE, R. RIVAS-LLAMAS, and A. V. SCHALLY: Response to LH – RH in Women before and after Treatment with Prednisone. Int. J. Fertil. **19**, 107 (1974).

62. HAUG, E., A. AAKVAAG, T. SAND, and P. A. TORJESEN: The Gonadotrophin Response to Synthetic Gonadotrophin-Releasing Hormone in Males in Relation to Age, Dose, and Basal Serum Levels of Testosterone, Oestradiol-17β and Gonadotrophins. Acta Endocr. **77**, 625 (1974).

63. JEPPSSON, S., G. RANNEVIK, and S. KULLANDER: Studies on the Descrased Gonadotropin Response after Administration of LH/FSH-Releasing Hormone during Pregnancy and the Puerperium. Am. J. Obstet. Gynecol. **120**, 1029 (1974).

64. JOB, J. C., P. E. GARNIER, J. L. CHAUSSAIN, R. SCHOLLER, J. E. TOUBLANC, and P. CANLORBE: Effect of Synthetic Luteinizing Hormone-Releasing Hormone (LH – RH) on the Release of Gonadotropins in Hypophyso-Gonadal Disorders of Children and Adolescents. V. Agonadism. J. Clin. Endocr. Metab. **38**, 1109 (1974).

65. JOB, J. C., P. E. GARNIER, J. L. CHAUSSAIN, J. E. TOUBLANC, and P. CANLORBE: Effect of Synthetic Luteinizing Hormone Releasing Hormone on the Release of Gonadotropins in Hypophysogonadal Disorders of Children and Adolescents. J. Pediatrics **84**, 371 (1974).

66. KASTIN, A. J., A. V. SCHALLY, D. GONZALEZ-BARCENA, D. H. COY, M. C. MILLER, H. PORIAS, and D. S. SCHALCH: Clinical Comparison of Natural LH – RH, Synthetic LH – RH, and Two Analogues of LH – RH. J. Clin. Endocr. Metab. **38**, 801 (1974).

67. KLEY, H. K., W. WIEGELMANN, E. NIESCHLAG, H. G. SOLBACH, H. ZIMMERMANN, and H. L. KRÜSKEMPER: LH, FSH and Testosterone in Plasma Following LH – RH Infusion: A Combined Test for Pituitary and Leydig Cell Function. Acta Endocr. **75**, 417 (1974).

68. MORTIMER, C. H., A. S. MCNEILLY, R. A. FISHER, M. A. F. MURRAY, and G. M. BESSER: Gonadotrophin-Releasing Hormone Therapy in Hypogonadal Males with Hypothalamic or Pituitary Dysfunction. Brit. Med. J. **4**, 617 (1974).

69. RANNEVIK, G., and J. THORELL: Effects of Synthetic LH/FSH-Releasing Hormone (LRH) on Plasma LH and FSH in Amenorrhoeic Women. Acta Endocr. **75**, 647 (1974).

70. RANNEVIK, G., S. JEPPSSON, S. KULLANDER, and J. THORELL: Effects of Synthetic Luteinizing Hormone-Releasing Hormone on Plasma Luteinizing Hormone and Follicle-Stimulating Hormone in Amenorrheic Women. Fertility Sterility **25**, 547 (1974).

71. ROBYN, C., H. SCHÖNDORF, O. JÜRGENSEN, J. SIOE ENG DERICKS-TAN, and H. D. TAUBERT: Oral Contraception Can Decrease the Pituitary Capacity to Release Gonadotrophins in Response to Synthetic LH-Releasing-Hormone. Arch. Gynäk. **216**, 73 (1974).

72. RÖMMLER, A., S. BAUMGARTEN und J. HAMMERSTEIN: Doppelstimulierung der Hypophyse mit synthetischem LH-Releasinghormon an drei aufeinanderfolgenden Tagen bei Männern sowie menstruierenden und amenorrhoischen Frauen. Geburtsh. und Frauenheilk. **34**, 842 (1974).

73. VANDENBERG, G., G. W. DE VANE, and S. S. C. YEN: Effects of Exogenous Estrogen and

Progestin on Pituitary Responsiveness to Synthetic Luteinizing Hormone-Releasing Factor. J. Clin. Invest. **53**, 1750 (1974).

74. WIEGELMANN, W., H. K. KLEY, H. G. SOLBACH und H. L. KRÜSKEMPER: Wachstumshormon, Gonadotropine und Cortisol im Plasma von Männern unter kombinierter Anwendung des Insulinhypoglykämie/LH – RH-Stimulationstestes. Klin. Wschr. **52**, 194 (1974).

75. GENNSER, G., P. LIEDHOLM, and J. THORELL: Pituitary Responses to Continuous Administration of LRH in Human Males with Oligozoospermia. Obstet. Gynecol. Survey **30**, 41 (1975).

76. HASHIMOTO, T., K. MIYAI, T. ONISHI, K. MATSUMOTO, and Y. KUMAHARA: Comparison of Short and Long-Term Treatment with Synthetic LH-Releasing Hormone and Clomiphene Citrate in Male Hypothalamic Hypogonadism. J. Clin. Endocr. Metab. **41**, 905 (1975).

77. DE KRETSER, D. M., H. G. BURGER, and R. DUMPYS: Serum LH and FSH Response in Four-Hour Infusions of Luteinizing Hormone-Releasing Hormone in Normal Men, Sertoli Cell Only Syndrome, and Klinefelter's Syndrome. J. Clin. Endocr. Metab. **41**, 876 (1975).

78. PALMER, R. L., A. H. CRISP, P. C. B. MACKINNON, M. FRANKLIN, J. BONNAR, and M. WHEELER: Pituitary Sensitivity to 50 µg LH/FSH – RH in Subjects with Anorexia Nervosa in Acute and Recovery Stages. Brit. Med. J. 179 (1975).

79. BOYAR, R. M.: Effects of Clomiphene Citrate on Pituitary FSH, FSH – RF, and Release of LH in Immature and Mature Rats. Endocrinol. **86**, 629 (1970).

80. JUTISZ, M., B. KERDELHUÉ, and A. BÉRAULT: Further Studies on Mechanism of Action of Luteinizing Hormone Releasing Factor Using in Vivo and in Vitro Techniques. In: The Human Testis. p. 221. New York and London: Plenum Press. 1970.

81. CRAMER, E. B., R. E. TRAUM, and S. A. D'ANGELO: Chronic Median Eminence Lesions and Ovariectomy: Effects on Follicle-Stimulating Hormone Secretion in the Rat. J. Endocr. **51**, 665 (1971).

82. JUTISZ, M.: On the Hypothalamic Regulation of the Adenohypophysial Gonadotropic Function. J. Neuro-Visc. Rel. Suppl. **X**, 22 (1971).

83. JUTISZ, M., B. KERLDELHUÉ, A. BÉRAULT, J. PELLETIER et A. TIXIER-VIDAL: Action in vitro du facteur hypothalamique de libération de l'hormone lutéinisante (LRF) sur l'antéhypophyse d'agnelle. Gen. Comp. Endocr. **17**, 22 (1971).

84. SCHALLY, A. V., A. J. KASTIN, and A. ARIMURA: Hypothalamic Follicle-Stimulating Hormone (FSH) and Luteinizing Hormone (LH)-Regulating Hormone: Structure, Physiology, and Clinical Studies. Fertility Sterility **22**, 703 (1971).

85. TIXIER-VIDAL, A., B. KERDELHUÉ, A. BÉRAULT, R. PICART et M. JUTISZ: Action in vitro du facteur hypothalamique de libération de l'hormone lutéinisante (LRF) sur l'antéhypophyse d'agnelle. Gen. Comp. Endocr. **17**, 33 (1971).

86. ARIMURA, A., L. DEBELJUK, and A. V. SCHALLY: Stimulation of FSH Release in Vivo by Prolonged Infusion of Synthetic LH – RH. Endocrinol. **91**, 529 (1972).

87. — — — LH Release by LH-Releasing Hormone in Golden Hamsters at Various Stages of Estrous Cycle. Proc. Soc. Exp. Biol. Med. **140**, 609 (1972).

88. DEBELJUK, L., A. ARIMURA, and A. V. SCHALLY: Effect of Testosterone and Estradiol on the LH and FSH Release Induced by LH-Releasing Hormone (LH – RH) in Intact Male Rats. Endocrinol. **90**, 1578 (1972).

89. HÉRY, M., B. KERDELHUÉ, E. PATTOU, M. JUTISZ, and C. KORDON: Changes in LH Levels in the Internal Jugular Vein Following Intracarotid Injection of Purified LH-Releasing Factor in Spayed Rats. Life Sci. **11**, 809 (1972).

90. JUTISZ, M., and B. KERDELHUÉ: In Vitro Studies on Synthetic LH – RH and its Assay in Plasma Using a Radioimmunological Method. Excerpta Media Intern. Congr. Ser. **263**, 98 (1972).

91. SANDOW, J., A. V. SCHALLY, H. G. SCHRÖDER, T. W. REDDING, W. HEPTNER, and H. G. VOGEL: Pharmacological Characteristics of a Synthetic Releasing Hormone LH/FSH – RH. Arzneim.-Forsch. 22, 1718 (1972).

92. ARIMURA, A., H. SATO, T. KUMASAKA, R. B. WOROBEC, L. DEBELJUK, J. DUNN, and A. V. SCHALLY: Production of Antiserum to LH-Releasing Hormone (LH – RH) Associated with Gonadal Atrophy in Rabbits: Development of Radioimmunoassays for LH – RH. Endocrinol. 93, 1092 (1973).

93. ARIMURA, A., L. DEBELJUK, M. SHIINO, E. G. RENNELS, and A. V. SCHALLY: Follicular Stimulation by Chronic Treatment with Synthetic LH-Releasing Hormone in Hypophysectomized Female Rats Bearing Pituitary Grafts. Endocrinol. 92, 1507 (1973).

94. CALAS, A., B. KERDELHUÉ, I. ASSENMACHER et M. JUTISZ: Les axones à LH–RH de l'éminence médiane. Mise en évidence chez le Canard par une technique immunocytochimique. Compt. Rend. Acad. Sci. 277, 2765 (1973).

95. CORBIN, A., and G. V. UPTON: Effect of Synthetic Luteinizing Hormone Releasing Hormone on Pituitary and Serum Levels of Luteinizing Hormone of Intact and Median Eminence Lesioned Male Rats. Experientia 29, 1025 (1973).

96. CRIGHTON, D. B., J. P. FOSTER, D. T. HOLLAND, and S. L. JEFFCOATE: Simultaneous Determination of Luteinizing Hormone and Luteinizing Hormone Releasing Hormone in the Jugular Venous Blood of the Sheep at Oestrus. J. Endocr. 59, 373 (1973).

97. FRASER, H. M., and A. GUNN: Effects of Antibodies to Luteinizing Hormone-Releasing Hormone in the Male Rabbit and on the Rat Oestrous Cycle. Nature 244, 160 (1973).

98. JUTISZ, M., and B. KERDELHUÉ: Immunological and Biological Specificity of Luteinizing Hormone Releasing Hormone (LH – RH) and its Pattern through the Estrus Cycle in two Species and in the Menstrual Cycle in Women. In: Recent Progess in Reproductive Endocrinology. p. 323. London: Academic Press. 1974.

99. PFAFF, D. W.: Luteinizing Hormone-Releasing Factor Potentiates Lordosis Behavior in Hypophysectomized Ovariectomized Female Rats. Science 182, 1148 (1973).

100. SCHALLY, A. V.: Hypothalamic LH and FSH Releasing Hormone; Physiological and Clinical Studies. Clin. Sci. 44, 1 (1973).

101. SCHALLY, A. V., T. W. REDDING, and A. ARIMURA: Effect of Sex Steroids on Pituitary Responses to LH- and FSH-Releasing Hormone in Vitro. Endocrinol. 93, 893 (1973).

102. TIXIER-VIDAL, A., B. KERDELHUÉ, and M. JUTISZ: Kinetics of Release of Luteinizing Hormone (LH) and Follicle Stimulating Hormone (FSH) by Primary Cultures of Dispersed Rat Anterior Pituitary Cells. Chronic Effect of Synthetic LH and FSH Realising Hormone. Life Sci. 12, 499 (1973).

103. AIYER, M. S., G. FINK, and F. GREIG: Changes in the Sensitivity of the Pituitary Gland to Luteinizing Hormone Releasing Factor During the Oestrous Cycle of the Rat. J. Endocr. 60, 47 (1974).

104. AIYER, M. S., and G. FINK: The Role of Sex Steroid Hormones in Modulating the Responsiveness of the Anterior Pituitary Gland to Luteinizing Hormone Releasing Factor in the Female Rat. J. Endocr. 62, 553 (1974).

105. AIYER, M. S., S. A. CHIAPPA, and G. FINK: A Priming Effect of Luteinizing Hormone Releasing Factor on the Anterior Pituitary Gland in the Female Rat. J. Endocr. 62, 573 (1974).

106. BORGEAT, P., F. LABRIE, J. COTÉ, F. RUEL, A. V. SCHALLY, D. H. COY, E. J. COY, and N. YANAIHARA: Parallel Stimulation of Cyclic AMP Accumulation and LH and FSH Release by Analogs of LH – RH in Vitro. Mol. Cell. Endocrinol. 1, 7 (1974).

107. CALAS, A., B. KERDELHUÉ, I. ASSENMACHER et M. JUTISZ: Les axones à LH – RH de l'Eminence Médiane. Etude ultrastructurale chez le Canard par une technique immunocytochimique. Compt. Rend. Acad. Sci. 278, 2557 (1974).

108. DUPONT, A., F. LABRIE, G. PELLETIER, R. PUVIANI, D. H. COY, E. J. COY, and A. V. SCHALLY: Organ Distribution of Radioactivity and Disappearance of Radioactivity

from Plasma after Administration of [³H] Luteinizing Hormone-Releasing Hormone to Mice and Rats. Neuroendocr. **16**, 65 (1974).

109. KORDON, C., B. KERDELHUÉ, E. PATTOU, and M. JUTISZ: Immunocytochemical Localization of LH — RH in Axons and Nerve Terminals of the Rat Median Eminence. Proc. Soc. Exp. Biol. Med. **147**, 122 (1974).

110. MAKINO, T., L. M. DEMERS, and R. O. GREEP: Factors Influencing Pituitary Glycogen Metabolism and Gonadotropic Hormone Release. I. Luteinizing Hormone-Releasing Hormone. Fertility Sterility **25**, 872 (1974).

111. MAREK, J., P. JAQUET, F. GOGAN, G. LOMBARDI, and J. VAGUE: Radioimmunoassay of Plasma LH in Rat: Basal Levels and Effects of Synthetic LH — RH. Endokrinol. **63**, 312 (1974).

112. SCHAMS, D., F. HÖFER, E. SCHALLENBERGER, M. HARTL, and H. KARG: Pattern of Luteinizing Hormone (LH) and Follicle Stimulating Hormone (FSH) in Bovine Blood Plasma after Injection of a Synthetic Gonadotropin-Releasing Hormone (Gn — RH). Theriogenology **1**, 137 (1974).

113. SIMONOVIC, I., M. MOTTA, and L. MARTINI: Acetylcholine and the Release of the Follicle-Stimulating Hormone-Releasing Factor. Endocrinol. **95**, 1373 (1974).

114. VILCHEZ-MARTINEZ, J. A., A. ARIMURA, and A. V. SCHALLY: Influence of Estradiol Benzoate on Pituitary Responsiveness to LH — RH at Different Stages of the Estrous Cycle in Rats. Proc. Soc. Exp. Biol. Med. **146**, 859 (1974).

115. VILCHEZ-MARTINEZ, J. A., A. V. SCHALLY, D. H. COY, E. J. COY, L. DEBELJUK, and A. ARIMURA: In Vivo Inhibition of LH Release by a Synthetic Antagonist of LH-Releasing Hormone (LH — RH). Endocrinol. **95**, 213 (1974).

116. BORGEAT, P., P. GARNEAU, and F. LABRIE: Calcium Requirement for Stimulation of Cyclic AMP Accumulation in Anterior Pituitary Gland by LH — RH. Mol. Cell. Endocrinol. **2**, 117 (1975).

117. CRIGHTON, D. B., J. P. FOSTER, W. HARESIGN, and S. A. SCOTT: Plasma LH and Progesterone Levels after Single or Multiple Injections of Synthetic LH — RH in Anoestrous Ewes and Comparison with Levels during the Oestrous Cycle. J. Reprod. Fert. **44**, 121 (1975).

118. FERLAND, L., P. BORGEAT, F. LABRIE, J. BERNARD, A. DE LEAN, and J. P. RAYNAUD: Changes of Pituitary Sensitivity to LH — RH during the Rat Estrous Cycle. Mol. Cell. Endocrinol. **2**, 107 (1975).

119. FINK, G.: The Responsiveness of the Anterior Pituitary Gland to Luteinizing Hormone Releasing Factor in Rats Exposed to Constant Light. J. Endocr. **65**, 439 (1975).

120. FRASER, H. M., S. L. JEFFCOATE, A. GUNN, and D. T. HOLLAND: Effect of Active Immunization to Luteinizing Hormone Releasing Hormone on Gonadotrophin Levels in Ovariectomized Rats. J. Endocr. **64**, 191 (1975).

121. HARESIGN, W., J. P. FOSTER, N. B. HAYNES, D. B. CRIGHTON, and G. E. LAMMING: Progesterone Levels Following Treatment of Seasonally Anoestrous Ewes with Synthetic LH-Releasing Hormone. J. Reprod. Fert. **43**, 269 (1975).

122. HARESIGN, W.: Ovarian Response to Synthetic LH — RH in Anoestrous Ewes. J. Reprod. Fert. **44**, 127 (1975).

123. KIZER, J. S., A. ARIMURA, A. V. SCHALLY, and M. J. BROWNSTEIN: Absence of Luteinizing Hormone-Releasing Hormone (LH — RH) from Catecholaminergic Neurons. Endocrinol. **96**, 523 (1975).

124. MOSS, R. L., and S. M. MCCANN: Action of Luteinizing Hormone-Releasing Factor (LRF) in the Initiation of Lordosis Behavior in the Estrone-Primed Ovariectomized Female Rat. Neuroendocr. **17**, 309 (1975).

125. SÉTÁLÓ, G., S. VIGH, A. V. SCHALLY, A. ARIMURA, and B. FLERKÓ: LH — RH-Containing Neural Elements in the Rat Hypothalamus. Endocrinol. **96**, 135 (1975).

126. SCHALLY, A. V., A. ARIMURA, A. J. KASTIN, H. MATSUO, Y. BABA, T. W. REDDING, R. M. G.

Nair, and L. Debeljuk: Gonadotropin Releasing Hormone: One Polypeptide Regulates Secretion of Luteinizing and Follicle-Stimulating Hormones. Science 173, 1036 (1971).

127. Arimura, A., and A. V. Schally: Hypothalamic LH- and FSH-Releasing Hormone: Reevaluation of the Concept that One Hypothalamic Hormone Controls the Release of LH and FSH. In: Biological Rhythms in Neuroendocrine Activity (M. Kawakami, ed.), p. 73. Tokyo: Igaku Shoin Ltd. 1974.

128. Zimmermann, E. A., K. C. Hsu, M. Ferin, and G. P. Kozlowski: Localization of Gonadotropin-Releasing Hormone (Gn – RH) in the Hypothalamus of the Mouse by Immunoperoxidase Technique. Endocrinol. 95, 1 (1974).

129. Corbin, A., E. L. Daniels, and J. E. Milmore: An "Internal" Feedback Mechanism Controlling Follicle Stimulating Hormone Releasing Factor. Endocrinol. 86, 735 (1970).

130. Barker, H. M., T. E. Isles, H. M. Fraser, and A. Gunn: Radioimmunoassay of Luteinizing Hormone Releasing Hormone. Nature 242, 527 (1973).

131. Kerdelhué, B., M. Jutisz, D. Gillessen, and R. O. Studer: Obtention of Antisera against a Hypothalamic Decapeptide (Luteinizing Hormone/Follicle Stimulating Hormone Releasing Hormone) which Stimulates the Release of Pituitary Gonadotropins and Development of its Radioimmunoassay. Biochim. Biophys. Acta 297, 540 (1973).

132. Bryce, G. F.: Development of a Radioimmunoassay for Luteotropin Releasing Hormone (LRH) and Thyrotropin Releasing Hormone (TRH), Immunochemistry 11, 507 (1974).

133. Faglia, G., P. Beck-Peccoz, P. Travaglini, A. Paracchi, A. Spada, and A. Lewin: Elevations in Plasma Growth Hormone Concentration after Luteinizing Hormone-Releasing Hormone (LRH) in Patients with Active Agromegaly. J. Clin. Endocr. Metab. 37, 338 (1973).

134. Kastin, A. J., D. Gonzalez-Barcena, H. G. Friesen, L. S. Jacobs, D. S. Schalch, A. Arimura, W. H. Daughaday, and A. V. Schally: Unaltered Plasma Prolactin Levels in Men After Administration of Synthetic LH-Releasing Hormone. J. Clin. Endocr. Metab. 36, 375 (1973).

135. Celis, M. E., S. Taleisnik, and R. Walter: Release of Pituitary Melanocyte-Stimulating Hormone by the Oxytocin Fragment, H – Cys – Tyr – Ilr – Gln – Asn – OH. Biochem. Biophys. Res. Commun. 45, 564 (1971).

136. Celis, M. E., and S. Taleisnik: In Vitro Formation of a MSH-Releasing Agent by Hypothalamic Extracts. Experientia 27, 1481 (1971).

137. Dular, R., F. S. LaBella, S. Vivian, and L. Eddie: Purification of Prolactin-Releasing and Inhibiting Factors from Beef. Endocrinol. 94, 563 (1974).

138. Hershman, J. M., and J. A. Pittman Jr.: Response to Synthetic Thyrotropin-Releasing Hormone in Man. J. Clin. Endocr. Metab. 31, 457 (1970).

139. Bangerter, S., S. Weiss, J. J. Staub, H. Bürgi, M. P. König und H. Studer: Effekt von peroral verabreichtem TRH auf die Radiojodaufnahme der Schilddrüse. Schweiz. med. Wschr. 101, 1269 (1971).

140. Beckers, C., A. Maskens et C. Cornette: Intérêt de la „Thyrotropin-Releasing Hormone" (TRH) en physiopathologie thyroïdienne. Ann. d'Endocr. 32, 865 (1971).

141. Bowers, C. Y., A. V. Schally, A. Kastin, A. Arimura, D. S. Schalch, C. Gual, E. Castineda, and K. Folkers: Synthetic Thyrotropin-Releasing Hormone. Activity in Men and Women, Specificity of Action, Inhibition by Triiodothyronine, and Activity Orally. J. Med. Chem. 14, 477 (1971).

142. Haigler, E. D., Jr., J. A. Pittman, Jr., J. M. Hershman, and C. M. Baugh: Direct Evaluation of Pituitary Thyrotropin Reserve Utilizing Synthetic Thyrotropin Releasing Hormone. J. Clin. Endocr. Metab. 33, 573 (1971).

143. Hershman, J. M., and J. A. Pittman, Jr.: Control of Thyrotropin Secretion in Man. N. Engl. J. Med. 285, 997 (1971).

144. Job, J. C., G. Milhaud, E. Binet, P. Rivaille et M. S. Moukhtar: Effet de l'hormone de libération de la thyréostimuline (TRH) sur le taux sanguin de thyréostimuline (TSH) chez l'enfant: enfants normaux, enfants atteints d'hypothyroïdie, d'hypopituitarisme, de goitre. Eur. J. Clin. Biol. Res. **16**, 537 (1971).

145. Labhart, A.: Die klinische Bedeutung der Neuroendokrinologie. Bull. schweiz. Akad. med. Wiss. **27**, 230 (1971).

146. Milhaud, G., P. Rivaille, M. S. Moukhtar, E. Binet, and J. C. Job: Response of Normal, Hypothyroid and Hypothalamo-Pituitary Insufficient Children to Synthetic Thyrotrophin-Releasing Hormone. J. Endocr. **51**, 483 (1971).

147. Pittman, J. A.: Hypopituitarism. In: The Thyroid, Third Edition (S. C. Werner, and S. H. Ingbar, eds.), p. 797. New York: Harper and Row. 1971.

148. Pittman, J. A., Jr., E. D. Haigler, Jr., J. M. Hershman, and C. S. Pittman: Hypothalamic Hypothyroidism. N. Engl. J. Med. **285**, 844 (1971).

149. Rothenbuchner, G., J. Birk, U. Loos, S. Raptis, E. F. Pfeiffer, J. Goldstein, L. Vanhaelst und B. E. Schreiber: Das Verhalten von thyreotropem Hormon (TSH), Trijodthyronin (T₃) und Thyroxin (T₄) nach Gabe von Thyreotropin-Releasing Hormon (TRH) bei stoffwechselgesunden Personen. Verh. Dtsch. Ges. inn. Med. **77**, 649 (1971).

150. Staub, J. J., S. Weiss, H. Kohler, H. Bürgi, M. P. König und H. Studer: Effekt von peroral verabreichtem „TSH releasing hormone" (TRH) auf die Thyroxinkonzentration des Blutes. Schweiz. med. Wschr. **101**, 1295 (1971).

151. Wagner, H., M. Hrubesch, H. Vosberg, K. Bökel, B. Brisse, G. Junge-Hülsing, and W. H. Hauss: Thyreotropin-Releasing Hormone and Serum Levels of H − TSH, HGH, Thyroxine, Insulin and Glucose. Horm. Metab. Res. **3**, 137 (1971).

152. Wagner, H., M. Hrubesch, K. Böckel, H. Vosberg, G. Junge-Hülsing und W. H. Hauss: Einfluß von synthetischem Thyreotropin-Releasing-Hormon auf den Serum-TSH-Spiegel des Menschen. In: Thyreotropin-Releasing-Hormon. (F. A. Horster, W. Wildmeister und F. E. Pausch, Herausgeber), S. 33. Stuttgart: F. K. Schattauer-Verlag. 1971.

153. Wagner, H., M. Hrubesch, K. Böckel, H. Vosberg und G. Junge-Hülsing: TSH-Spiegel im Serum bei Gesunden und Patienten mit Schilddrüsenfunktionsstörungen vor und nach intravenöser Gabe von synthetischem Thyreotropin-Releasing-Hormon. Med. Welt **22**, 1883 (1971).

154. Wagner, H., K. Böckel, M. Hrubesch, H. Vosberg, G. Jung-Hülsing und W. H. Hauss: Der TRH-Kurztest in der Schilddrüsenfunktionsdiagnostik. Nuc. Comp. **2**, 90 (1971).

155. Beckers, C., C. Cornette et M. Thalasso: Réactivité hypophysaire à la „Thyrotropin-Releasing Hormone" (TRH) chez des patients hyperthyroidiens considérés comme guéris cliniquement. Ann. d'Endocr. **33**, 642 (1972).

156. Bethge, H., W. Wildmeister und F. A. Horster: Plasma-Cortisol-Spiegel nach intravenöser Injektion von Thyreotropin-releasing Hormon (TRH) beim Menschen. Klin. Wschr. **50**, 211 (1972).

157. Faglia, G., P. Beck-Peccoz, B. Ambrosi, C. Ferrari, and P. Travaglini: The Effects of a Synthetic Thyrotrophin Releasing Hormone (TRH) in Normal and Endocrinopathic Subjects. Acta Endocr. **71**, 209 (1972).

158. Faglia, G., P. Beck-Peccoz, C. Ferrari, P. Travaglini, B. Ambrosi e A. Spada: Impiego del TRH nello studio della fisiopatologia del sistema ipotalamo-ipofiso-tiroideo nell'uomo. In: XIV Congresso nationale di endocrinologia, fisiologia e fisiopatologia ipotalomo ipofisaria. p. 78. Rom: Il Pensiero Scientifico. 1972.

159. Faglia, G., B. Ambrosi, P. Beck-Peccoz, P. Travaglini, and C. Ferrari: The Effect of Theophylline on Plasma Thyrotropin (HTSH) Response to Thyrotropin Releasing Factor (TRF) in Man. J. Clin. Endocr. **34**, 906 (1972).

160. Faglia, G., P. Beck-Peccoz, B. Ambrosi, C. Ferrari, and V. Neri: Prolonged and

Exaggerated Elevations in Plasma Thyrotropin (HTSH) after Thyrotropin Releasing Factor (TRF) in Patients with Pituitary Tumors. J. Clin. Endocr. **33**, 999 (1972).

161. FOLEY, T. P., Jr., L. S. JACOBS, W. HOFFMAN, W. H. DAUGHADAY, and R. M. BLIZZARD: Effects of Thyroid Hormone upon the Human Prolactin (HPr) Responses to Thyrotropin Releasing Hormone (TRH). Excerpta Medica Intern. Congr. Ser. **263**, 385 (1972).

162. HAIGLER, E. D., Jr., J. M. HERSHMAN, and J. A. PITTMAN, Jr.: Response to Orally Administered Synthetic Thyrotropin-Releasing Hormone in Man. J. Clin. Endocr. Metab. **35**, 631 (1972).

163. HORSTER, F. A., W. WILDMEISTER und F. E. PAUSCH: Thyreotropin-Releasing-Hormon. Stuttgart: F. K. Schattauer Verlag. 1972.

164. LOOS, U., G. ROTHENBUCHNER, J. BIRK, E. F. PFEIFFER, G. KNAPP, and B. E. SCHREIBER: Serum Levels of T_3, T_4 and TSH after Different Modes of TRH Administration. Excerpta Medica Intern. Congr. Ser. **263**, 136 (1972).

165. LUNDBERG, P. O., J. WÅLINDER, I. WERNER, and L. WIDE: Effects of Thyrotrophin-Releasing Hormone on Plasma Levels of TSH, FSH, LH and GH in Anorexia Nervosa. Europ. J. clin. Invest. **2**, 150 (1972).

166. PICKARDT, C. R., F. ERHARDT, K. HORN und P. C. SCRIBA: Kontrolle der Schilddrüsen-hormon-Behandlung der blanden Struma durch Bestimmung der Serum-TSH-Spiegel nach TRH-Belastung. Klin. Wschr. **50**, 1138 (1972).

167. PICKARDT, C. R., F. ERHARDT, R. FAHLBUSCH, J. GRÜNER, and P. C. SCRIBA: The Diagnostic Significance of the Stimulation of TSH Secretion by Administration of Thyrotropin Releasing Hormone (TRH) in Diseases of the Hypothalamus and Pituitary. Excerpta Medica Intern. Congr. Ser. **306**, 105 (1972).

168. PICKARDT, C. R., W. GEIGER, R. FAHLBUSCH und P. C. SCRIBA: Stimulation der TSH-Sekretion durch TRF-Belastung bei hypothalamischen und hypophysären Krankheitsbildern. Klin. Wschr. **50**, 42 (1972).

169. PITTMAN, J. A., Jr.: Hypothalamic Hyperthyroidism. N. Engl. J. Med. **287**, 356 (1972).

170. SHENKMAN, L., T. MITSUMA, B. SHOPSIN, and C. S. HOLLANDER: Pituitary and Thyroidal Responses to Thyrotropin Releasing Hormone (TRH). Excerpta Medica Intern. Congr. Ser. **263**, 338 (1972).

171. SHENKMAN, L., T. MITSUMA, A. SUPHAVAI, and C. S. HOLLANDER: Hypothalamic Hypothyroidism: J. Amer. Med. Assoc. **222**, 480 (1972).

172. TYSON, J. E., H. G. FRIESEN, and M. S. ANDERSON: Human Milk Secretion after TRH Induced Prolaction Release. Excerpta Medica Intern. Congr. Ser. **263**, 396 (1972).

173. — — — Human Lactational and Ovarian Response to Endogenous Prolactin Release. Science **177**, 897 (1972).

174. WAGNER, H., K. BÖCKEL, G. GROTE, M. HRUBESCH, H. VOSBERG, G. JUNGE-HÜLSING und W. H. HAUSS: Weitere Befunde zur Wirkung von synthetischem Thyreotropin-Releasing-Hormon. Med. Welt **23**, 1419 (1972).

175. WIEGELMANN, W., W. WILDMEISTER, F. A. HORSTER, and H. G. SOLBACH: Radioimmunoassay of HGH, ICSH and FSH in Plasma after i. v. Administration of Thyrotropin Releasing Hormone (TRH). Horm. Metab. Res. **4**, 482 (1972).

176. WILDMEISTER, W., H. DAWEKE, F. A. GRIES, D. GRÜNEKLEE, J. HESSING, and F. A. HORSTER: Influence of Synthetic TRH on Glucose, Free Fatty Acids and Insulin in Plasma of Healthy Persons. Horm. Metab. Res. **4**, 368 (1972).

177. CARLSON, H. E., L. S. JACOBS, and W. H. DAUGHADAY: Growth Hormone, Thyrotropin and Prolactin Responses to Thyrotropin-Releasing Hormone Following Diethylstilbestrol Pretreatment. J. Clin. Endocr. Metab. **37**, 488 (1973).

178. FAGLIA, G., G. FERRARI, P. BECK-PECCOZ, A. SPADA, P. TRAVAGLINI, and B. AMBROSI: Reduced Plasma Thyrotropin Response to Thyrotropin Releasing Hormone After Dexamethasone Administration in Normal Subjects. Horm. Metab. Res. **5**, 289 (1973).

179. FAGLIA, G., P. BECK-PECCOZ, C. FERRARI, B. AMBROSI, A. SPADA, and P. TRAVAGLINI:

Enhanced Plasma Thyrotrophin Response to Thyrotrophin-Releasing Hormone Following Oestradiol Administration in Man. Clin. Endocr. **2**, 207 (1973).

180. FAGLIA, G., P. BECK-PECCOZ, C. FERRARI, P. TRAVAGLINI, B. AMBROSI, and A. SPADA: Plasma Growth Hormone Response to Thyrotropin-Releasing Hormone in Patients with Active Acromegaly. J. Clin. Endocr. Metab. **36**, 1259 (1973).

181. FAGLIA, G., C. FERRARI, P. BECK-PECCOZ, B. AMBROSI, A. SPADA, P. TRAVAGLINI, and A. PARACCHI: Evaluation of Plasma Thyrotrophin Response to Thyrotrophin Releasing Hormone in Graves' Disease and in Hyperactive Thyroid Adenoma before and after Treatment. Folia Endocrinologica **26**, 245 (1973).

182. FAGLIA, G., P. BECK-PECCOZ, C. FERRARI, B. AMBROSI, A. SPADA, P. TRAVAGLINI, and S. PARACCHI: Plasma Thyrotropin Response to Thyrotropin-Releasing Hormone in Patients with Pituitary and Hypothalamic Disorders. J. Clin. Endocr. Metab. **37**, 595 (1973).

183. FAIRCLOUGH, P. D., R. J. CRYER, J. McALLISTER, L. HAWKINS, A. E. JONES, M. McKENDRICK, R. HALL, and G. M. BESSER: Serum TSH Responses to Intravenously and Orally Administered TRH in Man after Thyroidectomy for Carcinoma of the Thyroid. Clin. Endocr. **2**, 351 (1973).

184. GAUTVIK, K. M., B. D. WEINTRAUB, C. T. GRAEBER, F. MALOOF, J. E. ZUCKERMAN, and A. H. TASHJIAN, Jr.: Serum Prolactin and TSH: Effects of Nursing and pyroGlu – His – Pro–NH₂ Administration in Postpartum Women. J. Clin. Endocr. Metab. **37**, 135 (1973).

185. GORDIN, A., and B.-A. LAMBERG: Serum Thyrotrophin Response to Thyrotrophin Releasing Hormone and the Concentration of Free Thyroxine in Subacute Thyroiditis. Acta Endocr. **74**, 111 (1973).

186. JOB, J. C., J. L. CHAUSSAIN, L. VITRY, E. BINET et P. CANLORBE: Goitre simple juvénile et taux sanguin de thyréostimuline (TSH). Effects de l'hormone hypothalamique de libération de la thyréostimuline (TRH). Biomedicine **19**, 164 (1973).

187. KARLBERG, B. E.: Thyroid Nodule Autonomy: Its Demonstration by the Thyrotrophin Releasing Hormone (TRH) Stimulation Test. Acta Endocr. **73**, 689 (1973).

188. LUNDBERG, P. O., and L. WIDE: The Response to TRH, LH–RH, Metyrapone and Vasopressin in Patients with Hypothalamo-Pituitary Disorders. Europ. J. clin. Invest. **3**, 49 (1973).

189. MALVAUX, P., and C. BECKERS: Serum Thyrotrophin Response to Thyrotrophin-Releasing Hormone in Normal Children and in Patients with Short Stature and Various Endocrine or Genetic Diseases. Clin. Endocr. **2**, 219 (1973).

190. MITSUMA, T., L. SHENKMAN, A. SUPHAVAI, and C. S. HOLLANDER: Hypothalamic Hypothyroidism: Diminished Thyroidal Response to Thyrotropin-Releasing Hormone. Amer. J. Med. Sci. **265**, 315 (1973).

191. PICKARDT, C. R., F. ERHARDT, J. GRÜNER, H. G. HEINZE, K. HORN und P. C. SCRIBA: Stimulierbarkeit der TSH-Sekretion durch TRH bei autonomen Adenomen der Schilddrüse. Dtsch. med. Wschr. **98**, 152 (1973).

192. PRANGE, A. J., Jr., I. C. WILSON, P. P. LARA, J. F. WILBER, G. R. BREESE, L. B. ALLTOP, M. A. LIPTON: TRH (Lopremone): Psychobiological Responses of Normal Women. Arch. Gen. Psych. **29**, 28 (1973).

193. PRANGE, A. J., I. C. WILSON, and P. P. LARA: Pituitary and Antidepressant Responses of Women to TRH. Psychopharmacol. Bull. **9**, 28 (1973).

194. THORELL, J. I., and G. ADIELSSON: Antidepressive Effects of Electroconvulsive Therapy and Thyrotrophin-Releasing Hormone. Lancet 43 (1973).

195. TORJESEN, P. A., E. HAUG, and T. SAND: Effect of Thyrotrophin-Releasing Hormone on Serum Levels of Pituitary Hormones in Men and Women. Acta Endocr. **73**, 455 (1973).

196. WILSON, I. C., A. J. PRANGE, Jr., P. P. LARA, L. B. ALLTROP, R. A. STIKELEATHER, and M. A. LIPTON: TRH (Lopremone): Psychobiological Responses of Normal Women. Arch. Gen. Psych. **29**, 15 (1973).

197. BENKERT, O., A. GORDON, and D. MARTSCHKE: The Comparison of Thyrotropin Releasing Hormone. Luteinising Hormone-Releasing Hormone and Placebo in Depressive Patients Using a Double-Blind Cross-Over Technique. Psychopharmacol. **40**, 191 (1974).

198. DENIKER, P., D. GINESTET, H. LOO, E. ZARIFIAN et M.-J. COTTERAU: Etude préliminaire de l'action de la thyréostimuline hypothalamique (thyrotropine releasing hormone ou TRH) dans les états dépressifs. Ann. méd.-psych. **1**, 249 (1974).

199. DUSSAULT, J. H.: The Effect of Dexamethasone on TSH and Prolactin Secretion after TRH Stimulation. CMA J. **111**, 1195 (1974).

200. EHRENSING, R. H., A. J. KASTIN, D. S. SCHALCH, H. G. FRIESEN, J. R. VARGAS, and A. V. SCHALLY: Affective State and Thyrotropin and Prolactin Responses After Repeated Injections of Thyrotropin-Releasing Hormone in Depressed Patients. Am. J. Psych. **131**, 714 (1974).

201. FLASSE, C. et C. BECKERS: Importance de l'épreuve de la stimulation à la ,,Thyrotropin-Releasing Hormone" (TRH) dans les dysfonctionnements thyroidiens latents. Ann. d'Endocr. **35**, 195 (1974).

202. GAUTVIK, K. M., A. H. TASHJIAN, Jr., I. A. KOURIDES, B. D. WEINTRAUB, C. T. GRAEBER, F. MALOOF, K. SUZUKI, and J. E. ZUCKERMAN: Thyrotropin-Releasing Hormone is not the Sole Physiologic Mediator of Prolactin Release During Suckling. N. Engl. J. Med. **290**, 1162 (1974).

203. HOLLISTER, L. E., P. BERGER, F. L. OGLE, R. C. ARNOLD, and A. JOHNSON: Protirelin (TRH) in Depression. Arch. Gen. Psych. **31**, 468 (1974).

204. L'HERMITE, M., C. ROBYN, J. GOLSTEIN, G. ROTHENBUCHNER, J. BIRK, U. LOOS, M. BONNYNS, and L. VANHAELST: Prolactin and Thyrotropin in Thyroid Diseases: Lack of Evidence for a Physiological Role of Thyrotropin-Releasing Hormone in the Regulation of Prolactin Secretion. Horm. Metab. Res. **6**, 190 (1974).

205. LIUZZI, A., P. G. CHIODINI, L. BOTALLA, F. SILVESTRINI, and E. E. MÜLLER: Growth Hormone (GH)-Releasing Activity of TRH and GH-Lowering Effect of Dopaminergic Drugs in Acromegaly: Homogeneity in the Two Responses. J. Clin. Endocr. Metab. **39**, 871 (1974).

206. LOOS, U., G. ROTHENBUCHNER, J. BIRK, A. ISHIHARA und E. F. PFEIFFER: Die endokrine Ophthalmopathie — neue pathophysiologische Gesichtspunkte durch die Bestimmung von T_3, T_4 und TSH nach TRH-Stimulation. Verh. Dtsch. Ges. inn. Med. **80**, 1340 (1974).

207. MORTIMER, C. H., G. M. BESSER, D. J. GOLDIE, J. HOOK, and A. S. McNEILLY: The TSH, FSH and Prolactin Responses to Continuous Infusions of TRH and the Effects of Oestrogen Administration in Normal Males. Clin. Endocr. **3**, 97 (1974).

208. NILSSON, K. O., J. I. THORELL, and B. HÖKFELT: The Effect of Thyrotrophin Releasing Hormone on the Release of Thyrotrophin and other Pituitary Hormones in Man under Basal Conditions and Following Adrenergic Blocking Agents. Acta Endocr. **76**, 24 (1974).

209. OLIVER, C., J. P. CHARVET, J. L. CODACCIONI, J. VAGUE, and J. C. PORTER: TRH in Human C. S. F. Lancet **1974**, 873.

210. PITTMAN, Jr., J. A.: Thyrotropin-Releasing Hormone. Adv. Intern. Med. **19**, 303 (1974).

211. PRANGE, A. J., Jr., I. C. WILSON, P. P. LARA, and L. B. ALLTOP: Effects of Thyrotropin-Releasing Hormone in Depression. In: The Thyroid Axis, Drugs, and Behavior (A. J. PRANGE, ed.), p. 135. New York: Raven Press. 1974.

212. PRANGE, A. J., Jr., I. C. WILSON, P. P. LARA, J. F. WILBER, G. R. BREESE, L. B. ALLTOP, and M. A. LIPTON: Thyrotropin-Releasing Hormone: Psychobiological Responses of Normal Women. II. Pituitary-Thyroid Responses. In: The Thyroid Axis, Drugs, and Behavior. (A. J. PRANGE, ed.), p. 165. New York: Raven Press. 1974.

213. REFETOFF, S., V. S. FANG, B. RAPOPORT, and H. G. FRIESEN: Interrelationships in the Regulation of TSH and Prolactin Secretion in Man: Effects of L-Dopa, TRH and Thyroid Hormone in Various Combinations. J. Clin. Endocr. Metab. **38**, 450 (1974).

214. Rothenbuchner, G., D. A. Koutras, S. Raptis, J. Birk, U. Loos, G. Rigopoulos, and B. Malamos: The Effect of Thyrotrophin-Releasing Hormone on Serum TSH, T_4, and T_3 Levels in Endemic and Sporacid Nontoxic Goitre. Horm. Metab. Res. **6**, 501 (1974).

215. Snyder, P. J., L. S. Jacobs, M. M. Rabello, F. H. Sterling, R. N. Shore, R. D. Utiger, and W. H. Daughaday: Diagnostic Value of Thyrotrophin-Releasing Hormone in Pituitary and Hypothalamic Diseases. Ann. Intern. Med. **81**, 751 (1974).

216. Turek, I. S., and J. Rocha: Oral Thyrotropin-Releasing Hormone (TRH) in Depressive Illness. J. Clin. Pharmacol. **14**, 612 (1974).

217. Vanhaelst, L., J. Golstein, M. L'Hermite, C. Robyn, R. Leclercq, O. D. Bruno, and G. Copinschi: Effects of Dexamethasone and Secobarbital on the Pituitary Response to Thyrotropin Releasing Hormone (TRH) in Man: Synergistic Inhibition of Thyrotropin (TSH) Release. Neuroendocr. **16**, 282 (1974).

218. Weeke, J.: The Influence of the Circadian Thyrotropin Rhythm on the Thyrotropin Response to Thyrotropin-Releasing Hormone in Normal Subjects. Scand. J. clin. Lab. Invest. **33**, 17 (1974).

219. — Thyrotropin Response to Thyrotropin Releasing Hormone in Normal Subjects. Europ. J. clin. Invest. **4**, 29 (1974).

220. Weeke, J., and A. P. Hansen: Serum Thyrotropin during Daily Life and in Response to Thyrotropin Releasing Hormone in Normal Subjects and Juvenile Diabetics. Diabetologia **10**, 101 (1974).

221. Wenzel, K. W., H. Meinhold, M. Herpich, F. Adlkofer und H. Schleusener: TRH-Stimulationstest mit alters- und geschlechtsabhängigem TSH-Anstieg bei Normalpersonen. Klin. Wschr. **52**, 722 (1974).

222. Wenzel, K. W., H. Meinhold, M. Raffenberg, F. Adlkofer, and H. Schleusener: Classification of Hypothyroidism in Evaluating Patients after Radioiodine Therapy by Serum Cholesterol, T_3-Uptake. Total T_4, FT_4-Index, Total T_3, Basal TSH and TRH-Test. Europ. J. clin. Invest. **4**, 141 (1974).

223. Wenzel, K. W., H. Meinhold, H. Schleusener und H. Botsch: Verbesserte Beurteilungskriterien des autonomen Adenoms der Schilddrüse: Trijodthyronin-Konzentrationen im Serum, funktionelle Definition durch den TRH-Test. Dtsch. med. Wschr. **99**, 1465 (1974).

224. Gemsenjäger, E., J. J. Staub und J. Girard: Untersuchung des TSH vor und nach TRH-Stimulation bei blander Struma und Rezidivstruma. Helv. chir. Acta **42**, 81 (1975).

225. Maeda, K., Y. Kato, S. Ohgo, K. Chihara, Y. Yoshimoto, N. Yamaguchi, S. Kuromaru, and H. Imura: Growth Hormone and Prolactin Release After Injection of Thyrotropin-Releasing Hormone in Patients with Depression. J. Clin. Endocr. Metab. **40**, 501 (1975).

226. Pickardt, C. R., and A. von zur Mühlen: Established Applications for the TRH Stimulation Test. Acta Endocr. (Kbh.) Suppl. **193**, 178 (1975).

227. Prange, A. J., Jr., I. C. Wilson, G. R. Breese, and M. A. Lipton: Behavioral Effects of Hypothalamic Releasing Hormones in Animals and Men. Progr. Brain Res. **42**, 1 (1975).

228. Prange, A. J., Jr., and I. C. Wilson: Behavioral Effects of Thyrotropin Releasing Hormone in Animals and Man: A Generic Hypothesis. Psychopharmacol. Bull. **11**, 22 (1975).

229. Wagner, H., H. Vosberg, G. Grote, K. Böckel, M. Hrubesch und W. H. Hauss: Altersabhängige Abnahme der Stimulierbarkeit der TSH-Sekretion durch Thyreotropin-Releasing-Hormon bei Männern und Frauen. Z. Gerontol. **8**, 38 (1975).

230. Wenzel, K. W., H. Meinhold, and H. Schleusener: Different Effects of Oral Doses of Triiodothyronine or Thyroxine on the Inhibition of Thyrotrophin Releasing Hormone (TRH) Mediated Thyrotrophin (TSH) Response in Man. Acta Endocr. **80**, 42 (1975).

231. Wenzel, K. W.: Variationen der Thyrotropin-Releasing-Hormon (TRH) stimulierten Thyreotropin (TSH)-Antwort im Vergleich zum Schilddrüsensuppressionstest und den Trijodthyronin (T_3)- und Thyroxin (T_4)-Serumspiegeln bei sogenannter euthyreoter endokriner Ophthalmopathie. Endokrinol. Im Druck.

232. ZUPPINGER, K. A., E. E. JOSS, M. P. KÖNIG, J. J. STAUB, J. GIRARD, and H. EHREN-GRUBER: Effect of Oral Thyrotrophin-Releasing Hormone on Serum Thyroxine in Growth Hormone Deficient and Normal Children. Clin. Endocr. 4, 119 (1975).

233. AVERILL, R. L. W., and T. H. KENNEDY: Increased Thyrotrophic Hormone Secretion after Electrical Stimulation and Hypothalamic Extract Infusion. New Zealand Med. J. 65, 398 (1966).

234. BAUGH, C. M., C. L. KRUMDIECK, J. M. HERSHMAN, and J. A. PITTMAN, Jr.: Synthesis and Biological Activity of Thyrotropin-Releasing Hormone. Endocrinol. 87, 1015 (1970).

235. PITTMAN, J. A., Jr., E. DUBOVSKY, and R. J. BESCHI: Stimulation of Pituitary Glucose Oxidation by Thyrotropin-Releasing Hormone. Biochem. Biophys. Res. Commun. 5, 1246 (1970).

236. LABELLA, F. S., and S. R. VIVIAN: Effect of Synthetic TRF on Hormone Release from Bovine Anterior Pituitary in Vitro. Endocrinol. 88, 787 (1971).

237. WILDMEISTER, W., und F. A. HORSTER: Die Wirkung von synthetischem Thyrotrophin Releasing Hormon auf die Entwicklung eines experimentellen Exophthalmus beim Goldfisch. Acta Endocr. 68, 363 (1971).

238. HAUG, E., H. FREY, and T. SAND: Effect of Thyrotrophin Releasing-Hormone and Thyroid-Stimulating Hormone on Serum Protein-Bound ^{131}I. Acta Endocr. 70, 454 (1972).

239. PLOTNIKOFF, N. P., A. J. PRANGE, Jr., G. R. BREESE, M. S. ANDERSON, and I. C. WILSON: Thyrotropin Releasing Hormone: Enhancement of Dopa Activity by a Hypothalamic Hormone. Science 178, 417 (1972).

240. SCHAMS, D.: Prolactin Releasing Effects of TRH in the Bovine and their Depression by a Prolactin Inhibitor. Horm. Metab. Res. 4, 405 (1972).

241. STEINER, H., F. PIVA, G. GAVAZZI, R. O. STUDER, D. GILLESSEN, and L. MARTINI: Different Patterns of Thyrotropin (TSH)-Release, TSH-Resynthesis and of Corticosterone Depression after in Vivo Administration of Thyrotropin-Releasing Hormone (TRH) and of an Isosteric Analog of TRH. Horm. Metab. Res. 4, 484 (1972).

242. AVERILL, R. L. W., and J. S. EVANS: Thyrotrophin Secretion from Rat Pituitary Autografts Infused with Synthetic Thyrotrophin Releasing Factor. J. Endocr. 57, 539 (1973).

243. DANNIES, P. S., and A. H. TASHJIAN, Jr.: Effects of Thyrotropin-Releasing Hormone and Hydrocortisone on Synthesis and Degradation of Prolactin in a Rat Pituitary Cell Strain. J. Biol. Chem. 248, 6174 (1973).

244. DUBOVSKY, E. V., R. J. BESCHI, J. A. PITTMAN, and R. H. LINDSAY: Stimulation of TSH Release and Glucose Oxidation in Pituitaries from Thyroidectomized Rats by Thyrotropin Releasing Factor (TRF). Alabama J. Med. Sci. 10, 270 (1973).

245. GOURDJI, D., A. TIXIER-VIDAL, A. MORIN, P. PRADELLES, J.-L. MORGAT, P. FROMAGEOT, and B. KERDELHUÉ: Binding of a Tritiated Thyrotropin-Releasing Factor to a Prolactin Secreting Clonal Cell Line (GH 3). Exper. Cell Res. 82, 39 (1973).

246. HINKLE, P. M., and A. H. TASHJIAN, Jr.: Receptors for Thyrotropin-Releasing Hormone in Prolactin-Producing Rat Pituitary Cells in Culture. J. Biol. Chem. 248, 6180 (1973).

247. AVERILL, R. L. W.: Thyroid Activation in Rats after Intrapituitary Administration of Synthetic Thyrotrophin-Releasing Factor (TRF). Endocrinol. 94, 794 (1974).

248. AZIZI, F., A. G. VAGENAKIS, J. BOLLINGER, S. REICHLIN, J. E. BUSH, and L. E. BRAVERMAN: The Effect of a Single Large Dose of Thyrotropin-Releasing Hormone on Various Aspects of Thyroid Function in the Rat. Endocrinol. 95, 1767 (1974).

249. BREESE, G. R., B. R. COOPER, A. J. PRANGE, Jr., J. M. COTT, and M. A. LIPTON: Interactions of Thyrotropin-Releasing Hormone with Centrally Acting Drugs. In: The Thyroid Axis, Drugs, and Behavior (A. J. PRANGE, ed.), p. 115. New York: Raven Press. 1974.

250. Breese, G. R., J. M. Cott, B. R. Cooper, A. J. Prange, Jr., and M. A. Lipton: Antagonism of Ethanol Narcosis by Thyrotropin Releasing Hormone. Life Sci. **14,** 1053 (1974).

251. Brown, M. R., and G. A. Hedge: Effects of Glucocorticoids on TRH and TSH Secretion: Dose and Time Considerations. Amer. J. Physiol. **227,** 289 (1974).

252. — — In Vivo Effects of Prostaglandins on TRH-Induced TSH Secretion. Endocrinol. **95,** 1392 (1974).

253. Brunet, N., D. Gourdji, A. Tixier-Vidal, P. Pradelles, J. L. Morgat, and P. Fromageot: Chemical Evidence for Associated TRF with Subcellular Fractions after Incubation of Intact Rat Prolactin Cells (GH 3) with ^{3}H-Labelled TRF. FEBS Letters **38,** 129 (1974).

254. Dannies, P. S., and A. H. Tashjian, Jr.: Pyroglutamyl-Histidyl-Prolineamide (TRH) — A Neurohormone which Affects the Release and Synthesis of Prolactin and Thyrotropin. Israel J. Med. Sci. **10,** 1294 (1974).

255. Dyer, R. G., and R. E. J. Dyball: Evidence for a Direct Effect of LRF and TRF on Single Unit Activity in the Rostral Hypothalamus. Nature **252,** 486 (1974).

256. Hill-Samli, M., and R. M. MacLeod: Interaction of Thyrotropin-Releasing Hormone and Dopamine on the Release of Prolactin from the Rat Anterior Pituitary in Vitro. Endocrinol. **95,** 1189 (1974).

257. Horst, W. D., and N. Spirt: A Possible Mechanism for the Anti-Depressant Activity of Thyrotropin Releasing Hormone. Life Sci. **15,** 1073 (1974).

258. Jackson, I. M. D., and S. Reichlin: Thyrotropin Releasing Hormone (TRH): Distribution in the Brain, Blood and Urine of the Rat. Life Sci. **14,** 2259 (1974).

259. — — Thyrotropin-Releasing Hormone (TRH): Distribution in Hypothalamic and Extrahypothalamic Brain Tissues of Mammalian and Submammalian Chordates. Endocrinol. **95,** 854 (1974).

260. Oliver, C., R. L. Eskay, N. Ben-Jonathan, and J. C. Porter: Distribution and Concentration of TRH in the Rat Brain. Endocrinol. **95,** 540 (1974).

261. Oliver, C., A. Taurog, J. C. Porter: Physiologie de la sécrétion de la TRH. Nouv. Presse méd. **3,** 1941 (1974).

262. Plotnikoff, N. P., A. J. Prange, Jr., G. R. Breese, M. S. Anderson, and I. C. Wilson: The Effects of Thyrotropin-Releasing Hormone on DOPA Response in Normal, Hypophysectomized, and Thyroidectomized Animals. In: The Thyroid Axis, Drugs, and Behavior (A. J. Prange, ed.), p. 103. New York: Raven Press. 1974.

263. Plotnikoff, N. P., A. J. Prange, Jr., G. R. Breese, and I. C. Wilson: Thyrotropin Releasing Hormone: Enhancement of DOPA Activity in Thyroidectomized Rats. Life Sci. **14,** 1271 (1974).

264. Prange, A. J., Jr., G. R. Breese, J. M. Cott, B. R. Martin, B. R. Copper, I. C. Wilson, and N. P. Plotnikoff: Thyrotropin Releasing Hormone: Antagonism of Pentobarbital in Rodents. Life Sci. **14,** 447 (1974).

265. Scanes, C. G.: Some in Vitro Effects of Synthetic Thyrotropin Releasing Factor on the Secretion of Thyroid Stimulating Hormone from the Anterior Pituitary Gland of the Domestic Fowl. Neuroendocr. **15,** 1 (1974).

266. Steiner, H., M. Zanisi, and L. Martini: Antiovulatory Activity of the Thyrotropin Releasing Hormone in the Rat. Horm. Metab. Res. **6,** 432 (1974).

267. Steiner, H., H. Künzi, and R. O. Studer: Accumulation of Tritiated Thyrotropin Releasing Hormone in Different Organs, Especially in the Thyroid. Experientia **30,** 1096 (1974).

268. Takahara, J., A. Arimura, and A. V. Schally: Stimulation of Prolactin and Growth Hormone Release by TRH Infused into a Hypophysial Portal Vessel. Proc. Soc. Exp. Biol. Med. **146,** 831 (1974).

269. — — — Effect of Catecholamines on the TRH-Stimulated Release of Prolactin and Growth Hormone from Sheep Pituitaries in Vitro. Endocrinol. **95,** 1490 (1974).

270. TAUROG, A., C. OLIVER, R. L. ESKAY, J. C. PORTER, and J. M. McKENZIE: The Role of TRH in the Neoteny of the Mexican Axolotl (Ambystoma mexicanum). Gen. Comp. Endocr. 24, 267 (1974).

271. BREESE, G. R., J. M. COTT, B. R. COOPER, A. J. PRANGE, Jr., M. A. LIPTON, and N. P. PLOTNIKOFF: Effects of Thyrotropin-Releasing Hormone (TRH) on the Acitons of Pentobarbital and other Centrally Acting Drugs. J. Pharmacol. Exp. Therap. 193, 11 (1975).

272. CHEN, H. J., and J. MEITES: Effects of Biogenic Amines and TRH on Release of Prolactin and TSH in the Rat. Endocrinol. 96, 10 (1975).

273. DEIS, R. P., and N. ALONSO: Effect of Synthetic Thyrotrophin Releasing Factor on Prolactin and Luteinizing Hormone Secretion in Male and Female Rats During Various Reproductive States. J. Endocr. 67, 425 (1975).

274. HALL, T. R., A. CHADWICK, N. J. BOLTON, and C. G. SCANES: Prolactin Release in Vitro and in Vivo in the Pigeon and the Domestic Fowl Following Administration of Synthetic Thyrotrophin-Releasing Factor (TRF). Gen. Comp. Endocr. 25, 298 (1975).

275. HILL-SAMLI, M., and R. M. MACLEOD: Thyrotropin-Releasing Hormone Blockade of the Ergocryptine and Apomorphine Inhibition of Prolactin Release in Vitro. Proc. Soc. Exp. Biol. Med. 149, 511 (1975).

276. HUIDOBRO-TORO, J. P., A. SCOTTI DE CAROLIS, and V. G. LONGO: Intensification of Central Catecholaminergic and Serotonergic Processes by the Hypothalamic Factors MIF and TRF and by Angiotensin II. Pharmacol. Biochem. and Behavior 3, 235 (1975).

277. KATO, Y., K. CHIHARA, K. MAEDA, S. OHGO, Y. OKANISHI, and H. IMURA: Plasma Growth Hormone Responses to Thyrotropin-Releasing Hormone in the Urethane-Anesthetized Rat. Endocrinol. 96, 1114 (1975).

278. KRUSE, H.: Thyrotropin-Releasing Hormone: Interaction with Chlorpromazine in Mice, Rats, and Rabbits. J. Pharmacol. 6, 249 (1975).

279. NEMEROFF, C. B., A. J. PRANGE, Jr., G. BISSETTE, G. R. BREESE, and M. A. LIPTON: Thyrotropin-Releasing Hormone (TRH) and its β-Alanine Analogue: Potentiation of the Anticonvulsant Potency of Phenobarbital in Mice. Psychopharmacol. Commun. 1, 305 (1975).

280. PRANGE, A. J., G. R. BREESE, G. D. JAHNKE, B. R. MARTIN, B. R. COOPER, J. M. COTT, I. C. WILSON, L. B. ALLTOP, M. A. LIPTON, G. BISSETTE, C. B. NEMEROFF, and P. T. LOOSEN: Modification of Pentobarbital Effects by Natural and Synthetic Polypeptides: Dissociation of Brain and Pituitary Effects. Life Sci. 16, 1907 (1975).

281. WEI, E., S. SIGEL, H. LOH, and E. L. WAY: Thyrotropin-Releasing Hormone and Shaking Behaviour in Rat. Nature 253, 739 (1975).

282. PRANGE, A. J., Jr., I. C. WILSON, P. P. LARA, L. B. ALLTOP, and G. R. BREESE: Effects of Thyrotropin-Releasing Hormone in Depression. Lancet 1972, 999.

283. WILSON, I. C., A. J. PRANGE, Jr., P. P. LARA, L. B. ALLTOP, R. A. STIKELEATHER, and M. A. LIPTON: Thyrotropin-Releasing Hormone: Psychobiological Responses of Normal Women. I. Subjective Experiences. In: The Thyroid Axis, Drugs, and Behavior (A. J. PRANGE, Jr., ed.), p. 146. New York: Raven Press. 1974.

284. COPPEN, A., S. MONTGOMERY, M. PEET, J. BAILEY, V. MARKS, and P. WOODS: Thyrotrophin-Releasing Hormone in the Treatment of Depression. Lancet 433 (1974).

285. BENKERT, O.: Studies on Pituitary Hormones and Releasing Hormones in Depression and Sexual Impotence. Progr. Brain Res. 42, 25 (1975).

286. WILSON, I. C., P. P. LARA, and A. J. PRANGE, Jr.: Thyrotrophin-Releasing Hormone in Schizophrenia. Lancet 43 (1973).

287. METCALF, G.: TRH: A Possible Mediator of Thermoregulation. Nature 252, 310 (1974).

288. L'HERMITE, M., L. VANHAELST, G. COPINSCHI, R. LECLERCQ, J. GOLSTEIN, O. D. BRUNO, and C. ROBYN: Prolactin Release after Injection of Thyrotrophin-Releasing Hormone in Man. Lancet 763 (1972).

289. BOWERS, C. Y., H. G. FRIESEN, and K. FOLKERS: Thyrotropin Releasing Hormone and the Release of Prolactin. In: Biological Rhythms in Neuroendocrine Activity (M. KAWAKAMI, ed.), p. 102. Tokyo: Igaku Shoin Ltd. 1974.

290. BECKERS, C.: Immunoassay of TRH and its Problems. In: Frontiers of Hormone Research (M. MAROIS, ed.), p. 170. Basel: S. Karger. 1972.

291. OLIVER, C., J. P. CHARVET, J.-L. CODACCIONI, and J. VAGUE: Radioimmunoassay of Thyrotropin-Releasing Hormone (TRH) in Human Plasma and Urine. J. Clin. Endocr. Metab. **39**, 406 (1974).

292. DUPONT, A., F. LABRIE, L. LEVASSEUR, and A. V. SCHALLY: Characteristics of the Plasma Disappearance of [^{3}H]-Thyrotropin-Releasing Hormone in the Rat. Canad. J. Physiol. Pharmacol. **52**, 1012 (1974).

293. KENNEDY, J. F., C. J. GRAY, S. A. BARKER, L. ALBRIGHTON, C. Y. BOWERS, A. V. SCHALLY, and W. F. WHITE: Photo-Oxidation of Thyrotropin Releasing Hormone. Life Sci. **10**, 569 (1971).

294. SCHALLY, A. V., A. KUROSHIMA, Y. ISHIDA, A. ARIMURA, T. SAITO, C. Y. BOWERS, and S. L. STEELMAN: Purification of Growth Hormone-Releasing Factor from Beef Hypothalamus. Proc. Soc. Exp. Biol. Med. **122**, 821 (1966).

295. BORGEAT, P., F. LABRIE, G. POIRIER, G. CHAVANCY, and A. V. SCHALLY: Stimulation of Adenosine 3',5'-Cyclic Monophosphate Accumulation in Anterior Pituitary Gland in Vitro by Purified Growth Hormone-Releasing Hormone. Transact. Assoc. Amer. Phys. **86**, 284 (1973).

296. CURRIE, B. L., K. N.-G. JOHANSSON, T. GREIBROKK, K. FOLKERS, and C. Y. BOWERS: Identification and Purification of Factor A — GHRH from Hypothalami which Releases Growth Hormone. Biochem. Biophys. Res. Commun. **60**, 605 (1974).

297. JOHANSSON, K. N.-G., B. L. CURRIE, K. FOLKERS, and C. Y. BOWERS: Identification and Purification of Factor B — GHRH from Hypothalami which Releases Growth Hormone. Biochem. Biophys. Res. Commun. **60**, 610 (1974).

298. SCHALLY, A. V., Y. BABA, R. M. G. NAIR, and C. D. BENNETT: The Amino Acid Sequence of a Peptide with Growth Hormone-Releasing Activity Isolated from Porcine Hypothalamus. J. Biol. Chem. **246**, 6647 (1971).

299. SCHALLY, A. V., A. ARIMURA, I. WAKABAYASHI, T. W. REDDING, E. DICKERMAN, and J. MEITES: Biological Activity of a Synthetic Decapeptide Corresponding to the Proposed Growth Hormone-Releasing Hormone. Experientia **28**, 205 (1972).

300. FAUSZT, I., and S. BAJUSZ: Synthesis of a Proposed Porcine Growth Hormone Releasing Hormone (GH — RH) and the N-Terminal Decapeptide of the β-Chain of Human Hemoglobin. Acta Chim. Acad. Sci. Hung. **82**, 471 (1974).

301. GIORI, P., M. GUARNERI, D. MAZZOTTA, C. B. VICENTINI, and C. A. BENASSI: New Synthesis of Growth-Hormone Releasing Hormone by the Pyrazoline Active Ester Procedure. Eur. J. Med. Chem. **10**, 428 (1975).

302. YUDAEV, N., and Z. UTESHEVA: Purification and Some Properties of a Hypothalamic Factor Regulating the Release of Growth Hormone from Hypophysis. First International Congress of Biochemistry, Stockholm, **7**, 453 (1973).

303. SCHALLY, A. V., T. W. REDDING, J. TAKAHARA, D. H. COY, and A. ARIMURA: Lack of Growth Hormone-Releasing Activity of (Pyro)Glu — Ser — Gly — NH₂. Biochem. Biophys. Res. Commun. **55**, 556 (1973).

304. ZECH, K., and W. VOELTER: Synthese eines Peptids mit vermutlicher Wirkung des Wachstumshormon-freisetzenden Hormons (GH — RH). Z. Naturforsch. **29 b**, 818 (1974).

305. GONZALEZ-BARCENA, D., A. J. KASTIN, D. H. COY, S. GLICK, D. S. SCHALCH, L. A. LEE, J. P. ARZAC, and A. V. SCHALLY: Unaltered Plasma GH Levels in Acromegalics and Normal Men and Women after Administration of [Pyro]Glu — Ser — Gly — NH₂, a Proposed GH-Releasing Hormone. J. Clin. Endocr. Metab. **38**, 1134 (1974).

306. SCHÖNBERG, D., D. GUPTA, K. ZECH, and W. VOELTER: Synthesis and Biological Activity of a Peptide with Suggested Activity of Growth Hormone-Releasing Hormone (GH – RH). Abstracts Internat. Symp. on GH and Related Peptides Ricerca Sci. ed Educazione Permanente 2, Suppl. 1,57 (1975).

307. CELIS, M. E., S. TALEISNIK, and R. WALTER: Regulation of Formation and Proposed Structure of the Factor Inhibiting the Release of Melanocyte-Stimulating Hormone. Proc. Natl. Acad. Sci. 68, 1428 (1971).

308. PLOTNIKOFF, N. P., A. J. KASTIN, M. S. ANDERSON, and A. V. SCHALLY: DOPA Potentiation by a Hypothalamic Factor, MSH Release-Inhibiting Hormone (MIF). Life Sci. 10, 1279 (1971).

309. CELIS, M. E., S. HASE, and R. WALTER: Structure-Activity Studies of MSH-Release-Inhibiting Hormone. FEBS Letters 27, 327 (1972).

310. PLOTNIKOFF, N. P., A. J. KASTIN, M. S. ANDERSON, and A. V. SCHALLY: Oxotremorine Antagonism by a Hypothalamic Hormone, Melanocyte-Stimulating Hormone Release-Inhibiting Factor (MIF). Proc. Soc. Exp. Biol. Med. 140, 811 (1972).

311. FRIEDMAN, E., J. FRIEDMAN, and S. GERSHON: Dopamine Synthesis: Stimulation by a Hypothalamic Factor. Science 182, 831 (1973).

312. PLOTNIKOFF, N. P., A. J. KASTIN, M. S. ANDERSON, and A. V. SCHALLY: Deserpidine Antagonism by a Tripeptide, L-Prolyl-L-Leucylglycinamide. Neuroendocr. 11, 67 (1973).

313. WALTER, R., E. C. GRIFFITHS, and K. C. HOOPER: Production of MSH-Release-Inhibiting Hormone by a Particulate Preparation of Hypothalami: Mechanisms of Oxytocin Inactivation. Brain Res. 60, 449 (1973).

314. CASTENSSON, S., H. SIEVERTSSON, B. LINDEKE, and C. Y. SUM: Studies on the Inhibition of Oxotremorine Induced Tremor by a Melanocyte-Stimulating Hormone Release-Inhibiting Factor, Thyrotropin Releasing Hormone and Related Peptides. FEBS Letters 44, 101 (1974).

315. PLOTNIKOFF, N. P., and A. J. KASTIN: Pharmacological Studies with a Tripeptide, Prolyl-Leucyl Glycine Amide. Arch. Internat. Pharmocodyn. Thérapie 211, 211 (1974).

316. PLOTNIKOFF, N. P., and A. J. KASTIN: Oxotremorine Antagonism by Prolyl-Leucyl-Glycine-Amide Administered by Different Routes and with Several Anticholinergics. Pharmacol. Biochem. and Behavior 2, 417 (1974).

317. PLOTNIKOFF, N. P., F. N. MINARD, and A. J. KASTIN: DOPA Potentiation in Ablated Animals and Brain Levels of Biogenic Amines in Intact Animals after Prolyl-leucyl-glycinamide. Neuroendocr. 14, 271 (1974).

318. PLOTNIKOFF, N. P.: Prolyl-Leucyl-Glycine Amide (PLG) and Thyrotropin-Releasing Hormone (TRH): DOPA Potentiation and Biogenic Amine Studies. Progr. Brain Res. 42, 11 (1975).

319. WALTER, R., A. NEIDLE, and N. MARKS: Significant Differences in the Degradation of Pro – Leu – Gly – NH$_2$ by Human Serum and that of Other Species. Proc. Soc. Exp. Biol. Med. 148, 98 (1975).

320. BOWER, S. A., M. E. HADLEY, and V. J. HRUBY: Comparative MSH Release-Inhibiting Activities of Tocinoic Acid (the Ring of Oxytocin), and L – Pro – L – Leu – Gly – NH$_2$ (the Side Chain of Oxytocin). Biochem. Biophys. Res. Commun. 45, 1185 (1971).

321. NAIR, R. M. G., A. J. KASTIN, and A. V. SCHALLY: Isolation and Structure of Another Hypothalamic Peptide Possessing MSH-Release-Inhibiting Activity. Biochem. Biophys. Res. Commun. 47, 1420 (1972).

322. GRANT, N. H., D. E. CLARK, and E. I. ROSANOFF: Evidence that Pro – Leu – Gly – NH$_2$, Tocinoic Acid, and Des – Cys – Tocinoic Acid Do not Affect Secretion of Melanocyte Stimulating Hormone. Biochem. Biophys. Res. Commun. 51, 100 (1973).

323. DESLAURIERS, R., R. WALTER, and I. C. P. SMITH: Intramolecular Motion in Peptides Determined by ^{13}C NMR: A Spin-Lattice Relaxation Time-Study on MSH-Release-Inhibiting Factor. FEBS Letters 37, 27 (1973).

324. RALSTON, E., J.-L. DE COEN, and R. WALTER: Tertiary Structure of H−Pro−Leu−Gly−NH₂, the Factor that Inhibits Release of Melanocyte Stimulating Hormone, Derived by Conformational Energy Calculations. Proc. Natl. Acad. Sci. **71**, 1142 (1974).

325. LABELLA, F. S., R. DULAR, and S. VIVIAN: Anomalous Prolactin Release in Vitro in Response to Cold and Its Specific Blockade by a Purified Hypothalamic-Inhibiting Factor. Endocrinol. **92**, 1571 (1973).

326. GREIBROKK, T., B. L. CURRIE, K. N.-G. JOHANSSON, J. J. HANSEN, K. FOLKERS, and C. Y. BOWERS: Purification of a Prolactin Inhibiting Hormone and the Revealing of Hormone D−GHIH which Inhibits the Release of Growth Hormone. Biochem. Biophys. Res. Commun. **59**, 704 (1974).

327. TAKAHARA, J., A. ARIMURA, and A. V. SCHALLY: Suppression of Prolactin Release by a Purified Porcine PIF Preparation and Catecholamines Infused into a Rat Hypophysial Portal Vessel. Endocrinol. **95**, 462 (1974).

328. GREIBROKK, T., J. J. HANSEN, R. KNUDSEN, Y.-K. LAM, K. FOLKERS, and C. Y. BOWERS: On the Isolation of a Prolactin Inhibiting Factor (Hormone). Biochem. Biophys. Res. Commun. **67**, 338 (1975).

329. VALE, W., P. BRAZEAU, G. GRANT, A. NUSSEY, R. BURGUS, J. RIVIER, N. LING et R. GUILLEMIN: Premières observations sur le mode d'action de la somatostatine, un facteur hypothalamique qui inhibe la sécrétion de l'hormone de croissance. Compt. Rend. Acad. Sci. **275**, 2913 (1972).

330. ALBERTI, K. G. M. M., N. J. CHRISTENSEN, S. E. CHRISTENSEN, A. PRANGE HANSEN, J. IVERSEN, K. LUNDBAEK, K. SEYER-HANSEN, and H. ØRSKOV: Inhibition of Insulin Secretion by Somatostatin. Lancet 1299 (1973).

331. GUILLEMIN, R.: Blockersubstanz Somatostatin. Selecta **41**, 3726 (1973).

332. HALL, R., G. M. BESSER, A. V. SCHALLY, D. H. COY, D. EVERED, D. J. GOLDIE, A. J. KASTIN, A. S. MCNEILLY, C. H. MORTIMER, C. PHENEKOS, W. M. G. TUNBRIDGE, and D. WEIGHTMAN: Action of Growth-Hormone-Release Inhibitory Hormone in Healthy Men and in Acromegaly. Lancet 581 (1973).

333. PRANGE HANSEN, A., H. ØRSKOV, K. SEYER-HANSEN, and K. LUNDBAEK: Some Actions of Growth Hormone Release Inhibiting Factor. Brit. Med. J. **3**, 523 (1973).

334. BELANGER, A., F. LABRIE, P. BORGEAT, M. SAVARY, J. COTE, J. DROUIN, A. V. SCHALLY, D. H. COY, E. J. COY, H. U. IMMER, K. SESTANJ, V. R. NELSON, and M. GOTZ: Inhibition of Growth Hormone and Thyrotropin Release by Growth Hormone-Release Inhibiting Hormone. Mol. Cell. Endocrinol. **1**, 329 (1974).

335. BESSER, G. M., C. H. MORTIMER, D. CARR, A. V. SCHALLY, D. H. COY, D. EVERED, A. J. KASTIN, W. M. G. TUNBRIDGE, M. O. THORNER, and R. HALL: Growth Hormone Release Inhibiting Hormone in Acromegaly. Brit. Med. J. **1**, 352 (1974).

336. BESSER, G. M., C. H. MORTIMER, A. S. MCNEILLY, M. O. THORNER, G. A. BATISTONI, S. R. BLOOM, K. W. KASTRUP, K. F. HANSSEN, R. HALL, D. H. COY, A. J. KASTIN, and A. V. SCHALLY: Long-Term Infusion of Growth Hormone Release Inhibiting Hormone in Acromegaly: Effects on Pituitary and Pancreatic Hormones. Brit. Med. J. **4**, 622 (1974).

337. BLOOM, S. R., C. H. MORTIMER, M. O. THORNER, G. M. BESSER, R. HALL, A. GOMEZ-PAN, V. M. ROY, R. C. G. RUSSELL, D. H. COY, A. J. KASTIN, and A. V. SCHALLY: Inhibition of Gastrin and Gastric-Acid Secretion by Growth-Hormone Release-Inhibiting Hormone. Lancet 1106 (1974).

338. BORGEAT, P., F. LABRIE, J. DROUIN, A. BÉLANGER, H. U. IMMER, K. SESTANJ, V. NELSON, M. GOTZ, A. V. SCHALLY, D. H. COY, and E. J. COY: Inhibition of Adenosine 3',5'-Monophosphate Accumulation in Anterior Pituitary Gland in Vitro by Growth Hormone-Release Inhibiting Hormone. Biochem. Biophys. Res. Commun. **56**, 1052 (1974).

339. BRAZEAU, P., J. RIVIER, W. VALE, and R. Guillemin: Inhibition of Growth Hormone Secretion in the Rat by Synthetic Somatostatin. Endocrinol. **94**, 184 (1974).

340. BRAZEAU, P., and R. Guillemin: Somatostatin: Newcomer from the Hypothalamus. N. Engl. J. Med. **290**, 963 (1974).

341. CHRISTENSEN, S. E., A. P. HANSEN, J. IVERSEN, K. LUNDBAEK, H. ÖRSKOV, and K. SEYER-HANSEN: Somatostatin as a Tool in Studies of Basal Carbohydrate and Lipid Metabolism in Man: Modifications of Glucagon and Insulin Release. Scand. J. clin. Lab. Invest. **34**, 321 (1974).

342. COPINSCHI, G., E. VIRASORO, L. VANHAELST, R. LECLERCQ, J. GOLSTEIN, and M. L'HER-MITE: Specific Inhibition by Somatostatin of Growth Hormone Release after Hypogly-caemia in Normal Man. Clin. Endocr. **3**, 441 (1974).

343. CURRY, D. L., L. L. BENNETT, and C. H. LI: Direct Inhibition of Insulin Secretion by Synthetic Somatostatin. Biochem. Biophys. Res. Commun. **58**, 885 (1974).

344. CURRY, D. L., and L. L. BENNETT: Reversal of Somatostatin Inhibition of Insulin Secre-tion by Calcium. Biochem. Biophys. Res. Commun. **60**, 1015 (1974).

345. GERICH, J. E., M. LORENZI, V. SCHNEIDER, J. K. KARAM, J. RIVIER, R. GUILLEMIN, and P. H. FORSHAM: Effects of Somatostatin on Plasma Glucose and Glucagon Levels in Human Diabetes Mellitus. N. Engl. J. Med. **291**, 544 (1974).

346. GERICH, J. E., M. LORENZI, V. SCHNEIDER, C. W. KWAN, J. H. KARAM, R. GUILLEMIN, and P. H. FORSHAM: Inhibition of Pancreatic Glucagon Responses to Arginine by Somato-statin in Normal Man and in Insulin-Dependent Diabetics. Diabetes **23**, 876 (1974).

347. GIUSTINA, G., E. RESCHINI, M. PERACCHI, L. CANTALAMESSA, F. CAVAGNINI, M. PINTO, and P. BULGHERONI: Failure of Somatostatin to Suppress Thyrotropin Releasing Factor and Luteinizing Hormone Releasing Factor-Induced Growth Hormone Release in Acromegaly. J. Clin. Endocr. Metab. **38**, 906 (1974).

348. GRANT, N. H., D. SARANTAKIS, and J. P. YARDLEY: Action of Growth Hormone Release Inhibitory Hormone on Prolactin Release in Rat Pituitary Cell Cultures. J. Endocr. **61**, 163 (1974).

349. IVERSEN, J.: Inhibition of Pancreatic Glucagon Release by Somatostatin: In Vitro. Scand. J. clin. Lab. Invest. **33**, 125 (1974).

350. KATO, Y., K. CHIHARA, S. OHGO, and H. IMURA: Effects of Hypothalamic Surgery and Somatostatin on Chlorpromazine-Induced Growth Hormone Release in Rats. Endo-crinol. **95**, 1608 (1974).

351. — — — — Inhibiting Effect of Somatostatin on Growth Hormone Release Induced by Isoprenaline or Chlorpromazine in Rats. J. Endocr. **62**, 687 (1974).

352. LUFT, R., S. EFENDIC, T. HÖKFELT, O. JOHANSSON, and A. ARIMURA: Immunohisto-chemical Evidence for the Localization of Somatostatin-Like Immunoreactivity in a Cell Population of the Pancreatic Islets. Med. Biol. **52**, 428 (1974).

353. MORTIMER, C. H., W. M. G. TUNBRIDGE, D. CARR, L. YEOMANS, T. LIND, D. H. COY, S. R. BLOOM, A. KASTIN, C. N. MALLINSON, G. M. BESSER, A. V. SCHALLY, and R. HALL: Effects of Growth-Hormone Release-Inhibiting Hormone on Circulating Glucagon, Insulin, and Growth Hormone in Normal, Diabetic, Acromegalic, and Hypopituitary Patients. Lancet **1974**, 697.

354. PLOTNIKOFF, N. P., A. J. KASTIN, and A. V. SCHALLY: Growth Hormone Release Inhibiting Hormone: Neuropharmacological Studies. Pharmacol. Biochem. and Behavior **2**, 693 (1974).

355. SILER, T. M., S. S. C. YEN, W. VALE, and R. GUILLEMIN: Inhibition by Somatostatin on the Release of TSH Induced in Man by Thyrotropin-Releasing Factor. J. Clin. Endocr. Metab. **38**, 742 (1974).

356. WEEKE, J., A. PRANGE HANSEN, and K. LUNDBAEK: The Inhibition by Somatostatin of the Thyrotropin Response to Thyrotropin-Releasing Hormone in Normal Subjects. Scand. J. clin. Lab. Invest. **33**, 101 (1974).

357. YEN, S. S. C., T. M. SILER, and G. W. DEVANE: Effect of Somatostatin in Patients with Acromegaly. N. Engl. J. Med. **290**, 935 (1974).

358. ARIMURA, A., H. SATO, D. H. COY, and A. V. SCHALLY: Radioimmunoassay for GH-Release Inhibiting Hormone. Proc. Soc. Exp. Biol. Med. **148,** 784 (1975).

359. BENKER, G., und D. REINWEIN: Somatostatin. Wirkungen und mögliche klinische Bedeutung. Dtsch. med. Wschr. **100,** 961 (1975).

360. BESSER, G. M., A. M. PAXTON, S. A. N. JOHNSON, E. J. MOODY, C. H. MORTIMER, R. HALL, A. GOMEZ-PAN, A. V. SCHALLY, A. J. KASTIN, and D. H. COY: Impairment of Platelet Function by Growth-Hormone Release-Inhibiting Hormone. Lancet 1166 (1975).

361. CHRISTENSEN, S. E., A. PRANGE HANSEN, K. LUNDBAEK, H. ØRSKOV, and K. SEYER-HANSEN. Somatostatin and Insulinoma. Lancet 1426 (1975).

362. CHRISTENSEN, N. J., S. E. CHRISTENSEN, A. PRANGE HANSEN, and K. LUNDBAEK: The Effect of Somatostatin on Plasma Noradrenaline and Plasma Adrenaline Concentrations During Exercise and Hypoglycemia. Science Press, in press.

363. EFENDIĆ, S., and R. LUFT: Studies on the Inhibitory Effect of Somatostatin on Glucose Induced Insulin Release in the Isolated Perfused Rat Pancreas. Acta Endocr. **78,** 510 (1975).

364. — — Studies on the Mechanism of Somatostatin Action on Insulin Release in Man. Acta Endocr. **78,** 516 (1975).

365. NEUMAN, M.: Une nouvelle hormone hypothalamique, la somatostatine. Ars Medici **30,** 135 (1975).

366. PRANGE HANSEN, A., S. E. CHRISTENSEN, and K. LUNDBAEK: The Effect of Somatostatin on the Rise of Growth Hormone and Glucagon Secretion Induced by Arginine and L-Dopa in Diabetic Patients. Scand. J. clin. Lab. Invest. **35,** 205 (1975).

367. WAGNER, H., E. ZIERDEN, and W. H. HAUSS: Effects of Synthetic Somatostatin on Endotoxin-Induced Changes of Growth Hormone, Cortisol and Insulin in Plasma, Blood Sugar and Blood Leukocytes in Man. Klin. Wschr. **53,** 539 (1975).

368. WEEKE, J., A. PRANGE HANSEN, and K. LUNDBAEK: Inhibition by Somatostatin of Basal Levels of Serum Thyrotropin (TSH) in Normal Men. J. Clin. Endocr. Metab. **41,** 168 (1975).

369. WIEGELMANN, W., H. G. SOLBACH, H. K. KLEY, K. H. RUDORFF, J. HERRMANN, H. ZIMMERMANN und H. L. KRÜSKEMPER: Die Wirkung von synthetischem Somatostatin bei männlichen Normalpersonen und Akromegalen. Dtsch. med. Wschr. **100,** 331 (1975).

370. YEN, S. S. C., B. L. LASLEY, C. F. WANG, H. LEBLANC, and T. M. SILER: The Operating Characteristics of the Hypothalamic-Pituitary System during the Menstrual Cycle and Observations of Biological Action of Somatostatin. Recent Progr. Hormone Res. **31,** 321 (1975).

371. SCHALLY, A. V., C. Y. BOWERS, T. W. REDDING, and J. F. BARRETT: Isolation of Thyrotropin Releasing Factor (TRF) from Porcine Hypothalamus. Biochem. Biophys. Res. Commun. **25,** 165 (1966).

372. BURGUS, R., T. F. DUNN, D. DESIDERIO, D. N. WARD, W. VALE, and R. GUILLEMIN: Characterization of Ovine Hypothalamic Hypophysiotropic TSH-Releasing Factor. Nature **226,** 321 (1970).

373. SCHALLY, A. V., T. W. REDDING, C. Y. BOWERS, and J. F. BARRETT: Isolation and Properties of Porcine Thyrotropin-Releasing Hormone. J. Biol. Chem. **244,** 4077 (1969).

374. GUILLEMIN, R., and E. SAKIZ: A Proposal for a Reference Standard Preparation for the Hypothalamic Thyrotrophic Hormone Releasing Factor. Nature **207,** 297 (1965).

375. BURGUS, R., T. F. DUNN, D. M. DESIDERIO, D. N. WARD, W. VALE, R. GUILLEMIN, A. M. FELIX, D. GILLESSEN, and R. O. STUDER: Biological Activity of Synthetic Polypeptide Derivatives Related to the Structure of Hypothalamic TRF. Endocrinol. **86,** 573 (1970).

376. STUDER, R. O.: Chemistry of TRH. Frontiers Hormone Res. **1,** 4 (1972).

377. FOLKERS, K., F. ENZMANN, J. BØLER, C. Y. BOWERS, and A. V. SCHALLY: Discovery of Modification of the Synthetic Tripeptide-Sequence of the Thyrotropin Releasing Hormone Having Activity. Biochem. Biophys. Res. Commun. **37,** 123 (1969).

378. FOLKERS, K., J.-K. CHANG, B. L. CURRIE, C. Y. BOWERS, A. WEIL, and A. V. SCHALLY: Synthesis and Relationship of L-Glutaminyl-L-histidyl-L-prolinamide to the Thyrotropin Releasing Hormone. Biochem. Biophys. Res. Commun. 39, 110 (1970).
379. MERRIFIELD, R. B.: Solid Phase Peptide Synthese. I. The Synthesis of a Tetrapeptide. J. Amer. Chem. Soc. 85, 2149 (1963).
380. BRUNFELDT, K., J. HALSTRØM, and P. ROEPSTORFF: A Punched-Tape Controlled Peptide Synthesizer. Peptides 1968, 194.
381. BAYER, E., H. HAGENMAIER, G. JUNG, W. PARR, H. ECKSTEIN, P. HUNZIKER, and R. E. SIEVERS: The Problem of Failure Sequences in the Solid Phase Synthesis of Peptides. Peptides 1971, 65.
382. TOMETSKO, A. M., J. GARDEN, and J. TISCHIO: Automated Chemical Synthesis Controlled by Computer Generated Programs-Polypeptides. Rev. Sci. Instr. 42, 331 (1971).
383. VOELTER, W., K. ZECH und N. GRUBHOFER: Synthese biologisch aktiver Peptide mittels eines neuen Syntheseautomaten. Z. Naturforsch. 28 b, 625 (1973).
384. FLOURET, G. R.: Synthesis of Pyroglutamylhistidylprolineamide. J. Med. Chem. 13, 843 (1970).
385. GILLESSEN, D., A. M. FELIX, W. LERGIER und R. O. STUDER: Synthese des „Thyrotropin-releasing" Hormons (TRH) (Schaf) und verwandter Peptide. Helv. Chim. Acta 53, 63 (1970).
386. BØLER, J., J.-K. CHANG, F. ENZMANN, and K. FOLKERS: Synthesis of the Thyrotropin-Releasing Hormone. J. Med. Chem. 14, 475 (1971).
387. CHANG, J.-K., H. SIEVERTSSON, C. BOGENTOFT, B. CURRIE, K. FOLKERS, and G. D. DAVES, Jr.: Syntheses of Pyroglutamylhistidylprolinamide and Unusual Mass Fragmentation. J. Med. Chem. 14, 481 (1971).
388. CHANG, J.-K., H. SIEVERTSSON, B. L. CURRIE, K. FOLKERS, and C. BOWERS: Synthesis of Analogs of the Thyrotropin-Releasing Hormone and Structure-Activity Relationships. J. Med. Chem. 14, 484 (1971).
389. INOUYE, K., K. NAMBA, and H. OTSUKA: A Synthesis of Thyrotropin-Releasing Factor. Bull. Chem. Soc. Japan 44, 1689 (1971).
390. KÖNIG, W., und R. GEIGER: Eine neue Amid-Schutzgruppe. Chem. Ber. 103, 2041 (1970).
391. — — Pyroglutamylpeptide. Chem. Ber. 105, 2872 (1972).
392. KURATH, P., and A. M. THOMAS: N-Carbobenzoxy-L-pyroglutamyl-L-histidyl Peptides. Helv. Chim. Acta 56, 1656 (1973).
393. BAJUSZ, S., and I. FAUSZT: An Improved Method for the Synthesis of the Thyrotropin Releasing Hormone (TRH). Acta Chim. Acad. Sci. Hung. 75, 419 (1973).
394. HATANAKA, C., M. OBAYASHI, O. NISHIMURA, N. TOUKAI, and M. FUJINO: An Improved Synthesis of Thyrotropin Releasing Hormone (TRH) and Crystallization of the Tartrate. Biochem. Biophys. Res. Commun. 60, 1345 (1974).
395. BEYERMAN, H. C., P. KRANENBURG, and J. L. M. SYRIER: Synthesis of Thyrotropin-Releasing Hormone Pyroglutamyl-Histidyl-Proline Amide. Rec. Trav. Chim. 90, 791 (1971).
396. SYRIER, J. L. M., and H. C. BEYERMAN: On the Optical Purity of Synthetic Thyrotropin-Releasing Hormone. Rec. Trav. Chim. 93, 117 (1974).
397. BEYERMAN, H. C., J. HIRT, P. KRANENBURG, J. L. M. SYRIER, and A. VAN ZON: Excess Mixed Anhydride Peptide Synthesis with Histidine Derivatives. Rec. Trav. Chim. 93, 256 (1974).
398. RIVAILLE, P., and G. MILHAUD: An Efficient Synthesis of Thyrotropin Releasing Hormone (TRH). Helv. Chim. Acta 54, 355 (1971).
399. PIETTA, P. G., P. F. CAVALLO, K. TAKAHASHI, and G. R. MARSHALL: Preparation and Use of Benzhydrylamine Polymers in Peptide Synthesis, II. Syntheses of Thyrotropin Releasing Hormone, Thyrocalcitonin 26—32, and Eledoisin. J. Org. Chem. 39, 44 (1974).

400. Hofmann, K., and C. Y. Bowers: Polypeptides. XLVII. Effect of the Pyrazole-Imidazole Replacement on the Biological Activity of Thyrotropin-Releasing Hormone. J. Med. Chem. **13**, 1099 (1970).

401. Gillessen, D., F. Piva, H. Steiner und R. O. Studer: Über die Bedeutung des Histidins im „Thyrotropin-releasing" Hormon (TRH). Helv. Chim. Acta **54**, 1335 (1971).

402. Gillessen, D.: Persönliche Mitteilung. 1974.

403. Flouret, G. R., R. Morgan, R. Gendrich, J. Wilber, and M. Seibel: Synthesis of the Thyrotropin-Releasing Hormone Enantiomer and Some Diastereoisomers and in Vitro Studies of Their Biological Activity. J. Med. Chem. **16**, 1137 (1973).

404. Voelter, W., S. Fuchs und K. Zech: Synthese von D − His² − TRH, einem biologisch aktiven Analogon des Thyreotropin-freisetzenden Hormons (TRH). Tetrahedron Letters **1974**, 3975 (1974).

405. Voelter, W. und H. Horn: Synthese von [Tyr(OMe)²]Tyroliberin, einem neuen Analogon des Thyrotropin-freisetzenden Hormons. Hoppe-Seyler's Z. Physiol. Chem. **355**, 1466 (1974).

406. Zech, K., H. Horn und W. Voelter: Synthese von [2-Phe] − TRH, ein bioaktives Analogon des Thyreotropin-freisetzenden Hormons (TRH). Chemiker-Ztg. **98**, 209 (1974).

407. Sievertsson, H., S. Castensson, C. Y. Bowers, H. G. Friesen, and K. Folkers: Synthesis of Analogs of the Thyrotropin Releasing Hormone (TRH) and Bioassays for TRH- and TRH-Inhibiting Activities. Acta Pharm. Suecica **10**, 297 (1973).

408. Sievertsson, H., S. Castensson, O. Lindgren, and C. Y. Bowers: Studies on Tetrapeptides Related to Thyrotropin Releasing Hormone and Luteinizing Hormone Releasing Hormone. Acta Pharm. Suecica **11**, 67 (1974).

409. Lybeck, H., J. Leppäluoto, P. Virkkunen, D. Schafer, L. Carlsson, and J. Mulder: Suppression of TRH-Mediated Thyroidal Release of ¹³¹I by a Synthetic Analog. Neuroendocr. **12**, 366 (1973).

410. Seif, F. J., W. Klingler, K. Zech, and W. Voelter: Inhibition of TRH-Induced TSH Release by L-Thyroxine, L-Triiodothyronine, and their D-Isomers in Mice. Experientia **31**, 992 (1975).

411. Pradelles, P., J. L. Morgat, P. Fromageot, C. Oliver, P. Jacquet, D. Gourdji, and A. Tixier-Vidal: Preparation of Highly Labelled ³H-Thyreotropin Releasing Hormone (PGA − His − Pro(NH₂)) by Catalytic Hydrogenolysis. FEBS Letters **22**, 19 (1972).

412. Hinkle, P. M., E. L. Woroch, and A. H. Tashjian, Jr.: Receptor-Binding Affinities and Biological Activities of Analogs of Thyrotropin-Releasing Hormone in Prolactin-Producing Pituitary Cells in Culture. J. Biol. Chem. **249**, 3085 (1974).

413. Kalbacher, H., K. Zech und W. Voelter: Darstellung von (Pro-methylamid)³-TRH, einem Methylderivat des Thyreotropin-freisetzenden Hormons. Chemiker-Ztg. **99**, 460 (1975).

414. Kalbacher, H., W. A. König und W. Voelter: Synthese und spektroskopische Untersuchungen von N^{amid}Äthylthyroliberin. Hoppe-Seyler's Z. Physiol. Chem. **356**, 1827 (1975).

415. Klingler, W., H. Kalbacher, P. Göbel, F. J. Seif, K. Zech und W. Voelter: Zusammenhänge zwischen Struktur und biologischer Aktivität beim Thyroliberin (TRH). Hoppe-Seyler's Z. Physiol. Chem. **357**, 269 (1976).

416. Kalbacher, H., W. Voelter und M. Vajda: Synthese und ¹³C − NMR-Spektroskopie von (Pro − NH − propyl)³-Derivaten des Thyreotropin-freisetzenden Hormons (TRH). Acta Chim. Acad. Sci. Hung., im Druck.

417. Voelter, W., H. Kalbacher und K.-P. Zeller: Darstellung und massenspektroskopischer Strukturbeweis von (N-Butylprolinamid)³- und (N-Pentylprolinamid)³-TRH, drei neuen Derivaten des Thyreotropin-freisetzenden Hormons. Chemiker-Ztg. **100**, 238 (1976).

418. KALBACHER, H., W. KLINGLER und W. VOELTER: Synthese, spektroskopische Eigenschaften und biologische Aktivität von [Pro–alkylamid]³ – TRH-Derivaten. Chem. Ber., im Druck.

419. FUCHS, S.: Synthesen von biologisch aktiven TRH-Derivaten. Dissertation der Universität Tübingen. 1975.

420. FUCHS, S., W. KLINGLER und W. VOELTER: Synthese, spektroskopische Untersuchungen und biologische Aktivität von TRH-Derivaten der allgemeinen Form L – Pyr – L – His – L – Pro – NH – (CH₂)n – NH₂. Monatsh. Chem., im Druck.

421. SIEVERTSSON, H., J.-K. CHANG, K. FOLKERS, and C. Y. BOWERS: On the Role of the Histidine Moiety in the Structure of the Thyrotropin-Releasing Hormone. J. Med. Chem. 15, 219 (1972).

422. VALE, W., J. RIVIER, and R. BURGUS: Synthetic TRF (Thyrotropin Releasing Factor) Analogues: II. pGlu – N³ Me – His – Pro – NH₂: A Synthetic Analogue with Specific Activity Greater than that of TRF. Endocrinol. 89, 1485 (1971).

423. RIVIER, J., W. VALE, M. MONAHAN, N. LING, and R. BURGUS: Synthetic Thyrotropin-Releasing Factor Analogs. 3. Effect of Replacement or Modification of Histidine Residue on Biological Activity. J. Med. Chem. 15, 479 (1972).

424. SIEVERTSSON, H., J.-K. CHANG, K. FOLKERS, and C. Y. BOWERS: Synthesis of Di- and Tripeptides and Assay in Vivo for Activity in the Thyrotropin Releasing Hormone and the Luteinizing Releasing Hormone Systems. J. Med. Chem. 15, 8 (1971).

425. GÖBEL, P., W. KLINGLER, H. HORN und W. VOELTER: Über die biologische Wirkung eines Thyreotropin-„Releasing"-Hormon (TRH)-Derivates. Klin. Wschr. 52, 1128 (1974).

426. ZECH, K.: Hypothalamus-„Releasing"-Hormone; Synthesewege und Struktur. Dissertation der Universität Tübingen. 1973.

427. BIEMANN, K.: Mass Spectrometry, Organic Chemical Applications. New York: McGraw-Hill Book Co. 1962.

428. MCLAFFERTY, F. W.: Interpretation of Mass Spectra. New York: W. A. Benjamin. 1966.

429. SPITELLER, G.: Massenspektrometrische Strukturanalyse organischer Verbindungen. Weinheim: Verlag Chemie. 1966.

430. BUDZIKIEWICZ, H., C. DJERASSI, and D. H. WILLIAMS, Mass Spectrometry of Organic Compounds. San Francisco: Holden-Day. 1967.

431. KIENITZ, H.: Massenspektrometrie. Weinheim: Verlag Chemie. 1968.

432. BURGUS, R., T. F. DUNN, D. DESIDERIO et R. GUILLEMIN: Structure moléculaire du facteur hypothalamique hypophysiotrope TRF d'origine ovine: mise en évidence par spectrométrie de masse de la séquence PCA – His – Pro – NH₂. Compt. Rend. Acad. Sci. 269, 1870 (1969).

433. DESIDERIO, D. M., Jr., R. BURGUS, T. F. DUNN, W. VALE, R. GUILLEMIN, and D. N. WARD: The Elucidation of the Primary Structure of the Hypothalamic Thyroid Stimulating Hormone Releasing Factor of Ovine Origin by Means of Mass Spectrometry. Org. Mass. Spectrom. 5, 221 (1971).

434. ENZMANN, F., J. BØLER, K. FOLKERS, C. Y. BOWERS, and A. V. SCHALLY: Structure and Synthesis of the Thyrotropin-Releasing Hormone. J. Med. Chem. 14, 469 (1971).

435. LÜDERWALD, I., M. PRZYBYLSKI, H. RINGSDORF, D. SILBERHORN, K. ZECH, and W. VOELTER: Investigation of Hypothalamus Peptide Hormones and Related Derivatives by Field Desorption Mass Spectrometry. Zur Publikation eingereicht.

436. KÖNIG, W. A., S. FUCHS, K. ZECH und W. VOELTER: Massenspektrometrische Fragmentierung von Trimethylsilylderivaten (TMS) des Thyrotropin-„Releasing"-Hormons (TRH) und dessen Teilstrukturen. Z. Naturforsch. 28 b, 820 (1973).

437. VOELTER, W., K. ZECH und W. A. KÖNIG: Unveröffentlichte Ergebnisse.

438. FERMANDJIAN, S., P. PRADELLES, P. FROMAGEOT, and J.-J. DUNAND: Proton NMR Studies on Thyrotropin Releasing Factor. FEBS Letters 28, 156 (1972).

439. Gibbon, W. A., G. Némethy, A. Stern, and L. C. Craig: An Approach to Conformational Analysis of Peptides and Proteins in Solution Based on a Combination of Nuclear Magnetic Resonance Spectroscopy and Conformational Energy Calculations. Proc. Natl. Acad. Sci. **67**, 239 (1970).

440. Ramachandran, G. N., R. Chandrasekaran, and K. D. Kopple: Variation of the $NH - C^{\alpha}H$ Coupling Constant with Dehedral Angle in the NMR Spectra of Peptides. Biopolymers **10**, 2113 (1971).

441. Belle, J., M. Montagut et A.-M. Bellocq: Analyse conformationelle de l'hormone hypothalamique TRF de libération de la thyréostimuline. Compt. Rend. Acad. Sci. **275**, 471 (1972).

442. Boilot, J.-C., B. Clin, A.-M. Bellocq et B. Lemanceau: Analyse conformationnelle de l'hormone hypothalamique TRF par résonance magnétique nucléaire. Compt. Rend. Acad. Sci. **276**, 217 (1973).

443. Bellocq, A.-M., J.-C. Boilot, E. Dupart et M. Dubien: Analyse conformationnelle de l'hormone hypothalamique TRF par spectroscopie Raman. Compt. Rend. Acad. Sci. **276**, 423 (1973).

444. Donzel, B., J. Rivier, and M. Goodman: Conformational Studies on the Hypothalamic Thyrotropin Releasing Factor and Related Compounds by [1]H Nuclear Magnetic Resonance Spectroscopy. Biopolymers **13**, 2631 (1974).

445. Montagut, M., B. Lemanceau, and A.-M. Bellocq: Conformational Analysis of Thyrotropin Releasing Factor by Proton Magnetic Resonance Spectroscopy. Biopolymers **13**, 2615 (1974).

446. Sievertsson, H., S. Castensson, and C. Y. Bowers: On the Conformation of Thyrotropin Releasing Hormone. FEBS Letters **42**, 340 (1974).

447. Donzel, B., M. Goodman, J. Rivier, N. Ling, and W. Vale: Synthesis and Conformations of Hypothalamic Hormone Releasing Factors: Two-Analogues Containing Backbone N-Methyl Groups. Nature **256**, 750 (1975).

448. Breitmaier, E., G. Jung und W. Voelter: Impuls-Fourier-Transform-^{13}C-NMR-Spektroskopie, Grundlagen und Anwendungen. Angew. Chem. **83**, 659 (1971).

449. Stothers, J. B.: Carbon-[13]-NMR Spectroscopy. New York: Academic Press. 1972.

450. Levy, G. C., and G. L. Nelson: Carbon-13 Nuclear Magnetic Resonance for Organic Chemists. New York: Wiley-Interscience. 1972.

451. Breitmaier, E., and W. Voelter: ^{13}C–NMR Spectroscopy. Weinheim: Verlag Chemie. 1974.

452. Smith, I. C. P., R. Deslauriers, and R. Walter: A Carbon-13 Nuclear Magnetic Resonance Study of Neurohypophyseal Hormones and Related Oligopeptides. Chemistry and Biology of Peptides **29** (1972).

453. Deslauriers, R., R. Walter, and I. C. P. Smith: [13]C Nuclear Magnetic Resonance Studies of the Conformation of the X-Pro Bond in the Oligopeptide Hormones, Thyrotropin-Releasing Hormone, Luteinizing Hormone-Releasing Factor, Angiotensin and Melanocyte-Stimulating Hormone Release-Inhibiting Factor. Biochem. Biophys. Res. Commun. **53**, 244 (1973).

454. Deslauriers, R., C. Garrigou-Lagrange, A.-M. Bellocq, and I. C. P. Smith: Carbon-13 Nuclear Magnetic Resonance Studies on Thyrotropin-Releasing Factor and Related Peptides. FEBS Letters **31**, 59 (1973).

455. Voelter, W., O. Oster und Karl Zech: cis-trans-Isomerie des Thyreotropin-freisetzenden Hormons (TRH) in wäßriger Lösung. Angew. Chem. **86**, 46 (1974).

456. Deslauriers, R., W. H. McGregor, D. Sarantakis, and I. C. P. Smith: Carbon-13 Nuclear Magnetic Resonance Studies of Structure and Function in Thyrotropin-Releasing Factor. Determination of the Tautomeric Form of Histidine and Relationship to Biology Activity. Biochemistry **13**, 3443 (1974).

457. Voelter, W., G. Jung, E. Breitmaier und E. Bayer: ^{13}C-Chemische Verschiebungen von Aminosäuren und Peptiden. Z. Naturforsch. **26 b**, 213 (1971).

458. THOMAS, W. A., and M. K. WILLIAMS: [13]C Nuclear Magnetic Resonance Spectroscopy and cis/trans Isomerism in Dipeptides Containing Proline. Chem. Commun. 1972, 994.
459. DESLAURIERS, R., R. WALTER, and I. C. P. SMITH: A Carbon-13 Nuclear Magnetic Resonance Study of Oxytocin and its Oligopeptides. Biochem. Biophys. Res. Commun. 48, 854 (1972).
460. WÜTHRICH, K., A. TUN-KYI, and R. SCHWYZER: Manifestation in the ^{13}C – NMR Spectra of Two Different Molecular Conformations of a Cyclic Pentapeptide. FEBS Letters 25, 104 (1972).
461. VOELTER, W., und O. OSTER: Nachweis der cis-trans-Isometrie bei Prolinderivaten mit ^{13}C-NMR-Spektroskopie. Chemiker-Ztg. 96, 586 (1972).
462. — — The Influence of N-Protected Amino Acid Residues on the cis/trans Isomerism of Proline Moieties in Peptides. Org. Magn. Res. 5, 547 (1973).
463. VOELTER, W., O. OSTER und E. BREITMAIER: Nachweis der cis-trans-Isometrie bei Diprolinpeptiden mit ^{13}C-NMR-Spektroskopie. Z. Naturforsch. 28 b, 370 (1973).
464. OSTER, O.: Untersuchungen molekularer Strukturen von Proteinfragmenten. Dissertation der Universität Tübingen. 1973.
465. VOELTER, W., S. FUCHS, R. H. SEUFFER und K. ZECH: ^{13}C-NMR-Studien von geschützten Aminosäuren. Monatsh. Chem. 105, 1110 (1974).
466. OSTER, O., E. BREITMAIER, and W. VOELTER: ^{13}C NMR Studies of Proline Peptides and Carbohydrates. In: Nuclear Magnetic Resonance Spectroscopy of Nuclei Other than Protons (T. AXENROD and G. A. WEBB, eds.), p. 233. New York: John Wiley and Sons, Inc. 1974.
467. VOELTER, W., und K. ZECH: Unveröffentlichte Ergebnisse.
468. KALBACHER, H., W. VOELTER und M. VAJDA: Synthese und ^{13}C-NMR-Spektroskopie von [Pro – NH – Propyl]³-Derivaten des Thyreotropin-freisetzenden Hormons (TRH). Acta Chim. Acad. Sci. Hung., im Druck.
469. DESLAURIERS, R., R. WALTER, and I. C. P. SMITH: Intramolecular Motion in Peptides Determined by ^{13}C NMR: A Spin-Lattice Relaxation Time-Study on MSH-Release-Inhibiting Factor. FEBS Letters 37, 27 (1973).
470. DESLAURIERS, R., I. C. P. SMITH, and R. WALTER: Conformational Mobility of the Pyrrolidine Ring of Proline in Peptides and Peptide Hormones as Manifest in Carbon 13 Spin-Lattice Relaxation Times. J. Biol. Chem. 249, 7006 (1974).
471. HAAR, W., S. FERMANDJIAN, J. VICAR, K. BLAHA, and P. FROMAGEOT: ^{13}C-NMR Study of [85% ^{13}C-Enriched Proline]-TRF [13]C-[13]C Vicinal Coupling Constants and Conformation of the Proline Residue. Proc. Natl. Acad. Sci. 72, 4948 (1975).
472. VOELTER, W., K. ZECH, P. GÖBEL, O. OSTER und A. ATTANASIO: Circulardichroismusuntersuchungen des Thyreotropin-Releasinghormons (TRH). Z. Naturforsch. 30 b, 142 (1975).
473. PRADELLES, P., J. VICAR, J.-L. MORGAT, S. FERMANDJIAN, K. BLAHA, and P. FROMAGEOT: Influence of Organic Solvents on the Conformation of Thyreotropin Releasing Factor (TRF) Studied by Circular Dichroism. Persönliche Mitteilung.
474. VOELTER, W., H. HORN und J. BRUN: Unveröffentlichte Ergebnisse.
475. SCHALLY, A. V., A. ARIMURA, Y. BABA, R. M. G. NAIR, H. MATSUO, T. W. REDDING, L. DEBELJUK, and W. F. WHITE: Isolation and Properties of the FSH and LH-Releasing Hormone. Biochem. Biophys. Res. Commun. 43, 393 (1971).
476. AMOSS, M., R. BURGUS, R. E. BLACKWELL, W. VALE, R. FELLOWS, and R. GUILLEMIN: Purification, Amino Acid Composition and N-Terminus of the Hypothalamic Luteinizing Hormone Releasing Factor (LRF) of Ovine Origin. Biochem. Biophys. Res. Commun. 44, 205 (1971).
477. MATSUO, H., Y. BABA, R. M. G. NAIR, A. ARIMURA, and A. V. SCHALLY: Structure of the Porcine LH- and FSH-Releasing Hormone. I. The Proposed Amino Acid Sequence. Biochem. Biophys. Res. Commun. 43, 1334 (1971).

478. Currie, B. L., H. Sievertsson, C. Bogentoft, J.-K. Chang, and K. Folkers: Data on Structure of Bovine LRH by Inactivation. Biochem. Biophys. Res. Commun. **42,** 1180 (1971).

479. Bogentoft, C., B. L. Currie, H. Sievertsson, J.-K. Chang, and K. Folkers: On the Structure of the Hypothalamic Luteinizing Releasing Hormone. Evidence for the Presence of Arginine, Tyrosine, and Tryptophan by Inactivation. Biochem. Biophys. Res. Commun. **44,** 403 (1971).

480. Johansson, N. G., F. Hooper, H. Sievertsson, B. L. Currie, K. Folkers, and C. Y. Bowers: Biosynthesis in Vitro of the Luteinizing Releasing Hormone by Hypothalamic Tissue. Biochem. Biophys. Res. Commun. **49,** 656 (1972).

481. Johansson, K. N. G., B. L. Currie, K. Folkers, and C. Y. Bowers: Biosynthesis and Evidence for the Existence of the Follicle Stimulating Hormone Releasing Hormone. Biochem. Biophys. Res. Commun. **50,** 8 (1973).

482. — — — — Determination of Releasing Hormones in Subcellular Fractions from Porcine Hypothalamic Tissue. Biochem. Biophys. Res. Commun. **52,** 967 (1973).

483. — — — — Biosynthesis of the Luteinizing Hormone Releasing Hormone in Mitochondrial Preparations and by a Possible Pantetheine-Template Mechanism. Biochem. Biophys. Res. Commun. **53,** 502 (1973).

484. Wessels, P. L., J. Feeny, H. Gregory, and J. J. Gormley: High Resolution Nuclear Magnetic Resonance Studies of the Conformation of Luteinizing Hormone Releasing Hormone (LH–RH) and its Component Peptides. J. Chem. Soc. Perkin Trans. II **1973,** 1691.

485. Marche, P., J.-L. Morgat, and P. Fromageot: Solvent Effects on Luteinizing- and Follicle-Stimulating-Hormone Releasing Factor Polymorphism Studied by Circular Dichroism. Eur. J. Biochem. **40,** 513 (1973).

486. Geiger, R., W. König, H. Wissmann, K. Geisen, and F. Enzmann: Synthesis and Characterisation of a Decapeptide Having LH – RH/FSH – RH Activity. Biochem. Biophys. Res. Commun. **45,** 767 (1971).

487. Sievertsson, H., J. K. Chang, C. Bogentoft, B. L. Currie, K. Folkers, and C. Y. Bowers: Synthesis of the Luteinizing Releasing Hormone of the Hypothalamus and its Hormonal Activity. Biochem. Biophys. Res. Commun. **44,** 1566 (1971).

488. Beyerman, H. C., H. Hindriks, J. Hirt, E. W. B. de Leer, and A. van der Wiele: A Synthesis of the Decapeptide Sequence Proposed for the LH- and FSH-Releasing Hormone. Rec. Trav. Chim. **91,** 1239 (1972).

489. Rivier, J., M. Monahan, W. Vale, G. Grant, M. Amoss, R. E. Blackwell, R. Guillemin, and R. Burgus: Solid Phase Peptide Synthesis on a Benzhydrylamine Resin of LRF (Luteinizing Hormone Releasing Factor) and Analogues Including Antagonists. Chimia **26,** 300 (1972).

490. Schally, A. V., T. W. Redding, H. Matsuo, and A. Arimura: Stimulation of FSH and LH Release in Vitro by Natural and Synthetic LH and FSH Releasing Hormone. Endocrinol. **90,** 1561 (1972).

491. Sievertsson, H., J.-K. Chang, A. v. Klaudy, C. Bogentoft, B. L. Currie, K. Folkers, and C. Bowers: Two Syntheses of the Luteinizing Hormone Releasing Hormone of the Hypothalamus. J. Med. Chem. **15,** 222 (1972).

492. Voelter, W., und K. Zech: Synthese eines Dekapeptids mit der biologischen Aktivität des Schweine-LH-Releasing-Faktors. Chimia **26,** 313 (1972).

493. Crighton, D. B.: The Effect of Synthetic Gonadotrophin Releasing Factor on the Release of Luteinizing Hormone and Follicle-Stimulating Hormone from Ovine Pituitary Tissue in Vitro. J. Endocr. **58,** 387 (1973).

494. Fujino, M., T. Fukuda, S. Kobayashi, and M. Obayashi: Isobornyloxycarbonyl Function, a New Convenient Amino-Protecting Group in Peptide Synthesis. IV. Synthesis of Gonadotropin-Releasing Hormone (Gn – RH or LH – RH/FSH – RH). Chem. Pharm. Bull. **21,** 87 (1973).

495. FUJINO, M., and C. KITADA: Synthesis of Luteinizing Hormone-Releasing Hormone (LH−RH) by the Solid-Phase Method. J. Takeda Res. Laboratories **32**, 101 (1973).

496. MATSUEDA, R., H. MARUYAMA, E. KITAZAWA, H. TAKAHAGI, and T. MUKAIYAMA: Solid Phase Peptide Synthesis by Oxidation-Reduction Condensation. Synthesis of LH−RH by Fragment Condensation on Solid Support. Bull. Chem. Soc. Japan **46**, 3240 (1973).

497. SCHAFER, D. J., and A. D. BLACK: Synthesis of Luteinizing Hormone Releasing Hormone (LH−RH); A Simplified Approach to the Synthesis of Arginine Peptides. Tetrahedron Letters **1973**, 4071.

498. WHITE, W. F., M. T. HEDLUND, R. H. RIPPEL, W. ARNOLD, and G. R. FLOURET: Chemical and Biological Properties of Gonadotropin-Releasing Hormone Synthesized by the Solid-Phase Method. Endocrinol. **93**, 96 (1973).

499. YANAIHARA, N., C. YANAIHARA, M. SAKAGAMI, K. TSUJI, T. HASHIMOTO, T. KANEKO, H. OKA, A. V. SCHALLY, A. ARIMURA, and T. W. REDDING: Synthesis and Biological Evaluation of LH and FSH Releasing Hormone and Its Analogs. J. Med. Chem. **16**, 373 (1973).

500. FUJINO, M., S. KOBAYASHI, M. OBAYASHI, T. FUKUDA, S. SHINAGAWA, and O. NISHIMURA: The Use of N-Hydroxy-5-norbornene-2,3-dicarboximide Active Esters in Peptide Synthesis. Chem. Pharm. Bull. **22**, 1857 (1974).

501. KOCHMAN, K., B. KERDELHUÉ, U. ZOR, and M. JUTISZ: Studies of Enzymatic Degradation of Luteinizing Hormone-Releasing Hormone by Different Tissues. FEBS Letters **50**, 190 (1975).

502. C. Y. BOWERS, J.-K. CHANG, H. SIEVERTSSON, C. BOGENTOFT, B. L. CURRIE, and K. FOLKERS: Activity of a New Synthetic Tetrapeptide in Hypothalamic Luteinizing and Follicle Stimulating Releasing Hormone Assay Systems. Biochem. Biophys. Res. Commun. **44**, 414 (1971).

503. CHANG, J.-K., H. SIEVERTSSON, C. BOGENTOFT, B. L. CURRIE, K. FOLKERS, and C. Y. BOWERS: Discovery of a New Synthetic Tetrapeptide Having Luteinizing Releasing Hormone (LRH) Activity. Biochem. Biophys. Res. Commun. **44**, 409 (1971).

504. CHANG, J.-K., H. SIEVERTSSON, B. L. CURRIE, C. BOGENTOFT, K. FOLKERS, and C. Y. BOWERS: Synthesis of the Luteinizing-Releasing Hormone of the Hypothalamus and the 8-Lysine Analog. J. Med. Chem. **15**, 623 (1972).

505. FUJINO, M., S. KOBAYASHI, M. OBAYASHI, S. SHINAGAWA, T. FUKUDA, C. KITADA, R. NAKAYAMA, I. YAMAZAKI, W. F. WHITE, and R. H. RIPPEL: Structure-Activity Relationships in the C-Terminal Part of Luteinizing Hormone Releasing Hormone (LH−RH). Biochem. Biophys. Res. Commun. **49**, 863 (1972).

506. FUJINO, M., S. KOBAYASHI, M. OBAYASHI, T. FUKUDA, S. SHINAGAWA, I. YAMAZAKI, R. NAKAYAMA, W. F. WHITE, and R. H. RIPPEL: Syntheses and Biological Activities of Analogs of Luteinizing Hormone Releasing Hormone (LH−RH). Biochem. Biophys. Res. Commun. **49**, 698 (1972).

507. GEIGER, R., H. WISSMANN, W. KÖNIG, J. SANDOW, A. V. SCHALLY, T. W. REDDING, L. DEBELJUK, and A. ARIMURA: Synthesis and Biological Evaluation of 4-Alanine-Luteinizing Hormone-Releasing Hormone ([Ala⁴]LH−RH). Biochem. Biophys. Res. Commun. **49**, 1467 (1972).

508. GUILLEMIN, R., M. AMOSS, R. E. BLACKWELL, J. RIVIER, N. LING, and W. VALE: On the Biological Activities of the Synthetic Tetrapeptide Pyroglutamyl-tyrosyl-arginyl-tryptophanylamide. Biochem. Biophys. Res. Commun. **48**, 1093 (1972).

509. SCHALLY, A. V., A. ARIMURA, W. H. CARTER, T. W. REDDING, R. GEIGER, W. KÖNIG, H. WISSMAN, G. JAEGER, J. SANDOW, N. YANAIHARA, C. YANAIHARA, T. HASHIMOTO, and M. SAKAGAMI: Luteinizing Hormone-Releasing Hormone (LH−RH) Activity of Some Synthetic Polypeptides. I. Fragments Shorter than Decapeptide. Biochem. Biophys. Res. Commun. **48**, 366 (1972).

510. Coy, D. H., E. J. Coy, and A. V. Schally: Analogs of Luteinizing Hormone-Releasing Hormone Containing Derivatives of Phenylalanine in Place of Tyrosine. J. Med. Chem. 16, 827 (1973).

511. Fujino, M., S. Shinagawa, M. Obayashi, S. Kobayashi, T. Fukuda, I. Yamazaki, R. Nakayama, W. F. White, and R. H. Rippel: Further Studies on the Structure-Activity Relationships in the C-Terminal Part of Luteinizing Hormone-Releasing Hormone. J. Med. Chem. 16, 1144 (1973).

512. Fujino, M., S. Shinagawa, I. Yamazaki, S. Kobayashi, M. Obayashi, T. Fukuda, R. Nakayama, W. F. White, and R. H. Rippel: [Des – Gly – NH$_2^{10}$, Pro-ethylamide^9] – LH – RH: A Highly Potent Analog of Luteinizing Hormone Releasing Hormone. Arch. Biochem. Biophys. 154, 488 (1973).

513. Groom, G. V., and A. R. Boyns: Gonadotrophin Release from Human Foetal Pituitary Cultures Induced by Fragments of the Luteinizing Hormone-Releasing Hormone. FEBS Letters 33, 57 (1973).

514. Kennedy, J. F., C. J. Gray, S. Ramanvongse, L. Albrighton, and W. F. White: Chemico-Biological Relationships of Follicle Stimulating Hormone and Luteinizing Hormone-Releasing Hormone. Life Sci. 12, 533 (1973).

515. Rivaille, P., Lê Du et G. Milhaud: Intérêt de la synthèse de peptides en phase solide et préparation de décapeptides du type LH – RH. Compt. Rend. Acad. Sci. 277, 343 (1973).

516. Rippel, R. H., E. S. Johnson, W. F. White, M. Fujino, I. Yamazaki, and R. Nakayama: Ovulating and LH-Releasing Activity of a Highly Potent Analog of Synthetic Gonadotropin-Releasing Hormone. Endocrinol. 93, 1449 (1973).

517. Rivier, J., W. Vale, R. Burgus, N. Ling, M. Amoss, R. E. Blackwell, and R. Guillemin: Synthetic Luteinizing Hormone Releasing Factor Analogs. Series of Short-Chain Amide LRF Homologs Converging to the Amino Terminus. J. Med. Chem. 16, 545 (1973).

518. Yanaihara, N., C. Yanaihara, M. Sakagami, K. Tsuji, T. Hashimoto, T. Kaneko, H. Oka, A. V. Schally, A. Arimura, and T. W. Redding: Synthesis and Biological Evaluation of LH and FSH Releasing Hormone and its Analogs. J. Med. Chem. 16, 373 (1973).

519. Arimura, A., J. A. Vilchez-Martinez, D. H. Coy, E. J. Coy, Y. Hirotsu, and A. V. Schally: [D – Ala^6, Des – Gly – NH$_2^{10}$] – LH – RH – Ethylamide: A New Analogue with Unusually High LH – RH/FSH – RH Activity. Endocrinol. 95, 1174 (1974).

520. Arimura, A., J. A. Vilchez-Martinez, and A. V. Schally: In Vivo Comparison of LH – RH and FSH – RH Activities of [Des – Gly^{10}] [Pro^9 – Ethylamide] – LH – RH, [Des – Gly^{10}] [Pro^9 – Propylamide] – LH – RH, and LH – RH Using Immature Male Rats. Proc. Soc. Exp. Biol. Med. 146, 17 (1974).

521. Arnold, W., G. R. Flouret, R. Morgan, R. Rippel, and W. White: Synthesis and Biological Activity of Some Analogs of the Gonadotropin Releasing Hormone. J. Med. Chem. 17, 314 (1974).

522. Borvendég, J., S. Bajusz, I. Hermann, and A. Turán: Studies on the Structure-Activity Relationships of Synthetic LHRH Analogs. FEBS Letters 44, 233 (1974).

523. Coy, D. H., E. J. Coy, Y. Hirotsu, and A. V. Schally: Synthesis and Biological Properties of [2 – L – β – (Pyrazolyl – 3)alanine] – Luteinizing Hormone-Releasing Hormone. J. Med. Chem. 17, 140 (1974).

524. Coy, D. H., E. J. Coy, Y. Hirotsu, J. A. Vilchez-Martinez, A. V. Schally, J. W. van Nispen, and G. I. Tesser: Investigation of the Role of Tryptophan in the Luteinizing Hormone-Releasing Hormone. Biochemistry 13, 3550 (1974).

525. Fujino, M., T. Fukuda, S. Shinagawa, S. Kobayashi, I. Yamazaki, R. Nakayama, J. H. Seely, W. F. White, and R. H. Rippel: Synthetic Analogs of Luteinizing Hormone Releasing Hormone (LH – RH) Substituted in Position 6 and 10. Biochem. Biophys. Res. Commun. 60, 406 (1974).

526. Fujino, M., I. Yamazaki, S. Kobayashi, T. Fukuda, S. Shinagawa, R. Nakayama,

W. F. White, and R. H. Rippel: Some Analogs of Luteinizing Hormone Releasing Hormone (LH – RH) Having Intense Ovulation-Inducing Activity. Biochem. Biophys. Res. Commun. **57**, 1248 (1974).

527. Geiger, R., W. König, J. Sandow und A. V. Schally: [8-Homoarginin] (Luteinisierendes Hormon freisetzendes Hormon). Hoppe-Seyler's Z. Physiol. Chem. **355**, 1526 (1974).

528. Humphries, J., G. Fisher, Y.-P. Wan, K. Folkers, and C. Y. Bowers: Analogs of the Luteinizing Hormone-Releasing Hormone to Study Conformational Aspects of the Aromatic Amino Acid Moieties and Inhibition. J. Med. Chem. **17**, 569 (1974).

529. Humphries, J., Y.-P. Wan, G. Fisher, K. Folkers, and C. Y. Bowers: Acidic Analogs of the Luteinizing Hormone-Releasing Hormone and Conformational Aspects for Activity. Biochem. Biophys. Res. Commun. **57**, 675 (1974).

530. Immer, H. U., V. R. Nelson, C. Revesz, K. Sestanj, and M. Götz: Luteinizing Hormone-Releasing Hormone and Analogs. Synthesis and Biological Activity. J. Med. Chem. **17**, 1060 (1974).

531. Künzi, H., D. Gillessen, A. Trzeciak, R. O. Studer, B. Kerdelhué, M. Jutisz, and W. Lotz: Synthesis of Some Structural Analogues of LH – RH Modified in Position 5, their in Vivo and in Vitro Gonadotropin-Releasing Activity and Immunoreactivity. Helv. Chim. Acta **57**, 2131 (1974).

532. Rees, R. W. A., T. J. Foell, S.-Y. Chai, and N. H. Grant: Synthesis and Biological Activities of Analogs of the Luteinizing Hormone-Releasing Hormone (LH–RH) Modified in Position 2. J. Med. Chem. **17**, 1016 (1974).

533. Rivier, J., M. Amoss, C. Rivier, and W. Vale: Synthetic Luteinizing Hormone Releasing Factor. Short Chain Analogs. J. Med. Chem. **17**, 230 (1974).

534. Vilchez-Martinez, J. A., D. H. Coy, A. Arimura, E. J. Coy, Y. Hirotsu, and A. V. Schally: Synthesis and Biological Properties of [Leu – 6] – LH – RH and [D – Leu – 6, Des – Gly – NH$_2^{10}$] – LH – RH Ethylamide. Biochem. Biophys. Res. Commun. **59**, 1226 (1974).

535. Yabe, Y., K. Kitamura, C. Miura, and Y. Baba: Analogues of Luteinizing Hormone-Releasing Hormone with Modification in Position 8. Chem. Pharm. Bull. **22**, 2557 (1974).

536. Beattie, C. W., A. Corbin, T. J. Foell, V. Garsky, W. A. McKinley, R. W. A. Rees, D. Sarantakis, and J. P. Yardley: Luteinizing Hormone-Releasing Hormone. Antiovulatory Activity of Analogs Substituted in Position 2 and 6. J. Med. Chem. **18**, 1247 (1975).

537. Beddell, C. R., P. J. Fraser, D. Gilbert, P. J. Goodford, L. A. Lowe, and S. Wilkinson: Pseudosymmetry in the Structure of Luteinizing Hormone-Releasing Hormone Studies on a Series of Novel Analogs. J. Med. Chem. **18**, 417 (1975).

538. Coy, D. H., Y. Hirotsu, T. W. Redding, E. J. Coy, and V. A. Schally: Synthesis and Biological Properties of the 2–L–β–(Pyrazolyl–1)–alanine Analogs of Luteinizing Hormone-Releasing Hormone and Thyrotropin-Releasing Hormone. J. Med. Chem. **18**, 948 (1975).

539. Schafer, D. J., A. D. Black, J. D. Bower, and D. B. Crighton: Synthesis and Biological Activity of Luteinizing Hormone-Releasing Hormone and Related Peptides. J. Med. Chem. **18**, 613 (1975).

540. Shinagawa, S., and M. Fujino: Synthesis of a Highly Potent Analog of Luteinizing Hormone Releasing Hormone (LH – RH): [Des – Gly – NH$_2^{10}$, Pro – NH – Et^9] – LH – RH. Chem. Pharm. Bull. **23**, 229 (1975).

541. Yajima, H., M. Kurobe, I. Yo, N. Fujii, and Y. Baba: Studies on Peptides. LI. Application of the Solid Phase Synthesis for the Preparation of Pro-Analogues of LH and FSH Releasing Hormone. Chem. Pharm. Bull. **23**, 1622 (1975).

542. Brazeau, P., W. Vale, R. Burgus, N. Ling, M. Butcher, J. Rivier, and R. Guillemin:

Hypothalamic Polypeptide that Inhibits the Secretion of Immunoreactive Pituitary Growth Hormone. Science **179**, 77 (1973).

543. BURGUS, R., N. LING, M. BUTCHER, and R. GUILLEMIN: Primary Structure of Somatostatin, A Hypothalamic Peptide that Inhibits the Secretion of Pituitary Growth Hormone. Proc. Natl. Acad. Sci. **70**, 684 (1973).

544. HOLLADAY, L. A., and D. PUETT: Physicochemical Characteristics and a Proposed Conformation of Somatostatin. In: Peptides: Chemistry, Structure and Biology (R. WALTER and J. MEIENHOFER, eds.), p. 175. Ann. Arbor: Ann Arbor Science Publishers. 1975.

545. — — Somatostatin Conformation: Evidence for a Stable Intramolecular Structure from Circular Dichroism, Diffusion, and Sedimentation Equilibrium. Proc. Natl. Acad. Sci **73** (1976).

546. RIVIER, J., P. BRAZEAU, W. VALE, N. LING, R. BURGUS, C. GILON, J. YARDLEY et R. GUILLEMIN: Synthèse totale par phase solide d'un tétradécapeptide ayant les propriétés chimiques et biologiques de la somatostatine. Compt. Rend. Acad. Sci. **276**, 2737 (1973).

547. YAMASHIRO, D., and C. H. LI: Synthesis of a Peptide with Full Somatostatin Activity. Biochem. Biophys. Res. Commun. **54**, 882 (1973).

548. RIVIER, J. E. F.: Somatostatin. Total Solid Phase Synthesis. J. Amer. Chem. Soc. **96**, 2986 (1974).

549. SARANTAKIS, D., and W. A. MCKINLEY: Total Synthesis of Hypothalamic "Somatostatin". Biochem. Biophys. Res. Commun. **54**, 234 (1973).

550. IMMER, H. U., K. SESTANJ, V. R. NELSON, and M. GÖTZ: Synthesis of Somatostatin. Helv. Chim. Acta **57**, 730 (1974).

551. FUJII, N., and H. YAJIMA: Studies on Peptides. LII. Application of the Trifluoromethanesulphonic Acid Procedure to the Synthesis of a Peptide with Somatostatin Activity. Chem. Pharm. Bull. **23**, 1596 (1975).

(Received July 21, 1976)

Namenverzeichnis. Author Index

Sachverzeichnis. Subject Index

Von · By

A. SIEGEL, Wien

Satz: Austro-Filmsatz Richard Gerin, A-1020 Wien
Druck: Paul Gerin, A-1021 Wien

Fortschritte der Chemie organischer Naturstoffe
Progress in the Chemistry of Organic Natural Products

All Volumes and Cumulative Index 1—20 available / Alle Bände und Generalregister 1—20 lieferbar.

Price reduction for subscribers / Preisermäßigung für Subskribenten: 10%.

Special price reduction (20% of the list price) for the Vols. 1—20 plus Cumulative Index. / Vorzugspreis (20% Nachlaß) bei Bezug der Bände 1—20 inklusive Generalregister.

Volume 32: 31 figures. VIII, 560 pages. 1975.

Contents: W. K. SEIFERT, Carboxylic Acids in Petroleum and Sediments — P. G. SAMMES, Naturally Occurring 2,5-Dioxopiperazines and Related Compounds — R. J. HIGHET and E. A. SOKOLOSKI, Structural Investigations of Natural Products by Newer Methods of NMR Spectroscopy — P. M. SCOPES, Applications of the Chiroptical Techniques to the Study of Natural Products — H. C. VAN HUMMEL, Chemistry and Biosynthesis of Plant Galactolipids — H. KÖSSEL and H. SELIGER, Recent Advances in Polynucleotide Synthesis — Author Index — Subject Index.

Volume 33: 48 figures. VIII, 581 pages. 1976.

Contents: L. MINALE, G. CIMINO, S. DE STEFANO, and G. SODANO, Natural Products from Porifera — R. M. COATES, Biogenetic-Type Rearrangements of Terpenes — K. L. RINEHART, JR., and L. S. SHIELD, Chemistry of the Ansamycin Antibiotics — A. FONTANA and C. TONIOLO, The Chemistry of Tryptophan in Peptides and Proteins — P. HEMMERICH, The Present Status of Flavin and Flavocoenzyme Chemistry — Author Index — Subject Index.

Springer-Verlag Wien · New York